PROBABILITY AND
STATISTICAL INFERENCE

STATISTICS: Textbooks and Monographs

D. B. Owen, Founding Editor, 1972–1991

1. The Generalized Jackknife Statistic, *H. L. Gray and W. R. Schucany*
2. Multivariate Analysis, *Anant M. Kshirsagar*
3. Statistics and Society, *Walter T. Federer*
4. Multivariate Analysis: A Selected and Abstracted Bibliography, 1957–1972, *Kocherlakota Subrahmaniam and Kathleen Subrahmaniam*
5. Design of Experiments: A Realistic Approach, *Virgil L. Anderson and Robert A. McLean*
6. Statistical and Mathematical Aspects of Pollution Problems, *John W. Pratt*
7. Introduction to Probability and Statistics (in two parts), Part I: Probability; Part II: Statistics, *Narayan C. Giri*
8. Statistical Theory of the Analysis of Experimental Designs, *J. Ogawa*
9. Statistical Techniques in Simulation (in two parts), *Jack P. C. Kleijnen*
10. Data Quality Control and Editing, *Joseph I. Naus*
11. Cost of Living Index Numbers: Practice, Precision, and Theory, *Kali S. Banerjee*
12. Weighing Designs: For Chemistry, Medicine, Economics, Operations Research, Statistics, *Kali S. Banerjee*
13. The Search for Oil: Some Statistical Methods and Techniques, *edited by D. B. Owen*
14. Sample Size Choice: Charts for Experiments with Linear Models, *Robert E. Odeh and Martin Fox*
15. Statistical Methods for Engineers and Scientists, *Robert M. Bethea, Benjamin S. Duran, and Thomas L. Boullion*
16. Statistical Quality Control Methods, *Irving W. Burr*
17. On the History of Statistics and Probability, *edited by D. B. Owen*
18. Econometrics, *Peter Schmidt*
19. Sufficient Statistics: Selected Contributions, *Vasant S. Huzurbazar (edited by Anant M. Kshirsagar)*
20. Handbook of Statistical Distributions, *Jagdish K. Patel, C. H. Kapadia, and D. B. Owen*
21. Case Studies in Sample Design, *A. C. Rosander*
22. Pocket Book of Statistical Tables, *compiled by R. E. Odeh, D. B. Owen, Z. W. Birnbaum, and L. Fisher*
23. The Information in Contingency Tables, *D. V. Gokhale and Solomon Kullback*
24. Statistical Analysis of Reliability and Life-Testing Models: Theory and Methods, *Lee J. Bain*
25. Elementary Statistical Quality Control, *Irving W. Burr*
26. An Introduction to Probability and Statistics Using BASIC, *Richard A. Groeneveld*
27. Basic Applied Statistics, *B. L. Raktoe and J. J. Hubert*
28. A Primer in Probability, *Kathleen Subrahmaniam*
29. Random Processes: A First Look, *R. Syski*
30. Regression Methods: A Tool for Data Analysis, *Rudolf J. Freund and Paul D. Minton*
31. Randomization Tests, *Eugene S. Edgington*
32. Tables for Normal Tolerance Limits, Sampling Plans and Screening, *Robert E. Odeh and D. B. Owen*
33. Statistical Computing, *William J. Kennedy, Jr., and James E. Gentle*
34. Regression Analysis and Its Application: A Data-Oriented Approach, *Richard F. Gunst and Robert L. Mason*
35. Scientific Strategies to Save Your Life, *I. D. J. Bross*
36. Statistics in the Pharmaceutical Industry, *edited by C. Ralph Buncher and Jia-Yeong Tsay*
37. Sampling from a Finite Population, *J. Hajek*

Additional Volumes in Preparation

PROBABILITY AND STATISTICAL INFERENCE

Nitis Mukhopadhyay

University of Connecticut
Storrs, Connecticut

Routledge
Taylor & Francis Group

LONDON AND NEW YORK

First published 2000 by Marcel Dekker

Published 2020 by Routledge
2 Park Square, Milton Park, Abingdon, Oxon OX14 4RN
52 Vanderbilt Avenue, New York, NY 10017

First issued in paperback 2020

Routledge is an imprint of the Taylor & Francis Group, an informa business

Library of Congress Cataloging-in-Publication Data

Mukhopadhyay, Nitis.
    Probability and statistical inference / Nitis Mukhopadhyay.
        p. cm. – (Statistics, textbooks and monographs ; v. 162)
    Includes bibliographical references and index.
    ISBN 0-8247-0379-0 (alk. paper)
    1. Probabilities. 2. Mathematical statistics. I. Title. II. Series.

    QA273 .M85 2000
    519.2—dc21   00-022901

ISBN 13: 978-0-367-65949-3 (pbk)
ISBN 13: 978-0-8247-0379-0 (hbk)

With love and affection,
this book is dedicated to my parents

The late Mr. Manindra Chandra Mukherjee,

Mrs. Snehalata Mukherjee

It is my homage to the two best teachers I have ever known

Preface

This textbook aims to foster the theory of both probability and statistical inference for first-year graduate students in statistics or other areas in which a good understanding of statistical concepts is essential. It can also be used as a textbook in a junior/senior level course for statistics or mathematics/statistics majors, with emphasis on concepts and examples. The book includes the core materials that are usually taught in a two-semester or three-quarter sequence.

A distinctive feature of this book is its set of examples and exercises. These are essential ingredients in the total learning process. I have tried to make the subject come alive through many examples and exercises.

This book can also be immensely helpful as a supplementary text in a significantly higher level course (for example, Decision Theory and Advanced Statistical Inference) designed for second or third year graduate students in statistics.

The prerequisite is one year's worth of calculus. That should be enough to understand a major portion of the book. There are sections for which some familiarity with linear algebra, multiple integration and partial differentiation will be beneficial. I have reviewed some of the important mathematical results in Section 1.6.3. Also, Section 4.8 provides a selected review of matrices and vectors.

The first four chapters introduce the basic concepts and techniques in probability theory, *including* the calculus of probability, conditional probability, independence of events, Bayes's Theorem, random variables, probability distributions, moments and moment generating functions (mgf), probability generating functions (pgf), multivariate random variables, independence of random variables, standard probability inequalities, the exponential family of distributions, transformations and sampling distributions. Multivariate normal, t and F distributions have also been briefly discussed. Chapter 5 develops the notions of convergence in probability, convergence in distribution, the central limit theorem (CLT) for both the sample mean and sample variance, and the convergence of the density functions of the Chi-square, t and F distributions.

The remainder of the book systematically develops the concepts of statistical inference. It is my belief that the concept of "sufficiency" is the *heart* of statistical inference and hence this topic deserves appropriate care and respect in its treatment. I introduce the fundamental notions of sufficiency, Neyman factorization, information, minimal sufficiency, completeness, and ancillarity very early, in Chapter 6. Here, Basu's Theorem and the location, scale and location-scale families of distributions are also addressed.

The method of moment estimator, maximum likelihood estimator (MLE), Rao-Blackwell Theorem, Rao-Blackwellization, Cramér-Rao inequality, uniformly minimum variance unbiased estimator (UMVUE) and Lehmann-Scheffé Theorems are developed in Chapter 7. Chapter 8 provides the Neyman-Pearson theory of the most powerful (MP) and uniformly most powerful (UMP) tests of hypotheses as well as the monotone likelihood ratio (MLR) property. The concept of a UMP unbiased (UMPU) test is briefly addressed in Section 8.5.3. The confidence interval and confidence region methods are elaborated in Chapter 9. Chapter 10 is devoted entirely to the Bayesian methods for developing the concepts of the highest posterior density (HPD) credible intervals, the Bayes point estimators and tests of hypotheses.

Two-sided alternative hypotheses, likelihood ratio (LR) and other tests are developed in Chapter 11. Chapter 12 presents the basic ideas of large-sample confidence intervals and test procedures, including variance stabilizing transformations and properties of MLE. In Section 12.4, I explain how one arrives at the customary $\sin^{-1}(\sqrt{p})$, $\sqrt{\lambda}$, and $\tanh^{-1}(\rho)$ transformations in the case of Binomial(p), Poisson(λ), and the correlation coefficient ρ, respectively.

Chapter 13 introduces two-stage sampling methodologies for determining the required sample size needed to solve two simple problems in statistical inference for which, unfortunately, no fixed-sample-size solution exists. This material is included to emphasize that there is much more to explore beyond what is customarily covered in a standard one-year statistics course based on Chapters 1 -12.

Chapter 14 (Appendix) presents (i) a list of notation and abbreviations, (ii) short biographies of selected luminaries, and (iii) some of the standard statistical tables computed with the help of MAPLE. One can also find some noteworthy remarks and examples in the section on statistical tables. An extensive list of references is then given, followed by a detailed index.

In a two-semester sequence, probability theory is covered in the first part, followed by statistical inference in the second. In the first semester, the core material may consist of Chapters 1-4 and some parts of Chapter 5. In the second semester, the core material may consist of the remainder of Chapter 5 and Chapters 6-10 plus some selected parts of Chapters 11-13. The book covers more than enough ground to allow some flexibility in the selection of topics beyond the core. In a three-quarter system, the topics will be divided somewhat differently, but a year's worth of material taught in either a two-semester or three-quarter sequence will be similar.

Obviously there are competing textbooks at this level. What sets this book apart from the others? Let me *highlight* some of the novel features of this book:

1. The material is rigorous, both conceptually and mathematically, but I have adopted what may be called a "tutorial style." In Chapters 1-12, the reader will find numerous worked examples. Techniques and concepts are typically illustrated through a series of examples and related exercises, providing additional opportinities for absorption. It will be hard to find another book that has even one-half the number of worked examples!

2. At the end of each chapter, a long list of exercises is arranged according to the section of a concept's origin (for example, Exercise 3.4.2 is the second exercise related to the material presented in Section 3.4). Many exercises are direct follow-ups on the worked-out examples. Hints are frequently given in the exercises. This kind of drill helps to reinforce and emphasize important concepts as well as special mathematical techniques. I have found over the years that the ideas, principles, and techniques are appreciated more if the student solves similar examples and exercises. I let a reader build up his/her own confidence first and then challenge the individual to approach harder problems, with substantial hints when appropriate. I try to entice a reader to think through the examples and then do the problems.

3. I can safely remark that I often let the examples do the talking. After giving a series of examples or discussing important issues, I routinely summarize within a box what it is that has been accomplished or where one should go from here. This feature, I believe, should help a reader to focus on the topic just learned, and move on.

4. There are numerous figures and tables throughout the book. I have also used computer simulations in some instances. From the layout, it should be obvious that I have used the power of a computer very liberally.

I should point out that the book contains unique features throughout. Let me *highlight a few examples*:

a) In Section 2.4, the "moment problem" is discussed in an elementary fashion. The two given density functions plotted in Figure 2.4.1 have identical moments of all orders. This example is not new, but the two plots certainly should grab one's attention! Additionally, Exercise 2.4.6 guides a reader in the construction of other examples. Next, at this level, hardly any book discusses the role of a probability generating function. Section 2.5 does precisely that with the help of examples and exercises. Section 3.5 and related exercises show how easily one can construct examples of a collection of dependent random variables having certain independent subsets within the collection. With the help of interesting examples and discussions, Section 3.7 briefly unfolds the intricate relationship between "zero correlation" and "independence" for two random variables.

b) In Chapter 4, the Helmert transformation for a normal distribution, and the transformation involving the spacings for an exponential distribution, have both been developed thoroughly. The related remarks are expected to make many readers pause and think. Section 4.6 exposes readers to some continuous multivariate distributions other than the multivariate normal. Section 4.7 has special messages – in defining a random variable having the Student's t or F distribution, for example, one takes independent random variables in the numerator and denominator. But, what happens when the random variables in the numerator and denominator are dependent? Some possible answers are emphasized with the help of examples. Exercise 4.7.4 shows a way to construct examples where the distribution of a sample variance is a multiple of Chi-square even though the random samples do not come from a normal population!

c) The derivation of the central limit theorem for the sample variance (Theorem 5.3.6) makes clever use of several non-trivial ingredients from the theory of probability. In other words, this result reinforces the importance of many results taught in the preceding sections. That should be an important aspect of learning. No book at this level highlights this in the way I have. In Section 5.4, various convergence properties of the densities and percentage points of the Student's t and F distributions, for example, are laid out. The usefulness of such approximations is emphasized through computation. In no other book like this will one find such engaging discussions and comparisons.

d) No book covers the topics of Chapter 6, namely, sufficiency, information, and ancillarity, with nearly as much depth or breadth for the target audience. In particular, Theorem 6.4.2 helps in proving the sufficiency property of a statistic via its information content. The associated simple examples and exercises then drive the point home. One will discover out-of-the-ordinary remarks, ideas and examples throughout the book.

e) The history of statistics and statistical discoveries should not be separated from each other since neither can exist without the other. It may be noted that Folks (1981) first added some notable historical remarks within the material of his textbook written at the sophomore level. I have found that at all levels of instructions, students enjoy the history very much and they take more interest in the subject when the human element comes alive. Thus, I have added historical remarks liberally throughout the text. Additionally, in Section 14.2, I have given selected biographical notes on some of the exceptional contributors to the development of statistics. The biographies, in spite of some limitations/selection bias/exclusions, will hopefully inspire and energize the readers.

I assure the readers that a lot of effort has gone into this work. Several readers and reviewers have been very helpful. I remain eternally grateful to them. But, I alone am responsible for any remaining mistakes and errors. I will be absolutely delighted if the readers kindly point out errors of any kind or draw my attention to any part of the text requiring more explanation or improvement.

It has been a wonderful privilege on my part to teach and share my enthusiasm with the readers. I eagerly await to hear comments, criticisms, and suggestions (Electronic Mail: mukhop@uconnvm.uconn.edu). In the meantime,

Enjoy and Celebrate Statistics!!

I thank you, the readers, for considering my book and wish you all the very best.

Nitis Mukhopadhyay
January 1, 2000

Acknowledgments

A long time ago, in my transition from Salkia A. S. High School to Presidency College, Calcutta, followed by the Indian Statistical Institute, Calcutta, I had the good fortune of learning from many great teachers. I take this opportunity to express my sincerest gratitude to all my teachers, especially to Mr. Gobinda Bandhu Chowdhury, Professors Debabrata Basu, Biren Bose, Malay Ghosh, Sujit K. Mitra, and to Professor Anis C. Mukhopadhyay, who is my elder brother.

In good times and not so good times, I have been lucky to be able to count on my mentors, Professors P. K. Sen, Malay Ghosh, Bimal K. Sinha and Bikas K. Sinha, for support and guidance. From the bottom of my heart, I thank them for their kindness and friendship.

During the past twenty-five years, I have taught this material at a number of places, including Monash University in Melbourne, Australia, as well as the University of Minnesota-Minneapolis, the University of Missouri-Columbia, the Oklahoma State University-Stillwater and the University of Connecticut-Storrs. Any time a student asked me a question, I learned something. When a student did not ask questions whereas he/she perhaps should have, I have wondered why no question arose. From such soul searching, I learned important things, too. I have no doubt that the students have made me a better teacher. I thank all my students, both inside and outside of classrooms.

I am indebted tremendously to Dr. William T. Duggan. He encouraged me in writing this book since its inception and he diligently read several versions and caught many inconsistencies and errors. It is my delight to thank Bill for all his suggestions, patience, and valuable time.

My son, Shankha, has most kindly gone through the whole manuscript to "test" its flow and readability, and he did so during perhaps the busiest time of his life, just prior to his entering college. He suggested many stylistic changes and these have been very valuable. Shankha, thank you.

I thank Professor Tumulesh K. S. Solanky, for going through an earlier draft and for encouraging me throughout this project. I am also indebted to Professor Makoto Aoshima and I thank him for the valuable suggestions he gave me.

Without the support of the students, colleagues and staff of the Department of Statistics at the University of Connecticut-Storrs, this project could not have been completed. I remain grateful for this support, particularly to Professor Dipak K. Dey, the Head of the Department.

I am grateful to Mr. Greg Cicconetti, one of the graduate students in the Department of Statistics at the University of Connecticut-Storrs, for teaching me some tricks with computer graphics. He also enthusiastically helped me with some of the last minute details. Greg, thanks for the support.

I take this opportunity to especially thank my colleague, Professor Joe Glaz, who gave me constant moral support.

Those who know me personally may not believe that I have typed this book myself. I gathered that unbelievable courage because one special individual, Mrs. Cathy Brown, our department's administrative assistant, told me I could do it and that she would help me with Latex any time I needed help. She has helped me with Latex, and always with a smile, on innumerable occasions during the most frustrating moments. It is impossible for me to thank Cathy enough.

I remain grateful to the anonymous reviewers of the manuscript in various stages as well as to the editorial and production staff at Marcel Dekker, Inc. I am particularly indebted to Ms. Maria Allegra and Ms. Helen Paisner for their help and advice at all levels.

Last but not least, I express my heartfelt gratitude to my wife, Mahua, and our two boys, Shankha and Ranjan. During the past two years, we have missed a number of activities as a family. No doubt my family sacrificed much, but I did not hear many complaints. I was "left alone" to complete this project. I express my deepest appreciation to the three most sensible, caring, and loving individuals I know.

Contents

1

Notions of Probability

1.1 Introduction

In the study of the subject of *probability*, we first imagine an appropriate *random experiment*. A random experiment has three important components which are:

    a)   multiplicity of outcomes,
    b)   uncertainty regarding the outcomes, and
    c)   repeatability of the experiment in identical fashions.

Suppose that one tosses a regular coin up in the air. The coin has two sides, namely the head (H) and tail (T). Let us assume that the tossed coin will land on either H or T. Every time one tosses the coin, there is the possibility of the coin landing on its head or tail (*multiplicity of outcomes*). But, no one can say with absolute certainty whether the coin would land on its head, or for that matter, on its tail (*uncertainty regarding the outcomes*). One may toss this coin as many times as one likes under identical conditions (*repeatability*) provided the coin is not damaged in any way in the process of tossing it successively.

All three components are crucial ingredients of a random experiment. In order to contrast a random experiment with another experiment, suppose that in a lab environment, a bowl of pure water is boiled and the boiling temperature is then recorded. The first time this experiment is performed, the recorded temperature would read 100° Celsius (or 212° Fahrenheit). Under identical and perfect lab conditions, we can think of repeating this experiment several times, but then each time the boiling temperature would read 100° Celsius (or 212° Fahrenheit). Such an experiment will not fall in the category of a random experiment because the requirements of multiplicity and uncertainty of the outcomes are both violated here.

We interpret probability of an event as the *relative frequency* of the occurrence of that event in a number of independent and identical replications of the experiment. We may be curious to know the magnitude of the probability p of observing a head (H) when a particular coin is tossed. In order to gather valuable information about p, we may decide to toss the coin ten times, for example, and suppose that the following sequence of H and T is observed:

$$T\ HT\ HT\ HT\ T\ HTT\ H \tag{1.1.1}$$

Let n_k be the number of H's observed in a sequence of k tosses of the coin while n_k/k refers to the associated *relative frequency* of H. For the observed sequence in (1.1.1), the successive frequencies and relative frequencies of H are given in the accompanying Table 1.1.1. The observed values for n_k/k empirically provide a sense of what p may be, but admittedly this particular observed sequence of relative frequencies appears a little unstable. However, as the number of tosses increases, the oscillations between the successive values of n_k/k will become less noticeable. Ultimately n_k/k and p are expected to become indistinguishable in the sense that in the long haul n_k/k will be very close to p. That is, our instinct may simply lead us to interpret p as $\lim_{k \to \infty} (n_k/k)$.

Table 1.1.1. Behavior of the Relative Frequency of the H's

k	n_k	n_k/k	k	n_k	n_k/k
1	0	0	6	2	1/3
2	1	1/2	7	3	3/7
3	1	1/3	8	3	3/8
4	2	1/2	9	3	1/3
5	2	2/5	10	4	2/5

A random experiment provides in a natural fashion a list of *all* possible *outcomes*, also referred to as the *simple events*. These simple events act like "atoms" in the sense that the experimenter is going to observe only one of these simple events as a possible outcome when the particular random experiment is performed. A *sample space* is merely a set, denoted by **S**, which enumerates each and every possible simple event or outcome. Then, a probability scheme is generated on the subsets of **S**, including **S** itself, in a way which mimics the nature of the random experiment itself. Throughout, we will write $P(A)$ for the probability of a statement $A(\subseteq \mathbf{S})$. A more precise treatment of these topics is provided in the Section 1.3. Let us look at two simple examples first.

Example 1.1.1 Suppose that we toss a *fair coin* three times and record the outcomes observed in the first, second, and third toss respectively from left to right. Then the possible simple events are $HHH, HHT, HTH,$ HTT, THH, THT, TTH or TTT. Thus the sample space is given by

$$\mathbf{S} = \{HHH, HHT, HTH, HTT, THH, THT, TTH, TTT\}.$$

Since the coin is assumed to be fair, this particular random experiment generates the following probability scheme: $P(HHH) = P(HHT) = P(HTH) = P(HTT) = P(THH) = P(THT) = P(TTH) = P(TTT) = \frac{1}{8}$. ▲

Example 1.1.2 Suppose that we toss *two fair dice*, one red and the other yellow, at the same time and record the scores on their faces landing upward. Then, each simple event would constitute, for example, a pair ij where i is the number of dots on the face of the red die that lands up and j is the number of dots on the face of the yellow die that lands up. The sample space is then given by $\mathbf{S} = \{11, 12, ..., 16, 21, ..., 26, ..., 61, ..., 66\}$ consisting of exactly 36 possible simple events. Since both dice are assumed to be fair, this particular random experiment generates the following probability scheme: $P(ij) = \frac{1}{36}$ for all $i, j = 1, ..., 6$. ▲

Some elementary notions of set operations are reviewed in the Section 1.2. The Section 1.3 describes the setup for developing the formal *theory of probability*. The Section 1.4 introduces the concept of *conditional probability* followed by the notions such as the *additive rules, multiplicative rules,* and *Bayes's Theorem.* The Sections 1.5-1.6 respectively introduces the *discrete* and *continuous random variables,* and the associated notions of a *probability mass function* (pmf), *probability density function* (pdf) and the *distribution function* (df). The Section 1.7 summarizes some of the standard probability distributions which are frequently used in statistics.

1.2   About Sets

A set $\mathbf{S}$ is a collection of objects which are tied together with one or more common defining properties. For example, we may consider a set $\mathbf{S} = \{x : x \text{ is an integer}\}$ in which case $\mathbf{S}$ can be alternately written as $\{..., -2, -1, 0, 1, 2, ...\}$. Here the common defining property which ties in all the members of the set $\mathbf{S}$ is that they are integers.

Let us start with a set $\mathbf{S}$. We say that A is a *subset* of $\mathbf{S}$, denoted by $A \subseteq \mathbf{S}$, provided that each member of A is also a member of $\mathbf{S}$. Consider A and B which are both subsets of $\mathbf{S}$. Then, A is called a *subset* of B, denoted by $A \subseteq B$, provided that each member of A is also a member of B. We say that A is a *proper subset* of B, denoted by $A \subset B$, provided that A is a subset of B but there is at least one member of B which does *not* belong to A. Two sets A and B are said to be equal if and only if $A \subseteq B$ as well as $B \subseteq A$.

Example 1.2.1 Let us define $\mathbf{S} = \{1, 3, 5, 7, 9, 11, 13\}, A = \{1, 5, 7\}, B = \{1, 5, 7, 11, 13\}, C = \{3, 5, 7, 9\},$ and $D = \{11, 13\}$. Here, A, B, C and D are all proper subsets of **S**. It is obvious that A is a proper subset of B, but A is not a subset of either C or D. ▲

Suppose that A and B are two subsets of **S**. Now, we mention some customary set operations listed below:

Complement: $\quad A^c = \{x : x \in \mathbf{S} \text{ but } x \notin A\}$

Union: $\qquad\quad A \cup B = \{x : x \in A \text{ or } x \in B\}$

Intersection: $\quad A \cap B = \{x : x \in A \text{ and } x \in B\}$ $\qquad$ (1.2.1)

Symmetric Difference: $\quad A \triangle B = (A^c \cap B) \cup (A \cap B^c)$

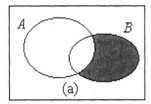 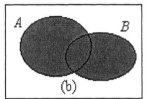

Figure 1.2.1. Venn Diagrams: Shaded Areas
Correspond to the Sets (a) $A^c \cap B$ (b) $A \cup B$

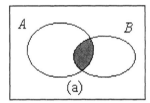 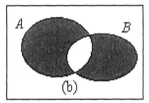

Figure 1.2.2. Venn Diagrams: Shaded Areas
Correspond to the Sets (a) $A \cap B$ (b) $A \triangle B$

The union and intersection operations also satisfy the following laws: For any subsets A, B, C of **S**, we have

Commutative Law: $\quad A \cup B = B \cup A, A \cap B = B \cap A$

Associative Law: $\quad (A \cup B) \cup C = A \cup (B \cup C);$
$\qquad\qquad\qquad\quad (A \cap B) \cap C = A \cap (B \cap C)$ $\qquad$ (1.2.2)

Distributive Law: $\quad (A \cup B) \cap C = (A \cap C) \cup (B \cap C)$
$\qquad\qquad\qquad\quad A \cup (B \cap C) = (A \cup B) \cap (A \cup C)$

We say that A and B are *disjoint* if and only if there is no common element between A and B, that is, if and only if $A \cap B = \varphi$, an empty set. Two disjoint sets A and B are also referred to as being *mutually exclusive*.

Example 1.2.2 (Example 1.2.1 Continued) One can verify that $A \cup B = \{1, 5, 7, 11, 13\}, B \cup C = \mathbf{S}$, but C and D are mutually exclusive. Also, A and D are mutually exclusive. Note that $A^c = \{3, 9, 11, 13\}, A \triangle C = \{1, 3, 9\}$ and $A \triangle B = \{11, 13\}$. ▲

Now consider $\{A_i; i \in I\}$, a collection of subsets of $\mathbf{S}$. This collection may be finite, countably infinite or uncountably infinite. We define

$$
\begin{array}{ll}
\textit{Union:} & \cup_{i \in I} A_i = \{x : x \in A_i \text{ for at least one } i \in I\}; \\
\textit{Intersection:} & \cap_{i \in I} A_i = \{x : x \in A_i \text{ for all } i \in I\}.
\end{array}
\qquad (1.2.3)
$$

The equation (1.2.3) lays down the set operations involving the union and intersection among arbitrary number of sets. When we specialize $I = \{1, 2, 3, ...\}$ in the definition given by (1.2.3), we can combine the notions of countably infinite number of unions and intersections to come up with some interesting sets. Let us denote

$$
B = \cap_{j=1}^{\infty} \cup_{i=j}^{\infty} A_i \text{ and } C = \cup_{j=1}^{\infty} \cap_{i=j}^{\infty} A_i. \qquad (1.2.4)
$$

Interpretation of the set B: Here the set B is the intersection of the collection of sets $\cup_{i=j}^{\infty} A_i, j \geq 1$. In other words, an element x will belong to B if and only if x belongs to $\cup_{i=j}^{\infty} A_i$ for each $j \geq 1$ which is equivalent to saying that there exists a sequence of positive integers $i_1 < i_2 < ... < i_k < ...$ such that $x \in A_{i_k}$ for all $k = 1, 2, ...$. That is, the set B corresponds to the elements which are *hit infinitely often* and hence B is referred to as the *limit* (as $n \to \infty$) *supremum* of the sequence of sets $A_n, n = 1, 2, ...$.

Interpretation of the set C: On the other hand, the set C is the union of the collection of sets $\cap_{i=j}^{\infty} A_i, j \geq 1$. In other words, an element x will belong to C if and only if x belongs to $\cap_{i=j}^{\infty} A_i$ for some $j \geq 1$ which is equivalent to saying that x belongs to $A_j, A_{j+1}, ...$ for some $j \geq 1$. That is, the set C corresponds to the elements which are *hit eventually* and hence C is referred to as the *limit* (as $n \to \infty$) *infimum* of the sequence of sets $A_n, n = 1, 2, ...$.

$$
\boxed{\cap_{j=1}^{\infty} \cup_{i=j}^{\infty} A_i = \limsup_{n \to \infty} A_n \text{ and } \cup_{j=1}^{\infty} \cap_{i=j}^{\infty} A_i = \liminf_{n \to \infty} A_n.}
$$

Theorem 1.2.1 (DeMorgan's Law) *Consider* $\{A_i; i \in I\}$, *a collection of subsets of* $\mathbf{S}$. *Then,*

$$
\left(\cup_{i \in I} A_i\right)^c = \cap_{i \in I}\left(A_i^c\right) \qquad (1.2.5)
$$

Proof Suppose that an element x belongs to the lhs of (1.2.5). That is, $x \in \mathbf{S}$ but $x \notin \cup_{i \in I} A_i$, which implies that x can not belong to any of the sets $A_i, i \in I$. Hence, the element x must belong to the set $\cap_{i \in I}(A_i^c)$. Thus, we have $\left(\cup_{i \in I} A_i\right)^c \subseteq \cap_{i \in I}(A_i^c)$

Suppose that an element x belongs to the rhs of (1.2.5). That is $x \in A_i^c$ for each $i \in I$, which implies that x can not belong to any of the sets $A_i, i \in I$. In other words, the element x can not belong to the set $\cup_{i \in I} A_i$, so that x must belong to the set $\left(\cup_{i \in I} A_i\right)^c$. Thus, we have $\left(\cup_{i \in I} A_i\right)^c \supseteq \cap_{i \in I}(A_i^c)$. The proof is now complete. ∎

Definition 1.2.1 *The collection of sets $\{A_i; i \in I\}$ is said to consist of disjoint sets if and only if no two sets in this collection share a common element, that is when $A_i \cap A_j = \varphi$ for all $i \neq j \in I$. The collection $\{A_i; i \in I\}$ is called a partition of* **S** *if and only if*

(i) $\{A_i; i \in I\}$ *consists of disjoint sets only, and*

(ii) $\{A_i; i \in I\}$ *spans the whole space* **S**, *that is $\cup_{i \in I} A_i = $* **S**.

Example 1.2.3 Let **S** $=(0, 1]$ and define the collection of sets $\{A_i; i \in I\}$ where $A_i = (\frac{1}{2^i}, \frac{1}{2^{i-1}}], i \in I = \{1, 2, 3, ...\}$. One should check that the given collection of intervals form a partition of $(0, 1]$. ▲

1.3   Axiomatic Development of Probability

The axiomatic theory of probability was developed by Kolmogorov in his 1933 monograph, originally written in German. Its English translation is cited as Kolmogorov (1950b). Before we describe this approach, we need to fix some ideas first. Along the lines of the examples discussed in the Introduction, let us focus on some random experiment in general and state a few definitions.

Definition 1.3.1 *A sample space is a set, denoted by* **S**, *which enumerates each and every possible outcome or simple event.*

In general an *event* is an *appropriate* subset of the sample space **S**, including the empty subset φ and the whole set **S**. In what follows we make this notion more precise.

Definition 1.3.2 *Suppose that $\mathcal{B} = \{A_i : A_i \subseteq $* **S**$, i \in I\}$ *is a collection of subsets of* **S**. *Then, $\mathcal{B}$ is called a Borel sigma-field or Borel sigma-algebra if the following conditions hold:*

(i)   *The empty set $\varphi \in \mathcal{B}$;*

(ii)   *If $A \in \mathcal{B}$, then $A^c \in \mathcal{B}$;*

(iii) *If $A_i \in \mathcal{B}$ for $i = 1, 2, ...$, then $\cup_{i=1}^{\infty} A_i \in \mathcal{B}$.*

In other words, the Borel sigma-field $\mathcal{B}$ is closed under the operations of complement and countable union of its members. It is obvious that the whole space $\mathbf{S}$ belongs to the Borel sigma-field $\mathcal{B}$ since we can write $\mathbf{S} = \varphi^c \in \mathcal{B}$, by the requirement (i)-(ii) in the Definition 1.3.2. Also if $A_i \in \mathcal{B}$ for $i = 1, 2, ..., k$, then $\cup_{i=1}^{k} A_i \in \mathcal{B}$, since with $A_i = \varphi$ for $i = k+1, k+2, ...$, we can express $\cup_{i=1}^{k} A_i$ as $\cup_{i=1}^{\infty} A_i$ which belongs to $\mathcal{B}$ in view of (iii) in the Definition 1.3.2. That is, $\mathcal{B}$ is obviously closed under the operation of finite unions of its members. See the Exercise 1.3.1 in this context.

Definition 1.3.3 *Suppose that a fixed collection of subsets $\mathcal{B} = \{A_i : A_i \subseteq \mathbf{S}, i \in I\}$ is a Borel sigma-field. Then, any subset A of $\mathbf{S}$ is called an event if and only if $A \in \mathcal{B}$.*

Frequently, we work with the Borel sigma-field which consists of all subsets of the sample space $\mathbf{S}$ but always it may not necessarily be that way. Having started with a fixed Borel sigma-field $\mathcal{B}$ of subsets of $\mathbf{S}$, a *probability scheme* is simply a way to assign numbers between zero and one to every event while such assignment of numbers must satisfy some general guidelines. In the next definition, we provide more specifics.

Definition 1.3.4 *A probability scheme assigns a unique number to a set $A \in \mathcal{B}$, denoted by $P(A)$, for every set $A \in \mathcal{B}$ in such a way that the following conditions hold:*

$$
\begin{aligned}
&(i) \quad P(\mathbf{S}) = 1; \\
&(ii) \quad P(A) \geq 0 \text{ for every } A \in \mathcal{B}; \\
&(iii) \quad P(\cup_{i=1}^{\infty} A_i) = \Sigma_{i=1}^{\infty} P(A_i) \text{ for all } A_i \in \mathcal{B} \text{ such that} \\
&\qquad A_i\text{'s are all pairwise disjoint, } i = 1, 2, ... \, .
\end{aligned}
\qquad (1.3.1)
$$

Now, we are in a position to claim a few basic results involving probability. Some are fairly intuitive while others may need more attention.

Theorem 1.3.1 *Suppose that A and B are any two events and recall that φ denotes the empty set. Suppose also that the sequence of events $\{B_i; i \geq 1\}$ forms a partition of the sample space $\mathbf{S}$. Then,*

$(i) \quad P(\varphi) = 0 \text{ and } P(A) \leq 1;$

$(ii) \quad P(A^c) = 1 - P(A);$

$(iii) \quad P(B \cap A^c) = P(B) - P(B \cap A);$

$(iv) \quad P(A \cup B) = P(A) + P(B) - P(A \cap B);$

$(v) \quad \text{If } A \subseteq B, \text{ then } P(A) \leq P(B);$

$(vi) \quad P(A) = \Sigma_{i=1}^{\infty} P(A \cap B_i).$

Proof (i) Observe that $\varphi \cup \varphi^c = \mathbf{S}$ and also φ, φ^c are disjoint events. Hence, by part (iii) in the Definition 1.3.4, we have $1 = P(\mathbf{S}) = P(\varphi \cup \varphi^c) = P(\varphi) + P(\varphi^c)$. Thus, $P(\varphi) = 1 - P(\varphi^c) = 1 - P(\mathbf{S}) = 1 - 1 = 0$, in view of part (i) in the Definition 1.3.4. The second part follows from part (ii). ◆

(ii) Observe that $A \cup A^c = \mathbf{S}$ and then proceed as before. Observe that A and A^c are disjoint events. ◆

(iii) Notice that $B = (B \cap A) \cup (B \cap A^c)$ where $B \cap A$ and $B \cap A^c$ are disjoint events. Hence by part (iii) in the Definition 1.3.4, we claim that

$$P(B) = P\{(B \cap A) \cup (B \cap A^c)\} = P(B \cap A) + P(B \cap A^c).$$

Now, the result is immediate. ◆

(iv) It is easy to verify that $A \cup B = (A \cap B^c) \cup (B \cap A^c) \cup (A \cap B)$ where the three events $A \cap B^c, B \cap A^c, A \cap B$ are also disjoint. Thus, we have

$$P(A \cup B)$$
$$= P(A \cap B^c) + P(B \cap A^c) + P(A \cap B)$$

in view of part (iii), Definition 1.3.4

$$= \{P(A) - P(A \cap B)\} + \{P(B) - P(A \cap B)\} + P(A \cap B),$$

which leads to the desired result. ◆

(v) We leave out its proof as the Exercise 1.3.4. ◆

(vi) Since the sequence of events $\{B_i; i \geq 1\}$ forms a partition of the sample space $\mathbf{S}$, we can write $A = A \cap \mathbf{S} = A \cap (\cup_{i=1}^{\infty} B_i) = \cup_{i=1}^{\infty} (A \cap B_i)$ where the events $A \cap B_i, i = 1, 2, \dots$ are also disjoint. Now, the result follows from part (iii) in the Definition 1.3.4. ■

Example 1.3.1 (Example 1.1.1 Continued) Let us define three events as follows:

$$A : \quad \text{We observe two heads}$$
$$B : \quad \text{We observe at least one tail} \qquad (1.3.2)$$
$$C : \quad \text{We observe no heads}$$

How can we obtain the probabilities of these events? First notice that as subsets of $\mathbf{S}$, we can rewrite these events as $A = \{HHT, HTH, THH\}, B = \{HHT, HTH, HTT, THH, THT, TTH, TTT\}$, and $C = \{TTT\}$. Now it becomes obvious that $P(A) = \frac{3}{8}, P(B) = \frac{7}{8}$, and $P(C) = \frac{1}{8}$. One can also see that $A \cap B = \{HHT, HTH, THH\}$ so that $P(A \cap B) = \frac{3}{8}$, whereas $A \cup C = \{HHT, HTH, THH, TTT\}$ so that $P(A \cup C) = \frac{4}{8} = \frac{1}{2}$. ▲

Example 1.3.2 (Example 1.1.2 Continued) Consider the following events:

$$D : \quad \text{The total from the red and yellow dice is 8}$$
$$E : \quad \text{The red die comes up with 2 more than the yellow die} \qquad (1.3.3)$$

Now, as subsets of the corresponding sample space $\mathbf{S}$, we can rewrite these events as $D = \{26, 35, 44, 53, 62\}$ and $E = \{31, 42, 53, 64\}$. It is now obvious that $P(D) = \frac{5}{36}$ and $P(E) = \frac{4}{36} = \frac{1}{9}$. ▲

Example 1.3.3 In a college campus, suppose that 2600 are women out of 4000 undergraduate students, while 800 are men among 2000 undergraduates who are under the age 25. From this population of undergraduate students if one student is selected at random, what is the probability that the student will be either a man or be under the age 25? Define two events as follows

A :   the selected undergradute student is male
B :   the selected undergradute student is under the age 25

and observe that $P(A) = 1400/4000, P(B) = 2000/4000, P(A \cap B) = 800/4000$. Now, apply the Theorem 1.3.3, part (iv), to write $P(A \cup B) = P(A) + P(B) - P(A \cap B) = (1400 + 2000 - 800)/4000 = 13/20.$ ▲

> Having a sample space **S** and appropriate events from a Borel sigma-field $\mathcal{B}$ of subsets of **S**, and a probability scheme satisfying (1.3.1), one can evaluate the probability of the legitimate events only. The members of $\mathcal{B}$ are the only legitimate events.

1.4   The Conditional Probability and Independent Events

Let us reconsider the Example 1.3.2. Suppose that the two fair dice, one red and the other yellow, are tossed in another room. After the toss, the experimenter comes out to announce that the event D has been observed. Recall that $P(E)$ was $\frac{1}{9}$ to begin with, but we know now that D has happened, and so the probability of the event E should be appropriately updated. Now then, how should one revise the probability of the event E, given the additional information?

The basic idea is simple: when we are told that the event D has been observed, then D should take over the role of the "sample space" while the original sample space **S** should be irrelevant at this point. In order to evaluate the probability of the event E in this situation, one should simply focus on the portion of E which is inside the set D. This is the fundamental idea behind the concept of *conditioning*.

Definition 1.4.1 *Let* **S** *and* $\mathcal{B}$ *be respectively the sample space and the Borel sigma-field. Suppose that A, B are two arbitrary events. The conditional probability of the event A given the other event B, denoted by $P(A \mid B)$, is defined as*

$$P(A \mid B) = \frac{P(A \cap B)}{P(B)} \quad \text{provided that } P(B) > 0. \tag{1.4.1}$$

In the same vein, we will write $P(B \mid A) = P(A \cap B)/P(A)$ provided that $P(A) > 0$.

Definition 1.4.2 *Two arbitrary events A and B are defined independent if and only if $P(A \mid B) = P(A)$, that is having the additional knowledge that B has been observed has not affected the probability of A, provided that $P(B) > 0$. Two arbitrary events A and B are then defined dependent if and only if $P(A \mid B) \neq P(A)$, in other words knowing that B has been observed has affected the probability of A, provided that $P(B) > 0$.*

In case the two events A and B are independent, intuitively it means that the occurrence of the event A (or B) does not affect or influence the probability of the occurrence of the other event B (or A). In other words, the occurrence of the event A (or B) yields no reason to alter the likelihood of the other event B (or A).

When the two events A, B are dependent, sometimes we say that B is *favorable* to A if and only if $P(A \mid B) > P(A)$ provided that $P(B) > 0$. Also, when the two events A, B are dependent, sometimes we say that B is *unfavorable* to A if and only if $P(A \mid B) < P(A)$ provided that $P(B) > 0$.

Example 1.4.1 (Example 1.3.2 Continued) Recall that $P(D \cap E) = P(53) = \frac{1}{36}$ and $P(D) = \frac{5}{36}$, so that we have $P(E \mid D) = P(D \cap E)/P(D) = \frac{1}{5}$. But, $P(E) = \frac{1}{9}$ which is different from $P(E \mid D)$. In other words, we conclude that D and E are two dependent events. Since $P(E \mid D) > P(E)$, we may add that the event D is favorable to the event E. ▲

The proof of the following theorem is left as the Exercise 1.4.1.

Theorem 1.4.1 *The two events B and A are independent if and only if A and B are independent. Also, the two events A and B are independent if and only if $P(A \cap B) = P(A)P(B)$.*

We now state and prove another interesting result.

Theorem 1.4.2 *Suppose that A and B are two events. Then, the following statements are equivalent:*

(i)   *The events A and B are independent;*
(ii)   *The events A^c and B are independent;*
(iii) *The events A and B^c are independent;*
(iv)   *The events A^c and B^c are independent.*

Proof It will suffice to show that part $(i) \Rightarrow$ part $(ii) \Rightarrow$ part $(iii) \Rightarrow$ part $(iv) \Rightarrow$ part (i).

(i) $\Rightarrow$ (ii) : Assume that A and B are independent events. That is, in view of the Theorem 1.4.1, we have $P(A \cap B) = P(A)P(B)$. Again in view of the Theorem 1.4.1, we need to show that $P(A^c \cap B) = P(A^c)P(B)$. Now, we apply parts (ii)-(iii) from the Theorem 1.3.1 to write

$$P(A^c \cap B) = P(B) - P(A \cap B) = P(B)\{1 - P(A)\} = P(B)P(A^c).$$

(ii) $\Rightarrow$ (iii) $\Rightarrow$ (iv) : These are left as the Exercise 1.4.2.

(iv) $\Rightarrow$ (i) : Assume that A^c and B^c are independent events. That is, in view of the Theorem 1.4.1, we have $P(A^c \cap B^c) = P(A^c)P(B^c)$. Again in view of the Theorem 1.4.1, we need to show that $P(A \cap B) = P(A)P(B)$. Now, we combine DeMorgan's Law from (1.2.5) as well as the parts (ii)-(iv) from the Theorem 1.3.1 to write

$$\begin{aligned}
P(A \cap B) \\
&= 1 - P(A^c \cup B^c) \\
&= 1 - \{P(A^c) + P(B^c) - P(A^c \cap B^c)\} \\
&= 1 - P(A^c) - P(B^c) + P(A^c)P(B^c) \\
&= \{1 - P(A^c)\}\{1 - P(B^c)\},
\end{aligned}$$

which is the same as $P(A)P(B)$, the desired claim. ∎

Definition 1.4.3 *A collection of events $A_1, ..., A_n$ are called mutually independent if and only if every sub-collection consists of independent events, that is*

$$P(\cap_{j=1}^{k} A_{i_j}) = \Pi_{j=1}^{k} P(A_{i_j})$$

for all $1 \leq i_1 < i_2 < ... < i_k \leq n$ and $2 \leq k \leq n$.

A collection of events $A_1, ..., A_n$ may be pairwise independent, that is, any two events are independent according to the Definition 1.4.2, but the whole collection of sets may not be mutually independent. See the Example 1.4.2.

Example 1.4.2 Consider the random experiment of tossing a fair coin twice. Let us define the following events:

A_1 :   Observe a head (H) on the first toss
A_2 :   Observe a head (H) on the second toss
A_3 :   Observe the same outcome on both tosses

The sample space is given by $\mathbf{S} = \{HH, HT, TH, TT\}$ with each outcome being equally likely. Now, rewrite $A_1 = \{HH, HT\}$, $A_2 = \{HH, TH\}$, $A_3 = \{HH, TT\}$. Thus, we have $P(A_1) = P(A_2) = P(A_3) = \frac{1}{2}$. Now, $P(A_1 \cap A_2) = P(HH) = \frac{1}{4} = P(A_1)P(A_2)$, that is the two events A_1, A_2 are independent. Similarly, one should verify that A_1, A_3 are independent, and so are also A_2, A_3. But, observe that $P(A_1 \cap A_2 \cap A_3) = P(HH) = \frac{1}{4}$ and it is not the same as $P(A_1)P(A_2)P(A_3)$. In other words, the three events A_1, A_2, A_3 are not mutually independent, but they are pairwise independent. ▲

1.4.1 Calculus of Probability

Suppose that A and B are two arbitrary events. Here we summarize some of the standard rules involving probabilities.

Additive Rule:

$$P(A \cup B) = P(A) + P(B) - P(A \cap B) \qquad (1.4.2)$$

Conditional Probability Rule:

$$P(A \mid B) = P(A \cap B)/P(B) \text{ provided that } P(B) > 0 \qquad (1.4.3)$$

Multiplicative Rule:

$$P(A \cap B) = P(A)P(B \mid A) \qquad (1.4.4)$$

The *additive rule* was proved in the Theorem 1.3.1, part (iv). The *conditional probability rule* is a restatement of (1.4.1). The *multiplicative rule* follows easily from (1.4.1). Sometimes the experimental setup itself may directly indicate the values of the various conditional probabilities. In such situations, one can obtain the probabilities of joint events such as $A \cap B$ by cross multiplying the two sides in (1.4.1). At this point, let us look at some examples.

Example 1.4.3 Suppose that a company publishes two magazines, M1 and M2. Based on their record of subscriptions in a suburb they find that sixty percent of the households subscribe only for M1, forty five percent subscribe only for M2, while only twenty percent subscribe for both M1 and M2. If a household is picked at random from this suburb, the magazines' publishers would like to address the following questions: What is the probability that the randomly selected household is a subscriber for (i) at least one of the magazines M1, M2, (ii) none of those magazines M1, M2, (iii) magazine M2 given that the same household subscribes for M1? Let us now define the two events

A : The randomly selected household subscribes for the magazine M1
B : The randomly selected household subscribes for the magazine M2

We have been told that $P(A) = .60, P(B) = .45$ and $P(A \cap B) = .20$. Then, $P(A \cup B) = P(A) + P(B) - P(A \cap B) = .85$, that is there is 85% chance that the randomly selected household is a subscriber for at least one of the magazines M1, M2. This answers question (i). In order to answer question (ii), we need to evaluate $P(A^c \cap B^c)$ which can simply be written as $1 - P(A \cup B) = .15$, that is there is 15% chance that the randomly selected household subscribes for none of the magazines M1, M2. Next, to answer question (iii), we obtain $P(B \mid A) = P(A \cap B)/P(A) = \frac{1}{3}$, that is there is one-in-three chance that the randomly selected household subscribes for the magazine M2 given that this household already receives the magazine M1. ▲

> In the Example 1.4.3, $P(A \cap B)$ was given to us, and so we could use (1.4.3) to find the conditional probability $P(B \mid A)$.

Example 1.4.4 Suppose that we have an urn at our disposal which contains eight green and twelve blue marbles, all of equal size and weight. The following random experiment is now performed. The marbles inside the urn are mixed and then one marble is picked from the urn at random, but we do not observe its color. This first drawn marble is not returned to the urn. The remaining marbles inside the urn are again mixed and one marble is picked at random. This kind of selection process is often referred to as *sampling without replacement*. Now, what is the probability that (i) both the first and second drawn marbles are green, (ii) the second drawn marble is green? Let us define the two events

A : The randomly selected first marble is green

B : The randomly selected second marble is green

Obviously, $P(A) = \frac{8}{20} = .4$ and $P(B \mid A) = \frac{7}{19}$. Observe that the experimental setup itself dictates the value of $P(B \mid A)$. A result such as (1.4.3) is not very helpful in the present situation. In order to answer question (i), we proceed by using (1.4.4) to evaluate $P(A \cap B) = P(A)P(B \mid A) = \frac{8}{20}\frac{7}{19} = \frac{14}{95}$. Obviously, $\{A, A^c\}$ forms a partition of the sample space. Now, in order to answer question (ii), using the Theorem 1.3.1, part (vi), we write

$$P(B) = P(A \cap B) + P(A^c \cap B), \qquad (1.4.5)$$

But, as before we have $P(A^c \cap B) = P(A^c)P(B \mid A^c) = \frac{12}{20}\frac{8}{19} = \frac{24}{95}$. Thus, from (1.4.5), we have $P(B) = \frac{14}{95} + \frac{24}{95} = \frac{38}{95} = \frac{2}{5} = .4$. Here, note that $P(B \mid A) \neq P(B)$ and so by the Definition 1.4.2, the two events A, B are dependent. One may *guess* this fact easily from the layout of the experiment itself. The reader should check that $P(A)$ would be equal to $P(B)$ whatever be the configuration of the urn. Refer to the Exercise 1.4.11. ▲

> In the Example 1.4.4, $P(B \mid A)$ was known to us, and so we could use (1.4.4) to find the joint probability $P(A \cap B)$.

1.4.2 Bayes's Theorem

We now address another type of situation highlighted by the following example.

Example 1.4.5 Suppose in another room, an experimenter has two urns at his disposal, urn #1 and urn #2. The urn #1 has eight green and twelve blue marbles whereas the urn #2 has ten green and eight blue marbles, all of same size and weight. The experimenter selects one of the urns at random with equal probability and from the selected urn picks a marble at random. It is announced that the selected marble in the other room turned out blue. What is the probability that the blue marble was chosen from the urn #2? We will answer this question shortly. ▲

The following theorem will be helpful in answering questions such as the one raised in the Example 1.4.5.

Theorem 1.4.3 (Bayes's Theorem) *Suppose that the events* $\{A_1, ..., A_k\}$ *form a partition of the sample space* **S** *and* B *is another event. Then,*

$$P(A_j \mid B) = \frac{P(A_j)P(B \mid A_j)}{\Sigma_{i=1}^{k} P(A_i)P(B \mid A_i)}, \text{ for fixed } j = 1, ..., k.$$

Proof Since $\{A_1, ..., A_k\}$ form a partition of **S**, in view of the Theorem 1.3.1, part (vi) we can immediately write

$$P(B) = \Sigma_{i=1}^{k} P(B \cap A_i) = \Sigma_{i=1}^{k} P(A_i)P(B \mid A_i), \tag{1.4.6}$$

by using (1.4.4). Next, using (1.4.4) once more, let us write

$$P(A_j \mid B) = P(A_j \cap B)/P(B). \tag{1.4.7}$$

The result follows by combining (1.4.6) and (1.4.7). ∎

This marvelous result and the ideas originated from the works of Rev. Thomas Bayes (1783). In the statement of the Theorem 1.4.3, note that the conditioning events on the rhs are $A_1, ..., A_k$, but on the lhs one has the conditioning event B instead. The quantities such as $P(A_i), i = 1, ..., k$ are often referred to as the *apriori* or *prior probabilities*, whereas $P(A_j \mid B)$ is referred to as the *posterior probability*. In Chapter 10, we will have more opportunities to elaborate the related concepts.

Example 1.4.6 (Example 1.4.5 Continued) Define the events

A_i : The urn #i is selected, $i = 1, 2$

B : The marble picked from the selected urn is blue

It is clear that $P(A_i) = \frac{1}{2}$ for $i = 1, 2$, whereas we have $P(\,B \mid A_1) = \frac{12}{20}$ and $P(\,B \mid A_2) = \frac{8}{18}$. Now, applying the Bayes Theorem, we have

$$P(A_2 \mid B)$$
$$= \{P(A_2)P(B \mid A_2)\}/\{P(A_1)P(B \mid A_1) + P(A_2)P(B \mid A_2)\}$$
$$= \{(\tfrac{1}{2})(\tfrac{8}{18})\}/\{(\tfrac{1}{2})(\tfrac{12}{20}) + (\tfrac{1}{2})(\tfrac{8}{18})\}$$

which simplifies to $\frac{20}{47}$. Thus, the chance that the randomly drawn blue marble came from the urn #2 was $\frac{20}{47}$ which is equivalent to saying that the chance of the blue marble coming from the urn #1 was $\frac{27}{47}$. ▲

> The Bayes Theorem helps in finding the conditional probabilities when the original conditioning events $A_1, ..., A_k$ and the event B reverse their roles.

Example 1.4.7 This example has more practical flavor. Suppose that 40% of the individuals in a population have some disease. The diagnosis of the presence or absence of this disease in an individual is reached by performing a type of blood test. But, like many other clinical tests, this particular test is not perfect. The manufacturer of the blood-test-kit made the accompanying information available to the clinics. If an individual has the disease, the test indicates the absence (*false negative*) of the disease 10% of the time whereas if an individual does not have the disease, the test indicates the presence (*false positive*) of the disease 20% of the time. Now, from this population an individual is selected at random and his blood is tested. The health professional is informed that the test indicated the presence of the particular disease. What is the probability that this individual does indeed have the disease? Let us first formulate the problem. Define the events

A_1 : The individual has the disease
A_1^c : The individual does not have the disease
B : The blood test indicates the presence of the disease

Suppose that we are given the following information: $P(A_1) = .4, P(A_1^c) = .6, P(B \mid A_1) = .9$, and $P(B \mid A_1^c) = .2$. We are asked to calculate the conditional probability of A_1 given B. We denote $A_2 = A_1^c$ and use the Bayes Theorem. We have

$$P(A_1 \mid B) = \frac{P(A_1)P(B \mid A_1)}{P(A_1)P(B \mid A_1) + P(A_2)P(B \mid A_2)} = \frac{(.4)(.9)}{(.4)(.9) + (.6)(.2)},$$

which is $\frac{3}{4}$. Thus there is 75% chance that the tested individual has the disease if we know that the blood test had indicated so. ▲

1.4.3   Selected Counting Rules

In many situations, the sample space **S** consists of only a finite number of equally likely outcomes and the associated Borel sigma-field $\mathcal{B}$ is the collection of all possible subsets of **S**, including the empty set φ and the whole set **S**. Then, in order to find the probability of an event A, it will be important to *enumerate* all the possible outcomes included in the sample space **S** and the event A. This section reviews briefly some of the customary counting rules followed by a few examples.

The Fundamental Rule of Counting: Suppose that there are k different tasks where the ith task can be completed in n_i ways, $i = 1, ..., k$. Then, the total number of ways these k tasks can be completed is given by $\Pi_{i=1}^{k} n_i$.

Permutations: The word *permutation* refers to *arrangements* of some objects taken from a collection of *distinct* objects.

> The *order* in which the objects are laid out *is* important here.

Suppose that we have n distinct objects. The number of ways we can arrange k of these objects, denoted by the symbol ^nP_k, is given by

$$^nP_k = n(n-1)(n-2)...(n-k+1) \text{ or } \Pi_{i=1}^{k}(n-i+1). \qquad (1.4.8)$$

The number of ways we can arrange all n objects is given by ^nP_n which is denoted by the special symbol

$$n! = n(n-1)...(3)(2)(1). \qquad (1.4.9)$$

We describe the symbol $n!$ as the "n factorial". We use the convention to interpret $0! = 1$.

Combinations: The word combination refers to the *selection* of some objects from a set of *distinct* objects without regard to the order.

> The *order* in which the objects are laid out *is not* important here.

Suppose that we have n distinct objects. The number of ways we can select k of these objects, denoted by the symbol $\binom{n}{k}$, is given by

$$\binom{n}{k} = \frac{n!}{k!(n-k)!}. \qquad (1.4.10)$$

But, observe that we can write $n! = n(n-1)...(n-k+1)(n-k)!$ and hence we can rewrite (1.4.10) as

$$\binom{n}{k} = \frac{n(n-1)(n-2)...(n-k+1)}{k!} = \frac{^nP_k}{k!}. \qquad (1.4.11)$$

We use the convention to interpret $\binom{n}{0} = 1$, whatever be the positive integer n.

The following result uses these combinatorial expressions and it goes by the name

Theorem 1.4.4 (Binomial Theorem) *For any two real numbers a, b and a positive integer n, one has*

$$(a + b)^n = \Sigma_{k=0}^n \binom{n}{k} a^k b^{n-k}. \tag{1.4.12}$$

Example 1.4.8 Suppose that a fair coin is tossed five times. Now the outcomes would look like $HHTHT, THTTH$ and so on. By the fundamental rule of counting we realize that the sample space **S** will consist of $2^5 (= 2 \times 2 \times 2 \times 2 \times 2)$, that is thirty two, outcomes each of which has five components and these are equally likely. How many of these five "dimensional" outcomes would include two heads? Imagine five distinct positions in a row and each position will be filled by the letter H or T. Out of these five positions, choose two positions and fill them both with the letter H while the remaining three positions are filled with the letter T. This can be done in $\binom{5}{2}$ ways, that is in $(5)(4)/2 = 10$ ways. In other words we have $P(\text{Two Heads}) = \binom{5}{2}/2^5 = 10/32 = 5/16$. ▲

Example 1.4.9 There are ten students in a class. In how many ways can the teacher form a committee of four students? In this selection process, naturally the order of selection is *not* pertinent. A committee of four can be chosen in $\binom{10}{4}$ ways, that is in $(10)(9)(8)(7)/4! = 210$ ways. The sample space **S** would then consist of 210 equally likely outcomes. ▲

Example 1.4.10 (Example 1.4.9 Continued) Suppose that there are six men and four women in the small class. Then what is the probability that a randomly selected committee of four students would consist of two men and two women? Two men and two women can be chosen in $\binom{6}{2}\binom{4}{2}$ ways, that is in 90 ways. In other words, $P(\text{Two men and two women are selected}) = \binom{6}{2}\binom{4}{2}/\binom{10}{4} = 90/210 = 3/7$. ▲

Example 1.4.11 John, Sue, Rob, Dan and Molly have gone to see a movie. Inside the theatre, they picked a row where there were exactly five empty seats next to each other. These five friends can sit in those chairs in exactly 5! ways, that is in 120 ways. Here, seating arrangement is naturally pertinent. The sample space **S** consists of 120 equally likely outcomes. But the usher does not know in advance that John and Molly must sit next to each other. If the usher lets the five friends take those seats randomly, what is the probability that John and Molly would sit next to each other? John and Molly may occupy the first two seats and other three friends may permute in 3! ways in the remaining chairs. But then John and Molly may occupy the second and third or the third and fourth or the fourth and fifth seats while the other three friends would take the remaining three seats in any order. One can also permute John and Molly in 2! ways. That is, we have P(John and Molly sit next to each other) $= (2!)(4)(3!)/5! = 2/5$. ▲

Example 1.4.12 Consider distributing a standard pack of 52 cards to four players (North, South, East and West) in a *bridge game* where each player gets 13 cards. Here again the ordering of the cards is *not* crucial to the game. The total number of ways these 52 cards can be distributed among the four players is then given by $n = \binom{52}{13}\binom{39}{13}\binom{26}{13}\binom{13}{13}$. Then, P(North will receive 4 aces and 4 kings) $= \binom{4}{4}\binom{4}{4}\binom{44}{5}\binom{39}{13}\binom{26}{13}\binom{13}{13}/n = \binom{44}{5}/\binom{52}{13} = \frac{11}{6431950} \approx 1.7102 \times 10^{-6}$, because North is given 4 aces and 4 kings plus 5 other cards from the remaining 44 cards while the remaining 39 cards are distributed equally among South, East and West. By the same token, P(South will receive exactly one ace) $= \binom{4}{1}\binom{48}{12}\binom{39}{13}\binom{26}{13}\binom{13}{13}/n = 4\binom{48}{12}/\binom{52}{13} = \frac{9139}{20825} \approx .43885$, since South would receive one out of the 4 aces and 12 other cards from 48 non-ace cards while the remaining 39 cards are distributed equally among North, East and West. ▲

1.5   Discrete Random Variables

A *discrete random variable*, commonly denoted by X, Y and so on, takes a finite or countably infinite number of possible values with specific probabilities associated with each value. In a collection agency, for example, the manager may look at the pattern of the number (X) of delinquent accounts. In a packaging plant, for example, one may be interested in studying the pattern of the number (X) of the defective packages. In a truckload of shipment, for example, the receiving agent may want to study the pattern of the number (X) of rotten oranges or the number (X) of times the shipment arrives late. One may ask: how many times (X) one must drill in a oil field in order to hit oil? These are some typical examples of *discrete* random variables.

In order to illustrate further, let us go back to the Example 1.1.2. Suppose that X is the total score from the red and yellow dice. The possible values of X would be any number from the set consisting of $2, 3, ..., 11, 12$. We had already discussed earlier how one could proceed to evaluate $P(X = x), x = 2, ..., 12$. In this case, the event $[X = 2]$ corresponds to the subset $\{11\}$, the event $[X = 3]$ corresponds to the subset $\{12, 21\}$, the event $[X = 4]$ corresponds to the subset $\{13, 22, 31\}$, and so on. Thus, $P(X = 2) = \frac{1}{36}, P(X = 3) = \frac{2}{36}, P(X = 4) = \frac{3}{36}$, and so on. The reader should easily verify the following entries:

$$
\begin{array}{lccccccccccc}
x: & 2 & 3 & 4 & 5 & 6 & 7 & 8 & 9 & 10 & 11 & 12 \\
P(X=x): & \frac{1}{36} & \frac{2}{36} & \frac{3}{36} & \frac{4}{36} & \frac{5}{36} & \frac{6}{36} & \frac{5}{36} & \frac{4}{36} & \frac{3}{36} & \frac{2}{36} & \frac{1}{36}
\end{array}
$$
(1.5.1)

Here, the set of the possible values for the random variable X happens to be *finite*.

On the other hand, when tossing a fair coin, let Y be the number of tosses of the coin required to observe the first head (H) to come up. Then, $P(Y = 1) = P(\text{The } H \text{ appears in the first toss itself}) = P(H) = \frac{1}{2}$, and $P(Y = 2) = P(\text{The first } H \text{ appears in the second toss}) = P(TH) = (\frac{1}{2})(\frac{1}{2}) = \frac{1}{4}$. Similarly, $P(Y = 3) = P(TTH) = \frac{1}{8}, ...$, that is

$$
P(Y = y) = (\tfrac{1}{2})^y, y = 1, 2, ...
$$

Here, the set of the possible values for the random variable Y is *countably infinite*.

1.5.1   Probability Mass and Distribution Functions

In general, a random variable X is a mapping (that is, a function) from the sample space $\mathbf{S}$ to a subset $\mathcal{X}$ of the real line $\Re$ which amounts to saying that the random variable X induces events $(\in \mathcal{B})$ in the context of $\mathbf{S}$. We may express this by writing $X : \mathbf{S} \to \mathcal{X}$. In the discrete case, suppose that X takes the possible values $x_1, x_2, x_3, ...$ with the respective probabilities $p_i = P(X = x_i), i = 1, 2, ...$. Mathematically, we evaluate $P(X = x_i)$ as follows:

$$
p_i = P(X = x_i) = \sum_{s \in \mathbf{S}: X(s) = x_i} P(s), i = 1, 2, ...
$$
(1.5.2)

In (1.5.1), we found $P(X = i)$ for $i = 2, 3, 4$ by following this approach whereas the space $\mathcal{X} = \{2, 3, ..., 12\}$ and $\mathbf{S} = \{11, ..., 16, 21, ..., 61, ..., 66\}$.

While assigning or evaluating these probabilities, one has to make sure that the following two conditions are satisfied:

$$
\begin{array}{ll}
i) & p_i \geq 0 \text{ for all } i = 1, 2, ..., \text{ and} \\
ii) & \Sigma_{i=1}^{\infty} p_i = 1.
\end{array}
$$
(1.5.3)

When both these conditions are met, we call an assignment such as

$$
\begin{array}{lccccccc}
X \text{ values}: & x_1 & x_2 & . & . & . & x_i & . & . & . \\
\text{Probabilities}: & p_1 & p_2 & . & . & . & p_i & . & . & .
\end{array}
$$
(1.5.4)

a *discrete distribution* or a *discrete probability distribution* of the random variable X.

The function $f(x) = P(X = x)$ for $x \in \mathcal{X} = \{x_1, x_2, ..., x_i, ...\}$ is customarily known as the *probability mass function* (pmf) of X.

We also define another useful function associated with a random variable X as follows:

$$F(x) = P\{X \leq x\} = P\{s : s \in \mathbf{S} \text{ such that } X(s) \leq x\}, \quad x \in \Re, \quad (1.5.5)$$

which is customarily called the *distribution function* (df) or the *cumulative distribution function* (cdf) of X. Sometimes we may instead write $F_X(x)$ for the df of the random variable X.

A *distribution function* (df) or *cumulative distribution function* (cdf) $F(x)$ for the random variable X is defined for all real numbers x

Once the pmf $f(x)$ is given as in case of (1.5.4), one can find the probabilities of events which are defined through the random variable X. If we denote a set $A(\subseteq \Re)$, then

$$P(X \in A) = \sum_{i:x_i \in A \cap \mathcal{X}} P(X = x_i). \qquad (1.5.6)$$

Example 1.5.1 Suppose that X is a discrete random variable having the following probability distribution:

$$\begin{array}{lccc} X \text{ values :} & 1 & 2 & 4 \\ \text{Probabilities :} & .2 & .4 & .4 \end{array} \qquad (1.5.7)$$

The associated df is then given by

$$F(x) = \begin{cases} 0 & \text{if} & x < 1 \\ .2 & \text{if} & 1 \leq x < 2 \\ .6 & \text{if} & 2 \leq x < 4 \\ 1.0 & \text{if} & 4 \leq x. \end{cases} \qquad (1.5.8)$$

This df looks like the step function in the Figure 1.5.1. From the Figure 1.5.1, it becomes clear that the jump in the value of $F(x)$ at the points $x = 1, 2, 4$ respectively amounts to $.2, .4$ and $.4$. These jumps obviously correspond to the assigned values of $P(X = 1), P(X = 2)$ and $P(X = 4)$. Also, the df is non-decreasing in x and it is discontinuous at the points

$x = 1, 2, 4$, namely at those points where the probability distribution laid out in (1.5.7) puts any positive mass. ▲

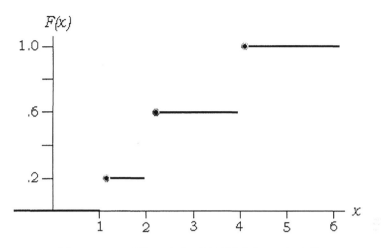

Figure 1.5.1. Plot of the DF $F(x)$ from (1.5.8)

Example 1.5.2 For the random variable X whose pmf is given by (1.5.1), the df is easily verified to be as follows:

x	$F(x)$	x	$F(x)$
$-\infty < x < 2$	0	$7 \leq x < 8$	21/36
$2 \leq x < 3$	1/36	$8 \leq x < 9$	26/36
$3 \leq x < 4$	3/36	$9 \leq x < 10$	30/36
$4 \leq x < 5$	6/36	$10 \leq x < 11$	33/36
$5 \leq x < 6$	10/36	$11 \leq x < 12$	35/36
$6 \leq x < 7$	15/36	$12 \leq x < \infty$	1

From this display, again few facts about the df $F(x)$ become clear. The df is non-decreasing in x and it is discontinuous at the points $x = 2, 3, ..., 12$, namely at those points where the probability distribution laid out in (1.5.1) puts any positive mass. Also, the amount of jump of the df $F(x)$ at the point $x = 2$ equals $\frac{1}{36}$ which corresponds to $P(X = 2)$, and the amount of jump of $F(x)$ at the point $x = 3$ equals $\frac{3}{36} - \frac{1}{36} = \frac{2}{36}$ which corresponds to $P(X = 3)$, and so on. ▲

In the two preceding examples, we gave tips on how the pmf could be constructed from the expression of the df $F(x)$. Suppose that we write

$$F(x-) = \lim F(h) \text{ as } h \uparrow x, \text{ for } x \in \Re. \qquad (1.5.9)$$

This $F(x-)$ is the limit of $F(h)$ as h converges to x from the left hand side of x. In the Example 1.5.1, one can easily verify that $F(1-) = 0, F(2-) = .2$ and $F(4-) = .6$. In other words, in this example, the *jump* of the df or the cdf $F(x)$ at the point

$$x = 1 \text{ amounts to } F(1) - F(1-) \text{ which is } P(X = 1) = .2,$$
$$x = 2 \text{ amounts to } F(2) - F(2-) \text{ which is } P(X = 2) = .4, \qquad (1.5.10)$$
$$x = 4 \text{ amounts to } F(4) - F(4-) \text{ which is } P(X = 4) = .4.$$

There is this natural correspondence between the jumps of a df, the left-limit of a df, and the associated pmf. A similar analysis in the case of the Example 1.5.2 is left out as the Exercise 1.5.4.

> For a discrete random variable X, one can obtain $P(X = x)$ as $F(x) - F(x-)$ where the left-limit $F(x-)$ comes from (1.5.9).

Example 1.5.3 (Example 1.5.2 Continued) For the random variable X whose pmf is given by (1.5.1), the probability that the total from the two dice will be smaller than 4 or larger than 9 can be found as follows. Let us denote an event $A = \{X < 4 \cup X > 9\}$ and we exploit (1.5.6) to write $P(A) = P(X < 4) + P(X > 9) = P(X = 2, 3) + P(X = 10, 11, 12) = \frac{9}{36} = \frac{1}{4}$. What is the probability that the total from the two dice will differ from 8 by at least 2? Let us denote an event $B = \{|X - 8| \geq 2\}$ and we again exploit (1.5.6) to write $P(B) = P(X \leq 6) + P(X \geq 10) = P(X = 2, 3, 4, 5, 6) + P(X = 10, 11, 12) = \frac{21}{36} = \frac{7}{12}$. ▲

Example 1.5.4 (Example 1.5.3 Continued) Consider two events A, B defined in the Example 1.5.3. One can obtain $P(A), P(B)$ alternatively by using the expression of the df $F(x)$ given in the Example 1.5.2. Now, $P(A) = P(X < 4) + P(X > 9) = F(3) + \{1 - F(9)\} = \frac{3}{36} + \{1 - \frac{30}{36}\} = \frac{9}{36} = \frac{1}{4}$. Also, $P(B) = P(X \leq 6) + P(X \geq 10) = F(6) + \{1 - F(9)\} = \frac{15}{36} + \{1 - \frac{30}{36}\} = \frac{21}{36} = \frac{7}{12}$. ▲

Next, let us summarize some of the important properties of a distribution function defined by (1.5.5) associated with an arbitrary discrete random variable.

Theorem 1.5.1 *Consider the df $F(x)$ with $x \in \Re$, defined by (1.5.5), for a discrete random variable X. Then, one has the following properties:*

(i)   *$F(x)$ is non-decreasing, that is $F(x) \leq F(y)$ for all $x \leq y$ where $x, y \in \Re$;*

(ii) $\lim_{x \to -\infty} F(x) = 0, \lim_{x \to \infty} F(x) = 1$;

(iii)   *$F(x)$ is right continuous, that is $F(x + h) \downarrow F(x)$ as $h \downarrow 0$, for all $x \in \Re$.*

This is a simple result to prove. We leave its proof as Exercise 1.5.6.

> It is true that any function $F(x), x \in \Re$, which satisfies the properties (i)-(iii) stated in the Theorem 1.5.1 is in fact a df of some uniquely determined random variable X.

1.6 Continuous Random Variables

A *continuous random variable*, commonly denoted by X, Y and so on, takes values only within subintervals of $\Re$ or within subsets generated by some appropriate subintervals of the real line $\Re$. At a particular location in a lake, for example, the manager of the local parks and recreation department may be interested in studying the pattern of the depth (X) of the water level. In a high-rise office building, for example, one may be interested to investigate the pattern of the waiting time (X) for an elevator on any of its floors. At the time of the annual music festival in an amusement park, for example, the resident manager at a nearby senior center may want to study the pattern of the loudness factor (X). These are some typical examples of *continuous* random variables.

1.6.1 *Probability Density and Distribution Functions*

We assume that the sample space **S** itself is a subinterval of $\Re$. Now, a continuous real valued random variable X is a mapping of **S** into the real line $\Re$. In this scenario, we no longer talk about $P(X = x)$ because this probability will be zero regardless of what specific value $x \in \Re$ one has in mind. For example, the waiting time (X) at a bus stop would often be postulated as a continuous random variable, and thus the probability of one's waiting *exactly* five minutes or *exactly* seven minutes at that stop for the next bus to arrive is simply zero.

In order to facilitate modeling a continuous stochastic situation, we start with a function $f(x)$ associated with each value $x \in \Re$ satisfying the following two properties:

$$
\begin{aligned}
&i) \quad f(x) \geq 0 \text{ for all } x \in \Re, \text{ and} \\
&ii) \quad \int_{\Re} f(x)dx = 1.
\end{aligned}
\tag{1.6.1}
$$

Observe that the interpretations of (1.5.3) and (1.6.1) are indeed very similar. The conditions listed in (1.6.1) are simply the continuous analogs of those in (1.5.3). Any function $f(x)$ satisfying (1.6.1) is called a *probability density function* (pdf).

> A *probability mass function* (pmf) or *probability density function* (pdf) $f(x)$ is respectively defined through (1.5.3) and (1.6.1).

Once the pdf $f(x)$ is specified, we can find the probabilities of various events defined in terms of the random variable X. If we denote a set $A(\subseteq \Re)$, then

$$P(X \in A) = \int_A f(x)dx, \tag{1.6.2}$$

where the convention is that we would integrate the function $f(x)$ only on that part of the set A wherever $f(x)$ is positive.

In other words, $P(X \in A)$ is given by the area under the curve $\{(x, f(x));$ for all $x \in A$ wherever $f(x) > 0\}$. In the Figure 1.6.1, we let the set A be the interval (a, b) and the shaded area represents the corresponding probability, $P(a < X < b)$.

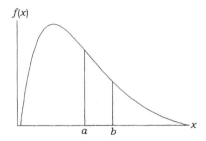

Figure 1.6.1. Shaded Area Under the PDF $f(x)$ Is $P(a < X < b)$

We define the *distribution function* (df) of a continuous random variable X by modifying the discrete analog from (1.5.5). We let

$$F(x) = P\{X \leq x\} = \int_{-\infty}^x f(y)dy, x \in \Re, \tag{1.6.3}$$

which also goes by the name, *cumulative distribution function* (cdf). Again, note that $F(x)$ is defined for all real values x. As before, sometimes we also write $F_X(x)$ for the df of the random variable X.

> Theorem 1.5.1 also holds for the distribution function $F(x)$ of any continuous random variable X. Indeed, the Theorem 1.5.1 and its converse hold for all random variables in general.

$$\tag{1.6.4}$$

Now, we state a rather important characteristic of a df without supplying its proof. The result is well-known. One will find its proof in Rao (1973, p. 85), among other places.

Theorem 1.6.1 *Suppose that $F(x), x \in \Re$, is the df of an arbitrary random variable X. Then, the set of points of discontinuity of the distribution function $F(x)$ is finite or at the most countably infinite.*

Example 1.6.1 Consider the following discrete random variable X first:

$$\begin{array}{llll} X \text{ value:} & -1 & 2 & 5 \\ \text{Probability:} & .1 & .3 & .6 \end{array} \qquad (1.6.5)$$

In this case, the df $F_X(x)$ is discontinuous at the finite number of points $x = -1, 2$ and 5. ▲

Example 1.6.2 Look at the next random variable Y. Suppose that

$$P(Y = y) = \left(\tfrac{1}{2}\right)^y, y = 1, 2, 3, \dots . \qquad (1.6.6)$$

In this case, the corresponding df $F_Y(y)$ is discontinuous at the countably infinite number of points $y = 1, 2, 3, \dots$. ▲

Example 1.6.3 Suppose that a random variable U has an associated *non-negative* function $f(u)$ given by

$$f(u) = \begin{cases} \left(\tfrac{1}{2}\right)^{u+1} & \text{if } u = 1, 2, 3, \dots \\ u & \text{if } 0 < u < 1 \\ 0 & \text{elsewhere.} \end{cases}$$

Observe that $\Sigma_{u=1}^{\infty}\left(\tfrac{1}{2}\right)^{u+1} + \int_{u=0}^{\infty} f(u)du = \tfrac{1}{2} + \tfrac{1}{2} = 1$ and thus $f(u)$ happens to be the distribution of U. This random variable U is *neither discrete nor continuous*. The reader should check that its df $F_U(u)$ is discontinuous at the countably infinite number of points $u = 1, 2, 3, \dots$. ▲

If a random variable X is continuous, then its df $F(x)$ turns out to be continuous at every point $x \in \Re$. However, we do not mean to imply that the df $F(x)$ will necessarily be differentiable at all the points. The df $F(x)$ may not be differentiable at a number of points. Consider the next two examples.

Example 1.6.4 Suppose that we consider a continuous random variable W with its pdf given by

$$f(w) = \begin{cases} \tfrac{3}{7}w^2 & \text{if } 1 < w < 2 \\ 0 & \text{if } w \le 1 \text{ or } w \ge 2. \end{cases} \qquad (1.6.7)$$

In this case, the associated df $F_W(w)$ is given by

$$F_W(w) = \begin{cases} 0 & \text{if } w \le 1 \\ \tfrac{1}{7}(w^3 - 1) & \text{if } 1 < w < 2 \\ 1 & \text{if } w \ge 2. \end{cases} \qquad (1.6.8)$$

The expression for $F_W(w)$ is clear for $w \leq 1$ since in this case we integrate the zero density. The expression for $F_W(w)$ is also clear for $w \geq 2$ because the associated integral can be written as $\int_{-\infty}^{1} f(v)dv + \int_{1}^{2} f(v)dv + \int_{2}^{w} f(v)dv = 0+1+0 = 1$. For $1 < w < 2$, the expression for $F_W(w)$ can be written as $\int_{-\infty}^{1} f(v)dv + \int_{1}^{w} f(v)dv = 0 + \int_{1}^{w} \frac{3}{7}v^2 dv = \frac{1}{7}(w^3 - 1)$. It should be obvious that the df $F_W(w)$ is continuous at all points $w \in \Re$ but $F_W(w)$ is *not differentiable* specifically at the two points $w = 1, 2$. In this sense, the df $F_W(w)$ may not be considered very smooth. ▲

The pmf in (1.6.6) gives an example of a *discrete random variable* whose df has countably infinite number of discontinuity points. The pdf in (1.6.9) gives an example of a *continuous random variable* whose df is not differentiable at countably infinite number of points.

Example 1.6.5 Consider the function $f(x)$ defined as follows:

$$f(x) = \begin{cases} 7(4)^{-i} & \text{whenever } x \in (\frac{1}{2^i}, \frac{1}{2^{i-1}}], i = 1, 2, 3, ... \\ 0 & \text{whenever } x \leq 0 \text{ or } x > 1 \end{cases} \qquad (1.6.9)$$

We leave it as the Exercise 1.6.4 to verify that (i) $f(x)$ is a genuine pdf, (ii) the associated df $F(x)$ is continuous at all points $x \in \Re$, and (iii) $F(x)$ is not differentiable at the points x belonging to the set $\{1, \frac{1}{2}, \frac{1}{2^2}, \frac{1}{2^3}, ...\}$ which is countably infinite. In the Example 1.6.2, we had worked with the point masses at countably infinite number of points. But, note that the present example is little different in its construction. ▲

In each case, observe that the set of the discontinuity points of the associated df $F(.)$ is at most countably infinite. This is exactly what one will expect in view of the Theorem 1.6.1. Next, we proceed to calculate probabilities of events.

The probability of an event or a set A is given by the area under the pdf $f(x)$ wherever $f(x)$ is positive, $x \in A$.

Example 1.6.6 (Example 1.6.4 Continued) For the continuous distribution defined by (1.6.7), suppose that the set A stands for the interval $(-1.5, 1.8)$. Then, $P(A) = \int_{-1.5}^{1} f(w)dw + \int_{1}^{1.8} f(w)dw = 0 + \int_{1}^{1.8} \frac{3}{7}w^2 dw = \frac{4.832}{7} \approx .69029$. Alternately, we may also apply the form of the df given by (1.6.8) to evaluate the probability, $P(A)$ as follows: $P(A) = P(1 < W < 1.8) = F_W(1.8) - F_W(1) = \frac{1}{7}\{(1.8)^3 - 1\} - 0 = \frac{4.832}{7} \approx .69029$. ▲

At all points $x \in \Re$ wherever the df $F(x)$ is differentiable, one must have the following result:

$$dF(x)/dx = f(x), \text{ the pdf of } X \text{ at the point } x. \qquad (1.6.10)$$

This is a simple restatement of the fundamental theorem of integral calculus.

Example 1.6.7 (Example 1.6.4 Continued) Consider the df $F_W(w)$ of the random variable W from (1.6.8). Now, $F_W(w)$ is not differentiable at the two points $w = 1, 2$. Except at the points $w = 1, 2$, the pdf $f(w)$ of the random variable W can be obtained from (1.6.8) as follows: $f(w) = \frac{d}{dw}F_W(w)$ which will coincide with zero when $-\infty < w < 1$ or $2 < w < \infty$, whereas for $1 < w < 2$ we would have $\frac{d}{dw}\{\frac{1}{7}(w^3 - 1)\} = \frac{3}{7}w^2$. This agrees with the pdf given by (1.6.7) except at the points $w = 1, 2$. ▲

For a continuous random variable X with its pdf $f(x)$ and its df $F(x), x \in \Re$, one has: (i) $P(X < a)$ or $P(X \leq a)$ is given by $F(a) = \int_{-\infty}^{a} f(y)dy$, (ii) $P(X > b)$ or $P(X \geq b)$ is given by $1 - F(b) = \int_{b}^{\infty} f(y)dy$, and (iii) $P(a < X < b)$ or $P(a \leq X < b)$ or $P(a < X \leq b)$ or $P(a \leq X \leq b)$ is given by $F(b) - F(a) = \int_{a}^{b} f(y)dy$. The general understanding, however, is that these integrals are carried out within the appropriate intervals wherever the pdf $f(x)$ is positive.

In the case of the Example 1.6.7, notice that the equation (1.6.10) does not quite lead to any specific expression for $f(w)$ at $w = 1, 2$, the points where $F_W(w)$ happens to be non-differentiable.. So, must $f(w)$ be defined at $w = 1, 2$ in exactly the same way as in (1.6.7)?

Let us write $I(.)$ for the indicator function of $(.)$. Since we only handle integrals when evaluating probabilities, without any loss of generality, the pdf given by (1.6.7) is considered equivalent to any of the pdf's such as $\frac{3}{7}w^2I(1 \leq w < 2)$ or $\frac{3}{7}w^2I(1 < w \leq 2)$ or $\frac{3}{7}w^2I(1 \leq w \leq 2)$. If we replace the pdf $f(w)$ from (1.6.7) by any of these other pdf's, there will be no substantive changes in the probability calculations.

From this point onward, when we define the various pieces of the pdf $f(x)$ for a *continuous* random variable X, we will not attach any importance on the locations of the equality signs placed to identify the boundaries of the pieces in the domain of the variable x.

1.6.2   The Median of a Distribution

Often a pdf has many interesting characteristics. One interesting characteristic is the position of the *median* of the distribution. In statistics, when comparing two income distributions, for example, one may simply look at the median incomes from these two distributions and compare them. Consider a *continuous* random variable X whose df is given by $F(x)$. We say that x_m is the median of the distribution if and only if $F(x_m) = \frac{1}{2}$, that is $P(X \leq x_m) = P(X \geq x_m) = \frac{1}{2}$. In other words, the median of a distribution is that value x_m of X such that 50% of the possible values of X are below x_m and 50% of the possible values of X are above x_m.

Example 1.6.8 (Example 1.6.4 Continued) Reconsider the pdf $f(w)$ from (1.6.7) and the df $F(w)$ from (1.6.8) for the random variable W. The median w_m of this distribution would be the solution of the equation $F(w_m) = \frac{1}{2}$. In view of (1.6.8) we can solve this equation exactly to obtain $u_m = (4.5)^{1/3} \approx 1.651$. ▲

1.6.3   Selected Reviews from Mathematics

A review of some useful results from both the differential and integral calculus as well as algebra and related areas are provided here. These are *not* laid out in the order of importance. Elaborate discussions and proofs are *not* included.

Finite Sums of Powers of Positive Integers:

$$
\begin{aligned}
1 + 2 + \ldots + n &= \tfrac{1}{2}n(n+1) \\
1^2 + 2^2 + \ldots + n^2 &= \tfrac{1}{6}n(n+1)(2n+1) \\
1^3 + 2^3 + \ldots + n^3 &= \left\{\tfrac{1}{2}n(n+1)\right\}^2 \\
1^4 + 2^4 + \ldots + n^4 &= \tfrac{1}{30}n(n+1)(2n+1)(3n^2 + 3n - 1)
\end{aligned}
\tag{1.6.11}
$$

Infinite Sums of Reciprocal Powers of Positive Integers: Let us write

$$
\zeta(p) = \Sigma_{n=1}^{\infty} n^{-p}
\tag{1.6.12}
$$

Then, $\zeta(p) = \infty$ if $p \leq 1$, but $\zeta(p)$ is finite if $p > 1$. It is known that $\zeta(2) = \frac{1}{6}\pi^2, \zeta(4) = \frac{1}{90}\pi^4$, and $\zeta(3) \approx 1.2020569$. Refer to Abramowitz and Stegun (1972, pp. 807-811) for related tables.

Results on Limits:

$$\lim_{n \to \infty} n^{1/n} = 1; \quad \lim_{n \to \infty} \left\{ 1 + \tfrac{a}{n} \right\} = 1, \text{ for fixed } a \in \Re;$$

$$\lim_{n \to \infty} \left\{ 1 + \tfrac{a}{n} \right\}^n = e^a, \text{ for fixed } a \in \Re. \tag{1.6.13}$$

The Big $O(.)$ and $o(.)$ Terms: Consider two terms a_n and b_n, both are real valued but depend on $n = 1, 2, \dots$. The term a_n is called $O(b_n)$ provided that $a_n/b_n \to c$, a constant, as $n \to \infty$. The term a_n is called $o(b_n)$ provided that $a_n/b_n \to 0$ as $n \to \infty$. Let $a_n = n - 1, b_n = 2n + \sqrt{n}$ and $d_n = 5n^{8/7} + n, n = 1, 2, \dots$. Then, one has, for example, $a_n = O(b_n), a_n = o(d_n), b_n = o(d_n)$. Also, one may write, for example, $a_n = O(n), b_n = O(n), d_n = O(n^{8/7}), d_n = o(n^2), d_n = o(n^{9/7})$.

Taylor Expansion: Let $f(.)$ be a real valued function having the finite n^{th} derivative $\frac{d^n}{dx^n} f(x)$, denoted by $f^{(n)}(.)$, everywhere in an open interval $(a, b) \subseteq \Re$ and assume that $f^{(n-1)}(.)$ is continuous on the closed interval $[a, b]$. Let $c \in [a, b]$. Then, for every $x \in [a, b], x \neq c$, there exists a real number ξ between the two numbers x and c such that

$$f(x) = f(c) + \Sigma_{i=1}^{n-1} \frac{(x - c)^i}{i!} f^{(i)}(c) + \frac{(x - c)^n}{n!} f^{(n)}(\xi). \tag{1.6.14}$$

Some Infinite Series Expansions:

$$
\begin{aligned}
e^x \quad &= 1 + \tfrac{x}{1!} + \tfrac{x^2}{2!} + \tfrac{x^3}{3!} + \dots \text{ for all } x \in \Re; \\
(1 - x)^{-m} &= 1 + mx + \tfrac{m(m+1)}{2!} x^2 + \tfrac{m(m+1)(m+2)}{3!} x^3 + \dots \\
&\quad \text{for all } x \in (-1, 1), m > 0; \\
\log(1 + x) &= x - \tfrac{1}{2} x^2 + \tfrac{1}{3} x^3 - \dots \text{ for all } x > -1.
\end{aligned}
\tag{1.6.15}
$$

Differentiation Under Integration (Leibnitz's Rule): Suppose that $f(x, \theta), a(\theta)$, and $b(\theta)$ are differentiable functions with respect to θ for $x \in \Re, \theta \in \Re$. Then,

$$
\begin{aligned}
\tfrac{d}{d\theta} \left[\int_{a(\theta)}^{b(\theta)} f(x, \theta) dx \right] &= f(b(\theta), \theta) \left(\tfrac{d}{d\theta} b(\theta) \right) - f(a(\theta), \theta) \left(\tfrac{d}{d\theta} a(\theta) \right) \\
&\quad + \int_{a(\theta)}^{b(\theta)} \left(\tfrac{\partial}{\partial \theta} f(x, \theta) \right) dx.
\end{aligned}
\tag{1.6.16}
$$

Differentiation Under Integration: Suppose that $f(x, \theta)$ is a differentiable function in θ, for $x \in \Re, \theta \in \Re$. Let there be another function $g(x, \theta)$ such that (i) $\left| \frac{\partial}{\partial \theta} f(x, \theta) \right|_{\theta = \theta_0} \right| \leq g(x, \theta)$ for all θ_0 belonging to some interval $(\theta - \varepsilon, \theta + \varepsilon)$, and (ii) $\int_{-\infty}^{\infty} g(x, \theta) dx < \infty$. Then

$$\frac{d}{d\theta} \left[\int_{-\infty}^{\infty} f(x, \theta) dx \right] = \int_{-\infty}^{\infty} \left(\frac{\partial}{\partial \theta} f(x, \theta) \right) dx. \tag{1.6.17}$$

Monotone Function of a Single Real Variable: Suppose that $f(x)$ is a real valued function of $x \in (a, b) \subseteq \Re$. Let us assume that $\frac{d}{dx} f(x)$ exists at each $x \in (a, b)$ and that $f(x)$ is continuous at $x = a, b$. Then,

(i) $f(x)$ is strictly increasing in (a, b) if $\frac{d}{dx} f(x) > 0$ for $x \in (a, b)$;

(ii) $f(x)$ is strictly decreasing in (a, b) if $\frac{d}{dx} f(x) < 0$ for $x \in (a, b)$.

$$\tag{1.6.18}$$

Gamma Function and Gamma Integral: The expression $\Gamma(\alpha)$, known as the *gamma function* evaluated at α, is defined as

$$\Gamma(\alpha) = \int_0^{\infty} e^{-x} x^{\alpha - 1} dx \text{ for } \alpha > 0. \tag{1.6.19}$$

The representation given in the rhs of (1.6.19) is referred to as the *gamma integral*. The gamma function has many interesting properties including the following:

$$\Gamma(\alpha + 1) = \alpha \Gamma(\alpha) \text{ with arbitrary } \alpha > 0; \Gamma(\tfrac{1}{2}) = \sqrt{\pi};$$

$$\text{and } \Gamma(n) = (n - 1)! \text{ for arbitrary } n = 1, 2, \dots . \tag{1.6.20}$$

Stirling's Approximation: From equation (1.6.19) recall that $\Gamma(\alpha) = \int_0^{\infty} e^{-x} x^{\alpha - 1} dx$, with $\alpha > 0$. Then,

$$\Gamma(\alpha) \backsim \sqrt{2\pi} e^{-\alpha} \alpha^{\alpha - \frac{1}{2}} \text{ for large values of } \alpha. \tag{1.6.21}$$

Writing $\alpha = n + 1$ where n is a positive integer, one can immediately claim that

$$n! \backsim \sqrt{2\pi} e^{-n} n^{n + \frac{1}{2}} \text{ for large values of } n. \tag{1.6.22}$$

The approximation for $n!$ given by (1.6.22) works particularly well even for n as small as five or six. The derivation of (1.6.22) from (1.6.21) is left as the Exercise 1.6.15.

Ratios of Gamma Functions and Other Approximations: Recall that $\Gamma(\alpha) = \int_0^\infty e^{-x} x^{\alpha-1} dx$, for $\alpha > 0$. Then, one has

$$\Gamma(c\alpha + d) \backsim \sqrt{2\pi} e^{-c\alpha}(c\alpha)^{c\alpha+d-\frac{1}{2}} \text{ for large values of } \alpha, \qquad (1.6.23)$$

with fixed numbers $c(> 0)$ and d, assuming that the gamma function itself is defined. Also, one has

$$\alpha^{d-c} \frac{\Gamma(\alpha+c)}{\Gamma(\alpha+d)} \backsim 1 + \frac{1}{2\alpha}(c-d)(c+d-1) \text{ for large values of } \alpha, \qquad (1.6.24)$$

with fixed numbers c and d, assuming that the gamma functions involved are defined.

Other interesting and sharper approximations for expressions involving the gamma functions and factorials can be found in the Section 6.1 in Abramowitz and Stegun (1972).

Beta Function and Beta Integral: The expression $b(\alpha, \beta)$, known as the *beta function* evaluated at α and β in that order, is defined as

$$b(\alpha, \beta) = \int_0^1 x^{\alpha-1}(1-x)^{\beta-1} dx \text{ for } \alpha > 0, \beta > 0. \qquad (1.6.25)$$

The representation given in the rhs of (1.6.25) is referred to as the *beta integral*. Recall the expression of $\Gamma(\alpha)$ from (1.6.19). We mention that $b(\alpha, \beta)$ can alternatively be expressed as follows:

$$b(\alpha, \beta) = \{\Gamma(\alpha)\Gamma(\beta)\}/\{\Gamma(\alpha+\beta)\} \text{ for } 0 < \alpha, \beta < \infty. \qquad (1.6.26)$$

Maximum and Minimum of a Function of a Single Real Variable: For some integer $n \geq 1$, suppose that $f(x)$ is a real valued function of a single real variable $x \in (a, b) \subseteq R$, having a continuous n^{th} derivative $\frac{d^n}{dx^n} f(x)$, denoted by $f^{(n)}(x)$, everywhere in the open interval (a, b). Suppose also that for some point $\xi \in (a, b)$, one has $f^{(1)}(\xi) = f^{(2)}(\xi) = \ldots = f^{(n-1)}(\xi) = 0$, but $f^{(n)}(\xi) \neq 0$. Then,

(i)   for n *even*, $f(x)$ has a local minimum at $x = \xi$ if $f^{(n)}(\xi) > 0$;

(ii)   for n *odd*, $f(x)$ has a local maximum at $x = \xi$ if $f^{(n)}(\xi) < 0$.
$$(1.6.27)$$

Maximum and Minimum of a Function of Two Real Variables: Suppose that $f(\mathbf{x})$ is a real valued function of a two-dimensional variable

$\mathbf{x} = (x_1, x_2) \in (a_1, b_1) \times (a_2, b_2) \subseteq \Re^2$. The process of finding where this function $f(\mathbf{x})$ attains its maximum or minimum requires knowledge of matrices and vectors. We briefly review some notions involving matrices and vectors in the Section 4.8. Hence, we defer to state this particular result from calculus in the Section 4.8. One should refer to (4.8.11)-(4.8.12) regarding this.

Integration by Parts: Consider two real valued functions $f(x), g(x)$ where $x \in (a, b)$, an open subinterval of $\Re$. Let us denote $\frac{d}{dx} f(x)$ by $f'(x)$ and the indefinite integral $\int g(x) dx$ by $h(x)$. Then,

$$\int_a^b f(x) g(x) dx = [f(x) h(x)]_{x=a}^{x=b} - \int_a^b f'(x) h(x) dx, \qquad (1.6.28)$$

assuming that all the integrals and $f'(x)$ are finite.

L'Hôpital's Rule: Suppose that $f(x)$ and $g(x)$ are two *differentiable* real valued functions of $x \in \Re$. Let us assume that $\lim_{x \to a} f(x) = 0$ and $\lim_{x \to a} g(x) = 0$ where a is a fixed real number, $-\infty$ or $+\infty$. Then,

$$\lim_{x \to a} \{f(x)/g(x)\} = \lim_{x \to a} \{f'(x)/g'(x)\}, \qquad (1.6.29)$$

where $f'(x) = df(x)/dx, g'(x) = dg(x)/dx$.

Triangular Inequality: For any two real numbers a and b, the following holds:

$$|a + b| \le |a| + |b|. \qquad (1.6.30)$$

From the triangular inequality it also follows that

$$|a - b| \ge |a| - |b|. \qquad (1.6.31)$$

One may use (1.6.30) and mathematical induction to immediately write:

$$|a_1 + a_2 + ... + a_k| \le |a_1| + |a_2| + ... + |a_k| \qquad (1.6.32)$$

where $a_1, a_2, ..., a_k$ are real numbers and $k \ge 2$.

1.7   Some Standard Probability Distributions

In this section we list a number of useful distributions. Some of these distributions will appear repeatedly throughout this book.

> As a convention, we often write down the pmf or the pdf $f(x)$
> only for those $x \in \mathcal{X}$ where $f(x)$ is positive.

1.7.1 Discrete Distributions

In this subsection, we include some standard discrete distributions. A few of these appear repeatedly throughout the text.

The Bernoulli Distribution: This is perhaps one of the simplest possible discrete random variables. We say that a random variable X has the Bernoulli(p) distribution if and only if its pmf is given by

$$f(x) = P(X = x) = p^x(1 - p)^{1-x} \text{ for } x = 0, 1, \qquad (1.7.1)$$

where $0 < p < 1$. Here, p is often referred to as a *parameter*. In applications, one may collect dichotomous data, for example simply record whether an item is defective ($x = 0$) or non-defective ($x = 1$), whether an individual is married ($x = 0$) or unmarried ($x = 1$), or whether a vaccine works ($x = 1$) or does not work ($x = 0$), and so on. In each situation, p stands for $P(X = 1)$ and $1 - p$ stands for $P(X = 0)$.

The Binomial Distribution: We say that a discrete random variable X has the Binomial(n, p) distribution if and only if its pmf is given by

$$f(x) = P(X = x) = \binom{n}{x} p^x(1 - p)^{n-x} \text{ for } x = 0, 1, ..., n, \qquad (1.7.2)$$

where $0 < p < 1$. Here again p is referred to as a *parameter*. Observe that the Bernoulli(p) distribution is same as the Binomial($1, p$) distribution.

The Binomial(n, p) distribution arises as follows. Consider repeating the Bernoulli experiment independently n times where each time one observes the outcome (0 or 1) where $p = P(X = 1)$ remains the same throughout. Let us obtain the expression for $P(X = x)$. Consider n distinct positions in a row where each position will be filled by the number 1 or 0. We want to find the probability of observing x many 1's, and hence additionally exactly $(n - x)$ many 0's. The probability of any such particular sequence, for example the first x many 1's followed by $(n - x)$ many 0's or the first 0 followed by x many 1's and then $(n - x - 1)$ many 0's, would each be $p^x(1 - p)^{n-x}$. But, then there are $\binom{n}{x}$ ways to fill the x positions each with the number 1 and $n - x$ positions each with the number 0, out of the total n positions. This verifies the form of the pmf in (1.7.2).

Recall the requirement in the part (ii) in (1.5.3) which demands that all the probabilities given by (1.7.2) must add up to one. In order to verify this directly, let us proceed as follows. We simply use the binomial expansion to write

$$\Sigma_{x=0}^{n} f(x) = \Sigma_{x=0}^{n} \binom{n}{x} p^x(1 - p)^{n-x} = [p + (1 - p)]^n = (1)^n = 1. \qquad (1.7.3)$$

Refer to the Binomial Theorem from (1.4.12). Next let us look at some examples.

Example 1.7.1 In a short multiple choice quiz, suppose that there are ten unrelated questions, each with five suggested choices as the possible answer. Each question has exactly one correct answer given. An unprepared student guessed all the answers in that quiz. Suppose that each correct (wrong) answer to a question carries one (zero) point. Let X stand for the student's quiz score. We can postulate that X has the Binomial($n = 10, p = \frac{1}{5}$) distribution. Then, $P(X = 0) = \binom{10}{0}(\frac{1}{5})^0(\frac{4}{5})^{10} = (\frac{4}{5})^{10} \approx .10737$. Also, $P(X \geq 8) = \binom{10}{8}(\frac{1}{5})^8(\frac{4}{5})^2 + \binom{10}{9}(\frac{1}{5})^9(\frac{4}{5})^1 + \binom{10}{10}(\frac{1}{5})^{10}(\frac{4}{5})^0 = 45(\frac{16}{5^{10}}) + 10(\frac{4}{5^{10}}) + (\frac{1}{5})^{10} = \frac{761}{5^{10}} \approx 7.7926 \times 10^{-5}$. In other words, the student may earn few points by using a strategy of plain guessing, but it will be hard to earn B or better in this quiz. ▲

Example 1.7.2 A study on occupational outlook reported that 5% of all plumbers employed in the industry are women. In a random sample of 12 plumbers, what is the probability that at most two are women? Since we are interested in counting the number of women among twelve plumbers, let us use the code one (zero) for a woman (man), and let X be the number of women in a random sample of twelve plumbers. We may assume that X has the Binomial($n = 12, p = .05$) distribution. Now, the probability that at most two are women is the same as $P(X \leq 2) = \binom{12}{0}(.05)^0(.95)^{12} + \binom{12}{1}(.05)^1(.95)^{11} + \binom{12}{2}(.05)^2(.95)^{10} \approx .98043$. ▲

The Poisson Distribution: We say that a discrete random variable X has the Poisson(λ) distribution if and only if its pmf is given by

$$f(x) = P(X = x) = e^{-\lambda}\lambda^x/x! \text{ for } x = 0, 1, 2, ..., \qquad (1.7.4)$$

where $0 < \lambda < \infty$. Here, λ is referred to as a *parameter*.

Recall the requirement in the part (ii) in (1.5.3) which demands that all the probabilities given by (1.7.4) must add up to one. In order to verify this directly, let us proceed as follows. We simply use the infinite series expansion of e^x from (1.6.15) to write

$$\Sigma_{x=0}^{\infty}f(x) = e^{-\lambda}\left\{1 + \frac{\lambda}{1!} + \frac{\lambda^2}{2!} + ...\right\} = e^{-\lambda}e^{\lambda} = 1. \qquad (1.7.5)$$

The Poisson distribution may arise in the following fashion. Let us re-consider the binomial distribution defined by (1.7.2) but pretend that we have a situation like this: we make $n \to \infty$ and $p \to 0$ in such a way that np remains a constant, say, $\lambda(> 0)$. Now then, we can rewrite the binomial probability for any fixed $x = 0, 1, 2, \ldots$,

$$
\begin{aligned}
f(x) &= \binom{n}{x} p^x (1-p)^{n-x} \\
&= \frac{n(n-1)\ldots(n-x+1)}{x!} \left(\frac{\lambda}{n}\right)^x \left(1 - \frac{\lambda}{n}\right)^{n-x} \\
&= \frac{1}{x!} \lambda^x \left(1 - \frac{1}{n}\right)\left(1 - \frac{2}{n}\right)\ldots\left(1 - \frac{x-1}{n}\right)\left(1 - \frac{\lambda}{n}\right)^{-x}\left(1 - \frac{\lambda}{n}\right)^n.
\end{aligned}
\tag{1.7.6}
$$

Observe that $(1 - \frac{k}{n}) \to 1$ as $n \to \infty$ for all fixed $k = 1, 2, \ldots, x-1, \lambda$. Also, $(1 - \frac{\lambda}{n})^n \to e^{-\lambda}$ as $n \to \infty$. See (1.6.13) as needed. Now, from (1.7.6) we can conclude that $\binom{n}{x} p^x (1-p)^{n-x} \to e^{-\lambda} \lambda^x / x!$ as $n \to \infty$.

Next let us look at some examples.

Example 1.7.3 In a certain manufacturing industry, we are told that minor accidents tend to occur independently of one another and they occur at a constant rate of three $(= \lambda)$ per week. A Poisson($\lambda = 3$) is assumed to adequately model the number of minor accidents occurring during a given week. Then, the probability that no minor accidents will occur during a week is given by $P(X = 0) = e^{-3} \approx 4.9787 \times 10^{-2}$. Also, the probability that more than two minor accidents will occur during a week is given by $P(X > 2) = 1 - P(X \le 2) = 1 - \{(e^{-3} 3^0 / 0!) + (e^{-3} 3 / 1!) + (e^{-3} 3^2 / 2!)\} = 1 - 8.5 e^{-3} \approx .57681$ ▲

Example 1.7.4 We are inspecting a particular brand of concrete slab specimens for any visible cracks. Suppose that the number (X) of cracks per concrete slab specimen has approximately a Poisson distribution with $\lambda = 2.5$. What is the probability that a randomly selected slab will have at least two cracks? We wish to evaluate $P(X \ge 2)$ which is the same as $1 - P(X \le 1) = 1 - [\{e^{-2.5}(2.5)^0 / 0!\} + \{e^{-2.5}(2.5)^1 / 1!\}] = 1 - (3.5)e^{-2.5} \approx .7127$. ▲

The Geometric Distribution: A discrete random variable X is said to have the Geometric(p) distribution if and only if its pmf is given by

$$
f(x) = P(X = x) = p(1-p)^{x-1} \text{ for } x = 1, 2, 3, \ldots,
\tag{1.7.7}
$$

where $0 < p < 1$. Here, p is referred to as a *parameter*.

The Geometric(p) distribution arises as follows. Consider repeating the Bernoulli experiment independently until we observe the value $x = 1$ for the first time. In other words, let X be number of trials needed for the independent run of Bernoulli data to produce the value 1 for the first time. Then, we have $P(X = x) = P(\text{Observing } x - 1 \text{ many 0's followed a single occurrence of 1 in the } x^{th} \text{ trial}) = p(1-p)^{x-1}, x = 1, 2, \ldots$.

Recall the requirement in the part (ii) in (1.5.3) which demands that all the probabilities given by (1.7.7) must add up to one. In order to verify this directly, let us proceed as follows. We denote $q \doteq 1 - p$ and simply use the expression for the sum of an infinite geometric series from (1.6.15), with $m = 1$, to write

$$\Sigma_{x=1}^{\infty} f(x) = p\{1 + q + q^2 + ...\} = p(1 - q)^{-1} = 1.$$

The expression of the df for the Geometric(p) random variable is also quite straight forward. One can easily verify that

$$F(x) = P(X \leq x) = 1 - q^x \text{ for any positive integer } x. \tag{1.7.8}$$

Example 1.7.5 Some geological exploration may indicate that a well drilled for oil in a region in Texas would strike oil with probability .3. Assuming that the oil strikes are independent from one drill to another, what is the probability that the first oil strike will occur on the sixth drill? Let X be the number of drills until the first oil strike occurs. Then, X is distributed as Geometric($p = .3$) so that one has $P(X = 6) = (.3)(.7)^5 \approx 5.0421 \times 10^{-2}$. ▲

Example 1.7.6 An urn contains six blue and four red marbles of identical size and weight. We reach in to draw a marble at random, and if it is red, we throw it back in the urn. Next, we reach in again to draw another marble at random, and if it is red then it is thrown back in the urn. Then, we reach in for the third draw and the process continues until we draw the first blue marble. Let X be total number of required draws. Then, this random variable X has the Geometric($p = .6$) distribution. What is the probability that we will need to draw marbles fewer than four times? Using the expression of the df $F(x)$ from (1.7.8) we immediately obtain $P(X < 4) = F(3) = 1 - (.4)^3 = .936$. ▲

The kind of sampling inspection referred to in the Example 1.7.6 falls under what is called *sampling with replacement*.

The Negative Binomial Distribution: A discrete random variable X is said to have the negative binomial distribution with μ and k, customarily denoted by NB(μ, k), if and only if its pmf is given by

$$f(x) = P(X = x) = \binom{k+x-1}{k-1} \left(\frac{\mu}{\mu+k}\right)^x \left(\frac{k}{\mu+k}\right)^k \text{ for } x = 0, 1, 2, ...,$$
$$\tag{1.7.9}$$

where $0 < \mu, k < \infty$. Here, μ and k are referred to as *parameters*. The parameterization given in (1.7.9) is due to Anscombe (1949). This form of the pmf is widely used in the areas such as entomology, plant science, and soil science. The parameters μ and k have physical interpretations in many applications in these areas.

Denoting $p = k/(\mu + k), q = \mu/(\mu + k)$, the pmf given by (1.7.9) can be rewritten in a more traditional way as follows:

$$f(x) = \binom{k+x-1}{k-1} q^x p^k \text{ for } x = 0, 1, 2, \ldots \qquad (1.7.10)$$

where $0 < p < 1, q = 1 - p$ and k *is a positive integer*. This form of the negative binomial distribution arises as follows. Suppose that we have the same basic setup as in the case of a geometric distribution, but instead we let X be the number of 0's observed before the kth occurrence of 1. Then, we have $P(X = 0) = P(\text{Observing } k \text{ many 1's right away}) = p^k$. Next, $P(X = 1) = P(\text{The last trial yields 1,but we observe } k - 1 \text{ many 1's and a single 0 before the occurrence of the 1 in the last trial}) = \binom{(k-1)+1}{k-1} qp^{k-1}p = \binom{k}{k-1} qp^k$. Also, $P(X = 2) = P(\text{The last trial yields 1, but we observe } k - 1$ many 1's and two 0's before the occurrence of the 1 in the last trial) $= \binom{(k-1)+2}{k-1} q^2 p^{k-1}p = \binom{k+1}{k-1} q^2 p^k$. Analogous arguments will eventually justify (1.7.10) in general.

Example 1.7.7 (Example 1.7.5 Continued) Some geological exploration indicates that a well drilled for oil in a region in Texas may strike oil with probability .3. Assuming that the oil strikes are independent from one drill to another, what is the probability that the third oil strike will occur on the tenth well drilled? Let X be the number of drilled wells until the third oil strike occurs. Then, using (1.7.10) we immediately get $P(X = 10) = \binom{9}{2}(.7)^7(.3)^3 \approx 8.0048 \times 10^{-2}$. ▲

The Discrete Uniform Distribution: Let X be a discrete random variable which takes the only possible values $x_1, \ldots, x_k$ each with the same probability $\frac{1}{k}$. Such X is said to have a discrete uniform distribution. We may write down its pmf as follows.

$$f(x) = \frac{1}{k} \text{ for } x = x_1, \ldots, x_k, \text{ with a fixed positive integer } k. \qquad (1.7.11)$$

Example 1.7.8 Suppose that we roll a fair die once and let X be the number of dots on the face of the die which lands up. Then, obviously $f(x) = \frac{1}{6}$ for $x = 1, \ldots, 6$ which corresponds to the pmf given by (1.7.11) with $k = 6$ and $x_1 = 1, x_2 = 2, \ldots, x_6 = 6$. ▲

1.7.2 Continuous Distributions

In this subsection we include some standard continuous distributions. A few of these appear repeatedly throughout the text.

The Uniform Distribution: A continuous random variable X has the uniform distribution on the interval (a, b), denoted by Uniform(a, b), if and only if its pdf is given by

$$f(x) = (b-a)^{-1} \text{ for } a < x < b, \qquad (1.7.12)$$

where $-\infty < a, b < \infty$. Here, a, b are referred to as *parameters*.

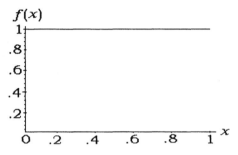

Figure 1.7.1. Uniform $(0, 1)$ Density

Let us ask ourselves: How can one directly check that $f(x)$ given by (1.7.12) is indeed a pdf? The function $f(x)$ is obviously non-negative for all $x \in \Re$. Next, we need to verify directly that the total integral is one. Let us write $\int_{-\infty}^{\infty} f(x)dx = \int_{a}^{b}(b-a)^{-1}dx = (b-a)^{-1}\int_{a}^{b}dx = (b-a)^{-1}[x]_{x=a}^{x=b} = (b-a)^{-1}(b-a) = 1$ since $b \neq a$. In other words, (1.7.12) defines a genuine pdf. Since this pdf puts equal weight at each point $x \in (a, b)$, it is called the Uniform(a, b) distribution. The pdf given by (1.7.12) when $a = 0, b = 1$ has been plotted in the Figure 1.7.1.

Example 1.7.9 The waiting time X at a bus stop, measured in minutes, may be uniformly distributed between zero and five. What is the probability that someone at that bus stop would wait more than 3.8 minutes for the bus? We have $P(X > 3.8) = \int_{3.8}^{5} f(x)dx = \int_{3.8}^{5} \frac{1}{5}dx = \frac{1}{5}[x]_{x=3.8}^{x=5} = \frac{1.2}{5} = .24$. ▲

The Normal Distribution: A continuous random variable X has the normal distribution with the *parameters* μ and σ^2, denoted by $N(\mu, \sigma^2)$, if and only if its pdf is given by

$$f(x) = \{\sigma\sqrt{2\pi}\}^{-1}exp\{-(x-\mu)^2/(2\sigma^2)\} \text{ for } x \in \Re, \qquad (1.7.13)$$

where $-\infty < \mu < \infty$ and $0 < \sigma < \infty$. Among all the continuous distributions, the normal distribution is perhaps the one which is most widely used in modeling data.

C. F. Gauss, the celebrated German mathematician of the eighteenth century, had discovered this distribution while analyzing the measurement errors in astronomy. Hence, the normal distribution is alternatively called a *Gaussian distribution*.

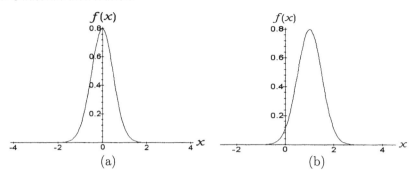

Figure 1.7.2. Normal Densities: (a) $N(0, .25)$ (b) $N(1, .25)$

Let us ask ourselves: How can one directly check that $f(x)$ given by (1.7.13) is indeed a pdf? The function $f(x)$ is obviously positive for all $x \in \Re$. Next, we need to verify directly that

$$\int_{-\infty}^{\infty} f(x)dx = 1. \tag{1.7.14}$$

Recall the gamma function $\Gamma(.)$ defined by (1.6.19). Let us substitute $u = (x - \mu)/\sigma, v = -u, w = \frac{1}{2}v^2$ successively, and rewrite the integral from (1.7.14) as

$$\int_{-\infty}^{\infty} f(x)dx$$

$$= \{\sqrt{2\pi}\}^{-1}\left[\int_{-\infty}^{0} exp\{-\tfrac{1}{2}u^2\}du + \int_{0}^{\infty} exp\{-\tfrac{1}{2}u^2\}du\right]$$

$$= \{\sqrt{2\pi}\}^{-1}\left[\int_{\infty}^{0} exp\{-\tfrac{1}{2}v^2\}(-dv) + \int_{0}^{\infty} exp\{-\tfrac{1}{2}u^2\}du\right] \tag{1.7.15}$$

$$= \{\sqrt{2\pi}\}^{-1}\left[\int_{0}^{\infty} exp\{-\tfrac{1}{2}v^2\}dv + \int_{0}^{\infty} exp\{-\tfrac{1}{2}u^2\}du\right].$$

In the last step in (1.7.15) since the two integrals are the same, we can claim that

$$\int_{-\infty}^{\infty} f(x)dx$$

$$= 2\{\sqrt{2\pi}\}^{-1}\int_{0}^{\infty} exp\{-\tfrac{1}{2}v^2\}dv$$

$$= \{\sqrt{\pi}\}^{-1}\int_{0}^{\infty} exp\{-w\}w^{-1/2}dw$$

$$= \{\sqrt{\pi}\}^{-1}\Gamma(\tfrac{1}{2})$$

$$= 1, \text{ since } \Gamma(\tfrac{1}{2}) = \sqrt{\pi}.$$

This proves (1.7.14). An alternative way to prove the same result using the polar coordinates has been indicated in the Exercise 1.7.20.

The pdf given by (1.7.13) is symmetric around $x = \mu$, that is we have $f(x - \mu) = f(x + \mu)$ for all fixed $x \in \Re$. In other words, once the curve $(x, f(x))$ is plotted, if we pretend to fold the curve around the vertical line $x = \mu$, then the two sides of the curve will lie exactly on one another. See the Figure 1.7.2.

The Standard Normal Distribution: The normal distribution with $\mu = 0, \sigma = 1$ is customarily referred to as the standard normal distribution and the standard normal random variable is commonly denoted by Z. The standard normal pdf and the df are respectively denoted by

$$\phi(z) = \{\sqrt{2\pi}\}^{-1} exp(-z^2/2) \text{ for } z \in \Re,$$
$$\Phi(z) = P(Z \le z) = \int_{-\infty}^{z} \phi(u)du \text{ for } z \in \Re. \tag{1.7.16}$$

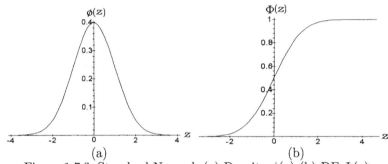

Figure 1.7.3. Standard Normal: (a) Density $\phi(z)$ (b) DF $\Phi(z)$

In the Figure 1.7.3 we have shown plots of the pdf $\phi(z)$ and the df $\Phi(z)$. In these plots, the variable z should stretch from $-\infty$ to ∞. But, the standard normal pdf $\phi(z)$, which is symmetric around $z = 0$, falls off very sharply. For all practical purposes, there is hardly any sizable density beyond the interval $(-3, 3)$. This is also reflected in the plot of the corresponding df $\Phi(z)$.

Unfortunately, there is no available simple analytical expression for the df $\Phi(z)$. The standard normal table, namely the Table 14.3.1, will facilitate finding various probabilities associated with Z. One should easily verify the following:

$$\Phi(-a) = 1 - \Phi(a) \text{ for all fixed } a > 0. \tag{1.7.17}$$

How is the df of the $N(\mu, \sigma^2)$ related to the df of the standard normal random variable Z? With any fixed $x \in \Re, v = (u - \mu)/\sigma$, observe that we can proceed as in (1.7.15) and write

$$
\begin{aligned}
P\{X \leq x\} \\
&= \{\sigma\sqrt{2\pi}\}^{-1} \int_{-\infty}^{x} exp\{-(u-\mu)^2/(2\sigma^2)\}du \\
&= \{\sqrt{2\pi}\}^{-1} \int_{-\infty}^{(x-\mu)/\sigma} exp\{-\tfrac{1}{2}v^2\}dv \\
&= \Phi\left(\tfrac{x-\mu}{\sigma}\right),
\end{aligned}
\tag{1.7.18}
$$

with the $\Phi(.)$ function defined in (1.7.16).

In the Figure 1.7.2, we plotted the pdf's corresponding to the $N(0,.25)$ and $N(1,.25)$ distributions. By comparing the two plots in the Figure 1.7.2 we see that the shapes of the two pdf's are exactly same whereas in (b) the curve's point of symmetry $(x = 1)$ has moved by a unit on the right hand side. By comparing the plots in the Figures 1.7.2(a)-1.7.3(a), we see that the $N(0,.25)$ pdf is more concentrated than the standard normal pdf around their points of symmetry $x = 0$.

In (1.7.13), we added earlier that the parameter μ indicates the point of symmetry of the pdf. When σ is held fixed, this pdf's shape remains intact whatever be the value of μ. But when μ is held fixed, the pdf becomes more (less) concentrated around the fixed center μ as σ becomes smaller (larger).

Example 1.7.10 Suppose that the scores of female students on the recent Mathematics Scholastic Aptitude Test were normally distributed with $\mu = 520$ and $\sigma = 100$ points. Find the proportion of female students taking this exam who scored (i) between 500 and 620, (ii) more than 650. Suppose that $F(.)$ stands for the df of X. To answer the first part, we find $P(500 < X < 620) = F(620) - F(500) = \Phi(\frac{620-520}{100}) - \Phi(\frac{500-520}{100}) = \Phi(1) - \Phi(-.2) = \Phi(1) + \Phi(.2) - 1$, using (1.7.18). Thus, reading the entries from the standard normal table (see Chapter 14), we find that $P(500 < X < 620) = .84134 + .57926 - 1 = .4206$. Next, we again use (1.7.18) and the standard normal table to write $P(X > 650) = 1 - F(650) = 1 - \Phi(\frac{650-520}{100}) = 1 - \Phi(1.3) = 1 - .90320 = .0968$. ▲

Theorem 1.7.1 *If X has the $N(\mu, \sigma^2)$ distribution, then $W = (X-\mu)/\sigma$ has the standard normal distribution.*

Proof Let us first find the df of W. For any fixed $w \in \Re$, in view of (1.7.18), we have

$$
P(W \leq w) = P(X \leq \mu + \sigma w) = \Phi\left(\frac{(\mu+\sigma w)-\mu}{\sigma}\right) = \Phi(w). \tag{1.7.19}
$$

Since $\Phi(w)$ is differentiable for all $w \in \Re$, using (1.6.10) we can claim that the pdf of W would be given by $d\Phi(w)/dw = \phi(w)$ which is the pdf of the standard normal random variable. ∎

The Gamma Distribution: The expression $\Gamma(\alpha)$ was introduced in (1.6.19). We say that a positive continuous random variable X has the gamma distribution involving α and β, denoted by Gamma(α, β), if and only if its pdf is given by

$$f(x) = \{\beta^\alpha \Gamma(\alpha)\}^{-1} e^{-x/\beta} x^{\alpha-1} \text{ for } 0 < x < \infty, \qquad (1.7.20)$$

where $0 < \alpha, \beta < \infty$. Here, α and β are referred to as *parameters*. By varying the values of α and β, one can generate interesting shapes for the associated pdf.

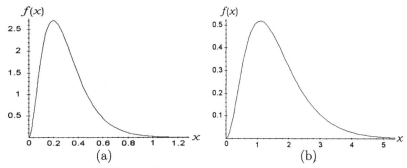

Figure 1.7.4. (a) Gamma$(3, .1)$ Density (b) Gamma$(3.2, .5)$ Density

In the Figure 1.7.4, the two pdf's associated with the Gamma$(\alpha = 3, \beta = .1)$ and Gamma$(\alpha = 3.2, \beta = .5)$ distributions have been plotted. The gamma distribution is known to be *skewed* to the right and this feature is apparent from the plots provided. As $\alpha\beta$ increases, the point where the pdf attains its maximum keeps moving farther to the right hand side. This distribution appears frequently as a statistical model for data obtained from reliability and survival experiments as well as clinical trials.

Let us ask ourselves: How can one directly check that $f(x)$ given by (1.7.20) is indeed a pdf? The $f(x)$ is obviously non-negative for all $x \in \Re^+$. Next, we need to verify directly that

$$\int_0^\infty f(x)dx = 1. \qquad (1.7.21)$$

Let us substitute $u = x/\beta$ which is one-to-one and rewrite the integral from (1.7.21) as

$$\{\beta^\alpha \Gamma(\alpha)\}^{-1} \int_0^\infty e^{-x/\beta} x^{\alpha-1} dx$$
$$= \{\beta^\alpha \Gamma(\alpha)\}^{-1} \int_0^\infty e^{-u} (\beta u)^{\alpha-1} (\beta du) \qquad (1.7.22)$$
$$= \{\Gamma(\alpha)\}^{-1} \int_0^\infty e^{-u} u^{\alpha-1} du,$$

which verifies (1.7.21) since $\Gamma(\alpha) = \int_0^\infty e^{-u} u^{\alpha-1} du$.

The exponential Distribution: This is a very special distribution which coincides with the Gamma$(1, \beta)$ distribution where $\beta(> 0)$ is referred to as a *parameter*. This distribution is widely used in reliability and survival analyses. The associated pdf is given by

$$f(x) = \beta^{-1} e^{-x/\beta} \text{ for } 0 < x < \infty, \tag{1.7.23}$$

where $0 < \beta < \infty$. This distribution is also skewed to the right.

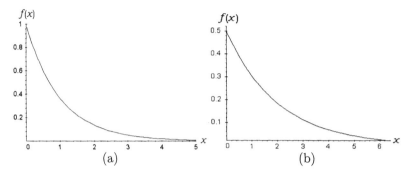

Figure 1.7.5. Exponential Densities: (a) $\beta = 1$ (b) $\beta = 2$

The pdf given by (1.7.23) has been plotted in the Figure 1.7.5 with $\beta = 1, 2$. The plot (a) corresponds to the pdf of the *standard exponential* variable. The plot (b) corresponds to $\beta = 2$ and it is clear that this pdf has moved to the rhs compared with the standard exponential pdf.

> The *standard exponential* pdf refers to (1.7.23) when we fix $\beta = 1$. $\qquad (1.7.24)$

As the parameter β increases, the pdf given by (1.7.23) will move farther to the right and as a result the concentration in the right hand vicinity of zero will decrease. In other words, as β increases, the probability of failure of items in the early part of a life test would decrease if X represents the life expectancy of the tested items.

The reader can easily check that the corresponding df is given by

$$F(x) = \begin{cases} 0 & \text{if} \quad -\infty < x \leq 0 \\ 1 - exp(-x/\beta) & \text{if} \quad x > 0. \end{cases} \tag{1.7.25}$$

In the context of a reliability experiment, suppose that X represents the length of life of a particular component in a machine and that X has the exponential distribution defined by (1.7.23). With any two fixed positive numbers a and b, we then have

$$P\{X > a + b \mid X > a\}$$
$$= P\{X > a + b \cap X > a\}/P\{X > a\}$$
$$= P\{X > a + b\}/P\{X > a\} \qquad (1.7.26)$$
$$= exp\,(-b/\beta)\,,\ \text{using (1.7.25)}$$
$$= P\{X > b\},\ \text{which does not involve } a.$$

This conveys the following message: Given that a component has lasted up until the time a, the conditional probability of its surviving beyond the time $a + b$ is same as $P(X > b\}$, regardless of the magnitude of a. In other words, the life of the component ignores the aging process regardless of its own age. This interesting feature of the exponential distribution is referred to as its *memoryless property*. The recently edited volume of Balakrishnan and Basu (1995) gives a synthesis of the gamma, exponential, and other distributions.

The Chi-square Distribution: We say that a positive continuous random variable X has the Chi-square distribution with ν degrees of freedom denoted by χ_ν^2, with $\nu = 1, 2, 3, ...$, if X has the Gamma($\frac{1}{2}\nu, 2$) distribution. Here, the *parameter* ν is referred to as the *degree of freedom*. By varying the values of ν, one can generate interesting shapes for the associated pdf.

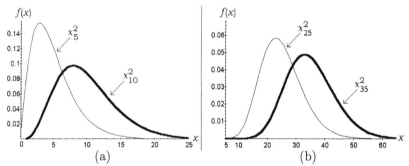

Figure 1.7.6. PDF's: (a) χ_5^2 Thin; χ_{10}^2 Thick (b) χ_{25}^2 Thin; χ_{35}^2 Thick

A Chi-square random variable is derived from the Gamma family and so it should not be surprising to learn that Chi-square distributions are skewed to the right too. In the Figure 1.7.6, we have plotted the pdf's corresponding of the χ_ν^2 random variable when $\nu = 5, 10, 25, 30$. From these figures, it should be clear that as the degree of freedom ν increases, the pdf tends to move more toward the rhs. From the Figure 1.7.6 (b) it appears that the shape of the pdf resembles more like that of a symmetric distribution when

$\nu = 25$ and 35. In the Section 5.4.1, the reader will find a more formal statement of this empirical observation for large values of ν. One may refer to (5.4.2) for a precise statement of the relevant result.

The Lognormal Distribution: We say that a positive continuous random variable X has the lognormal distribution if and only if its pdf is given by

$$f(x) = \{\sigma\sqrt{2\pi}\}^{-1}x^{-1}exp[-\{log(x) - \mu\}^2/(2\sigma^2)] \text{ for } 0 < x < \infty, \quad (1.7.27)$$

where $-\infty < \mu < \infty$ and $0 < \sigma < \infty$ are referred to as *parameters*. The pdf given by (1.7.27) when $\mu = 0$ and $\sigma = 1$ has been plotted in the Figure 1.7.8 and it also looks fairly skewed to the right. We leave it as the Exercise 1.7.15 to verify that

$$P(X \leq x) = \Phi(\{log(x) - \mu\}/\sigma), \text{ for all } x > 0. \quad (1.7.28)$$

We may immediately use (1.7.28) to claim that

$$P\{log(X) \leq x\} = P\{X \leq e^x\} = \Phi(\{x - \mu\}/\sigma), \text{ for all } x \in \Re. \quad (1.7.29)$$

That is, the pdf of $log(X)$ must coincide with that of the $N(\mu, \sigma^2)$ random variable. Thus, the name "lognormal" appears quite natural in this situation.

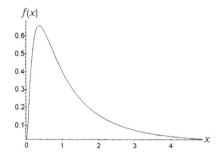

Figure 1.7.7. Lognormal Density: $\mu = 0, \sigma = 1$

The Student's t Distribution: The pdf of a Student's t random variable with ν degrees of freedom, denoted by t_ν, is given by

$$f(x) = a(1 + \tfrac{1}{\nu}x^2)^{-\frac{1}{2}(\nu+1)} \quad \text{for } -\infty < x < \infty, \quad (1.7.30)$$

with $a \equiv a(\nu) = \{\sqrt{\nu\pi}\}^{-1}\Gamma(\tfrac{1}{2}(\nu + 1))\{\Gamma(\tfrac{1}{2}\nu)\}^{-1}, \nu = 1, 2, 3, \dots$. One can easily verify that this distribution is symmetric about $x = 0$. Here, the *parameter* ν is referred to as the *degree of freedom*. This distribution plays a key role in statistics. It was discovered by W. S. Gosset under the pseudonym "Student" which was published in 1908. By varying the values of ν, one can generate interesting shapes for the associated pdf.

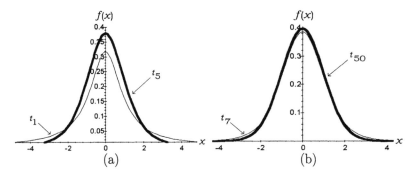

Figure 1.7.8. Student's t PDF's: (a) t_1 and t_5 (b) t_7 and t_{50}

In the Figure 1.7.8, we have plotted the Student's t_ν variable's pdf given by (1.7.30) when $\nu = 1, 5, 7$ and 50. As the degree of freedom ν increases, one can see that the pdf has less spread around the point of symmetry $x = 0$. If one compares the Figure 1.7.8 (b) with $\nu = 50$ and the Figure 1.7.3 (a) for the pdf of the standard normal variable, the naked eyes may not find any major differences. In the Section 5.4.2, the reader will find a more formal statement of this empirical observation for large values of ν. One may refer to (5.4.3) for a precise statement.

Table 1.7.1. Comparison of the Tail Probabilities for the
Student's t_ν and the Standard Normal Distributions

	$P(Z > 1.5)$: 6.6807×10^{-2}	$P(Z > 1.96)$: 2.4998×10^{-2}	$P(Z > 2.5)$: 6.2097×10^{-3}	$P(Z > 5)$: 2.8665×10^{-7}
ν	$P(t_\nu > 1.5)$	$P(t_\nu > 1.96)$	$P(t_\nu > 2.5)$	$P(t_\nu > 5)$
10	8.2254×10^{-2}	3.9218×10^{-2}	1.5723×10^{-2}	2.6867×10^{-4}
15	7.7183×10^{-2}	3.4422×10^{-2}	1.2253×10^{-2}	7.9185×10^{-5}
30	7.2033×10^{-2}	2.9671×10^{-2}	9.0578×10^{-3}	1.1648×10^{-5}
100	6.8383×10^{-2}	2.6389×10^{-2}	7.0229×10^{-3}	1.2251×10^{-6}
500	6.7123×10^{-2}	2.5275×10^{-2}	6.3693×10^{-3}	3.973×10^{-7}
1000	6.6965×10^{-2}	2.5137×10^{-2}	6.2893×10^{-3}	3.383×10^{-7}

It is true, however, that the tails of the t distributions are "heavier" than those of the standard normal distribution. In order to get a feeling for this, one may look at the entries given in the Table 1.7.1. It is clear that $P(t_\nu > 1.5)$ decreases as ν is successively assigned the value $10, 15, 30, 100, 500$

and 1000. Even when $\nu = 1000$, we have $P(t_\nu > 1.5) = .066965$ whereas $P(Z > 1.5) = .066807$, that is $P(t_\nu > 1.5) > P(Z > 1.5)$. In this sense, the tail of the t_ν distribution is *heavier* than that of the standard normal pdf. One may also note that the discrepancy between $P(t_\nu > 5)$ and $P(Z > 5)$ stays large even when $\nu = 1000$.

The Cauchy Distribution: It corresponds to the pdf of a Student's t_ν random variable with $\nu = 1$, denoted by Cauchy$(0, 1)$. In this special case, the pdf from (1.7.30) simplifies to

$$f(x) = \pi^{-1}(1 + x^2)^{-1} \quad \text{for } -\infty < x < \infty. \tag{1.7.31}$$

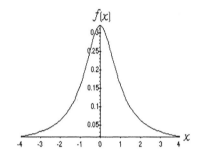

Figure 1.7.9. Cauchy$(0, 1)$ Density

The pdf from (1.7.31) has been plotted in the Figure 1.7.9 which naturally coincides with the Figure 1.7.8 (a) with $\nu = 1$. The pdf from (1.7.31) is obviously symmetric about $x = 0$.

Let us ask ourselves: How can one directly check that $f(x)$ given by (1.7.31) is indeed a pdf? Obviously, $f(x)$ is positive for all $x \in \Re$. Next, let us evaluate the whole integral and write

$$\pi^{-1} \int_{-\infty}^{\infty} (1 + x^2)^{-1} dx = \pi^{-1} [\arctan(x)]_{-\infty}^{\infty} = \pi^{-1} [\tfrac{1}{2}\pi - (-\tfrac{1}{2}\pi)] = 1. \tag{1.7.32}$$

The df of the Cauchy distribution is also simple to derive. One can verify that

$$F(x) = P(X \le x) = \tfrac{1}{\pi} \arctan(x) + \tfrac{1}{2} \quad \text{for all } x \in \Re. \tag{1.7.33}$$

The Cauchy distribution has heavier tails compared with those of the standard normal distribution. From (1.7.31) it is clear that the Cauchy pdf $f(x) \approx \tfrac{1}{\pi} x^{-2}$ for large values of $|x|$. It can be argued then that $\tfrac{1}{\pi} x^{-2} \to 0$ slowly compared with $(\sqrt{2\pi})^{-1} e^{-x^2/2}$ when $|x| \to \infty$.

The F Distribution: We say that a positive continuous random variable X has the F distribution with ν_1, ν_2 degrees of freedom, in that order and denoted by F_{ν_1,ν_2}, if and only if its pdf is given by

$$f(x) = kx^{\frac{1}{2}(\nu_1-2)}\{1 + (\nu_1/\nu_2)x\}^{-\frac{1}{2}(\nu_1+\nu_2)} \quad \text{for } 0 < x < \infty, \quad (1.7.34)$$

with $k \equiv k(\nu_1, \nu_2) = (\nu_1/\nu_2)^{\frac{1}{2}\nu_1}\Gamma((\nu_1 + \nu_2)/2)\{\Gamma(\nu_1/2)\Gamma(\nu_2/2)\}^{-1}$. Here, ν_1 and ν_2 are referred to as the *parameters*. By varying the values of ν_1 and ν_2, one can generate interesting shapes for this pdf.

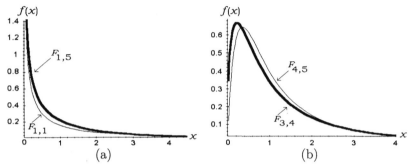

Figure 1.7.10. F Densities: (a) $F_{1,1}$ and $F_{1,5}$

(b) $F_{4,5}$ and $F_{3,4}$

The pdf from (1.7.34) has been plotted in the Figure 1.7.10 when we fix $(\nu_1, \nu_2) = (1,1), (1,5), (4,5), (3,4)$. One realizes that the F distribution is skewed to the right.

The Beta Distribution: Recall the expression $\Gamma(\alpha)$, the beta function $b(\alpha, \beta)$, and that $b(\alpha, \beta) = \Gamma(\alpha)\Gamma(\beta)\{\Gamma(\alpha + \beta)\}^{-1}$ respectively from (1.6.19), (1.6.25)-(1.6.26). A continuous random variable X, defined on the interval $(0, 1)$, has the beta distribution with *parameters* α and β, denoted by Beta(α, β), if and only if its pdf is given by

$$f(x) = \{b(\alpha, \beta)\}^{-1}x^{\alpha-1}(1 - x)^{\beta-1} \text{ for } 0 < x < 1, \quad (1.7.35)$$

where $0 < \alpha, \beta < \infty$. By varying the values of α and β, one can generate interesting shapes for this pdf. In general, the Beta distributions are fairly skewed when $\alpha \neq \beta$. The beta pdf from (1.7.35) has been plotted in the Figure 1.7.11 for $(\alpha, \beta) = (2,5), (4,5)$. The pdf in the Figure 1.7.11 (b), however, looks almost symmetric. It is a simple matter to verify that a random variable distributed as Beta$(1, 1)$ is equivalent to the Uniform$(0, 1)$

random variable.

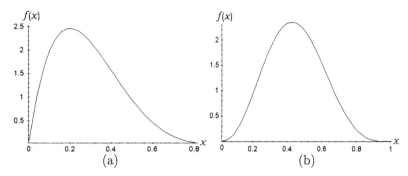

Figure 1.7.11. Beta Densities: (a) $\alpha = 2, \beta = 5$ (b) $\alpha = 4, \beta = 5$

The Negative Exponential Distribution: We say that a continuous random variable X has the negative exponential distribution involving γ and β, if and only if its pdf is given by

$$f(x) = \beta^{-1} e^{-(x-\gamma)/\beta} \text{ for } \gamma < x < \infty, \qquad (1.7.36)$$

where $-\infty < \gamma < \infty, 0 < \beta < \infty$. Here, γ and β are referred to as *parameters*. This distribution is widely used in modeling data arising from experiments in reliability and life tests. When the minimum *threshold* parameter γ is assumed zero, one then goes back to the exponential distribution introduced earlier in (1.7.23).

In the Exercise 1.7.23, we have asked to plot this pdf for different values of β and γ. Based on these plots, for some fixed number a, the reader should think about the possible monotonicity property of $P_{\gamma,\beta}\{X > a\}$ as (i) a function of β when γ is kept fixed, or (ii) a function of γ when β is kept fixed.

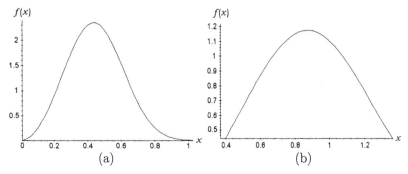

Figure 1.7.12. Weibull Densities: (a) $\alpha = 3, \beta = .5$ (b) $\alpha = 3, \beta = 1$

The Weibull Distribution: We say that a positive continuous random variable X has the Weibull distribution if and only if its pdf is given by

$$f(x) = \alpha\beta^{-\alpha}x^{\alpha-1}\exp\left(-[x/\beta]^{\alpha}\right)I(x > 0), \qquad (1.7.37)$$

where $\alpha(> 0)$ and $\beta(> 0)$ are referred to as *parameters*. This pdf is also skewed to the right. By varying the values of α and β, one can generate interesting shapes for the associated pdf. The Weibull pdf from (1.7.37) has been plotted in the Figure 1.7.12 for $(\alpha, \beta) = (3, .5), (3, 1)$. These figures are skewed to the right too.

The Rayleigh Distribution We say that a positive continuous random variable X has the Rayleigh distribution if and only if its pdf is given by

$$f(x) = 2\theta^{-1}x\exp(-x^2/\theta) \text{ for } 0 < x < \infty, \qquad (1.7.38)$$

where $\theta(> 0)$ is referred to as a *parameter*. In reliability studies and related areas, this distribution is used frequently.

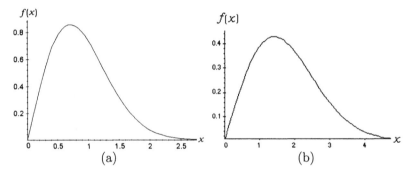

Figure 1.7 13. Raleigh Densities: (a) $\theta = 1$ (b) $\theta = 4$

The pdf from (1.7.38) has been plotted in the Figure 1.7.13 for $\theta = 1, 4$. This pdf is again skewed to the right. The tail in the rhs became heavier as θ increased from 1 to 4.

1.8   Exercises and Complements

1.1.1 (Example 1.1.1 Continued) In the set up of the Example 1.1.1, find the probability of observing no heads or two heads.

1.1.2 (Example 1.1.2 Continued) In the set up of the Example 1.1.2, find the probability of observing a difference of one between the scores which come up on the two dice. Also, find the probability of observing the red die scoring higher than that from the yellow die.

1.1.3 Five equally qualified individuals consisting of four men and one woman apply for two identical positions in a company. The two positions are filled by selecting two from this applicant pool at random.

(*i*)   Write down the sample space **S** for this random experiment;
(*ii*) Assign probabilities to the simple events in the sample space **S**;
(*iii*) Find the probability that the woman applicant is selected for a position.

1.2.1 One has three fair dice which are red, yellow and brown. The three dice are rolled on a table at the same time. Consider the following two events:

A : The sum total of the scores from all three dice is 10
B : The sum total of the scores from the red and brown dice exceeds 8

Find $P(A), P(B)$ and $P(A \cap B)$.

1.2.2 Prove the set relations given in (1.2.2).

1.2.3 Suppose that A, B, C are subsets of **S**. Show that
(*i*) $A \triangle C \subseteq (A \triangle B) \cup (B \triangle C)$;
(*ii*) $(A \triangle B) \cup (B \triangle C)$
        $= [(A \cup B) \cap (A \cap B)^c] \cup [(B \cup C) \cap (B \cap C)^c].$
{*Hint*: Use the Venn Diagram.}

1.2.4 Suppose that $A_1, ..., A_n$ are Borel sets, that is they belong to $\mathcal{B}$. Define the following sets: $B_1 = A_1, B_2 = A_2 \cap A_1^c, B_3 = A_3 \cap (A_1 \cup A_2)^c, ..., B_n = A_n \cap (A_1 \cup ... \cup A_{n-1})^c$. Show that
(*i*) $B_1, ..., B_n$ are Borel sets;
(*ii*) $B_1, ..., B_n$ are disjoint sets;
(*iii*) $\cup_{i=1}^n A_i = \cup_{i=1}^n B_i.$

1.2.5 Suppose that $S = \{(x,y) : x^2 + y^2 \le 1\}$. Extend the ideas from the Example 1.2.3 to obtain a partition of the circular disc, **S**. {*Hint:* How about considering $A_i = \{(x,y) : \frac{1}{2^i} < x^2 + y^2 \le \frac{1}{2^{i-1}}\}, i = 1, 2, ...?$ Any other possibilities?}

1.3.1 Show that the Borel sigma-field $\mathcal{B}$ is closed under the operation of (i) finite intersection, (ii) countably infinite intersection, of its members. {*Hint*: Can DeMorgan's Law (Theorem 1.2.1) be used here?}

1.3.2 Suppose that A and B are two arbitrary Borel sets. Then, show that $A \triangle B$ is also a Borel set.

1.3.3 Suppose that $A_1, ..., A_k$ are disjoint Borel sets. Then show that $P(\cup_{i=1}^{k} A_i) = \Sigma_{i=1}^{k} P(A_i)$. Contrast this result with the requirement in part (iii) of the Definition 1.3.4.

1.3.4 Prove part (v) in the Theorem 1.3.1.

1.3.5 Consider a sample space **S** and suppose that $\mathcal{B}$ is the Borel sigma-field of subsets of **S**. Let A, B, C be events, that is these belong to $\mathcal{B}$. Now, prove the following statements:

(i) $P(A \cup B) \leq P(A) + P(B)$;

(ii) $P(A \cap B) \geq P(A) + P(B) - 1$;

(iii) $P(A \cap C) \leq \min\{P(A), P(C)\}$;

(iv) $P(A \cup B \cup C) = P(A) + P(B) + P(C) - P(A \cap B)$
       $- P(A \cap C) - P(B \cap C) + P(A \cap B \cap C)$.

{*Hint*: Use the Theorem 1.3.1 to prove Parts (i)-(iii). Part (iv) should follow from the Theorem 1.3.1, part (iv). The result in part (ii) is commonly referred to as the *Bonferroni inequality*. See the Theorem 3.9.10.}

1.3.6 (Exercise 1.3.5 Continued) Suppose that $A_1, ..., A_n$ are Borel sets, that is they belong to $\mathcal{B}$. Show that

$$P(A_1 \cup ... \cup A_n) = \Sigma_{i=1}^{n} P(A_i) - \Sigma\Sigma_{j>i=1}^{n} P(A_i \cap A_j)$$

$$+ \Sigma\Sigma\Sigma_{k>j>i=1}^{n} P(A_i \cap A_j \cap A_k) - ... + (-1)^{n-1} P(A_1 \cap ... \cap A_n).$$

{*Hint*: Use mathematical induction and part (iv) in the Exercise 1.3.5.}

1.3.7 (Exercise 1.3.5 Continued) Suppose that $A_1, ..., A_n$ are Borel sets, that is they belong to $\mathcal{B}$. Show that

$$P(A_1 \cap ... \cap A_n) \geq \Sigma_{i=1}^{n} P(A_i) - (n-1).$$

This is commonly referred to as the *Bonferroni inequality*. {*Hint*: Use mathematical induction and part (ii) in the Exercise 1.3.5. See the Theorem 3.9.10.}

1.3.8 Suppose that A_1, A_2 are Borel sets, that is they belong to $\mathcal{B}$. Show that

$$P(A_1 \Delta A_2) = P(A_1) + P(A_2) - 2P(A_1 \cap A_2),$$

where recall from (1.2.1) that $A_1 \Delta A_2$ stands for the symmetric difference of A_1, A_2.

1.3.9 Suppose that A_1, A_2, A_3 are Borel sets, that is they belong to $\mathcal{B}$. Recall from (1.2.1) that $A_i \Delta A_j$ stands for the symmetric difference of A_i, A_j. Show that

(i) $A_1 \Delta A_3 \subseteq (A_1 \Delta A_2) \cup (A_2 \Delta A_3)$;

(ii) $P(A_1 \Delta A_3) \leq P(A_1 \Delta A_2) + P(A_2 \Delta A_3)$.

1.4.1 Prove the Theorem 1.4.1.

1.4.2 In the Theorem 1.4.2, prove that (ii) $\Rightarrow$ (iii) $\Rightarrow$ (iv).

1.4.3 Consider a sample space **S** and suppose that A, B are two events which are mutually exclusive, that is, $A \cap B = \varphi$. If $P(A) > 0, P(B) > 0$, then can these two events A, B be independent? Explain.

1.4.4 Suppose that A and B are two events such that $P(A) = .7999$ and $P(B) = .2002$. Are A and B mutually exclusive events? Explain.

1.4.5 A group of five seniors and eight first year graduate students are available to fill the vacancies of local news reporters at the radio station in a college campus. If four students are to be randomly selected from this pool for interviews, find the probability that at least two first year graduate students are among the chosen group.

1.4.6 Four cards are drawn at random from a standard playing pack of fifty two cards. Find the probability that the random draw will yield

    (i) an ace and three kings;

    (ii) the ace, king, queen and jack of clubs;

    (iii) the ace, king, queen and jack from the same suit;

    (iv) the four queens.

1.4.7 In the context of the Example 1.1.2, let us define the three events:

    E : The sum total of the scores from the two up faces exceeds 8

    F : The score on the red die is twice that on the yellow die

    G : The red die scores one point more than the yellow die

Are the events E, F independent? Are the events E, G independent?

1.4.8 (Example 1.4.7 Continued) In the Example 1.4.7, the prevalence of the disease was 40% in the population whereas the diagnostic blood test had the 10% false negative rate and 20% false positive rate. Instead assume that the prevalence of the disease was $100p\%$ in the population whereas the diagnostic blood test had the $100\alpha\%$ false negative rate and $100\beta\%$ false positive rate, $0 < p, \alpha, \beta < 1$. Now, from this population an individual is selected at random and his blood is tested. The health professional is informed that the test indicated the presence of the particular disease. Find the expression of the probability, involving p, α and β, that this individual does indeed have the disease. {*Hint*: Try and use the Bayes's Theorem.}

1.4.9 Let us generalize the scenario considered in the Examples 1.4.5-1.4.6. The urn #1 contains eight green and twelve blue marbles whereas the urn #2 has ten green and eight blue marbles. Suppose that we also have the urn #3 which has just five green marbles. These marbles have the same size and weight. Now, the experimenter first randomly selects an urn, with equal probability, and from the selected urn draws one marble at random. Given that the selected marble was green, find the probability that

(i) the urn #1 was selected;
(ii) the urn #2 was selected;
(iii) the urn #3 was selected.

{*Hint*: Try and use the Bayes Theorem.}

1.4.10 (Exercise 1.2.1 Continued) One has three fair dice which are red, yellow and brown. The three dice are rolled on a table at the same time. Consider the following events:

A :   The total score from all three dice is 10
B :   The total score from the red and brown dice exceeds 8
Are A, B independent events?

1.4.11 (Example 1.4.4 Continued) Show that $P(A) = P(B)$ assuming that there are respectively m, n green and blue marbles in the urn to begin with. This shows that the result $P(A) = P(B)$ we had in the Example 1.4.4 was not just a coincidence after all.

1.4.12 Show that $\Sigma_{r=0}^{n} \binom{n}{r} = 2^n$. {*Hint*: Can the binomial theorem, namely $(a + b)^n = \Sigma_{r=0}^{n} \binom{n}{r} a^r b^{n-r}$, be used here?}

1.4.13 Suppose that we have twenty beads of which eight are red, seven are green and five are blue. If we set them up in a row on a table, how many different patterns are possible?

1.4.14 Suppose that four boys and four girls are waiting to occupy a chair each, all placed in a row adjacent to each other. If they are seated randomly, what is the probability that the boys and girls will alternate? {*Hint*: The total possible arrangement is 8!. Note that each arrangement could start with a boy or a girl.}

1.4.15 Suppose that we have n different letters for n individuals as well as n envelopes correctly addressed to those n individuals. If these letters are randomly placed in these envelopes so that exactly one letter goes in an envelope, then what is the probability that at least one letter will go in the correct envelope? Obtain the expression of this probability when we let n go to ∞. {*Hint*: Let A_i be the event that the ith letter is stuffed in its correct envelope, $i = 1, ..., n$. We are asked to evaluate $P(A_1 \cup ... \cup A_n)$. Apply the result from the Exercise 1.3.6. Observe also that $P(A_i) = \frac{(n-1)!}{n!}, P(A_i \cap A_j) = \frac{(n-2)!}{n!}, P(A_i \cap A_j \cap A_k) = \frac{(n-3)!}{n!}, ..., P(A_1 \cap ... \cap A_n) = \frac{1}{n!}$, for $i \neq j \neq k...$. Now all there is left is to count the number of terms in the single, double, triple sums and so on. The answer for $P(A_1 \cup ... \cup A_n)$ should simplify to $1 - \frac{1}{2!} + \frac{1}{3!} - ... + (-1)^{n-1} \frac{1}{n!}$ which is approximately $1 - e^{-1}$ for very large n. See (1.6.13).}

1.4.16 Consider the digits $0, 1, ..., 9$. Use these digits at random to form a four (five) digit number. Then, find the probability of forming

   (*i*)   a four digit random number, not starting with a zero, which would be an even number while each digit appears exactly once;

   (*ii*)   a four digit random number which would be an even number where the digits can be repeated, and starting with a zero is allowed;

   (*iii*)   a five digit random number, not starting with a zero, which would be divisible by the number five while each digit appears only once.

1.4.17 In a twin engine plane, we are told that the two engines ($\#1, \#2$) function independently. We are also told that the plane flies just fine when at least one of the two engines are working. During a flying mission, individually the engine $\#1$ and $\#2$ respectively may fail with probability .001 and .01. Then, during a flying mission, what is the probability that the plane would crash? The plane would complete its mission?

1.4.18 Suppose that $A_1, ..., A_k$ are disjoint events. Let B be another event. Then, show that

$$P\{\cup_{i=1}^k A_i \mid B\} = \Sigma_{i=1}^k P\{A_i \mid B\}.$$

1.4.19 Suppose that A_1, A_2 are events. Then, show that

$$P(A_1) = P(A_2) \text{ if and only if } P(A_1 \cap A_2^c) = P(A_1^c \cap A_2).$$

1.4.20 Suppose that A_1, A_2 are disjoint events. Then, show that

   (*i*) $P(A_1 \mid A_2^c) = P(A_1)/\{1 - P(A_2)\}$ if $P(A_2) \neq 1$;

   (*ii*) $P(A_1 \mid A_1 \cup A_2) = P(A_1)/\{P(A_1) + P(A_2)\}$.

1.5.1 Suppose that a random variable X has the following pmf:

X values:	-2	0	1	3	8
Probabilities:	.2	p	.1	$2p$	.4

where $p \in (0, .1]$.

   (*i*)   Is it possible to determine p uniquely?

   (*ii*)   Find $P\{|X - .5| > 2\}$ and $P\{|X - .5| \geq 2.5\}$.

1.5.2 Suppose that a random variable X has the following pmf:

X values:	-2	0	1	3	8
Probabilities:	.2	p	.1	$.3 - p$	.4

where $p \in (0, .3)$. Is it possible to determine p uniquely? With a fixed but arbitrary $p \in (0, .3)$, find $P\{|X - .5| > 2\}, P\{X^2 < 4\}$ and $P\{|X - .3| \geq 1.7\}$.

1.5.3 Suppose that a random variable X has the following pmf:

X values:	-4	-2	1	3	6
Probabilities:	$.3 - p^2$	p^2	$.1$	$2p$	$.4$

where $p^2 \leq .3$. Is it possible to determine p uniquely?

1.5.4 (Example 1.5.2 Continued) Use the form of the df $F(x)$ from the Example 1.5.2 and perform the types of calculations carried out in (1.5.10).

1.5.5 A large envelope has twenty cards of same size. The number two is written on ten cards, the number four is written on six cards, and the number five is written on the remaining four cards. The cards are mixed inside the envelope and we go in to take out one card at random. Let X be the number written on the selected card.

(i)   Derive the pmf $f(x)$ of the random variable X;
(ii)   Derive the df $F(x)$ of the random variable X;
(iii)   Plot the df $F(x)$ and check its points of discontinuities;
(iv)   Perform the types of calculations carried out in (1.5.10).

1.5.6 Prove the Theorem 1.5.1.

1.5.7 An urn contains m red and n blue marbles of equal size and weight. The marbles are mixed inside the urn and then the experimenter selects four marbles at random. Suppose that the random variable X denotes the number of red marbles selected. Derive the pmf of X when initially the respective number of marbles inside the urn are given as follows:

(i) $m = 5, n = 3$;
(ii) $m = 3, n = 5$;
(iii) $m = 4, n = 2$.

In each case, watch carefully the set of possible values of the random variable X wherever the pmf is positive. Next write down the pmf of X when m and n are arbitrary. This situation is referred to as *sampling without replacement* and the corresponding distribution of X is called the *Hypergeometric distribution.*

1.5.8 A fair die is rolled n times independently while we count the number of times (X) the die lands on a three or six. Derive the pmf $f(x)$ and the df $F(x)$ of the random variable X. Evaluate $P(X < 3)$ and $P(|X - 1| > 1)$.

1.6.1 Let c be a positive constant and consider the pdf of a random variable X given by

$$f(x) = \begin{cases} c(2-x) & \text{if } 0 \le x \le 1 \\ 0 & \text{elsewhere.} \end{cases}$$

The Figure 1.8.1 gives a plot of $f(x)$ when $c = \frac{4}{3}$. It is implicit that $x, f(x)$ are respectively plotted on the horizontal and vertical axes.

Find the correct value of c. Find the df $F(x)$ and plot it. Does $F(x)$ have any points of discontinuity? Find the set of points, if non-empty, where $F(x)$ is not differentiable. Calculate $P(-.5 < X \le .8)$. Find the median of this distribution.

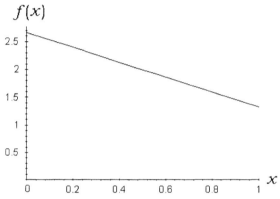

Figure 1.8.1. The PDF $f(x)$ from the Exercise 1.6.1 with $c = \frac{4}{3}$

1.6.2 Let c be a constant and consider the pdf of a random variable X given by

$$f(x) = \begin{cases} c + \frac{1}{4}x & \text{if } -2 \le x \le 0 \\ c - \frac{1}{4}x & \text{if } 0 \le x \le 2 \\ 0 & \text{elsewhere.} \end{cases}$$

Find the value of c. Find the df $F(x)$ and plot it. Does $F(x)$ have any points of discontinuity? Find the set of points, if non-empty, where $F(x)$ is not differentiable. Calculate $P(-1.5 < X \le 1.8)$. Find the median of this distribution.

1.6.3 Let c be a positive constant and consider the pdf of a random variable X given by

$$f(x) = \begin{cases} x & \text{if } 0 \le x \le 1 \\ cx^2 & \text{if } 2 \le x \le 3 \\ 0 & \text{elsewhere.} \end{cases}$$

Find the value of c. Find the df $F(x)$ and plot it. Does $F(x)$ have any points of discontinuity? Find the set of points, if nonempty, where $F(x)$ is not differentiable. Calculate $P(-1.5 < X \le 1.8)$. Find the median of this distribution.

1.6.4 (Example 1.6.5 Continued) Consider the function $f(x)$ defined as follows:

$$f(x) = \begin{cases} 7(4)^{-i} & \text{whenever } x \in (\frac{1}{2^i}, \frac{1}{2^{i-1}}], i = 1, 2, 3, \dots \\ 0 & \text{whenever } x \le 0 \text{ or } x > 1. \end{cases}$$

Let us denote the set $A = \{1, \frac{1}{2}, \frac{1}{2^2}, \frac{1}{2^3}, \dots\}$. Show that

(i)   the function $f(x)$ is a genuine pdf;
(ii)   the associated df $F(x)$ is continuous at all points $x \in A$;
(iii)   Find the set of points, if nonempty, where $F(x)$ is not differentiable.

1.6.5 Along the lines of the construction of the specific pdf $f(x)$ given in the Exercise 1.6.4, examine how one can find other examples of pdf's so that the associated df's are non-differentiable at countably infinite number of points.

1.6.6 Is $f(x) = x^{-2}I(1 < x < \infty)$ a genuine pdf? If so, find $P\{2 < X \le 3\}, P\{|X - 1| \le .5\}$ where X is the associated random variable.

1.6.7 Consider a random variable X having the pdf $f(x) = c(x - 2x^2 + x^3)I(0 < x < 1)$ where c is a positive number and $I(.)$ is the indicator function. Find c first and then evaluate $P(X > .3)$. Find the median of this distribution.

1.6.8 Suppose that a random variable X has the Rayleigh distribution with

$$f(x) = 2\theta^{-1}x\exp(-x^2/\theta)I(x > 0),$$

where $\theta(> 0)$ is referred to as a *parameter*. First show directly that $f(x)$ is indeed a pdf. Then derive the expression of the df explicitly. Find the median of this distribution. {*Hint:* Try substitution $u = x^2/\theta$ during the integration.}

1.6.9 Suppose that a random variable X has the Weibull distribution with

$$f(x) = \alpha^{-1}\beta x^{\beta-1}\exp(-x^\beta/\alpha)I(x > 0),$$

where $\alpha(> 0)$ and $\beta(> 0)$ are referred to as *parameters*. First show directly that $f(x)$ is indeed a pdf. Then derive the expression of the df explicitly. Find the median of this distribution. {*Hint:* Try the substitution $v = x^\beta/\alpha$ during the integration.}

1.6.10 Consider the function $f(x) = e^{-x^2}(1+x^2)$ for $x \in \Re$. Use (1.6.18) to examine the monotonicity of $f(x)$ in x.

1.6.11 Consider the function $f(x) = exp\{x - x^{\frac{1}{2}} - \frac{1}{2}log(x)\}$ for $x > 0$. Use (1.6.18) to show that $f(x)$ is increasing (decreasing) when $\left|x^{1/2} + \frac{1}{2}\right| < (>)\frac{3}{4}$.

1.6.12 Consider the function $f(x) = e^{-|x|}(1+x^2)$ for $x \in \Re$. Use (1.6.18) to examine the monotonicity of $f(x)$ in x.

1.6.13 Use the method of integration by parts from (1.6.28) to evaluate

(i) $\int_0^\infty xe^{-\frac{1}{3}x}dx$;

(ii) $\int_0^\infty x^2e^{-\frac{1}{3}x}dx$;

(iii) $\int_0^\infty xe^{-\frac{1}{3}x^2}dx$;

(iv) $\int_1^2 xlog(x)dx$.

1.6.14 By the appropriate substitutions, express the following in the form of a gamma integral as in (1.6.19). Then evaluate these integrals in terms of the gamma functions.

(i) $\int_0^\infty xe^{-\frac{1}{3}x}dx$;

(ii) $\int_0^\infty x^2e^{-\frac{1}{3}x}dx$;

(iii) $\int_0^\infty xe^{-\frac{1}{3}x^2}dx$.

1.6.15 Use *Stirling's approximation*, namely that $\Gamma(\alpha) \backsim \sqrt{2\pi}e^{-\alpha}\alpha^{\alpha-\frac{1}{2}}$ for large values of α, to prove: $n! \backsim \sqrt{2\pi}e^{-n}n^{n+\frac{1}{2}}$ for large positive integral values of n. {*Hint*: Observe that $n! = \Gamma(n+1) \backsim \sqrt{2\pi}e^{-(n+1)}(n+1)^{n+\frac{1}{2}}$. But, one can rewrite the last expression as $\sqrt{2\pi}e^{-n}n^{n+\frac{1}{2}}\,e^{-1}(1+n^{-1})^n\,(1+n^{-1})^{\frac{1}{2}}$ and then appeal to (1.6.13).}

1.7.1 Consider the random variable which has the following *discrete uniform distribution*:

$$P(X = i) = \frac{1}{n} \text{ for } i = 1, ..., n.$$

Derive the explicit expression of the associated df $F(x), x \in \Re$. Evaluate $\Sigma_{i=1}^n iP(X = i)$ and $\Sigma_{i=1}^n i^2 P(X = i)$. {*Hint*: Use (1.6.11).}

1.7.2 The probability that a patient recovers from a stomach infection is .9. Suppose that ten patients are known to have contracted this infection. Then

(i)   what is the probability that exactly seven will recover?

(ii)   what is the probability that at least five will recover?

(iii) what is the probability that at most seven will recover?

1.7.3 Suppose that a random variable X has the Binomial(n, p) distribution, $0 < p < 1$. Show that

$$P\{X > 1 \mid X \geq 1\} = \frac{1 - (1-p)^n - np(1-p)^{n-1}}{1 - (1-p)^n}.$$

{*Hint*: Use (1.4.3), (1.7.2) and direct calculations.}

1.7.4 Suppose that a random variable X has the Binomial(n, p) distribution, $0 < p < 1$. For all $x = 0, 1, ..., n$, show that

$$\frac{P\{X = x\}}{P\{X = x+1\}} = \frac{(x+1)(1-p)}{(n-x)p}.$$

This recursive relation helps enormously in computing the binomial probabilities successively for all n, particularly when n is large. {*Hint*: Use (1.7.2) and direct calculations.}

1.7.5 (Exercise 1.7.4 Continued) Suppose that a random variable X has the Binomial(n, p) distribution, $0 < p < 1$. Find the value of x at which $P(X = x)$ is maximized. {*Hint*: Use the result from the Exercise 1.7.4 to order the probabilities $P(X = x)$ and $P(X = x+1)$ for each x first.}

1.7.6 The switchboard rings at a service desk according to a Poisson distribution on the average five $(= \lambda)$ times in a ten minute interval. What is the probability that during a ten minute interval, the service desk will receive

(*i*)   no more than three calls?
(*ii*)   at least two calls?
(*iii*) exactly five calls?

1.7.7 Suppose that a random variable X has the Poisson(λ) distribution, $0 < \lambda < \infty$. Show that

$$P\{X > 1 \mid X \geq 1\} = \frac{1 - e^{-\lambda} - \lambda e^{-\lambda}}{1 - e^{-\lambda}}.$$

{*Hint*: Use (1.4.3), (1.7.4) and direct calculations.}

1.7.8 Suppose that a random variable X has the Poisson(λ) distribution, $0 < \lambda < \infty$. For all $x = 0, 1, 2, ...$, show that

$$\frac{P\{X = x\}}{P\{X = x+1\}} = \frac{(x+1)}{\lambda}.$$

This recursive relation helps enormously in computing the Poisson probabilities successively for all x, particularly when x is large. {*Hint*: Use (1.7.4) and direct calculations.}

1.7.9 In this exercise, you are given the expressions of the pdf of different random variables. In each case, identify the random variable by its standard name and specify the values of the associated parameters. Also, find the value of c in each case.

(*i*) $f(x) = cexp(-\pi x)I(x > 0)$, c is a positive constant;

(*ii*) $f(x) = cexp(-\pi x^2)$, $x \in \Re$, and c is a positive constant;

(*iii*) $f(x) = cexp(-x^2 - \frac{1}{4}x)$, $x \in \Re$, and c is a positive constant;

(*iv*) $f(x) = 4x^c exp(-2x)I(x > 0)$, c is a positive constant;

(*v*) $f(x) = \frac{128}{3}x^4 exp(-cx)I(x > 0)$, c is a positive constant;

(*vi*) $f(x) = 105x^4(1 - x)^c I(0 < x < 1)$, c is a positive constant.

1.7.10 Suppose that a random variable X has the Gamma$(2, 1)$ distribution with its pdf given by (1.7.20). Find the expressions for $P(X < a), P(X > b)$ in the simplest form where a, b are positive numbers. {*Hint:* Observe that $P(X < a) = \int_0^a xe^{-x}dx$ and evaluate this integral by parts.}

1.7.11 Suppose that a random variable X has the Gamma$(3, 2)$ distribution with its pdf given by (1.7.20). Find the expressions for $P(X < a), P(X > b)$ in the simplest form where a, b are positive numbers. {*Hint:* Observe that $P(X < a) = \frac{1}{16}\int_0^a x^2 e^{-x/2}dx$ and evaluate this integral by parts.}

1.7.12 Suppose that a random variable X has the χ_2^2 distribution. Derive the expression of the associated df explicitly. Also, find $P(X > a + b \mid X > a)$ where a, b are positive numbers. Does this answer depend on the number a? Explain.

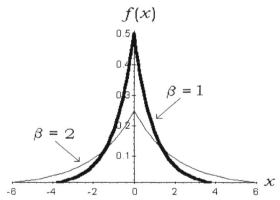

Figure 1.8.2. The PDF $f(x)$ from the Exercise 1.7.13 with $\theta = 0$: Thick $\beta = 1$; Thin $\beta = 2$

1.7.13 Suppose that a random variable X has the following pdf involving

θ and β:

$$f(x) = \tfrac{1}{2\beta} exp\{-|x - \theta|/\beta\} \text{ for all } x \in \Re,$$

where $\theta \in \Re$ and $\beta \in \Re^+$. Here, θ and β are referred to as *parameters*. In the statistical literature, this is known as the *Laplace* or *double exponential* distribution. The Figure 1.8.2 gives a plot of this pdf when $\theta = 0$ and $\beta = 1, 2$. It is implicit that $x, f(x)$ are respectively plotted on the horizontal and vertical axes.

(i)   Show that the pdf $f(x)$ is symmetric about $x = \theta$;

(ii)   Derive the expression of the df explicitly;

(iii)   Show that $P\{|X - \theta| > a + b \mid |X - \theta| > a\}$ does not depend on a, for any two positive numbers a and b;

(iv)   Let $\theta = 0$ and $\beta = 1$. Obtain an expression for the right hand tail area probability, $P(X > a), a > 0$. Then compare tail area probability with that of the Cauchy distribution from (1.7.31). Which distribution has "heavier" tails? Would this answer change if $\theta = 0$ and $\beta = 2$ instead? {*Hint*: Recall the types of discussions we had around the Table 1.7.1.}

1.7.14 (Exercise 1.6.7 Continued) Consider a random variable X having the pdf $f(x) = c(x - 2x^2 + x^3)I(0 < x < 1)$ where c is a positive number and $I(.)$ is the indicator function. Find the value of c. {*Hint*: In order to find c, should you match this $f(x)$ up against the Beta density?}

1.7.15 Show that $f(x) = \{\sigma\sqrt{2\pi}\}^{-1}x^{-1}exp[-\{log(x) - \mu\}^2/(2\sigma^2)]I(x > 0)$ given by (1.7.27) is a bona-fide pdf with $\mu \in \Re, \sigma \in \Re^+$. Also verify that

(i) $P(X \leq x) = \Phi(\{log(x) - \mu\}/\sigma)$, for all $x \in \Re^+$;

(ii) $P\{log(X) \leq x\} = \Phi(\{x - \mu\}/\sigma)$, for all $x \in \Re$.

{*Hint:* Show that (a) $f(x)$ is always positive and (b) evaluate $\int_0^\infty f(x)dx$ by making the substitution $u = log(x)$.}

1.7.16 The neighborhood fishermen know that the change of depth $(X,$ measured in feet) at a popular location in a lake from one day to the next is a random variable with the pdf given below:

$$f(x) = cI(-3 < x < 2),$$

where c is a positive number. Then,

(i)   find the value of the constant c;

(ii)   determine the expression of the df;

(iii)   find the median of this distribution.

1.7.17 The neighborhood fishermen know that the change of depth $(X,$ measured in feet) at a popular location in a lake from one day to the next is a random variable with the pdf given below:

$$f(x) = kI(-3 < x < a),$$

where k is a positive number and a is some number. But, it is known that $P(X \leq 2) = .9$. Then,

(i)   find the values of the constants k and a;
(ii)   determine the expression of the df;
(iii)   find the median of this distribution.

1.7.18 Suppose that the shelf life (X, measured in hours) of certain brand of bread has the exponential pdf given by (1.7.23). We are told that 90% of this brand of bread stays suitable for sale at the most for three days from the moment they are put on the shelf. What percentage of this brand of bread would last for sale at the most for two days from the moment they are put on the shelf?

1.7.19 Let X be distributed as $N(\mu, \sigma^2)$ where $\mu \in \Re, \sigma \in \Re^+$.

(i)   Suppose that $P(X \leq 10) = .25$ and $P(X \leq 50) = .75$. Can μ and σ be evaluated uniquely?
(ii)   Suppose that median of the distribution is 50 and $P(X > 100) = .025$. Can μ and σ be evaluated uniquely?

1.7.20 Let $\phi(z) = \{\sqrt{2\pi}\}^{-1} e^{-z^2/2}, z \in \Re$. Along the lines of (1.7.15)-(1.7.16) we claim that $\phi(z)$ is a valid pdf. One can alternately use the polar coordinates to show that $\int_{-\infty}^{\infty} \phi(z)dz = 1$ by going through the following steps.

(i)   Show that it is enough to prove: $\int_0^{\infty} e^{-z^2/2}dz = \sqrt{\pi/2}$;
(ii)   Verify: $\left[\int_0^{\infty} e^{-z^2/2}dz\right]^2 = \int_0^{\infty} e^{-u^2/2}du \int_0^{\infty} e^{-v^2/2}dv$. Hence,

show that $\int_{t=0}^{\infty} \int_{s=0}^{\infty} e^{-(u^2+v^2)/2}dudv = \frac{1}{2}\pi$;
(iii)   In the double integral found in part (ii), use the substitutions $u = r\cos(\theta), v = r\sin(\theta), 0 < r < \infty$ and $0 < \theta < 2\pi$. Then,

rewrite $\int_{t=0}^{\infty} \int_{s=0}^{\infty} e^{-(u^2+v^2)/2}dudv = \int_{\theta=0}^{2\pi}\left[\int_{r=0}^{\infty} e^{-r^2/2}rdr\right]d\theta$;
(iv)   Evaluate explicitly to show that $\int_{r=0}^{\infty} e^{-r^2/2}rdr = 1$;
(v)   Does part (iii) now lead to part (ii)?

1.7.21 A soft-drink dispenser can be adjusted so that it may fill μ ounces of the drink per cup. Suppose that the ounces of fill (X) are normally distributed with parameters μ ounce and $\sigma = .25$ ounce. Obtain the setting for μ so that 8-ounce cups will overflow only 1.5% of the time.

1.7.22 Consider an arbitrary random variable Y which may be discrete or continuous. Let A be an arbitrary event (Borel set) defined through the random variable Y. For example, the event A may stand for the set where $Y \geq 2$ or $|Y| > 4 \cup |Y| \leq \frac{1}{2}$ or one of the many other possibilities. Define a new random variable $X = I(A)$, the indicator variable of the set A, that is:

$$X = \begin{cases} 0 & \text{if } A \text{ occurs} \\ 1 & \text{if } A^c \text{ occurs.} \end{cases}$$

Argue that X is a Bernoulli variable, defined in (1.7.1), with $p = P(A)$.

1.7.23 Consider the negative exponential pdf

$$f(x) = \beta^{-1} e^{-(x-\gamma)/\beta} I(\gamma < x < \infty)$$

from (1.7.36) where $\beta \in \Re^+, \gamma \in \Re$. Plot the pdf for several values of β and γ. Answer the following questions by analyzing these plots.

(i)   If β is held fixed, will $P_{\gamma=1}\{X > 3\}$ be larger than $P_{\gamma=2}\{X > 3\}$?

(ii)   If γ is held fixed, will $P_{\beta=2}\{X > 4\}$ be larger than $P_{\beta=3}\{X > 4\}$?

2

Expectations of Functions of Random Variables

2.1 Introduction

In Chapter 1 we introduced the notions of discrete and continuous random variables. We start Chapter 2 by discussing the concept of the *expected value* of a random variable. The expected value of a random variable is sometimes judged as the "center" of the probability distribution of the variable. The *variance* of a random variable then quantifies the average squared deviation of a random variable from its "center". The Section 2.2 develops these concepts. The Section 2.3 introduces the notion of the expected value of a general function of a random variable which leads to the related notions of *moments* and *moment generating functions* (mgf) of random variables. In Section 2.4 we apply a powerful result involving the mgf which says that a finite mgf determines a probability distribution uniquely. We also give an example which shows that the finiteness of all moments alone may not determine a probability distribution uniquely. The Section 2.5 briefly touches upon the notion of a *probability generating function* (pgf) which consequently leads to the idea of *factorial moments* of a random variable.

2.2 Expectation and Variance

Let us consider playing a game. The house will roll a fair die. The player will win \$8 from the house whenever a six comes up but the player will pay \$2 to the house anytime a face other than the six comes up. Suppose, for example, that from ten successive turns we observed the following up faces on the rolled die: $4, 3, 6, 6, 2, 5, 3, 6, 1, 4$. At this point, the player is ahead by \$10(= \$24 − \$14) so that the player's average win per game thus far has been exactly one dollar. But in the long run, what is expected to be the player's win per game? Assume that the player stays in the game k times in succession and by that time the face six appears n_k times while a non-six face appears $k - n_k$ times. At this point, the player's win will then amount to $8n_k - 2(k - n_k)$ so that the player's win per game (W_k) should be $\{8n_k - 2(k - n_k)\}/k$ which is rewritten as follows:

$$W_k = (8)\left(k^{-1}n_k\right) + (-2)\{1 - (k^{-1}n_k)\}. \qquad (2.2.1)$$

What will W_k amount to when k is very large? Interpreting probabilities as the limiting relative frequencies we can say that $\lim_{k\to\infty} \frac{n_k}{k}$ should coincide with the probability of seeing the face six in a single roll of a fair die which is nothing but $\frac{1}{6}$. Hence, the player's ultimate win per game is going to be

$$\lim_{k\to\infty} W_k = (8)(1/6) + (-2)(5/6) = -1/3. \qquad (2.2.2)$$

In other words, in this game the player will lose one-third of a dollar in the long run. It is seen readily from (2.2.2) that we multiplied the possible value of the win with its probability and added up these terms.

In (2.2.2), the final answer is really the weighted average of the two possible values of the win where the weights are the respective probabilities. This is exactly how we interpret this player's *expected* win (per game) in the long haul. The process is intrinsically a limiting one and in general we will proceed to define the expected value of a random variable X as simply the weighted average of all possible values of the random variable. More precise statements would follow.

Let us begin with a random variable X whose pmf or pdf is $f(x)$ for $x \in \mathcal{X} \subseteq \Re$. We use the following *convention*:

In a discrete case, we denote the space $\mathcal{X} = \{x_1, x_2, ...\}$ where $f(x_i)$ is positive for all $i = 1, 2, ...$ and $\Sigma_{i=1}^{\infty} f(x_i) = 1$. In a continuous case, the space $\mathcal{X}$ consists of union of sub-intervals so that $f(x)$ is positive for all $x \in \mathcal{X}$ and $\int_{\mathcal{X}} f(x)dx = 1$. Such a space $\mathcal{X}$ is called the *support of the distribution* of X.

$$(2.2.3)$$

In some examples when X is a discrete random variable, the space $\mathcal{X}$ will consist of finitely many points. One may recall (1.5.1) as an example. On the other hand, in a continuous case in general, the space $\mathcal{X}$ will be the union of subintervals of the real line $\Re$. However, in many examples the space $\mathcal{X}$ will be $\Re, \Re^+$ or the interval $(0,1)$. In general, $\mathcal{X}$ is called the *support of the distribution* of X whether it is a discrete random variable or a continuous random variable.

Definition 2.2.1 *The expected value of the random variable X, denoted by $E(X), E\{X\}$ or $E[X]$, is defined as:*

$$\Sigma_{i:x_i \in \mathcal{X}} \, x_i f(x_i) \qquad \text{when X is discrete, or}$$
$$\int_{\mathcal{X}} \, x f(x)dx \qquad \text{when X is continuous,}$$

where $\mathcal{X}$ is the support of X. The expected value is also called the mean of the distribution and is frequently assigned the symbol μ.

In the Definition 2.2.1, the term $\Sigma_{i:x_i\in\mathcal{X}}\, x_i f(x_i)$ is interpreted as the infinite sum, $x_1 f(x_1) + x_2 f(x_2) + x_3 f(x_3) + \dots$ provided that $x_1, x_2, x_3, \dots$ belong to $\mathcal{X}$, the support of the random variable X.

In statistics, the expected value of X is often referred to as the *center* of the probability distribution $f(x)$. Let us look at the following discrete probability distributions for the two random variables X and Y respectively.

X values: -1 1   3   5   7
Probabilities: .1   .2   .4   .2   .1

Y values: -1 1   3   5   7
Probabilities: .2   .15   .3   .15   .2

(2.2.4)

Here we have $\mathcal{X} = \mathcal{Y} = \{-1, 1, 3, 5, 7\}$. Right away one can verify that

$$E(X) = (-1)(.1) + (1)(.2) + 3(.4) + 5(.2) + 7(.1) = 3,$$
$$E(Y) = (-1)(.2) + (1)(.15) + 3(.3) + 5(.15) + 7(.2) = 3.$$

(2.2.5)

We may summarize the same information by saying that $\mu_X = \mu_Y = 3$.

In a continuous case the technique is not vastly different. Consider a random variable X whose pdf is given by

$$f(x) = \begin{cases} \frac{3}{8}x^2 & \text{if } 0 < x < 2 \\ 0 & \text{elsewhere.} \end{cases}$$

(2.2.6)

Here we have $\mathcal{X} = (0, 2)$. The corresponding expected value can be found as follows:

$$E(X) = \int_0^2 xf(x)dx = \int_0^2 \frac{3}{8}x^3 dx = \frac{3}{8}\left[\frac{1}{4}x^4\right]_{x=0}^{x=2} = \frac{3}{2}.$$

(2.2.7)

We will give more examples shortly. But before doing so, let us return to the two random variables X and Y defined via (2.2.4). Both can take the same set of possible values with positive probabilities but they have different probability distributions. In (2.2.5) we had shown that their expected values were same too. In other words the mean by itself may not capture all the interesting features in a distribution.

We may ask ourselves: Between X and Y, which one is more variable? In order to address this question, we need an appropriate measure of *variation* in a random variable.

Definition 2.2.2 *The variance of the random variable X, denoted by $V(X), V\{X\}$ or $V[X]$, is defined to be $E\{(X - \mu)^2\}$ which is the same as:*

$$\Sigma_{i:x_i\in\mathcal{X}}\, \{x_i - \mu\}^2 f(x_i) \quad \text{when } X \text{ is discrete, or}$$
$$\int_{\mathcal{X}} \{x - \mu\}^2 f(x)dx \quad \text{when } X \text{ is continuous;}$$

where $\mathcal{X}$ is the support of X and $\mu = E(X)$. The variance is frequently assigned the symbol σ^2. The positive square root of σ^2, namely σ, is called the standard deviation of X.

The two quantities μ and σ^2 play important roles in statistics. The variance measures the average squared distance of a random variable X from the center μ of its probability distribution. Loosely speaking, a large value of σ^2 indicates that on the average X has a good chance to stray away from its mean μ, whereas a small value of σ^2 indicates that on the average X has a good chance to stay close to its mean μ.

Next, we state a more general definition followed by a simple result. Then, we provide an alternative formula to evaluate σ^2.

Definition 2.2.3 *Start with a random variable X and consider a function $g(X)$ of the random variable. Suppose that $f(x)$ is the pmf or pdf of X where $x \in \mathcal{X}$. Then, the expected value of the random variable $g(X)$, denoted by $E(g(X)), E\{g(X)\}$ or $E[g(X)]$, is defined as:*

$$\Sigma_{i:x_i \in \mathcal{X}} \, g(x_i)f(x_i) \quad \text{when } X \text{ is discrete, or}$$
$$\int_{\mathcal{X}} g(x)f(x)dx \quad \text{when } X \text{ is continuous,}$$

where $\mathcal{X}$ is the support of X.

Theorem 2.2.1 *Let X be a random variable. Suppose that we also have real valued functions $g_i(x)$ and constants $a_i, i = 0, 1, ..., k$. Then, we have*

$$E\{a_0 + \Sigma_{i=1}^{k} a_i g_i(X)\} = a_0 + \Sigma_{i=1}^{k} a_i E\{g_i(X)\}$$

as long as all the expectations involved are finite. That is, the expectation is a linear operation.

Proof We supply a proof assuming that X is a continuous random variable with its pdf $f(x), x \in \mathcal{X}$. In the discrete case, the proof is similar. Let us write

$$E\{a_0 + \Sigma_{i=1}^{k} a_i g_i(X)\}$$

$$= \int_{\mathcal{X}} \left[a_0 + \{\Sigma_{i=1}^{k} a_i g_i(x)\} \right] f(x)dx$$

$$= \int_{\mathcal{X}} a_0 f(x)dx + \int_{\mathcal{X}} \{\Sigma_{i=1}^{k} a_i g_i(x)\} f(x)dx$$

$$= a_0 \int_{\mathcal{X}} f(x)dx + \Sigma_{i=1}^{k} \int_{\mathcal{X}} a_i g_i(x)f(x)dx, \because a_0 \text{ is a}$$

constant and using property of the integral operations.

Hence, we have

$$E\{a_0 + \Sigma_{i=1}^{k} a_i g_i(X)\} = a_0 + \Sigma_{i=1}^{k} a_i \int_{\mathcal{X}} g_i(x)f(x)dx, \qquad (2.2.8)$$

since a_i's are all constants and $\int_{\mathcal{X}} f(x)dx = 1$. Now, we have the desired result. ∎

Theorem 2.2.2 *Let X be a random variable. Then, we have*

$$V(X) = E(X^2) - E^2(X). \tag{2.2.9}$$

Proof We prove this assuming that X is a continuous random variable with its pdf $f(x), x \in \mathcal{X}$. In a discrete case, the proof is similar. Note that μ, which is $E(X)$, happens to be a constant. So, from the Definition 2.2.2 we have

$$V(X) = \int_{\mathcal{X}} (x - \mu)^2 f(x)dx = E(X^2 - 2\mu X + \mu^2).$$

Hence, in view of the Theorem 2.2.1, we have

$$V(X) = E(X^2) - 2\mu E(X) + \mu^2 = E(X^2) - 2\mu^2 + \mu^2 = E(X^2) - \mu^2, \tag{2.2.10}$$

which is the desired result. ∎

Example 2.2.1 We now go back to the two discrete random variables X, Y defined in (2.2.4). Recall that $\mu_X = \mu_Y = 3$. Now, using the Definition 2.2.3 we note that

$$\begin{aligned} E(X^2) &= (1)(.1) + (1)(.2) + 9(.4) + 25(.2) + 49(.1) = 13.8, \\ E(Y^2) &= (1)(.2) + (1)(.15) + 9(.3) + 25(.15) + 49(.2) = 16.6. \end{aligned} \tag{2.2.11}$$

Then using the Theorem 2.2.2, we have the corresponding variances $\sigma_X^2 = 4.8$ and $\sigma_Y^2 = 7.6$. We find that the random variable Y is more variable than the random variable X. Incidentally, the associated standard deviations are $\sigma_X \approx 2.19, \sigma_Y \approx 2.76$ respectively. ▲

Example 2.2.2 Next we go back to the continuous random variable X defined in (2.2.6). Recall from (2.2.7) that X had its mean equal 1.5. Next, using the Definition 2.2.3 we note that

$$E(X^2) = \int_0^2 x^2 f(x)dx = \tfrac{3}{8} \int_0^2 x^4 dx = \tfrac{3}{8} \left[\tfrac{1}{5} x^5 \right]_{x=0}^{x=2} = \tfrac{12}{5}.$$

Thus, using the Theorem 2.2.2, we have $V(X) = \sigma_X^2 = \tfrac{12}{5} - (\tfrac{3}{2})^2 = \tfrac{3}{20} = .15$ and the associated standard deviation is $\sigma_X \approx .387$. ▲

Next, we state two simple but useful results. Proofs of Theorems 2.2.3-2.2.4 have respectively been included as Exercises 2.2.18-2.2.19 with some hints.

Theorem 2.2.3 *Let X and Y be random variables. Then, we have*

(i) $a \le E(X) \le b$ if the support $\mathcal{X}$ of X is the interval $[a, b]$;

(ii) $E(X) \le E(Y)$ if $X \le Y$ w.p.1.

Theorem 2.2.4 *Let X be a random variable. Then, we have*

$$V(aX + b) = a^2 V(X),$$

where a and b are any two fixed real numbers.

We now consider another result which provides an interesting perspective by expressing the mean of a *continuous* random variable X in terms of its tail area probabilities when X is assumed *non-negative*.

Theorem 2.2.5 *Let X be a non-negative continuous random variable with its distribution function $F(x)$. Suppose that $\lim_{x \to \infty} x\{1 - F(x)\} = 0$. Then, we have:*

$$E(X) = \int_0^\infty \{1 - F(x)\} dx$$

Proof We have assumed that $X \ge 0$ w.p.1 and thus

$$
\begin{aligned}
E(X) &= \int_0^\infty x f(x) dx \\
&= \int_0^\infty x dF(x), \because dF(x)/dx = f(x) \text{ from (1.6.10)} \\
&= -\int_0^\infty x d\{1 - F(x)\} \\
&\quad - \left\{ [x\{1 - F(x)\}]_{x=0}^{x=\infty} - \int_0^\infty \{1 - F(x)\} dx \right\}, \\
&\quad \text{using integration by parts from (1.6.28)} \\
&= \int_0^\infty \{1 - F(x)\} dx \text{ since } \lim_{x \to \infty} x\{1 - F(x)\} \\
&\quad \text{is assumed to be zero.}
\end{aligned}
$$

The proof is now complete. ■

We can write down a discrete analog of this result as well. Look at the following result.

Theorem 2.2.6 *Let X be a positive integer valued random variable with its distribution function $F(x)$. Then, we have*

$$E(X) = \Sigma_{x=0}^\infty \{1 - F(x)\}.$$

Proof Recall that $F(x) = P(X \le x)$ so that $1 - F(x) = P(X > x)$ for $x = 1, 2, \ldots$. Let us first verify that

$$X = \Sigma_{x=0}^\infty I(X > x) \text{ w.p.1,} \tag{2.2.12}$$

where as usual $I(X > x)$ stands for the indicator function of the event $X > x$. If X happens to be one, then $I(X > 0) = 1$, but $I(X > x) = 0$ for all $x = 1, 2, \ldots$. Hence, the rhs of (2.2.12) is also one. In other words when $X = 1$, the two sides of (2.2.12) agree. Now if X happens to be two, then $I(X > 0) = I(X > 1) = 1$, but $I(X > x) = 0$ for all $x = 2, 3, \ldots$. Hence, the rhs of (2.2.12) is also then two. In other words when $X = 2$, the two sides of (2.2.12) agree. The reader should check that (2.2.12) holds whatever be the value of X observed from the set $\{1, 2, \ldots\}$.

For any fixed $x \in \{0, 1, 2, \ldots\}$ the random variable $I(X > x)$ takes only one of the values 0 or 1. Hence from the Definition 2.2.1 of the expectation or using the Example 2.2.3 which follows shortly, we can claim that $E[I(X > x)] = P(X > x)$ and this equals $1 - F(x)$.

Now the desired result will follow provided that the expectation operation "E" can be taken inside the summation in order to write $E[\Sigma_{x=0}^{\infty} I(X > x)] = \Sigma_{x=0}^{\infty} E[I(X > x)] = \Sigma_{x=0}^{\infty} P(X > x)$. Since $I(X > x) \geq 0$ w.p.1, indeed by applying the Monotone Convergence Theorem (refer to the Exercise 2.2.24) one can justify exchanging the expectation and the infinite summation. The proof is now complete. ∎

The Exercise 2.2.21 provides an alternate sufficient condition for the conclusion of the Theorem 2.2.5 to hold. The Exercises 2.2.22-2.2.23 give interesting applications of the Theorems 2.2.5-2.2.6.

At this point we consider some of the standard distributions from Section 1.7 and derive the expressions for their means and variances. The exact calculation of the mean and variance needs special care and attention to details. In what follows, we initially emphasize this.

2.2.1   The Bernoulli Distribution

From (1.7.1) recall that a Bernoulli(p) random variable X takes the values 1 and 0 with probability p and $1 - p$ respectively. Then according to the Definitions 2.2.1-2.2.3 and using (2.2.9), we have

$$\mu = E(X) = (1)(p) + (0)(1 - p) = p,$$

$$E(X^2) = (1)^2(p) + (0)^2(1 - p) = p, \qquad (2.2.13)$$

$$\sigma^2 = V(X) = E(X^2) - E^2(X) = p - p^2 = p(1 - p).$$

Example 2.2.3 (Exercise 1.7.22 Continued) Consider an arbitrary random variable Y which may be discrete or continuous. Let A be an arbitrary event (Borel set) defined through the random variable Y. For example, the event A may stand for the set where $Y \geq 2$ or $|Y| > 4 \cup |Y| \leq \frac{1}{2}$ or one of

the many other possibilities. Define a new random variable $X = I(A)$, the
indicator variable of the set A, that is:

$$X = \begin{cases} 0 & \text{if } A^c \text{ occurs} \\ 1 & \text{if } A \text{ occurs.} \end{cases}$$

Then X is a Bernoulli random variable, defined in (1.7.1), with $p = P(A)$.
Hence, applying (2.2.13) we conclude that $\mu = P(A)$ and $\sigma^2 = P(A)\{1 - P(A)\}$. Consider selecting a random digit from $0, 1, 2, ..., 9$ with equal probability and let A be the event that the random digit is divisible by 4. Then,
$P(A) = \frac{2}{10} = .2$ so that for the associated Bernoulli random variable X
one concludes that $\mu = .2$ and $\sigma^2 = .16$. ▲

2.2.2   The Binomial Distribution

Suppose that X has the Binomial(n, p) distribution with its pmf $f(x) = \binom{n}{x} p^x (1 - p)^{n-x}$ where $x = 0, 1, ..., n$, $0 < p < 1$, given by (1.7.2). Now,
observe that $x! = x(x - 1)!, n! = n(n - 1)!$ for $x \geq 1, n \geq 1$ and so we can
write

$$\Sigma_{x=0}^n x \binom{n}{x} p^x (1 - p)^{n-x} = n \sum_{x=1}^n \frac{(n-1)!}{(x-1)!(n-x)!} p^x (1 - p)^{n-x}.$$

Thus, using the Binomial Theorem, we obtain

$$E(X) = np\Sigma_{x=1}^n \binom{n-1}{x-1} p^{x-1}(1 - p)^{n-x} = np\{p + (1 - p)\}^{n-1} = np.$$
$$(2.2.14)$$

In order to evaluate the variance, let us use the Theorems 2.2.1-2.2.2 and
note that

$$V(X) = E\{X(X - 1) + X\} - \mu^2 = E\{X(X - 1)\} + \mu - \mu^2. \quad (2.2.15)$$

We now proceed along the lines of (2.2.14) by omitting some of the intermediate steps and write

$$E\{X(X - 1)\} = n(n - 1)p^2\Sigma_{x=2}^n \binom{n-2}{x-2} p^{x-2}(1 - p)^{n-x} = n(n - 1)p^2.$$
$$(2.2.16)$$

Next, we combine (2.2.15)-(2.2.16) to obtain

$$V(X) = n(n - 1)p^2 + np - (np)^2 = np(1 - p).$$

In other words, for the Binomial(n, p) variable, one has

$$\mu = np \text{ and } \sigma^2 = np(1 - p). \quad (2.2.17)$$

2.2.3 The Poisson Distribution

Suppose that X has the Poisson(λ) distribution with its pmf $f(x) = e^{-\lambda}\lambda^x/x!$ where $x = 0, 1, ...,$ $0 < \lambda < \infty$, given by (1.7.4). In order to find the mean and variance of this distribution, one may proceed in the same way as in the binomial case. After the dust settles, one has to find the sums of the following two infinite series:

$$I = \Sigma_{x=1}^{\infty} x e^{-\lambda}\lambda^x (x!)^{-1} \text{ and } II = \Sigma_{x=2}^{\infty} x(x-1)e^{-\lambda}\lambda^x (x!)^{-1}.$$

Now, $I = \lambda e^{-\lambda} \Sigma_{x=1}^{\infty} \lambda^{x-1}\{(x-1)!\}^{-1} = \lambda e^{-\lambda}e^{\lambda} = \lambda$ and similarly $II = \lambda^2$. Here, the exponential series expansion from (1.6.15) helps. The details are left out as the Exercise 2.2.7. Finally, one has

$$\mu = \lambda \text{ and } \sigma^2 = \lambda. \tag{2.2.18}$$

2.2.4 The Uniform Distribution

Suppose that X has the Uniform(α, β) distribution with its pdf $f(x) = (\beta - \alpha)^{-1}$ for $\alpha < x < \beta$, given by (1.7.12). Here α, β are two real numbers. In order to find the mean and variance of this distribution, we proceed as follows. Let us write

$$E(X) = \int_\alpha^\beta x f(x)dx = (\beta - \alpha)^{-1} \left[\tfrac{1}{2}x^2\right]_{x=\alpha}^{x=\beta} = \tfrac{1}{2}(\alpha + \beta). \tag{2.2.19}$$

The preceding answer should not be surprising. The uniform distribution puts the same density or weight, uniformly on each point $x \in (\alpha, \beta)$, so that we should expect the midpoint of the interval (α, β) to be designated as the mean or the center of this distribution. Next, we write

$$E(X^2) = \int_\alpha^\beta x^2 f(x)dx = (\beta - \alpha)^{-1} \left[\tfrac{1}{3}x^3\right]_{x=\alpha}^{x=\beta} = \tfrac{1}{3}(\beta^2 + \alpha\beta + \alpha^2). \tag{2.2.20}$$

Now we combine the Theorem 2.2.2 and (2.2.19)-(2.2.20) to claim that

$$V(X) = E(X^2) - \mu^2 = \tfrac{1}{3}(\beta^2 + \alpha\beta + \alpha^2) - \tfrac{1}{4}(\alpha + \beta)^2 = \tfrac{1}{12}(\beta - \alpha)^2.$$

Finally, we summarize:

$$\mu = \tfrac{1}{2}(\alpha + \beta) \text{ and } \sigma^2 = \tfrac{1}{12}(\beta - \alpha)^2. \tag{2.2.21}$$

2.2.5 The Normal Distribution

First, let us consider the standard normal random variable Z having its pdf $\phi(z) = \{\sqrt{2\pi}\}^{-1}exp(-z^2/2)$ for $-\infty < z < \infty$, given by (1.7.16).

In order to find the mean and variance of this distribution, we make the one-to-one substitution $u = -z$ along the lines of (1.7.15) when $z < 0$, and write

$E(Z)$

$$= \int_{-\infty}^{\infty} z\phi(z)dz$$

$$= \{\sqrt{2\pi}\}^{-1} \left[\int_{-\infty}^{0} z\exp(-\tfrac{1}{2}z^2)dz + \int_{0}^{\infty} z\exp(-\tfrac{1}{2}z^2)dz \right] \qquad (2.2.22)$$

$$= \{\sqrt{2\pi}\}^{-1} \left[-\int_{0}^{\infty} u\exp(-\tfrac{1}{2}u^2)du + \int_{0}^{\infty} z\exp(-\tfrac{1}{2}z^2)dz \right]$$

$$= -I_1 + J_1, \text{ say.}$$

But, the two integrals I_1, J_1 are equal. Hence, $E(Z) = 0$. Next, in order to evaluate $E(Z^2)$ we first write

$$E(Z^2) = \{\sqrt{2\pi}\}^{-1} \left[\int_{-\infty}^{0} z^2 \exp(-\tfrac{1}{2}z^2)dz + \int_{0}^{\infty} z^2 \exp(-\tfrac{1}{2}z^2)dz \right].$$

Then, we proceed as before with the one-to-one substitution $u = -z$ when $z < 0$, and rewrite

$E(Z^2)$

$$= \{\sqrt{2\pi}\}^{-1} \left[\int_{\infty}^{0} (-u)^2 \exp(-\tfrac{1}{2}u^2)(-du) + \int_{0}^{\infty} z^2 \exp(-\tfrac{1}{2}z^2)dz \right]$$

$$= \{\sqrt{2\pi}\}^{-1} \left[\int_{0}^{\infty} u^2 \exp(-\tfrac{1}{2}u^2)du + \int_{0}^{\infty} z^2 \exp(-\tfrac{1}{2}z^2)dz \right]$$

$$= 2\{\sqrt{2\pi}\}^{-1} \int_{0}^{\infty} u^2 \exp(-\tfrac{1}{2}u^2)du$$

since the two integrals in the previous step are equal.

$$(2.2.23)$$

Now, in order to evaluate the last integral from (2.2.23), we further make the one-to-one substitution $v = \tfrac{1}{2}u^2$ when $u > 0$, and proceed as follows:

$$\{\sqrt{2\pi}\}^{-1} \int_{0}^{\infty} u^2 \exp(-\tfrac{1}{2}u^2)du$$

$$= \{\sqrt{\pi}\}^{-1} \int_{0}^{\infty} v^{1/2} \exp(-v)dv$$

$$= \{\sqrt{\pi}\}^{-1}\Gamma(\tfrac{3}{2}), \text{ where } \Gamma(.) \text{ comes from (1.6.19)} \qquad (2.2.24)$$

$$= \{\sqrt{\pi}\}^{-1}\tfrac{1}{2}\Gamma(\tfrac{1}{2}),$$

which reduces to $\tfrac{1}{2}$ since $\Gamma(\tfrac{1}{2}) = \sqrt{\pi}$. Refer to (1.6.20). Now, combining (2.2.23)-(2.2.24) we see clearly that $E(Z^2) = 1$. In other words, the mean and variance of the standard normal variable Z are given by

$$\mu_Z = 0 \text{ and } \sigma_Z^2 = 1. \qquad (2.2.25)$$

The result in (2.2.22), namely that $E(Z) = 0$, may not surprise anyone because the pdf $\phi(z)$ is symmetric about $z = 0$ and $E(Z)$ is finite. Refer to the Exercise 2.2.13 in this context.

Example 2.2.4 Now, how should one evaluate, for example, $E[(Z - 2)^2]$? Use the Theorem 2.2.1 and proceed directly to write $E[(Z - 2)^2] = E[Z^2] - 4E[Z] + 4 = 1 - 0 + 4 = 5$. How should one evaluate, for example, $E[(Z - 2)(2Z + 3)]$? Again use the Theorem 2.2.1 and proceed directly to write $E[(Z - 2)(2Z + 3)] = 2E[Z^2] - E[Z] - 6 = 2 - 0 - 6 = -4$. ▲

Next suppose that X has the $N(\mu, \sigma^2)$ distribution with its pdf $f(x) = \{\sigma\sqrt{2\pi}\}^{-1} exp\{-(x - \mu)^2/(2\sigma^2)\}$ for $-\infty < x < \infty$, given by (1.7.13). Here we recall that $(\mu, \sigma^2) \in \Re \times \Re^+$.

Let us find the mean and variance of this distribution. One may be tempted to use a result that $(X - \mu)/\sigma$ is standard normal and exploit the expressions summarized in (2.2.25). But, the reader may note that we have not yet derived fully the distribution of $(X - \mu)/\sigma$. We came incredibly close to it in (1.7.18). This matter is delegated to Chapter 4 for a fuller treatment through transformations and other techniques.

Instead we give a direct approach. In order to calculate the mean and variance of X, let us first evaluate $E\{(X - \mu)/\sigma\}$ and $E\{(X - \mu)^2/\sigma^2\}$ by making the one-to-one substitution $w = (x - \mu)/\sigma$ in the respective integrals where $x \in \Re$:

$$
\begin{aligned}
E\{(X &- \mu)/\sigma\} \\
&= \{\sigma\sqrt{2\pi}\}^{-1} \int_{-\infty}^{\infty} [(x - \mu)/\sigma] exp\{-(x - \mu)^2/(2\sigma^2)\}dx \\
&= \{\sqrt{2\pi}\}^{-1} \int_{-\infty}^{\infty} w\, exp\{-w^2/2\}dw \\
&= 0, \text{ since } E(Z) = 0 \text{ in view of (2.2.25).}
\end{aligned}
\tag{2.2.26}
$$

Now, by appealing to the Theorem 2.2.1, we can see that

$$
0 = E\{(X - \mu)/\sigma\} = [E(X)/\sigma] - (\mu/\sigma) \Rightarrow E(X) = \mu,
\tag{2.2.27}
$$

since $\sigma > 0$. Again, with the same substitution we look at

$$
\begin{aligned}
E\{(X &- \mu)^2/\sigma^2\} \\
&= \{\sigma\sqrt{2\pi}\}^{-1} \int_{-\infty}^{\infty} w^2 exp\{-w^2/2\}(\sigma dw) \\
&= \{\sqrt{2\pi}\}^{-1} \int_{-\infty}^{\infty} w^2 exp\{-w^2/2\}dw \\
&= 1, \text{ since } E(Z^2) = 1 \text{ in view of (2.2.25).}
\end{aligned}
\tag{2.2.28}
$$

Next, by appealing to the Theorem 2.2.1 again, we can write

$$
1 = \{E(X^2) - \mu^2\}/\sigma^2 = V(X)/\sigma^2, \text{ since } \mu = E(X) \text{ from (2.2.27).}
\tag{2.2.29}
$$

We combine (2.2.27), (2.2.29) and summarize by saying that for the random variable X distributed as $N(\mu, \sigma^2)$, we have

$$
\mu_X = \mu \text{ and } \sigma_X^2 = \sigma^2.
\tag{2.2.30}
$$

2.2.6 The Laplace Distribution

Suppose that a random variable X has the *Laplace* or the *double exponential* pdf $f(x) = \frac{1}{2\beta} exp\{-|x-\theta|/\beta\}$ for all $x \in \Re$ where $\theta \in \Re$ and $\beta \in \Re^+$. This pdf is symmetric about $x = \theta$. In order to evaluate $E(X)$, let us write

$$\int_{-\infty}^{\infty} xf(x)dx$$
$$= \theta + \frac{1}{2\beta} \int_{-\infty}^{\infty} (x-\theta)e^{-|x-\theta|/\beta}dx$$
$$= \theta + \frac{\beta}{2} \int_{-\infty}^{\infty} ue^{-|u|}du \text{ where } (x-\theta)/\beta = u \qquad (2.2.31)$$
$$= \theta + \frac{\beta}{2} \left[\int_{\infty}^{0} ve^{-v}dv + \int_{0}^{\infty} ue^{-u}du \right] \text{ with } v = -u.$$

This reduces to θ since $\int_{\infty}^{0} ve^{-v}dv = -\int_{0}^{\infty} ue^{-u}du$. Next, the variance, $V(X)$ is given by

$$\frac{\beta^2}{2} \int_{-\infty}^{\infty} u^2 e^{-|u|}du \text{ where } (x-\theta)/\beta = u$$
$$= \frac{\beta^2}{2} \left[-\int_{\infty}^{0} v^2 e^{-v}dv + \int_{0}^{\infty} u^2 e^{-u}du \right] \text{ with } v = -u. \qquad (2.2.32)$$

One can see that $\int_{0}^{\infty} v^2 e^{-v}dv = -\int_{\infty}^{0} v^2 e^{-v}dv$ and hence the last step of (2.2.32) can be rewritten as

$$V(X) = \beta^2 \int_{0}^{\infty} u^2 e^{-u}du = \beta^2 \Gamma(3) = 2\beta^2 \qquad (2.2.33)$$

with $\Gamma(.)$ defined in (1.6.19). For the Laplace distribution, we then summarize our findings as follows:

$$\mu_X = \theta \text{ and } \sigma_X^2 = 2\beta^2. \qquad (2.2.34)$$

2.2.7 The Gamma Distribution

We consider a random variable X which has the Gamma(α, β) distribution with its pdf $f(x) = \{\beta^\alpha \Gamma(\alpha)\}^{-1}e^{-x/\beta}x^{\alpha-1}$ for $0 < x < \infty$, given by (1.7.20). Here, we have $(\alpha, \beta) \in \Re^+ \times \Re^+$. In order to derive the mean and variance of this distribution, we proceed with the one-to-one substitution $u = x/\beta$ along the lines of (1.7.22) where $x > 0$, and express $E(X)$ as

$$\{\beta^\alpha \Gamma(\alpha)\}^{-1} \int_{0}^{\infty} xe^{-x/\beta}x^{\alpha-1}dx = \beta\{\Gamma(\alpha)\}^{-1} \int_{0}^{\infty} e^{-u}u^\alpha du. \qquad (2.2.35)$$

In other words, the mean of the distribution simplifies to

$$\beta\{\Gamma(\alpha)\}^{-1}\Gamma(\alpha+1) = \alpha\beta, \quad \because \Gamma(\alpha+1) = \alpha\Gamma(\alpha) \text{ from } (1.6.20).$$
$$(2.2.36)$$

Then, we express $E(X^2)$ analogously as

$$\{\beta^\alpha \Gamma(\alpha)\}^{-1} \int_0^\infty x^2 e^{-x/\beta} x^{\alpha-1} dx = \beta^2 \{\Gamma(\alpha)\}^{-1} \int_0^\infty e^{-u} u^{\alpha+1} du.$$
$$(2.2.37)$$

In other words, $E(X^2)$ simplifies to

$$\beta^2 \{\Gamma(\alpha)\}^{-1} \Gamma(\alpha+2) = (\alpha+1)\alpha\beta^2, \text{ since } \Gamma(\alpha+2)$$
$$(\alpha+1)\alpha\Gamma(\alpha) \text{ from (1.6.20)}.$$
$$(2.2.38)$$

Hence, $V(X) = E(X^2) - E^2(X) = (\alpha+1)\alpha\beta^2 - \alpha^2\beta^2 = \alpha\beta^2$. In summary, for the random variable X distributed as Gamma(α, β), we have

$$\mu_X = \alpha\beta \text{ and } \sigma_X^2 = \alpha\beta^2.$$
$$(2.2.39)$$

2.3 The Moments and Moment Generating Function

Start with a random variable X and consider a function of the random variable, $g(X)$. Suppose that $f(x)$ is the pmf or pdf of X where $x \in \mathcal{X}$ is the support of the distribution of X. Now, recall the Definition 2.2.3 for the *expected value* of the random variable $g(X)$, denoted by $E[g(X)]$.

We continue to write $\mu = E(X)$. When we specialize $g(x) = x - \mu$, we obviously get $E[g(X)] = 0$. Next, if we let $g(x) = (x - \mu)^2$, we get $E[g(X)] = \sigma^2$. Now these notions are further extended by considering two other special choices of functions, namely, $g(x) = x^r$ or $(x - \mu)^r$ for *fixed* $r = 1, 2, \dots$.

Definition 2.3.1 *The rth moment of a random variable X, denoted by η_r, is given by $\eta_r = E[X^r]$, for fixed $r = 1, 2, \dots$. The first moment η_1 is the mean or the expected value μ of the random variable X.*

Definition 2.3.2 *The rth central moment of a random variable X around its mean μ, denoted by μ_r, is given by $\mu_r = E[(X - \mu)^r]$ with fixed $r = 1, 2, \dots$.*

> Recall that the first central moment μ_1 is zero and the second central moment μ_2 turns out to be the variance σ^2 of X, assuming that μ and σ^2 are finite.

Example 2.3.1 It is known that the infinite series $\Sigma_{i=1}^\infty i^{-p}$ converges if $p > 1$, and it diverges if $p \le 1$. Refer back to (1.6.12). With some fixed

$p > 1$, let us write $\Sigma_{i=1}^{\infty} i^{-p} = \zeta(p)$, a positive and finite real number, and define a random variable X which takes the values $i \in \mathcal{X} = \{1, 2, 3, ...\}$ such that $P(X = i) = \{\zeta(p)\}^{-1} i^{-p}, i = 1, 2, ...$. This is obviously a discrete probability mass function. Now, let us fix $p = 2$. Since $\zeta(1) = \Sigma_{i=1}^{\infty} i^{-1}$ is not finite, it is clear that η_1 or $E(X)$ is not finite for this random variable X. This example shows that it is fairly simple to construct a discrete random variable X for which even the mean μ is not finite. ▲

> The Example 2.3.1 shows that the moments of a random variable may not be finite.

Now, for any random variable X, the following conclusions should be fairly obvious:

(i) η_r is finite $\Rightarrow \mu_r$ is finite;

(ii) η_r is finite $\Rightarrow \eta_s$ is finite for all $s = 1, ..., r - 1$;   (2.3.1)

(iii) η_r is finite $\Rightarrow \mu_s$ is finite for all $s = 1, ..., r - 1$.

The part (iii) in (2.3.1) follows immediately from the parts (i) and (ii).

> The finiteness of the rth moment η_r of X does not necessarily imply the finiteness of the sth moment η_s of X when $s > r$. Look at the Example 2.3.2.

Example 2.3.2 (Example 2.3.1 Continued) For the random variable X defined in the Example 2.3.1, the reader should easily verify the claims made in the adjoining table:

Table 2.3.1. Existence of Few Lower Moments
But Non-Existence of Higher Moments

p	Finite η_r	Infinite η_r
2	none	$r = 1, 2, ...$
3	$r = 1$	$r = 2, 3, ...$
4	$r = 1, 2$	$r = 3, 4, ...$
.	.	.
k	$r = 1, ..., k - 2$	$r = k - 1, k, ...$

The Table 2.3.1 shows that it is a simple matter to construct discrete random variables X for which μ may be finite but its variance σ^2 may not be finite, or μ and σ^2 both could be finite but the third moment η_3 may not be finite, and so on. ▲

It will be instructive to find simple examples of continuous random variables and other discrete random variables with interesting features analogous to those cited in the Example 2.3.2. These are left as Exercises 2.3.8-2.3.10.

When the Definition 2.2.3 is applied with $g(x) = e^{tx}$, one comes up with a very useful and special function in statistics. Look at the following definition.

Definition 2.3.3 *The moment generating function (mgf) of a random variable X, denoted by $M_X(t)$, is defined as*

$$M_X(t) = E[e^{tX}],$$

provided that the expectation is finite for $|t| < a$ with some $a > 0$.

As usual, the exact expression of the mgf $M_X(t)$ would then be derived analytically using one of the following expressions:

$$\begin{matrix} \Sigma_{i:x_i \in \mathcal{X}} \, e^{tx_i} f(x_i) & \text{when } X \text{ is discrete, or} \\ \int_{\mathcal{X}} \, e^{tx} f(x) dx & \text{when } X \text{ is continuous.} \end{matrix} \qquad (2.3.2)$$

The function $M_X(t)$ bears the name mgf because one can derive *all* the moments of X by starting from its mgf. In other words, *all* moments of X can be *generated* from its mgf provided that the mgf itself is finite.

Theorem 2.3.1 *If a random variable X has a finite mgf $M_X(t)$, for $|t| < a$ with some $a > 0$, then the rth moment η_r of X, given in the Definition 2.3.1, is the same as $d^r M_X(t)/dt^r$ when evaluated at $t = 0$.*

Proof Let us first pretend that X is a continuous random variable so that $M_X(t) = \int_{\mathcal{X}} e^{tx} f(x) dx$. Now, assume that the differentiation operator of $M_X(t)$ with respect to t can be taken inside the integral with respect to x. One may refer to (1.6.16)-(1.6.17) for situations where such interchanges are permissible. We write

$$dM_X(t)/dt = \tfrac{d}{dt} \int_{\mathcal{X}} e^{tx} f(x) dx = \int_{\mathcal{X}} \tfrac{\partial}{\partial t} \{e^{tx} f(x)\} \, dx = \int_{\mathcal{X}} x e^{tx} f(x) dx, \qquad (2.3.3)$$

and then it becomes clear that $dM_X(t)/dt$ when evaluated at $t = 0$ will coincide with $\int_{\mathcal{X}} x f(x) dx$ which is $\eta_1 (= \mu)$. Similarly let us use (2.3.3) to claim that

$$d^2 M_X(t)/dt^2 = \int_{\mathcal{X}} \tfrac{\partial}{\partial t} \{x e^{tx} f(x)\} \, dx = \int_{\mathcal{X}} x^2 e^{tx} f(x) dx. \qquad (2.3.4)$$

Hence, $d^2 M_X(t)/dt^2$ when evaluated at $t = 0$ will coincide with $\int_{\mathcal{X}} x^2 f(x) dx$ which is η_2. The rest of the proof proceeds similarly upon successive differentiation of the mgf $M_X(t)$. A discrete scenario can be handled by replacing the integral with a sum. ∎

> A finite mgf $M_X(t)$ determines a unique infinite sequence
> of moments of a random variable X.

Remark 2.3.1 Incidentally, the sequence of moments $\{\eta_r = E(X^r) : r = 1, 2, ...\}$ is sometimes referred to as the sequence of *positive integral moments* of X. The Theorem 2.3.1 provides an explicit tool to restore all the positive integral moments of X by successively differentiating its mgf and then letting $t = 0$ in the expression. It is interesting to note that the negative integral moments of X, that is when $r = -1, -2, ...$ in the Definition 2.3.1, are also hidden inside the same mgf. These negative moments of X can be restored by implementing a process of *successive integration* of the mgf, an operation which is viewed as the opposite of differentiation. Precise statements of the regularity conditions under which the negative moments of X can be derived with this approach are found in Cressie et al. (1981).

> The positive integral moments of X can be found by
> successively differentiating its mgf with respect to t
> and letting $t = 0$ in the final derivative.

The following result is very useful and simple. We leave its proof as an exercise. There will be ample opportunities to use this theorem in the sequel.

Theorem 2.3.2 *Suppose that a random variable X has the mgf $M_X(t)$, for $\mid t \mid < a$ with some $a > 0$. Let $Y = cX + d$ be another random variable where c, d are fixed real numbers. Then, the mgf $M_Y(t)$ of Y is related to the mgf $M_X(t)$ as follows:*

$$M_Y(t) = e^{td} M_X(tc)$$

In the following subsections, we show the derivations of the mgf in the case of a few specific distributions. We also exhibit some immediate applications of the Theorem 2.3.1.

2.3.1   The Binomial Distribution

Suppose that X has the Binomial(n, p) distribution with its pmf $f(x) = \binom{n}{x} p^x (1-p)^{n-x}$ for $x = 0, 1, ..., n$ and $0 < p < 1$, given by (1.7.2). Here, *for all* fixed $t \in \Re$ we can express $M_X(t)$ as

$$\begin{aligned}
&\Sigma_{x=0}^n \binom{n}{x} \left(pe^t\right)^x (1-p)^{n-x} \\
&= \{(1-p) + pe^t\}^n, \text{ using the Binomial Theorem.}
\end{aligned} \tag{2.3.5}$$

Observe that $log(M_X(t)) = nlog\{(1 - p) + pe^t\}$ so that we obtain

$$dM_X(t)/dt = npe^t\{(1 - p) + pe^t\}^{-1}M_X(t). \qquad (2.3.6)$$

Next, using the chain rule of differentiation, from (2.3.6) we can write

$$d^2 M_X(t)/dt^2 = npe^t\{(1 - p) + pe^t\}^{-1}M_X(t) - n(pe^t)^2\{(1 - p)$$
$$+pe^t\}^{-2}M_X(t) + npe^t\{(1 - p) + pe^t\}^{-1}\{dM_X(t)/dt\}. \qquad (2.3.7)$$

From (2.3.7) it is obvious that $dM_X(t)/dt$, when evaluated at $t = 0$, reduces to

$$\eta_1 = np\{(1 - p) + p\}^{-1}M_X(0) = np, \text{ since } M_X(0) = 1. \qquad (2.3.8)$$

We found the mean of the distribution in (2.2.14). Here we have another way to check that the mean is np.

Also, the Theorem 2.3.1 would equate the expression of $E(X^2)$ with $d^2 M_X(t)/dt^2$ evaluated at $t = 0$. From (2.3.6) it is obvious that $d^2 M_X(t)/dt^2$, when evaluated at $t = 0$, should lead to η_2, that is

$$E(X^2)$$
$$= np\{(1 - p) + p\}^{-1}M_X(0) - np^2\{(1 - p) + p\}^{-2}M_X(0)$$
$$+ np\{(1 - p) + p\}^{-1}\{dM_X(t)/dt\}\,|_{t=0} \qquad (2.3.9)$$
$$= np - np^2 + n^2p^2, \text{ since } M_X(0) = 1 \text{ and}$$
$$\{dM_X(t)/dt\}\,|_{t=0} = np.$$

Hence, one has $V(X) = E(X^2) - \mu^2 = np - np^2 + n^2p^2 - (np)^2 = np - np^2 = np(1 - p)$. This matches with the expression of σ^2 given in (2.2.17).

2.3.2   The Poisson Distribution

Suppose that X has the Poisson(λ) distribution with its pmf $f(x) = e^{-\lambda}\lambda^x/x!$ for $x = 0, 1, ...$ and $0 < \lambda < \infty$, given by (1.7.4). Here, for *all* fixed $t \in \Re$ we can express $M_X(t)$ as

$$e^{-\lambda}\Sigma_{x=0}^{\infty}(\lambda e^t)^x\frac{1}{x!}$$
$$= exp\{-\lambda + \lambda e^t\}, \text{ using the exponential series expansion} \qquad (2.3.10)$$
$$e^a = 1 + \frac{a}{1!} + \frac{a^2}{2!} + ... \text{ from (1.6.15).}$$

Observe that $log(M_X(t)) = -\lambda + \lambda e^t$ so that one has

$$dM_X(t)/dt = \lambda e^t M_X(t). \qquad (2.3.11)$$

Next, using the chain rule of differentiation, from (2.3.11) one obtains

$$d^2 M_X(t)/dt^2 = \lambda e^t M_X(t) + (\lambda e^t)^2 M_X(t). \tag{2.3.12}$$

Since $M_X(0) = 1$, from (2.3.11) it is obvious that $dM_X(t)/dt$, when evaluated at $t = 0$, reduces to $\eta_1 = \lambda$ which matches with the expression of μ found earlier in (2.2.18). Here is another way to check that the mean of the distribution is λ.

Also, the Theorem 2.3.1 would equate the expression of η_2 or $E(X^2)$ with $d^2 M_X(t)/dt^2$, evaluated at $t = 0$. From (2.3.12) it should become obvious that $d^2 M_X(t)/dt^2$, evaluated at $t = 0$, should be $\eta_2 = \lambda + \lambda^2$ since $M_X(0) = 1$. Hence, we have $V(X) = E(X^2) - \mu^2 = \lambda + \lambda^2 - \lambda^2 = \lambda$. This again matches with the expression of σ^2 given in (2.2.18).

2.3.3 The Normal Distribution

Let us first suppose that Z has the standard normal distribution with its pdf $\phi(z) = \{\sqrt{2\pi}\}^{-1} exp(-z^2/2)$ for $-\infty < z < \infty$, given by (1.7.16). Now, *for all* fixed $t \in \Re$ we can express the mgf as

$$
\begin{aligned}
M_Z(t) &= \int_{-\infty}^{\infty} e^{tz} \phi(z) dz \\
&= \{\sqrt{2\pi}\}^{-1} \int_{-\infty}^{\infty} exp(tu - \tfrac{1}{2}u^2) du \\
&= exp(\tfrac{1}{2}t^2) \int_{-\infty}^{\infty} h(u) du \text{ where} \\
&h(u) = \{\sqrt{2\pi}\}^{-1} exp\{-\tfrac{1}{2}(u - t)^2\} \text{ for } u \in \Re.
\end{aligned} \tag{2.3.13}
$$

Observe that the form of the function $h(u)$ used in the last step in (2.3.13) resembles the pdf of a random variable having the $N(t, 1)$ distribution for all fixed $t \in \Re$. So, we must have $\int_{-\infty}^{\infty} h(u) du = 1$ for all fixed $t \in \Re$. In other words, (2.3.13) leads to the following conclusion:

$$M_Z(t) = exp\{\tfrac{1}{2}t^2\} \text{ where } Z \text{ is } N(0,1), \text{ for all } t \in \Re. \tag{2.3.14}$$

Now, suppose that X is distributed as $N(\mu, \sigma^2)$. We can write $X = \sigma Y + \mu$ where $Y = (X - \mu)/\sigma$. We can immediately use the Theorem 2.3.2 and claim that

$$M_X(t) = e^{t\mu} M_Y(t\sigma). \tag{2.3.15}$$

Now, by substituting $y = (x - \mu)/\sigma$ in the integral, the expression of $M_Y(t)$ can be found as follows:

$$
\begin{aligned}
&M_Y(t) \\
&= \{\sigma\sqrt{2\pi}\}^{-1} \int_{-\infty}^{\infty} exp\left\{t\left(\tfrac{x-\mu}{\sigma}\right)\right\} exp\left\{-\tfrac{1}{2}\left(\tfrac{x-\mu}{\sigma}\right)^2\right\} dx \\
&= \{\sqrt{2\pi}\}^{-1} \int_{-\infty}^{\infty} exp(ty) exp\{-\tfrac{1}{2}y^2\} dy,
\end{aligned}
$$

which was exactly the same integral evaluated in (2.3.13). In other words, from (2.3.14) we can immediately claim that $M_Y(t) = exp(\frac{1}{2}t^2)$. Thus, one can rewrite (2.3.15) as follows:

$$M_X(t) = exp(t\mu + \tfrac{1}{2}t^2\sigma^2) \text{ where } X \text{ is } N(\mu, \sigma^2), \text{ for all } t \in \Re. \quad (2.3.16)$$

Now, $log(M_X(t)) = t\mu + \frac{1}{2}t^2\sigma^2$ so that $dM_X(t)/dt = (\mu + t\sigma^2)M_X(t)$. Hence, $dM_X(t)/dt$, when evaluated at $t = 0$, reduces to $\eta_1 = \mu$ because one has $M_X(0) = 1$.

Also, using the chain rule of differentiation we have $d^2M_X(t)/dt^2 = \sigma^2 M_X(t) + (\mu + t\sigma^2)^2 M_X(t)$ so that $d^2M_X(t)/dt^2$, when evaluated at $t = 0$, reduces to $\eta_2 = \sigma^2 + \mu^2$. In view of the Theorem 2.3.1, we can say that μ is the mean of X and $V(X) = E(X^2) - \mu^2 = \sigma^2 + \mu^2 - \mu^2 = \sigma^2$. These answers were verified earlier in (2.2.30).

One can also easily evaluate higher order derivatives of the mgf. In view of the Theorem 2.3.1 one can claim that $d^k M_X(t)/dt^k$ with $k = 3$ and 4, when evaluated at $t = 0$, will reduce to η_3 or $E(X^3)$ and η_4 or $E(X^4)$ respectively. Then, in order to obtain the *third* and *fourth central moments* of X, one should proceed as follows:

$$\mu_3 = E\{X^3 - 3\mu X^2 + 3\mu^2 X - \mu^3\} = \eta_3 - 3\mu\eta_2 + 2\mu^3, \quad (2.3.17)$$

by the Theorem 2.2.1. Similarly,

$$\begin{aligned} \mu_4 &= E\{X^4 - 4\mu X^3 + 6\mu^2 X^2 - 4\mu^3 X + \mu^4\} \\ &= \eta_4 - 4\mu\eta_3 + 6\mu^2\eta_2 - 3\mu^4, \text{ by the Theorem 2.2.1.} \end{aligned} \quad (2.3.18)$$

Look at the related Exercises 2.3.1-2.3.2.

Example 2.3.3 Suppose that Z is the standard normal variable. How should one directly derive the expression for $E(|Z|)$? Here, the mgf of Z from (2.3.14) is not going to be of much help. Let us apply the Definition 2.2.3 and proceed as follows. As before, we substitute $u = -z$ when $z < 0$, and express $E(|Z|)$ as

$$\begin{aligned} &\{\sqrt{2\pi}\}^{-1} \int_{-\infty}^{\infty} |z| exp\{-z^2/2\}dz \\ &= \{\sqrt{2\pi}\}^{-1} \left[-\int_{\infty}^{0} u\,exp\{-u^2/2\}du + \int_{0}^{\infty} z\,exp\{-z^2/2\}dz \right]. \end{aligned} \quad (2.3.19)$$

In the last step of (2.3.19), one immediately realizes that the two integrals are really the same that is, $\int_{0}^{\infty} z\,exp\{-z^2/2\}dz = -\int_{\infty}^{0} u\,exp\{-u^2/2\}du$ and hence one has

$$E(|Z|) = 2\{\sqrt{2\pi}\}^{-1} \int_{0}^{\infty} z\,exp\{-z^2/2\}dz. \quad (2.3.20)$$

Next, along the lines of (2.2.22)-(2.2.24) we can rewrite the integral on the rhs of (2.3.20) as a gamma integral to finally claim that

$$E(|Z|) = \sqrt{(2/\pi)}. \tag{2.3.21}$$

Also look at the closely related Exercises 2.3.3-2.3.4. ▲

Example 2.3.4 Suppose that Z is the standard normal variable. How should one directly derive the expression for $E(e^{|Z|})$? Here, the mgf of Z from (2.3.14) is not going to be of much help. Let us apply the Definition 2.2.3 and proceed as follows:

$$
\begin{aligned}
E(e^{|Z|}) &= \{\sqrt{2\pi}\}^{-1} \int_{-\infty}^{\infty} exp\{|z| - (z^2/2)\}dz \\
&= 2\{\sqrt{2\pi}\}^{-1} \int_{0}^{\infty} exp\{z - (z^2/2)\}dz \\
&= 2\sqrt{e} \int_{0}^{\infty} h(z)dz \text{ where } h(z) = \\
&\quad \{\sqrt{2\pi}\}^{-1} exp\{-\tfrac{1}{2}(z-1)^2\}, z \in \Re \\
&= 2\sqrt{e}\Phi(1), \text{ since } h(z) \text{ is a normal pdf in } z.
\end{aligned}
$$

Look at the related Exercise 2.3.16 where we ask for the mgf of $|Z|$. ▲

2.3.4   The Gamma Distribution

Let us suppose that a random variable X has the Gamma(α, β) distribution with its pdf $f(x) = \{\beta^{\alpha}\Gamma(\alpha)\}^{-1}e^{-x/\beta}x^{\alpha-1}$ where $0 < x < \infty$ and $(\alpha, \beta) \in \Re^{+} \times \Re^{+}$, given by (1.7.20). Now, let us denote $\beta^{*} = \beta(1 - \beta t)^{-1}$ for all $t < \beta^{-1}$ so that β^{*} is positive. Thus, we can express $M_X(t)$ as

$$
\begin{aligned}
\int_{0}^{\infty} e^{tx}f(x)dx \\
= \{\beta^{\alpha}\Gamma(\alpha)\}^{-1}\{\beta^{*\alpha}\Gamma(\alpha)\} \int_{0}^{\infty} h(u)du \text{ where} \\
h(u) = \{\beta^{*\alpha}\Gamma(\alpha)\}^{-1}exp\{-u/\beta^{*}\}u^{\alpha-1} \text{ for } u \in \Re^{+}.
\end{aligned}
\tag{2.3.22}
$$

Observe that the function $h(u)$ used in the last step in (2.3.22) resembles the pdf of a random variable having the Gamma(α, β^{*}) distribution for $u \in \Re^{+}$ where β^{*} is positive. So, we must have $\int_{-\infty}^{\infty} h(z)dz = 1$. In other words, (2.3.22) leads to the following conclusion:

$$M_X(t) = (1 - \beta t)^{-\alpha} \text{ where } X \text{ is Gamma}(\alpha, \beta), \text{ for all } t < \beta^{-1}. \tag{2.3.23}$$

Now, $log(M_X(t)) = -\alpha log(1 - \beta t)$ so that one can immediately have $dM_X(t)/dt = \alpha\beta(1 - \beta t)^{-1}M_X(t)$. Hence, $dM_X(t)/dt$, when evaluated at $t = 0$, reduces to $\alpha\beta$ because $M_X(0) = 1$. Next, we use the chain rule of differentiation in order to write $d^2 M_X(t)/dt^2 = \alpha(1+\alpha)\beta^2(1-\beta t)^{-2}M_X(t)$

so that $d^2M_X(t)/dt^2$, when evaluated at $t = 0$, reduces to $\alpha(1 + \alpha)\beta^2$. In view of the Theorem 2.3.1, we can say that $\alpha\beta$ is the mean of X and $V(X) = E(X^2) - \mu^2 = \alpha(1 + \alpha)\beta^2 - \mu^2 = \alpha\beta^2$. These same answers were derived earlier in (2.2.39).

How should one derive the expression for $E(X^r)$ when r is arbitrary? We no longer restrict r to be a positive integer. Now, the expression of the mgf of X given by (2.3.23) may not be of immediate help. One may pursue a direct approach along the lines of (2.2.35)-(2.2.38). Let us write

$$E(X^r) = \{\beta^\alpha \Gamma(\alpha)\}^{-1} \int_0^\infty e^{-x/\beta} x^{(\alpha+r)-1} dx, \tag{2.3.24}$$

and observe that the integrand is the kernel (that is the part involving x only) of the gamma pdf provided that $\alpha + r > 0$. In other words, we must have

$$\int_0^\infty e^{-x/\beta} x^{(\alpha+r)-1} dx = \beta^{\alpha+r} \Gamma(\alpha+r) \text{ when } \alpha + r > 0. \tag{2.3.25}$$

Next, we combine (2.3.24)-(2.3.25) to conclude that

$$E(X^r) = \beta^r \Gamma(\alpha+r)\{\Gamma(\alpha)\}^{-1} \text{ when } \alpha > -r, \tag{2.3.26}$$

regardless of what r is. When r is positive, the expression for $E(X^r)$ stays the same, but no additional condition on α is warranted. The result quoted in (2.3.26) would be very helpful in the sequel.

Suppose that X is distributed as Gamma(α, β). Then, $E(X^r)$ is finite if and only if $\alpha > -r$. Look at (2.3.24)-(2.3.26).

Special Case 1: The Exponential Distribution

The exponential distribution was defined in (1.7.23). This random variable X has the pdf $f(x) = \beta^{-1} e^{-x/\beta} I(x > 0)$ which is equivalent to saying that X is distributed as Gamma$(1, \beta)$ with $\beta \in \Re^+$. In this special situation, we can summarize the following results.

The mgf is $M_X(t) = (1 - \beta t)^{-1}$ for all $t < \beta^{-1}$;

$$E(X) = \beta, E(X^2) = 2\beta^2, V(X) = \beta^2. \tag{2.3.27}$$

These can be checked out easily using (2.3.23) as well as (2.3.26).

Special Case 2: The Chi-square Distribution

The Chi-square distribution was also introduced in the Section 1.7. This random variable X has the pdf $f(x) = \{2^{\nu/2} \Gamma(\nu/2)\}^{-1} e^{-x/2} x^{(\nu/2)-1} I(x > 0)$ which is equivalent to saying that X is distributed as Gamma$(\frac{\nu}{2}, 2)$ with positive integral values of ν. Recall that ν is referred to as the degree of freedom and X is often denoted by χ^2_ν. Then, in this special situation, we can summarize the following results.

The mgf of χ^2_ν is $(1 - 2t)^{-\nu/2}$ for all $t < \frac{1}{2}$;

$$E(X) = \nu, E(X^2) = \nu(\nu + 2), V(X) = 2\nu. \qquad (2.3.28)$$

These can again be checked out easily from (2.3.23) as well as (2.3.26).

2.4   Determination of a Distribution via MGF

Next, we emphasize the role of a moment generating function in uniquely determining the probability distribution of a random variable. We state an important and useful result below. Its proof is out of scope for this book. We will have ample opportunities to rely upon this result in the sequel.

Theorem 2.4.1 *Let $M(t)$ be a finite mgf for $\mid t \mid < a$ with some $a > 0$. Then $M(t)$ corresponds to the mgf associated with the probability distribution of a uniquely determined random variable.*

This simple looking result however has deep implications. Suppose that U is a discrete random variable and assume that somehow one knows the expression of its mgf $M_U(t)$. Now if $M_U(t)$, for example, looks exactly like the mgf of a Binomial random variable, then we can conclude that U indeed has the Binomial distribution. But on the other hand if $M_U(t)$ looks exactly like the mgf of a Poisson random variable, then again we will conclude that U has the Poisson distribution, and so on. If U has a continuous distribution instead and $M_U(t)$, for example, looks exactly like the mgf of a Gamma random variable, then U indeed must have a Gamma distribution. These conclusions are supported by the Theorem 2.4.1. We will exploit such implications in the sequel.

Example 2.4.1 Suppose that a random variable X takes the possible values $0, 1$ and 2 with the respective probabilities $\frac{1}{8}, \frac{1}{4}$ and $\frac{5}{8}$. The mgf of X is obviously given by $M_X(t) = \frac{1}{8} + \frac{1}{4}e^t + \frac{5}{8}e^{2t} = \frac{1}{8}(1 + 2e^t + 5e^{2t})$. Observe how the mgf of a discrete random variable is formed and how easy it is to identify the probability distribution by inspecting the appearance of the terms which together build up the function $M_X(t)$. Now suppose that we have a random variable U whose mgf $M_U(t) = \frac{1}{5}(1 + e^t + 3e^{2t})e^{-t}$. What is the probability distribution of U? Let us rewrite this mgf as $.2e^{-t} + .2 + .6e^t$. We can immediately claim that U takes the possible values $-1, 0$ and 1 with the respective probabilities $.2, .2$ and $.6$. In view of the Theorem 2.4.1, we can claim that the random variable U and its distribution we have indicated are unique. ▲

Example 2.4.2 Suppose that U is a random variable such that $M_U(t) = \frac{1}{16}(1 + e^t)^4$. We can rewrite $M_U(t) = (\frac{1}{2} + \frac{1}{2}e^t)^4$ which agrees with the expression of $M_X(t)$ given by (2.3.5) where $n = 4$ and $p = \frac{1}{2}$. Hence U must be distributed as Binomial$(4, \frac{1}{2})$ by the Theorem 2.4.1. ▲

Example 2.4.3 Suppose that U is a random variable such that $M_U(t) = exp\{\pi t^2\}$ which agrees with the expression of $M_X(t)$ given by (2.3.16) with $\mu = 0, \sigma^2 = 2\pi$. Hence, U must be distributed as $N(0, 2\pi)$. ▲

A finite mgf determines the distribution uniquely.

Before we move on, let us attend to one other point involving the moments. A finite mgf uniquely determines a probability distribution, but on the other hand, the moments by themselves alone may not be able to identify a unique random variable associated with all those moments. Consider the following example.

Example 2.4.4 Rao (1973, p.152) had mentioned the construction of the following two pdf's, originally due to C. C. Heyde. Consider two positive continuous random variables X and Y whose pdf's are respectively given by

$$
\begin{aligned}
f(x) &= (2\pi)^{-1/2}x^{-1}exp[-\tfrac{1}{2}(log(x))^2] && \text{for } x > 0, \\
g(y) &= f(y)[1 + c\sin(2\pi log(y))] && \text{for } y > 0,
\end{aligned}
\tag{2.4.1}
$$

where c is a fixed number, $-1 \leq c \leq 1$ and $c \neq 0$. We leave it as the Exercise 2.4.4 to show that $E[X^r] = E[Y^r]$ for all $r = 1, 2, \dots$. But, certainly we can

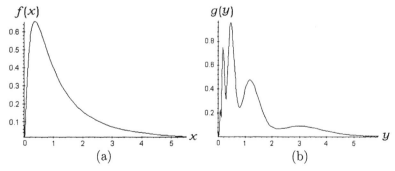

Figure 2.4.1. The Two PDF's from (2.4.1): (a) $f(x)$ (b) $g(y)$
Where $c = \frac{1}{2}$

claim that X and Y have different probability distributions because their two pdf's are obviously different when $c \neq 0$. In the Figure 2.4.1, these two pdf's have been plotted when $c = \frac{1}{2}$. Here all the moments of X are finite, but in the Exercise 2.4.5 one would verify that the mgf for the random variable X is not finite. In another related Exercise 2.4.6, examples of other pairs of random variables with analogous characteristics can be found. ▲

> Finite moments alone may not determine a distribution uniquely. The Example 2.4.4 highlights this point.

Depending on the particular situation, however, an infinite sequence of all finite moments may or may not characterize a probability distribution uniquely. Any elaborate discussion of such issues would be out of scope of this book. The readers may additionally consult the Section 3, Chapter 7 in Feller (1971) on "Moment Problems." Chung (1974) also includes relevant details.

2.5   The Probability Generating Function

We have seen that a mgf generates the moments. Analogously, there is another function which generates the probabilities. Suppose that X is a non-negative random variable. A *probability generating function* (pgf) of X is defined by

$$P_X(t) = E[t^X] \text{ with } t > 0. \tag{2.5.1}$$

The explicit form of a pgf is often found when the random variable X is integer-valued. From this point onward, let us include *only non-negative integer-valued random variables* in this discussion.

Why this function $P_X(t)$ is called a pgf should become clear once we write it out fully as follows:

$$P_X(t) = p_0 + p_1 t + p_2 t^2 + p_3 t^3 + \dots \tag{2.5.2}$$

where $p_i = P(X = i)$ for $i = 1, 2, \dots$. We see immediately that the coefficient of t^i is p_i, the probability that $X = i$ for $i = 1, 2, \dots$.

> In statistics, $E\{X(X-1)\dots(X-k+1)\}$ is referred to as the kth order factorial moment of the random variable X.   (2.5.3)

In the light of the Theorem 2.3.1 one can justify the following result:

$d^k P_X(t)/dt^k$, evaluated at $t = 1$, will reduce to the kth order factorial moment of X, for $k = 1, 2, \dots$.   (2.5.4)

One should verify that

$$P_X(t) = M_X(log(t)) \text{ with } t > 0, \tag{2.5.5}$$

where $P_X(t)$ is the pgf of X defined in (2.5.1).

2.6   Exercises and Complements

2.2.1 A game consists of selecting a three-digit number. If one guesses the number correctly, the individual is paid $800 for each dollar the person bets. Each day there is a new winning number. Suppose that an individual bets $1 each day for a year, then how much money can this individual be expected to win?

2.2.2 An insurance company sells a life insurance policy with the face value of $2000 and a yearly premium of $30. If 0.4% of the policy holders can be expected to die in the course of a year, what would be the company's expected earnings per policyholder in any year?

2.2.3 (Exercise 1.5.1 Continued) Suppose that a random variable X has the following pmf:

X values:	-2	0	1	3	8
Probabilities:	.2	p	.1	$2p$	.4

where $p \in (0, .1]$. Evaluate μ and σ for the random variable X.

2.2.4 (Exercise 1.5.2 Continued) Suppose that a random variable X has the following pmf:

X values:	-2	0	3	5	8
Probabilities:	.2	p	.1	$.3 - p$	.4

where $p \in (0, .3)$. With a fixed but arbitrary $p \in (0, .3)$, find the expressions of μ and σ for the random variable X depending upon p. Is there some $p \in (0, .3)$ for which the $V(X)$ is minimized?

2.2.5 Consider a random variable X which has the following discrete uniform distribution along the lines of (1.7.11):

$$P(X = i) = n^{-1} \text{ for } i = 1, ..., n$$

Derive the explicit expressions of μ and σ for this distribution. {*Hint*: Use the first two expressions from (1.6.11).}

2.2.6 Let X be a discrete random variable having the Geometric(p) distribution whose pmf is $f(x) = p(1 - p)^{x-1}$ for $x = 1, 2, 3, ...$ and $0 < p < 1$, given by (1.7.7). Derive the explicit expressions of μ and σ for this distribution. An alternative way to find μ has been provided in the Exercise 2.2.23. {*Hint*: Use the expressions for sums of the second and third infinite series expansions from (1.6.15).}

2.2.7 Suppose that X has the Poisson(λ) distribution with $0 < \lambda < \infty$. Show that $\mu = \sigma^2 = \lambda$ which would verify (2.2.18). See Section 2.2.3 for some partial calculations.

2.2.8 For the negative binomial pmf defined by (1.7.10), find the expressions for the mean and variance.

2.2.9 Let c be a positive constant and consider the pdf of a random variable X given by

$$f(x) = \begin{cases} c(2 - x) & \text{if } 0 \le x \le 1 \\ 0 & \text{elsewhere.} \end{cases}$$

(i)   Explicitly evaluate c, μ and σ;
(ii)   Evaluate $E\{(1 - X)^{-1/2}\}$;
(iii)   Evaluate $E\{X^3(1 - X)^{3/2}\}$.
{*Hints*: In parts (ii)-(iii) first derive the expressions using direct integration techniques. Alternatively reduce each expected value in parts (ii)-(iii) as a combination of the appropriate beta functions defined in (1.6.25).}

2.2.10 Let c be a constant and consider the pdf of a random variable X given by

$$f(x) = \begin{cases} c + \frac{1}{4}x & \text{if } -2 \le x \le 0 \\ c - \frac{1}{4}x & \text{if } 0 \le x \le 2 \\ 0 & \text{elsewhere.} \end{cases}$$

Explicitly evaluate c, μ and σ.

2.2.11 Suppose that X has the Uniform(0, 1) distribution. Evaluate
(i) $E[X^{-1/2}(1 - X)^{10}]$;
(ii) $E[X^{3/2}(1 - X)^{5/2}]$;
(iii) $E[(X^2 + 2X^{1/2} - 3X^{5/2})(1 - X)^{10}]$.
{*Hint*: First express these expectations as the integrals using the Definition 2.2.3. Can these integrals be evaluated using the forms of beta integrals defined in (1.6.25)?}

2.2.12 Consider a random variable X having the pdf $f(x) = c(x - 2x^2 + x^3)I(0 < x < 1)$ where c is a positive number. Explicitly evaluate μ and σ for this distribution. {*Hint*: In order to find c, should one match this $f(x)$ with a beta density?}

2.2.13 Suppose that X is a continuous random variable having its pdf $f(x)$ with the support $\mathcal{X} = (a, b), -\infty \leq a < b \leq \infty$. Additionally suppose that $f(x)$ is symmetric about the point $x = c$ where $a < c < b$. Then, show that both the mean and median of this distribution will coincide with c as long as the mean is finite. {*Hint*: First show that the point of symmetry c for $f(x)$ must be the midpoint between a and b. Consider splitting the relevant integrals as the sum of the integrals over the intervals $(-\infty, a), (a, c), (c, b)$ and (b, ∞). A word of caution may be in order. A pdf may be symmetric about c, but that may not imply $\mu = c$ simply because μ may not even be finite. Look at the Exercise 2.3.8.}

2.2.14 Suppose that a random variable X has the Rayleigh distribution, that is its pdf is given by

$$f(x) = 2\theta^{-1}x\,exp(-x^2/\theta)I(x > 0),$$

where $\theta(> 0)$. Evaluate μ and σ for this distribution. {*Hint*: Try the substitution $u = x^2/\theta$ during the integration. Is this substitution one-to-one?}

2.2.15 Suppose that a random variable X has the Weibull distribution, that is its pdf is given by

$$f(x) = \alpha\beta^{-\alpha}x^{\alpha-1}\,exp(-[x/\beta]^{\alpha})I(x > 0),$$

where $\alpha(> 0)$ and $\beta(> 0)$. Evaluate μ and σ for this distribution. {*Hint*: Try the substitution $u = [x/\beta]^{\alpha}$ during the integration. Is this substitution one-to-one?}

2.2.16 Suppose that we have a random variable X which has the lognormal pdf $f(x) = \{\sqrt{2\pi}\}^{-1}x^{-1}\,exp[-\{log(x)\}^2/2]$ for $0 < x < \infty$, given by (1.7.27). Derive the expression of the mean and variance. {*Hint*: While considering the integrals over $(0, \infty)$, try the substitution $y = log(x)$ and look at the integrands as functions of the new variable y which varies between $(-\infty, \infty)$. Try to express these integrands in the variable y to resemble the pdf's of normal random variables.}

2.2.17 For the negative exponential distribution with its pdf $f(x) = \beta^{-1}e^{-(x-\gamma)/\beta}$ for $\gamma < x < \infty$, defined in (1.7.36), show that $\mu = \gamma + \beta$ and $\sigma = \beta$.

2.2.18 Prove Theorem 2.2.3. {*Hint*: First suppose that $f(x)$ is the pdf of X with its support $\mathcal{X} = [a, b]$. To prove part (i), multiply each x throughout the interval $[a, b]$ by $f(x)$ to write $af(x) \leq xf(x) \leq bf(x)$ and then integrate all sides (with respect to x) from a to b. The result will follow since $\int_a^b f(x)dx = 1$. The discrete case can be handled analogously. In part (ii), note that $Y - X \geq 0$ w.p.1 and hence the result follows from part (i).}

2.2.19 Prove Theorem 2.2.4. {*Hint*: By the Theorem 2.2.1, note that $E(aX+b) = aE(X)+b$. Next, apply the Definition 2.2.2 and write $V(aX+b) = E\{(aX+b)-[aE(X)+b]\}^2 = E\{[aX-aE(X)]^2\} = a^2E\{[X-E(X)]^2\}$ which simplifies to $a^2V(X)$.}

2.2.20 Suppose that X is a random variable whose second moment is finite. Let $g(a) = E[(X - a)^2], a \in \Re$. Show that $g(a)$ is minimized with respect to a when $a = \mu$, the mean of X. {*Hint*: Write out $E[(X - a)^2] = E[X^2] - 2aE[X] + a^2$. Treat this as a function of a single variable a and minimize this with respect to a. Refer to (1.6.27).}

2.2.21 In the Theorem 2.2.5, we had *assumed* that $\lim_{x\to\infty} x\{1-F(x)\} = 0$ for the df $F(x)$ of a non-negative continuous random variable X. Write $x\{1 - F(x)\} = \{1 - F(x)\}/x^{-1}$ and then it is clear that $\lim_{x\to\infty} x\{1 - F(x)\}$ unfortunately takes the form of 0/0. Hence, by applying the L'Hôpital's rule (refer to (1.6.29)) show that we may assume instead $\lim_{x\to\infty} x^2 f(x) = 0$ where $f(x)$ is the associated pdf.

2.2.22 (Exercise 2.2.21 Continued) Suppose that X has the exponential distribution with its pdf $f(x) = \beta^{-1}e^{-x/\beta}$ for $x > 0, \beta > 0$. Obtain the expression of μ by applying the integral expression found in the Theorem 2.2.5. Is the sufficient condition $\lim_{x\to\infty} x\{1 - F(x)\} = 0$ satisfied here? How about the sufficient condition $\lim_{x\to\infty} x^2 f(x) = 0$?

2.2.23 This exercise provides an application of the Theorem 2.2.6. Suppose that X has the Geometric(p) distribution, defined in (1.7.7), with its pmf $f(x) = pq^{x-1}, x = 1, 2, ..., 0 < p < 1, q = 1 - p$.

(i)   Show that $F(x) = 1 - q^x, x = 1, 2, ...$;

(ii)   Perform the infinite summation $\Sigma_{x=0}^{\infty}\{1 - F(x)\}$ using the expression for $\Sigma_{x=0}^{\infty}q^x$, and hence find μ.

A direct approach using the Definition 2.2.1 to find μ was sought out in the Exercise 2.2.6.

2.2.24 (**Monotone Convergence Theorem**) Let $\{X_n; n \geq 1\}$ be a sequence of *non-negative* random variables. Then, show that

$$E[\Sigma_{n=1}^{\infty}X_n] = \Sigma_{n=1}^{\infty}E[X_n].$$

{*Note*: We state this result here for completeness. It has been utilized in the proof of the Theorem 2.2.6. It may be hard to prove this result at the level of this book. See also the Exercise 5.2.22 (a).}

2.3.1 Suppose that Z has the standard normal distribution. By directly evaluating the appropriate integrals along the lines of (2.2.22)-(2.2.24) and Example 2.3.4 show that

(i) $E(Z^r) = 0$ for any odd integer $r > 0$;
(ii) $E(Z^r) = \pi^{-1/2}2^{r/2}\Gamma(\frac{1}{2}r + \frac{1}{2})$ for any even integer $r > 0$;
(iii) $\mu_4 = 3$ using part (ii).

2.3.2 (Exercise 2.3.1 Continued) Suppose that a random variable X has the $N(\mu, \sigma^2)$ distribution. By directly evaluating the relevant integrals, show that

(i) $E\{(X - \mu)^r\} = 0$ for all positive odd integer r;
(ii) $E\{(X - \mu)^r\} = \pi^{-1/2}\sigma^r 2^{r/2}\Gamma\left(\frac{1}{2}r + \frac{1}{2}\right)$ for all positive
         even integer r;
(iii)   the fourth central moment μ_4 reduces to $3\sigma^4$.

2.3.3 Suppose that Z has the standard normal distribution. Along the lines of the Example 2.3.4, that is directly using the techniques of integration, show that

$$E(|Z|^r) = \pi^{-1/2}2^{r/2}\Gamma(\frac{1}{2}r + \frac{1}{2})$$

for any number $r > 0$.

2.3.4 (Exercise 2.3.3 Continued) Why is it that the answer in the Exercise 2.3.1, part (ii) matches with the expression of $E(|Z|^r)$ given in the Exercise 2.3.3? Is it possible to somehow connect these two pieces?

2.3.5 (Exercise 2.2.13 Continued) Suppose that X is a continuous random variable having its pdf $f(x)$ with the support $\mathcal{X} = (a, b), -\infty \leq a < b \leq \infty$. Additionally suppose that $f(x)$ is symmetric about the point $x = c$ where $a < c < b$. Then, show that $E\{(X - c)^r\} = 0$ for all positive odd integer r as long as the rth order moment is finite. {*Hint*: In principle, follow along the derivations in the part (i) in the Exercise 2.3.2.}

2.3.6 (Exercise 2.3.1 Continued) Derive the same results stated in the Exercise 2.3.1 by successively differentiating the mgf of Z given by (2.3.14).

2.3.7 (Exercise 2.3.2 Continued) Suppose that X has the $N(\mu, \sigma^2)$ distribution. Show that $\mu_3 = 0$ and $\mu_4 = 3\sigma^4$ by successively differentiating the mgf of X given by (2.3.16).

2.3.8 Consider the Cauchy random variable X, defined in (1.7.31), whose pdf is given by $f(x) = \pi^{-1}(1 + x^2)^{-1}$ for $x \in \Re$. Show that $E(X)$ does not exist.

2.3.9 Give an example of a continuous random variable, other than Cauchy, for which the mean μ is infinite. {*Hint*: Try a continuous version of the Example 2.3.1. Is $f(x) = x^{-2}I(1 < x < \infty)$ a genuine pdf? What happens to the mean of the corresponding random variable?}

2.3.10 Give different examples of continuous random variables for which

(i)   the mean μ is finite, but the variance σ^2 is infinite;

(ii)   the mean μ and variance σ^2 are both finite, but μ_3 is infinite;

(iii) μ_{10} is finite, but μ_{11} is infinite.

{Hint: Extend the ideas from the Example 2.3.2 and try other pdf's similar to that given in the "hint" for the Exercise 2.3.9.}

2.3.11 For the binomial and Poisson distributions, derive the third and fourth central moments by the method of successive differentiation of their respective mgf's from (2.3.5) and (2.3.10).

2.3.12 Consider the expression of the mgf of an exponential distribution given in (2.3.27). By successively differentiating the mgf, find the third and fourth central moments of the exponential distribution.

2.3.13 Consider the expression of the mgf of a Chi-square distribution given in (2.3.28). By successively differentiating the mgf, find the third and fourth central moments of the Chi-square distribution.

2.3.14 (Exercise 2.2.16 Continued) Suppose that X has the lognormal pdf $f(x) = \{\sqrt{2\pi}\}^{-1}x^{-1}exp[-\{log(x)\}^2/2]$ for $0 < x < \infty$, given by (1.7.27). Show that the rth moment η_r is finite and find its expression for each $r = 1, 2, 3, \dots$. {Hint: Substitute $y = log(x)$ in the integrals and see that the integrals would resemble the mgf with respect to the normal pdf for some appropriate values t.}

2.3.15 Suppose that a random variable X has the Laplace or the double exponential pdf $f(x) = \frac{1}{2\beta}exp\{-|x|/\beta\}$ for all $x \in \Re$ where $\beta \in \Re^+$. For this distribution,

(i)   derive the expression of its mgf;

(ii)   derive the expressions of the third and fourth central moments.

{Hint: Look at the Section 2.2.6. Evaluate the relevant integrals along the lines of (2.2.31)-(2.2.33).}

2.3.16 Suppose that Z has the standard normal distribution. Along the lines of the Example 2.3.4, derive the expression of the mgf $M_{|Z|}(t)$ of the random variable $|Z|$ for t belonging to $\Re$.

2.3.17 Prove the Theorem 2.3.2.

2.3.18 (Exercise 2.2.14 Continued) Suppose that we have a random variable X which has the Rayleigh distribution, that is its pdf is given by $f(x) = 2\theta^{-1}x exp(-x^2/\theta)I(x > 0)$ where $\theta(> 0)$. Evaluate $E(X^r)$ for any arbitrary but fixed $r > 0$. {Hint: Try the substitution $u = x^2/\theta$ during the integration.}

2.3.19 (Exercise 2.2.15 Continued) Suppose that we have a random variable X which has the Weibull distribution, that is its pdf is given by

$f(x) = \alpha\beta^{-\alpha}x^{\alpha-1}exp(-[x/\beta]^{\alpha})I(x>0)$ where $\alpha(>0)$ and $\beta(>0)$. Evaluate $E(X^r)$ for any arbitrary but fixed $r>0$. {*Hint:* Try the substitution $u = [x/\beta]^{\alpha}$ during the integration.}

2.4.1 In this exercise, you are given the expressions of the mgf of different random variables. In each case, (i) identify the random variable either by its standard name or by explicitly writing down its pmf or the pdf, (ii) find the values of both μ and σ for the random variable X.

(i) $M_X(t) = e^{5t}$, for $t \in \Re$;

(ii) $M_X(t) = 1$, for $t \in \Re$;

(iii) $M_X(t) = \frac{1}{2}(1+e^t)$, for $t \in \Re$;

(iv) $M_X(t) = \frac{1}{3}(e^{-2t}+1+e^t)$, for $t \in \Re$;

(v) $M_X(t) = \frac{1}{10}(e^{2t}+3+6e^{4t})$, for $t \in \Re$;

(vi) $M_X(t) = \frac{1}{2401}(3e^t+4)^4$, for $t \in \Re$.

{*Hint:* Think of a discrete random variable and how one actually finds its mgf. Then, use Theorem 2.4.1.}

2.4.2 In this exercise, you are given the expressions for the mgf of a random variable X. In each case, (i) identify the random variable X either by its standard name or by explicitly writing down its pmf or the pdf, (ii) find the values of both μ and σ for the random variable X.

(i) $E[e^{tX}] = e^{25t^2}$, for $t \in \Re$;

(ii) $E[e^{tX}] = e^{t^2}$, for $t \in \Re$;

(iii) $M_X(t) = (1-6t+9t^2)^{-2}$, for $t < \frac{1}{3}$.

{*Hint:* Think of a continuous random variable from the Section 1.7 and see if that helps in each part. Then, use Theorem 2.4.1.}

2.4.3 A random variable X has its mgf given by

$$M_X(t) = \frac{1}{5}\left\{\frac{1+2e^{2t}}{e^{4t}} + \frac{e^{3t}+e^{6t}}{e^{5t}}\right\} \text{ for } t \in \Re.$$

Find μ and σ^2. Can the distribution of X be identified?

2.4.4 (Example 2.4.4 Continued) Show that $E(X^r) = E(Y^r)$ for all $r = 1, 2, ...$ where the pdf of X is $f(x) = (2\pi)^{-1/2}x^{-1}exp[-\frac{1}{2}(log(x))^2]I(x>0)$ and that of Y is $g(y) = f(y)[1+c\sin(2\pi log(y))]I(y>0)$. Here, c is a fixed number, $-1 \le c \le 1$ and $c \ne 0$. {*Hint:* Note that the pdf $f(x)$ actually matches with the lognormal density from (1.7.27). When handling $g(y)$, first show that the multiplier of $f(y)$ is always positive. Then while evaluating $\int_0^{\infty} g(y)dy$, all one needs to show is that $\int_0^{\infty} \sin(2\pi log(x))f(x)dx = 0$. In the process of evaluating the relevant integrals, try the substitution $u = log(x)$.}

2.4.5 (Exercise 2.4.4 Continued) Suppose a random variable X has the lognormal pdf $f(x) = (2\pi)^{-1/2}x^{-1}exp[-\frac{1}{2}(log(x))^2]I(x > 0)$. Show that the mgf of X does not exist. {*Hint*: In the process of evaluating the relevant integral, one may try the substitution $u = log(x)$.}

2.4.6 (Exercise 2.4.4 Continued) Consider the two pdf's $f(x)$ and $g(y)$ with $x > 0, y > 0$, as defined in the Exercise 2.4.4. Let $a(x)$ with $x > 0$ be any other pdf which has all its (positive integral) moments finite. For example, $a(x)$ may be the pdf corresponding to the Gamma(α, β) distribution. On the other hand, there is no need for $a(x)$ to be positive for all $x > 0$. Consider now two non-negative random variables U and V with the respective pdf's $f_0(u)$ and $g_0(v)$ with $u > 0, v > 0$ where

$$f_0(u) = pa(u) + (1-p)f(u),$$

$$g_0(v) = pa(v) + (1-p)g(v) \text{ with fixed } p, 0 < p < 1.$$

Show that U and V have the same infinite sequence of moments but they naturally have different distributions. We plotted the two pdf's $f_0(u)$ and $g_0(v)$ with $c = p = \frac{1}{2}$ and $a(x) = \frac{1}{10}e^{-x/10}$. In comparison with the plots given in the Figure 2.4.1, the two present pdf's appear more skewed to the right. The reader should explore different shapes obtainable by using different choices of the function $a(x)$ and the numbers c and p. In view of the Exercise 2.4.5, one should check that neither U nor V has a finite mgf.

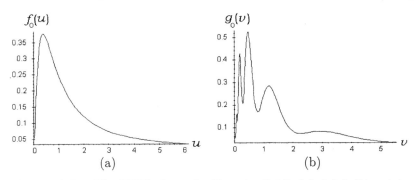

Figure 2.6.1. The PDF's from the Exercise 2.4.6: (a) $f_0(u)$ (b) $g_0(v)$
Where $c = p = \frac{1}{2}$, $a(x) = \frac{1}{10}exp(-x/10)$

2.4.7 (Exercise 2.2.5 Continued) Consider a random variable X which has the following discrete uniform distribution along the lines of (1.7.11):

$$P(X = i) = n^{-1} \text{ for } i = 1, ..., n$$

Derive the explicit expression of the mgf of X and hence derive μ and σ for this distribution. {*Hint*: Note that $E(e^{tX}) = \Sigma_{i=1}^{n} e^{ti} \frac{1}{n} = \frac{1}{n} e^{t}[1 + e^{t} + e^{2t} + ... + e^{(n-1)t}]$. Can this be simplified using the geometric progression?}

2.4.8 A random variable X has its pdf

$$f(x) = c\, exp\{-x^2 - dx\} \text{ for } x \in \Re,$$

where $c(> 0)$ and d are appropriate constants. We are also told that $E(X) = -\frac{3}{2}$.

    (*i*)   Determine c and d;

    (*ii*)   Identify the distribution by its name;

    (*iii*)   Evaluate the third and fourth central moments of X.

2.4.9 A random variable X has its pdf

$$f(x) = c\, exp\{-x/3\}x^d \text{ for } x \in \Re^{+},$$

where $c(> 0)$ and $d(> 0)$ are appropriate constants. We are also told that $E(X) = 10.5$ and $V(X) = 31.5$.

    (*i*)   Determine c and d;

    (*ii*)   Identify the distribution by its name;

    (*iii*)   Evaluate the third and fourth moments, η_3 and η_4, for X.

2.4.10 A random variable X has its mgf given by

$$M_X(t) = e^{t}(5 - 4e^{t})^{-1} \text{ for } t < .223.$$

Evaluate $P(X = 4 \text{ or } 5)$. {*Hint*: What is the mgf of a geometric random variable?}

2.5.1 Verify (2.5.4).

2.5.2 Let X be a discrete random variable having the Geometric(p) distribution whose pmf is $f(x) = P(X = x) = p(1-p)^{x-1}$ for $x = 1, 2, 3, ...$ and $0 < p < 1$, given by (1.7.7). Show that pgf of X is given by

$$P_X(t) = pt\{1 - (1 - p)t\}^{-1} \text{ with } t > 0.$$

Evaluate the first two factorial moments of X and hence find the expressions of μ and σ.

2.5.3 Let X be a discrete random variable having the Poisson(λ) distribution with $0 < \lambda < \infty$. Derive the expression of the pgf of X. Hence, evaluate the kth order factorial moment of X and show that it reduces to λ^k.

2.5.4 Let X be a discrete random variable having the Binomial(n, p) distribution with $0 < p < 1$. Derive the expression of the pgf of X. Hence, evaluate the kth order factorial moment of X and show that it reduces to $n(n-1)...(n-k+1)p^k$.

2.5.5 Consider the pgf $P_X(t)$ defined by (2.5.1). Suppose that $P_X(t)$ is twice differentiable and assume that the derivatives with respect to t and expectation can be interchanged. Let us denote $P_X'(t) = \frac{d}{dt}P_X(t)$ and $P_X''(t) = \frac{d^2}{dt^2}P_X(t)$. Then, show that

    (i) $E[X] = P_X'(1)$;

    (ii) $V[X] = P_X''(1) + P_X'(1) - \{P_X'(1)\}^2$.

2.5.6 Suppose that a random variable X has the Beta$(3,5)$ distribution. Evaluate the third and fourth factorial moments of X. {*Hint*: Observe that $E[X(X-1)(X-2)] = E[X(X-1)^2 - X(X-1)] = E[X(X-1)^2] - E[X(X-1)]$. Write each expectation as a beta integral and evaluate the terms accordingly to come up with the third factorial moment. The other part can be handled by extending this idea.}

3

Multivariate Random Variables

3.1 Introduction

Suppose that we draw two random digits, each from the set $\{0, 1, ..., 9\}$, with equal probability. Let X_1, X_2 be respectively the sum and the difference of the two selected random digits. Using the sample space technique, one can easily verify that $P(X_1 = 0) = \frac{1}{100}, P(X_1 = 1) = \frac{2}{100}, P(X_1 = 2) = \frac{3}{100}$ and eventually obtain the distribution, namely $P(X_1 = i)$ for $i = 0, 1, ..., 18$. Similarly, one can also evaluate $P(X_2 = j)$ for $j = -9, -8, ..., 8, 9$. Now, suppose that one has observed $X_1 = 3$. Then there is no chance for X_2 to take a value such as 2. In other words, the *bivariate discrete* random variable (X_1, X_2) varies together in a certain way. This sense of *joint variation* is the subject matter of the present chapter.

Generally speaking, we may have $k(\geq 2)$ real valued random variables $X_1, ..., X_k$ which vary individually as well as jointly. For example, during the "health awareness week" we may consider a population of college students and record a randomly selected student's height (X_1), weight (X_2), age (X_3) and blood pressure (X_4). Each individual random variable may be assumed to follow some appropriate probability distribution in the population, but it may also be quite reasonable to assume that these four variables together follow a certain joint probability distribution. This example falls in the category of *multivariate continuous* random variables.

The Section 3.2 introduces the ideas of *joint, marginal and conditional distributions* in a discrete situation including the *multinomial distribution*. The Section 3.3 introduces analogous topics in the continuous case. We begin with the bivariate scenario for simplicity. The notion of a *compound distribution* and the evaluation of the associated mean and variance are explored in the Examples 3.3.6-3.3.8. Multidimensional analysis is included in the Section 3.3.2. The notions of the *covariances* and *correlation coefficients* between random variables are explored in the Section 3.4. We briefly revisit the multinomial distribution in the Section 3.4.1. The Section 3.5 introduces the concept of the *independence* of random variables. The celebrated *bivariate normal distribution* is discussed in the Section 3.6 and the associated marginal as well as the conditional pdf's are found in this special situation. The relationships between the zero correlation and possible independence are explained in the Section 3.7. The Section 3.8 summarizes both the *one-* and *multi-parameter exponential families* of distributions. This chapter draws to an end with some of the standard inequalities which are widely used in statistics. The Sections 3.9.1-3.9.4 include more details than Sections 3.9.5-3.9.7.

3.2   Discrete Distributions

Let us start with an example of *bivariate discrete* random variables.

Example 3.2.1 We toss a fair coin twice and define $U_i = 1$ or 0 if the ith toss results in a head (H) or tail (T), $i = 1, 2$. We denote $X_1 = U_1 + U_2$ and $X_2 = U_1 - U_2$. Utilizing the techniques from the previous chapters one can verify that X_1 takes the values $0, 1$ and 2 with the respective probabilities $\frac{1}{4}, \frac{1}{2}$ and $\frac{1}{4}$, whereas X_2 takes the values $-1, 0$ and 1 with the respective probabilities $\frac{1}{4}, \frac{1}{2}$ and $\frac{1}{4}$. When we think of the distribution of X_i alone, we do not worry much regarding the distribution of $X_j, i \neq j = 1, 2$. But how about studying the random variables (X_1, X_2) together? Naturally, the pair (X_1, X_2) takes one of the nine $(= 3^2)$ possible pairs of values: $(0, -1), (0, 0), (0, 1), (1, -1), (1, 0), (1, 1), (2, -1), (2, 0)$, and $(2, 1)$.

Table 3.2.1. Joint Distribution of X_1 and X_2

		X_1 values			Row Total
		0	1	2	
	−1	0	.25	0	.25
X_2 values	0	.25	0	.25	.50
	1	0	.25	0	.25
Col. Total		.25	.50	.25	1.00

Note that $X_1 = 0$ implies that we must have observed TT, in other words $U_1 = U_2 = 0$, so that $P(X_1 = 0 \cap X_2 = 0) = P(TT) = \frac{1}{4}$. Whereas $P(X_1 = 0 \cap X_2 = -1) = P(X_1 = 0 \cap X_2 = 1) = 0$. Other entries in the Table 3.2.1 can be verified in the same way. The entries in the Table 3.2.1 provide the joint probabilities, $P(X_1 = i \cap X_2 = j)$ for all $i = 0, 1, 2$ and $j = -1, 0, 1$. Such a representation is provides what is known as the *joint distribution* of X_1, X_2. The column and row totals respectively line up exactly with the individual distributions of X_1, X_2 respectively which are also called the *marginal distributions* of X_1, X_2. ▲

3.2.1   The Joint, Marginal and Conditional Distributions

Suppose that we have $k(\geq 2)$ discrete random variables $X_1, ..., X_k$ where X_i takes one of the possible values x_i belonging to its support $\mathcal{X}_i$, $i = 1, ..., k$. Here, $\mathcal{X}_i$ can be at most countably infinite. The *joint probability mass function* (pmf) of $\mathbf{X} = (X_1, ..., X_k)$ is then given by

$$f(\mathbf{x}) = P\{X_1 = x_1, ..., X_k = x_k\}, \mathbf{x} = (x_1, ..., x_k),$$
$$\text{for all } x_i \in \mathcal{X}_i, i = 1, ..., k. \tag{3.2.1}$$

A function such as $f(\mathbf{x})$ would be a genuine *joint pmf* if and only if the following two conditions are met:

(i) $f(\mathbf{x}) \geq 0$ for all $\mathbf{x} = (x_1, ..., x_k) \in \Pi_{i=1}^{k} \mathcal{X}_i = \mathcal{X}$, say, and

(ii) $\sum\limits_{\text{all } \mathbf{x} \in \mathcal{X}} f(\mathbf{x}) = 1.$ $\tag{3.2.2}$

These are direct multivariable extensions of the requirements laid out earlier in (1.5.3) in the case of a single real valued random variable.

The *marginal distribution* of X_i corresponds to the marginal probability mass function defined by

$$f_i(x_i) = \sum\limits_{\mathbf{x}: x_j \in \mathcal{X}_j \text{ for all } j \neq i} f(\mathbf{x}), x_i \in \mathcal{X}_i, i = 1, ..., k. \tag{3.2.3}$$

Example 3.2.2 (Example 3.2.1 Continued) We have $\mathcal{X}_1 = \{0, 1, 2\}, \mathcal{X}_2 = \{-1, 0, 1\}$, and the joint pmf may be summarized as follows: $f(x_1, x_2) = 0$ when $(x_1, x_2) = (0, -1), (0, 1), (1, 0), (2, -1), (2, 1)$, but $f(x_1, x_2) = .25$ when $(x_1, x_2) = (0, 0), (1, -1), (1, 1), (2, 0)$. Let us apply (3.2.3) to obtain the marginal pmf of X_1.

$$f_1(0) = P\{X_1 = 0\} = \sum\limits_{x_2 \in \mathcal{X}_2} P\{X_1 = 0 \cap X_1 = x_2\} = 0 + .25 + 0 = .25,$$

$$f_1(1) = P\{X_1 = 1\} = \sum\limits_{x_2 \in \mathcal{X}_2} P\{X_1 = 1 \cap X_1 = x_2\} = .25 + 0 + .25 = .50,$$

$$f_1(2) = P\{X_1 = 2\} = \sum\limits_{x_2 \in \mathcal{X}_2} P\{X_1 = 2 \cap X_1 = x_2\} = 0 + .25 + 0 = .25,$$

which match with the respective column totals in the Table 3.2.1. Similarly, the row totals in the Table 3.2.1 will respectively line up exactly with the marginal distribution of X_2. ▲

In the case of k-dimensions, the notation becomes cumbersome in defining the notion of the *conditional* probability mass functions. For simplicity, we explain the idea only in the bivariate case.

For $i \neq j$, the *conditional pmf* of X_i at the point x_i given that $X_j = x_j$, denoted by $f_{i|j}(x_i)$, is defined by

$$f_{i|j}(x_i) = \frac{f(\mathbf{x})}{f_j(x_j)} \text{ for all } x_i \in \mathcal{X}_i \text{ where } \mathbf{x} = (x_i, x_j), f_j(x_j) > 0.$$

(3.2.4)

Now, suppose that $P(X_j = x_j)$, which is equivalent to $f_j(x_j)$, is positive. Then, using the notion of the conditional probability from (1.4.1) we can express $f_{i|j}(x_i)$ as

$$P\{X_i = x_i \mid X_j = x_j\} = \frac{P\{X_i = x_i \cap X_j = x_j\}}{P(X_j = x_j)} = \frac{f(x_i, x_j)}{f_j(x_j)}$$

for $x_i \in \mathcal{X}_i, i \neq j$. In the two-dimensional case, we have two conditional pmf's, namely $f_{1|2}(x_1)$ and $f_{2|1}(x_2)$ which respectively represent the conditional pmf of X_1 given $X_2 = x_2$ and that of X_2 given $X_1 = x_1$.

Once the conditional pmf $f_{1|2}(x_1)$ of X_1 given $X_2 = x_2$ and the conditional pmf $f_{2|1}(x_2)$ of X_2 given $X_1 = x_1$ are found, the *conditional mean* of X_i given $X_j = x_j$, denoted by $E[X_i \mid X_j = x_j]$, is defined as follows:

$$E[X_i \mid X_j = x_j] = \sum_{x_i \in \mathcal{X}_i} x_i f_{i|j}(x_i) \text{ with any fixed } x_j \in \mathcal{X}_j,$$

$$\text{for } i \neq j = 1, 2.$$

(3.2.5)

This is not really any different from how we interpreted the expected value of a random variable in the Definition 2.2.1. To find the conditional expectation of X_i given $X_j = x_j$, we simply multiply each value x_i by the conditional probability $P\{X_i = x_i \mid X_j = x_j\}$ and then sum over all possible values $x_i \in \mathcal{X}_i$ with any fixed $x_j \in \mathcal{X}_j$, for $i \neq j = 1, 2$.

Example 3.2.3 (Example 3.2.2 Continued) One can verify that given $X_2 = -1$, the conditional pmf $f_{1|2}(x_1)$ corresponds to 0, 1 and 0 respectively when $x_1 = 0, 1, 2$. Other conditional pmf's can be found analogously. ▲

Example 3.2.4 Consider two random variables X_1 and X_2 whose joint distribution is given as follows:

Table 3.2.2. Joint Distribution of X_1 and X_2

		X₁ values			Row Total
		-1	2	5	
	1	.12	.18	.25	.55
X_2 values					
	2	.20	.09	.16	.45
Col. Total		.32	.27	.41	1.00

One has $\mathcal{X}_1 = \{-1, 2, 5\}, \mathcal{X}_2 = \{1, 2\}$. The marginal pmf's of X_1, X_2 are respectively given by $f_1(-1) = .32, f_1(2) = .27, f_1(5) = .41$, and $f_2(1) = .55, f_2(2) = .45$. The conditional pmf of X_1 given that $X_2 = 1$, for example, is given by $f_{1|2}(-1) = \frac{12}{55}, f_{1|2}(2) = \frac{18}{55}, f_{1|2}(5) = \frac{25}{55}$. We can apply the notion of the conditional expectation from (3.2.5) to evaluate $E[X_1 \mid X_2 = 1]$ and write

$$E[X_1 \mid X_2 = 1] = (-1)(\tfrac{12}{55}) + (2)(\tfrac{18}{55}) + (5)(\tfrac{25}{55}) = \tfrac{149}{55} \approx 2.7091.$$

That is, the conditional average of X_1 given that $X_2 = 1$ turns out to be approximately 2.7091. Earlier we had found the marginal pmf of X_1 and hence we have

$$E[X_1] = (-1)(.32) + (2)(.27) + (5)(.41) = 2.27.$$

Obviously, there is a conceptual difference between the two expressions $E[X_1]$ and $E[X_1 \mid X_2 = 1]$. ▲

If we have a function $g(x_1, x_2)$ of two variables x_1, x_2 and we wish to evaluate $E[g(X_1, X_2)]$, then how should we proceed? The approach involves a simple generalization of the Definition 2.2.3. One writes

$$E[g(X_1, X_2)] = \underset{x_2 \in \mathcal{X}_2}{\Sigma} \underset{x_1 \in \mathcal{X}_1}{\Sigma} g(x_1, x_2) f(x_1, x_2). \qquad (3.2.6)$$

Example 3.2.5 (Example 3.2.4 Continued) Suppose that $g(x_1, x_2) = x_1 x_2$ and let us evaluate $E[g(X_1, X_2)]$. We then obtain

$$E[g(X_1, X_2)] = (-1)(1)(.12) + (-1)(2)(.20) + (2)(1)(.18)$$
$$+ (2)(2)(.09) + (5)(1)(.25) + (5)(2)(.16) = 3.05.$$
$$(3.2.7)$$

Instead, if we had $h(x_1, x_2) = x_1^2 x_2$, then one can similarly check that $E[h(X_1, X_2)] = 16.21$. ▲

3.2.2   The Multinomial Distribution

Now, we discuss a special multivariate discrete distribution which appears in the statistical literature frequently and it is called the *multinomial distribution*. It is a direct generalization of the binomial setup introduced earlier in (1.7.2). First, let us look at the following example.

Example 3.2.6 Suppose that we have a fair die and we roll it twenty times. Let us define $X_i =$ the number of times the die lands up with the face having i dots on it, $i = 1, ..., 6$. It is not hard to see that individually

X_i has the Binomial $(20, \frac{1}{6})$ distribution, for every fixed $i = 1, ..., 6$. It is clear, however, that $\Sigma_{i=1}^{6} X_i$ must be exactly twenty in this example and hence we conclude that $X_1, ..., X_6$ are not all free-standing random variables. Suppose that one has observed $X_1 = 2, X_2 = 4, X_3 = 4, X_5 = 2$ and $X_6 = 3$, then X_4 must be 5. But, when we think of X_4 alone, its marginal distribution is Binomial $(20, \frac{1}{6})$ and so by itself it can take any one of the possible values $0, 1, 2, ..., 20$. On the other hand, if we assume that $X_1 = 2, X_2 = 4, X_3 = 4, X_5 = 2$ and $X_6 = 3$, then X_4 has to be 5. That is to say that there is certainly some kind of dependence among these random variables $X_1, ..., X_6$. What is the exact nature of this joint distribution? We discuss it next. ▲

We denote a vector valued random variable by $\mathbf{X} = (X_1, ..., X_k)$ in general. A discrete random variable $\mathbf{X}$ is said to have a *multinomial distribution* involving n and $p_1, ..., p_k$, denoted by $\text{Mult}_k(n, p_1, ..., p_k)$, if and only if the joint pmf of $\mathbf{X}$ is given by

$$f(\mathbf{x}) = \frac{n!}{\Pi_{i=1}^{k}(x_i!)} \Pi_{i=1}^{k} p_i^{x_i} \text{ for } x_i = 0, 1, ..., n, 0 < p_i < 1,$$

$$i = 1, ..., k, \ \Sigma_{i=1}^{k} x_i = n \text{ and } \Sigma_{i=1}^{k} p_i = 1. \tag{3.2.8}$$

We may think of the following experiment which gives rise to the multinomial pmf. Let us visualize k boxes lined up next to each other. Suppose that we have n marbles which are tossed into these boxes in such a way that each marble will land in one and only one of these boxes. Now, let p_i stand for the probability of a marble landing in the ith box, with $0 < p_i < 1, i = 1, ..., k, \Sigma_{i=1}^{k} p_i = 1$. These p's are assumed to remain same for each tossed marble. Now, once these n marbles are tossed into these boxes, let $X_i = $ the number of marbles which land inside the box $\#i, i = 1, ..., k$. The joint distribution of $\mathbf{X} = (X_1, ..., X_k)$ is then characterized by the pmf described in (3.2.8).

How can we verify that the expression given by (3.2.8) indeed corresponds to a genuine pmf? We must check the requirements stated in (3.2.2). The first requirement certainly holds. In order to verify the requirement (ii), we need an extension of the Binomial Theorem. For completeness, we state and prove the following result.

Theorem 3.2.1 (Multinomial Theorem) *Let $a_1, ..., a_k$ be arbitrary real numbers and n be a positive integer. Then, we have*

$$(a_1 + ... + a_k)^n = \sum_{\text{all } \mathbf{x} \in \Pi_{i=1}^{k} \mathcal{X}_i} \frac{n!}{\Pi_{i=1}^{k}(x_i!)} \Pi_{j=1}^{k} a_j^{x_j}, \tag{3.2.9}$$

where $x_i \in \mathcal{X}_i = \{0, 1, ..., n\}, \mathbf{x} = (x_1, ..., x_n)$ and $\Sigma_{i=1}^{k} x_i = n$.

Proof First consider the case $k = 2$. By the Binomial Theorem, we get

$$(a_1 + a_2)^n = \Sigma_{x=0}^n \frac{n!}{x!(n-x)!} a_1^x a_2^{n-x} = \sum_{\text{all } \mathbf{x} \in \Pi_{i=1}^2 \mathcal{X}_i} \frac{n!}{\Pi_{i=1}^2 (x_i!)} \Pi_{j=1}^2 a_j^{x_j},$$

$$(3.2.10)$$

which is the same as (3.2.9). Now, in the case $k = 3$, by using (3.2.10) repeatedly let us write

$$(a_1 + a_2 + a_3)^n$$
$$= \Sigma_{x=0}^n \frac{n!}{x!(n-x)!} (a_1 + a_2)^x a_3^{n-x} \qquad (3.2.11)$$
$$= \Sigma_{x=0}^n \Sigma_{y=0}^x \frac{n!}{y!(x-y)!(n-x)!} a_1^y a_2^{x-y} a_3^{n-x},$$

which is the same as (3.2.9). Next, let us assume that the expansion (3.2.9) holds for all $n \leq r$, and let us prove that the same expansion will then hold for $n = r + 1$. By mathematical induction, the proof will then be complete. Let us write

$$(a_1 + ... + a_r + a_{r+1})^n$$
$$= \Sigma_{x=0}^n \frac{n!}{x!(n-x)!} (a_1 + ... + a_r)^x a_{r+1}^{n-x},$$
by using the binomial expansion (3.2.10)

$$= \Sigma_{x=0}^n \frac{n!}{x!(n-x)!} \left\{ \sum_{\text{all } \mathbf{x} \in \Pi_{i=1}^r \mathcal{Y}_i} \frac{x!}{\Pi_{i=1}^r (x_i!)} \Pi_{j=1}^r a_j^{x_j} \right\} a_{r+1}^{n-x}$$

with $\mathcal{Y}_i = \{0, 1, ..., x\}, \mathbf{x} = (x_1, ..., x_r), \Sigma_{i=1}^r x_i = x,$ (3.2.12)

$x = 0, 1, ..., n$, by using (3.2.8) with $k = r$

$$= \sum_{\text{all } \mathbf{x} \in \Pi_{i=1}^{r+1} \mathcal{X}_i} \frac{n!}{\Pi_{i=1}^{r+1} (x_i!)} \Pi_{j=1}^{r+1} a_j^{x_j}$$

with $\mathcal{X}_i = \{0, 1, ..., x\}, \mathbf{x} = (x_1, ..., x_r), \Sigma_{i=1}^r x_i = x,$

$x = 0, 1, ..., n,$

which is the same as (3.2.9). The proof is now complete. ∎

The fact that the expression in (3.2.8) satisfies the requirement (ii) stated in (3.2.2) follows immediately from the Multinomial Theorem. That is, the function $f(\mathbf{x})$ given by (3.2.8) is a genuine multivariate pmf.

> The *marginal distribution* of the random variable X_i is Binomial(n, p_i) for each fixed $i = 1, ..., k$.

In the case of the multinomial distribution (3.2.8), suppose that we focus on the random variable X_i alone for some arbitrary but otherwise fixed

$i = 1, ..., k$. The question is this: Out of the n marbles, how many (that is, X_i) would land in the ith box? We are simply counting how many marbles would fall in the ith box and how many would fall outside, that is in any one of the other $k-1$ boxes. This is the typical binomial situation and hence we observe that the random variable X_i has the Binomial(n, p_i) distribution for each fixed $i = 1, ..., k$. Hence, from our discussions on the binomial distribution in Section 2.2.2 and (2.2.17), it immediately follows that

$$E(X_i) = np_i \text{ and } V(X_i) = np_i(1 - p_i)$$
$$\text{for each fixed but otherwise arbitrary } i = 1, ..., k. \tag{3.2.13}$$

Example 3.2.7 (Example 3.2.6 Continued) In the die rolling example, $\mathbf{X} = (X_1, ..., X_k)$ has the Mult$_k(n, p_1, ..., p_k)$ distribution where $n = 20, k = 6$, and $p_1 = ... = p_6 = \frac{1}{6}$. ▲

> The derivation of the moment generating function (mgf) of the
> multinomial distribution and some of its applications are
> highlighted in Exercise 3.2.8.

The following theorems are fairly straightforward to prove. We leave their proofs as the Exercises 3.2.5-3.2.6.

Theorem 3.2.2 *Suppose that the random vector* $\mathbf{X} = (X_1, ..., X_k)$ *has the Mult$_k(n, p_1, ..., p_k)$ distribution. Then, any subset of the X variables of size r, namely* $(X_{i_1}, ..., X_{i_r})$ *has a multinomial distribution in the sense that* $(X_{i_1}, ..., X_{i_r}, X_{r+1}^*)$ *is Mult$_{r+1}(n, p_{i_1}, ..., p_{i_r}, 1 - \Sigma_{j=1}^r p_{i_j})$ where* $1 \leq i_1 < i_2 < ... < i_r \leq k$ *and* $X_{r+1}^* = n - \Sigma_{j=1}^r X_{i_j}$.

Theorem 3.2.3 *Suppose that the random vector* $\mathbf{X} = (X_1, ..., X_k)$ *has the Mult$_k(n, p_1, ..., p_k)$ distribution. Consider any subset of the X variables of size r, namely* $(X_{i_1}, ..., X_{i_r})$. *The conditional joint distribution of* $(X_{i_1}, ..., X_{i_r})$ *given all the remaining X's is also multinomial with its conditional pmf*

$$P\{X_{i_1} = x_{i_1}, X_{i_2} = x_{i_2}, ..., X_{i_r} = x_{ir} \mid \sum_{l \neq i_1, i_2, ..., i_r}^{k} X_l = t\}$$

$$= \frac{(n-t)!}{\Pi_{j=1}^r x_{i_j}!} \Pi_{j=1}^r (p_{i_j}')^{x_{i_j}}, \quad x_{i_j} = 0, 1, ..., n - t, \Sigma_{j=1}^r x_{i_j} = n - t,$$

with $0 \leq t \leq n$, $p_{i_j}' = p_{i_j}/\{\Sigma_{j=1}^r p_{i_j}\}$ *where* $1 \leq i_1 < i_2 < ... < i_r \leq k$.

Example 3.2.8 (Example 3.2.7 Continued) In the die rolling example, suppose that we are simply interested in counting how many times the faces with the numbers 1 and 5 land up. In other words, our focus is on the three dimensional random variable (X_1, X_5, X_3^*) where $X_3^* = n - (X_1 + X_5)$. By the Theorem 3.2.2, the three dimensional variable of interest (X_1, X_5, X_3^*) has the $\text{Mult}_3(n, \frac{1}{6}, \frac{1}{6}, \frac{2}{3})$ distribution where $n = 20$. ▲

3.3   Continuous Distributions

Generally speaking, we may be dealing with $k(\geq 2)$ real valued random variables $X_1, ..., X_k$ which vary individually as well as jointly. We may go back to the example cited earlier in the introduction. Consider a population of college students and record a randomly selected student's height (X_1), weight (X_2), age (X_3), and blood pressure (X_4). It may be quite reasonable to assume that these four variables together has some *multivariate continuous* joint probability distribution in the population.

The notions of the *joint, marginal,* and *conditional distributions* in the case of multivariate continuous random variables are described in the Section 3.3.1. This section begins with the bivariate scenario for simplicity. The notion of a *compound distribution* and the evaluation of the associated mean and variance are explored in the Examples 3.3.6-3.3.8. Multidimensional analysis is included in the Section 3.3.2.

3.3.1   The Joint, Marginal and Conditional Distributions

For the sake of simplicity, we begin with the bivariate continuous random variables. Let X_1, X_2 be continuous random variables taking values respectively in the spaces $\mathcal{X}_1, \mathcal{X}_2$ both being subintervals of the real line $\Re$. Consider a function $f(x_1, x_2)$ defined for $(x_1, x_2) \in \mathcal{X}_1 \times \mathcal{X}_2$. Throughout we will use the following convention:

> Let us presume that $\mathcal{X}_1 \times \mathcal{X}_2$ is the support of the probability distribution in the sense that $f(x_1, x_2) > 0$ for all $(x_1, x_2) \in \mathcal{X}_1 \times \mathcal{X}_2$ and $f(x_1, x_2) = 0$ for all $(x_1, x_2) \in \Re^2 - (\mathcal{X}_1 \times \mathcal{X}_2)$.

Such a function $f(x_1, x_2)$ would be called a *joint pdf* of the random vector $\mathbf{X} = (X_1, X_2)$ if the following condition is satisfied:

$$\int_{\mathcal{X}_2} \int_{\mathcal{X}_1} f(x_1, x_2) dx_1 dx_2 = 1. \tag{3.3.1}$$

Comparing (3.3.1) with the requirement (ii) in (3.2.2) and (1.6.1), one will notice obvious similarities. In the case of one-dimensional random variables, the integral in (1.6.1) was interpreted as the *total area* under the density curve $y = f(x)$. In the present situation involving two random variables, the integral in (3.3.1) will instead represent the *total volume* under the density surface $z = f(x_1, x_2)$.

The *marginal distribution* of one of the random variables is obtained by integrating the joint pdf with respect to the remaining variable. The *marginal pdf* of X_i is then formally given by

$$f_i(x_i) = \int_{\mathcal{X}_j} f(x_1, x_2) dx_j \text{ for } x_i \in \mathcal{X}_i \text{ with fixed } i = 1, 2, \ j \neq i.$$
(3.3.2)

Visualizing the notion of the *conditional distribution* in a continuous scenario is little tricky. In the discrete case, recall that $f_i(x_i)$ was simply interpreted as $P(X_i = x_i)$ and hence the conditional pmf was equivalent to the corresponding conditional probability given by (3.2.4) as long as $P(X_i = x_i) > 0$. In the continuous case, however, $P(X_i = x_i) = 0$ for all $x_i \in \mathcal{X}_i$ with $i = 1, 2$. So, conceptually how should one proceed to define a *conditional pdf*?

Let us first derive the conditional df of X_1 given that $x_2 \leq X_2 \leq x_2 + h$ where $h(> 0)$ is a small number. Assuming that $P(x_2 \leq X_2 \leq x_2 + h) > 0$, we have

$$P\{X_1 \leq x_1 \mid x_2 \leq X_2 \leq x_2 + h\}$$
$$= \frac{P\{X_1 \leq x_1 \cap x_2 \leq X_2 \leq x_2 + h\}}{P\{x_2 \leq X_2 \leq x_2 + h\}}$$
$$= \frac{\int_{t=x_2}^{x_2+h} \int_{s=-\infty}^{x_1} f(s,t) ds dt}{\int_{t=x_2}^{x_2+h} f_2(t) dt}.$$
(3.3.3)

From the last step in (3.3.3) it is clear that as $h \downarrow 0$, the *limiting* value of this conditional probability takes the form of $0/0$. Thus, by appealing to the L'Hôpital's rule from (1.6.29) we can conclude that

$$\lim_{h \downarrow 0} P\{X_1 \leq x_1 \mid x_2 \leq X_2 \leq x_2 + h\}$$
$$= \lim_{h \downarrow 0} \frac{\frac{d}{dh} \left[\int_{t=x_2}^{x_2+h} \int_{s=-\infty}^{x_1} f(s,t) ds dt \right]}{\frac{d}{dh} \left[\int_{t=x_2}^{x_2+h} f_2(t) dt \right]}$$
$$= \lim_{h \downarrow 0} \frac{\int_{s=-\infty}^{x_1} f(s, x_2 + h) ds}{f_2(x_2 + h)}.$$
(3.3.4)

Next, by differentiating the last expression in (3.3.4) with respect to x_1, one obtains the expression for the conditional pdf of X_1 given that $X_2 = x_2$.

Hence, the *conditional pdf* of X_1 given that $X_2 = x_2$ is given by

$$f_{1|2}(x_1) = \frac{f(x_1, x_2)}{f_2(x_2)} \text{ for all } (x_1, x_2) \in \mathcal{X}_1 \times \mathcal{X}_2 \text{ such that } f_2(x_2) > 0,$$

(3.3.5)

whereas the *conditional pdf* of X_2 given $X_1 = x_1$ is analogously given by

$$f_{2|1}(x_2) = \frac{f(x_1, x_2)}{f_1(x_1)} \text{ for all } (x_1, x_2) \in \mathcal{X}_1 \times \mathcal{X}_2 \text{ such that } f_1(x_1) > 0.$$

(3.3.6)

Once these conditional pdf's are obtained, one can legitimately start talking about the *conditional moments* of one of the random variables given the other. For example, in order to find the conditional mean and the conditional variance of X_1 given that $X_2 = x_2$, one should proceed as follows:

$$\mu_{1|2} = E[X_1 \mid X_2 = x_2] = \int_{\mathcal{X}_1} x_1 f_{1|2}(x_1) dx_1;$$

$$E[X_1^2 \mid X_2 = x_2] = \int_{\mathcal{X}_1} x_1^2 f_{1|2}(x_1) dx_1; \qquad (3.3.7)$$

$$\sigma_{1|2}^2 = E[X_1^2 \mid X_2 = x_2] - \mu_{1|2}^2.$$

It should be clear that $\mu_{1|2}$ and $\sigma_{1|2}^2$ respectively from (3.3.7) would correspond to the *conditional mean* and the *conditional variance* of X_1 given that $X_2 = x_2$. In general, we can write

$$E[g(X_1, X_2)] = \int_{\mathcal{X}_2} \int_{\mathcal{X}_1} g(x_1, x_2) f(x_1, x_2) dx_1 dx_2;$$

$$E[h(X_i) \mid X_j = x_j] = \int_{\mathcal{X}_i} h(x_i) f_{i|j}(x_i) dx_i \text{ with } x_j \in \mathcal{X}_j \qquad (3.3.8)$$

$$\text{whenever } f_j(x_j) > 0 \text{ for } i \neq j = 1, 2 \,.$$

Example 3.3.1 Consider two random variables X_1 and X_2 whose joint continuous distribution is given by the following pdf:

$$f(x_1, x_2) = \begin{cases} 6(1 - x_2) & \text{if } 0 < x_1 < x_2 < 1 \\ 0 & \text{elsewhere} \end{cases} \qquad (3.3.9)$$

Obviously, one has $\mathcal{X}_1 = \mathcal{X}_2 = (0, 1)$. Next, we integrate this joint pdf with respect to x_2 to obtain the marginal pdf of X_1 and write

$$f_1(x_1) = \int_{x_1}^1 f(x_1, x_2) dx_2 = \int_{x_1}^1 6(1 - x_2) dx_2 = 3(1 - x_1)^2, \qquad (3.3.10)$$

for $0 < x_1 < 1$. In the same fashion, we obtain the marginal pdf of X_2 as follows:

$$f_2(x_2) = \int_0^{x_2} f(x_1, x_2) dx_1 = \int_0^{x_2} 6(1 - x_2) dx_1 = 6x_2(1 - x_2), \qquad (3.3.11)$$

for $0 < x_2 < 1$. Next, we can simply use the marginal pdf $f_2(x_2)$ of X_2 to obtain

$$E(X_2) = 6 \int_0^1 x_2^2(1 - x_2)dx_2 = \tfrac{1}{2},$$
$$E(X_2^2) = 6 \int_0^1 x_2^3(1 - x_2)dx_2 = \tfrac{3}{10}, \qquad (3.3.12)$$

so that

$$V(X_2) = E(X_2^2) - E^2(X_2) = \tfrac{3}{10} - \tfrac{1}{4} = \tfrac{1}{20}. \qquad (3.3.13)$$

The reader may similarly evaluate $E(X_1)$ and $V(X_1)$. ▲

Example 3.3.2 (Example 3.3.1 Continued) One should easily verify that

$$f_{1|2}(x_1) = 1/x_2 \text{ if } 0 < x_1 < x_2 \text{ and } 0 < x_2 < 1;$$
$$f_{2|1}(x_2) = 2(1 - x_2)(1 - x_1)^{-2} \text{ if } x_1 < x_2 \le 1 \text{ and } 0 < x_1 < 1; \qquad (3.3.14)$$

are respectively the two conditional pdf's. For fixed $0 < x_1 < 1$, the conditional expectation of X_2 given that $X_1 = x_1$ is obtained as follows:

$$\begin{aligned}
\mu_{2|1} &= 2(1 - x_1)^{-2} \int_{x_1}^1 x_2(1 - x_2)dx_2 \\
&= 2(1 - x_1)^{-2} \left[\tfrac{1}{2}x_2^2 - \tfrac{1}{3}x_2^3\right]_{x_2=x_1}^{x_2=1} \qquad (3.3.15) \\
&= \tfrac{1}{3}(1 - x_1)^{-1}(1 + x_1 - 2x_1^2).
\end{aligned}$$

One can evaluate the expression of $E[X_2^2 \mid X_1 = x_1]$ as follows:

$$\begin{aligned}
E[X_2^2 \mid X_1 = x_1] \\
= 2(1 - x_1)^{-2} \int_{x_1}^1 x_2^2(1 - x_2)dx_2 \qquad (3.3.16) \\
= \tfrac{1}{6}(1 - x_1)^{-1}(1 + x_1 + x_1^2 - 3x_1^3),
\end{aligned}$$

so that using (3.3.14) and after considerable simplifications, we find:

$$\sigma_{2|1}^2 = V[X_2 \mid X_1 = x_1] = E[X_2^2 \mid X_1 = x_1] - \mu_{2|1}^2 = \tfrac{1}{18}(1 - x_1)^2. \quad (3.3.17)$$

The details are left out as Exercise 3.3.1. ▲

Example 3.3.3 Consider two random variables X_1 and X_2 whose joint continuous distribution is given by the following pdf:

$$f(x_1, x_2) = \begin{cases} 2(1 - x_1) & \text{if } 0 < x_1, x_2 < 1 \\ 0 & \text{elsewhere} \end{cases} \qquad (3.3.18)$$

Obviously, one has $\mathcal{X}_1 = \mathcal{X}_2 = (0, 1)$. One may check the following expressions for the marginal pdf's easily:

$$f_1(x_1) = 2(1 - x_1) \text{ if } 0 < x_1 < 1;$$
$$f_2(x_2) = 1 \text{ if } 0 < x_2 < 1. \qquad (3.3.19)$$

One should derive the conditional pdf's and the associated conditional means and variances. See the Exercise 3.3.2. ▲

> In a continuous bivariate distribution, the joint, marginal, and conditional pdf's were defined in (3.3.1)-(3.3.2) and (3.3.5)-(3.3.6).

In the case of a two-dimensional random variable, anytime we wish to calculate the probability of an event $A(\subseteq \Re^2)$, we may use an approach which is a generalization of the equation (1.6.2) in the univariate case: One would write

$$P(A) = \underset{A\cap(\mathcal{X}_1\times\mathcal{X}_2)}{\int\int} f(x_1, x_2)dx_1 dx_2. \tag{3.3.20}$$

We emphasize that the convention is to integrate $f(x_1, x_2)$ only on that part of the set A where $f(x_1, x_2)$ is positive.

If we wish to evaluate the conditional probability of an event $B(\subseteq \Re)$ given, say, $X_1 = x_1$, then we should integrate the conditional pdf $f_{2|1}(x_2)$ of X_2 given $X_1 = x_1$ over that part of the set B where $f_{2|1}(x_2)$ is positive. That is, one has

$$P(B \mid X_1 = x_1) = \underset{B\cap\mathcal{X}_2}{\int} f_{2|1}(x_2)dx_2. \tag{3.3.21}$$

Example 3.3.4 (Example 3.3.2 Continued) In order to appreciate the essence of what (3.3.20) says, let us go back for a moment to the Example 3.3.2. Suppose that we wish to find the probability of the set or the event A where $A = \{X_1 \le .2 \cap .3 < X_2 \le .8\}$. Then, in view of (3.3.20), we obtain $P(A) = \int_{x_1=0}^{.2}\int_{x_2=.3}^{.8} 6(1 - x_2)dx_2 dx_1 = 6\int_{x_1=0}^{.2}\left\{\left[x_2 - \frac{1}{2}x_2^2\right]_{.3}^{.8}\right\} dx_1 = 1.2(.845 - .62) = .27.$ ▲

Example 3.3.5 Consider two random variables X_1 and X_2 whose joint continuous distribution is given by the following pdf:

$$f(x_1, x_2) = \begin{cases} 3x_1 & \text{if } 0 < x_2 < x_1 < 1 \\ 0 & \text{elsewhere} \end{cases} \tag{3.3.22}$$

Obviously, one has $\mathcal{X}_1 = \mathcal{X}_2 = (0, 1)$. One may easily check the following expressions for the marginal pdf's:

$$\begin{aligned} f_1(x_1) &= 3x_1^2 \text{ if } 0 < x_1 < 1; \\ f_2(x_2) &= \tfrac{3}{2}(1 - x_2^2) \text{ if } 0 < x_2 < 1. \end{aligned} \tag{3.3.23}$$

Now, suppose that we wish to compute the conditional probability that $X_2 < .2$ given that $X_1 = .5$. We may proceed as follows. We have

$$f_{2|1}(x_2) = \frac{f(.5, x_2)}{f_1(.5)} = \frac{1.50}{0.75} = 2 \text{ for } 0 < x_2 < .5, \tag{3.3.24}$$

and hence using (3.3.24), one can write

$$P\{X_2 < .2 \mid X_1 = .5\} = \int_0^{.2} f_{2|1}(x_2)dx_2 = \int_0^{.2} 2dx_2 = .4. \qquad (3.3.25)$$

One may, for example, evaluate $\mu_{2|1}$ too given that $X_1 = .5$. We leave it out as the Exercise 3.3.3. ▲

We defined the conditional mean and variance of a random variable X_1 given the other random variable X_2 in (3.3.7). The following result gives the tool for finding the unconditional mean and variance of X_1 utilizing the expressions of the conditional mean and variance.

Theorem 3.3.1 *Suppose that* $\mathbf{X} = (X_1, X_2)$ *has a bivariate pmf or pdf, namely* $f(x_1, x_2)$ *for* $x_i \in \mathcal{X}_i$, *the support of* $X_i, i = 1, 2$. *Let* $E_1[.]$ *and* $V_1[.]$ *respectively denote the expectation and variance with respect to the marginal distribution of* X_1. *Then, we have*

(i) $E[X_2] = E_1[E_{2|1}\{X_2 \mid X_1\}];$

(ii) $V(X_2) = V_1[E_{2|1}\{X_2 \mid X_1\}] + E_1[V_{2|1}\{X_2 \mid X_1\}].$

Proof (i) Note that

$$h(x_1) = E_{2|1}\{X_2 \mid X_1 = x_1\} = \int_{\mathcal{X}_2} x_2 \frac{f(x_1, x_2)}{f_1(x_1)} dx_2. \qquad (3.3.26)$$

Next, we can rewrite

$$
\begin{aligned}
E[X_2] & \\
&= \int_{\mathcal{X}_1} \left\{ \int_{\mathcal{X}_2} x_2 [f(x_1, x_2)/f_1(x_1)]dx_2 \right\} f_1(x_1)dx_1 \\
&= \int_{\mathcal{X}_1} h(x_1)f_1(x_1)dx_1, \text{ from (3.3.26)} \\
&= E[h(X_1)],
\end{aligned}
\qquad (3.3.27)
$$

which is the desired result.

(ii) In view of part (i), we can obviously write

$$V[X_2] = E_1[g(X_1)] \qquad (3.3.28)$$

where $g(x_1) = E\{(X_2 - \mu_2)^2 \mid X_1 = x_1\}$. As before, denoting $E\{X_2 \mid X_1 = x_1\}$ by $\mu_{2|1}$, let us rewrite the function $g(x_1)$ as follows:

$$
\begin{aligned}
g(x_1) & \\
&= E[\{(X_2 - \mu_{2|1}) - (\mu_{2|1} - \mu_1)\}^2 \mid X_1 = x_1] \\
&= E[(X_2 - \mu_{2|1})^2 \mid X_1 = x_1] + E[(\mu_{2|1} - \mu_2)^2 \mid X_1 = x_1] \\
&\quad - 2E[(X_2 - \mu_{2|1})(\mu_{2|1} - \mu_2) \mid X_1 = x_1] \\
&= V[X_2 \mid X_1 = x_1] + (\mu_{2|1} - \mu_2)^2 \\
&\quad - 2E[(X_2 - \mu_{2|1})(\mu_{2|1} - \mu_2) \mid X_1 = x_1].
\end{aligned}
\qquad (3.3.29)
$$

But, observe that the third term in the last step (3.3.29) can be simplified as follows:

$$E[(X_2 - \mu_{2|1})(\mu_{2|1} - \mu_2) \mid X_1 = x_1]$$

$$= (\mu_{2|1} - \mu_2)E[(X_2 - \mu_{2|1}) \mid X_1 = x_1] \qquad (3.3.30)$$

$$= (\mu_{2|1} - \mu_2)(\mu_{2|1} - \mu_{2|1})$$

which is zero. Now, we combine (3.3.29)-(3.3.30) and obtain

$$E\{(X_2 - \mu_2)^2 \mid X_1 = x_1\} = V[X_2 \mid X_1 = x_1] + (\mu_{2|1} - \mu_2)^2. \qquad (3.3.31)$$

At this point, (3.3.31) combined with (3.3.29) then leads to the desired result stated in part (ii). ∎

> The next three Examples 3.3.6-3.3.8, while applying the Theorem 3.3.1, introduce what are otherwise known in statistics as the *compound distributions*.

Example 3.3.6 Suppose that conditionally given $X_1 = x_1$, the random variable X_2 is distributed as $N(\beta_0 + \beta_1 x_1, x_1^2)$ for any fixed $x_1 \in \Re$. Here, β_0, β_1 are two fixed real numbers. Suppose also that marginally, X_1 is distributed as $N(3, 10)$. How should we proceed to find $E[X_2]$ and $V[X_2]$? The Theorem 3.3.1 (i) will immediately imply that

$$E[X_2] = E_1[E_{2|1}\{X_2 \mid X_1\}] = E[\beta_0 + \beta_1 X_1] = \beta_0 + \beta_1 E[X_1], \qquad (3.3.32)$$

which reduces to $\beta_0 + 3\beta_1$. Similarly, the Theorem 3.3.1 (ii) will imply that

$$V[X_2] = V[\beta_0 + \beta_1 X_1] + E[X_1^2] = \beta_1^2 V[X_1] + \{V[X_1] + E^2[X_1]\}, \qquad (3.3.33)$$

which reduces to $10\beta_1^2 + 19$. Refer to the Section 2.3.3 as needed. In a situation like this, the marginal distribution of X_2 is referred to as a *compound distribution*. Note that we have been able to derive the expressions of the mean and variance of X_2 without first identifying the marginal distribution of X_2. ▲

Example 3.3.7 Suppose that conditionally given $X_1 = x_1$, the random variable X_2 is distributed as Poisson(x_1) for any fixed $x_1 \in \Re^+$. Suppose also that marginally, X_1 is distributed as Gamma$(\alpha = 3, \beta = 10)$. How should we proceed to find $E[X_2]$ and $V[X_2]$? The Theorem 3.3.1 (i) will immediately imply that

$$E[X_2] = E_1[E_{2|1}\{X_2 \mid X_1\}] = E[X_1] = \alpha\beta = 30. \qquad (3.3.34)$$

Similarly, the Theorem 3.3.1 (ii) will imply that

$$
\begin{aligned}
V[X_2] \\
&= V[X_1] + E[X_1], \text{ because in a Poisson distribution,} \\
&\quad \text{the mean and variance happen to be equal} \\
&= \alpha\beta^2 + \alpha\beta,
\end{aligned}
\tag{3.3.35}
$$

which reduces to 330. Refer to the Sections 2.3.2 and 2.3.4 as needed. In a situation like this, again the marginal distribution of X_2 is referred to as a *compound distribution*. Note that we have been able to derive the expressions of the mean and variance of X_2 without first identifying the marginal distribution of X_2. ▲

Example 3.3.8 Suppose that conditionally given $X_1 = x_1$, the random variable X_2 is distributed as Binomial(n, x_1) for any fixed $x_1 \in (0, 1)$. Suppose also that marginally, X_1 is distributed as Beta$(\alpha = 4, \beta = 6)$. How should we proceed to find $E[X_2]$ and $V[X_2]$? The Theorem 3.3.1 (i) will immediately imply that

$$
E[X_2] = E_1[E_{2|1}\{X_2 \mid X_1\}] = nE[X_1] = \frac{n\alpha}{\alpha+\beta} = \frac{4}{10}n.
\tag{3.3.36}
$$

Similarly, the Theorem 3.3.1 (ii) will imply that

$$
V[X_2] = V[nX_1] + E[nX_1(1 - X_1)] = n^2 V[X_1] + nE[X_1(1 - X_1)].
\tag{3.3.37}
$$

Refer to the Sections 2.3.1 and equation (1.7.35) as needed. One can verify that $V[X_1] = \frac{\alpha\beta}{(\alpha+\beta)^2(\alpha+\beta+1)} = \frac{24}{1100}$. Also, we have $E[X_1(1 - X_1)] = \frac{1}{b(4,6)} \int_0^1 x(1-x)x^3(1-x)^5 dx = \frac{1}{b(4,6)} \int_0^1 x^4(1-x)^6 dx = \frac{b(5,7)}{b(4,6)} = \frac{24}{110}$. Hence, we have

$$
V[X_2] = \frac{24}{1100}n^2 + \frac{24}{110}n = \frac{12}{55}n(1 + \frac{1}{10}n).
\tag{3.3.38}
$$

In a situation like this, again the marginal distribution of X_2 is referred to as a *compound distribution*. Note that we have been able to derive the expressions of the mean and variance of X_2 without first identifying the marginal distribution of X_2. ▲

Example 3.3.9 (Example 3.3.2 Continued) Let us now apply the Theorem 3.3.1 to reevaluate the expressions for $E[X_2]$ and $V[X_2]$. Combining (3.3.15) with the Theorem 3.3.1 (i) and using direct integration with respect to the marginal pdf $f_1(x_1)$ of X_1 from (3.3.10), we get

$$
\begin{aligned}
E[X_2] \\
&= E\left\{\tfrac{1}{3}(1 - X_1)^{-1}(1 + X_1 - 2X_1^2)\right\} \\
&= \tfrac{1}{3}E\left\{(1 - X_1)^{-1}[3(1 - X_1) - 2(1 - X_1)^2]\right\} \\
&= 1 - (\tfrac{2}{3})(\tfrac{3}{4}), \text{ since } E[(1 - X_1)] = \tfrac{3}{4} \\
&= \tfrac{1}{2},
\end{aligned}
\tag{3.3.39}
$$

which matches with the answer found earlier in (3.3.12). Next, combining (3.3.15) and (3.3.17) with the Theorem 3.3.1 (ii), we should then have

$$
\begin{aligned}
V[X_2] \\
&= E\left\{\tfrac{1}{18}(1 - X_1)^2\right\} + V\left\{\tfrac{1}{3}(1 - X_1)^{-1}(1 + X_1 - 2X_1^2)\right\} \\
&= \tfrac{1}{18}E\left\{(1 - X_1)^2\right\} + \tfrac{1}{9}V\left\{(1 - X_1)^{-1}[3(1 - X_1) - 2(1 - X_1)^2]\right\} \\
&= \tfrac{1}{18}E\left\{(1 - X_1)^2\right\} + \tfrac{4}{9}V\left\{(1 - X_1)\right\}.
\end{aligned}
\tag{3.3.40}
$$

Now, we combine (3.3.39) and (3.3.40). We also use the marginal pdf $f_1(x_1)$ of X_1 from (3.3.10) and direct integration to write

$$
V[X_2] = (\tfrac{1}{18})(\tfrac{3}{5}) + \tfrac{4}{9}[\tfrac{3}{5} - (\tfrac{3}{4})^2] = \tfrac{1}{20},
\tag{3.3.41}
$$

since $E[(1 - X_1)] = \tfrac{3}{4}$ and $E[(1 - X_1)^2] = \tfrac{3}{5}$. The answer given earlier in (3.3.13) and the one found in (3.3.41) match exactly. ▲

3.3.2 Three and Higher Dimensions

The ideas expressed in (3.3.2) and (3.3.5)-(3.3.6) extend easily in a *multivariate* situation too. For example, let us suppose that a random vector $\mathbf{X} = (X_1, X_2, X_3, X_4)$ has the joint pdf $f(x_1, x_2, x_3, x_4)$ where $x_i \in \mathcal{X}_i$, the support of $X_i, i = 1, 2, 3, 4$. The marginal pdf of (X_1, X_3), that is the joint pdf of X_1 and X_3, for example, can be found as follows:

$$
f_{1,3}(x_1, x_3) = \int_{\mathcal{X}_4} \int_{\mathcal{X}_2} f(x_1, x_2, x_3, x_4) dx_2 dx_4.
\tag{3.3.42}
$$

The conditional pdf of (X_2, X_4) given that $X_1 = x_1, X_3 = x_3$ will be of the form

$$
f_{2,4|1,3}(x_2, x_4) = \frac{f(x_1, x_2, x_3, x_4)}{f_{1,3}(x_1, x_3)} \text{ when } f_{1,3}(x_1, x_3) > 0,
\tag{3.3.43}
$$

and the x's belong to the $\mathcal{X}$'s. The notions of expectations and conditional expectations can also be generalized in a natural fashion in the case of a multidimensional random vector $\mathbf{X}$ by extending the essential ideas from (3.3.7)-(3.3.8).

If one has a k-dimensional random variable $\mathbf{X} = (X_1, ..., X_k)$, the joint pdf would then be written as $f(\mathbf{x})$ or $f(x_1, ..., x_k)$. The joint pdf of any subset of random variables, for example, X_1 and X_3, would then be found by integrating $f(x_1, ..., x_k)$ with respect to the remaining variables $x_2, x_4, ..., x_k$. One can also write down the expressions of the associated conditional pdf's of any subset of random variables from $\mathbf{X}$ given the values of any other subset of random variables from $\mathbf{X}$.

Theorem 3.3.2 *Let* $\mathbf{X} = (X_1, ..., X_k)$ *be any* k-*dimensional discrete or continuous random variable. Suppose that we also have real valued functions* $h_i(\mathbf{x})$ *and constants* $a_i, i = 0, 1, ..., p$. *Then, we have*

$$E\{a_0 + \Sigma_{i=1}^{p} a_i h_i(\mathbf{X})\} = a_0 + \Sigma_{i=1}^{p} a_i E\{h_i(\mathbf{X})\}$$

as long as all the expectations involved are finite. That is, the expectation is a linear operation.

Proof Let us write $\mathcal{X} = \Pi_{i=1}^{k} \mathcal{X}_i$ and hence we have

$$E\{a_0 + \Sigma_{i=1}^{p} a_i h_i(\mathbf{X})\}$$

$$= \int \cdots \int_{\mathcal{X}} \{a_0 + \Sigma_{i=1}^{p} a_i h_i(\mathbf{X})\} f(\mathbf{x}) \Pi_{i=1}^{k} dx_i$$

$$= a_0 \int \cdots \int_{\mathcal{X}} f(\mathbf{x}) \Pi_{i=1}^{k} dx_i + \Sigma_{i=1}^{p} a_i \int \cdots \int_{\mathcal{X}} h_i(\mathbf{x}) f(\mathbf{x}) \Pi_{i=1}^{k} dx_i$$

$$= a_0 + \Sigma_{i=1}^{p} a_i E\{h_i(\mathbf{X})\}, \text{ since } \int \cdots \int_{\mathcal{X}} f(\mathbf{x}) \Pi_{i=1}^{k} dx_i = 1$$

$$\text{and } \int \cdots \int_{\mathcal{X}} h_i(\mathbf{x}) f(\mathbf{x}) \Pi_{i=1}^{k} dx_i = E[h_i(\mathbf{X})].$$

$$(3.3.44)$$

Now, the proof is complete. ∎

Next, we provide two specific examples.

Example 3.3.10 Let us denote $\mathcal{X}_1 = \mathcal{X}_3 = (0, 1)$, $\mathcal{X}_2 = (0, 2)$ and define

$$a_1(x_1) = \begin{cases} 2x_1 & \text{if } 0 < x_1 < 1 \\ 0 & \text{otherwise} \end{cases} \qquad a_2(x_2) = \begin{cases} \frac{3}{8}x_2^2 & \text{if } 0 < x_2 < 2 \\ 0 & \text{otherwise} \end{cases}$$

$$(3.3.45)$$

$$a_3(x_3) = \begin{cases} x_3 + 2x_3^3 & \text{if } 0 < x_3 < 1 \\ 0 & \text{otherwise} \end{cases} \qquad (3.3.46)$$

Note that these are non-negative functions and $\int_{\mathcal{X}_i} a_i(x_i) dx_i = 1$ for all

$i = 1, 2, 3$. With $\mathbf{x} = (x_1, x_2, x_3)$ and $\mathcal{X} = \Pi_{i=1}^3 \mathcal{X}_i$, let us denote

$$f(\mathbf{x}) = \begin{cases} \frac{1}{3}a_1(x_1)a_2(x_2) + \frac{1}{3}a_2(x_2)a_3(x_3) \\ \qquad + \frac{1}{6}a_3(x_3)a_1(x_1) & \text{if } \mathbf{x} \in \mathcal{X} \qquad (3.3.47) \\ 0 & \text{otherwise} \end{cases}$$

Now, one can easily see that

$$\int_{x_3=0}^1 \int_{x_2=0}^2 \int_{x_1=0}^1 a_1(x_1)a_2(x_2)\Pi_{i=1}^3 dx_i$$
$$= \int_{x_3=0}^1 dx_3 \int_{x_2=0}^2 a_2(x_2)dx_2 \int_{x_1=0}^1 a_1(x_1)dx_1 = 1,$$

$$\int_{x_3=0}^1 \int_{x_2=0}^2 \int_{x_1=0}^1 a_2(x_2)a_3(x_3)\Pi_{i=1}^3 dx_i$$
$$= \int_{x_3=0}^1 a_3(x_3)dx_3 \int_{x_2=0}^2 a_2(x_2)dx_2 \int_{x_1=0}^1 dx_1 = 1,$$

and

$$\int_{x_3=0}^1 \int_{x_2=0}^2 \int_{x_1=0}^1 a_3(x_3)a_1(x_1)\Pi_{i=1}^3 dx_i$$
$$= \int_{x_3=0}^1 a_3(x_3)dx_3 \int_{x_2=0}^2 dx_2 \int_{x_1=0}^1 a_1(x_1)dx_1 = 2.$$

Hence, for the function $f(\mathbf{x})$ defined in (3.3.47) we have

$$\int_{x_3=0}^1 \int_{x_2=0}^2 \int_{x_1=0}^1 f(\mathbf{x})\Pi_{i=1}^3 dx_i = 1, \qquad (3.3.48)$$

and also observe that $f(\mathbf{x})$ is always non-negative. In other words, $f(\mathbf{x})$ is a pdf of a three dimensional random variable, say, $\mathbf{X} = (X_1, X_2, X_3)$. The joint pdf of X_1, X_3 is then given by

$$\begin{aligned} g(x_1, x_3) \\ &= \int_{x_2=0}^2 f(\mathbf{x})dx_2 \\ &= \frac{1}{3}\int_{x_2=0}^2 a_1(x_1)a_2(x_2)dx_2 + \frac{1}{3}\int_{x_2=0}^2 a_2(x_2)a_3(x_3)dx_2 \qquad (3.3.49) \\ &\quad + \frac{1}{6}\int_{x_2=0}^2 a_3(x_3)a_1(x_1)dx_2 \\ &= \frac{1}{3}\{a_1(x_1) + a_3(x_3) + a_3(x_3)a_1(x_1)\}, x_i \in \mathcal{X}_i, i = 1, 3. \end{aligned}$$

One may directly check by double integration that $g(x_1, x_3)$ is a genuine pdf with its support being $\mathcal{X}_1 \times \mathcal{X}_3$. For any $0 < x_2 < 2$, with $g(x_1, x_3)$ from (3.3.49), the conditional pdf of X_2 given that $X_1 = x_1, X_3 = x_3$ is then given by

$$f_{2|1,3}(x_2) = f(\mathbf{x})/g(x_1, x_3) \text{ with fixed } 0 < x_1, x_3 < 1. \qquad (3.3.50)$$

Utilizing (3.3.49), it also follows that for any $0 < x_3 < 1$, the marginal pdf of X_3 is given by

$$
\begin{aligned}
g_3(x_3) &= \int_{x_1=0}^{1} g(x_1, x_3) dx_1 \\
&= \tfrac{1}{3} \int_{x_1=0}^{1} \{a_1(x_1) + a_3(x_3) + a_3(x_3)a_1(x_1)\} dx_1 \\
&= \tfrac{1}{3} \{1 + 2a_3(x_3)\} \text{ for any } 0 < x_3 < 1.
\end{aligned}
\tag{3.3.51}
$$

Utilizing the expressions of $g(x_1, x_3), g_3(x_3)$ from (3.3.49) and (3.3.51), for any $0 < x_1 < 1$, we can write down the conditional pdf of X_1 given that $X_3 = x_3$ as follows:

$$
g_{1|3}(x_1) = g(x_1, x_3)/g_3(x_3) \text{ for fixed } 0 < x_3 < 1. \tag{3.3.52}
$$

After obtaining the expression of $g(x_1, x_3)$ from (3.3.49), one can easily evaluate, for example $E(X_1 X_3)$ or $E(X_1 X_3^2)$ respectively as the double integrals $\int_{x_3=0}^{1} \int_{x_1=0}^{1} x_1 x_3 g(x_1, x_3) dx_1 dx_3$ or $\int_{x_3=0}^{1} \int_{x_1=0}^{1} x_1 x_3^2 g(x_1, x_3) dx_1 dx_3$. Look at the Exercises 3.3.6-3.3.7. ▲

> In the two Examples 3.3.10-3.3.11, make a special note of how the joint densities $f(\mathbf{x})$ of more than two continuous random variables have been constructed. The readers should create a few more of their own. Is there any special role of the specific forms of the functions $a_i(x_i)$'s? Look at the Exercise 3.3.8.

Example 3.3.11 Let us denote $\mathcal{X}_1 = \mathcal{X}_3 = \mathcal{X}_4 = \mathcal{X}_5 = (0,1)$, $\mathcal{X}_2 = (0,2)$ and recall the functions $a_1(x_1), a_2(x_2)$ and $a_3(x_3)$ from (3.3.45)-(3.3.46). Additionally, let us denote

$$
a_4(x_4) = \begin{cases} 1 & \text{if } 0 < x_4 < 1 \\ 0 & \text{otherwise} \end{cases} \qquad
a_5(x_5) = \begin{cases} 4x_5^3 & \text{if } 0 < x_5 < 1 \\ 0 & \text{otherwise} \end{cases}
$$

Note that these are non-negative functions and also one has $\int_{\mathcal{X}_i} a_i(x_i) dx_i = 1$ for all $i = 1, ..., 5$. With $\mathbf{x} = (x_1, x_2, x_3, x_4, x_5)$ and $\mathcal{X} = \Pi_{i=1}^{5} \mathcal{X}_i$, let us denote

$$
f(\mathbf{x}) = \begin{cases} \tfrac{1}{5}a_1(x_1)a_2(x_2) + \tfrac{1}{5}a_2(x_2)a_3(x_3) \\ \qquad + \tfrac{1}{10}\{a_3(x_3)a_1(x_1) + a_4(x_4) + a_4(x_4)a_5(x_5)\} \text{ if } \mathbf{x} \in \mathcal{X} \\ 0, \qquad \text{otherwise.} \end{cases}
\tag{3.3.53}
$$

It is a fairly simple matter to check that $\int \int \int \int \int_{\mathcal{X}} a_1(x_1)a_2(x_2)\Pi_{i=1}^{5} dx_i = 1$, $\int \int \int \int \int_{\mathcal{X}} a_2(x_2)a_3(x_3)\Pi_{i=1}^{5} dx_i = 1$, $\int \int \int \int \int_{\mathcal{X}} a_3(x_3)a_1(x_1)\Pi_{i=1}^{5} dx_i =$

$2, \int \int \int \int \int\limits_{\mathcal{X}} a_4(x_4)\Pi_{i=1}^5 dx_i = 2$, and $\int \int \int \int \int\limits_{\mathcal{X}} a_4(x_4)a_5(x_5)\Pi_{i=1}^5 dx_i = 2$.

Hence, one verifies that $\int \int \int \int \int\limits_{\mathcal{X}} f(\mathbf{x})\Pi_{i=1}^5 dx_i = 1$ so that (3.3.53) does

provide a genuine pdf since $f(\mathbf{x})$ is non-negative for all $\mathbf{x} \in \mathcal{X}$. Along the lines of the example 3.3.10, one should be able to write down (i) the marginal joint pdf of (X_2, X_3, X_5), (ii) the conditional pdf of X_3, X_5 given the values of X_1, X_4, (iii) the conditional pdf of X_3 given the values of X_1, X_4, (iv) the conditional pdf of X_3 given the values of X_1, X_2, X_4, among other similar expressions. Look at the Exercises 3.3.9-3.3.10. Look at the Exercise 3.3.11 too for other possible choices of a_i's. ▲

3.4   Covariances and Correlation Coefficients

Since the random variables $X_1, ..., X_k$ vary together, we may look for a measure of the strength of such interdependence among these variables. The *covariance* aims at capturing this sense of *joint dependence* between two real valued random variables.

Definition 3.4.1 *The covariance between any two discrete or continuous random variables X_1 and X_2, denoted by $Cov(X_1, X_2)$, is defined as*

$$Cov(X_1, X_2) = E\left[(X_1 - \mu_1)(X_2 - \mu_2)\right], \tag{3.4.1}$$

where $\mu_i = E(X_i), i = 1, 2$.

On the rhs of (3.4.1) we note that it is the expectation of a specific function $g(X_1, X_2) = (X_1 - \mu_1)(X_2 - \mu_2)$ of the two random variables X_1, X_2. By combining the ideas from (3.2.6) and (3.3.8) we can easily rewrite (3.4.1). Then, using the Theorem 3.3.2 we obtain

$$
\begin{aligned}
Cov&(X_1, X_2) \\
&= E(X_1 X_2) - \mu_2 E(X_1) - \mu_1 E(X_2) + \mu_1 \mu_2 \\
&= E(X_1 X_2) - \mu_2 \mu_1 - \mu_1 \mu_2 + \mu_1 \mu_2 \\
&= E(X_1 X_2) - \mu_1 \mu_2.
\end{aligned} \tag{3.4.2}
$$

We now rewrite (3.4.2) in its customary form, namely

$$Cov(X_1, X_2) = E(X_1 X_2) - E(X_1)E(X_2), \tag{3.4.3}$$

whether X_1, X_2 are discrete or continuous random variables, provided that the expectations $E(X_1 X_2), E(X_1)$ and $E(X_2)$ are finite. From (3.4.3) it also becomes clear that

$$Cov(X_1, X_2) = Cov(X_2, X_1), \tag{3.4.4}$$

where X_1, X_2 are any two arbitrary discrete or continuous random variables, that is the *covariance measure is symmetric* in the two variables. From (3.4.3) it obviously follows that

$$Cov(X_1, X_1) = V(X_1), \qquad (3.4.5)$$

where X_1 is any arbitrary random variable having a finite second moment.

Theorem 3.4.1 *Suppose that $X_i, Y_i, i = 1, 2$ are any discrete or continuous random variables. Then, we have*

(i) $Cov(X_1, c) = 0$ *where $c \in \Re$ is fixed, if $E(X_1)$ is finite;*

(ii) $Cov(X_1 + X_2, Y_1 + Y_2) = Cov(X_1, Y_1) + Cov(X_1, Y_2)$
$+ Cov(X_2, Y_1) + Cov(X_2, Y_2)$ *provided that $Cov(X_i, Y_j)$
is finite for $i, j = 1, 2$.*

In other words, the covariance is a bilinear operation which means that it is a linear operation in both coordinates.

Proof (i) Use (3.4.3) where we substitute $X_2 = c$ w.p.1. Then, $E(X_2) = c$ and hence we have

$$Cov(X_1, c) = E(cX_1) - E(X_1)E(c) = cE(X_1) - E(X_1)c = 0.$$

(ii) First let us evaluate $Cov(X_1 + X_2, Y_1)$. One gets

$$\begin{aligned}
Cov(X_1 + X_2, Y_1) \\
&= E\{(X_1 + X_2)Y_1\} - E\{(X_1 + X_2)\}E(Y_1), \text{ by } (3.4.3) \\
&= E(X_1 Y_1) + E(X_2 Y_1) - \{E(X_1) + E(X_2)\}E(Y_1),
\end{aligned}$$

by the Theorem 3.3.2

$$\begin{aligned}
&= E(X_1 Y_1) + E(X_2 Y_1) - E(X_1)E(Y_1) - E(X_2)E(Y_1) \\
&= \{E(X_1 Y_1) - E(X_1)E(Y_1)\} + \{E(X_2 Y_1) - E(X_2)E(Y_1)\} \\
&= Cov(X_1, Y_1) + Cov(X_2, Y_1), \text{ by } (3.4.3).
\end{aligned}$$

$$(3.4.6)$$

Next, we exploit (3.4.6) repeatedly to obtain

$$\begin{aligned}
Cov(X_1 + X_2, Y_1 + Y_2) \\
&= Cov(X_1, Y_1 + Y_2) + Cov(X_2, Y_1 + Y_2) \\
&= Cov(Y_1 + Y_2, X_1) + Cov(Y_1 + Y_2, X_2)
\end{aligned}$$

by using (3.4.4)

$$\begin{aligned}
&= Cov(Y_1, X_1) + Cov(Y_2, X_1) + Cov(Y_1, X_2) \\
&\quad + Cov(Y_2, X_2).
\end{aligned}$$

$$(3.4.7)$$

This leads to the final conclusion by appealing again to (3.4.4). ∎

Example 3.4.1 (Examples 3.2.4 and 3.2.5 Continued) We already know that $E(X_1) = 2.27$ and also $E(X_1 X_2) = 3.05$. Similarly, we can find $E(X_2) = (1)(.55) + 2(.45) = 1.45$ and hence, $Cov(X_1, X_2) = 3.05 - (2.27)(1.45) = -0.2415$. ▲

Example 3.4.2 (Example 3.3.3 Continued) It is easy to see that $E(X_1) = \int_0^1 x_1 f_1(x_1) dx_1 = 2 \int_0^1 (x_1 - x_1^2) dx_1 = 2 \left[\frac{1}{2} x_1^2 - \frac{1}{3} x_1^3 \right]_{x_1=0}^{x_1=1} = \frac{1}{3}$ and similarly $E(X_2) = \frac{1}{2}$. Also, one has $E(X_1 X_2) = \int_{x_2=0}^1 \int_{x_1=0}^1 x_1 x_2 f(x_1, x_2) dx_1 dx_2 = 2 \int_{x_2=0}^1 \int_{x_1=0}^1 (x_1 - x_1^2) x_2 dx_1 dx_2 = 2 \left\{ \int_{x_1=0}^1 (x_1 - x_1^2) dx_1 \right\} \left\{ \int_{x_2=0}^1 x_2 dx_2 \right\} = (\frac{1}{3})(\frac{1}{2}) = \frac{1}{6}$. Next, we appeal to (3.4.3) and observe that $Cov(X_1, X_2) = E(X_1 X_2) - E(X_1)E(X_2) = \frac{1}{6} - (\frac{1}{3})(\frac{1}{2}) = 0$. ▲

Example 3.4.3 (Example 3.3.5 Continued) It is easy to see that $E(X_1) = \int_0^1 x_1 f_1(x_1) dx_1 = 3 \int_0^1 x_1^3 dx_1 = \frac{3}{4}$ and $E(X_2) = \int_0^1 x_2 f_2(x_2) dx_2 = \frac{3}{2} \int_0^1 (x_2 - x_2^3) dx_2 = \frac{3}{2} \left[\frac{1}{2} x_2^2 - \frac{1}{4} x_2^4 \right]_{x_2=0}^{x_2=1} = \frac{3}{8}$. Again, one similarly has $E(X_1 X_2) = \int_{x_2=0}^1 \int_{x_1=0}^1 x_1 x_2 f(x_1, x_2) dx_1 dx_2 = 3 \int_{x_2=0}^1 \int_{x_1=x_2}^1 x_1^2 x_2 dx_1 dx_2$. This reduces to $3 \int_{x_2=0}^1 x_2 \left\{ \int_{x_1=x_2}^1 x_1^2 dx_1 \right\} dx_2 = \int_{x_2=0}^1 x_2 (1 - x_2^3) dx_2 = \frac{3}{10}$. Next, we appeal to (3.4.3) and observe that $Cov(X_1, X_2) = E(X_1 X_2) - E(X_1)E(X_2) = \frac{3}{10} - (\frac{3}{4})(\frac{3}{8}) = \frac{3}{160}$. ▲

The term $Cov(X_1, X_2)$ can be any real number, positive, negative or zero. However, if we simply look at the value of $Cov(X_1, X_2)$, it will be hard for us to say very much about the strength of the dependence between the two random variables X_1 and X_2 under consideration. It should be apparent that the term $Cov(X_1, X_2)$ has the unit of measurement which is same as that for the variable $X_1 X_2$, the product of X_1 and X_2. If X_1, X_2 are both measured in inches, then $Cov(X_1, X_2)$ would have to be recorded in square inches.

A standardized version of the covariance term, commonly known as the *correlation coefficient,* was made popular by Karl Pearson. Refer to Stigler (1989) for the history of the invention of correlation. We discuss related historical matters later.

Definition 3.4.2 *The correlation coefficient between any two random variables X_1 and X_2, denoted by ρ_{X_1, X_2}, is defined as follows:*

$$\rho_{X_1, X_2} = \frac{Cov(X_1, X_2)}{\sigma_1 \sigma_2}, \qquad (3.4.8)$$

whenever one has $-\infty < Cov(X_1, X_2) < \infty, 0 < \sigma_1^2 = V(X_1) < \infty$ and $0 < \sigma_2^2 = V(X_2) < \infty$.

> When we consider two random variables X_1 and X_2 only, we may supress the subscripts from ρ_{X_1,X_2} and simply write ρ instead.

Definition 3.4.3 *Two random variables X_1, X_2 are respectively called negatively correlated, uncorrelated, or positively correlated if and only if ρ_{X_1,X_2} is negative, zero or positive.*

Before we explain the role of a correlation coefficient any further, let us state and prove the following result.

Theorem 3.4.2 *Consider any two discrete or continuous random variables X_1 and X_2 for which we can assume that $-\infty < Cov(X_1, X_2) < \infty, 0 < V(X_1) < \infty$ and $0 < V(X_2) < \infty$. Let ρ_{X_1,X_2}, defined by (3.4.8), stand for the correlation coefficient between X_1 and X_2. We have the following results:*

(i)   *Let $Y_i = c_i + d_i X_i$ where $-\infty < c_i < \infty$ and $0 < d_i < \infty$ are fixed numbers, $i = 1, 2$. Then, $\rho_{Y_1,Y_2} = \rho_{X_1,X_2}$;*

(ii) $|\rho_{X_1,X_2}| \le 1$;

(iii)   *In part (ii), the equality holds, that is ρ_{X_1,X_2} is $+1$ or -1, if and only if X_1 and X_2 are linearly related. In other words, ρ_{X_1,X_2} is $+1$ or -1 if and only if $X_1 = a + bX_2$ w.p.1 for some real numbers a and b.*

Proof (i) We apply the Theorem 3.3.2 and Theorem 3.4.1 to claim that

$$Cov(Y_1, Y_2) = d_1 d_2 Cov(X_1, X_2). \qquad (3.4.9)$$

Also, we have

$$V(Y_1) = d_1^2 V(X_1) \text{ and } V(Y_2) = d_2^2 V(X_2). \qquad (3.4.10)$$

Next we combine (3.4.8)-(3.4.10) to obtain

$$
\begin{aligned}
\rho_{Y_1,Y_2} \\
&= Cov(Y_1, Y_2)/\sqrt{V(Y_1)V(Y_2)} \\
&= d_1 d_2 Cov(X_1, X_2)/\sqrt{d_1^2 V(X_1) d_2^2 V(X_2)} \\
&= Cov(X_1, X_2)/\sqrt{V(X_1)V(X_2)} \\
&= \rho_{X_1,X_2}.
\end{aligned}
\qquad (3.4.11)
$$

(ii) We apply the Cauchy-Schwarz inequality (Theorem 3.9.5) or directly the covariance inequality (Theorem 3.9.6) from the Section 3.9 and immediately note that

$$Cov^2(X_1, X_2)$$
$$= E^2\left[(X_1 - \mu_1)(X_2 - \mu_2)\right] \text{ where } \mu_i = E(X_i), i = 1, 2$$
$$\leq E\left[(X_1 - \mu_1)^2\right] E\left[(X_2 - \mu_2)^2\right] \qquad (3.4.12)$$
$$= V(X_1)V(X_2).$$

From (3.4.12) we conclude that

$$\rho^2_{X_1, X_2} = Cov^2(X_1, X_2)/\{V(X_1)V(X_2)\} \leq 1. \qquad (3.4.13)$$

Hence, one has $\left|\rho_{X_1, X_2}\right| \leq 1$.

(iii) We conclude that $\rho^2_{X_1, X_2} = 1$ provided that we have equality through-out (3.4.13). The covariance inequality (Theorem 3.9.6) dictates that we can have equality in (3.4.13) *if and only if* the two random variables $(X_1 - \mu_1)$ and $(X_2 - \mu_2)$ are linearly related w.p.1. In other words, one will have $\rho^2_{X_1, X_2} = 1$ *if and only if* $(X_1 - \mu_1) = c(X_2 - \mu_2) + d$ w.p.1 where c and d are two real numbers. The result follows with $a = \mu_1 - c\mu_2 + d$ and $b = c$. One may observe that $\rho_{X_1, X_2} = +1$ or -1 according as b is positive or negative. ∎

From the Definition 3.4.3, recall that in the case when $\rho_{X_1, X_2} = 0$, the two random variables X_1, X_2 are called *uncorrelated*. The case of zero cor-relation is addressed in more detail in the Section 3.7.

If the correlation coefficient ρ is one in magnitude, then the two random variables are linearly related with probability one. The zero correlation is referred to as uncorrelation.

Example 3.4.4 (Example 3.4.1 Continued) One can check that $V(X_1) = 6.4971$ and $V(X_2) = .6825$. We also found earlier that $Cov(X_1, X_2) = -0.2415$. Thus, using (3.4.8) we get $\rho_{X_1, X_2} = Cov(X_1, X_2)/(\sigma_1\sigma_2) = -0.2415/\sqrt{(6.4971)(.6825)} \approx -.11468.$ ▲

Example 3.4.5 (Example 3.4.2 Continued) We already had shown that $Cov(X_1, X_2) = 0$. One can also check that both $V(X_1)$ and $V(X_2)$ are finite, and hence $\rho_{X_1, X_2} = 0.$ ▲

Example 3.4.6 (Example 3.4.3 Continued) We already had shown that $Cov(X_1, X_2) = \frac{3}{160}$ whereas $E(X_1) = \frac{3}{4}, E(X_2) = \frac{3}{8}$. Proceeding analo-gously, one can check that $E(X_1^2) = \int_0^1 3x_1^4 dx_1 = \frac{3}{5}$ and $E(X_2^2) = \frac{3}{2}\int_0^1(x_2^2 - x_2^4)dx_2 = \frac{1}{5}$. Thus, $V(X_1) = \frac{3}{5} - (\frac{3}{4})^2 = \frac{3}{80}$ and $V(X_2) = \frac{1}{5} - (\frac{3}{8})^2 = \frac{19}{320}$. Hence, $\rho_{X_1, X_2} = (\frac{3}{160})/\sqrt{(\frac{3}{80})(\frac{19}{320})} \approx .39736.$ ▲

Next, we summarize some more useful results for algebraic manipulations involving expectations, variances, and covariances.

Theorem 3.4.3 *Let us write* $a_1, ..., a_k$ *and* $b_1, ..., b_k$ *for arbitrary but fixed real numbers. Also recall the arbitrary k-dimensional random variable* $\mathbf{X} = (X_1, ..., X_k)$ *which may be discrete or continuous. Then, we have*

(i) $E\left\{\Sigma_{i=1}^{k} a_i X_i\right\} = \Sigma_{i=1}^{k} a_i E\{X_i\};$

(ii) $Cov(X_i, X_j) = E[X_i X_j] - E[X_i]E[X_j]$ *for all* $i, j;$

(iii) $V\left\{\Sigma_{i=1}^{k} a_i X_i\right\} = \Sigma_{i=1}^{k} a_i^2 V\{X_i\} + 2 \underset{1 \le i < j \le k}{\Sigma\Sigma} a_i a_j Cov(X_i, X_j);$

(iv) $Cov\left(\Sigma_{i=1}^{k} a_i X_i, \Sigma_{j=1}^{k} b_j X_j\right) = \Sigma_{i=1}^{k} \Sigma_{j=1}^{k} a_i b_j Cov(X_i, X_j).$

Proof The parts (i) and (ii) follow immediately from the Theorem 3.3.2 and (3.4.3). The parts (iii) and (iv) follow by successively applying the bilinear property of the covariance operation stated precisely in the Theorem 3.4.1, part (ii). The details are left out as the Exercise 3.4.5. ∎

> One will find an immediate application of Theorem 3.4.3
> in a multinomial distribution.

3.4.1 The Multinomial Case

Recall the multinomial distribution introduced in the Section 3.2.2. Suppose that $\mathbf{X} = (X_1, ..., X_k)$ has the $\text{Mult}_k(n, p_1, ..., p_k)$ distribution with the associated pmf given by (3.2.8). We had mentioned that X_i has the Binomial(n, p_i) distribution for all fixed $i = 1, ..., k$. How should we proceed to derive the expression of the covariance between X_i and X_j for all fixed $i \ne j = 1, ..., k$?

Recall the way we had motivated the multinomial pmf given by (3.2.8). Suppose that we have n marbles and these are tossed so that each marble lands in one of the k boxes. The probability that a marble lands in box #i is, say, $p_i, i = 1, ..., k, \Sigma_{i=1}^{k} p_i = 1$. Then, let X_l be the number of marbles landing in the box #$l, l = i, j$. Define

$$Y_l = \begin{cases} 1 & \text{if } l^{th} \text{ marble lands in box \#}i \\ 0 & \text{otherwise,} \end{cases}$$

$$(3.4.14)$$

$$Z_l = \begin{cases} 1 & \text{if } l^{th} \text{ marble lands in box \#}j \\ 0 & \text{otherwise.} \end{cases}$$

Then, utilizing (3.4.14) we can write

$$X_i = \Sigma_{l=1}^{n} Y_l \text{ and } X_j = \Sigma_{l=1}^{n} Z_l, \qquad (3.4.15)$$

so that we have

$Cov(X_i, X_j)$
$= Cov(\Sigma_{l=1}^n Y_l, \Sigma_{l=1}^n Z_l)$, using (3.4.15)
$= \Sigma_{l=1}^n \Sigma_{m=1}^n Cov(Y_l, Z_m)$, using Theorem 3.4.3, (iv)
$= \Sigma_{l=1}^n Cov(Y_l, Z_l)$, since the marbles land independently
in the boxes, that is $Cov(Y_l, Z_m) = 0$ if $l \neq m$ $\hspace{2em}$ (3.4.16)
$= nCov(Y_1, Z_1)$, since the covariances in the preceding

$\hspace{2em}$ step are all equal
$= n\{E[Y_1 Z_1] - E[Y_1]E[Z_1]\}.$

It is not hard to see that $E[Y_1 Z_1] = 0, E[Y_1] = p_i, E[Z_1] = p_j$, and hence we can rewrite (3.4.16) to claim that

$$Cov(X_i, X_j) = n\{0 - p_i p_j\} = -n p_i p_j, \hspace{2em} (3.4.17)$$

for all fixed $i \neq j = 1, ..., k$.

This negative covariance should intuitively make sense because out of the n marbles, large number of marbles in box #i would necessarily force the number of marbles in the box #j to be small. Next, one can simply obtain

$$\begin{aligned}\rho_{X_i, X_j} &= Cov(X_i, X_j)/\sqrt{V(X_i)V(X_j)} \\ &= -n p_i p_j / \{\sqrt{n p_i(1-p_i)}\sqrt{n p_j(1-p_j)}\} \hspace{2em} (3.4.18) \\ &= -\sqrt{p_i/(1-p_i)}\sqrt{p_j/(1-p_j)},\end{aligned}$$

for all fixed $i \neq j = 1, ..., k$.

3.5 Independence of Random Variables

Suppose that we have a k-dimensional random variable $\mathbf{X} = (X_1, ..., X_k)$ whose joint *pmf or pdf* is written as $f(\mathbf{x})$ or $f(x_1, ..., x_k)$ with $x_i \in \mathcal{X}_i(\subseteq \Re), i = 1, ..., k$. Here $\mathcal{X}_i$ is the support of the random variable $X_i, i = 1, ..., k$, where these random variables can be *discrete or continuous*.

Definition 3.5.1 *Let $f_i(x_i)$ denote the marginal pmf or pdf of $X_i, i = 1, ..., k$. We say that $X_1, ..., X_k$ form a collection of independent random variables if and only if*

$$f(x_1, ..., x_k) = \Pi_{i=1}^k f_i(x_i) \text{ for each } x_i \in \mathcal{X}_i, i = 1, ..., k. \hspace{2em} (3.5.1)$$

Also, $X_1, ..., X_k$ form a collection of dependent random variables if and only if these are not independent.

> In order to show that $X_1, ..., X_k$ are *dependent*, one simply needs
> to show the existence of a particular set of values $x_1, ..., x_k$,
> with $x_i \in \mathcal{X}_i, i = 1, ..., k$ such that $f(x_1, ..., x_k) \neq \Pi_{i=1}^k f_i(x_i)$.

Example 3.5.1 (Example 3.2.4 Continued) We immediately note that $P(X_1 = -1 \cap X_2 = 1) = .12, P(X_1 = -1) = .32$ and $P(X_2 = 1) = .55$. But, $P(X_1 = -1)P(X_2 = 1) = .176$ which is different from $P(X_1 = -1 \cap X_2 = 1)$. We have shown that when $x_1 = -1, x_2 = 1$ are chosen, then, $f(x_1, x_2) \neq f_1(x_1)f_2(x_2)$. Thus, the two random variables X_1, X_2 are *dependent*. ▲

Example 3.5.2 (Example 3.3.1 Continued) From (3.3.9)-(3.3.11) we have $f(x_1, x_2) = 6(1 - x_2)$ for $0 < x_1 < x_2 < 1, \Pi_{i=1}^2 f_i(x_i) = 18x_2(1 - x_1)^2(1 - x_2)$ for $0 < x_1, x_2 < 1$. For $x_1 = .4, x_2 = .5$, we have $f(x_1, x_2) = 3$, but $\Pi_{i=1}^2 f_i(x_i) = 1.62 \neq f(x_1, x_2)$. Thus, the two random variables X_1, X_2 are *dependent*. ▲

Example 3.5.3 Consider two random variables X_1, X_2 having their joint pdf given by

$$f(x_1, x_2) = \begin{cases} 4x_1x_2 & \text{if } 0 < x_1, x_2 < 1 \\ 0 & \text{elsewhere.} \end{cases}$$

First, one can easily show that $f_i(x_i) = 2x_i$ for $0 < x_i < 1, i = 1, 2$. Since one obviously has $f(x_1, x_2) = f_1(x_1) f_2(x_2)$ for all $0 < x_1, x_2 < 1$, it follows that X_1, X_2 are *independent*. ▲

> It is important to understand, for example, that in order for the
> k random variables $X_1, X_2, ..., X_k$ to be independent, their
> joint and the marginal pmf's or the pdf's must satisfy (3.5.1).
> Look at the next example for the case in point.

Example 3.5.4 It is easy to construct examples of random variables X_1, X_2, X_3 such that (i) X_1 and X_2 are independent, (ii) X_2 and X_3 are independent, and (iii) X_1 and X_3 are also independent, but (iv) X_1, X_2, X_3 together are dependent. Suppose that we start with the non-negative integrable functions $f_i(x_i)$ for $0 < x_i < 1$ which are *not identically equal to unity* such that we have $\int_{x_i=0}^1 f_i(x_i)dx_i = 1, i = 1, 2, 3$. Let us then define

$$g(x_1, x_2, x_3) = \tfrac{1}{4}\{f_1(x_1)f_2(x_2) + f_2(x_2)f_3(x_3) + f_1(x_1)f_3(x_3) + 1\} \tag{3.5.2}$$

for $0 < x_i < 1, i = 1, 2, 3$. It is immediate to note that $g(x_1, x_2, x_3)$ is a bona fide pdf. Now, focus on the three dimensional random variable $\mathbf{X} = (X_1, X_2, X_3)$ whose joint pdf is given by $g(x_1, x_2, x_3)$. The marginal pdf of the random variable X_i is given by $g_i(x_i) = \frac{1}{2}\{f_i(x_i) + 1\}$ for $0 < x_i < 1, i = 1, 2, 3$. Also the joint pdf of (X_i, X_j) is given by $g_{i,j}(x_i, x_j) = \frac{1}{4}\{f_i(x_i)f_j(x_j) + f_i(x_i) + f_j(x_j) + 1\}$ for $i \neq j = 1, 2, 3$. Notice that $g_{i,j}(x_i, x_j) = g_i(x_i)g_j(x_j)$ for $i \neq j = 1, 2, 3$, so that we can conclude: X_i and X_j are independent for $i \neq j = 1, 2, 3$. But, obviously $g(x_1, x_2, x_3)$ does not match with the expression of the product $\Pi_{i=1}^3 g_i(x_i)$ for $0 < x_i < 1, i = 1, 2, 3$. Refer to the three Exercises 3.5.4-3.5.6 in this context. ▲

In the Exercises 3.5.7-3.5.8 and 3.5.17, we show ways to construct a four-dimensional random vector $\mathbf{X} = (X_1, X_2, X_3, X_4)$ such that X_1, X_2, X_3 are independent, but X_1, X_2, X_3, X_4 are dependent.

For a set of *independent* vector valued random variables $\mathbf{X}_1, ..., \mathbf{X}_p$, not necessarily all of the same dimension, an important consequence is summarized by the following result.

Theorem 3.5.1 *Suppose that* $\mathbf{X}_1, ..., \mathbf{X}_p$ *are independent vector valued random variables. Consider real valued functions* $g_i(\mathbf{x}_i)$, $i = 1, ..., p$. *Then, we have*

$$E[\Pi_{i=1}^p g_i(\mathbf{X}_i)] = \Pi_{i=1}^p E[g_i(\mathbf{X}_i)]$$

as long as $E[g_i(\mathbf{X}_i)]$ *is finite, where this expectation corresponds to the integral with respect to the marginal distribution of* $\mathbf{X}_i$, $i = 1, ..., p$. *Here, the* $\mathbf{X}_i$*'s may be discrete or continuous.*

Proof Let the random vector $\mathbf{X}_i$ be of dimension k_i and let its support be $\mathcal{X}_i \subseteq \Re^{k_i}, i = 1, ..., k$. Since $\mathbf{X}_1, ..., \mathbf{X}_p$ are independent, their joint pmf or pdf is given by

$$f(\mathbf{x}_1, ..., \mathbf{x}_p) = \Pi_{i=1}^p f_i(\mathbf{x}_i) \text{ for } x_i \in \mathcal{X}_i. \tag{3.5.3}$$

Then, we have

$$
\begin{aligned}
E[\Pi_{i=1}^p g_i(\mathbf{X}_i)] \\
= \int_{\mathcal{X}_p} \cdots \int_{\mathcal{X}_1} \Pi_{i=1}^p g_i(\mathbf{x}_i) \Pi_{i=1}^p f_i(\mathbf{x}_i) \Pi_{i=1}^p d\mathbf{x}_i \\
= \Pi_{i=1}^p \int_{\mathcal{X}_i} g_i(\mathbf{x}_i) f_i(\mathbf{x}_i) d\mathbf{x}_i \\
= \Pi_{i=1}^p E[g_i(\mathbf{X}_i)],
\end{aligned}
\tag{3.5.4}
$$

which is the desired result. ∎

> One important consequence of this theorem is this: if two real valued
> random variables are *independent*, then their covariance, *when finite*,
> is necessarily zero. This in turn then implies that the correlation
> coefficient between those two random variables is indeed zero.

In other words, *independence* between the random variables X_1 and X_2
would lead to the *zero correlation* between X_1 and X_2 as long as ρ_{X_1,X_2} is
finite. We state this more generally as follows.

Theorem 3.5.2 *Suppose that* $\mathbf{X}_1, \mathbf{X}_2$ *are two independent vector valued
random variables, not necessarily of the same dimension. Then,*

    (i) $Cov(g_1(\mathbf{X}_1), g_2(\mathbf{X}_2)) = 0$ *where* $g_1(.), g_2(.)$ *are real valued
functions, if* $E[g_1 g_2], E[g_1]$ *and* $E[g_2]$ *are all finite;*

    (ii) $\mathbf{g}_1(\mathbf{X}_1), \mathbf{g}_2(\mathbf{X}_2)$ *are independent where* $\mathbf{g}_1(.), \mathbf{g}_2(.)$ *do not need
to be restricted as only real valued.*

Proof (i) Let us assume that $E[g_1 g_2], E[g_1]$ and $E[g_2]$ are all finite.
Then, we appeal to the Theorem 3.5.1 to assert that $E[g_1(\mathbf{X}_1)g_2(\mathbf{X}_2)] = E[g_1(\mathbf{X}_1)]E[g_2(\mathbf{X}_2)]$. But, since $Cov(g_1(\mathbf{X}_1), g_2(\mathbf{X}_2))$ is $E[g_1(\mathbf{X}_1)g_2(\mathbf{X}_2)] - E[g_1(\mathbf{X}_1)]E[g_2(\mathbf{X}_2)]$, we thus have $Cov(g_1(\mathbf{X}_1), g_2(\mathbf{X}_2))$ obviously reducing
to zero.

(ii) We give a sketch of the main ideas. Let A_i be any Borel set in the
range space of the function $\mathbf{g}_i(.), i = 1, 2$. Now, let $B_i = \{\mathbf{x}_i \in \mathcal{X}_i : \mathbf{g}_i(\mathbf{x}_i) \in A_i\}$. Now,

$$P\{\mathbf{g}_1(\mathbf{X}_1) \in A_1 \cap \mathbf{g}_2(\mathbf{X}_2) \in A_2\}$$
$$= P\{\mathbf{X}_1 \in B_1 \cap \mathbf{X}_2 \in B_2\}$$
$$= P\{\mathbf{X}_1 \in B_1\}P\{\mathbf{X}_2 \in B_2\}, \qquad (3.5.5)$$
$$\text{since } \mathbf{X}_1, \mathbf{X}_2 \text{ are independent}$$
$$= P\{\mathbf{g}_1(\mathbf{X}_1) \in A_1\}P\{\mathbf{g}_2(\mathbf{X}_2) \in A_2\},$$

which is the desired result. ∎

> In the Theorem 3.5.2, one may be tempted to prove part (i) using the
> result from part (ii) plus the Theorem 3.5.1, and wonder whether the
> requirements of the finite moments stated therein are crucial for the
> conclusion to hold. See the following example for the case in point.

Example 3.5.5 Suppose that X_1 is distributed as the standard normal
variable with its pdf $f_1(x_1) = \{\sqrt{2\pi}\}^{-1} exp(-x_1^2/2) = \phi(x_1)$ defined in
(1.7.16) and X_2 is distributed as the Cauchy variable with its pdf $f_2(x_2) = \pi^{-1}(1 + x_2^2)^{-1}$ defined in (1.7.31), $-\infty < x_1, x_2 < \infty$. Suppose also that

these two random variables are independent. But, we can not claim that $Cov(X_1, X_2) = 0$ because $E[X_2]$ is *not* finite. ▲

> Two random variables X_1 and X_2 may be independent,
> but that may not necessarily imply that $Cov(X_1, X_2) = 0$.
> This has been emphasized in the Example 3.5.5.

Example 3.5.6 What will happen to the conclusions in the Theorem 3.3.1, parts (i)-(ii), when the two random variables X_1 and X_2 are independent? Let X_1 and X_2 be independent. Then, $E_{2|1}\{X_2 \mid X_1\} = E[X_2]$ which is a constant, so that $E_1[E\{X_2\}] = E[X_2]$. Also, $V_1[E_{2|1}\{X_2 \mid X_1\}] = V_1[E\{X_2\}] = 0$ since the variance of a constant is zero. On the other hand, $V_{2|1}\{X_2 \mid X_1\} = E_{2|1}\{X_2^2 \mid X_1\} - E_{2|1}^2\{X_2 \mid X_1\} = E[X_2^2] - E^2[X_2] = V[X_2]$ which is a constant. Thus, $E_1[V_{2|1}\{X_2 \mid X_1\}] = E_1[V\{X_2\}] = V[X_2]$. Hence, $V_1[E_{2|1}\{X_2 \mid X_1\}] + E_1[V_{2|1}\{X_2 \mid X_1\}]$ simplifies to $V[X_2]$. ▲

We have more to say on similar matters again in the Section 3.7. We end this section with yet another result which will come in very handy in verifying whether some jointly distributed random variables are independent. In some situations, applying the Definition 3.5.1 may be a little awkward because (3.5.1) requires one to check whether the joint pmf or the pdf is the same as the product of the marginal pmf's or pdf's. But, then one may not yet have identified all the marginal pmf's or pdf's explicitly! What is the way out? Look at the following result. Its proof is fairly routine and we leave it as the Exercise 3.5.9.

Suppose that we have the random variables $X_1, ..., X_k$ where $\mathcal{X}_i$ is the support of $X_i, i = 1, ..., k$. Next, we formalize the notion of the supports $\mathcal{X}_i$'s being *unrelated to each other*.

Definition 3.5.2 *We say that the supports $\mathcal{X}_i$'s are unrelated to each other if the following condition holds: Consider any subset of the random variables, $X_{i_j}, j = 1, ..., p, p = 1, ..., k - 1$. Conditionally, given that $X_{i_j} = x_{i_j}$, the support of the remaining random variable X_l stays the same as $\mathcal{X}_l$ for any $l \neq i_j, j = 1, ..., p, p = 1, ..., k - 1$.*

Theorem 3.5.3 *Suppose that the random variables $X_1, ..., X_k$ have the joint pmf or pdf $f(x_1, ..., x_k), x_i \in \mathcal{X}_i$, the support of $X_i, i = 1, ..., k$. Assume that these supports are unrelated to each other according to the Definition 3.5.2. Then, $X_1, ..., X_k$ are independent if and only if*

$$f(x_1, ..., x_k) = \Pi_{j=1}^k h_j(x_j) \text{for all } x_i \in \mathcal{X}_i, i = 1, ..., k \qquad (3.5.6)$$

for some functions $h_1(x_1), ..., h_k(x_k)$.

Example 3.5.7 Suppose that the joint pdf of X_1, X_2 is given by

$$f(x_1, x_2) = k\,exp\{-\theta x_1 - \tfrac{1}{\theta}x_2\}, x_1 > 0, x_2 > 0 \qquad (3.5.7)$$

for some $\theta > 0$ where $k(> 0)$ is a constant. Obviously, we can write $f(x_1, x_2) = kh_1(x_1)h_2(x_2)$ for all positive numbers x_1, x_2 where $h_1(x_1) = exp\{-\theta x_1\}$ and $h_2(x_2) = exp\{-\tfrac{1}{\theta}x_2\}$. Also the supports $\mathcal{X}_1 = \Re^+, \mathcal{X}_2 = \Re^+$ are *unrelated* to each other. Hence, by appealing to the Theorem 3.5.3 we can claim that X_1 and X_2 are independent random variables. One should note two interesting facts here. The two chosen functions $h_1(x_1)$ and $h_2(x_2)$ are not even probability densities to begin with. The other point is that the choices of these two functions are not unique. One can easily replace these by $kdh_1(x_1)$ and $kd^{-1}h_2(x_2)$ respectively where $d(> 0)$ is any arbitrary number. Note that we did not have to determine k in order to reach our conclusion that X_1 and X_2 are independent random variables. ▲

Example 3.5.8 (Example 3.5.7 Continued) One can check that $f_1(x_1) = \theta\,exp\{-\theta x_1\}$ and $f_2(x_2) = \theta^{-1}exp\{-\tfrac{1}{\theta}x_2\}$, and then one may apply (3.5.1) instead to arrive at the same conclusion. In view of the Theorem 3.5.3 it becomes really simple to write down the marginal pdf's in the case of independence. ▲

Example 3.5.9 Suppose that the joint pdf of X_1, X_2, X_3 is given by

$$f(x_1, x_2, x_3) = c\,exp\{-2x_1 - \pi x_3^2\}/[\sqrt{x_1}(1 + x_2^2)]$$
$$\text{for } x_1 \in \Re^+, (x_2, x_3) \in \Re^2 \qquad (3.5.8)$$

where $c(> 0)$ is a constant. One has $f(x_1, x_2, x_3) = ch_1(x_1)h_2(x_2)h_3(x_3)$ with $h_1(x_1) = \{\sqrt{x_1}\}^{-1}\,exp\{-2x_1\}, h_2(x_2) = (1 + x_2^2)^{-1}$ and $h_3(x_3) = exp\{-\pi x_3^2\}$, for all values of $x_1 \in \Re^+, (x_2, x_3) \in \Re^2$. Also, the supports $\mathcal{X}_1 = \Re^+, \mathcal{X}_2 = \Re, \mathcal{X}_3 = \Re$ are *unrelated* to each other. Hence, by appealing to the Theorem 3.5.3 we can claim that X_1, X_2 and X_3 are independent random variables. One should again note the interesting facts. The three chosen functions $h_1(x_1), h_2(x_2)$ and $h_3(x_3)$ are not even probability densities to begin with. The other point is that the choices of these three functions are not unique. ▲

Example 3.5.10 (Example 3.5.9 Continued) One should check these out: By looking at $h_1(x_1)$ one can immediately guess that X_1 has the Gamma$(\alpha = \tfrac{1}{2}, \beta = \tfrac{1}{2})$ distribution so that its normalizing constant $\sqrt{2/\pi}$. By looking at $h_2(x_2)$ one can immediately guess that X_2 has the Cauchy distribution so that its normalizing constant $1/\pi$. Similarly, by looking at $h_3(x_3)$ one can immediately guess that X_3 has the $N(0, \tfrac{1}{2\pi})$ distribution so that its normalizing constant is unity. Hence, $c = (\sqrt{2/\pi})(1/\pi)(1) =$

$\sqrt{2/\pi^3}$. Refer to (1.7.13), (1.7.20) and (1.7.31) as needed. Note that the Theorem 3.5.3 really makes it simple to write down the marginal pdf's in the case of independence. ▲

> It is crucial to note that the supports $\mathcal{X}_i$'s are assumed to be *unrelated to each other* so that the representation given by (3.5.6) may lead to the conclusion of *independence* among the random variables X_i's. Look at the Example 3.5.11 to see what may happen when the supports are related.

Example 3.5.11 (Examples 3.3.1 and 3.5.2 Continued) Consider two random variables X_1 and X_2 whose joint continuous distribution is given by the following pdf:

$$f(x_1, x_2) = \begin{cases} 6(1 - x_2) & \text{if } 0 < x_1 < x_2 < 1 \\ 0 & \text{elsewhere} \end{cases} \tag{3.5.9}$$

Obviously, one has $\mathcal{X}_1 = \mathcal{X}_2 = (0, 1)$. One may be tempted to denote, for example, $h_1(x_1) = 6, h_2(x_2) = (1 - x_2)$. At this point, one may be tempted to claim that X_1 and X_2 are independent. But, that will be wrong! On the whole space $\mathcal{X}_1 \times \mathcal{X}_2$, one can not claim that $f(x_1, x_2) = h_1(x_1)h_2(x_2)$. One may check this out by simply taking, for example, $x_1 = \frac{1}{2}, x_2 = \frac{1}{4}$ and then one has $f(x_1, x_2) = 0$ whereas $h_1(x_1)h_2(x_2) = \frac{9}{2}$. But, of course the relationship $f(x_1, x_2) = h_1(x_1)h_2(x_2)$ holds in the subspace where $0 < x_1 < x_2 < 1$. From the Example 3.5.2 one will recall that we had verified that in fact the two random variables X_1 and X_2 were *dependent*. ▲

3.6 The Bivariate Normal Distribution

Let (X_1, X_2) be a two-dimensional continuous random variable with the following joint pdf:

$$f(x_1, x_2) = c \, exp\left[-\tfrac{1}{2}(1 - \rho^2)^{-1}\{u_1^2 - 2\rho u_1 u_2 + u_2^2\}\right] \tag{3.6.1}$$

with

$$u_1 = (x_1 - \mu_1)/\sigma_1, u_2 = (x_2 - \mu_2)/\sigma_2,$$
$$c = \{2\pi\sigma_1\sigma_2(1 - \rho^2)^{\frac{1}{2}}\}^{-1}, \quad -\infty < x_1, x_2 < \infty, \tag{3.6.2}$$
$$-\infty < \mu_1, \mu_2 < \infty, \quad 0 < \sigma_1, \sigma_2 < \infty, \quad -1 < \rho < 1.$$

The pdf given by (3.6.1) is known as the *bivariate normal* or *two-dimensional normal density*. Here, $\mu_1, \mu_2, \sigma_1, \sigma_2$ and ρ are referred to as the parameters of the distribution. A random variable (X_1, X_2) having the pdf given by (3.6.1) is denoted by $N_2(\mu_1, \mu_2, \sigma_1^2, \sigma_2^2, \rho)$. The pdf $f(x_1, x_2)$ from (3.6.1) is centered around the point (μ_1, μ_2) in the x_1, x_2 plane. The bivariate normal pdf given by (3.6.1) has been plotted in the Figures 3.6.1-3.6.2.

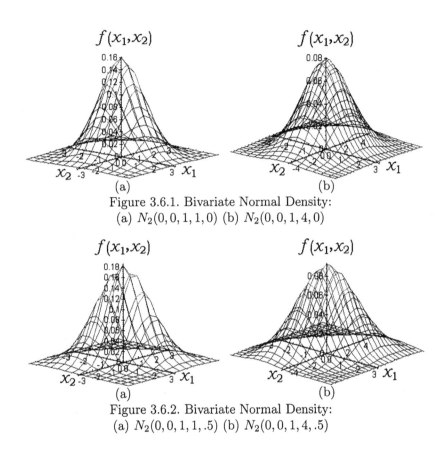

Figure 3.6.1. Bivariate Normal Density:
(a) $N_2(0, 0, 1, 1, 0)$ (b) $N_2(0, 0, 1, 4, 0)$

Figure 3.6.2. Bivariate Normal Density:
(a) $N_2(0, 0, 1, 1, .5)$ (b) $N_2(0, 0, 1, 4, .5)$

The Figure 3.6.2 (a) appears more concentrated around its center than its counterpart in the Figure 3.6.1 (a). This is due to the fact that we have taken $\rho = .5$ to draw the former picture in contrast with the value $\rho = 0$ in the latter picture, while the other parameters are held fixed. Clearly, the ordinate at the center in the Figure 3.6.1 (a) happens to be $\frac{1}{2\pi}$ ($\approx$ .15915) whereas the ordinate at the center in the Figure 3.6.2 (a) happens to be $\left(\frac{1}{2\pi}\right)\left(\frac{2}{\sqrt{3}}\right)$ ($\approx$ .18378) which is larger. This justifies the preceding claim. One observes a similar feature when the Figures 3.6.1 (b)-3.6.2 (b) are visually compared.

> How can one show directly that $f(x_1, x_2)$ from (3.6.1) is indeed a genuine pdf? The derivation follows shortly.

The function $f(x_1, x_2)$ is always positive. So, we merely need to verify that the double integral of $f(x_1, x_2)$ over the whole space $\Re^2$ is unity. With u_1, u_2 from (3.6.2), let us then rewrite

$$
\begin{aligned}
u_1^2 - 2\rho u_1 u_2 + u_2^2 &= \{u_1^2 - 2\rho u_1 u_2 + \rho^2 u_2^2\} + (1 - \rho^2) u_2^2 \\
&= (u_1 - \rho u_2)^2 + (1 - \rho^2) u_2^2.
\end{aligned} \tag{3.6.3}
$$

Hence, with c defined in (3.6.2) we obtain

$$f(x_1, x_2)$$

$$
= c\,exp \left[-\frac{1}{2(1-\rho^2)} \left\{ \frac{(x_1 - \mu_1)}{\sigma_1} - \frac{\rho(x_2 - \mu_2)}{\sigma_2} \right\}^2 - \frac{(x_2 - \mu_2)^2}{2\sigma_2^2} \right]. \tag{3.6.4}
$$

Now, for all fixed $x_2 \in \Re$, let us denote

$$
g(x_1, x_2) = exp \left[-\frac{1}{2(1-\rho^2)\sigma_1^2} \left\{ x_1 - \left(\mu_1 + \frac{\rho \sigma_1 (x_2 - \mu_2)}{\sigma_2} \right) \right\}^2 \right],
$$

$$
h(x_2) = exp \left[-\frac{(x_2 - \mu_2)^2}{2\sigma_2^2} \right], \tag{3.6.5}
$$

so that with $a = \{2\pi(1 - \rho^2)\sigma_1^2\}^{-1/2}$ we obtain

$$\int_\Re \int_\Re f(x_1, x_2) dx_1 dx_2 = ca^{-1} \int_\Re h(x_2) \left\{ \int_\Re a g(x_1, x_2) dx_1 \right\} dx_2. \tag{3.6.6}$$

Next, look at the expression of $ag(x_1, x_2)$ obtained from (3.6.5) and note that for all fixed x_2, it resembles the pdf of a univariate normal variable with mean $\mu_1 + \dfrac{\rho \sigma_1 (x_2 - \mu_2)}{\sigma_2}$ and variance $(1 - \rho^2)\sigma_1^2$ at the point $x_1 \in \Re$. Hence, we must have

$$\int_\Re ag(x_1, x_2) dx_1 = 1 \tag{3.6.7}$$

for all fixed $x_2 \in \Re$. Now, we can combine (3.6.6)-(3.6.7) to claim that

$$\int_\Re \int_\Re f(x_1, x_2) dx_1 dx_2 = ca^{-1} \int_\Re h(x_2) dx_2. \tag{3.6.8}$$

Again note that with $b = \{2\pi\sigma_2^2\}^{-1/2}$, the expression $bh(x_2)$ obtained from (3.6.5) happens to be the pdf of a normal variable with mean μ_2 and variance σ_2^2 at the point $x_2 \in \Re$. Hence, we must have

$$\int_\Re bh(x_2)dx_2 = 1, \qquad (3.6.9)$$

so that (3.6.8) can be rewritten as

$$\int_\Re \int_\Re f(x_1, x_2)dx_1 dx_2 = ca^{-1}b^{-1} = 1, \qquad (3.6.10)$$

by the definition of c from (3.6.2). Thus, we have directly verified that the function $f(x_1, x_2)$ given by (3.6.1) is indeed a genuine pdf of a two-dimensional random variable with its support $\Re^2$.

Theorem 3.6.1 *Suppose that* (X_1, X_2) *has the* $N_2(\mu_1, \mu_2, \sigma_1^2, \sigma_2^2, \rho)$ *distribution with its pdf* $f(x_1, x_2)$ *given by (3.6.1). Then,*

(i)   *the marginal distribution of* X_i *is given by* $N(\mu_i, \sigma_i^2), i = 1, 2$;

(ii)   *the conditional distribution of* $X_1 \mid X_2 = x_2$ *is normal with mean* $\mu_1 + \dfrac{\rho\sigma_1}{\sigma_2}(x_2 - \mu_2)$ *and variance* $\sigma_1^2(1 - \rho^2)$, *for all fixed* $x_2 \in \Re$;

(iii)   *the conditional distribution of* $X_2 \mid X_1 = x_1$ *is normal with mean* $\mu_2 + \dfrac{\rho\sigma_2}{\sigma_1}(x_1 - \mu_1)$ *and variance* $\sigma_2^2(1 - \rho^2)$, *for all fixed* $x_1 \in \Re$.

Proof (i) We simply show the derivation of the marginal pdf of the random variable X_2. Using (3.3.2) one gets for any fixed $x_2 \in \Re$,

$$f_2(x_2)$$
$$= \int_\Re f(x_1, x_2)dx_1$$
$$= c\int_\Re g(x_1, x_2)h(x_2)dx_1, \text{ using } (3.6.4)\text{-}(3.6.5)$$
$$= ch(x_2)\int_\Re g(x_1, x_2)dx_1,$$

which can be expressed as

$$ca^{-1}h(x_2)\int_\Re ag(x_1, x_2)dx_1 \text{ with } a = \{2\pi(1 - \rho^2)\sigma_1^2\}^{-1/2}$$
$$= ca^{-1}h(x_2), \text{ since } \int_\Re ag(x_1, x_2)dx_1 = 1 \text{ from } (3.6.7) \qquad (3.6.11)$$
$$= \{\sigma_2\sqrt{2\pi}\}^{-1}exp\{-(x_2 - \mu_2)^2/(2\sigma_2^2)\}.$$

This shows that X_2 is distributed as $N(\mu_2, \sigma_2^2)$. The marginal pdf of X_1 can be found easily by appropriately modifying (3.6.5) first. We leave this as the Exercise 3.6.1.

(ii) Let us denote $M = \left\{x_1 - \left(\mu_1 + \rho\sigma_1\sigma_2^{-1}(x_2 - \mu_2)\right)\right\}^2$ and $a = \{2\pi(1-\rho^2)\sigma_1^2\}^{-1/2}$. Utilizing (3.3.5), (3.6.4)-(3.6.5) and (3.6.11), the conditional pdf of $X_1 \mid X_2 = x_2$ is given by

$$
\begin{aligned}
f_{1|2}(x_1) \\
&= f(x_1, x_2)/f_2(x_2) \\
&= cg(x_1, x_2)h(x_2)/\{ca^{-1}h(x_2)\} \\
&= \{2\pi(1 - \rho^2)\sigma_1^2\}^{-1/2} exp\left[-\{2(1 - \rho^2)\sigma_1^2\}^{-1}M\right].
\end{aligned}
\tag{3.6.12}
$$

The expression of $f_{1|2}(x_1)$ found in the last step in (3.6.12) resembles the pdf of a normal random variable with mean $\mu_1 + \rho\sigma_1\sigma_2^{-1}(x_2 - \mu_2)$ and variance $(1 - \rho^2)\sigma_1^2$.

(iii) Its proof is left as the Exercise 3.6.2. ∎

Example 3.6.1 Suppose that (X_1, X_2) is distributed as $N_2(0, 0, 4, 1, \rho)$ where $\rho = \frac{1}{2}$. From the Theorem 3.6.1 (i) we already know that $E[X_2] = 0$ and $V[X_2] = 1$. Utilizing part (iii), we can also say that $E[X_2 \mid X_1 = x_1] = \frac{1}{4}x_1$ and $V[X_2 \mid X_1 = x_1] = \frac{3}{4}$. Now, one may apply the Theorem 3.3.1 to find indirectly the expressions of $E[X_2]$ and $V[X_2]$. We should have $E[X_2] = E\{E[X_2 \mid X_1 = x_1]\} = E[\frac{1}{4}X_1] = 0$, whereas $V[X_2] = V\{E[X_2 \mid X_1 = x_1]\} + E\{V[X_2 \mid X_1 = x_1]\} = V[\frac{1}{4}X_1] + E[\frac{3}{4}] = \frac{1}{16}V[X_1] + \frac{3}{4} = \frac{1}{16}(4) + \frac{3}{4} = 1$. Note that in this example, we did not fully exploit the form of the *conditional distribution* of $X_2 \mid X_1 = x_1$. We merely used the expressions of the conditional mean and variance of $X_2 \mid X_1 = x_1$. ▲

> In the Example 3.6.2, we exploit more fully the form of the *conditional distributions*.

Example 3.6.2 Suppose that (X_1, X_2) is distributed as $N_2(0, 0, 1, 1, \rho)$ where $\rho = \frac{1}{2}$. We wish to evaluate $E\{exp[\frac{1}{2}X_1X_2]\}$. Using the Theorem 3.3.1 (i), we can write

$$
\begin{aligned}
E\{exp[\tfrac{1}{2}X_1X_2]\} &= E\{E(exp[\tfrac{1}{2}X_1X_2] \mid X_1 = x_1)\} \\
&= E\{E(exp[\tfrac{1}{2}x_1X_2] \mid X_1 = x_1)\}.
\end{aligned}
\tag{3.6.13}
$$

But, from the Theorem 3.6.1 (iii) we already know that the conditional distribution of $X_2 \mid X_1 = x_1$ is normal with mean $\frac{1}{2}x_1$ and variance $\frac{3}{4}$. Now, $E(exp[\frac{1}{2}x_1X_2] \mid X_1 = x_1)$ can be viewed as the conditional mgf of $X_2 \mid X_1 = x_1$. Thus, using the form of the mgf of a univariate normal random variable from (2.3.16), we obtain

$$
E(exp[\tfrac{1}{2}x_1X_2] \mid X_1 = x_1) = exp\{\tfrac{11}{32}x_1^2\}.
\tag{3.6.14}
$$

From the Theorem 3.6.1 (i), we also know that marginally X_1 is distributed as $N(0,1)$. Thus, we combine (3.6.13)-(3.6.14) and get

$$
\begin{aligned}
E\{exp[\tfrac{1}{2}X_1X_2]\} &= E[exp\{\tfrac{11}{32}X_1^2\}] \\
&= \{\sqrt{2\pi}\}^{-1} \int_{\Re} exp\{\tfrac{11}{32}u^2\}exp\{-\tfrac{1}{2}u^2\}du \qquad (3.6.15) \\
&= \{\sqrt{2\pi}\}^{-1} \int_{\Re} exp\{-\tfrac{5}{32}u^2\}du.
\end{aligned}
$$

With $\sigma^2 = \tfrac{16}{5}$, let us denote $h(u) = \{\sigma\sqrt{2\pi}\}^{-1} exp\{-\tfrac{5}{32}u^2\}$, $u \in \Re$. Then, $h(u)$ is the pdf of a random variable having the $N(0, \sigma^2)$ distribution so that $\int_{\Re} h(u)du = 1$. Hence, from (3.6.15) we have

$$
E\{exp[\tfrac{1}{2}X_1X_2]\} = \sigma \int_{\Re} h(u)du = \sigma = \sqrt{16/5} \approx 1.7889. \qquad (3.6.16)
$$

In the same fashion one can also derive the mgf of the random variable X_1X_2, that is the expression for the $E\{exp[tX_1X_2]\}$ for some appropriate range of values of t. We leave this as the Exercise 3.6.3. ▲

The reverse of the conclusion given in the Theorem 3.6.1, part (i) is not necessarily true. That is, the marginal distributions of both X_1 and X_2 can be univariate normal, but this does not imply that (X_1, X_2) is jointly distributed as N_2. Look at the next example.

Example 3.6.3 In the bivariate normal distribution (3.6.1), each random variable X_1, X_2 individually has a normal distribution. But, it is easy to construct two dependent continuous random variables X_1 and X_2 such that marginally each is normally distributed whereas jointly (X_1, X_2) is not distributed as N_2.

Let us temporarily write $f(x_1, x_2; \mu_1, \mu_2, \sigma_1^2, \sigma_2^2, \rho)$ for the pdf given in (3.6.1). Next, consider any arbitrary $0 < \alpha, \rho < 1$ and fix them. Let us now define

$$
\begin{aligned}
&g(x_1, x_2; \rho) \\
&= \alpha\, f(x_1, x_2; 0, 0, 1, 1, \rho) + (1 - \alpha)\, f(x_1, x_2; 0, 0, 1, 1, -\rho)
\end{aligned} \qquad (3.6.17)
$$

for $-\infty < x_1, x_2 < \infty$. Since the non-negative functions $f(x_1, x_2; 0, 0, 1, 1, \rho)$ and $f(x_1, x_2; 0, 0, 1, 1, -\rho)$ are both pdf's on $\Re^2$, we must have

$$
\begin{aligned}
&\iint_{\Re^2} f(x_1, x_2; 0, 0, 1, 1, \rho)dx_1dx_2 \\
&= \iint_{\Re^2} f(x_1, x_2; 0, 0, 1, 1, -\rho)dx_1dx_2 = 1.
\end{aligned}
$$

Hence, we can express $\int\int\limits_{\Re^2} g(x_1, x_2; \rho) dx_1 dx_2$ as

$$\alpha \int\int\limits_{\Re^2} f(x_1, x_2; 0, 0, 1, 1, \rho) dx_1 dx_2$$

$$+(1 - \alpha) \int\int\limits_{\Re^2} f(x_1, x_2; 0, 0, 1, 1, -\rho) dx_1 dx_2 \qquad (3.6.18)$$

$$= 1.$$

Also, $g(x_1, x_2; \rho)$ is non-negative for all $(x_1, x_2) \in \Re^2$. Thus, $g(x_1, x_2)$ is a genuine pdf on the support $\Re^2$.

Let (X_1, X_2) be the random variables whose joint pdf is $g(x_1, x_2; \rho)$ for all $(x_1, x_2) \in \Re^2$. By direct integration, one can verify that marginally, both X_1 and X_2 are indeed distributed as the standard normal variables.

The joint pdf $g(x_1, x_2; \rho)$ has been plotted in the Figures 3.6.3 (a) and (b) with $\alpha = .5, .1$ respectively and $\rho = .5$. Comparing these figures visually with those plotted in the Figures 3.6.1-3.6.2, one may start wondering whether $g(x_1, x_2; \rho)$ may correspond to some bivariate normal pdf after all!

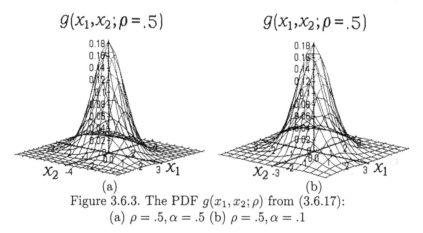

Figure 3.6.3. The PDF $g(x_1, x_2; \rho)$ from (3.6.17):
(a) $\rho = .5, \alpha = .5$ (b) $\rho = .5, \alpha = .1$

But, the fact of the matter is that the joint pdf $g(x_1, x_2; \rho)$ from (3.6.17) does not quite match with the pdf of any bivariate normal distribution. Look at the next example for some explanations. ▲

> How can one prove that the joint pdf $g(x_1, x_2; \rho)$ from (3.6.17)
> can not match with the pdf of any bivariate normal distribution?
> Look at the Example 3.6.4.

Example 3.6.4 (Example 3.6.3 Continued) Consider, for example, the situation when $\rho = .5, \alpha = .5$. Using the Theorem 3.6.1 (ii), one can check that

$$E[X_1 \mid X_2 = x_2] = \alpha\{\tfrac{1}{2}x_2\} + (1-\alpha)\{-\tfrac{1}{2}x_2\} = 0, \qquad (3.6.19)$$

whatever be fixed $x_2 \in \Re$, since $\alpha = .5$. Suppose that it is possible for the pair (X_1, X_2) to be distributed as the bivariate normal variable, $N_2(0,0,1,1,\rho^*)$ with some $\rho^* \in (-1,1)$. But, then $E[X_1 \mid X_2 = x_2]$ must be $\rho^* x_2$ which has to match with the answer zero obtained in (3.6.19), for all $x_2 \in \Re$. In other words, ρ^* must be zero. Hence, for all $(x_1, x_2) \in \Re^2$ we should be able to write

$$
\begin{aligned}
g(x_1, x_2; \rho = .5) \\
= \tfrac{1}{2\pi\sqrt{3}} e^{-\frac{2}{3}(x_1^2 + x_2^2)} \{ e^{\frac{2}{3}x_1 x_2} + e^{-\frac{2}{3}x_1 x_2} \} \\
\equiv \tfrac{1}{2\pi} e^{-\frac{1}{2}(x_1^2 + x_2^2)}, \text{ the joint pdf of } N_2(0,0,1,1,0) \\
= h(x_1, x_2), \text{ say.}
\end{aligned}
\qquad (3.6.20)
$$

Now, $g(0,0; \rho = .5) = \tfrac{1}{\pi\sqrt{3}}$, but $h(0,0) = \tfrac{1}{2\pi}$. It is obvious that $g(0,0; \rho = .5) \neq h(0,0)$. Hence, it is impossible for the random vector (X_1, X_2) having the pdf $g(x_1, x_2; \rho = .5)$ to be matched with any bivariate normal random vector. ▲

> The Exercise 3.6.8 gives another pair of random variables X_1 and X_2 such that marginally each is normally distributed whereas jointly (X_1, X_2) is not distributed as N_2.

For the sake of completeness, we now formally define what is known as the *regression function* in statistics.

Definition 3.6.1 *Suppose that (X_1, X_2) has the $N_2(\mu_1, \mu_2, \sigma_1^2, \sigma_2^2, \rho)$ distribution with its pdf $f(x_1, x_2)$ given by (3.6.1). The conditional mean of $X_1 \mid X_2 = x_2$, that is, $\mu_1 + \dfrac{\rho\sigma_1}{\sigma_2}(x_2 - \mu_2)$ is known as the regression function of X_1 on X_2. Analogously, the conditional mean of $X_2 \mid X_1 = x_1$, that is, $\mu_2 + \dfrac{\rho\sigma_2}{\sigma_1}(x_1 - \mu_1)$ is known as the regression function of X_2 on X_1.*

Even though linear regression analysis is out of scope for this textbook, we simply mention that it plays a very important role in statistics. The readers have already noted that the regression functions in the case of a bivariate normal distribution turn out to be straight lines.

We also mention that Tong (1990) had written a whole book devoted entirely to the multivariate normal distribution. It is a very valuable resource, particularly because it includes the associated tables for the percentage points of the distribution.

3.7 Correlation Coefficient and Independence

We begin this section with a result which clarifies the role of the zero correlation in a bivariate normal distribution.

Theorem 3.7.1 *Suppose that (X_1, X_2) has the bivariate normal distribution $N_2(\mu_1, \mu_2, \sigma_1^2, \sigma_2^2, \rho)$ with the joint pdf given by (3.6.1) where $-\infty < \mu_1, \mu_2 < \infty, 0 < \sigma_1, \sigma_2 < \infty$ and $-1 < \rho < 1$. Then, the two random variables X_1 and X_2 are independent if and only if the correlation coefficient $\rho = 0$.*

Proof We first verify the "necessary part" followed by the "sufficiency part".

Only if part: Suppose that X_1 and X_2 are independent. Then, in view of the Theorem 3.5.2 (i), we conclude that $Cov(X_1, X_2) = 0$. This will imply that $\rho = 0$.

If part: From (3.6.1), let us recall that the joint pdf of (X_1, X_2) is given by

$$f(x_1, x_2) = c \; exp \left[-\tfrac{1}{2}(1 - \rho^2)^{-1} \{u_1^2 - 2\rho u_1 u_2 + u_2^2 \} \right] \qquad (3.7.1)$$

with

$$u_1 = (x_1 - \mu_1)/\sigma_1, u_2 = (x_2 - \mu_2)/\sigma_2,$$

$$c = \{ 2\pi\sigma_1\sigma_2(1 - \rho^2)^{\frac{1}{2}} \}^{-1}, \; -\infty < x_1, x_2 < \infty,$$

$$-\infty < \mu_1, \mu_2 < \infty, \; 0 < \sigma_1, \sigma_2 < \infty, \; -1 < \rho < 1.$$

But, when $\rho = 0$, this joint pdf reduces to

$$f(x_1, x_2) = c \; exp \left[-\tfrac{1}{2} \{u_1^2 + u_2^2 \} \right] = h(x_1)h(x_2), \qquad (3.7.2)$$

where $h(x_i) = \{\sigma_i\sqrt{2\pi}\}^{-1}\exp\{-\tfrac{1}{2}(x_i - \mu_i)^2/\sigma_i^2\}, i = 1, 2$. By appealing to the Theorem 3.5.3 we conclude that X_1 and X_2 are independent. ∎

> However, the zero correlation coefficient between two arbitrary random variables does not necessarily imply that these two variables are independent. Examples 3.7.1-3.7.2 emphasize this point.

Example 3.7.1 Suppose that X_1 is $N(0, 1)$ and let $X_2 = X_1^2$. Then, $Cov(X_1, X_2) = E(X_1 X_2) - E(X_1)E(X_2) = E(X_1^3) - E(X_1)E(X_2) = 0$, since $E(X_1) = 0$ and $E(X_1^3) = 0$. That is, the correlation coefficient ρ_{X_1, X_2} is zero. But the fact that X_1 and X_2 are dependent can be easily verified as follows. One can claim that $P\{X_2 > 4\} > 0$, however, the conditional probability, $P\{X_2 > 4 \mid -2 \le X_1 \le 2\}$ is same as $P\{X_1^2 > 4 \mid -2 \le X_1 \le 2\}$

which happens to be zero. Thus, we note that $P\{X_2 > 4 \mid -2 \le X_1 \le 2\} \ne P\{X_2 > 4\}$. Hence, X_1 and X_2 are dependent variables. ▲

> One can easily construct similar examples in a discrete situation. Look at the Exercises 3.7.1-3.7.3.

Example 3.7.2 Suppose that Θ is distributed uniformly on the interval $[0, 2\pi)$. Let us denote $X_1 = \cos(\Theta), X_2 = \sin(\Theta)$. Now, one has $E[X_1] = \int_0^{2\pi} \cos(\theta) \frac{1}{2\pi} d\theta = \frac{1}{2\pi} [\sin(\theta)]_{\theta=0}^{\theta=2\pi} = 0$. Also, one can write $E[X_1 X_2] = \int_0^{2\pi} \cos(\theta) \sin(\theta) \frac{1}{2\pi} d\theta = \frac{1}{4\pi} \int_0^{2\pi} \sin(2\theta) d\theta = \frac{1}{8\pi} [-\cos(2\theta)]_{\theta=0}^{\theta=2\pi} = 0$. Thus, $Cov(X_1, X_2) = E(X_1 X_2) - E(X_1)E(X_2) = 0 - 0 = 0$. That is, the correlation coefficient ρ_{X_1, X_2} is zero. But the fact that X_1 and X_2 are dependent can be easily verified as follows. One observes that $X_1^2 + X_2^2 = 1$ and hence conditionally given $X_1 = x_1$, the random variable X_2 can take one of the possible values, $-\sqrt{1 - x_1^2}$ or $+\sqrt{1 - x_1^2}$ with probability $\frac{1}{2}$ each. Suppose that we fix $x_1 = \frac{\sqrt{3}}{2}$. Then, we argue that $P\{-\frac{1}{4} < X_2 < \frac{1}{4} \mid X_1 = \frac{\sqrt{3}}{2}\} = 0$, but obviously $P\{-\frac{1}{4} < X_2 < \frac{1}{4}\} > 0$. So, the random variables X_1 and X_2 are dependent. ▲

> Theorem 3.7.1 mentions that $\rho_{X_1, X_2} = 0$ implies independence between X_1 and X_2 when their joint distribution is N_2. But, $\rho_{X_1, X_2} = 0$ may sometimes imply independence between X_1 and X_2 even when their joint distribution is different from the bivariate normal. Look at the Example 3.7.3.

Example 3.7.3 The zero correlation coefficient implies *independence* not merely in the case of a bivariate normal distribution. Consider two random variables X_1 and X_2 whose joint probability distribution is given as follows: Each expression in the Table 3.7.1 involving the p's is assumed positive and smaller than unity.

Table 3.7.1. Joint Probability Distribution of X_1 and X_2

		X_1 values		Row
		0	1	Total
X_2 values	0	$1 - p_1$ $-p_2 + p$	$p_1 - p$	$1 - p_2$
	1	$p_2 - p$	p	p_2
Col. Total		$1 - p_1$	p_1	1

Now, we have $Cov(X_1, X_2) = E(X_1 X_2) - E(X_1)E(X_2) = P\{X_1 = 1 \cap X_2 = 1\} - P\{X_1 = 1\}P\{X_2 = 1\} = p - p_1 p_2$, and hence the zero correlation coefficient between X_1 and X_2 will amount to saying that $p = p_1p_2$ where so far p, p_1 and p_2 have all been assumed to lie between $(0, 1)$ but they are otherwise arbitrary. Now, we must have then $P(X_1 = 1 \cap X_2 = 0) = p_1 - p$, and $P(X_1 = 0 \cap X_2 = 0) = 1 - p_1 - p_2 + p$, and $P(X_1 = 0 \cap X_2 = 1) = p_2 - p$. But, now $P(X_1 = 0 \cap X_2 = 1) = p_2 - p = p_2 - p_1p_2 = p_2(1 - p_1) = P(X_1 = 0)P(X_2 = 1)$; $P(X_1 = 1 \cap X_2 = 0) = p_1 - p = p_1 - p_1p_2 = p_1(1 - p_2) = P(X_1 = 1)P(X_2 = 0)$; and $P(X_1 = 0 \cap X_2 = 0) = 1 - p_1 - p_2 + p = 1 - p_1 - p_2 + p_1p_2 = (1 - p_1)(1 - p_2) = P(X_1 = 0)P(X_2 = 0)$. Hence, the two such random variables X_1 and X_2 are independent. Here, the zero correlation coefficient implied independence, in other words the property that "the zero correlation coefficient implies independence" is not a unique characteristic property of a bivariate normal distribution. ▲

Example 3.7.4 There are other simple ways to construct a pair of random variables with the zero correlation coefficient. Start with two random variables U_1, U_2 such that $V(U_1) = V(U_2)$. Let us denote $X_1 = U_1 + U_2$ and $X_2 = U_1 - U_2$. Then, use the *bilinear property* of the covariance function which says that the covariance function is linear in both components. This property was stated in the Theorem 3.4.3, part (iv). Hence, $Cov(X_1, X_2) = Cov(U_1 + U_2, U_1 - U_2) = Cov(U_1, U_1) - Cov(U_1, U_2) + Cov(U_2, U_1) - Cov(U_2, U_2) = V(U_1) - V(U_2) = 0$. ▲

3.8   The Exponential Family of Distributions

The *exponential family of distributions* happens to be very rich when it comes to statistical modeling of datasets in practice. The distributions belonging to this family enjoy many interesting properties which often attract investigators toward specific members of this family in order to pursue statistical studies. Some of those properties and underlying data reduction principles, such as *sufficiency* or *minimal sufficiency*, would impact significantly in Chapter 6 and others. To get an idea, one may simply glance at the broad ranging results stated as Theorems 6.2.2, 6.3.3 and 6.3.4 in Chapter 6. In this section, we discuss briefly both the *one-parameter* and *multi-parameter* exponential families of distributions.

3.8.1   One-parameter Situation

Let X be a random variable with the pmf or pdf given by $f(x; \theta)$, $x \in \mathcal{X} \subseteq \Re$, $\theta \in \Theta \subseteq \Re$. Here, θ is the single parameter involved in the expression of $f(x; \theta)$ which is frequently referred to as a *statistical model*.

Definition 3.8.1 *We say that $f(x;\theta)$ belongs to the one-parameter exponential family if and only if we can express*

$$f(x;\theta) = a(\theta)g(x)exp\{b(\theta)R(x)\} \; for \; x \in \mathcal{X} \; and \; \theta \in \Theta, \qquad (3.8.1)$$

with appropriate forms of real valued functions $a(\theta) \geq 0, b(\theta), g(x) \geq 0$ and $R(x)$, where $\mathcal{X} \subseteq \Re$ and Θ is a subinterval of the real line $\Re$. It is crucial to note that the expressions of $a(\theta)$ and $b(\theta)$ can not involve x, while the expressions of $g(x)$ and $R(x)$ can not involve θ.

Many standard distributions, including several listed in Section 1.7, belong to this rich class. Let us look at some examples.

Example 3.8.1 Consider the Bernoulli distribution from (1.7.1). Let us rewrite the pmf $f(x;p)$ as

$$p^x(1-p)^{1-x} = (1-p)\{p(1-p)^{-1}\}^x = (1-p)exp[x\log\{p(1-p)^{-1}\}], \qquad (3.8.2)$$

which now resembles (3.8.1) where $\theta = p, a(\theta) = 1 - \theta, g(x) = 1, b(\theta) = \log\{\theta(1-\theta)^{-1}\}, R(x) = x$, and $\Theta = (0,1), \mathcal{X} = \{0,1\}$. ▲

Example 3.8.2 Let X be distributed as Poisson(λ), defined in (1.7.4) where $\lambda(> 0)$. With $\mathcal{X} = \{0,1,2,...\}$, $\theta = \lambda$, and $\Theta = (0,\infty)$, the pmf $f(x;\theta) = e^{-\theta}\theta^x/x!$ has the same representation given in (3.8.1) where $g(x) = (x!)^{-1}, a(\theta) = exp\{-\theta\}, b(\theta) = \log(\theta)$ and $R(x) = x$. ▲

Example 3.8.3 Let X be distributed as $N(\mu,1)$, defined in (1.7.13) where $\mu \in \Re$. With $\mathcal{X} = \Re$, $\theta = \mu$, and $\Theta = \Re$, the pdf $f(x;\theta) = \{\sqrt{2\pi}\}^{-1} exp\{-\frac{1}{2}(x-\theta)^2\}$ has the same representation given in (3.8.1) where $g(x) = exp\{-\frac{1}{2}x^2\}, a(\theta) = \{\sqrt{2\pi}\}^{-1}exp\{-\frac{1}{2}\theta^2\}, b(\theta) = \theta$ and $R(x) = x$. ▲

One should not however expect that every possible distribution would necessarily belong to this one-parameter exponential family. There are many examples where the pmf or the pdf does not belong to the class of distributions defined via (3.8.1).

> All one-parameter distributions do not necessarily belong to the exponential family (3.8.1). Refer to the Examples 3.8.4-3.8.6.

In the Table 3.8.1, we explicitly show the correspondence between some of the standard distributions mentioned in Section 1.7 and the associated representations showing their memberships in the one-parameter exponential family. While verifying the entries given in the Table 3.8.1, the reader will immediately realize that the expressions for $a(\theta), b(\theta)$, $g(x)$ and $R(x)$ are not really unique.

Example 3.8.4 Consider a random variable X having a negative exponential pdf, defined in (1.7.36). The pdf is $\theta^{-1}exp\{-(x-\theta)/\theta\}I(x>\theta)$ with $\theta > 0$, where $I(.)$ is an indicator function. Recall that $I(A)$ is 1 or 0 according as the set A or A^c is observed. The term $I(x>\theta)$ can not be absorbed in the expressions for $a(\theta), b(\theta)$, $g(x)$ or $R(x)$, and hence this distribution does not belong to a one-parameter exponential family. ▲

Table 3.8.1. Selected Members of the One-parameter
Exponential Family (3.8.1)

θ	$a(\theta)$	$b(\theta)$	$g(x)$	$R(x)$
		Bernoulli(p)		
p	$1-\theta$	$log\left(\frac{\theta}{1-\theta}\right)$	1	x
		Binomial(n,p)		
p	$(1-\theta)^n$	$log\left(\frac{\theta}{1-\theta}\right)$	$\binom{n}{x}$	x
		Poisson(λ)		
λ	$e^{-\theta}$	$log(\theta)$	$(x!)^{-1}$	x
		NB$(\mu,k), k$ known		
μ	$\left(\frac{k}{\theta+k}\right)^k$	$log\left(\frac{\theta}{\theta+k}\right)$	$\binom{k+x-1}{k-1}$	x
		Geometric(p)		
p	$\theta(1-\theta)^{-1}$	$log(1-\theta)$	1	x
		N$(\mu,\sigma^2), \mu$ known		
σ	$\frac{1}{\theta}exp\left\{-\frac{\mu^2}{2\theta^2}\right\}$	θ^{-2}	$\{2\pi\}^{-\frac{1}{2}}$	$-\frac{1}{2}x^2$ $+\mu x$
		N$(\mu,\sigma^2), \sigma$ known		
μ	$\frac{1}{\sqrt{2\pi}}exp\left\{-\frac{\theta^2}{2\sigma^2}\right\}$	$\frac{\theta}{\sigma^2}$	$\frac{1}{\sigma}exp\left\{-\frac{x^2}{2\sigma^2}\right\}$	x
		Gamma$(\alpha,\beta), \alpha$ known		
β	$\{\theta^\alpha\Gamma(\alpha)\}^{-1}$	$-\theta^{-1}$	$x^{\alpha-1}$	x
		Gamma$(\alpha,\beta), \beta$ known		
α	$\{\beta^\theta\Gamma(\theta)\}^{-1}$	θ	$\frac{1}{x}exp\{-\frac{x}{\beta}\}$	$log(x)$

Example 3.8.5 Suppose that a random variable X has the uniform distribution, defined in (1.7.12), on the interval $(0,\theta)$ with $\theta > 0$. The pdf can be rewritten as $f(x;\theta) = \theta^{-1}I(0<x<\theta)$. Again, the term $I(x>\theta)$ can not be absorbed in the expressions for $a(\theta), b(\theta)$, $g(x)$ or $R(x)$, and hence this distribution does not belong to a one-parameter exponential family. ▲

> In the two Examples 3.8.4-3.8.5, the support $\mathcal{X}$ depended
> on the single parameter θ. But, even if the support $\mathcal{X}$ does
> not depend on θ, in some cases the pmf or the pdf *may not*
> *belong* to the *exponential family* defined via (3.8.1).

Example 3.8.6 Suppose that a random variable X has the $N(\theta, \theta^2)$
distribution where $\theta(> 0)$ is the *single parameter*. The corresponding pdf
$f(x; \theta)$ can be expressed as

$$\{\theta\sqrt{2\pi}\}^{-1} exp\{-\tfrac{1}{2}(x - \theta)^2/\theta^2\} = \{\sqrt{2\pi e}\}^{-1}\theta^{-1} exp\left\{-\frac{x^2}{2\theta^2} + \frac{x}{\theta}\right\},$$

(3.8.3)

which does not have the same form as in (3.8.1). In other words, this
distribution does not belong to a one-parameter exponential family. ▲

> A distribution such as $N(\theta, \theta^2)$ with $\theta(> 0)$ is said to belong to a
> *curved exponential family,* introduced by Efron (1975, 1978).

3.8.2 Multi-parameter Situation

Let X be a random variable with the pmf or pdf given by $f(x; \boldsymbol{\theta})$, $x \in$
$\mathcal{X} \subseteq \Re$, $\boldsymbol{\theta} = (\theta_1, ..., \theta_k) \in \Theta \subseteq \Re^k$. Here, $\boldsymbol{\theta}$ is a vector valued parameter
having k components involved in the expression of $f(x; \boldsymbol{\theta})$ which is again
referred to as a *statistical model.*

Definition 3.8.2 *We say that $f(x; \boldsymbol{\theta})$ belongs to the k-parameter expo-*
nential family if and only if one can express

$$f(x; \boldsymbol{\theta}) = a(\boldsymbol{\theta})g(x)exp\{\Sigma_{i=1}^{k}b_i(\boldsymbol{\theta})R_i(x)\} \qquad (3.8.4)$$

with some appropriate forms for $g(x) \geq 0$, $a(\boldsymbol{\theta}) \geq 0, b_i(\boldsymbol{\theta})$ and $R_i(x)$,
$i = 1, ..., k$. It is crucial to note that the expressions of $a(\boldsymbol{\theta})$ and $b_i(\boldsymbol{\theta}), i =$
$1, ..., k$, can not involve x, while the expressions of $g(x)$ and $R_1(x), ..., R_k(x)$
can not involve $\boldsymbol{\theta}$.

Many standard distributions, including several listed in Section 1.7, belong to this rich class. In order to involve only statistically meaningful
reparameterizations while representing $f(x; \boldsymbol{\theta})$ in the form given by (3.8.4),
one would **assume** that the following regulatory conditions are satisfied:

Condition 1: Neither the b_i's nor the R_i's satisfy a linear constraint;

Condition 2: The parameter space contains a k-dimensional rectangle.

(3.8.5)

In the contexts of both the one-parameter and multi-parameter exponential families, there are such notions referred to as the *natural parameterization*, the *natural parameter space*, and the *natural exponential family*. A serious discussion of these topics needs substantial mathematical depth beyond the assumed prerequisites. Elaborate discussions of intricate issues and related references are included in Chapter 2 of both Lehmann (1983, 1986) and Lehmann and Casella (1998), as well as Barndorff-Nielson (1978). Let us again consider some examples.

Example 3.8.7 Let X be distributed as $N(\mu, \sigma^2)$, with $k = 2, \theta = (\mu, \sigma) \in \Re \times \Re^+$ where μ and σ are both treated as parameters. Then, the corresponding pdf has the form given in (3.8.4) where $x \in \Re$, $\theta_1 = \mu$, $\theta_2 = \sigma$, $R_1(x) = x, R_2(x) = x^2$, and $a(\boldsymbol{\theta}) = \{2\pi\theta_2^2\}^{-\frac{1}{2}} exp\{-\theta_1^2(2\theta_2^2)^{-1}\}, g(x) = 1, b_1(\boldsymbol{\theta}) = \theta_1\theta_2^{-2}$, and $b_2(\boldsymbol{\theta}) = -(2\theta_2^2)^{-1}$. ▲

Example 3.8.8 Let X be distributed as Gamma(α, β) where both $\alpha(> 0)$ and $\beta(> 0)$ are treated as parameters. The pdf of X is given by (1.7.20) so that $f(x; \alpha, \beta) = \{\beta^\alpha \Gamma(\alpha)\}^{-1} exp(-x/\beta) x^{\alpha-1}$, where we have $k = 2$, $\boldsymbol{\theta} = (\alpha, \beta) \in \Re^+ \times \Re^+$, $x \in \Re^+$. We leave it as an exercise to verify that this pdf is also of the form given in (3.8.4). ▲

The regularity conditions stated in (3.8.5) may sound too mathematical. But, the major consolation is that many standard and useful distributions in statistics belong to the exponential family and that the mathematical conditions stated in (3.8.5) are routinely satisfied.

3.9   Some Standard Probability Inequalities

In this section, we develop some inequalities which are frequently encountered in statistics. The introduction to each inequality is followed by a few examples. These inequalities apply to both discrete and continuous random variables.

3.9.1   Markov and Bernstein-Chernoff Inequalities

Theorem 3.9.1 (Markov Inequality) *Suppose that W is a real valued random variable such that $P(W \geq 0) = 1$ and $E(W)$ is finite. Then, for any fixed $\delta (> 0)$, one has:*

$$P(W \geq \delta) \leq \delta^{-1} E(W). \tag{3.9.1}$$

Proof Suppose that the event A stands for the set $[W \geq \delta]$, and then we can write

$$W \geq \delta \, I_A, \qquad\qquad (3.9.2)$$

where recall that I_A is the indicator function of A. One can check the validity of (3.9.2) as follows. On the set A, since one has $I_A = 1$, the rhs of (3.9.2) is δ and it is true then that $W \geq \delta$. On the set A^c, however, one has $I_A = 0$, so that the rhs of (3.9.2) is zero, but we have assumed that $W \geq 0$ w.p.1. Next, observe that I_A is a Bernoulli random variable with $p = P(A)$. Refer to the Example 2.2.3 as needed. Since, $W - \delta I_A \geq 0$ w.p.1, we have $E(W - \delta I_A) \geq 0$. But, $E(W - \delta I_A) = E(W) - \delta P(A)$ so that $E(W) - \delta P(A) \geq 0$, which implies that $P(A) \leq \delta^{-1} E(W)$. ∎

Example 3.9.1 Suppose that X is distributed as Poisson($\lambda = 2$). From the Markov inequality, we can claim, for example, that $P\{X \geq 1\} \leq (1)(2) = 2$. But this bound is useless because we know that $P\{X \geq 1\}$ lies between 0 and 1. Also, the Markov inequality implies that $P\{X \geq 2\} \leq (\frac{1}{2})(2) = 1$, which is again a useless bound. Similarly, the Markov inequality implies that $P\{X \geq 10\} \leq (\frac{1}{10})(2) = .2$, whereas the true probability, $P\{X \geq 10\} = 1 - .99995 = .00005$. There is a serious discrepancy between the actual value of $P\{X \geq 10\}$ and its upper bound. However, this discussion may not be very fair because after all the Markov inequality provides an upper bound for $P(W \geq \delta)$ without assuming much about the exact distribution of W. ▲

Example 3.9.2 Consider a random variable with its distribution as follows:

X values:	$-\frac{1}{7}$	1	10
Probabilities:	.7	.1	.2

One may observe that $E[X] = (-\frac{1}{7})(.7) + (1)(.1) + (10)(.2) = 2$ and $P\{X \geq 10\} = .2$. In this case, the upper bound for $P\{X \geq 10\}$ obtained from (3.9.1) is also .2, which happens to match with the exact value of $P\{X \geq 10\}$. But, that should not be the key issue. The point is this: The upper bound provided by the Markov inequality is distribution-free so that it works for a broad range of unspecified distributions. ▲

> Note that the upper bound for $P(W \geq \delta)$ given by (3.9.1) is useful only when it is smaller than unity.

> The upper bound given by (3.9.1) may appear crude, but even so, the Markov inequality will work like a charm in some derivations. The next Theorem 3.9.2 highlights one such application.

Theorem 3.9.2 (Bernstein-Chernoff Inequality) *Suppose that X is a real valued random variable whose mgf $M_X(t)$ is finite for some $t \in T \subseteq$*

$(0, \infty)$. *Then, for any fixed real number* a, *one has*

$$P\{X \geq a\} \leq \underset{t \in \mathcal{T}}{Inf} \left\{ e^{-ta} M_X(t) \right\} \qquad (3.9.3)$$

Proof Observe that for any fixed $t > 0$, the three sets $[X \geq a], [tX \geq ta]$ and $[e^{tX} \geq e^{ta}]$ are equivalent. Hence, for $t \in \mathcal{T}$, we can rewrite $P\{X \geq a\}$ as

$$P\{tX \geq ta\} = P\{e^{tX} \geq e^{ta}\} \leq e^{-ta} E\{e^{tX}\} = e^{-ta} M_X(t). \qquad (3.9.4)$$

using (3.9.1) with $W = e^{tX}, \delta = e^{ta}$. Now, since we are looking for an upper bound, it makes sense to pick the smallest upper bound available in this class. Thus, we take $\underset{t \in \mathcal{T}}{Inf} \left\{ e^{-ta} M_X(t) \right\}$ as the upper bound. ■

Example 3.9.3 Suppose that the random variable X has the exponential distribution with its pdf $f(x) = e^{-x} I(x > 0)$. The corresponding mgf is given by $M_X(t) = (1 - t)^{-1}$ for $t \in \mathcal{T} = (-\infty, 1)$. For any fixed $a > 1$, the Theorem 3.9.2 implies the following:

$$P\{X \geq a\} \leq \underset{0 < t < 1}{Inf} \left\{ e^{-ta} (1 - t)^{-1} \right\}. \qquad (3.9.5)$$

Now, let us denote $g(t) = e^{-ta}(1 - t)^{-1}$ and $h(t) = log(g(t))$ for $0 < t < 1$. It is easy to check that $h'(t) = -a + (1 - t)^{-1}, h''(t) = (1 - t)^{-2}$ so that $h'(t) = 0$ when $t \equiv t_0 = 1 - a^{-1}$ which belongs to $(0, 1)$. Also, $h''(t_0)$ is positive so that the function $h(t)$ attains the minimum at the point $t = t_0$. Refer to (1.6.27) as needed. Since $g(t)$ is a one-to-one monotone function of $h(t)$, we conclude that the function $g(t)$ attains the minimum at the point $t = t_0$. In other words, we can rewrite (3.9.5) as

$$P\{X \geq a\} \leq e^{-t_0 a} (1 - t_0)^{-1} = ae^{1-a}. \qquad (3.9.6)$$

Does the upper bound in (3.9.6) lie between 0 and 1? In order to answer this question, next look at the function $m(x) = (x - 1) - log(x)$ for $x \geq 1$. Obviously, $m(1) = 0$ and $m'(x) = 1 - x^{-1} \geq 0$, where equality holds if and only if $x = 1$. That is, for all $x > 1$, we have an increasing function $m(x)$. But, since $m(1) = 0$, we claim that $m(x) > 0$ for $x > 1$. In other words, for $a > 1$, we have $log(a) < a - 1$ so that $a < e^{a-1}$. Hence, the upper bound given by (3.9.6) is a number which lies between zero and one. ▲

> The types of bounds given by (3.9.5) are customarily used in studying the rate of convergence of tail area probabilities.

Example 3.9.4 Suppose that X is a random variable whose mgf $M_X(t)$ is finite for some $t \in \mathcal{T} \subseteq (-\infty, 0)$. Then, it follows from the Theorem 3.9.2 that for any fixed real number a, one can claim: $P\{X \leq a\} \leq \underset{t \in (-\infty, 0)}{Inf}$ $\{e^{-ta} M_X(t)\}$. Its verification is left as the Exercise 3.9.1. ▲

The following section provides yet another application of the Markov inequality.

3.9.2 Tchebysheff's Inequality

This inequality follows from a more general inequality (Theorem 3.9.4) which is stated and proved a little later. We take the liberty to state the simpler version separately for its obvious prominence in the statistical literature.

Theorem 3.9.3 (Tchebysheff's Inequality) *Suppose that X is a real valued random variable with the finite second moment. Let us denote its mean μ and variance $\sigma^2(> 0)$. Then, for any fixed real number $\varepsilon(> 0)$, one has*

$$P\{|X - \mu| \geq \varepsilon\} \leq \sigma^2/\varepsilon^2. \qquad (3.9.7)$$

We know that $P\{|X - \mu| < k\sigma\} = 1 - P\{|X - \mu| \geq k\sigma\}$. Thus, with $k > 0$, if we substitute $\varepsilon = k\sigma$ in (3.9.7), we can immediately conclude:

$$P\{|X - \mu| < k\sigma\} \geq 1 - k^{-2}. \qquad (3.9.8)$$

In statistics, sometimes (3.9.8) is also referred to as the Tchebysheff's inequality. Suppose we denote $p_k = P\{|X - \mu| < k\sigma\}$. Again, (3.9.7) or equivalently (3.9.8) provide distribution-free bounds for some appropriate probability. Yet, let us look at the following table:

Table 3.9.1. Values of p_k and the Tchebysheff's Lower Bound (3.9.8)

	$k = 1$	$k = 2$	$k = 3$	$k = 4$
Tchebysheff's Bound	0	$\frac{3}{4} = .75$	$\frac{8}{9} \approx .88889$	$\frac{15}{16} = .9375$
p_k: X is $N(0,1)$	.68268	.95450	.99730	.99994

In the case of the standard normal distribution, the Tchebysheff's lower bound for p_k appears quite reasonable for $k = 3, 4$. In the case $k = 1$, the Tchebysheff's inequality provides a trivial bound whatever be the distribution of X.

Theorem 3.9.4 *Suppose that X is a real valued random variable such that with some $r > 0$ and $\tau \in T(\subseteq \Re)$, one has $\psi_r = E\{|X - \tau|^r\}$ which is finite. Then, for any fixed real number $\varepsilon(> 0)$, one has*

$$P\{|X - \tau| \geq \varepsilon\} \leq \psi_r/\varepsilon^r. \tag{3.9.9}$$

Proof Note that $P\{|X - \tau| \geq \varepsilon\} = P\{W \geq \varepsilon^r\}$ where $W = |X - \tau|^r$. Now, the inequality (3.9.9) follows immediately by invoking the Markov inequality. ∎

> The Tchebysheff's inequality follows immediately
> from (3.9.9) by substituting $r = 2$ and $\tau = \mu$.

3.9.3   Cauchy-Schwarz and Covariance Inequalities

If we have independent random variables X_1 and X_2, then we know from the Theorem 3.5.1 that $E[X_1X_2] = E[X_1]E[X_2]$. But, if X_1 and X_2 are dependent, then it is not always so simple to evaluate $E[X_1X_2]$. The Cauchy-Schwarz inequality allows us to split $E[X_1X_2]$ in the form of an upper bound having two separate parts, one involving only X_1 and the other involving only X_2.

Theorem 3.9.5 (Cauchy-Schwarz Inequality) *Suppose that we have two real valued random variables X_1 and X_2, such that $E[X_1^2], E[X_2^2]$ and $E[X_1X_2]$ are all finite. Then, we have*

$$E^2[X_1X_2] \leq E[X_1^2]E[X_2^2]. \tag{3.9.10}$$

In (3.9.10), the equality holds if and only if $X_1 = kX_2$ w.p.1 for some constant k.

Proof First note that if $E[X_2^2] = 0$, then $X_2 = 0$ w.p.1 so that both sides of (3.9.10) will reduce to zero. In other words, (3.9.10) holds when $E[X_2^2] = 0$.

Now we assume that $E[X_2^2] > 0$. Let λ be any real number. Then we can write

$$E[(X_1 + \lambda X_2)^2]$$

$$= \left\{ \lambda\sqrt{E[X_2^2]} + \frac{E[X_1X_2]}{\sqrt{E[X_2^2]}} \right\}^2 + \frac{E[X_1^2]E[X_2^2] - E^2[X_1X_2]}{E[X_2^2]}. \tag{3.9.11}$$

But note that $(X_1 + \lambda X_2)^2$ is a non-negative random variable whatever be λ, and so $E[(X_1 + \lambda X_2)^2] \geq 0$ whatever be λ. If we substitute $\lambda \equiv \lambda_0 =$

$-E[X_1 X_2]/E[X_2^2]$ in the last step of (3.9.11), we get

$$0 \le E[(X_1 + \lambda_0 X_2)^2] = \frac{E[X_1^2]E[X_2^2] - E^2[X_1 X_2]}{E[X_2^2]},$$

that is $E[X_1^2]E[X_2^2] - E^2[X_1 X_2] \ge 0$ since $E[X_2^2] > 0$. Thus, (3.9.10) has been proven.

Next, let us suppose that equality holds in (3.9.10), that is, $E[X_1^2]E[X_2^2] - E^2[X_1 X_2] = 0$. Then, from (3.9.10) we conclude that

$$E[(X_1 + \lambda X_2)^2] = \left\{ \lambda\sqrt{E[X_2^2]} + \frac{E[X_1 X_2]}{\sqrt{E[X_2^2]}} \right\}^2 \quad \text{for all } \lambda \in \Re. \quad (3.9.12)$$

It should be clear from (3.9.12) that with $\lambda \equiv \lambda_0 = -E[X_1 X_2]/E[X_2^2]$, we have

$$E[(X_1 + \lambda_0 X_2)^2] = 0,$$

and again since $(X_1 + \lambda_0 X_2)^2$ is a non-negative random variable w.p.1, one can conclude that $X_1 + \lambda_0 X_2 = 0$ w.p.1. Hence, $X_1 = kX_2$ w.p.1 with some real number $k = -\lambda_0$.

Let us suppose that $X_1 = kX_2$ w.p.1 so that $E[X_1 X_2] = kE[X_2^2]$ and $E[X_1^2]E[X_2^2] = k^2 E^2[X_2^2]$. Hence, one immediately has $E^2[X_1 X_2] - E[X_1^2]E[X_2^2] = \{kE[X_2^2]\}^2 - k^2 E^2[X_2^2] = 0$. Now the proof is complete. ∎

Theorem 3.9.6 (Covariance Inequality) *Suppose that we have two real valued random variables X_1, X_2, such that $E[X_1^2], E[X_2^2]$ and $E[X_1 X_2]$ are all finite. Then, we have*

$$Cov^2(X_1, X_2) \le V[X_1]V[X_2]. \quad (3.9.13)$$

In (3.9.13), the equality holds if and only if $X_1 = a + bX_2$ w.p.1 for some constants a and b.

Proof Recall from (3.4.1) that $Cov(X_1, X_2) = E\{(X_1 - \mu_1)(X_2 - \mu_2)\}$ where $\mu_i = E[X_i], i = 1, 2$. Let us denote $U_i = X_i - \mu_i, i = 1, 2$, and then by the Cauchy-Schwarz inequality (3.9.10) we have

$$Cov^2(X_1, X_2) = E^2(U_1 U_2) \le E(U_1^2)E(U_2^2) = V[X_1]V[X_2], \quad (3.9.14)$$

which verifies (3.9.13). The covariance inequality will become an equality *if and only if* we have equality throughout (3.9.14), that is *if and only if* $E^2(U_1 U_2) = E(U_1^2)E(U_2^2)$. From the Cauchy-Schwarz inequality it then follows that we will have equality in (3.9.13) *if and only if* $U_1 = kU_2$ w.p.1 with some constant k, which will be equivalent to the claim that $X_1 = a + bX_2$ w.p.1. where $a = \mu_1 - k\mu_2$ and $b = k$. ∎

Earlier we had seen applications of this covariance inequality in proving the parts (ii)-(iii) in the Theorem 3.4.2 where we had shown that $\left|\rho_{X_1,X_2}\right| \le$ 1 whereas $\left|\rho_{X_1,X_2}\right| = 1$ if and only if $X_1 = a + bX_2$ w.p.1.

Example 3.9.5 (Examples 3.2.4, 3.2.5, 3.4.1 Continued) For the two random variables on hand, recall that we had $V(X_1) = 6.4971$ and $V(X_2) = .6825$. Thus, $Cov^2(X_1, X_2) \le (6.4971)(.6825) \approx 4.4343$. But, we knew that $Cov(X_1, X_2) = -0.2415$. ▲

Example 3.9.6 (Example 3.4.3, 3.4.6 Continued) For the two random variables on hand, recall that we had $V(X_1) = \frac{3}{80}$ and $V(X_2) = \frac{19}{320}$. Thus, $Cov^2(X_1, X_2) \le (\frac{3}{80})(\frac{19}{320}) = \frac{57}{25600} \approx .00223$. But, we had found earlier that $Cov(X_1, X_2) = \frac{3}{160} \approx .01875$. ▲

Example 3.9.7 Suppose that X_2 is distributed as $N(0, 5)$ and let $X_1 = 2X_2 - 10^{-3}X_2^2 + 1$. Note that $V[X_2] = 5$, but

$$
\begin{aligned}
V[X_1] \\
&= 4V[X_2] + 10^{-6}V[X_2^2] - 4 \times 10^{-3}Cov(X_2, X_2^2), \\
&\quad \text{using part (iii) from Theorem 3.4.3} \\
&= 4V[X_2] + 10^{-6}\{E[X_2^4] - E^2[X_2^2]\}, \text{ since} \qquad (3.9.15)\\
&\quad Cov(X_2, X_2^2) = E[X_2^3] - E[X_2]E[X_2^2] = 0 \\
&= 4(5) + 10^{-6}\{3(5)^2 - (5)^2\} \\
&= 20.00005.
\end{aligned}
$$

The covariance inequality (3.9.13) would imply that

$$Cov^2(X_1, X_2) \le (5)(20.00005) = 100.00025.$$

However, we know that $Cov(X_1, X_2) = Cov(X_2, X_1)$ and hence, by appealing to the Theorem 3.4.3 (ii), we can express $Cov(X_2, X_1)$ as

$$
\begin{aligned}
Cov(X_2, 2X_2 &- 10^{-3}X_2^2 + 1) \\
&= 2Cov(X_2, X_2) - 10^{-3}Cov(X_2, X_2^2) \\
&= 2V[X_2], \text{ since } Cov(X_2, X_2^2) = 0 \text{ from } (3.9.15) \\
&= 10.
\end{aligned}
$$

The upper bound for $Cov^2(X_1, X_2)$ given by the covariance inequality was 100.00025 whereas the exact value of $Cov^2(X_1, X_2)$ is 100. This upper bound and the actual value of $Cov^2(X_1, X_2)$ are almost indistinguishable because, for all practical purposes, X_1 is a linear function of X_2. ▲

3.9.4 Jensen's and Lyapunov's Inequalities

Consider a real valued random variable X with finite variance. It is clear that

$$E[X^2] - E^2[X] = V[X] \geq 0$$

which would imply that $E[X^2] \geq E^2[X]$. That is, if we let $f(x) = x^2$, then we can claim

$$E[f(X)] \geq f(E[X]). \tag{3.9.16}$$

The question is this: Can we think of a large class of functions $f(.)$ for which the inequality in (3.9.16) would hold? In order to move in that direction, we start by looking at the rich class of *convex functions* $f(x)$ of a single real variable x.

Definition 3.9.1 *Consider a function* $f : \Re \to \Re$. *The function* f *is called convex if and only if*

$$f(\alpha u + (1 - \alpha)v) \leq \alpha f(u) + (1 - \alpha)f(v) \tag{3.9.17}$$

for all $u, v \in \Re$ *and* $0 \leq \alpha \leq 1$. *The function* f *is called concave if and only if* $-f$ *is convex.*

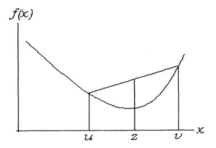

Figure 3.9.1. A Convex Function $f(x)$

The situation has been depicted in the Figure 3.9.1. The definition of the *convexity* of a function demands the following geometric property: Start with *any* two arbitrary points u, v on the real line and consider the *chord* formed by joining the two points $(u, f(u))$ and $(v, f(v))$. Then, the function f evaluated at any intermediate point z, such as $\alpha u + (1 - \alpha)v$ with $0 \leq$

$\alpha \le 1$, will lie under the chord.

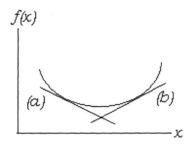

Figure 3.9.2. A Convex Function $f(x)$ and Tangent Lines $(a), (b)$

We may also look at a convex function in a slightly different way. A function f is convex if and only if the curve $y = f(x)$ never goes below the tangent of the curve drawn at the point $(z, f(z))$, for any $z \in \Re$. This has been depicted in the Figure 3.9.2 where we have drawn two tangents labelled as (a) and (b).

In the Figure 3.9.3, we have shown two curves, $y = f(x)$ and $y = g(x), x \in \Re$. The function $f(x)$ is convex because it lies above its tangents at any point whatsoever. But, the function $g(x)$ is certainly not convex because a

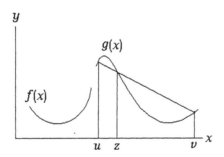

Figure 3.9.3. Two Functions: $f(x)$ Is Convex, $g(x)$ Is Not Convex

part of the curve $y = g(x)$ lies below the chord as shown which is a violation of the requirement set out in (3.9.17). In the Figure 3.9.3, *if* we drew the tangent to the curve $y = g(x)$ at the point $(u, g(u))$, then a part of the curve will certainly lie below this tangent!

Example 3.9.8 Consider the following functions: (i) $f(x) = x^2$ for $x \in \Re$; (ii) $f(x) = e^x$ for $x \in \Re$; (iii) $f(x) = \max(0, x)$ for $x \in \Re$; and (iv) $f(x) = log(x)$ for $x \in \Re^+$. The reader should verify geometrically or otherwise that the first three functions are convex, while the fourth one is concave. ▲

Suppose that $f(x)$ is twice differentiable at all points $x \in \Re$. Then, $f(x)$ is *convex* (*concave*) if $d^2 f(x)/dx^2$ is *positive* (*negative*) for all $x \in \Re$. This is a *sufficient condition*. Look at the Example 3.9.9.

$$(3.9.18)$$

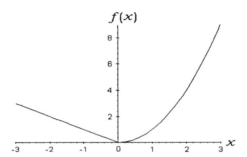

Figure 3.9.4. Plot of $f(x)$ Defined in (3.9.19)

Example 3.9.9 The convexity or concavity of the functions cited in parts (i), (ii) and (iv) of the Example 3.9.8 can be easily checked by verifying the *sufficient condition* given in (3.9.18). But, a function $f(x)$ may not even be differentiable at all points $x \in \Re$, and yet it may be convex (or concave). The function cited in part (iii) of the Example 3.9.8 is of this type. Let us define another function

$$f(x) = \begin{cases} -x & \text{if } x \le 0 \\ x^2 & \text{if } x > 0, \end{cases} \qquad (3.9.19)$$

which has been plotted in the Figure 3.9.4. One should verify that at the point $x = 0$, the left hand derivative of $f(x)$ is -1 whereas the right hand derivative of $f(x)$ is zero. In other words, the function $f(x)$ is not differentiable at the point $x = 0$, that is the condition laid out in (3.9.18) is not useful here.

But, after examining the plot of $f(x)$ in the Figure 3.9.4, it becomes apparent that $f(x)$ is indeed a convex function. One should, however, check this claim more formally. ▲

Theorem 3.9.7 (Jensen's Inequality) *Suppose that X is a real valued random variable and let $f(x), x \in \Re$ be a convex function. Assume that $E[X]$ is finite. Then, one has*

$$E[f(X)] \ge f(E[X]). \qquad (3.9.20)$$

Proof Note that if $E[f(X)]$ is infinite, then the required inequality certainly holds. So, let us assume now that $E[f(X)]$ is finite. Since $f(x)$ is convex, the curve $y = f(x)$ must lie above the tangents at all points $(u, f(u))$ for any $u \in \Re$. With $u = E[X]$, consider specifically the point $(u, f(u))$ at which the equation of the tangent line would be given by $y = f(E[X]) + b\{x - E[X]\}$ for some appropriate b. But, then we can claim that

$$f(x) \geq f(E[X]) + b\{x - E[X]\} \text{ for all } x \in \Re. \tag{3.9.21}$$

Thus, using (3.9.21), we have

$$E[f(X)] \geq E\left[f(E(X)) + b\{X - E(X)\}\right] = f(E(X)) + b\{E(X) - E(X)\},$$

which reduces to $f(E(X))$. Thus, we have the desired result. ∎

> In the statement of the Jensen's inequality (3.9.20), equality holds only when $f(x)$ is linear in $x \in \Re$. Also, the inequality in (3.9.20) gets reversed when $f(x)$ is concave.

Example 3.9.10 Suppose that X is distributed as Poisson$(\lambda), \lambda > 0$. Then, using the Jensen's inequality, for example, we immediately claim that $E[X^3] > \lambda^3$. This result follows because the function $f(x) = x^3, x \in \Re^+$ is convex and $E[X] = \lambda$. ▲

Example 3.9.11 Suppose that X is distributed as $N(-1, \sigma^2), \sigma > 0$. Then, for example, we have $E[|X|] > 1$. This result follows immediately from the Jensen's inequality because the function $f(x) = |x|, x \in \Re$ is convex and $|E[X]| = |-1| = 1$. ▲

Example 3.9.12 Suppose that X is distributed as $N(\mu, \sigma^2), -\infty < \mu < \infty, \sigma > 0$. Then, for example, we have $E[(X - \mu)^4] \geq \left\{E[(X - \mu)^2]\right\}^2$, by the Jensen's inequality with $f(x) = x^2, x \in \Re^+$ so that one can claim $E[(X - \mu)^4] \geq (\sigma^2)^2 = \sigma^4$. ▲

Example 3.9.13 Suppose that one is proctoring a makeup examination for two students who got started at the same time. Let X_i be the time to complete the examination for the ith student, $i = 1, 2$. Let $X = \max(X_1, X_2)$, the duration of time the proctor must be present in the room. Then, one has

$$\begin{aligned}
E[X] \\
&= E[\max(0, X_2 - X_1) + X_1] \\
&= E[\max(0, X_2 - X_1)] + E[X_1] \\
&\geq \max(0, E[X_2 - X_1]) + E[X_1], \text{ in view of the Jensen's} \\
&\qquad \text{inequality, since } f(x) = \max(0, x), x \in \Re \text{ is convex} \\
&= \max(0, E[X_2] - E[X_1]) + E[X_1],
\end{aligned}$$

so that $E[X] \geq \max(E[X_1], E[X_2])$. ▲

Example 3.9.14 Suppose that X is exponentially distributed with mean $\lambda > 0$. Then, what can we say about $E[X^{1/2}]$? Look at the function $f(x) = \sqrt{x}$ for $x > 0$. This function is *concave* because $f''(x) = -\frac{1}{4}x^{-3/2} < 0$. Hence, from the Jensen's inequality we can immediately claim that $E[X^{1/2}] \leq \sqrt{E[X]} = \sqrt{\lambda}$. Similarly, one can also show that $E[log(X)] \leq log(E[X]) = log(\lambda)$. ▲

What follows is yet another nice application of the Jensen's inequality.

Theorem 3.9.8 (Lyapunov's Inequality) *Suppose that X is a real valued random variable. Let $m_r = E^{1/r}[|X|^r]$. Then, m_r is increasing for $r \geq 1$.*

Proof We want to show that

$$m_r \geq m_s \text{ if } r > s \geq 1. \tag{3.9.22}$$

Observe that

$$E[|X|^r] = E\left[\{|X|^s\}^{r/s}\right] \geq [E\{|X|^s\}]^{r/s},$$

using Jensen's inequality since $f(x) = x^{r/s}, x > 0$, is convex. Thus, one has

$$E^{1/r}[|X|^r] \geq E^{1/s}[|X|^s].$$

which is the desired result. ∎

3.9.5 Hölder's Inequality

In the Cauchy-Schwarz inequality (3.9.10), the upper bound consisted of the product of $E[X^2]$ and $E[Y^2]$. However, what if we have a situation like this: For the random variable X, higher than second moments may be assumed finite, but for the random variable Y, lower than second moments may be assumed finite. Under this scenario, the Cauchy-Schwarz inequality does not help much to obtain an upper bound for $E^2[X_1 X_2]$. The following inequality is more flexible. For brevity, we do not supply its proof.

Theorem 3.9.9 (Hölder's Inequality) *Let X and Y be two real valued random variables. Then, with $r > 1, s > 1$ such that $r^{-1} + s^{-1} = 1$, one has*

$$|E[XY]| \leq E^{1/r}[|X|^r]E^{1/s}[|Y|^s],$$

provided that $E[|X|^r]$ and $E[|Y|^s]$ are finite.

Example 3.9.15 Suppose that we have two *dependent* random variables X and Y where X is distributed as $N(0,1)$, but the pdf of Y is given by

$g(y) = \frac{5}{3}y^{-8/3}I(1 < y < \infty)$. Now, by direct integration, one can check that $E[Y^{4/3}]$ is finite but $E[Y^2]$ is not. Suppose that we want to get an upper bound for $|E[XY]|$. The Cauchy-Schwarz inequality does not help because $E[Y^2]$ is not finite even though $E[X^2]$ is. Let us try and apply the Hölder's inequality with $r = 4$ and $s = \frac{4}{3}$ in order to obtain

$$|E[XY]| \leq E^{1/4}[|X|^4]E^{3/4}[|Y|^{4/3}] = 5^{3/4} \approx 3.3437.$$

Note that this upper bound is finite whatever be the nature of dependence between these two random variables X and Y. ▲

3.9.6 Bonferroni Inequality

We had mentioned this inequality in the Exercise 1.3.5, part (ii), for two events only. Here we state it more generally.

Theorem 3.9.10 (Bonferroni Inequality) *Consider a sample space* **S** *and suppose that* $\mathcal{B}$ *is the Borel sigma-field of subsets of* **S**. *Let* $A_1, ..., A_k$ *be events, that is these belong to* $\mathcal{B}$. *Then,*

$$P\{\cap_{i=1}^k A_i\} \geq \Sigma_{i=1}^k P(A_i) - (k-1).$$

Proof In Chapter 1, we had proved that

$$P(A_1 \cup A_2) = P(A_1) + P(A_2) - P(A_1 \cap A_2),$$

so that we can write

$$
\begin{aligned}
P(A_1 \cap A_2) &= P(A_1) + P(A_2) - P(A_1 \cup A_2) \\
&\geq P(A_1) + P(A_2) - 1.
\end{aligned}
\tag{3.9.23}
$$

This is valid because $P(A_1 \cup A_2) \leq 1$. Similarly, we can write

$$
\begin{aligned}
P(A_1 \cap A_2 \cap A_3) \\
\geq P(A_1 \cap A_2) + P(A_3) - 1, \text{ using (3.9.23)} \\
\geq P(A_1) + P(A_2) - 1 + P(A_3) - 1, \\
\text{using (3.9.23) again} \\
= P(A_1) + P(A_2) + P(A_3) - 2.
\end{aligned}
\tag{3.9.24}
$$

So, we first verified the desired result for $k = 2$ and assuming this, we then proved the same result for $k = 3$. The result then follows by the mathematical induction. ∎

The Exercise 3.9.10 gives a nice application combining the Tchebysheff's inequality and Bonferroni inequality.

3.9.7 Central Absolute Moment Inequality

Let X be a real valued random variable from some population having its pmf or pdf $f(x)$. Suppose that $E[X]$, that is the population mean, is μ. In statistics, $E\left[\mid X - \mu \mid^k\right]$ is customarily called the kth central absolute moment of $X, k = 1, 2, \ldots$. For example, suppose that X has the following distribution:

X values:	-1	0	2	2.5
Probabilities:	.1	.2	.4	.3

One can check that $E[X] = -.1 + 0 + .8 + .75 = 1.45$ which is μ. Now let us fix $k = 3$ and evaluate the 3rd central absolute moment of X. We have

$$E\left[\mid X - 1.45 \mid^3\right]$$
$$= \mid -1 - 1.45 \mid^3 (.1) + \mid 0 - 1.45 \mid^3 (.2)$$
$$+ \mid 2 - 1.45 \mid^3 (.4) + \mid 2.5 - 1.45 \mid^3 (.3)$$
$$= 1.4706125 + .609725 + .06655 + .3472875$$
$$= 2.494175.$$

Now, one may ask: how is the 3rd central absolute moment of X any different from the 3rd central moment of X? One should check that

$$E\left[\{X - 1.45\}^3\right] = (-1 - 1.45)^3(.1) + (0 - 1.45)^3(.2)$$
$$+ (2 - 1.45)^3(.4) + (2.5 - 1.45)^3(.3)$$
$$= -1.6665.$$

One thing should be clear. The kth central absolute moment of X should always be non-negative for any $k = 1, 2, \ldots$.

Now consider this scenario. Let $X_1, \ldots, X_n$ be real valued random samples from a population having its pmf or pdf $f(x)$. Let the population mean be μ. One may be tempted to evaluate the average *magnitude* of the discrepancy between the sample mean $\overline{X}(= n^{-1}\Sigma_{i=1}^n X_i), n \geq 1$, and the population mean μ which will amount to $E\left[\mid \overline{X} - \mu \mid\right]$. One may even like to evaluate $E\left[\mid \overline{X} - \mu \mid^k\right]$ for $k = 1, 2, \ldots$. But, then we need the *exact distribution* of $\overline{X}$ to begin with. From Chapter 4 it will be clear that the determination of the exact distribution of $\overline{X}$ can be hard, even for some of the simplest looking population pmf's or pdf's $f(x)$.

The beauty and importance of the next inequality will be appreciated more once we realize that it provides an upper bound for $E\left[\mid \overline{X} - \mu \mid^k\right]$

even though we may not know the exact distribution of $\overline{X}, k = 1, 2, \ldots$. The following result actually gives an upper bound for $E\left[|\overline{X} - \mu|^{2\xi}\right]$ with $\xi \geq \frac{1}{2}$.

Theorem 3.9.11 (Central Absolute Moment Inequality) *Suppose that we have iid real valued random variables $X_1, X_2, \ldots$ having the common mean μ. Let us also assume that $E[|X_1|^{2\xi}] < \infty$ for some $\xi \geq \frac{1}{2}$. Then, we have*

$$E\left[|\overline{X} - \mu|^{2\xi}\right] \leq kn^{-\tau},$$

where k does not depend on n, and $\tau = 2\xi - 1$ or ξ according as $\frac{1}{2} \leq \xi < 1$ or $\xi \geq 1$ respectively.

The methods for exact computations of central moments for the sample mean $\overline{X}$ were systematically developed by Fisher (1928). The classic textbook of Cramér (1946a) pursued analogous techniques extensively. The particular inequality stated here is the special case of a more general large deviation inequality obtained by Grams and Serfling (1973) and Sen and Ghosh (1981) in the case of Hoeffding's (1948) *U-statistics*. A sample mean $\overline{X}$ turns out to be one of the simplest U-statistics.

3.10   Exercises and Complements

3.2.1 (Example 3.2.2 Continued) Evaluate $f_2(i)$ for $i = -1, 0, 1$.

3.2.2 (Example 3.2.4 Continued) Evaluate $E[X_1 \mid X_2 = x_2]$ where $x_2 = 1, 2$.

3.2.3 (Example 3.2.5 Continued) Check that $E[X_1^2 X_2] = 16.21$. Also evaluate $E[X_1^2 X_2^2], E[|X_1| X_2^3], E[|X_1|^{1/2} X_2^3]$ and $E[X_1^2 X_2(1 - X_2)]$.

3.2.4 Suppose that the random variables X_1 and X_2 have the following joint distribution.

X_2 values	X_1 values		
	0	1	2
0	0	3/15	3/15
1	2/15	6/15	0
2	1/15	0	0

Using this joint distribution,

(i)   find the marginal pmf's for X_1 and X_2;
(ii)   find the conditional pmf's $f_{1|2}(x_1)$ and $f_{2|1}(x_2)$;
(iii)   evaluate $E[X_1 \mid X_2 = x_2]$ and $E[X_2 \mid X_1 = x_1]$;
(iv)   evaluate $E[X_1^2 X_2], E[X_1^2 X_2^2], E[X_1 X_2^3]$ and $E[X_1^2 X_2(1 - X_2)]$.

3.2.5 Prove the Theorem 3.2.2.

3.2.6 Prove the Theorem 3.2.3.

3.2.7 Consider an urn containing sixteen marbles of same size and weight of which two are red, five are yellow, three are green, and six are blue. The marbles are all mixed and then randomly one marble is picked from the urn and its color is recorded. Then this marble is returned to the urn and again all the marbles are mixed, followed by randomly picking a marble from the urn and its color recorded. Again this marble is also returned to the urn and the experiment continues in this fashion. This process of selecting marbles is called *sampling with replacement*. After the experiment is run n times, suppose that one looks at the number of red (X_1), yellow (X_2), green (X_3) and blue (X_4) marbles which are selected.

(i)   What is the joint distribution of (X_1, X_2, X_3, X_4)?
(ii)   What is the mean and variance of X_3?
(iii)   If $n = 10$, what is the joint distribution of (X_1, X_2, X_3, X_4)?
(iv)   If $n = 15$, what is the conditional distribution of (X_1, X_2, X_3) given that $X_4 = 5$?
(v)   If $n = 10$, calculate $P\{X_1 = 1 \cap X_2 = 3 \cap X_3 = 6 \cap X_4 = 0\}$.

{*Hint*: In parts (i) and (v), try to use (3.2.8). In part (ii), can you use (3.2.13)? For part (iii)-(iv), use the Theorem 3.2.2-3.2.3 respectively.}

3.2.8 The mgf of a random vector $\mathbf{X} = (X_1, ..., X_k)$, denoted by $M_{\mathbf{X}}(\mathbf{t})$ with $\mathbf{t} = (t_1, ..., t_k)$, is defined as $E[exp\{t_1 X_1 + ... + t_k X_k\}]$. Suppose that $\mathbf{X}$ is distributed as $\text{Mult}_k(n, p_1, ..., p_k)$. Then,

(i)   show that $M_{\mathbf{X}}(\mathbf{t}) = \{p_1 e^{t_1} + ... + p_k e^{t_k}\}^n$;
(ii)   find the mgf $M_U(t)$ of $U = X_1 + X_3$ by substituting $t_1 = t_3 = t$ and $t_i = 0$ for all $i \neq 1, 3$ in the part (i). What is the distribution of U?
(iii)   use the procedure from part (ii) to find the mgf of the sum of any subset of random variables from $\mathbf{X}$.

3.2.9 Suppose that the random variables X_1 and X_2 have the following joint pmf:

$$f(x_1, x_2) = \binom{4}{x_1}\binom{3}{x_2}\binom{2}{3 - x_1 - x_2} / \binom{9}{3},$$

where x_1, x_2 are integers, $0 \leq x_1 \leq 3, 0 \leq x_2 \leq 3$, but $1 \leq x_1 + x_2 \leq 3$. Then,

(i)   find the marginal pmf's for X_1 and X_2;
(ii)   find the conditional pmf's $f_{1|2}(x_1)$ and $f_{2|1}(x_2)$;
(iii)   evaluate $E[X_1 \mid X_2 = x_2]$ and $E[X_2 \mid X_1 = x_1]$;
(iv)   evaluate $E[X_1 X_2], E[X_1^2 X_2^2]$, and $E[X_1^2 X_2(1 - X_2)]$.

3.2.10 Suppose that $\mathbf{X} = (X_1, ..., X_5)$ has the $\mathrm{Mult}_5(n, p_1, ..., p_5)$ with $0 < p_i < 1, i = 1, ..., 5, \Sigma_{i=1}^5 p_i = 1$. What is the distribution of $\mathbf{Y} = (Y_1, Y_2, Y_3)$ where $Y_1 = X_1 + X_2, Y_2 = X_3, Y_3 = X_4 + X_5$? {*Hint*: Use the Theorem 3.2.2.}

3.2.11 Suppose that $\mathbf{X} = (X_1, X_2, X_3)$ has the $\mathrm{Mult}_3(n, p_1, p_2, p_3)$ with $0 < p_i < 1, i = 1, 2, 3, \Sigma_{i=1}^3 p_i = 1$. Show that the conditional distribution of X_1 given that $X_1 + X_2 = r$ is $\mathrm{Bin}(r, p_1/(p_1 + p_2))$. {*Hint*: Use the Theorem 3.2.3.}

3.2.12 Suppose that X_1 and X_2 have the joint pmf

$$f(x_1, x_2) = p^2 q^{x_2}, x_1 = 0, 1, 2, ..., x_2 \text{ and } x_2 = 0, 1, 2, ...$$

with $0 < p < 1, q = 1 - p$. Find the marginal distributions of X_1 and X_2 and evaluate $P\{X_2 - X_1 \le 1\}$.

3.2.13 Suppose that X_1 and X_2 have the joint pmf

$$f(x_1, x_2) = \tfrac{1}{21}(x_1 + x_2), x_1 = 1, 2, 3 \text{ and } x_2 = 1, 2$$

(i)   Show that $f_1(x_1) = \frac{1}{21}(2x_1 + 3)$ for $x_1 = 1, 2, 3$;
(ii)   Show that $f_2(x_2) = \frac{1}{7}(x_2 + 2)$ for $x_2 = 1, 2$;
(iii)   Evaluate $P\{X_1 \le 2\}$ and $P\{X_2 < 2\}$.

3.2.14 Suppose that X_1 and X_2 have the joint pdf given by

$$f(x_1, x_2) = \begin{cases} 1 & \text{if } 0 < x_1, x_2 < 1 \\ 0 & \text{elsewhere.} \end{cases}$$

The surface represented by this pdf is given in the Figure 3.10.1. Show that $P\{X_1 X_2 > a\} = 1 - a + a \log(a)$ for any $0 < a < 1$. {*Hint*: Note that

$$P\{X_1 X_2 > a\} = \int_{x_1=a}^{1} \int_{x_2=a/x_1}^{1} f(x_1, x_2) dx_2 dx_1.\}$$

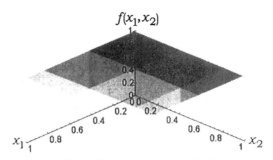

Figure 3.10.1. Plot of the PDF from the Exercise 3.2.14

3.3.1 (Examples 3.3.1-3.3.2 Continued) Evaluate $E(X_1)$ and $V(X_1)$. Also, evaluate $\mu_{1|2}$ and $\sigma_{1|2}^2$.

3.3.2 (Example 3.3.3 Continued) Answer the following questions.
(i)   Evaluate $E(X_i)$ and $V(X_i), i = 1, 2$ using (3.3.19);
(ii)   Evaluate $E[X_1(1 - X_1)], E[X_1^2(1 - X_2)^3]$ and $E[(X_1 + X_2)^2]$;
(iii)   Show that $f_{i|j}(x_i) = f_i(x_i)$ for $i \neq j = 1, 2$.

3.3.3 (Example 3.3.5 Continued) Evaluate $E(X_i)$ and $V(X_i), i = 1, 2$. Evaluate $\mu_{i|j}$ and $\sigma_{i|j}^2$ for $i \neq j, i, j = 1, 2$. Also evaluate

$$E[X_1(1 - X_1)], E[X_1^2(1 - X_2)^3] \text{ and } E[(X_1 + X_2)^2].$$

3.3.4 Let c be a positive constant such that X_1 and X_2 have the joint pdf given by

$$f(x_1, x_2) = \begin{cases} cx_1 x_2^2 & \text{if } 0 < x_1, x_2 < 2 \\ 0 & \text{elsewhere.} \end{cases}$$

Find the value of c. Derive the expressions of $f_1(x_1), f_2(x_2), f_{1|2}(x_1)$, and $f_{2|1}(x_2)$. Evaluate $E(X_i)$ and $V(X_i), i = 1, 2$. Evaluate $\mu_{i|j}$ and $\sigma_{i|j}^2$ for $i \neq j, i, j = 1, 2$. Also evaluate $E[X_1(1 - X_1)], E[X_1^2(1 - X_2)^3]$ and $E[(X_1 +$

$X_2)^2]$. The surface represented by this pdf is given in the Figure 3.10.2.

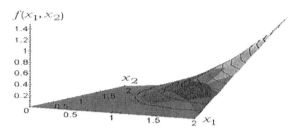

Figure 3.10.2. Plot of the PDF from the Exercise 3.3.4 with $c = \frac{3}{16}$

3.3.5 Let c be a positive constant such that X_1 and X_2 have the joint pdf given by

$$f(x_1, x_2) = \begin{cases} cexp\{-2x_1 - \frac{1}{2}x_2\} & \text{if } 0 < x_1, x_2 < \infty \\ 0 & \text{elsewhere.} \end{cases}$$

Find the value of c. Derive the expressions of $f_1(x_1), f_2(x_2), f_{1|2}(x_1)$, and $f_{2|1}(x_2)$. Does either of X_1, X_2 have a standard distribution which matches with one of those distributions listed in Section 1.7? Evaluate $E(X_i)$ and $V(X_i), i = 1, 2$. Evaluate $\mu_{i|j}$ and $\sigma^2_{i|j}$ for $i \neq j, i, j = 1, 2$. Also evaluate $E[X_1(1 - X_1)], E[X_1^2(1 - X_2)^3]$ and $E[(X_1 + X_2)^2]$. The surface represented by this pdf is given in the Figure 3.10.3.

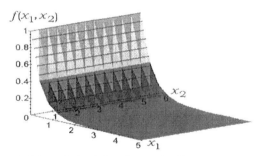

Figure 3.10.3. Plot of the PDF from the Exercise 3.3.5 with $c = 1$

3.3.6 (Example 3.3.10 Continued) Evaluate $E[X_1 X_2]$ and $E[X_1 X_3^2]$. Also evaluate $E[X_1(1 - X_1)], E[X_1 X_2 X_3^2]$ and $E[(X_1 + X_2 + X_3)^2]$.

3.3.7 (Example 3.3.10 Continued) Find the expressions for $E[X_1 \mid X_2 = x_2, X_3 = x_3]$ and $E[X_1 X_3^2 \mid X_2 = x_2]$ with $0 < x_2 < 2, 0 < x_3 < 1$.

{*Hints*: First find the conditional pdf's of X_1 given $X_2 = x_2, X_3 = x_3$ and that of (X_1, X_3) given $X_2 = x_2$. Next, express the expected values as the respective integrals and then evaluate them.}

3.3.8 (Example 3.3.10 Continued) Choose any three specific pdf's on three finite intervals as $a_i(x_i), i = 1, 2, 3$. Then, construct an appropriate combination of these a_i's to define a pdf $f(\mathbf{x})$ along the line of (3.3.47). Try putting these components in a certain order and adjust the coefficients so that the whole integral is one. {*Hint*: Examine closely why we had the success in the equation (3.3.48).}

3.3.9 (Example 3.3.11 Continued) Evaluate $E[X_1 X_2 X_5]$ and $E[X_4^2 X_5]$. Also evaluate $E[X_1(1 - X_5)], E[X_1 X_2 (2 - X_3)^2]$ and $E[(X_1 + X_2 + X_3 + X_4 + X_5)^2]$.

3.3.10 (Example 3.3.11 Continued) Find the expressions for $E[X_1(1 - X_5) \mid X_2 = x_2, X_4 = x_4]$ and $E[X_1 X_3^2 \mid X_5 = x_5]$ with $0 < x_2 < 2, 0 < x_4, x_5 < 1$. {*Hints*: First find the conditional pdf's of (X_1, X_5) given $X_2 = x_2, X_4 = x_4$ and that of (X_1, X_3) given $X_5 = x_5$. Next, express the expected values as the respective integrals and then evaluate them.}

3.3.11 (Example 3.3.11 Continued) Choose any five specific pdf's on five finite intervals as $a_i(x_i), i = 1, ..., 5$. Then, construct an appropriate combination of these a_i's to define a pdf $f(\mathbf{x})$ along the line of (3.3.53). Try putting these components in a certain order and adjust the coefficients so that the whole integral is one. {*Hint*: Examine closely how we had defined the function f in the equation (3.3.53).}

3.3.12 Let c be a positive constant such that X_1, X_2 and X_3 have the joint pdf given by

$$f(x_1, x_2, x_3) = \begin{cases} c\,exp\{-2x_1 - 4x_2^2 - \frac{1}{2}x_3\} & \text{if } 0 < x_1, x_3 < \infty, \\ & -\infty < x_2 < \infty \\ 0 & \text{elsewhere.} \end{cases}$$

Find the value of c. Derive the marginal pdf's of X_1, X_2 and X_3. Does either of X_1, X_2, X_3 have a standard distribution which matches with one of those distributions listed in Section 1.7? Evaluate $E[X_1 X_2]$ and $E[X_1 X_3^2]$. Also evaluate $E[X_1(1 - X_1)], E[X_1 X_2 X_3^2]$ and $E[(X_1 + X_2 + X_3)^2]$.

3.4.1 (Example 3.2.1 Continued) In the case of the random variables X_1, X_2 whose joint pdf was defined in the Table 3.2.1, calculate ρ, the correlation coefficient between X_1 and X_2.

3.4.2 (Exercise 3.2.4 Continued) In the case of the random variables X_1, X_2 whose joint pdf was defined in the Exercise 3.2.4, calculate ρ, the correlation coefficient between X_1 and X_2.

3.4.3 (Example 3.3.10 Continued) Evaluate $Cov(X_i, X_j), i \neq j = 1, 2, 3$. Then, evaluate $\rho_{X_i, X_j}, i \neq j = 1, 2, 3$.

3.4.4 (Example 3.3.11 Continued) Evaluate $Cov(X_i, X_j), i \neq j = 1, ..., 5$. Then, evaluate $\rho_{X_i, X_j}, i \neq j = 1, ..., 5$.

3.4.5 Prove the Theorem 3.4.3.

3.4.6 With any two random variables X_1 and X_2, show that

$$Cov(X_1 + X_2, X_1 - X_2) = V(X_1) - V(X_2),$$

provided that $Cov(X_1, X_2), V(X_1)$, and $V(X_2)$ are finite.

3.4.7 (Exercise 3.4.6 Continued) Consider any two random variables X_1 and X_2 for which one has $V(X_1) = V(X_2)$. Then, for this pair of random variables X_1 and X_2 we must have $Cov(X_1, X_2) = 0$. Next, using this ammunition, construct several pairs of random variables which are uncorrelated.

3.4.8 (Examples 3.2.6-3.2.7 Continued) Consider the multinomial random variable $\mathbf{X} = (X_1, ..., X_6)$ defined in the Example 3.2.6. Evaluate ρ_{X_i, X_j} for all $i \neq j = 1, ..., 6$.

3.4.9 (Example 3.2.8 Continued) Consider the multinomial random variable $\mathbf{X} = (X_1, X_5, X_3^*)$ defined in the Example 3.2.6 with $X_3^* = n - (X_1 + X_5)$. Evaluate $\rho_{X_1, X_5}, \rho_{X_1, X_3^*}$ and ρ_{X_5, X_3^*}.

3.4.10 (Exercise 3.2.7 Continued) Consider the multinomial random variable $\mathbf{X} = (X_1, X_2, X_3, X_4)$ defined in the Exercise 3.2.7. Evaluate ρ_{X_i, X_j} for all $i \neq j = 1, 2, 3, 4$.

3.4.11 (Exercise 3.2.10 Continued) Recall that $\mathbf{X} = (X_1, ..., X_5)$ has the $\text{Mult}_5(n, p_1, ..., p_5)$ with $0 < p_i < 1, i = 1, ..., 5, \Sigma_{i=1}^5 p_i = 1$ whereas $\mathbf{Y} = (Y_1, Y_2, Y_3)$ where $Y_1 = X_1 + X_2, Y_2 = X_3, Y_3 = X_4 + X_5$. Evaluate ρ_{Y_i, Y_j} for all $i \neq j = 1, 2, 3$.

3.4.12 Suppose that X_1 and X_2 have the joint pdf given by

$$f(x_1, x_2) = \begin{cases} x_1 + x_2 & \text{if } 0 < x_1, x_2 < 1 \\ 0 & \text{elsewhere.} \end{cases}$$

(i)   Show that $f_1(x_1) = x_1 + \frac{1}{2}, f_2(x_2) = x_2 + \frac{1}{2}$ for $0 < x_1, x_2 < 1$;

(ii) Evaluate $P\{\frac{1}{2} \leq X_1 \leq \frac{3}{4} \mid \frac{1}{3} \leq X_2 \leq \frac{2}{3}\}$.

3.4.13 Suppose that X_1 and X_2 have the joint pdf given by

$$f(x_1, x_2) = \begin{cases} 3(x_1^2 x_2 + x_1 x_2^2) & \text{if } 0 < x_1, x_2 < 1 \\ 0 & \text{elsewhere.} \end{cases}$$

Show that $\rho_{X_1, X_2} = -\frac{5}{139}$.

3.5.1 (Exercise 3.3.4 Continued) Let c be a positive constant such that X_1 and X_2 have the joint pdf given by

$$f(x_1, x_2) = \begin{cases} cx_1 x_2^2 & \text{if } 0 < x_1, x_2 < 2 \\ 0 & \text{elsewhere.} \end{cases}$$

Prove whether or not X_1 and X_2 are independent. Solve this exercise first by directly applying the Definition 3.5.1. Then, repeat this exercise by applying the Theorem 3.5.3.

3.5.2 (Exercise 3.3.5 Continued) Let c be a positive constant such that X_1 and X_2 have the joint pdf given by

$$f(x_1, x_2) = \begin{cases} cexp\{-2x_1 - \frac{1}{2}x_2\} & \text{if } 0 < x_1, x_2 < \infty \\ 0 & \text{elsewhere.} \end{cases}$$

Prove whether or not X_1 and X_2 are independent. Solve this exercise first by directly applying the Definition 3.5.1. Then, repeat this exercise by applying the Theorem 3.5.3.

3.5.3 (Exercise 3.3.12 Continued) Let c be a positive constant such that X_1, X_2 and X_3 have the joint pdf given by

$$f(x_1, x_2, x_3) = \begin{cases} cexp\{-2x_1 - 4x_2^2 - \frac{1}{2}x_3\} & \text{if } 0 < x_1, x_3 < \infty \text{ and} \\ & \qquad -\infty < x_2 < \infty \\ 0 & \text{elsewhere.} \end{cases}$$

Prove whether or not X_1, X_2 and X_3 are independent. Solve this exercise first by directly applying the Definition 3.5.1. Then, repeat this exercise by applying the Theorem 3.5.3.

3.5.4 (Example 3.5.4 Continued) Verify all the steps in the Example 3.5.4.

3.5.5 (Example 3.5.4 Continued) In the Example 3.5.4, suppose that we fix $f_1(x_1) = 2x_1, f_2(x_2) = 3x_2^2$ and $f_3(x_3) = \frac{5}{2}x_3^{3/2}, 0 < x_1, x_2, x_3 < 1$. Then, form the function $g(x_1, x_2, x_3)$ as in (3.5.2).

    (i)   Directly by integration, check that $g(x_1, x_2, x_3)$ is a pdf;

    (ii)   Directly by integration, find the expressions of all pairwise marginal pdf's $g_{i,j}(x_i, x_j)$ and single marginal pdf's $g_i(x_i)$ for $i \neq j = 1, 2, 3$;

    (iii)   Show directly that the X's are pairwise independent, but X_1, X_2, X_3 are not independent.

3.5.6 (Example 3.5.4 Continued) Construct some other specific functions $f_i(x_i), i = 1, 2, 3$ satisfying the conditions laid out in the Example 3.5.4 and observe what happens in some of those special situations.

3.5.7 Let us start with the non-negative integrable functions $f_i(x_i)$ for $0 < x_i < 1$ which are *not identically equal to unity* such that we can claim: $\int_{x_i=0}^{1} f_i(x_i) dx_i = 1, i = 1, 2, 3, 4$. With $\mathbf{x} = (x_1, x_2, x_3, x_4)$, let us then define

$$g(\mathbf{x}) = \tfrac{1}{8}\{f_1(x_1)f_2(x_2)f_3(x_3)f_4(x_4) + f_1(x_1)f_2(x_2) + f_2(x_2)f_3(x_3)$$
$$f_1(x_1)f_3(x_3) + f_1(x_1) + f_2(x_2) + f_3(x_3) + f_4(x_4)\}$$

for $0 < x_i < 1, i = 1, 2, 3, 4$.

(i)   Directly by integration, check that $g(x_1, x_2, x_3, x_4)$ is a pdf;
(ii)   Directly by integration, find the expression of the marginal pdf $g_{1,2,3}(x_1, x_2, x_3)$;
(iii)   Show directly that X_1, X_2, X_3 are independent;
(iv)   Consider the random variables X_1, X_2, X_3, X_4 whose joint pdf is given by $g(x_1, x_2, x_3, x_4)$. Show that X_1, X_2, X_3, X_4 are not independent.

3.5.8 (Exercise 3.5.7 Continued) Along the lines of the Exercise 3.5.6, construct some specific functions $f_i(x_i), i = 1, 2, 3, 4$ satisfying the conditions laid out in the Exercise 3.5.7 and observe what happens in some of those special situations.

3.5.9 Prove the Theorem 3.5.3.

3.5.10 Suppose that X_1 and X_2 have the joint pmf

$$f(x_1, x_2) = (2^{x_1-1}3^{x_2})^{-1}, \text{ for } x_1, x_2 = 1, 2, \dots .$$

Prove whether or not X_1 and X_2 are independent.

3.5.11 Suppose that X_1 and X_2 have the joint pmf

$$f(x_1, x_2) = 2/\{n(n+1)\}, \text{ for } x_2 = 1, \dots, x_1; x_1 = 1, 2, \dots, n.$$

Prove that X_1 and X_2 are dependent.

3.5.12 Suppose that X_1 and X_2 have the joint pmf

$$f(x_1, x_2) = \frac{e^{-(a+b)}a^{x_1}b^{x_2-x_1}}{x_1!(x_2-x_1)!}, \text{ for } x_1 = 0, 1, 2, \dots, x_2; \text{ and}$$

$$x_2 = 0, 1, 2, \dots; \text{ with positive numbers } a \text{ and } b.$$

Prove whether or not X_1 and X_2 are independent.

3.5.13 Suppose that X_1 and X_2 have the joint pmf such that $P\{X_1 = 2 \cap X_2 = 3\} = \frac{1}{3}$, $P\{X_1 = 2 \cap X_2 = -1\} = a$, $P\{X_1 = -1 \cap X_2 = 3\} = b$ and $P\{X_1 = -1 \cap X_2 = -1\} = \frac{1}{6}$ where a and b are appropriate numbers. Determine a and b when X_1 and X_2 are independent.

3.5.14 Suppose that X_1 and X_2 have the joint pdf

$$f(x_1, x_2) = \begin{cases} ke^{-(x_1+x_2)} & \text{if } 0 < x_2 < x_1 < \infty \\ 0 & \text{elsewhere,} \end{cases}$$

where k is a positive constant.

(i)   Show that $k = 2$;

(ii)   Show that $f_1(x_1) = 2\{e^{-x_1} - e^{-2x_1}\}I(0 < x_1 < \infty)$ and $f_2(x_2) = 2e^{-2x_2}I(0 < x_2 < \infty)$. Find $P(X_2 \geq 3)$;

(iii) Show that $f_{1|2}(x_1) = e^{-(x_1-x_2)}I(x_2 < x_1 < \infty)$ and $f_{2|1}(x_2) = e^{-x_2}(1 - e^{-x_1})^{-1}I(0 < x_2 < x_1)$;

(iv)   Prove whether or not X_1 and X_2 are independent.

3.5.15 Suppose that X_1 and X_2 have the joint pdf

$$f(x_1, x_2) = \begin{cases} \frac{1}{8}(6 - x_1 - x_2) & \text{if } 0 < x_1 < 2, 2 < x_2 < 4 \\ 0 & \text{elsewhere.} \end{cases}$$

(i)   Show that $f_1(x_1) = \frac{1}{4}(3 - x_1)I(0 < x_1 < 2)$ and $f_2(x_2) = \frac{1}{4}(5 - x_2)I(2 < x_2 < 4)$;

(ii)   Show that $f_{1|2}(x_1) = \dfrac{6 - x_1 - x_2}{2(5 - x_2)}I(0 < x_1 < 2, 2 < x_2 < 4)$ and $f_{2|1}(x_2) = \dfrac{6 - x_1 - x_2}{2(3 - x_1)}I(0 < x_1 < 2, 2 < x_2 < 4)$;

(iii) Prove whether or not X_1 and X_2 are independent.

3.5.16 Suppose that X_1 and X_2 have the joint pdf

$$f(x_1, x_2) = \begin{cases} \frac{1}{8}\pi^2 \sin(\frac{1}{2}\pi\{x_1 + x_2\}) & \text{if } 0 < x_1, x_2 < 1 \\ 0 & \text{elsewhere.} \end{cases}$$

The surface represented by this pdf is given in the Figure 3.10.4.

(i)   Show that $f_1(x_1) = \frac{1}{4}\pi\{\cos(\frac{1}{2}\pi x_1) + \sin(\frac{1}{2}\pi x_1)\}$ for $0 < x_1$
      < 1, and $f_2(x_2) = \frac{1}{4}\pi\{\cos(\frac{1}{2}\pi x_2) + \sin(\frac{1}{2}\pi x_2)\}$ for
      $0 < x_2 < 1$;

(ii) Prove whether or not X_1 and X_2 are independent.

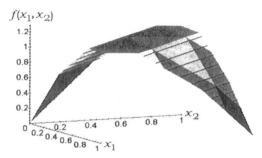

Figure 3.10.4. Plot of the PDF from the Exercise 3.5.16

3.5.17 Suppose that the random vector $\mathbf{X} = (X_1, X_2, X_3, X_4)$ has its
joint pdf given by

$$f(x_1, x_2, x_3, x_4) = \frac{1}{16\pi^2} \, exp\{-\frac{1}{2}u\}$$

where $u = \frac{1}{2}\{x_1^2 + x_2^2 + x_3^2 + x_1 x_2 + x_1 x_3 + x_2 x_3\} + 4x_4^2$

$$-2\{x_1 x_4 + x_2 x_4 + x_3 x_4\} \text{ for } (x_1, x_2, x_3, x_4) \in \Re^4.$$

(i)   Find the marginal pdf's $f_i(x_i)$, $i = 1, 2, 3, 4$. Show that each
       X_i is distributed as $N(0,4)$, $i = 1, 2, 3$, but X_4 is distributed
       as $N(0,1)$;

(ii)   By integrating $f(x_1, x_2, x_3, x_4)$ with respect to x_4 alone,
       obtain the joint pdf of (X_1, X_2, X_3);

(iii) Combine the parts (i) and (ii) to verify that X_1, X_2, X_3
       form a set of independent random variables;

(iv)   Show that X_1, X_2, X_3, X_4 do not form a set of independent
       random variables.

3.5.18 (Exercise 3.5.17 Continued) Consider the random vector $\mathbf{X} = (X_1, X_2, X_3, X_4)$ with its joint pdf given in the Exercise 3.5.17. Using this random vector $\mathbf{X}$, find few other four-dimensional random vectors $\mathbf{Y} = (Y_1, Y_2, Y_3, Y_4)$ where Y_1, Y_2, Y_3 are independent, but Y_1, Y_2, Y_3, Y_4 are not.

3.5.19 Suppose that X has the following pdf with $-\infty < \mu < \infty$, $0 < \sigma < \infty$:

$$f(x) = \begin{cases} \{\sigma\sqrt{2\pi}\}^{-1} x^{-1} exp \left\{ -\frac{1}{2\sigma^2} (log \, (x) - \mu)^2 \right\} & \text{if } x > 0 \\ 0 & \text{elsewhere} \end{cases}$$

One will recall from (1.7.27) that this pdf is known as the lognormal density and the corresponding X is called a lognormal random variable. Suppose that X_1, X_2 are iid having the common pdf $f(x)$. Let r and s be arbitrary, but fixed real numbers. Then, obtain the expression for $E[X_1^r X_2^s]$.

3.6.1 Derive the marginal pdf of X_1 in the Theorem 3.6.1, part (i).

3.6.2 Prove Theorem 3.6.1, part (iii).

3.6.3 (Example 3.6.2 Continued) Suppose that (X_1, X_2) is distributed as $N_2(0, 0, 1, 1, \rho)$ where one has $\rho \in (-1, 1)$. Find the expression of the mgf of the random variable $X_1 X_2$, that is $E\{exp[tX_1 X_2]\}$ for t belonging to an appropriate subinterval of $\Re$.

3.6.4 Suppose that the joint pdf of (X_1, X_2) is given by

$$f(x_1, x_2) = \{2\pi\sqrt{3}\}^{-1} exp\{-\tfrac{1}{6}[(2x_1 - x_2)^2 + 2x_1 x_2]\}$$

for $-\infty < x_1, x_2 < \infty$. Evaluate $E[X_i], V[X_i]$ for $i = 1, 2$, and ρ_{X_1, X_2}.

3.6.5 Suppose that the joint pdf of (X_1, X_2) is given by

$$f(x_1, x_2) = k\, exp\{-\tfrac{1}{216}[16(x_1 - 2)^2 - 12(x_1 - 2)(x_2 + 3)$$
$$+9(x_2 + 3)^2]\}$$

for $-\infty < x_1, x_2 < \infty$ where k is a positive number. Evaluate $E[X_i], V[X_i]$ for $i = 1, 2$, and ρ_{X_1, X_2}.

3.6.6 Suppose that (X_1, X_2) is distributed as $N_2(3, 1, 16, 25, \tfrac{2}{5})$. Evaluate $P\{3 < X_2 < 8 \mid X_1 = 7\}$ and $P\{-3 < X_1 < 3 \mid X_2 = -4\}$.

3.6.7 Suppose that X_1 is distributed as $N(\mu, \sigma^2)$ and conditionally the distribution of X_2 given that $X_1 = x_1$ is $N(x_1, \sigma^2)$. Then, show that the joint distribution of (X_1, X_2) is given by $N_2(\mu, \mu, \sigma^2, 2\sigma^2, 1/\sqrt{2})$.

3.6.8 Suppose that the joint pdf of (X_1, X_2) is given by

$$f(x_1, x_2) = (2\pi)^{-1} exp\{-\tfrac{1}{2}[x_1{}^2 + x_2{}^2]\}(1 + x_1 x_2\, exp\{-\tfrac{1}{2}[x_1{}^2$$
$$+x_2^2 - 2]\})$$

for $-\infty < x_1, x_2 < \infty$.

(i)   By direct integration, verify that $f(x_1, x_2)$ is a genuine pdf;

(ii)   Show that the marginal distributions of X_1, X_2 are both univariate normal;

(iii)   Does this $f(x_1, x_2)$ match with the density given by (3.6.1)?

3.6.9 Suppose that the joint pdf of (X_1, X_2, X_3) is given by

$$f(x_1, x_2, x_3) = (2\pi)^{-3/2} exp\{-\tfrac{1}{2}[x_1{}^2 + x_2{}^2 + x_3^2]\}(1+$$
$$x_1 x_2 x_3\, exp\{-\tfrac{1}{2}[x_1{}^2 + x_2^2 + x_3^2 - 3]\})$$

for $-\infty < x_1, x_2, x_3 < \infty$. Show that X_1, X_2, X_3 are dependent random variables. Also show that each pair $(X_1, X_2), (X_2, X_3)$ and (X_1, X_3) is distributed as a bivariate normal random vector.

3.7.1 This exercise provides a discrete version of the Example 3.7.1. Suppose that a random variable X_1 has the following probability distribution.

X_1 values:	-5	-2	0	2	5
Probabilities:	.25	.2	.1	.2	.25

Define $X_2 = X_1^2$. Then, show that

(i) $E[X_1] = E[X_1^3] = 0$;

(ii) $Cov(X_1, X_2) = 0$, that is the random variables X_1 and X_2 are uncorrelated;

(iii) X_1 and X_2 are dependent random variables.

3.7.2 (Exercise 3.7.1 Continued) From the Exercise 3.7.1, it becomes clear that a statement such as "the zero correlation need not imply independence" holds not merely for the specific situations handled in the Examples 3.7.1-3.7.2. Find a few other examples of the type considered in the Exercise 3.7.1.

3.7.3 In view of the Exercises 3.7.1-3.7.2, consider the following generalization. Suppose that X_1 has an arbitrary discrete distribution, symmetric about zero and having its third moment finite. Define $X_2 = X_1^2$. Then, show that

(i) $E[X_1] = E[X_1^3] = 0$;

(ii) $Cov(X_1, X_2) = 0$, that is the random variables X_1 and X_2 are uncorrelated;

(iii) X_1 and X_2 are dependent random variables.

3.7.4 (Example 3.7.1 Continued) In the Example 3.7.1, we used nice tricks with the standard normal random variable. Is it possible to think of another example by manipulating some other continuous random variable to begin with? What if one starts with X_1 that has a continuous symmetric distribution about zero? For example, suppose that X_1 is distributed uniformly on the interval $(-1, 1)$ and let $X_2 = X_1^2$. Are X_1 and X_2 uncorrelated, but dependent? Modify this situation to find other examples, perhaps with yet some other continuous random variable X_1 defined on the whole real line $\Re$ and $X_2 = X_1^2$.

3.7.5 In the types of examples found in the Exercises 3.7.1-3.7.4, what if someone had defined $X_2 = X_1^4$ or $X_2 = X_1^6$ instead? Would X_1 and X_2 thus defined, for example, be uncorrelated, but dependent in the earlier exercises?

3.7.6 (Example 3.7.3 Continued) The zero correlation between two random variables sometimes does indicate independence without bivariate normality. The Example 3.7.3 had dealt with a situation like this. In that example, will the same conclusion hold if both X_1 and X_2 were allowed to take two arbitrary values other than 0 and 1? {*Hint*: First recall from the Theorem 3.4.2, part (i) that the correlation coefficient between X_1 and X_2 would be the same as that between Y_1 and Y_2 where we let $Y_i = (X_i - a_i)/b_i$, with $a_i \in \Re, b_i \in \Re^+$ being any fixed numbers. Then, use this result to reduce the given problem to a situation similar to the one in the Example 3.7.3.}

3.7.7 Suppose that X_1 and X_2 have the joint pdf

$$f(x_1, x_2) = \begin{cases} k\{(x_1 + x_2) - (x_1^2 + x_2^2)\} & \text{if } 0 < x_1, x_2 < 1 \\ 0 & \text{elsewhere,} \end{cases}$$

where k is some positive constant. Show that X_1 and X_2 are uncorrelated, but these are dependent random variables.

3.7.8 Suppose that X_1 and X_2 have the joint pdf

$$f(x_1, x_2) = (4\pi)^{-1}[x_1^2 + x_2^2] exp\{-\tfrac{1}{2}[x_1{}^2 + x_2{}^2]\}$$

for $-\infty < x_1, x_2 < \infty$. Show that X_1 and X_2 are uncorrelated, but these are dependent random variables.

3.7.9 (Example 3.7.4 Continued) Suppose that (U_1, U_2) is distributed as $N_2(5, 15, 8, 8, \rho)$ for some $\rho \in (-1, 1)$. Let $X_1 = U_1 + U_2$ and $X_2 = U_1 - U_2$. Show that X_1 and X_2 are uncorrelated.

3.8.1 Verify the entries given in the Table 3.8.1.

3.8.2 Consider the Beta(α, β) distribution defined by (1.7.35). Does the pdf belong to the appropriate (that is, one- or two-parameter) exponential family when

(i) α is known, but β is unknown?
(ii) β is known, but α is unknown?
(iii) α and β are both unknown?

3.8.3 Consider the Beta(α, β) distribution defined by (1.7.35). Does the pdf belong to the appropriate (that is, one- or two-parameter) exponential family when

(i) $\alpha = \beta = \theta$, but $\theta(> 0)$ is unknown?
(ii) $\alpha = \theta, \beta = 2\theta$, but $\theta(> 0)$ is unknown?

3.8.4 Suppose that X has the uniform distribution on the interval $(-\theta, \theta)$ with $\theta(> 0)$ unknown. Show that the corresponding pdf does not belong to the one-parameter exponential family defined by (3.8.1). {*Hint*: Use ideas similar to those from the Example 3.8.5.}

3.8.5 Consider two random variables X_1, X_2 whose joint pdf is given by

$$f(x_1, x_2) = exp\{-\theta x_1 - \theta^{-1}x_2\}I(x_1 > 0 \cap x_2 > 0)$$

with $\theta(> 0)$ unknown.

    (*i*)   Are X_1, X_2 independent?

    (*ii*)   Does this pdf belong to the one-parameter exponential family?

3.8.6 Does the multinomial pmf with unknown parameters $p_1, ..., p_k$ defined in (3.2.8) belong to the multi-parameter exponential family? Is the number of parameters k or $k - 1$?

3.8.7 Express the bivariate normal pdf defined in (3.6.1) in the form of an appropriate member of the one-parameter or multi-parameter exponential family in the following situations when the pdf involves

    (*i*)   all the parameters $\mu_1, \mu_2, \sigma_1, \sigma_2, \rho$;

    (*ii*) $\mu_1 = \mu_2 = 0$, and the parameters σ_1, σ_2, ρ;

    (*iii*) $\sigma_1 = \sigma_2 = 1$, and the parameters μ_1, μ_2, ρ;

    (*iv*) $\sigma_1 = \sigma_2 = \sigma$, and the parameters $\mu_1, \mu_2, \sigma, \rho$;

    (*v*) $\mu_1 = \mu_2 = 0, \sigma_1 = \sigma_2 = 1$, and the parameter ρ.

3.8.8 Consider the Laplace or the double exponential pdf defined as

$$f(x; \theta) = \tfrac{1}{2}exp\{-|x - \theta|\}$$

for $-\infty < x, \theta < \infty$ where θ is referred to as a parameter. Show that $f(x; \theta)$ does not belong to the one-parameter exponential family.

3.8.9 Suppose that a random variable X has the Rayleigh distribution with

$$f(x; \theta) = 2\theta^{-1}x exp(-x^2/\theta)I(x > 0),$$

where $\theta(> 0)$ is referred to as a *parameter*. Show that $f(x; \theta)$ belongs to the one-parameter exponential family.

3.8.10 Suppose that a random variable X has the Weibull distribution with

$$f(x; \alpha) = \alpha^{-1}\beta x^{\beta-1} exp(-x^\beta/\alpha)I(x > 0),$$

where $\alpha(> 0)$ is referred to as a *parameter*, while $\beta(> 0)$ is assumed known. Show that $f(x; \alpha)$ belongs to the one-parameter exponential family.

3.9.1 (Example 3.9.4 Continued) Prove the claim made in the Example 3.9.4 which stated the following: Suppose that X is a random variable whose mgf $M_X(t)$ is finite for some $t \in \mathcal{T} \subseteq (-\infty, 0)$. Then, it follows from the Theorem 3.9.2 that for any fixed real number a, one has $P\{X \leq a\} \leq \underset{t\in(-\infty,0)}{Inf}\ \{e^{-ta}M_X(t)\}$.

3.9.2 Suppose that X has the Gamma$(4, \frac{1}{2})$ distribution.

(i)   Use Markov inequality to get an upper bound for $P\{X \geq 12\}$;

(ii)   Use Bernstein-Chernoff inequality to get an upper bound for $P\{X \geq 12\}$;

(iii) Use Tchebysheff's inequality to obtain an upper bound for $P\{X \geq 12\}$.

3.9.3 (Exercise 3.6.6 Continued) Suppose that (X_1, X_2) is distributed as $N_2(3, 1, 16, 25, \frac{3}{5})$.

(i)   Use Tchebysheff's inequality to get a lower bound for $P\{-2 < X_2 < 10 \mid X_1 = 7\}$;

(ii)   Use Cauchy-Schwarz inequality to obtain an upper bound for (a) $E[|X_1 X_2|]$ and (b) $E[X_1^2 X_2^2]$.

{*Hint*: In part (i), work with the conditional distribution of $X_2 \mid X_1 = 7$ and from X_2, subtract the right conditional mean so that the Tchebysheff's inequality can be applied. In part (ii), apply the Cauchy-Schwarz inequality in a straightforward manner.}

3.9.4 Suppose that X is a positive integer valued random variable and denote $p_k = P(X = k), k = 1, 2, \dots$. Assume that the sequence $\{p_k; k \geq 1\}$ is non-increasing. Show that $p_k \leq 2k^{-2}E[X]$ for any $k = 1, 2, \dots$. {*Hint*: $E[X] \geq \Sigma_{i=1}^k i p_i \geq \Sigma_{i=1}^k i p_k = \frac{1}{2}k(k+1)p_k \geq \frac{1}{2}k^2 p_k$. Refer to (1.6.11).}

3.9.5 Suppose that X is a discrete random variable such that $P(X = -a) = P(X = a) = \frac{1}{8}$ and $P(X = 0) = \frac{3}{4}$ where $a(> 0)$ is some fixed number. Evaluate μ, σ and $P(|X| \geq 2\sigma)$. Also, obtain Tchebysheff's upper bound for $P(|X| \geq 2\sigma)$. Give comments.

3.9.6 Verify the convexity or concavity property of the following functions.

(i) $h(x) = x^5$ for $-1 < x < 0$;

(ii) $h(x) = |x|^3$ for $x \in \Re$;

(iii) $h(x) = x^4 e^{-2x}$ for $x \in \Re^+$;

(iv) $h(x) = e^{-\frac{1}{2}x^2}$ for $x \in \Re$;

(v) $h(x) = x^{-1}$ for $x \in \Re^+$.

3.9.7 Suppose that $0 < \alpha, \beta < 1$ are two *arbitrary* numbers. Show that

$$\alpha log\{(1 - \beta)/\alpha\} + (1 - \alpha)log\{\beta/(1 - \alpha)\} < 0.$$

{*Hint*: Is $log(x)$ with $x > 0$ convex or concave?}

3.9.8 First show that $f(x) = x^{-1}$ for $x \in \Re^+$ is a convex function. Next, suppose that X is a real valued random variable such that $P(X > 0) = 1$ and $E[X]$ is finite. Show that

(i) $E[1/X] > 1/E[X]$;

(ii) $Cov(X, 1/X) < 0$.

3.9.9 Suppose that X and Y are two random variables with finite second moments. Also, assume that $P(X + Y = 0) < 1$. Show that

$$\left[E\{(X+Y)^2\}\right]^{1/2} \le \left[E(X^2)\right]^{1/2} + \left[E(Y^2)\right]^{1/2}.$$

{*Hint*: Observe that $(X + Y)^2 \le |X|\{|X + Y|\} + |Y|\{|X + Y|\}$. Take expectations throughout and then apply the Cauchy-Schwarz inequality on the rhs.}

3.9.10 Suppose that $X_1, ..., X_n$ are arbitrary random variables, each with zero mean and unit standard deviation. For arbitrary $\varepsilon(> 1)$, show that

$$P\{\cap_{i=1}^n |X_i| \le \varepsilon\sqrt{n}\} \ge 1 - \varepsilon^{-2}.$$

{*Hint*: Let A_i be the event that $|X_i| \le \varepsilon\sqrt{n}, i = 1, ..., n$. By Tchebysheff's inequality, $P(A_i) \ge 1 - (n\varepsilon^2)^{-1}, i = 1, ..., n$. Then apply Bonferroni inequality, namely, $P\{\cap_{i=1}^n A_i\} \ge \Sigma_{i=1}^n P(A_i) - (n-1) \ge n\{1 - (n\varepsilon^2)^{-1}\} - (n-1) = 1 - \varepsilon^{-2}.\}$

3.9.11 Suppose that Y is a random variable for which $E[Y] = 3$ and $E[Y^2] = 13$. Show that $P\{-2 < Y < 8\} > \frac{21}{25}$. {*Hint*: Check that $V[Y] = 4$ so that $P\{-2 < Y < 8\} = P\{|Y - 3| < 5\} > 1 - \frac{4}{25} = \frac{21}{25}$, by Tchebysheff's inequality.}

4

Functions of Random Variables and Sampling Distribution

4.1 Introduction

It is often the case when we need to determine the distribution of a function of one or more random variables. There is, however, no one unique approach to achieve this goal. In a particular situation, one or more approaches given in this chapter may be applicable. Which method one would ultimately adopt in a particular situation may largely depend on one's taste. For completeness, we include standard approaches to handle various types of problems involving transformations and sampling distributions.

In Section 4.2 we start with a technique involving distribution functions. This approach works well in the case of discrete as well as continuous random variables. Section 4.2 also includes distributions of *order statistics* in the continuous case. Next, Section 4.3 demonstrates the usefulness of a moment generating function (mgf) in deriving distributions. Again, this approach is shown to work well in the case of discrete as well as continuous random variables. Section 4.4 exclusively considers continuous distributions and transformations involving one, two, and several random variables. One highlight in this section consists of the *Helmert (Orthogonal) Transformation* in the case of a normal distribution, and another consists of a transformation involving the spacings between the successive order statistics in the case of an exponential distribution. In both these situations, we deal with one-to-one transformations from n random variables to another set of n random variables, followed by discussions on Chi-square, Student's t, and F distributions. Section 4.5 includes sampling distributions for both *one-sample* and *two-sample* problems. Section 4.6 briefly touches upon the multivariate normal distributions and provides the distribution of the *Pearson correlation coefficient* in the bivariate normal case. Section 4.7 shows, by means of several examples, the importance of independence of various random variables involved in deriving *sampling distributions* such as the Student's t and F distributions as well as in the reproductive properties of normal and Chi-square random variables.

The following example shows the immediate usefulness of results which evolve via transformations and sampling distributions. We first solve the

177

problem using the direct calculation, and then provide an alternative quick and painless way to solve the same problem by applying a result derived later in this chapter.

Example 4.1.1 Let us consider a popular game of "hitting the bull's eye" at the point of intersection of the horizontal and vertical axes. From a fixed distance, one aims a dart at the center and with the motion of the wrist, lets it land on the game board, say at the point with the *rectangular coordinates* (Z_1, Z_2). Naturally then, the distance (from the origin) of the point on the game board where the dart lands is $(Z_1^2 + Z_2^2)^{1/2}$. A smaller distance would indicate better performance of the player. Suppose that Z_1 and Z_2 are two independent random variables both distributed as $N(0, 1)$. We wish to calculate $P\{Z_1^2 + Z_2^2 \leq a\}$ where $a > 0$, that is the probability that the thrown dart lands within a distance of $\sqrt{a}$ from the aimed target. The joint pdf of (Z_1, Z_2) would obviously amount to $f(z_1, z_2) = (2\pi)^{-1} exp\{-\frac{1}{2}(z_1^2 + z_2^2)\}$ with $-\infty < z_1, z_2 < \infty$. We then go back to the heart of (3.3.20) as we proceed to evaluate the double integral,

$$\iint\limits_{z_1^2 + z_2^2 \leq a} (2\pi)^{-1} exp\left\{-\tfrac{1}{2}(z_1^2 + z_2^2)\right\} dz_1 dz_2. \tag{4.1.1}$$

Let us substitute

$$z_1 = \sqrt{r}\sin(\theta), z_2 = \sqrt{r}\cos(\theta) \text{ with } r > 0 \text{ and } 0 < \theta < 2\pi, \tag{4.1.2}$$

which transform a point (z_1, z_2) on the plane in the rectangular coordinates system to the point (r, θ) on the same plane, but in the *polar coordinates* system. Now, the matrix of the first partial derivatives of z_1 and z_2 with respect to θ and r would be

$$A = \begin{pmatrix} \frac{\partial z_1}{\partial \theta} & \frac{\partial z_1}{\partial r} \\ \frac{\partial z_2}{\partial \theta} & \frac{\partial z_2}{\partial r} \end{pmatrix} = \begin{pmatrix} \sqrt{r}\cos(\theta) & \frac{1}{2\sqrt{r}}\sin(\theta) \\ -\sqrt{r}\sin(\theta) & \frac{1}{2\sqrt{r}}\cos(\theta) \end{pmatrix}$$

so that its determinant, $det(A) = \frac{1}{2}$. Now, we can rewrite the double integral from (4.1.1) as

$$(2\pi)^{-1} \int_{r=0}^{a} \int_{\theta=0}^{2\pi} exp\{-r/2\} det(A) d\theta dr$$

$$= \tfrac{1}{2}(2\pi)^{-1} \int_{\theta=0}^{2\pi} d\theta \int_{r=0}^{a} exp\{-r/2\} dr \tag{4.1.3}$$

$$= \tfrac{1}{2}(2\pi)^{-1}(2\pi)2\{1 - e^{-\frac{1}{2}a}\}$$

$$= 1 - e^{-\frac{1}{2}a}.$$

But, later in this chapter (Example 4.3.6), we show that the distribution of $W = Z_1^2 + Z_2^2$ is indeed χ_2^2, that is a Chi-square with two degrees of freedom. Thus the pdf of W is given by $g(w) = \frac{1}{2}exp\{-\frac{1}{2}w\}I(w > 0)$. Refer to Section 1.7 as needed. The required probability, $P(W \leq a)$ is simply then

$$\int_0^a exp\{-w/2\}d(w/2) = [exp\{-w/2\}]_{w=0}^{w=a} = 1 - e^{-\frac{1}{2}a}, \qquad (4.1.4)$$

which matches with the answer found earlier in (4.1.3). This second approach appears much simpler. ▲

In the Exercise 4.1.1, we have proposed a direct approach to evaluate $P(Z_1^2 < bZ_2^2)$, with some fixed but arbitrary $b > 0$, where Z_1, Z_2 are iid standard normal. The Exercise 4.5.4, however, provides an easier way to evaluate the same probability by considering the sampling distribution of the random variable Z_1/Z_2 which just happens (Exercise 4.5.1 (iii)) to have the Cauchy distribution defined in (1.7.31). The three Exercises 4.5.7-4.5.9 also fall in the same category.

4.2 Using Distribution Functions

Suppose that we have *independent* random variables $X_1, ..., X_n$ and let us denote $Y = g(X_1, ..., X_n)$, a real valued function of these random variables. Our goal is to derive the distribution of Y.

If $X_1, ..., X_n$ or Y happen to be discrete random variables, we set out to evaluate the expression of $P(Y = y)$ for all appropriate $y \in \mathcal{Y}$, and then by identifying the form of $P(Y = y)$, we are often led to one of the standard distributions.

In the continuous case, the present method consists of first obtaining the distribution function of the new random variable Y, denoted by $F(y) = P(Y \leq y)$ for all appropriate values of $y \in \mathcal{Y}$. If $F(y)$ is a differentiable function of y, then by differentiating $F(y)$ with respect to y, one can obtain the pdf of Y at the point y. In what follows, we discuss the discrete and continuous cases separately.

4.2.1 Discrete Cases

Here we show how some discrete situations can be handled. The exact techniques used may vary from one problem to another.

Example 4.2.1 Suppose that X_1, X_2 are two independent random variables having respectively the following probability distributions:

X_1 values:	0	1	3
Probabilities:	.2	.3	.5

X_2 values:	-1	0	2	2.5
Probabilities:	.1	.2	.4	.3

Now, $P(X_1 = x_1 \cap X_2 = x_2) = P(X_1 = x_1)P(X_2 = x_2)$ for all $x_1 = 0,1,3$ and $x_2 = -1,0,2,2.5$. Suppose that we want to obtain the pmf for the random variable $Y = X_1 + X_2$. The possible values y of Y belong to the set $\mathcal{Y} = \{-1,0,1,2,2.5,3,3.5,5,5.5\}$. Now $P(Y = 0) = P(X_1 = 0 \cap X_2 = 0) + P(X_1 = 1 \cap X_2 = -1) = (.2)(.2) + (.3)(.1) = .07$. Also, $P(Y = 2) = P(X_1 = 0 \cap X_2 = 2) + P(X_1 = 3 \cap X_2 = -1) = (.2)(.4) + (.5)(.1) = .13$, and this way the pmf function $g(y) = P(Y = y)$ can be obtained for all $y \in \mathcal{Y}$. ▲

Example 4.2.2 Let X_1 and X_2 be independent random variables, where X_1 is Binomial(n_1,p) and X_2 is Binomial(n_2,p) with $0 < p < 1$. Let us find the distribution of $Y = X_1 + X_2$. With $0 \le y \le n_1 + n_2$, one can express $P(Y = y)$ as

$$
\begin{aligned}
P(X_1 + X_2 = y) &= \Sigma_{a=0}^{n_1} P(X_1 = a \cap X_2 = y - a), \\
&\quad \text{since the events are mutually exclusive} \\
&= \Sigma_{a=0}^{n_1} P(X_1 = a)P(X_2 = y - a), \\
&\quad \text{since } X_1 \text{ and } X_2 \text{ are independent} \\
&= \Sigma_{a=0}^{n_1} \binom{n_1}{a} p^a (1-p)^{n_1-a} \binom{n_2}{y-a} p^{y-a}(1-p)^{n_2-y+a} \\
&= p^y (1-p)^{n_1+n_2-y} \Sigma_{a=0}^{n_1} \binom{n_1}{a}\binom{n_2}{y-a}.
\end{aligned}
$$
$$(4.2.1)$$

In order to evaluate the summation in the last step in (4.2.1), one needs to compare the coefficients of $p^y(1-p)^{n_1+n_2-y}$ on both sides of the identity:

$$\{p + (1-p)\}^{n_1}\{p + (1-p)\}^{n_2} \equiv \{p + (1-p)\}^{n_1+n_2}.$$

Hence, $\Sigma_{a=0}^{n_1} \binom{n_1}{a}\binom{n_2}{y-a}$ can be replaced by $\binom{n_1+n_2}{y}$, a fact that follows from the Binomial Theorem (see (1.4.12)). That is, we can simply rewrite

$$P(Y = y) = \binom{n_1+n_2}{y} p^y(1-p)^{n_1+n_2-y}, \quad y = 0,1,...,n_1 + n_2,$$

which is the pmf of the Binomial$(n_1 + n_2, p)$ variable. Hence, the random variable Y has the Binomial $(n_1 + n_2, p)$ distribution. ▲

Example 4.2.3 Suppose that $X_1, ..., X_k$ are independent, X_i is distributed as the Binomial$(n_i, p), 0 < p < 1, i = 1, ..., k$. Obviously $X_1 + X_2 + X_3 = \{X_1 + X_2\} + X_3$, and of course $X_1 + X_2$ and X_3 are *independently distributed* as the binomials with the same p. Hence, using the Example 4.2.2, we conclude that $X_1 + X_2 + X_3$ is Binomial$(\{n_1 + n_2\} + n_3, p)$, because we are simply adding two independent Binomial random variables respectively with parameters $(n_1 + n_2, p)$ and (n_3, p). Next, one may use mathematical induction to claim that $\Sigma_{i=1}^{k} X_i$ is thus distributed as the Binomial$(\Sigma_{i=1}^{k} n_i, p)$. ▲

4.2.2 Continuous Cases

The distribution function approach works well in the case of continuous random variables. Suppose that X is a continuous real valued random variable and let Y be a real valued function of X. The basic idea is first to express the distribution function $G(y) = P(Y \le y)$ of Y in the form of the probability of an appropriate event defined through the original random variable X. Once the expression of $G(y)$ is found in a closed form, one would obviously obtain $dG(y)/dy$, whenever $G(y)$ is differentiable, as the pdf of the transformed random variable Y in the appropriate domain space for y. This technique is explained with examples.

Example 4.2.4 Suppose that a random variable X has the pdf

$$f(x) = \begin{cases} \frac{3}{7}x^2 & \text{if } 1 < x < 2 \\ 0 & \text{elsewhere.} \end{cases}$$

Let $Y = X^2$ and we first obtain $G(y) = P(Y \le y)$ for all y in the real line. Naturally, $G(y) = 0$ if $y \le 1$ and $G(y) = 1$ if $y \ge 4$. But, for $1 < y < 4$, we have $G(y) = P(X \le y^{\frac{1}{2}}) = \int_1^{\sqrt{y}} \frac{3}{7}x^2 dx = \frac{1}{7}(y^{\frac{3}{2}} - 1)$. Hence,

$$g(y) = \frac{dG(y)}{dy} = \begin{cases} \frac{3}{14}y^{\frac{1}{2}} & \text{if } 1 < y < 4 \\ 0 & \text{if } y \le 1 \text{ or } y \ge 4, \end{cases}$$

which is the pdf of Y. ▲

> In a continuous case, for the transformed variable $Y = g(X)$, first find the df $G(y) = P(Y \le y)$ of Y in the appropriate space for y. Then, $\frac{d}{dy}G(y)$, whenever $G(y)$ is differentiable, would be pdf of Y for the appropriate y values.

Example 4.2.5 Let X have an arbitrary continuous distribution with its pdf $f(x)$ and the df $F(x)$ on the interval $(a, b) \subseteq \Re$. We first find the pdf of the random variable $F(X)$ and denote $W = F(X)$. Let $F^{-1}(.)$ be the inverse function of $F(.)$, and then one has for $0 < w < 1$:

$$G_1(w) = P\{W \le w\} = P\left(X \le F^{-1}(w)\right) = F\left(F^{-1}(w)\right) = w,$$

and thus the pdf of W is given by

$$g_1(w) = \frac{dG_1(w)}{dw} = \begin{cases} 1 & \text{if } 0 < w < 1 \\ 0 & \text{elsewhere.} \end{cases}$$

That is, W has a Uniform distribution on the interval $(0,1)$. Next, define $U = - \, log(F(X))$ and $V = -2log(F(X))$. The domain space of U and V are both $(0,\infty)$. Now, for $0 < u < \infty$, one has the df of U,

$$G(u) = P(U \leq u) = P(W \geq e^{-u}) = \int_{e^{-u}}^{\infty} g_1(w)dw = 1 - e^{-u},$$

so that U has the pdf $f(u) = \frac{d}{du}G(u) = e^{-u}$, which corresponds to the pdf of a standard exponential random variable which is defined as Gamma$(1,1)$. Refer to $(1.7.24)$. Similarly, for $0 < v < \infty$, we write the df of V,

$$H(v) = P(V \leq v) = P(W \geq e^{-\frac{1}{2}v}) = 1 - e^{-\frac{1}{2}v},$$

so that V has the pdf $g(v) = \frac{d}{dv}H(v) = \frac{1}{2}e^{-\frac{1}{2}v}$, which matches with the pdf of a Chi-square random variable with two degrees of freedom. Again refer back to Section 1.7 as needed. ▲

Example 4.2.6 Suppose that Z has a standard normal distribution and let $Y = Z^2$. Denote $\phi(.)$ and $\Phi(.)$ respectively for the pdf and df of Z. Then, for $0 < y < \infty$, we have the df of Y,

$$G(y) = P(Y \leq y) = P(|Z| \leq y^{\frac{1}{2}}) = 2\Phi(y^{\frac{1}{2}}) - 1,$$

and hence the pdf of Y will be given by

$$g(y) = \frac{d}{dy}G(y) = 2\phi(y^{\frac{1}{2}})\frac{1}{2}y^{-\frac{1}{2}} = \frac{1}{\sqrt{2\pi}}e^{-\frac{1}{2}y}y^{-\frac{1}{2}}.$$

Now $g(y)$ matches with the pdf of a Chi-square random variable with one degree of freedom. That is, Z^2 is distributed as a Chi-square with one degree of freedom. Again, refer back to Section 1.7. ▲

Suppose that X has a continuous random variable with its df $F(x)$. Then, $F(X)$ is Uniform on the interval $(0,1)$, $-log\{F(X)\}$ is standard exponential, and $-2log\{F(X)\}$ is χ_2^2.

4.2.3 The Order Statistics

Let us next turn to the distributions of *order-statistics*. Consider independent and identically distributed (iid) continuous random variables $X_1, ..., X_n$ having the common pdf $f(x)$ and the df $F(x)$. Suppose that $X_{n:1} \leq X_{n:2} \leq ... \leq X_{n:n}$ stand for the corresponding ordered random variables where $X_{n:i}$ is referred to as the ith order statistic. Let us denote $Y_i = X_{n:i}$ for $i = 1, ..., n$ and $\mathbf{Y} = (y_1, ..., y_n)$. The joint pdf of $Y_1, ..., Y_n$, denoted by $f_{\mathbf{Y}}(y_1, ..., y_n)$, is given by

$$f_{\mathbf{Y}}(y_1, ..., y_n) = \begin{cases} n!\Pi_{i=1}^{n}f(y_i) & \text{if } -\infty < y_1 \leq ... \leq y_n < \infty \\ 0 & \text{elsewhere.} \end{cases} \tag{4.2.2}$$

In (4.2.2), the multiplier $n!$ arises because $y_1, ..., y_n$ can be arranged among themselves in $n!$ ways and the pdf for any such single arrangement amounts to $\Pi_{i=1}^{n} f(y_i)$. Often we are specifically interested in the *smallest* and *largest* *order statistics*. For the largest order statistic Y_n, one can find the distribution as follows:

$$\begin{aligned} P(Y_n \leq y) &= P\,(X_i \leq y \text{ for each } i = 1, ..., n) \\ &= \Pi_{i=1}^{n} P\,(X_i \leq y), \because \ X\text{'s are independent} \qquad (4.2.3) \\ &= \{F(y)\}^n \end{aligned}$$

and hence the pdf of Y_n would be given by

$$g(y) = \tfrac{d}{dy} P(Y_n \leq y) = \tfrac{d}{dy}[\{F(y)\}^n] = n\{F(y)\}^{n-1} f(y) \qquad (4.2.4)$$

in the appropriate space for the y values. In the same fashion, for the smallest order statistic Y_1, we can write:

$$\begin{aligned} P(Y_1 > y) &= P\,(X_i > y \text{ for each } i = 1, ..., n) \\ &= \Pi_{i=1}^{n} P\,(X_i > y), \because \ X\text{'s are independent} \qquad (4.2.5) \\ &= \{1 - F(y)\}^n \end{aligned}$$

and thus the pdf of Y_1 would be given by

$$h(y) = \tfrac{d}{dy} P(Y_1 \leq y) = \tfrac{d}{dy}[1 - \{1 - F(y)\}^n] = n\{1 - F(y)\}^{n-1} f(y) \qquad (4.2.6)$$

in the appropriate space for the y values.

In the Exercise 4.2.5, we have indicated how one can find the joint pdf of any two order statistics $Y_i = X_{n:i}$ and $Y_j = X_{n:j}$.

Using the Exercise 4.2.5, one can derive the joint pdf of Y_1 and Y_n. In order to write down the joint pdf of (Y_1, Y_n) at a point (y_1, y_n) quickly, we adopt the following *heuristic* approach. Since y_1, y_n are assumed fixed, each of the remaining $n - 2$ order statistics can be anywhere between y_1 and y_n, while these could be any $n - 2$ of the original n random X's. Now, $P\{y_1 < X_i < y_n\} = F(y_n) - F(y_1)$, for each $i = 1, ..., n$. Hence, with $k = \frac{n!}{(n-2)!}$, the joint pdf of (Y_1, Y_n) would be given by:

$$\begin{aligned} &f(y_1, y_n) \\ &= \begin{cases} k\{F(y_n) - F(y_1)\}^{n-2} f(y_1) f(y_n) & \text{if } -\infty < y_1 < y_n < \infty \\ 0 & \text{elsewhere.} \end{cases} \end{aligned}$$

$$(4.2.7)$$

In the same fashion, one can easily write down the joint pdf of any subset of the order statistics. Now, let us look at some examples.

> In the Exercises 4.2.7-4.2.8, we show how one can find the pdf of the *range*, $X_{n:n} - X_{n:1}$ when one has the random samples $X_1, ..., X_n$ from the *Uniform* distribution on the interval $(0, \theta)$ with $\theta \in \Re^+$ or on the interval $(\theta, \theta + 1)$ with $\theta \in \Re$ respectively.

Example 4.2.7 Suppose that $X_1, ..., X_n$ are iid random variables distributed as Uniform on the interval $(0, \theta)$ with $\theta > 0$. Consider the largest and smallest order statistics Y_n and Y_1, respectively. Note that

$$f(x) = \begin{cases} \theta^{-1} & \text{if } 0 < x < \theta \\ 0 & \text{elsewhere} \end{cases} \qquad F(x) = \begin{cases} 0 & \text{if } x \leq 0 \\ \frac{x}{\theta} & \text{if } 0 < x < \theta \\ 1 & \text{if } x \geq \theta \end{cases}$$

and in view of (4.2.4) and (4.2.6), the marginal pdf of Y_n and Y_1 will be respectively given by $g(y) = ny^{n-1}\theta^{-n}$ and $h(y) = n(\theta - y)^{n-1}\theta^{-n}$ for $0 < y < \theta$. In view of (4.2.7), the joint pdf of (Y_1, Y_n) would be given by $f(y_1, y_n) = n(n-1)(y_n - y_1)^{n-2}\theta^{-n}$ for $0 < y_1 < y_n < \theta$. ▲

> The next example points out the modifications needed in (4.2.4) and (4.2.6)-(4.2.7) in order to find the distributions of various order statistics from a set of *independent, but not identically distributed continuous random variables.*

Example 4.2.8 Consider independent random variables X_1 and X_2 where their respective pdf's are given by $f_1(x) = \frac{1}{3}x^2 I(-1 < x < 2)$ and $f_2(x) = \frac{5}{33}x^4 I(-1 < x < 2)$. Recall that here and elsewhere $I(.)$ stands for the indicator function of $(.)$. One has the distribution functions $F_1(x) = \int_{-1}^{x} \frac{1}{3}t^2 dt = \frac{1}{9}(x^3 + 1)$, $F_2(x) = \int_{-1}^{x} \frac{5}{33}t^4 dt = \frac{1}{33}(x^5 + 1)$ for $-1 < x < 2$. Hence, for $-1 < y < 2$, the distribution function $F(y)$ of $Y_2 = X_{2:2}$, the larger order statistic, can be found as follows:

$$P(Y_2 \leq y) = P(X_1 \leq y \cap X_2 \leq y) = F_1(y)F_2(y) = \frac{1}{297}(y^3 + 1)(y^5 + 1), \tag{4.2.8}$$

since the X's are independent. By differentiating $F(y)$ with respect to y, one can immediately find the pdf of $X_{2:2}$. Similarly one can handle the distribution of $X_{2:1}$, the smaller order statistic. The same idea easily extends for n *independent*, but *not identically distributed* continuous random variables. ▲

4.2.4 The Convolution

We start with what is also called the *convolution theorem*. This result will sometimes help to derive the distribution of sums of independent random variables. In this statement, we assume that the support of (X_1, X_2) is $\Re^2$. When the support is a proper subset of $\Re^2$, the basic result would still hold except that the range of the integral on the rhs of (4.2.9) should be appropriately adjusted on a case by case basis.

Theorem 4.2.1 (Convolution Theorem) *Suppose that X_1 and X_2 are independent continuous random variables with the respective pdf's $f_1(x_1)$ and $f_2(x_2)$, for $(x_1, x_2) \in \Re^2$. Let us denote $U = X_1 + X_2$. Then, the pdf of U is given by*

$$g(u) = \int_{x_1=-\infty}^{\infty} f_1(x_1) f_2(u - x_1) dx_1 \qquad (4.2.9)$$

for $u \in \Re$.

Proof First let us obtain the df of the random variable U. With $u \in \Re$, we write

$$
\begin{aligned}
G(u) &= P\{U \le u\} \\
&= P\{X_1 + X_2 \le u\} \\
&= \int_{x_1=-\infty}^{\infty} \left[\int_{x_2=-\infty}^{u-x_1} f(x_1, x_2) dx_2 \right] dx_1 \\
&= \int_{x_1=-\infty}^{\infty} \left[\int_{x_2=-\infty}^{u-x_1} f_1(x_1) f_2(x_2) dx_2 \right] dx_1, \\
&\qquad \text{since } X_1 \text{ and } X_2 \text{ are independent} \\
&= \int_{x_1=-\infty}^{\infty} f_1(x_1) \left[\int_{x_2=-\infty}^{u-x_1} f_2(x_2) dx_2 \right] dx_1.
\end{aligned}
\qquad (4.2.10)
$$

Thus, by differentiating the df $G(u)$ from (4.2.10) with respect to u, we get the pdf of U and write

$$
\begin{aligned}
g(u) &= \tfrac{d}{du} G(u) \\
&= \int_{x_1=-\infty}^{\infty} \tfrac{d}{du} \left\{ f_1(x_1) \left[\int_{x_2=-\infty}^{u-x_1} f_2(x_2) dx_2 \right] \right\} dx_1, \\
&\qquad \text{by the Leibnitz's rule (1.6.16)} \\
&= \int_{x_1=-\infty}^{\infty} f_1(x_1) \tfrac{d}{du} \left[\int_{x_2=-\infty}^{u-x_1} f_2(x_2) dx_2 \right] dx_1 \\
&= \int_{x_1=-\infty}^{\infty} f_1(x_1) f_2(u - x_1) dx_1 \text{ by (1.6.16),}
\end{aligned}
\qquad (4.2.11)
$$

which proves the result. ∎

> If X_1 and X_2 are independent continuous random variables, then the pdf of U given by (4.2.9) can be equivalently written as $\int_{x_2=-\infty}^{\infty} f_2(x_2) f_1(u - x_2) dx_2$ instead.

Example 4.2.9 Let X_1 and X_2 be independent exponential random variables, with their respective pdf's given by $f_1(x_1) = e^{-x_1}I(x_1 \in \Re^+)$ and $f_2(x_2) = e^{-x_2}I(x_2 \in \Re^+)$. What is the pdf of $U = X_1 + X_2$? It is obvious that the pdf $g(u)$ will be zero when $u \leq 0$, but it will be positive when $u > 0$. Using the convolution theorem, for $u > 0$, we can write $g(u) = \int_{x_1=0}^{u} e^{-x_1} e^{-(u-x_1)} dx_1 = \int_{x_1=0}^{u} e^{-u} dx_1 = e^{-u} \int_{x_1=0}^{u} dx_1 = ue^{-u}$ which coincides with the gamma pdf in (1.7.20) with $\alpha = 2$ and $\beta = 1$. In other words, the random variable U is distributed as Gamma(2, 1). See the related Exercise 4.2.19. ▲

A result analogous to the Theorem 4.2.1 for the product of two continuous independent random variables X_1 and X_2 has been set aside as the Exercise 4.2.13.

Example 4.2.10 Suppose that X_1 and X_2 are independent standard normal random variables with the respective pdf's $f_1(x_1) = \phi(x_1)$ and $f_2(x_2) = \phi(x_2)$ for $(x_1, x_2) \in \Re^2$. What is the pdf of $U = X_1 + X_2$? It is obvious that the pdf $g(u)$ will be positive for all $u \in \Re$. Using the convolution theorem, we can write $g(u) = \int_{-\infty}^{\infty} \phi(x_1)\phi(u - x_1)dx_1$. Now, we can simplify the expression of $g(u)$ as follows:

$$g(u) = \tfrac{1}{2\pi} \int_{-\infty}^{\infty} e^{-x^2/2} e^{-(u^2 - 2ux + x^2)/2} dx = \tfrac{1}{2\sqrt{\pi}} e^{-u^2/4} \int_{-\infty}^{\infty} h(x)dx,$$
$$(4.2.12)$$

where we let $h(x) = \frac{1}{\sqrt{\pi}} e^{-(x - \frac{1}{2}u)^2}$. But, note that for fixed u, the function $h(x)$ is the pdf of a $N(\frac{1}{2}u, \frac{1}{2})$ variable. Thus, $\int_{-\infty}^{\infty} h(x)dx$ must be one. Hence, from (4.2.12), we can claim that $g(u) = \frac{1}{2\sqrt{\pi}} e^{-u^2/4}$ for $u \in \Re$. But, $g(u)$ matches with the pdf of a $N(0, 2)$ variable and hence U is distributed as $N(0, 2)$. See the related Exercise 4.2.11. ▲

In the next example, X_1 and X_2 are *not independent*, but the convolution or the distribution function approach still works.

Example 4.2.11 Suppose that X_1 and X_2 have their joint pdf given by

$$f(x_1, x_2) = \begin{cases} 3x_1 & \text{if } 0 < x_2 < x_1 < 1 \\ 0 & \text{elsewhere.} \end{cases}$$

What is the pdf of $U = X_1 - X_2$? We first obtain the df $G(u)$ of U.

Obviously, $G(u) = 0$ for $u \leq 0$ or $u \geq 1$. With $0 < u < 1$, we write

$$
\begin{aligned}
G(u) \\
&= P\{X_1 \leq X_2 + u\} \\
&= \int_{x_1=0}^{u} \int_{x_2=0}^{x_1} 3x_1 dx_2 dx_1 + \int_{x_1=u}^{1} \int_{x_2=x_1-u}^{x_1} 3x_1 dx_2 dx_1 \\
&= 3 \int_{x_1=0}^{u} x_1^2 dx_1 + 3u \int_{x_1=u}^{1} x_1 dx_1 \\
&= \tfrac{1}{2}(3u - u^3).
\end{aligned}
$$

Thus, the pdf of U is given by

$$
g(u) = \tfrac{d}{du} G(u) = \begin{cases} \tfrac{3}{2}(1 - u^2) & \text{for } 0 < u < 1 \\ 0 & \text{for } u \leq 0 \text{ and } u \geq 1. \end{cases}
$$

We leave out some of the intermediate steps as the Exercise 4.2.12. ▲

The convolution approach leads to the distribution of $U = X_1 + X_2$ where X_1, X_2 are iid Uniform$(0, 1)$. The random variable U is often said to have the *triangular* distribution. See the Exercise 4.2.9.

4.2.5 The Sampling Distribution

Suppose that we consider a large population of all adult men in a city and we wish to gather information regarding the *distribution* of heights of these individuals. Let X_j stand for the height of the jth individual selected randomly from the whole population, $j = 1, ..., n$. Here, n stands for the sample size. Since the population is large, we may assume from a practical point of view, that these n individuals are selected independently of each other. The simple random sampling *with replacement* (see the Example 1.7.6) would certainly generate independence. What is the *sampling distribution* of, say, the sample mean, $\overline{X} = n^{-1}\Sigma_{j=1}^{n} X_j$? How can we proceed to understand what this theoretical distribution of $\overline{X}$ may possibly mean to us in real life? A simple minded approach may consist of randomly selecting n individuals from the relevant population and record each individual's height. The data would look like n numbers $x_{1,1}, ..., x_{1,n}$ which will give rise to an observed value $\overline{X}$, namely $\overline{x}_1 = n^{-1}\Sigma_{i=1}^{n} x_{1,i}$. From the same population, if we contemplate selecting another batch of n individuals, drawn independently from the first set, and record their heights, we will end up with another data of n numbers $x_{2,1}, ..., x_{2,n}$ leading to another observed value of the sample mean $\overline{X}$, namely $\overline{x}_2 = n^{-1}\Sigma_{i=1}^{n} x_{2,i}$. Though we do not physically have to go through this process of independent resampling with the same sample of size n, we may think of such a hypothetical process of the identical replication, infinitely many times. Through these replications, we will finally arrive at the specific observed values $\overline{x}_i, i = 1, 2, \ldots$ for the sample mean $\overline{X}$ based on a sample of size n. In real life, however, it will be impossible to replicate the process infinitely many times, but suppose instead that we replicate one hundred times thereby coming up with the observed values $\overline{x}_i, i = 1, 2, \ldots, 100$. One can easily draw a *relative frequency histogram* for this data consisting of the observed values $\overline{x}_i, i = 1, 2, \ldots, 100$. The shape of this histogram will give us important clues regarding the nature of the theoretical distribution of the sample mean, $\overline{X}$. If we generated more than one hundred $\overline{x}_i$ values, then the observed relative frequency histogram and the true pmf or pdf of $\overline{X}$ would have more similarities. With this perception of the independent resampling again and again, the adjective "sampling" is attached when we talk about the "distribution" of $\overline{X}$. In statistical applications, the sample mean $\overline{X}$ frequently arises and its theoretical distribution, as told by its pmf or the pdf, is customarily referred to as the *sampling distribution* of $\overline{X}$. In the same vein, one can think about the sampling distributions of many other characteristics, for example, the sample median, sample standard deviation or the sample maximum in a random sample of size n from an appropriate population under consideration. In practice, for example, it is not uncommon to hear about the sampling distribution of the median income in a population or the sampling distribution of the record rainfall data in a particular state over a period. We would use both phrases, the sampling distribution and distribution, quite interchangeably.

Example 4.2.12 (Example 4.2.4 Continued) Let X be a random variable with its pdf

$$f(x) = \begin{cases} \frac{3}{7}x^2 & \text{if } 1 < x < 2 \\ 0 & \text{elsewhere.} \end{cases} \qquad (4.2.13)$$

Now, suppose that we wish to select 1000 random samples from this population. How can we accomplish this? Observe that the df of X is

$$F(x) = \begin{cases} \frac{1}{7}(x^3 - 1) & \text{if } 1 < x < 2 \\ 0 & \text{elsewhere} \end{cases} \qquad (4.2.14)$$

and we know that $F(X)$ must be distributed as the Uniform$(0, 1)$ random variable. Using the MINITAB Release 12.1 and its uniform distribution generator, we first obtain 1000 observed values $u_1, u_2, \ldots, u_{1000}$ from the Uniform$(0, 1)$ distribution. Then, we let

$$F(x_i) = \tfrac{1}{7}(x_i^3 - 1) = u_i \Rightarrow x_i = (7u_i + 1)^{1/3}, i = 1, \ldots, 1000. \qquad (4.2.15)$$

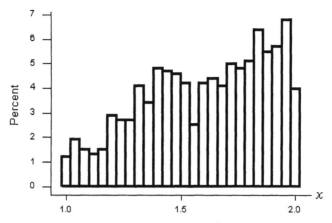

Figure 4.2.1. The Histogram of 1000 Sample Values Drawn
According to the PDF $f(x)$ from (4.2.13)

These x_i's form a random sample of size 1000 from a population with
its pdf $f(x)$ given by (4.2.13). Next, using the MINITAB Release 12.1, a
histogram was constructed from these 1000 sample values $x_1, x_2, ..., x_{1000}$
which is presented in the Figure 4.2.1. This histogram provides a sample-
based picture approximating the true pdf given by (4.2.13).

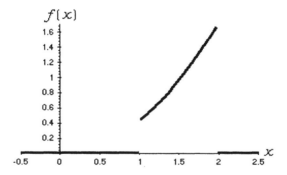

Figure 4.2.2. The Plot of the PDF $f(x)$ from (4.2.13)

In this situation, we know the exact form of the pdf $f(x)$ which is plotted between -0.5 and 2.5 in the Figure 4.2.2 with the help of MINITAB,
Release 12.1. Visual examinations reveal that there are remarkable similarities between the histogram from the Figure 4.2.1 and the plot of $f(x)$
from the Figure 4.2.2. In practice, a sample-based histogram is expected to
be a good approximation of the population distribution, particularly if the
sample size is not too small. ▲

4.3 Using the Moment Generating Function

The moment generating function (mgf) of a random variable was introduced in the Section 2.3. In the Section 2.4 we had emphasized that a finite mgf indeed pinpoints a unique distribution (Theorem 2.4.1). The implication we wish to reinforce is that a finite mgf corresponds to the probability distribution of a uniquely determined random variable. For this reason alone, the finite mgf's can play important roles in deriving the distribution of functions of random variables.

One may ponder over what kinds of functions of random variables we should be looking at in the first place. We may point out that in statistics one frequently faces the problem of finding the distribution of a very special type of function, namely a linear function of independent and identically distributed (iid) random variables. The sample mean and its distribution, for example, fall in this category. In this vein, the following result captures the basic idea. It also provides an important tool for future use.

Theorem 4.3.1 *Consider a real valued random variable X_i having its mgf $M_{X_i}(t) = E(e^{tX_i})$ for $i = 1, ..., n$. Suppose that $X_1, ..., X_n$ are independent. Then, the mgf of $U = \Sigma_{i=1}^{n} a_i X_i$ is given by $\Pi_{i=1}^{n} M_{X_i}(ta_i)$, where $a_1, ..., a_n$ are any arbitrary but otherwise fixed real numbers.*

Proof The proof merely uses the Theorem 3.5.1 which allows us to split the expected value of a product of n *independent* random variables as the product of the n individual expected values. We write

$$
\begin{aligned}
M_U(t) \;\; &= E[exp(tU)] \\
&= E[exp\{t\Sigma_{i=1}^{n} a_i X_i\}] \\
&= \Pi_{i=1}^{n} E[exp\{ta_i X_i\}], \because \; X\text{'s are independent and} \\
& \qquad \text{using Theorem 3.5.1 with } g_i(x_i) = exp(ta_i x_i) \\
&= \Pi_{i=1}^{n} M_{X_i}(ta_i),
\end{aligned}
$$

which completes the proof. ∎

> First, find the mgf $M_U(t)$. Visually match this mgf with that of one of the standard distributions. Then, U has that same distribution.

A simple looking result such as the Theorem 4.3.1 has deep implications. First, the mgf $M_U(t)$ of U is determined by invoking Theorem 4.3.1. In view of the Theorem 2.4.1, since the mgf of U and the distribution of U correspond uniquely to each other, all we have to do then is to match the form of this mgf $M_U(t)$ with that of one of the standard distributions. This way we will identify the distribution of U. There are many situations where this simple approach works just fine.

Example 4.3.1 (Example 4.2.3 Continued) Let $X_1, ..., X_k$ be independent and let X_i be Binomial(n_i, p), $i = 1, ..., k$. Write $U = \Sigma_{i=1}^k X_i$, $q = 1-p$, $n = \Sigma_{i=1}^k n_i$. Then invoking Theorem 4.3.1 and using the expression of the mgf of the Binomial (n_i, p) random variable from (2.3.5), one has $M_U(t) = \Pi_{i=1}^k (q + pe^t)^{n_i} = (q + pe^t)^{\Sigma_{i=1}^k n_i} = (q + pe^t)^n$. The final expression of $M_U(t)$ matches with the expression for the mgf of the Binomial (n, p) random variable found in (2.3.5). Thus, we can claim that the sampling distribution of U is Binomial with parameters n and p since the correspondence between a distribution and its finite mgf is unique. ▲

Example 4.3.2 Suppose that we roll a fair die n times and let X_i denote the score, that is the number on the face of the die which lands upward on the ith toss, $i = 1, ..., n$. Let $U = \Sigma_{i=1}^n X_i$, the total score from the n tosses. What is the probability that $U = 15$? The technique of full enumeration will be very tedious. On the other hand, let us start with the mgf of any of the X_i's which is given by $M_{X_i}(t) = \frac{1}{6}e^t(1 + e^t + ... + e^{5t})$ and hence $M_U(t) = \frac{1}{6^n}e^{nt}(1 + e^t + ... + e^{5t})^n$. In the full expansion of $M_U(t)$, the coefficient of e^{kt} must coincide with $P(X = k)$ for $k = n, n+1, ..., 6n-1, 6n$. This should be clear from the way the mgf of any discrete distribution is constructed. Hence, if $n = 4$, we have $P(X = 15) = \frac{1}{6^4}\{$coefficient of e^{11t} in the expansion of $(1 + e^t + ... + e^{5t})^4\} = \frac{140}{6^4}$. But, if $n = 10$, we have $P(X = 15) = \frac{1}{6^{10}}\{$coefficient of e^{5t} in the expansion of $(1 + e^t + ... + e^{5t})^{10}\} = \frac{2002}{6^{10}}$. One can find the pmf of U as an exercise. ▲

Example 4.3.3 Let $X_1, ..., X_n$ be independent random variables, X_i distributed as $N(\mu_i, \sigma_i^2)$, $i = 1, ..., n$. Write $U = \Sigma_{i=1}^n X_i$, $\mu = \Sigma_{i=1}^n \mu_i$ and $\sigma^2 = \Sigma_{i=1}^n \sigma_i^2$. Then invoking Theorem 4.3.1 and using the expression for the mgf of the $N(\mu_i, \sigma_i^2)$ random variable from (2.3.16), one has $M_U(t) = \Pi_{i=1}^n \{e^{t\mu_i + \frac{1}{2}t^2\sigma_i^2}\} = e^{t\Sigma_{i=1}^n \mu_i + \frac{1}{2}t^2\Sigma_{i=1}^n \sigma_i^2} = e^{t\mu + \frac{1}{2}t^2\sigma^2}$. The final expression of $M_U(t)$ matches with the expression for the mgf of the $N(\mu, \sigma^2)$ random variable found in (2.3.16). Thus, we can claim that the sampling distribution of U is normal with parameters μ and σ^2 since the correspondence between a distribution and its finite mgf is unique. ▲

Example 4.3.4 Let $X_1, ..., X_n$ be independent random variables, X_i having a Gamma distribution with the pdf $f_i(x) = c_i e^{-x/\beta} x^{\alpha_i - 1}$ with $c_i = \{\beta^{\alpha_i} \Gamma(\alpha_i)\}^{-1}$, for $0 < x, \alpha_i, \beta < \infty$, $i = 1, ..., n$. Let us write $U = \Sigma_{i=1}^n X_i$, $\alpha = \Sigma_{i=1}^n \alpha_i$. Then, for $0 < t < \beta^{-1}$, as before one has $M_U(t) = \Pi_{i=1}^n (1-\beta t)^{-\alpha_i} = (1-\beta t)^{-\Sigma_{i=1}^n \alpha_i} = (1-\beta t)^{-\alpha}$. Here we used the expression for the mgf of the Gamma(α_i, β) random variable from (2.3.23). Thus, we can claim that the sampling distribution of U is Gamma(α, β) since the correspondence between a distribution and its finite mgf is unique. ▲

Example 4.3.5 Let $X_1, ..., X_n$ be independent random variables, X_i be distributed as $N(\mu_i, \sigma_i^2)$, $i = 1, ..., n$. With fixed but otherwise arbitrary real numbers $a_1, ..., a_n$, we write $U = \Sigma_{i=1}^n a_i X_i$ and then denoting $\mu = \Sigma_{i=1}^n a_i \mu_i$ and $\sigma^2 = \Sigma_{i=1}^n a_i^2 \sigma_i^2$, we claim along the lines of the Example 4.3.3 that the sampling distribution of U turns out to be $N(\mu, \sigma^2)$. It is left as the Exercise 4.3.3. ▲

For the record, we now state the following results. Each part has already been verified in one form or another. It will, however, be instructive to supply direct proofs using the mgf technique under these special situations. These are left as the Exercise 4.3.8.

Theorem 4.3.2 (Reproductive Property of Independent Normal, Gamma and Chi-square Distributions) *Let $X_1, ..., X_n$ be independent random variables. Write $U = \Sigma_{i=1}^n X_i$ and $\overline{X} = n^{-1} U$. Then, one can conclude the following:*

(i)   *If X_i's have the common $N(\mu, \sigma^2)$ distribution, then U is distributed as $N(n\mu, n\sigma^2)$ and hence $\overline{X}$ is distributed as $N(\mu, \frac{1}{n}\sigma^2)$;*

(ii)   *If X_i's have the common* Gamma(α, β) *distribution, then U is distributed as* Gamma$(n\alpha, \beta)$;

(iii)   *If X_i has a* Gamma$(\frac{1}{2}\nu_i, 2)$ *distribution, that is a* Chi-square *distribution with ν_i degrees of freedom for $i = 1, ..., n$, then U is distributed as* Gamma$(\frac{1}{2}\nu, 2)$ *with $\nu = \Sigma_{i=1}^n \nu_i$, which is a* Chi-square *distribution with ν degrees of freedom.*

Example 4.3.6 (Example 4.1.1 Continued) Now, let us briefly go back to the Example 4.1.1. There, we had Z_1, Z_2 iid $N(0, 1)$ and in view of the Example 4.2.6 we can claim that Z_1^2, Z_2^2 are iid χ_1^2. Thus, by using the reproductive property of the independent Chi-squares, we note that the random variable $W = Z_1^2 + Z_2^2$ is distributed as the χ_2^2 random variable. The pdf of W happens to be $g(w) = \frac{1}{2}e^{-w/2}I(w > 0)$. Hence, in retrospect, (4.1.4) made good sense. ▲

4.4   A General Approach with Transformations

This is a more elaborate methodology which can help us to derive distributions of functions of random variables. We state the following result without giving its proof.

Theorem 4.4.1 *Consider a real valued random variable X whose pdf is $f(x)$ at the point x belonging to some subinterval $\mathcal{X}$ of the real line $\Re$. Suppose that we have a one-to-one function $g : \mathcal{X} \to \Re$ and let $g(x)$ be differentiable with respect to $x(\in \mathcal{X})$. Define the transformed random variable*

$Y = g(X)$. *Suppose that the domain space* $\mathcal{Y}$ *of* Y *is also a subinterval of* $\Re$. *Then, the pdf of* Y *is given by*

$$h(y) = f(g^{-1}(y))\left|\frac{d}{dy}g^{-1}(y)\right|, \tag{4.4.1}$$

for $y \in \mathcal{Y}$.

That is, in the expression of $f(x)$, the x variable is first replaced by the new variable $g^{-1}(y)$ everywhere and then we multiply the resulting expression by the absolute value of the Jacobian of transformation which is $\left|\frac{d}{dy}g^{-1}(y)\right|$. Incidentally, note that $\left|\frac{d}{dy}g^{-1}(y)\right|$ is evaluated as the absolute value of $\frac{dx}{dy}$ and then replacing the variable $x \in \mathcal{X}$ in terms of the new variable $y \in \mathcal{Y}$.

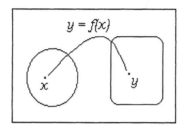

Figure 4.4.1. A Mapping of x to y

Recall the convention that when a pdf is written down with its support, it is understood that the density is zero elsewhere.

Example 4.4.1 (Example 4.2.5 Continued) We write $W = F(X)$ where we have $g(x) = F(x)$ for $x \in (a, b)$, so that $x = F^{-1}(w)$ for $w \in (0, 1)$. This transformation from x to w is one-to-one. Now, $\frac{dw}{dx} = f(x) = f(F^{-1}(w))$. Then, using (4.4.1) the pdf of W becomes

$$g_1(w) = f(F^{-1}(w))\left|1/\{f(F^{-1}(w))\}\right| = 1, \text{ for } 0 < w < 1.$$

In other words, $W = F(X)$ is distributed uniformly on the interval $(0, 1)$. Next, we have $U = q(W)$ where $q(w) = -log(w)$ for $0 < w < 1$. This transformation from w to u is also one-to-one. Using (4.4.1) again, the pdf of U is obtained as follows:

$$g(u) = \left|\frac{d}{du}q^{-1}(u)\right| = e^{-u}, \text{ for } 0 < u < \infty$$

which shows that U is distributed as a standard exponential variable or Gamma$(1, 1)$. ▲

Example 4.4.2 Let U be a Uniform$(0,1)$ variable, and define $V = -log(U)$, $W = -2log(U)$. From the steps given in the Example 4.4.1, it immediately follows that V has the *standard exponential distribution*. Using (4.4.1) again, one should verify that the pdf of W is given by $g(w) = \frac{1}{2}e^{-w/2}I(w > 0)$ which coincides with the pdf of a Chi-square random variable with 2 degrees of freedom. ▲

In the case when $g(.)$ considered in the statement of the Theorem 4.4.1 is *not* one-to-one, the result given in the equation (4.4.1) needs some minor adjustments. Let us briefly explain the necessary modifications. We begin by partitioning the $\mathcal{X}$ space into $A_1, ..., A_k$ in such a way that the individual transformation $g : A_i \rightarrow \mathcal{Y}$ becomes one-to-one for each $i = 1, ..., k$. That is, when we restrict the mapping g on $A_i \rightarrow \mathcal{Y}$, given a specific value $y \in \mathcal{Y}$, we can find a unique corresponding x in A_i which is mapped into y, and suppose that we denote $x = g_i^{-1}(y), i = 1, ..., k$. Then one essentially applies the Theorem 4.4.1 on each set in the partition where g_i, which is the same as g restricted on $A_i \rightarrow \mathcal{Y}$, is one-to-one. The pdf $h(y)$ of Y is then obtained as follows:

$$h(y) = \Sigma_{i=1}^k f(g_i^{-1}(y)) \left| \frac{d}{dy} g_i^{-1}(y) \right| \text{ for } y \in \mathcal{Y}. \qquad (4.4.2)$$

Example 4.4.3 Suppose that Z is a standard normal variable and $Y = Z^2$. This transformation from z to y is *not* one-to one, but the two pieces of transformations from $z \in [0, \infty)$ to $y \in \Re^+$ and from $z \in (-\infty, 0)$ to $y \in \Re^+$ are individually one-to-one. Also recall that the pdf of Z, namely, $\phi(z)$ is symmetric about $z = 0$. Hence, for $0 < y < \infty$, we use (4.4.2) to write down the pdf $h(y)$ of Y as follows:

$$h(y) = \phi(y^{\frac{1}{2}}) \left| \frac{1}{2} y^{-\frac{1}{2}} \right| + \phi(-y^{\frac{1}{2}}) \left| -\frac{1}{2} y^{-\frac{1}{2}} \right| = \phi(y^{\frac{1}{2}}) y^{-\frac{1}{2}} = \frac{1}{\sqrt{2\pi}} e^{-\frac{1}{2}y} y^{-\frac{1}{2}},$$

which coincides with the pdf of a Chi-square variable with one degree of freedom. Recall the earlier Example 4.2.6 in this context where the same pdf was derived using a different technique. See also the Exercises 4.4.3-4.4.4. ▲

Example 4.4.4 Suppose that X has the Laplace or double exponential distribution with its pdf given by $f(x) = \frac{1}{2}e^{-|x|}$ for $x \in \Re$. Let the transformed variable be $Y = |X|$. But, $g(x) = |x| : \Re \rightarrow \Re^+$ is *not one-to-one*. Hence, for $0 < y < \infty$, we use (4.4.2) to write down the pdf $h(y)$ of Y as follows:

$$h(y) = \frac{1}{2}e^{-y} |+1| + \frac{1}{2}e^{-y} |-1| = e^{-y},$$

which coincides with the pdf of a standard exponential variable. See also the Exercise 4.4.5. ▲

4.4.1   Several Variable Situations

In this subsection, we address the case of several variables. We start with the machinery available for one-to-one transformations in a general case. Later, we include an example when the transformation is not one-to-one. First, we clarify the techniques in the two-dimensional case. One may review the Section 4.8 for some of the details about matrices.

Suppose that we start with real valued random variables $X_1, ..., X_n$ and we have the transformed real valued random variables $Y_i = g_i(X_1, ..., X_n)$, $i = 1, ..., n$ where the transformation from $(x_1, ..., x_n) \in \mathcal{X}$ to $(y_1, ..., y_n) \in \mathcal{Y}$ is assumed to be *one-to one*. For example, with $n = 3$, we may have on hand three transformed random variables $Y_1 = X_1 + X_2, Y_2 = X_1 - X_2$, and $Y_3 = X_1 + X_2 + X_3$.

We start with the joint pdf $f(x_1, ..., x_n)$ of $X_1, ..., X_n$ and first replace all the variables $x_1, ..., x_n$ appearing within the expression of $f(x_1, ..., x_n)$ in terms of the transformed variables $g_i^{-1}(y_1, ..., y_n)$, $i = 1, ..., n$. In other words, since the transformation $(x_1, ..., x_n)$ to $(y_1, ..., y_n)$ is one-to one, we can theoretically think of *uniquely* expressing x_i in terms of $(y_1, ..., y_n)$ and write $x_i \equiv b_i(y_1, ..., y_n), i = 1, ..., n$. Thus, $f(x_1, ..., x_n)$ will be replaced by $f(b_1, ..., b_n)$. Then, we multiply $f(b_1, ..., b_n)$ by the *absolute value of the determinant of the Jacobian matrix* of transformation, written exclusively involving the transformed variables $y_1, ..., y_n$ only. Let us define the matrix

$$J = \begin{pmatrix} \frac{\partial x_1}{\partial y_1} & \frac{\partial x_1}{\partial y_2} & \cdot & \cdot & \frac{\partial x_1}{\partial y_n} \\ \frac{\partial x_2}{\partial y_1} & \frac{\partial x_2}{\partial y_2} & \cdot & \cdot & \frac{\partial x_2}{\partial y_n} \\ \cdot & \cdot & \cdot & \cdot & \cdot \\ \cdot & \cdot & \cdot & \cdot & \cdot \\ \frac{\partial x_n}{\partial y_1} & \frac{\partial x_n}{\partial y_2} & \cdot & \cdot & \frac{\partial x_n}{\partial y_n} \end{pmatrix}. \tag{4.4.3}$$

The understanding is that each x_i variable is replaced by $b_i(y_1, ..., y_n), i = 1, ..., n$, while forming the matrix J. Let $det(J)$ stand for the *determinant* of the matrix J and $| det(J) |$ stand for the absolute value of $det(J)$. Then the joint pdf of the transformed random variables $Y_1, ..., Y_n$ is given by

$$g(y_1, ..., y_n) = f\left(b_1(y_1, ..., y_n), ..., b_n(y_1, ..., y_n)\right) | det(J) | \tag{4.4.4}$$

for $y \in \mathcal{Y}$.

On the surface, this approach may appear complicated but actually it is not quite so. The steps involved are explained by means of couple of examples.

Example 4.4.5 Let X_1, X_2 be independent random variables and X_i be distributed as Gamma$(\alpha_i, \beta), \alpha_i > 0, \beta > 0$, $i = 1, 2$. Define the transformed variables $Y_1 = X_1 + X_2, Y_2 = X_1/(X_1 + X_2)$. Then one can *uniquely* express $x_1 = y_1 y_2$ and $x_2 = y_1(1 - y_2)$. It is easy to verify that

$$J = \begin{pmatrix} y_2 & y_1 \\ 1 - y_2 & -y_1 \end{pmatrix},$$

so that $det(J) = -y_1 y_2 - y_1(1 - y_2) = -y_1$. Now, writing the constant c instead of the expression $\{\beta^{\alpha_1 + \alpha_2} \Gamma(\alpha_1) \Gamma(\alpha_2)\}^{-1}$, the joint pdf of X_1 and X_2 can be written as

$$f(x_1, x_2) = c \, e^{-(x_1 + x_2)/\beta} x_1^{\alpha_1 - 1} x_2^{\alpha_2 - 1}$$

for $0 < x_1, x_2 < \infty$. Hence using (4.4.4) we can rewrite the joint pdf of Y_1 and Y_2 as

$$
\begin{aligned}
g(y_1, y_2) &= c \, e^{-y_1/\beta} (y_1 y_2)^{\alpha_1 - 1} \{y_1(1 - y_2)\}^{\alpha_2 - 1} \, | -y_1 | \\
&= c \, e^{-y_1/\beta} y_1^{\alpha_1 + \alpha_2 - 1} y_2^{\alpha_1 - 1} (1 - y_2)^{\alpha_2 - 1}
\end{aligned}
\qquad (4.4.5)
$$

for $0 < y_1 < \infty, 0 < y_2 < 1$. The terms involving y_1 and y_2 in (4.4.5) factorize and also either variable's domain does not involve the other variable. Refer back to the Theorem 3.5.3 as needed. It follows that Y_1 and Y_2 are independent random variables, Y_1 is distributed as Gamma($\alpha_1 + \alpha_2, \beta$) and Y_2 is distributed as Beta(α_1, α_2) since we can rewrite c as $\{\beta^{\alpha_1 + \alpha_2} \Gamma(\alpha_1 + \alpha_2)\}^{-1} \{b(\alpha_1, \alpha_2)\}^{-1}$ where $b(\alpha_1, \alpha_2)$ stands for the *beta function*, that is

$$b(\alpha_1, \alpha_2) = \{\Gamma(\alpha_1) \Gamma(\alpha_2)\}/\Gamma(\alpha_1 + \alpha_2).$$

One may refer to (1.6.19) and (1.6.25)-(1.6.26) to review the gamma and beta functions. ▲

Example 4.4.6 (Example 4.4.5 Continued) Suppose that $X_1, ..., X_n$ are independent random variables where X_i is Gamma(α_i, β), $i = 1, ..., n$. In view of the Example 4.4.5, we can conclude that $X_1 + X_2 + X_3 = (X_1 + X_2) + X_3$ is then Gamma($\alpha_1 + \alpha_2 + \alpha_3, \beta$) whereas $X_1/(X_1 + X_2 + X_3)$ is Beta($\alpha_1, \alpha_2 + \alpha_3$) and these are independent random variables. Thus, by the mathematical induction one can claim that $\Sigma_{i=1}^n X_i$ is Gamma($\Sigma_{i=1}^n \alpha_i, \beta$), $X_1/\{\Sigma_{i=1}^n X_i\}$ is Beta($\alpha_1, \Sigma_{i=2}^n \alpha_i$), whereas these are also distributed independently. ▲

> In the next two examples one has X_1 and X_2 dependent.
> Also, the transformed variables Y_1 and Y_2 are dependent.

Example 4.4.7 Suppose that X_1 and X_2 have their joint pdf given by

$$f(x_1, x_2) = \begin{cases} 2e^{-(x_1 + x_2)} & \text{if } 0 < x_1 < x_2 < \infty \\ 0 & \text{elsewhere.} \end{cases}$$

Let $Y_1 = X_1$ and $Y_2 = X_1 + X_2$. We first wish to obtain the joint pdf of Y_1 and Y_2. Then, the goal is to derive the marginal pdf's of Y_1, Y_2.

The one-to-one transformation $(x_1, x_2) \rightarrow (y_1, y_2)$ leads to the inverse: $x_1 = y_1, x_2 = y_2 - y_1$ so that $|det(J)| = 1$. Now, $x_2 > 0$ implies that $0 < 2y_1 < y_2 < \infty$ since $y_1 < y_2 - y_1$. Thus, (4.4.4) leads to the following joint pdf of Y_1 and Y_2 :

$$h(y_1, y_2) = 2e^{-y_2} I(0 < 2y_1 < y_2 < \infty).$$

The marginal pdf's of Y_1, Y_2 can be easily verified as the following:

$$h_1(y_1) = \int_{2y_1}^{\infty} h(y_1, y_2) dy_2 = 2e^{-2y_1} \text{ for } 0 < y_1 < \infty,$$

$$h_2(y_2) = \int_0^{y_2/2} h(y_1, y_2) dy_1 = y_2 e^{-y_2} \text{ for } 0 < y_2 < \infty.$$

We leave out some of the intermediate steps as the Exercise 4.4.6. ▲

Example 4.4.8 Suppose that X_1 and X_2 have their joint pdf given by

$$f(x_1, x_2) = \begin{cases} 3(x_1 + x_2) & \text{if } 0 < x_1, x_2 < 1, 0 < x_1 + x_2 < 1 \\ 0 & \text{elsewhere.} \end{cases}$$

Let $Y_1 = X_1 + X_2$ and $Y_2 = X_1 - X_2$. We first wish to obtain the joint pdf of Y_1 and Y_2. Then, the goal is to derive the marginal pdf's of Y_1, Y_2.

The one-to-one transformation $(x_1, x_2) \rightarrow (y_1, y_2)$ leads to the inverse: $x_1 = \frac{1}{2}(y_1 + y_2), x_2 = \frac{1}{2}(y_1 - y_2)$ so that $|det(J)| = \frac{1}{2}$. Observe that: $0 < x_1 < 1 \Rightarrow 0 < y_1 + y_2 < 2$; $0 < x_2 < 1 \Rightarrow 0 < y_1 - y_2 < 2$; $0 < x_1 + x_2 < 1 \Rightarrow 0 < y_1 < 1$. Let $\mathcal{Y} = \{(y_1, y_2) \in \Re : 0 < y_1 < 1, 0 < y_1 + y_2 < 2, 0 < y_1 - y_2 < 2\}$. The joint pdf of Y_1 and Y_2 is then given by

$$h(y_1, y_2) = \frac{3}{2} y_1 I((y_1, y_2) \in \mathcal{Y}).$$

The marginal pdf's of Y_1, Y_2 can be easily verified as the following:

$$h_1(y_1) \quad = \int_{-y_1}^{y_1} h(y_1, y_2) dy_2 = 3y_1^2 \text{ for } 0 < y_1 < 1;$$

$$h_2(y_2) \quad = \begin{cases} \int_{-y_2}^1 h(y_1, y_2) dy_1 = \frac{3}{4}(1 - y_2^2) \text{ for } -1 < y_2 < 0 \\ \int_{y_2}^1 h(y_1, y_2) dy_1 = \frac{3}{4}(1 - y_2^2) \text{ for } 0 < y_2 < 1 \end{cases}$$

$$= \frac{3}{4}(1 - y_2^2) \text{ for } -1 < y_2 < 1.$$

We leave out some of the intermediate steps as the Exercise 4.4.7. ▲

Example 4.4.9 The Helmert Transformation: This consists of a very special kind of *orthogonal transformation* from a set of n iid $N(\mu, \sigma^2)$

random variables $X_1, ..., X_n$, with $n \geq 2$, to a new set of n random variables $Y_1, ..., Y_n$ defined as follows:

$$
\begin{aligned}
Y_1 &= \tfrac{1}{\sqrt{n}}(X_1 + ... + X_n) \\
Y_2 &= \tfrac{1}{\sqrt{2}}(X_1 - X_2) \\
Y_3 &= \tfrac{1}{\sqrt{6}}(X_1 + X_2 - 2X_3)
\end{aligned}
$$

(4.4.6)

$$
Y_n = \tfrac{1}{\sqrt{n(n-1)}}\{X_1 + ... + X_{n-1} - (n-1)X_n\}.
$$

$Y_1, ..., Y_n$ so defined are referred to as the *Helmert variables.*

Let us denote the matrix

$$
A = \begin{pmatrix}
\frac{1}{\sqrt{n}} & \frac{1}{\sqrt{n}} & \frac{1}{\sqrt{n}} & \cdot\;\cdot & \frac{1}{\sqrt{n}} \\
\frac{1}{\sqrt{2}} & -\frac{1}{\sqrt{2}} & 0 & \cdot\;\cdot & 0 \\
\frac{1}{\sqrt{6}} & \frac{1}{\sqrt{6}} & -\frac{2}{\sqrt{6}} & \cdot\;\cdot & 0 \\
\cdot & \cdot & \cdot & \cdot & \cdot \\
\cdot & \cdot & \cdot & \cdot & \cdot \\
\frac{1}{\sqrt{n(n-1)}} & \frac{1}{\sqrt{n(n-1)}} & \frac{1}{\sqrt{n(n-1)}} & \cdot\;\cdot & -\frac{n-1}{\sqrt{n(n-1)}}
\end{pmatrix}. \quad (4.4.7)
$$

Then, one has $\mathbf{y} = A\mathbf{x}$ where $\mathbf{x}' = (x_1, ..., x_n)$ and $\mathbf{y}' = (y_1, ..., y_n)$.

$A_{n \times n}$ is an orthogonal matrix. So, A' is the inverse of A. This implies that $\Sigma_{i=1}^{n} x_i^2 = \mathbf{x}'\mathbf{x} = \mathbf{x}'A'A\mathbf{x} = \mathbf{y}'\mathbf{y} = \Sigma_{i=1}^{n} y_i^2$. In $\Re^n$, a sphere in the $\mathbf{x}$-coordinates continues to look like a sphere in the $\mathbf{y}$-coordinates when the $\mathbf{x}$ axes are rotated orthogonally to match with the new $\mathbf{y}$ axes.

Observe that the matrix J defined in (4.4.3) coincides with the matrix A' in the present situation and hence one can immediately write $| \det(J) | = | \det(A') | = | \{\det(AA')\}^{\frac{1}{2}} | = 1$.

Now, the joint pdf of $X_1, ..., X_n$ is given by

$$
f(x_1, ..., x_n) = \{\sigma\sqrt{2\pi}\}^{-n} exp\left\{-\tfrac{1}{2\sigma^2}\Sigma_{i=1}^{n}(x_i^2 - 2\mu x_i + \mu^2)\right\}
$$

for $-\infty < x_1, ..., x_n < \infty$, and thus using (4.4.4) we obtain the joint pdf of $Y_1, ..., Y_n$ as follows:

$$
\begin{aligned}
g(y_1, ..., y_n) \\
= \{\sigma\sqrt{2\pi}\}^{-n} \; exp\left\{-\tfrac{1}{2\sigma^2}(\Sigma_{i=1}^{n}y_i^2 - 2\mu\sqrt{n}y_1 + n\mu^2)\right\} \\
= \{\sigma\sqrt{2\pi}\}^{-n} \; exp\left\{-\tfrac{1}{2\sigma^2}(\Sigma_{i=2}^{n}y_i^2 + (y_1 - \sqrt{n}\mu)^2)\right\},
\end{aligned}
$$

(4.4.8)

where $-\infty < y_1, ..., y_n < \infty$. In (4.4.8), the terms involving $y_1, ..., y_n$ factorize and none of the variables' domains involve the other variables. Thus it is clear (Theorem 3.5.3) that the random variables $Y_1, ..., Y_n$ are *independently distributed*. Using such a factorization, we also observe that Y_1 is distributed as $N(\sqrt{n}\mu, \sigma^2)$ whereas $Y_2, ..., Y_n$ are iid $N(0, \sigma^2)$. ▲

The following result plays a crucial role in many statistical analyses. Incidentally, one should note that sometimes a population is also referred to as a *universe* in the statistical literature. The result we are about to mention is also indispensable in much of the distribution theory. These points will become clear in the sequel.

Theorem 4.4.2 (Joint Sampling Distribution of $\overline{X}$ and S^2 from a Normal Universe) *Suppose that $X_1, ..., X_n$ are iid $N(\mu, \sigma^2)$ random variables, $n \geq 2$. Define the sample mean $\overline{X} = n^{-1}\Sigma_{i=1}^n X_i$, and the sample variance $S^2 = (n-1)^{-1}\Sigma_{i=1}^n (X_i - \overline{X})^2$. Then, we have:*

(i) *The sample mean $\overline{X}$ is distributed independently of the sample variance S^2;*

(ii) *The sampling distribution of $\overline{X}$ is $N(\mu, \frac{1}{n}\sigma^2)$ and that of $(n-1)S^2/\sigma^2$ is Chi-square with $(n-1)$ degrees of freedom.*

Proof (i) Using the Helmert transformation from (4.4.6), we can rewrite $\overline{X} = n^{-\frac{1}{2}}Y_1$. Next, observe that

$$(n-1)S^2 = \Sigma_{i=1}^n X_i^2 - n\overline{X}^2 = \Sigma_{i=1}^n Y_i^2 - Y_1^2 = \Sigma_{i=2}^n Y_i^2, \qquad (4.4.9)$$

using the same Helmert variables. It is clear that $\overline{X}$ is a function of Y_1 alone, whereas from (4.4.9) we note that S^2 depends functionally on $(Y_2, ..., Y_n)$ only. But, since $Y_1, ..., Y_n$ are all independent, we conclude that $\overline{X}$ is distributed independently of S^2. Refer to the Theorem 3.5.2, part (ii) as needed.

(ii) Recall that $\overline{X} = n^{-\frac{1}{2}}Y_1$ where Y_1 is distributed as $N(\sqrt{n}\mu, \sigma^2)$ and so the sampling distribution of $\overline{X}$ follows immediately. From (4.4.9) again, it is clear that $(n-1)S^2/\sigma^2 = \Sigma_{i=2}^n Y_i^2/\sigma^2$ which is the sum of $(n-1)$ independent Chi-square random variables each having one degree of freedom. Hence, using the reproductive property of independent Chi-square variables (Theorem 4.3.2, part (iii)), it follows that $(n-1)S^2/\sigma^2$ has a Chi-square distribution with $(n-1)$ degrees of freedom. ∎

Remark 4.4.1 Let us reconsider the setup in the Example 4.4.9 and Theorem 4.4.2. It is important to note that the sample variance S^2 is the average of $Y_2^2, ..., Y_n^2$ and each Helmert variable *independently* contributes one degree of freedom toward the total $(n-1)$ degrees of freedom. Having n observations $X_1, ..., X_n$, the decomposition in (4.4.9) shows how exactly S^2 can be split up into $(n-1)$ *independent and identically distributed* components. This is the central idea which eventually leads to the *Analyses of Variance* techniques, used so widely in statistics.

> The Exercise 4.6.7 gives another transformation which proves that $\overline{X}$ and S^2 are independent when the X's are iid $N(\mu, \sigma^2)$.

Remark 4.4.2 Let us go back one more time to the setup considered in the Example 4.4.9 and Theorem 4.4.2. Another interesting feature can be noticed among the Helmert variables $Y_2, ..., Y_n$. Only the Helmert variable Y_n functionally depends on the last observed variable X_n. This particular feature has an important implication. Suppose that we have an additional observation X_{n+1} at our disposal beyond $X_1, ..., X_n$. Then, with $\overline{X}_{new} = (n+1)^{-1}\Sigma_{i=1}^{n+1}X_i$, the new sample variance S_{new}^2 and its decomposition would be expressed as

$$S_{new}^2 = n^{-1}\Sigma_{i=1}^{n+1}(X_i - \overline{X}_{new})^2 \Rightarrow nS_{new}^2 = \Sigma_{i=2}^{n}Y_i^2 + Y_{n+1}^2, \quad (4.4.10)$$

where $Y_{n+1} = \{X_1 + ... + X_n - nX_{n+1}\}/\sqrt{(n+1)n}$. Here, note that $Y_2, ..., Y_n$ based on $X_1, ..., X_n$ alone remain exactly same as in (4.4.6). In other words, the Helmert transformation shows exactly how the sample variance is affected in a sequential setup when we let n increase successively.

Remark 4.4.3 By extending the two-dimensional *polar transformation* mentioned in (4.1.2) to the n-dimension, one can supply an alternative proof of the Theorem 4.4.2. Indeed in Fisher's writings, one often finds derivations using the n-dimensional polar transformations and the associated geometry. We may also add that we could have used one among many choices of orthogonal matrices instead of the specific Helmert matrix A given by (4.4.7) in proving the Theorem 4.4.2. If we did that, then we would be hard pressed to claim useful interpretations like the ones given in our Remarks 4.4.1-4.4.2 which guide the readers in the understanding of some of the deep-rooted ideas in statistics.

> Suppose that $X_1, ..., X_n$ are iid random variables with $n \geq 2$. Then, $\overline{X}$ and S^2 are independent $\Rightarrow X_i$'s are normally distributed.

Remark 4.4.4 Suppose that from the iid random samples $X_1, ..., X_n$, one obtains the sample mean $\overline{X}$ and the sample variance S^2. If it is then assumed that $\overline{X}$ and S^2 are independently distributed, then effectively one is not assuming any less than normality of the original iid random variables. In other words, the independence of $\overline{X}$ and S^2 is a *characteristic property* of the normal distribution alone. This is a deep result in probability theory. For a proof of this fundamental characterization of a normal distribution and other historical notes, one may refer to Zinger (1958), Lukacs (1960), and Ramachandran (1967, Section 8.3), among a host of other sources. Look at the Exercises 4.4.22-4.4.23 which can be *indirectly* solved using this characterization.

Example 4.4.10 The Exponential Spacings: Let $X_1, ..., X_n$ be iid with the common pdf given by

$$f(x) = \beta^{-1} exp(-x/\beta) I(x > 0), \ \beta > 0. \qquad (4.4.11)$$

Define the order statistics $X_{n:1} \leq X_{n:2} \leq ... \leq X_{n:n}$ and let us write $Y_i = X_{n:i}, \ i = 1, ..., n$. The X_i's and $X_{n:i}$'s may be interpreted as the *failure times* and the *ordered failure times* respectively. Let us denote

$$U_1 = Y_1 \quad U_2 = Y_2 - Y_1 \quad U_3 = Y_3 - Y_2 \quad ... \quad U_n = Y_n - Y_{n-1} \quad (4.4.12)$$

and these are referred to as the *spacings* between the successive order statistics or failure times. In the context of a life-testing experiment, U_1 corresponds to the waiting time for the first failure, and U_i corresponds to the time between the $(i-1)^{th}$ and the ith failures, $i = 2, ..., n$. Here, we have a one-to-one transformation on hand from the set of n variables $(y_1, ..., y_n)$ to another set of n variables $(u_1, ..., u_n)$. Now, in view of (4.4.4), the joint pdf of the order statistics $Y_1, ..., Y_n$ is given by

$$g(y_1, ..., y_n)$$
$$= n! \beta^{-n} exp\left(-\Sigma_{i=1}^n y_i/\beta\right) \text{ if } 0 < y_1 \leq y_2 \leq ... \leq y_n < \infty. \qquad (4.4.13)$$

Note that $y_i = \Sigma_{j=1}^i u_j, \Sigma_{i=1}^n (n-i+1)u_i = \Sigma_{i=1}^n y_i, \ , i = 1, ..., n$. One can also verify that $| \ det(J) |= 1$, and thus using (4.4.4) and (4.4.13), the joint pdf of $U_1, .., U_n$ can be written as

$$h(u_1, ..., u_n)$$
$$= n! \beta^{-n} exp\left\{-[nu_1 + (n-1)u_2 + ... + 2u_{n-1} + u_n]/\beta\right\} \qquad (4.4.14)$$

for $0 < u_1, ..., u_n < \infty$. Next, (4.4.14) can be easily rewritten as

$$h(u_1, ..., u_n) = \Pi_{i=1}^n \{(n-i+1)\beta^{-1} \ exp\left[-(n-i+1)u_i/\beta\right]\} \qquad (4.4.15)$$

for $0 < u_1, ..., u_n < \infty$. In (4.4.15), the terms involving $u_1, ..., u_n$ factorize and none of the variables' domains involve any other variables. Thus it is clear (Theorem 3.5.3) that the random variables $U_1, ..., U_n$ are *independently distributed*. Using such a factorization, we also observe that U_i has an exponential distribution with the mean $(n-i+1)^{-1}\beta, i = 1, ..., n$. ▲

Example 4.4.11 (Example 4.4.10 Continued) Suppose that the X's are iid and the common pdf is the same as the one in (4.4.11). For all $k = 1, ..., n - 1$, observe that

$$X_{n:k+1} - X_{n:k} = Y_{k+1} - Y_k = U_{k+1},$$

and hence

$$E[X_{n:k+1} - X_{n:k}] = E[U_{k+1}] = (n - k)^{-1}\beta.$$

Since $X_{n:k} = \Sigma_{j=1}^k U_j$, we have

$$E[X_{n:k}] = \Sigma_{j=1}^k E[U_j] = \beta\left\{\frac{1}{n} + \frac{1}{n-1} + ... + \frac{1}{n-k+1}\right\}. \qquad (4.4.16)$$

It is worthwhile to note that we have succeeded in deriving an expression for the expected value of the kth order statistic $X_{n:k}$ without finding the marginal distribution of $X_{n:k}$. These techniques are particularly useful in the areas of reliability and survival analyses. ▲

If the X's are iid normal, the Helmert transformation provides a natural way to consider intricate properties of $\overline{X}$ and S^2. If the X's are iid exponential or negative exponential, the transformation involving the spacings between the successive order statistics is a natural one to consider instead.

Example 4.4.12 The Negative Exponential Distribution: Suppose that $X_1, ..., X_n$ are iid random variables having a common pdf given by

$$f(x) = \sigma^{-1}exp\left\{-(x - \mu)/\sigma\right\}I(x > \mu)$$

$$\text{for } -\infty < \mu < \infty, 0 < \sigma < \infty, \qquad (4.4.17)$$

where μ is the location parameter and σ is the scale parameter. In reliability applications, μ is often referred to as the minimum guarantee time or the minimum threshold, and hence μ is assumed positive in such applications. Refer back to (1.7.36) as needed. We will, however, continue to assume that μ is an arbitrary real number. Let us consider the two random variables

$$X_{n:1} \quad \text{and} \quad T = \Sigma_{i=1}^n (X_i - X_{n:1}) \qquad (4.4.18)$$

and look at their distributions. Denote $W_i = X_i - \mu$ and then it is clear that $W_1, ..., W_n$ are iid having the common pdf given by in (4.4.11) with β replaced by σ. Following the Example 4.4.10, we write $U_1 = W_{n:1}, U_2 =$

$W_{n:2} - W_{n:1}, ..., U_n = W_{n:n} - W_{n:n-1}$, and realize that $n(X_{n:1} - \mu)/\sigma$ is same as nU_1/σ, which has the standard exponential distribution. That is, $n(X_{n:1} - \mu)/\sigma$ is distributed as the standard exponential distribution which is the same as Gamma $(1,1)$. Also, recall that $U_1, ..., U_n$ must be *independent* random variables. Next, one notes that

$$
\begin{aligned}
T &= \Sigma_{i=1}^n (X_i - X_{n:1}) \\
&= \Sigma_{i=1}^n (W_i - W_{n:1}) \\
&= \Sigma_{i=1}^n (W_{n:i} - W_{n:1}), \quad \because \Sigma_{i=1}^n W_i = \Sigma_{i=1}^n W_{n:i} \quad (4.4.19) \\
&= \Sigma_{i=1}^n (n - i + 1)U_i - nU_1 \\
&= \Sigma_{i=2}^n (n - i + 1)U_i.
\end{aligned}
$$

Now, it is clear that T is distributed independently of $X_{n:1}$ because T functionally depends only on $(U_2, ..., U_n)$ whereas $X_{n:1}$ functionally depends only on U_1. But, U_1 is independent of $(U_2, ..., U_n)$. Also, note that

$$
2T\sigma^{-1} = \Sigma_{i=2}^n 2(n - i + 1)U_i\sigma^{-1} = \Sigma_{i=2}^n Z_{in}, \quad (4.4.20)
$$

where $Z_{2n}, ..., Z_{nn}$ are iid random variables having the Chi-square distribution with two degrees of freedom. Thus, using the reproductive property of independent Chi-square variables (Theorem 4.3.2, part (iii)), we conclude that $2\Sigma_{i=1}^n (X_i - X_{n:1})/\sigma$ has the Chi-square distribution with $2(n - 1)$ degrees of freedom. ▲

> Suppose that $X_1, ..., X_n$ are iid random variables. If their common distribution is negative exponential, then $X_{n:1}$ and $\Sigma_{i=1}^n (X_i - X_{n:1})$ are independent.

Remark 4.4.5 If we compare (4.4.20) with the representation given in (4.4.9), it may appear that in principle, the basic essence of the Remark 4.4.1 holds in this case too. It indeed does, but only partially. One realizes fast that in the present situation, one is forced to work with the spacings between the successive order statistics. Thus, the decomposition of $2T\sigma^{-1}$ into *unit independent components* consists of $(n - 1)$ random terms, each depending on the sample size n. This is fundamentally different from what we had observed in (4.4.9) and emphasized earlier in the Remark 4.4.2.

> In the next example X_1 and X_2 are independent, but the transformed variables Y_1 and Y_2 are dependent.

Example 4.4.13 (Example 4.4.5 Continued) Suppose that X_1 and X_2 are iid standard exponential random variables. Thus,

$$
f(x_1, x_2) = e^{-(x_1+x_2)} \text{ for } 0 < x_1, x_2 < \infty.
$$

We denote $y_1 = x_1 + x_2, y_2 = x_2$ so that for this one-to-one transformation we have $x_1 = y_1 - y_2, x_2 = y_2$ where $0 < y_2 < y_1 < \infty$. One can verify that $|det(J)| = 1$, and hence the joint pdf of Y_1 and Y_2 would become

$$g(y_1, y_2) = e^{-y_1} \text{ for } 0 < y_2 < y_1 < \infty. \qquad (4.4.21)$$

Then, from (4.4.21) we obtain the marginal pdf of Y_1 as

$$g_1(y_1) = \int_{y_2=0}^{y_1} e^{-y_1} dy_2 = y_1 e^{-y_1} \text{ for } 0 < y_1 < \infty.$$

In other words, $Y_1 = X_1 + X_2$ has the Gamma$(2, 1)$ distribution. We leave out the intermediate steps as the Exercise 4.4.16. ▲

> In the next example X_1, X_2 and X_3 are independent, but the transformed variables Y_1, Y_2 and Y_3 are dependent.

Example 4.4.14 (Example 4.4.13 Continued) Suppose that X_1, X_2 and X_3 are iid standard exponential random variables. Thus,

$$f(x_1, x_2, x_3) = e^{-(x_1+x_2+x_3)} \text{ for } 0 < x_1, x_2, x_3 < \infty.$$

We denote $y_1 = x_1 + x_2 + x_3, y_2 = x_2, y_3 = x_3$ so that for this one-to-one transformation we have $x_1 = y_1 - y_2 - y_3, x_2 = y_2, x_3 = y_3$ where $0 < y_2 < y_1 < \infty, 0 < y_3 < y_1 < \infty$ and $y_2 + y_3 < y_1$. One can verify that $|det(J)| = 1$, and hence the joint pdf of Y_1, Y_2 and Y_3 would become

$$g(y_1, y_2, y_3) = e^{-y_1} \text{ for } 0 < y_2 < y_1 < \infty, 0 < y_3 < y_1 < \infty, \\ \text{and } y_2 + y_3 < y_1. \qquad (4.4.22)$$

Then, from (4.4.22) we obtain the marginal pdf of Y_1 as

$$g_1(y_1) = \int_{y_3=0}^{y_1} \int_{y_2=0}^{y_1-y_3} e^{-y_1} dy_2 dy_3 = \tfrac{1}{2} y_1^2 e^{-y_1} \text{ for } 0 < y_1 < \infty.$$

In other words, $Y_1 = X_1 + X_2 + X_3$ has the Gamma$(3, 1)$ distribution. We leave out the intermediate steps as the Exercise 4.4.17. ▲

> In the next example X_1 and X_2 are dependent, but the transformed variables Y_1 and Y_2 are independent.

Example 4.4.15 Suppose that the random vector (X_1, X_2) has the bivariate normal distribution, $N_2(0, 0, \sigma^2, \sigma^2, \rho)$ with $0 < \sigma < \infty, -1 < \rho < 1$.

Let us derive the joint pdf of $Y_1 = X_1 + X_2$ and $Y_2 = X_1 - X_2$. Refer to the Section 3.6 as needed. With $c = \{2\pi\sigma^2(1 - \rho^2)^{-1/2}\}^{-1}$, we start with

$$f(x_1, x_2) = c\; exp\left\{-\frac{1}{2\sigma^2(1 - \rho^2)}\left[x_1^2 - 2\rho x_1 x_2 + x_2^2\right]\right\},$$

where $-\infty < x_1, x_2 < \infty$. For the one-to-one transformation on hand, we have $x_1 = \frac{1}{2}(y_1 + y_2), x_2 = \frac{1}{2}(y_1 - y_2)$ with $0 < y_1, y_2 < \infty$. One can easily verify that $|det(J)| = \frac{1}{2}$, and hence the joint pdf of Y_1 and Y_2 would become

$$g(y_1, y_2) = c\; exp\left\{-\frac{1}{4\sigma^2(1 - \rho^2)}\left[(1 - \rho)y_1^2 + (1 + \rho)y_2^2\right]\right\} \qquad (4.4.23)$$

for $-\infty < y_1, y_2 < \infty$. In (4.4.23), the terms involving y_1, y_2 factorize and none of the variables' domains involve the other variable. Thus it is clear (Theorem 3.5.3) that the random variables Y_1, Y_2 are *independently distributed*. Using such a factorization, we also observe that $Y_1 = X_1 + X_2$ is distributed as $N(0, 2\sigma^2(1 + \rho))$ whereas $Y_2 = X_1 - X_2$ is distributed as $N(0, 2\sigma^2(1 - \rho))$. ▲

Suppose that the transformation from $(X_1, ..., X_n) \to (Y_1, ..., Y_n)$ is *not* one-to-one. Then, the result given in the equation (4.4.4) will need slight adjustments. Let us briefly explain the necessary modifications. We begin by partitioning the $\mathcal{X}$ space, that is the space for $(x_1, ..., x_n)$, into $A_1, ..., A_k$ in such a way that the associated transformation from $A_i \to \mathcal{Y}$, that is the space for $(y_1, ..., y_n)$, becomes separately one-to-one for each $i = 1, ..., k$. In other words, when we restrict the original mapping of $(x_1, ..., x_n) \to (y_1, ..., y_n)$ on $A_i \to \mathcal{Y}$, given a specific $(y_1, ..., y_n) \in \mathcal{Y}$, we can find a unique $(x_1, ..., x_n)$ in A_i which is mapped into $(y_1, ..., y_n)$. Given $(y_1, ..., y_n)$, let the associated x_j be expressed as $x_j = b_{ij}(y_1, ..., y_n), j = 1, ..., n, i = 1, ..., k$. Suppose that the corresponding Jacobian matrix is denoted by $J_i, i = 1, ..., k$. Then one essentially applies (4.4.4) on each piece $A_1, ..., A_k$ and the pdf $g(y_1, ..., y_n)$ of Y is obtained as follows:

$$g(y_1, ..., y_n) = \Sigma_{i=1}^k f\left(b_{i1}(y_1, ..., y_n), ..., b_{in}(y_1, ..., y_n)\right) \mid det(J_i) \mid \quad (4.4.24)$$

for $y \in \mathcal{Y}$. For a clear understanding, let us look at the following example.

Example 4.4.16 Suppose that X_1 and X_2 are iid $N(0, 1)$ variables. Let us denote $Y_1 = X_1/X_2$ and $Y_2 = X_2^2$. Obviously, the transformation $(x_1, x_2) \to (y_1, y_2)$ is *not* one-to-one. Given (y_1, y_2), the inverse solution would be $(y_1\sqrt{y_2}, \sqrt{y_2})$ or $(-y_1\sqrt{y_2}, -\sqrt{y_2})$. We should take $A_1 = \{(x_1, x_2) \in \Re^2 : x_2 > 0\}, A_2 = \{(x_1, x_2) \in \Re^2 : x_2 < 0\}$. Now, we apply

(4.4.24) directly as follows: With $-\infty < y_1 < \infty, 0 < y_2 < \infty$, we get

$g(y_1, y_2)$

$$= \frac{1}{2\pi} e^{-\{(y_1\sqrt{y_2})^2 + (\sqrt{y_2})^2\}/2} \left| det \begin{pmatrix} \sqrt{y_2} & \frac{y_1}{2\sqrt{y_2}} \\ 0 & \frac{1}{2\sqrt{y_2}} \end{pmatrix} \right|$$

$$+ \frac{1}{2\pi} e^{-\{(-y_1\sqrt{y_2})^2 + (-\sqrt{y_2})^2\}/2} \left| det \begin{pmatrix} -\sqrt{y_2} & -\frac{y_1}{2\sqrt{y_2}} \\ 0 & \frac{1}{2\sqrt{y_2}} \end{pmatrix} \right|$$

$$= \frac{1}{2\pi} e^{-\{1+y_1^2\}y_2/2}.$$

$$(4.4.25)$$

With $-\infty < y_1 < \infty$ and the substitution $u = \{1 + y_1^2\}y_2/2$, the marginal pdf $g_1(y_1)$ of Y_1 is then obtained from (4.4.25) as

$$\frac{1}{2\pi} \int_{y_2=0}^{\infty} e^{-\{1+y_1^2\}y_2/2} dy_2 = \frac{1}{2\pi} \frac{2}{\{1+y_1^2\}} \int_{u=0}^{\infty} e^{-u} du = \frac{1}{\pi\{1+y_1^2\}},$$

which coincides with the Cauchy pdf defined in (1.7.31). That is, X_1/X_2 has the Cauchy distribution.

Next, with the substitution $h(y_1) = \frac{1}{\sqrt{2\pi}}(y_2)^{-1/2} e^{-y_1^2/(2y_2^{-1})}$ for $0 < y_2 < \infty$, the marginal pdf $g_2(y_2)$ of Y_2 is then obtained from (4.4.25) as

$$\frac{1}{2\pi} \int_{y_1=-\infty}^{\infty} e^{-\{1+y_1^2\}y_2/2} dy_1 = \frac{1}{2\pi} e^{-y_2/2} \sqrt{2\pi}(y_2)^{-1/2} \int_{y_1=-\infty}^{\infty} h(y_1) dy_1.$$

$$(4.4.26)$$

But, note that $h(y_1)$ resembles the pdf of the $N(0, y_2^{-1})$ variable for any fixed $0 < y_2 < \infty$. Hence, $\int_{y_1=-\infty}^{\infty} h(y_1) dy_1$ must be one for any fixed $0 < y_2 < \infty$. That is, for all fixed $0 < y_2 < \infty$, we get

$$g_2(y_2) = \frac{1}{\sqrt{2\pi}} e^{-y_2/2}(y_2)^{-1/2},$$

which coincides with the pdf of the χ_1^2 variable. In other words, X_2^2 has the χ_1^2 distribution. ▲

The Exercise 4.4.18 provides a substantial generalization in the sense that X_1/X_2 can be claimed to have a Cauchy distribution even when (X_1, X_2) has the $N_2(0, 0, \sigma^2, \sigma^2, \rho)$ distribution. The transformation used in the Exercise 4.4.18 is also different from the one given in the Example 4.4.16.

4.5   Special Sampling Distributions

In the Section 1.7, we had listed some of the standard distributions by including their pmf or pdf. We recall some of the continuous distributions, particularly the normal, gamma, Chi-square, Student's t, the F and beta distributions. At this point, it will help to remind ourselves some of the techniques used earlier to find the moments of normal, Chi-square or gamma variates.

In this section, we explain how one constructs the Student's t and F variables. Through examples, we provide *some justifications* for the importance of both the Student's t and F variates. When we move to the topics of statistical inference, their importance will become even more convincing.

4.5.1   The Student's t Distribution

Definition 4.5.1 *Suppose that X is a* standard normal *variable, Y is a* Chi-square *variable with ν degrees of freedom, and that X and Y are independently distributed. Then, the random variable $W = \dfrac{X}{\sqrt{\{Y\nu^{-1}\}}}$ is said to have the Student's t distribution, or simply the t distribution, with ν degrees of freedom.*

Theorem 4.5.1 *The pdf of the random variable W mentioned in the Definition 4.5.1, and distributed as t_ν, is given by*

$$h(w) = a(1 + \tfrac{1}{\nu}w^2)^{-\frac{1}{2}(\nu+1)} \quad for \ -\infty < w < \infty,$$

with $a \equiv a(\nu) = \{\sqrt{\nu\pi}\}^{-1}\Gamma(\tfrac{1}{2}(\nu+1))\{\Gamma(\tfrac{1}{2}\nu)\}^{-1}, \nu = 1,2,3,\dots$.

Proof The joint pdf of X and Y is given by

$$f(x,y) = ce^{-x^2/2}e^{-y/2}y^{(\nu-2)/2} \tag{4.5.1}$$

for $-\infty < x < \infty, 0 < y < \infty$ where $c = \{\sqrt{2\pi}2^{\nu/2}\Gamma(\nu/2)\}^{-1}$. Let us denote $u = y$ and $w = x/\sqrt{\{y\nu^{-1}\}}$, so that the inverse transformation is given by $x = w\sqrt{u/\nu}$ and $y = u$. Note that $|J| = \sqrt{u/\nu}$. From (4.5.1), the joint pdf of U and W can be written as

$$g(u,w) = c\nu^{-1/2}e^{-u(1+w^2\nu^{-1})/2}u^{(\nu-1)/2}$$

for $0 < u < \infty, -\infty < w < \infty$. Thus, for $-\infty < w < \infty$ and with the substitution $s = u(1 + w^2\nu^{-1})/2$, the pdf $h(w)$ of W is given by

$$\int_{u=0}^{\infty} g(u,w)du$$
$$= c\nu^{-1/2}2^{(\nu+1)/2}(1 + w^2\nu^{-1})^{-(\nu+1)/2}\int_{s=0}^{\infty} e^{-s}s^{(\nu-1)/2}ds$$
$$= c\nu^{-1/2}2^{(\nu+1)/2}(1 + w^2\nu^{-1})^{-(\nu+1)/2}\Gamma((\nu+1)/2),$$

which matches with the intended result. Some of the details are left out as the Exercise 4.5.1. ∎

Once the pdf $h(w)$ is simplified in the case $\nu = 1$, we find that $h(w) = \pi^{-1}(1+w^2)^{-1}$ for $-\infty < w < \infty$. In other words, when $\nu = 1$, the Student's t distribution coincides with the Cauchy distribution. Verification of this is left as the Exercise 4.5.2.

In some related problems we can go quite far without looking at the pdf of the Student's t variable. This point is emphasized next.

A case in point: note that $-W$ thus defined can simply be written as $(-X)/\sqrt{(Y\nu^{-1})}$. Since (i) $-X$ is distributed as standard normal, (ii) $-X$ is distributed independently of Y, we can immediately conclude that W and $-W$ have identical distributions, that is the Student's t distribution is symmetric around zero. In other words, the pdf $h(w)$ is symmetric around $w = 0$. To conclude this, it is not essential to look at the pdf of W. On the other hand, one may arrive at the same conclusion by observing that the pdf $h(w)$ is such that $h(w) = h(-w)$ for all $w > 0$.

How about finding the moments of the Student's t variable? Is the pdf $h(w)$ essential for deriving the moments of W? The answer is: we do not really need it. For any positive integer k, observe that

$$E[W^k] = \nu^{k/2}E[X^kY^{-k/2}] = \nu^{k/2}E[X^k]E[Y^{-k/2}] \qquad (4.5.2)$$

as long as $E[Y^{-k/2}]$ is finite. We could split the expectation because X and Y are assumed independent. But, it is clear that the expression in (4.5.2) will lead to finite entities provided that appropriate negative moments of a Chi-square variable exist. We had discussed similar matters for the gamma distributions in (2.3.24)-(2.3.26).

By appealing to (2.3.26) and (4.5.2), we claim that for the Student's t variable W given in the Definition 4.5.1, we have $E(W)$ finite if $\frac{1}{2}\nu > -(-\frac{1}{2})$ that is if $\nu > 1$, whereas $E(W^2)$ is finite, that is $V(W)$ is finite when $\frac{1}{2}\nu > -(-1)$ or $\nu > 2$. One should verify the following claims:

$$E(W) = \nu^{\frac{1}{2}}E(X)E\left(Y^{-\frac{1}{2}}\right) = 0 \text{ if } \nu > 1; \qquad (4.5.3)$$

and also,

$$V(W) = \nu E(X^2)E(Y^{-1}) = \tfrac{1}{2}\nu\Gamma(\tfrac{1}{2}\nu - 1)\{\Gamma(\tfrac{1}{2}\nu)\}^{-1} = \nu/(\nu - 2) \text{ if } \nu > 2. \qquad (4.5.4)$$

Example 4.5.1 The One-Sample Problem: Suppose that $X_1, ..., X_n$ are iid $N(\mu, \sigma^2)$, $-\infty < \mu < \infty$, $0 < \sigma < \infty$, $n \geq 2$. Let us recall that $\sqrt{n}(\overline{X} - \mu)/\sigma$ is distributed as $N(0, 1)$, $(n - 1)S^2/\sigma^2$ is χ^2_{n-1}, and these two are also independent (Theorem 4.4.2). Then, we rewrite

$$\sqrt{n}(\overline{X} - \mu)/S = \left\{\sqrt{n}(\overline{X} - \mu)/\sigma\right\} / \left\{(n - 1)S^2/[\sigma^2(n - 1)]\right\}^{\frac{1}{2}}. \qquad (4.5.5)$$

In other words, $\sqrt{n}(\overline{X} - \mu)/S$ has the same representation as that of W with $\nu = n - 1$. Hence, we can claim that

$$
\begin{array}{c}
\sqrt{n}(\overline{X} - \mu)/S \text{ has the Student's } t \text{ distribution} \\
\text{with } \nu = (n - 1) \text{ degrees of freedom.}
\end{array} \tag{4.5.6}
$$

A standardized variable such as $\sqrt{n}(\overline{X} - \mu)/S$ is widely used in the sequel when the population mean μ and variance σ^2 are both unknown.

W. S. Gosset was a pioneer in the development of statistical methods for design and analysis of experiments. He is perhaps better known under the pseudonym "Student" than under his own name. In most of his papers, he preferred to use the pseudonym "Student" instead of his given name. His path-breaking 1908 paper gave the foundation of this t-distribution. ▲

Example 4.5.2 The Two-Sample Problem: Suppose that the random variables $X_{i1}, ..., X_{in_i}$ are iid $N(\mu_i, \sigma^2), i = 1, 2$, and that the X_{1j}'s are independent of the X_{2j}'s. With $n_i \geq 2$, let us denote

$$
\begin{array}{c}
\overline{X}_i = n_i^{-1}\Sigma_{j=1}^{n_i} X_{ij}, \ S_i^2 = (n_i - 1)^{-1}\Sigma_{j=1}^{n_i}(X_{ij} - \overline{X}_i)^2 \\
S_P^2 = (n_1 + n_2 - 2)^{-1}\left\{(n_1 - 1)S_1^2 + (n_2 - 1)S_2^2\right\}
\end{array} \tag{4.5.7}
$$

for $i = 1, 2$.

$$\boxed{S_P^2 \text{ is called the } pooled \ sample \ variance.}$$

Now, $(n_i - 1)S_i^2/\sigma^2$ is $\chi^2_{n_i-1}, i = 1, 2$, and these are also independent. Using the reproductive property of independent Chi-squares (Theorem 4.3.2, part (iii)) we claim that $\left\{(n_1 - 1)S_1^2 + (n_2 - 1)S_2^2\right\}\sigma^{-2}$ has a Chi-square distribution with $(n_1 + n_2 - 2)$ degrees of freedom. Also, $(\overline{X}_1, \overline{X}_2)$ and S_P^2 are independent. Along the lines of the Example 4.5.1, we can claim that

$$
\begin{array}{c}
\{n_1^{-1} + n_2^{-1}\}^{-\frac{1}{2}}[(\overline{X}_1 - \overline{X}_2) - (\mu_1 - \mu_2)]S_P^{-1} \text{ has the Student's } t \\
\text{distribution with } \nu = (n_1 + n_2 - 2) \text{ degrees of freedom.}
\end{array}
$$

$$(4.5.8)$$

This two-sample Student's t distribution is widely used in the statistical literature. ▲

4.5.2   The F Distribution

Definition 4.5.2 *Let X, Y be independent Chi-square random variables distributed respectively with ν_1 and ν_2 degrees of freedom. Then, the random variable $U = (X/\nu_1) \div (Y/\nu_2)$ is said to have the F distribution with degrees of freedom ν_1, ν_2, in that order.*

Theorem 4.5.2 *The pdf of the random variable U mentioned in the Definition 4.5.2, and distributed as F_{ν_1,ν_2}, is given by*

$$h(u) = ku^{\frac{1}{2}(\nu_1-2)}\{1 + (\nu_1/\nu_2)u\}^{-\frac{1}{2}(\nu_1+\nu_2)} \quad for\ 0 < u < \infty,$$

with $k \equiv k(\nu_1, \nu_2) = (\nu_1/\nu_2)^{\frac{1}{2}\nu_1}\Gamma((\nu_1 + \nu_2)/2)\{\Gamma(\nu_1/2)\Gamma(\nu_2/2)\}^{-1}$ and $\nu_1, \nu_2 = 1, 2, \dots$.

Proof The joint pdf of X and Y is given by

$$f(x, y) = ce^{-(x+y)/2}x^{(\nu_1-2)/2}y^{(\nu_2-2)/2} \tag{4.5.9}$$

for $0 < x, y < \infty$ where $c = \{2^{(\nu_1+\nu_2)/2}\Gamma(\nu_1/2)\Gamma(\nu_2/2)\}^{-1}$. Let us denote $u = \frac{\nu_2}{\nu_1}\frac{x}{y}$ and $v = y$, so that the inverse transformation is given by $x = \frac{\nu_1}{\nu_2}uv$ and $y = v$. Note that $|J| = \frac{\nu_1}{\nu_2}v$. From (4.5.9), the joint pdf of U and V can be written as

$$h(u, v) = c(\frac{\nu_1}{\nu_2})^{\nu_1/2}exp\{-v(1 + \frac{\nu_1}{\nu_2}u)/2\}v^{(\nu_1+\nu_2-2)/2}u^{(\nu_1-2)/2}$$

for $0 < u, v < \infty$. Thus, for $0 < u < \infty$, the pdf $h(u)$ of U is given by

$$c(\tfrac{\nu_1}{\nu_2})^{\nu_1/2}u^{(\nu_1-2)/2}\int_{v=0}^{\infty} exp\{-v(1 + \tfrac{\nu_1}{\nu_2}u)/2\}v^{(\nu_1+\nu_2-2)/2}dv$$

$$= c(\tfrac{\nu_1}{\nu_2})^{\nu_1/2}u^{(\nu_1-2)/2}\left\{2/(1 + \tfrac{\nu_1}{\nu_2}u)\right\}^{(\nu_1+\nu_2)/2}\Gamma((\nu_1 + \nu_2)/2),$$

by viewing the integrand as a gamma pdf,

which matches with the intended result. ∎

In some related problems we can go quite far without looking at the pdf of the F variable. This point is emphasized next.

From the Definition 4.5.1 of the Student's t random variable W, it is clear that $W^2 = X^2(Y/\nu)^{-1}$. Since (i) X^2, Y are distributed as Chi-squares respectively with one and ν degrees of freedom, (ii) X^2 and Y are also independent, we can conclude that W^2 has the same form as in the Definition 4.5.2 for U. Thus, the pdf of U is not essential to arrive at the conclusion:

t_ν^2 has the same distribution as that of $F_{1,\nu}$

The F distribution is *not symmetric* in the same sense as the t distribution is. But, from the Definition 4.5.2, it follows immediately though that $1/U$, which is the same as $(Y/\ \nu_2) \div (X/\ \nu_1)$, should be distributed as F_{ν_2,ν_1}. That is, $1/F$ has a F distribution too. This feature may be intuitively viewed as the "symmetry property" of the pdf $h(u)$. The explicit form of the pdf has not played any crucial role in this conclusion.

How about finding the moments of the F_{ν_1,ν_2} variable? The pdf $h(u)$ is not essential for deriving the moments of U. For any positive integer k, observe that

$$E[U^k] = (\nu_2/\nu_1)^k E[X^k Y^{-k}] = (\nu_2/\nu_1)^k E[X^k] E[Y^{-k}] \qquad (4.5.10)$$

as long as $E[Y^{-k}]$ is finite. We can split the expectation in (4.5.10) because X and Y are assumed independent. But, it is clear that the expression in (4.5.10) will lead to finite entities provided that appropriate negative moment of a Chi-square variable exists. We had discussed similar matters for the gamma distributions in (2.3.24)-(2.3.26).

By appealing to (2.3.26) and (4.5.10), we claim that for the F_{ν_1,ν_2} variable given in the Definition 4.5.2, we have $E(U)$ finite if $\nu_2 > 2$, whereas $E(U^2)$ is finite, that is $V(U)$ is finite when $\nu_2 > 4$. One should verify the following claims:

$$E[U] = (\nu_2/\nu_1)E[X]E[Y^{-1}] = \nu_2/(\nu_2 - 2) \text{ if } \nu_2 > 2; \qquad (4.5.11)$$

and also

$$\begin{aligned} V[U] &= (\nu_2/\nu_1)^2 E[X^2]E[Y^{-2}] - E^2[U] \\ &= [2\nu_2^2(\nu_1 + \nu_2 - 2)]/[\nu_1(\nu_2 - 2)^2(\nu_2 - 4)] \text{ if } \nu_2 > 4. \end{aligned} \qquad (4.5.12)$$

Example 4.5.3 The Two-Sample Problem: Let $X_{i1}, ..., X_{in_i}$ be iid $N(\mu_i, \sigma_i^2)$, $i = 1, 2$, and that the X_{1j}'s are independent of the X_{2j}'s. For $n_i \geq 2$, we denote $\overline{X}_i = n_i^{-1}\Sigma_{j=1}^{n_i}X_{ij}$, $S_i^2 = (n_i - 1)^{-1}\Sigma_{j=1}^{n_i}(X_{ij} - \overline{X}_i)^2$ as in the Example 4.5.2. Now, (i) $(n_i - 1)S_i^2/\sigma_i^2$ is $\chi_{n_i-1}^2$, $i = 1, 2$, (ii) they are also independent, and hence in view of the Definition 4.5.2, the random variable $U = (S_1/\sigma_1)^2 \div (S_2/\sigma_2)$ has the F distribution with degrees of freedom $\nu_1 = n_1 - 1, \nu_2 = n_2 - 1$. ▲

4.5.3 The Beta Distribution

Recall the random variable U mentioned in the Definition 4.5.2. With $k \equiv k(\nu_1,\nu_2) = (\nu_1/\nu_2)^{\frac{1}{2}\nu_1}\Gamma((\nu_1+\nu_2)/2)\{\Gamma(\nu_1/2)\Gamma(\nu_2/2)\}^{-1}$, the pdf of U was

$$h(u) = ku^{\frac{1}{2}(\nu_1-2)}\{1 + (\nu_1/\nu_2)u\}^{-\frac{1}{2}(\nu_1+\nu_2)} \qquad (4.5.13)$$

for $0 < u < \infty$. Now, let us define $Z = [(\nu_1/\nu_2)U]/[1 + (\nu_1/\nu_2)U]$. The transformation from $u \to z$ is one-to-one and we have $(\nu_1/\nu_2)u = z(1 -$

$z)^{-1}, 0 < u < \infty, 0 < z < 1$. One may also check that $(\nu_1/\nu_2)\frac{du}{dz} = (1-z)^{-2}$. Thus, combining (4.4.1) and (4.5.13) in a straightforward fashion, we can write down the pdf of Z as follows: For $0 < z < 1$, we have

$$g(z) = (\nu_2/\nu_1)^{\frac{1}{2}\nu_1} k z^{\frac{1}{2}\nu_1 - 1}(1 - z)^{\frac{1}{2}\nu_2 - 1}$$

which simplifies to

$$g(z) = \Gamma((\nu_1 + \nu_2)/2)\{\Gamma(\nu_1/2)\Gamma(\nu_2/2)\}^{-1} z^{\frac{1}{2}\nu_1 - 1}(1 - z)^{\frac{1}{2}\nu_2 - 1}. \quad (4.5.14)$$

It coincides with the beta pdf defined in (1.7.35) where $\alpha = \frac{1}{2}\nu_1$ and $\beta = \frac{1}{2}\nu_2$. That is,

$$\boxed{[(\nu_1/\nu_2)F_{\nu_1,\nu_2}]/[1 + (\nu_1/\nu_2)F_{\nu_1,\nu_2}] \text{ has Beta}(\tfrac{1}{2}\nu_1, \tfrac{1}{2}\nu_2) \text{ distribution.}}$$

4.6   Special Continuous Multivariate Distributions

We now include some interesting aspects of the *multivariate* normal, t, and F distributions. It will become clear shortly that both the multivariate t and F distributions are close associates of the multivariate normal distribution. One may review the Section 4.8 for some of the details about matrices.

Tong's (1990) book is devoted to the multivariate normal distributions and includes valuable tables. It briefly discusses the multivariate t and F distributions too. The references to the tables and other features for the multivariate normal, t and F distributions can be found in Johnson and Kotz (1972).

We included important properties of a bivariate normal distribution in the Section 3.6. The sampling distributions in the context of a bivariate normal population, however, is included in the present section.

4.6.1   The Normal Distribution

The bivariate normal density was given in (3.6.1). The general multivariate normal density is more involved. But, without explicitly referring to the pdf, one can derive many interesting and useful properties. The following broad definition of the *p-dimensional normality* can be found in Rao (1973, p. 518). The advantage of adopting this definition over another relying explicitly on the multivariate pdf will be clear from the Examples 4.6.1-4.6.2.

Definition 4.6.1 *A $p(\geq 1)$ random vector* $\mathbf{X} = (X_1, ..., X_p)$ *is said to have a p-dimensional normal distribution, denoted by N_p, if and only if* *each linear function $\Sigma_{i=1}^{p} a_i X_i$ has the univariate* normal *distribution for all fixed, but arbitrary real numbers* $a_1, ..., a_p$.

Example 4.6.1 Suppose that $X_1, ..., X_n$ are iid $N(\mu, \sigma^2)$, $-\infty < \mu < \infty$, $0 < \sigma < \infty$. Consider the sample mean $\overline{X}$, and let us look at the two-dimensional joint distribution of $(X_1, \overline{X})$. Obviously, $E(X_1) = E(\overline{X}) = \mu$, $V(X_1) = \sigma^2$ and $V(\overline{X}) = \frac{1}{n}\sigma^2$. Also, we can write $Cov(X_1, \overline{X}) = Cov\left(X_1, \frac{1}{n}\{X_1 + ...X_n\}\right) = \Sigma_{i=1}^{n} Cov(X_1, \frac{1}{n}X_i)$. Refer to the Theorem 3.4.3, part (iii) as needed. But, the covariance between X_1 and X_j is zero for $1 < j \le n$. That is, $Cov(X_1, \overline{X}) = \frac{1}{n}Cov(X_1, X_1) = \frac{1}{n}\sigma^2$ and hence $\rho_{X_1, \overline{X}} = 1/\sqrt{n}$. Any linear function L of X_1 and $\overline{X}$ is also a linear function of n original iid normal random variables $X_1, ..., X_n$, and hence L itself is distributed normally. This follows from the reproductive property of independent normal variables (Theorem 4.3.2, part (i)). So, by the Definition 4.6.1 it follows that $(X_1, \overline{X})$ is jointly distributed as a bivariate normal variable. Thus, the conditional distribution of X_1 given $\overline{X} = \overline{x}$ is $N\left(\overline{x}, \sigma^2(1 - \frac{1}{n})\right)$ which provides the following facts: $E(X_1 \mid \overline{X} = \overline{x}) = \overline{x}$ and $V(X_1 \mid \overline{X} = \overline{x}) = \sigma^2(1 - \frac{1}{n})$. Refer to the Theorem 3.6.1 as needed. Look at the Exercise 4.6.1 ▲

Example 4.6.2 (Example 4.6.1 Continued) One can easily find the conditional expectation of $aX_1 + bX_2$ given $\overline{X} = \overline{x}$ in the following way where a and b are fixed non-zero real numbers. Observe that $E(aX_1 + bX_2 \mid \overline{X} = \overline{x}) = E(aX_1 \mid \overline{X} = \overline{x}) + E(bX_2 \mid \overline{X} = \overline{x}) = (a + b)\overline{x}$. Also note, for example, that $E(X_1^2 \mid \overline{X} = \overline{x}) = V(X_1 \mid \overline{X} = \overline{x}) + \{E(X_1 \mid \overline{X} = \overline{x})\}^2 = \sigma^2(1 - \frac{1}{n}) + \overline{x}^2$. One can exploit similar techniques to obtain other expected values in this context. Look at the Exercise 4.6.2. ▲

Example 4.6.3 (Example 4.6.1 Continued) A result such as $E(X_1 \mid \overline{X} = \overline{x}) = \overline{x}$ can be alternately derived as follows. Obviously, $E(\overline{X} \mid \overline{X} = \overline{x}) = \overline{x}$, and this can be rewritten as $\overline{x} = \frac{1}{n}E\{\Sigma_{i=1}^{n} X_i \mid \overline{X} = \overline{x}\} = \frac{1}{n}nE(X_1 \mid \overline{X} = \overline{x}) = E(X_1 \mid \overline{X} = \overline{x})$ which was the intended result in the first place. This argument works here because $E\{X_i \mid \overline{X} = \overline{x}\} = E(X_1 \mid \overline{X} = \overline{x})$ for each $i = 1, ..., n$. Observe that this particular approach merely used the fact that the X_i's are iid, but it did not exploit the fact that the X_i's are iid normal to begin with. ▲

Example 4.6.4 (Example 4.6.2 Continued) A result such as $E(X_1^2 \mid \overline{X} = \overline{x}) = \sigma^2(1 - \frac{1}{n}) + \overline{x}^2$ can be alternately derived as follows. Obviously, $\sigma^2 = E(S^2) = E(S^2 \mid \overline{X} = \overline{x})$ since S^2 and $\overline{X}$ are independent. Here, we have indeed used the full power of the assumption of normality. Recall the Remark 4.4.4 in this context. So one can write $(n-1)\sigma^2 = E\{\Sigma_{i=1}^{n}(X_i - \overline{X})^2 \mid \overline{X} = \overline{x}\} = E\{\Sigma_{i=1}^{n}X_i^2 - n\overline{X}^2 \mid \overline{X} = \overline{x}\} = E\{\Sigma_{i=1}^{n}X_i^2 \mid \overline{X} = \overline{x}\} - nE\{\overline{X}^2 \mid \overline{X} = \overline{x}\} = nE\{X_1^2 \mid \overline{X} = \overline{x}\} - n\overline{x}^2$ and hence $nE\{X_1^2 \mid \overline{X} = $

$\bar{x}\} = n\bar{x}^2 + (n-1)\sigma^2$. This leads to the desired conclusion: $E\{X_1^2 \mid \overline{X} = \bar{x}\} = \bar{x}^2 + (1 - \frac{1}{n})\sigma^2$. ▲

For completeness, we now give the pdf of the p-dimensional normal random vector. To be specific, let us denote a p-dimensional *column vector* $\mathbf{X}$ whose *transpose* is $\mathbf{X'} = (X_1, ..., X_p)$, consisting of the real valued random variable X_i as its ith component, $i = 1, ..., p$. We denote $E[X_i] = \mu_i, V[X_i] = \sigma_{ii}$, and $Cov(X_i, X_j) = \sigma_{ij}, 1 \le i \ne j \le p$. Suppose that we denote the mean vector $\boldsymbol{\mu}$ where $\boldsymbol{\mu'} = (\mu_1, ..., \mu_p)$ and then write down σ_{ij} as the $(i, j)^{th}$ element of the $p \times p$ matrix $\boldsymbol{\Sigma}, 1 \le i \ne j \le p$. Then, $\boldsymbol{\Sigma} = (\sigma_{ij})_{p \times p}$ is referred to as the *variance-covariance matrix* or the *dispersion matrix* of the random vector $\mathbf{X}$.

Assume that the matrix $\boldsymbol{\Sigma}$ has the *full rank* which is equivalent to saying that $\boldsymbol{\Sigma}^{-1}$ exists. Then, the random vector $\mathbf{X}$ has the p-dimensional normal distribution with mean vector $\boldsymbol{\mu}$ and dispersion matrix $\boldsymbol{\Sigma}$, denoted by $N_p(\boldsymbol{\mu}, \boldsymbol{\Sigma})$, provided that the joint pdf of $X_1, ..., X_p$ is given by

$$f(\mathbf{x}) = c \; exp\left\{-\tfrac{1}{2}(\mathbf{x} - \boldsymbol{\mu})'\boldsymbol{\Sigma}^{-1}(\mathbf{x} - \boldsymbol{\mu})\right\}, \text{ with } \mathbf{x} \in \Re^p$$
$$\text{and } c = \left\{(2\pi)^{p/2}\left[det\,(\boldsymbol{\Sigma})\right]^{1/2}\right\}^{-1}, \tag{4.6.1}$$

where $\mathbf{x} = (x_1, ..., x_p)$. One should check that the bivariate normal pdf from (3.6.1) can be written in this form too. We leave it as the Exercise 4.6.3.

C. F. Gauss originally derived the density function given in (4.6.1) from that of linear functions of independent normal variables around 1823-1826. In many areas, including the sciences and engineering, the normal distributions are also frequently referred to as the *Gaussian distributions*.

| Suppose that $X_1, ..., X_p$ are jointly distributed as $N_p(\boldsymbol{\mu}, \boldsymbol{\Sigma})$. Then, $X_1, ..., X_p$ are independent $\Leftrightarrow$ $\boldsymbol{\Sigma}$ is a diagonal matrix. | (4.6.2) |

| Let $X_1, ..., X_p$ be jointly distributed as $N_p(\boldsymbol{\mu}, \boldsymbol{\Sigma})$. Then, any sub-vector $X_{i_1}, X_{i_2}, ..., X_{i_k}$ are jointly distributed as $N_k(\boldsymbol{\theta}, \mathbf{A})$ where $\boldsymbol{\theta'} = (\mu_{i_1}, \mu_{i_2}, ..., \mu_{i_k})$ and the matrix $\mathbf{A}_{k \times k}$ is the part of $\boldsymbol{\Sigma}$ corresponding to the rows/columns numbered $i_1, i_2, ..., i_k, 1 \le i_1 < i_2 < ... < i_k \le p$. | (4.6.3) |

We leave the proofs of (4.6.2)-(4.6.3) as the Exercise 4.6.4.

Example 4.6.5 (Exercise 3.5.17 Continued) Suppose that the random vector $\mathbf{X} = (X_1, X_2, X_3, X_4)$ has the 4-dimensional normal distribution namely, $N_4(\boldsymbol{\mu}, \boldsymbol{\Sigma})$ with $\boldsymbol{\mu} = \mathbf{0}$ and

$$\boldsymbol{\Sigma} = \begin{pmatrix} 4 & 0 & 0 & 1 \\ 0 & 4 & 0 & 1 \\ 0 & 0 & 4 & 1 \\ 1 & 1 & 1 & 1 \end{pmatrix} \text{ which is partitioned into } \begin{pmatrix} P & Q \\ Q' & S \end{pmatrix}$$

with the matrices and vectors $P_{3\times 3} = 4I, Q'_{1\times 3} = (111)$ and $S_{1\times 1} = 1$. The matrix $\boldsymbol{\Sigma}$ is p.d. and it is easy to check that $det(\boldsymbol{\Sigma}) = 16$.

Let us denote the matrices

$$E_{1\times 1} = S - Q'P^{-1}Q = 1 - \tfrac{1}{4}\,\mathbf{1}'I_{3\times 3}\mathbf{1} = 1 - \tfrac{3}{4} = \tfrac{1}{4},$$

$$F_{3\times 1} = P^{-1}Q = \tfrac{1}{4}I_{3\times 3}\mathbf{1} = \tfrac{1}{4}\mathbf{1}.$$

Thus, applying (4.8.10), we have

$$\boldsymbol{\Sigma}^{-1} = \begin{pmatrix} 1/2 & 1/4 & 1/4 & -1 \\ 1/4 & 1/2 & 1/4 & -1 \\ 1/4 & 1/4 & 1/2 & -1 \\ -1 & -1 & -1 & 4 \end{pmatrix}$$

and hence

$$\mathbf{x}'\boldsymbol{\Sigma}^{-1}\mathbf{x} = \tfrac{1}{2}\{x_1^2 + x_2^2 + x_3^2 + x_1x_2 + x_1x_3 + x_2x_3\} + 4x_4^2$$

$$-2\{x_1x_4 + x_2x_4 + x_3x_4\} = u, \text{ say.}$$

In other words, in this particular situation, the pdf from (4.6.1) will reduce to

$$f(x_1, x_2, x_3, x_4) = \tfrac{1}{16\pi^2}\, exp\{-\tfrac{1}{2}u\}$$

for $(x_1, x_2, x_3, x_4) \in \mathfrak{R}^4$. Thus, in the Exercise 3.5.17, the given random vector $\mathbf{X}$ actually had this particular 4-dimensional normal distribution even though at that time we did not explicitly say so. ▲

Sampling Distributions : The Bivariate Normal Case

For the moment, let us focus on the bivariate normal distribution. Suppose that $(X_1, Y_1), ..., (X_n, Y_n)$ are iid $N_2(\mu_1, \mu_2, \sigma_1^2, \sigma_2^2, \rho)$ where $-\infty < \mu_1, \mu_2 < \infty, 0 < \sigma_1^2, \sigma_2^2 < \infty$ and $-1 < \rho < 1, n \geq 2$. Let us denote

$$\overline{X} = n^{-1}\Sigma_{i=1}^n X_i \qquad\qquad \overline{Y} = n^{-1}\Sigma_{i=1}^n Y_i$$

$$S_1^2 = (n-1)^{-1}\Sigma_{i=1}^n(X_i - \overline{X})^2 \qquad S_2^2 = (n-1)^{-1}\Sigma_{i=1}^n(Y_i - \overline{Y})^2$$

$$S_{12} = (n-1)^{-1}\Sigma_{i=1}^n(X_i - \overline{X})(Y_i - \overline{Y}) \qquad r = S_{12}/(S_1 S_2).$$

$$(4.6.4)$$

> Here, r is defined as the *Pearson correlation coefficient*
> or simply the *sample correlation coefficient*.

By separately using the marginal distributions of $X_1, ..., X_n$ and $Y_1, ..., Y_n$, we can right away claim the following sampling distributions:

$$\overline{X} \text{ is } N(\mu_1, \tfrac{1}{n}\sigma_1^2) \qquad \overline{Y} \text{ is } N(\mu_2, \tfrac{1}{n}\sigma_2^2)$$

$$(n-1)S_1^2\sigma_1^{-2} \text{ is } \chi_{n-1}^2 \qquad (n-1)S_2^2\sigma_2^{-2} \text{ is } \chi_{n-1}^2. \qquad (4.6.5)$$

The joint distribution of $(\overline{X}, \overline{Y})$ is not very difficult to obtain by means of transformations. We leave this out as the Exercise 4.6.8.

Alternately, note, however, that any linear function of $\overline{X}$ and $\overline{Y}$ is also a linear function of the original iid random vectors (X_i, Y_i), $i = 1, ..., n$. Then, by appealing to the Definition 4.6.1 of a multivariate normal distribution, we can immediately claim the following result:

$$(\overline{X}, \overline{Y}) \text{ is distributed as } N_2(\mu_1, \mu_2, n^{-1}\sigma_1^2, n^{-1}\sigma_2^2, \rho^*) \text{ with some } \rho^*. \qquad (4.6.6)$$

How does one find ρ^*? Invoking the bilinear property of covariance from Theorem 3.4.3, part (iii), one can express $Cov(\overline{X}, \overline{Y})$ as

$$n^{-2}\Sigma_{i=1}^n\Sigma_{j=1}^n Cov\,(X_i, Y_j) = n^{-2}\Sigma_{i=1}^n Cov\,(X_i, Y_i) = n^{-1}\rho\sigma_1\sigma_2,$$

so that the population correlation coefficient between $\overline{X}$ and $\overline{Y}$ is simplified to $n^{-1}\rho\sigma_1\sigma_2 / \left(\sqrt{n^{-2}\sigma_1^2\sigma_2^2}\right) = \rho$.

The distribution of the *Pearson correlation coefficient*

$$r = S_{12}/(S_1 S_2) \qquad (4.6.7)$$

is quite complicated, particularly when $\rho \neq 0$. Without explicitly writing down the pdf of r, it is still simple enough to see that the distribution of r can not depend on the values of μ_1, μ_2, σ_1^2 and σ_2^2. To check this claim, let us denote $U_i = (X_i - \mu_1)/\sigma_1$, $V_i = (Y_i - \mu_2)/\sigma_2$, and then observe that the random vectors (U_i, V_i), $i = 1, ..., n$, are distributed as iid $N_2(0, 0, 1, 1, \rho)$. But, r can be equivalently expressed in terms of the U_i's and V_i's as follows:

$$r = \frac{\Sigma_{i=1}^n(U_i - \overline{U})(V_i - \overline{V})}{\sqrt{\Sigma_{i=1}^n(U_i - \overline{U})^2}\sqrt{\Sigma_{i=1}^n(V_i - \overline{V})^2}}. \qquad (4.6.8)$$

> From (4.6.8), it is clear that the distribution of r depends *only* on ρ.

Francis Galton introduced a numerical measure, r, which he termed "reversion" in a lecture at the Royal Statistical Society on February 9, 1877 and later called "regression". The term "cor-relation" or "correlation" probably appeared first in Galton's paper to the Royal Statistical Society on December 5, 1888. At that time, "correlation" was defined in terms of deviations from the median instead of the mean. Karl Pearson gave the definition and calculation of correlation as in (4.6.7) in 1897. In 1898, Pearson and his collaborators discovered that the standard deviation of r happened to be $(1 - \rho^2)/\sqrt{n}$ when n was large. "Student" derived the "probable error of a correlation coefficient" in 1908. Soper (1913) gave large sample approximations for the mean and variance of r which were better than those proposed earlier by Pearson. Refer to DasGupta (1980) for some of the historical details.

The unsolved problem of finding the exact pdf of r for normal variates came to R. A. Fisher's attention via Soper's 1913 paper. The pdf of r was published in the year 1915 by Fisher for all values of $\rho \in (-1, 1)$. Fisher, at the age of 25, brilliantly exploited the n-dimensional geometry to come up with the solution, reputedly within one week. Fisher's genius immediately came into limelight. Following the publication of Fisher's results, however, Karl Pearson set up a major cooperative study of the correlation. One will notice that in the team formed for this cooperative project [Soper et al. (1917)] studying the distribution of the sample correlation coefficient, the young Fisher was not included. This happened in spite of the fact that Fisher was right there and he already earned quite some fame. Fisher felt hurt as he was left out of this project. One thing led to another. R. A. Fisher and Karl Pearson continued criticizing each other even more as each held on to his philosophical stand.

We will merely state the pdf of r when $\rho = 0$. This pdf is given by

$$f(r) = c \ (1 - r^2)^{\frac{1}{2}(n-4)} \ for \ -1 < r < 1, \tag{4.6.9}$$

where $c = \Gamma \left(\frac{1}{2}(n - 1) \right) \left\{ \sqrt{\pi} \ \Gamma \left(\frac{1}{2}(n - 2) \right) \right\}^{-1}$ for $n \geq 3$. Using (4.6.9) and some simple transformation techniques, one can easily derive the following result:

$$r(n - 2)^{1/2}(1 - r^2)^{-1/2} \text{ has the Student's } t \text{ distribution}$$
$$\text{with } (n - 2) \text{ degrees of freedom when } \rho = 0. \tag{4.6.10}$$

The verification of the claim in (4.6.10) is left as the Exercise 4.6.9. Fisher (1915) also gave the exact pdf of r in the form of an infinite power series for all values of $\rho \in (-1, 0) \cup (0, 1)$.

Sampling Distributions : The Multivariate Normal Case

Now, we briefly touch upon some aspects of sampling distributions in the context of a multivariate normal population. Suppose that $\mathbf{X}_1, ..., \mathbf{X}_n$ are iid $N_p(\boldsymbol{\mu}, \boldsymbol{\Sigma})$ where $\boldsymbol{\mu} \in \Re^p$ and $\boldsymbol{\Sigma}$ is a $p \times p$ p.d. matrix, $n \geq 2$. Let us denote

$$\overline{\mathbf{X}} = n^{-1}\Sigma_{i=1}^n \mathbf{X}_i \text{ and } \mathbf{W} = \Sigma_{i=1}^n (\mathbf{X}_i - \overline{\mathbf{X}})(\mathbf{X}_i - \overline{\mathbf{X}})'. \qquad (4.6.11)$$

Observe that $\overline{\mathbf{X}}$ is a p-dimensional column vector whereas $\mathbf{W}$ is a $p \times p$ matrix, both functionally depending on the random samples $\mathbf{X}_1, ..., \mathbf{X}_n$.

Theorem 4.6.1 *Suppose that* $\mathbf{X}_1, ..., \mathbf{X}_n$ *are iid* $N_p(\boldsymbol{\mu}, \boldsymbol{\Sigma})$ *where* $\boldsymbol{\mu} \in \Re^p$ *and* $\boldsymbol{\Sigma}$ *is a* $p \times p$ *p.d. matrix,* $n \geq 2$. *Then, we have the following sampling distributions*:

- (i) $\overline{\mathbf{X}}$ *is distributed as* $N_p(\boldsymbol{\mu}, n^{-1}\boldsymbol{\Sigma})$;
- (ii) $n(\overline{\mathbf{X}} - \boldsymbol{\mu})'\boldsymbol{\Sigma}^{-1}(\overline{\mathbf{X}} - \boldsymbol{\mu})$ *is distributed as* χ_p^2;
- (iii) *For any fixed vector* $\mathbf{a} \in \Re^p - \{\mathbf{0}\}, \mathbf{a}'\mathbf{W}\mathbf{a}/\mathbf{a}'\boldsymbol{\Sigma}\mathbf{a}$ *is distributed as* χ_{n-1}^2;
- (iv) $\overline{\mathbf{X}}$ *and* $\mathbf{W}$ *are distributed independently.*

The part (i) in this theorem is easily proved by applying the Definition 4.6.1. Also, the part (ii) can be easily proved when $p = 2$. We leave their verifications out in the Exercise 4.6.10. Proofs of parts (ii)-(iv) in their fullest generality are, however, out of scope for this book. The readers, however, should exploit these results to avoid laborious calculations whenever possible.

4.6.2   The t Distribution

This distribution comes up frequently in the areas of multiple comparisons and selection and ranking. Let the random vector $\mathbf{X}$, where $\mathbf{X}' = (X_1, ..., X_p)$, have a p-dimensional normal distribution with the mean vector $\mathbf{0}$ and the $p \times p$ dispersion matrix $\boldsymbol{\Sigma}$. We assume that each diagonal entry in $\boldsymbol{\Sigma}$ is 1. That is, the random variables $X_1, ..., X_p$ have each been standardized. In other words, the $(i, j)^{th}$ entry in the matrix $\boldsymbol{\Sigma}$ corresponds to the population correlation coefficient ρ_{ij} between the two variables $X_i, X_j, 1 \leq i \neq j \leq p$. In this special situation, one also refers to $\boldsymbol{\Sigma}$ as a *correlation matrix*. Suppose that Y is a positive real valued random variable distributed as χ_ν^2. It is also assumed that Y and $(X_1, ..., X_p)$ are independent. Let us denote p new random variables

$$T_i = X_i \div (Y/\nu)^{1/2}, i = 1, ..., p. \qquad (4.6.12)$$

The *marginal* distribution of each T_i has the Student's t distribution with ν degrees of freedom. This follows from the definition of the Student's t random variable. Jointly, however, $(T_1, ..., T_p)$ is said to have the p-dimensional t distribution which depends on the correlation matrix Σ, denoted by $Mt_p(\nu, \Sigma)$. One may refer to Johnson and Kotz (1972, p. 134). If we assume that Σ^{-1} exists, then the *joint pdf* of $T_1, ..., T_p$, distributed as $Mt_p(\nu, \Sigma)$, is given by

$$f(t_1, ..., t_p)$$
$$= \frac{\Gamma((\nu + p)/2)}{(\pi\nu)^{p/2}\Gamma(\nu/2)\left(det(\Sigma)\right)^{1/2}} \left(1 + \nu^{-1}\mathbf{t}'\Sigma^{-1}\mathbf{t}\right)^{-(\nu+p)/2}, \quad \mathbf{t} \in \Re^p$$

(4.6.13)

with $\mathbf{t}' = (t_1, ..., t_p)$. The derivation of this joint pdf is left as the Exercise 4.6.11.

Fundamental works in this area include references to Cornish (1954, 1962), Dunnett (1955), and Dunnett and Sobel (1954, 1955), among others. One may also refer to Johnson and Kotz (1972) and Tong (1990) to find other sources.

Example 4.6.6 Suppose that $U_{i1}, ..., U_{in_i}$ are iid $N(\mu_i, \sigma^2), i = 1, 2, 3$ as in the setup of a one-way analysis of variance. Let the three treatments be independent too. From the ith treatment, we obtain the sample mean $\overline{U}_i$ and the sample variance $S_i^2, i = 1, 2, 3$. Since the population variances are all equal to σ^2, we obtain the *pooled* sample variance

$$S_P^2 = \{n_1 + n_2 + n_3 - 3\}^{-1}\Sigma_{i=1}^3(n_i - 1)S_i^2. \tag{4.6.14}$$

Let $X_i = (\overline{U}_{i+1} - \overline{U}_1 - \mu_{i+1} + \mu_1)/\{\sigma\sqrt{n_{i+1}^{-1} + n_1^{-1}}\}, i = 1, 2$. Then, (X_1, X_2) is distributed as $N_2(0, \Sigma)$ with $\Sigma_{2\times2} = \begin{pmatrix} \sigma_{11} & \sigma_{12} \\ \sigma_{21} & \sigma_{22} \end{pmatrix}$ where $\sigma_{11} = \sigma_{22} = 1$ and $\sigma_{12} = \sigma_{21} = n_1^{-1}/\{\sqrt{n_2^{-1} + n_1^{-1}}\sqrt{n_3^{-1} + n_1^{-1}}\}$. Also, $Y = \{n_1 + n_2 + n_3 - 3\}S_P^2/\sigma^2$ is distributed as a Chi-square random variable with the degree of freedom $\nu = n_1 + n_2 + n_3 - 3$. Next, the random variable Y and the random vector (X_1, X_2) are independently distributed. Thus, with $T_i = \sqrt{\nu}X_i Y^{-1/2}, i = 1, 2$, the corresponding random vector (T_1, T_2) is distributed as $Mt_2(\nu, \Sigma)$. ▲

4.6.3   The F Distribution

This distribution also arises frequently in the areas of analysis of variance, multiple comparisons, and selection and ranking. Suppose that we have independent random variables $X_0, X_1, ..., X_p$ where X_i is distributed as $\chi^2_{\nu_i}, i = 0, 1, ..., p$. This is often the situation in the case of analysis of variance, where these X's customarily stand for the sums of squares due to the error and treatments. We denote the new random variables

$$U_i = (X_i/\nu_i) \div (X_0/\nu_0), i = 1, ..., p. \qquad (4.6.15)$$

The *marginal* distribution of the random variable U_i is F_{ν_i,ν_0}. This follows from the definition of a F random variable. Jointly, however, $(U_1, ..., U_p)$ is said to have the p-dimensional F distribution, denoted by $MF_p(\nu_0, \nu_1, ..., \nu_p)$. One may refer to Johnson and Kotz (1972, p. 240). The *joint pdf* of $U_1, ..., U_p$, distributed as $MF_p(\nu_0, \nu_1, ..., \nu_p)$, is given by

$$f(u_1, ..., u_p) = \frac{\Gamma(\nu/2)\Pi^p_{i=0}\nu_i^{\nu_i/2}}{\Pi^p_{i=0}\Gamma(\nu_i/2)} \frac{\Pi^p_{i=1}u_i^{(\nu_i-2)/2}}{\left(\nu_0 + \Sigma^p_{i=1}\nu_i u_i\right)^{\nu/2}}, \ \mathbf{u} \in \Re^{+p}, \qquad (4.6.16)$$

with $\mathbf{u}' = (u_1, ..., u_p)$ and $\nu = \Sigma^p_{i=0}\nu_i$. The derivation of this joint pdf is left as the Exercise 4.6.13.

Fundamental works in this area include references to Finney (1941) and Kimbal (1951), among others. One may also refer to Johnson and Kotz (1972) and Tong (1990) to find other sources.

4.7   Importance of Independence in Sampling Distributions

In the reproductive property of the normal distributions and Chi-square distributions (Theorem 4.3.2, parts (i) and (iii)) or in the definitions of the Student's t and F distributions, we *assumed independence* among various random variables. Now, we address the following important question. If those independence assumptions are violated, would one have similar standard distributions in the end? In general, the answer is "no" in many circumstances. In this section, some of the ramifications are discussed in a very simple way.

4.7.1   *Reproductivity of Normal Distributions*

In the Theorem 4.3.2, part (i), we learned that *independent* normal variables add up to another normal random variable. According to the Definition 4.6.1, as long as $(X_1, ..., X_n)$ is distributed as the n-dimensional normal N_n, this reproductive property continues to hold. But, as soon as we stray away from the structure of the multivariate normality for $(X_1, ..., X_n)$, the marginals of $X_1, ..., X_n$ can each be normal in a particular case, but their sum need not be distributed as a normal variable. We recall an earlier example.

Example 4.7.1 (Example 3.6.3 Continued) Let us rewrite the bivariate normal pdf given in (3.6.1) as $f(x_1, x_2; \mu_1, \mu_2, \sigma_1^2, \sigma_2^2, \rho)$. Next, consider any arbitrary $0 < \alpha, \rho < 1$ and fix them. We recall that we had

$$g(x_1, x_2; \rho)$$
$$= \alpha\, f(x_1, x_2; 0, 0, 1, 1, \rho) + (1 - \alpha)\, f(x_1, x_2; 0, 0, 1, 1, -\rho) \qquad (4.7.1)$$

for $-\infty < x_1, x_2 < \infty$. Here, (X_1, X_2) has a bivariate *non-normal* distribution, but marginally both X_1, X_2 have the standard normal distribution. However one can show that $X_1 + X_2$ is *not* a univariate normal variable. These claims are left as the Exercises 4.7.1-4.7.2. ▲

4.7.2 Reproductivity of Chi-square Distributions

Suppose that X_1 and X_2 are independent random variables which are respectively distributed as χ_m^2 and χ_n^2. From the Theorem 4.3.2, part (iii), we can then claim that $X_1 + X_2$ will be distributed as χ_{m+n}^2. But, will the same conclusion necessarily hold if X_1 and X_2 are respectively distributed as χ_m^2 and χ_n^2, but X_1 and X_2 are dependent? The following simple example shows that when X_1 and X_2 are dependent, but X_1 and X_2 are both Chi-squares, then $X_1 + X_2$ need not be Chi-square.

Example 4.7.2 Suppose that (U_1, V_1) is distributed as $N_2(0, 0, 1, 1, \rho)$. We assume that $-1 < \rho < 1$. Let $X_1 = U_1^2$, $X_2 = V_1^2$ and thus the marginal distributions of X_1, X_2 are both χ_1^2. Let us investigate the distribution of $Y = X_1 + X_2$. Observe that

$$Y = \tfrac{1}{2}\left\{ (U_1 + V_1)^2 + (U_1 - V_1)^2 \right\}, \qquad (4.7.2)$$

where the joint distribution of $(U_1 + V_1, U_1 - V_1)$ is actually N_2, because any linear function of $U_1 + V_1, U_1 - V_1$ is ultimately a linear function of U_1 and V_1, and is thus univariate normal. This follows (Definition 4.6.1) from the fact that (U_1, V_1) is assumed N_2. Now, we can write

$$Cov(U_1 + V_1, U_1 - V_1) = V(U_1) - V(V_1) = 0,$$

and hence from our earlier discussions (Theorem 3.7.1), it follows that $U_1 + V_1$ and $U_1 - V_1$ are independent random variables. Also, $U_1 + V_1$ and $U_1 - V_1$ are respectively distributed as $N(0, 2(1 + \rho))$ and $N(0, 2(1 - \rho))$.

Thus, with $0 < t < \min\left\{(2(1+\rho))^{-1}, (2(1-\rho))^{-1}\right\}$, the mgf $M_Y(t)$ of Y is given by

$$E\left\{e^{tY}\right\} = E\left\{exp\left[\tfrac{1}{2}t\{(U_1 + V_1)^2 + (U_1 - V_1)^2\}\right]\right\}$$
$$= E\left\{exp\left[\tfrac{1}{2}t(U_1 + V_1)^2\right]\right\} E\left\{exp\left[\tfrac{1}{2}t(U_1 - V_1)^2\right]\right\}$$

since $U_1 + V_1$ and $U_1 - V_1$ are independent.

Hence, we can express $M_Y(t)$ as

$$\{1 - 2(1+\rho)t\}^{-\frac{1}{2}}\{1 - 2(1-\rho)t\}^{-\frac{1}{2}} = \left\{(1-2t)^2 - 4\rho^2 t^2\right\}^{-\frac{1}{2}},$$
$$(4.7.3)$$

which will coincide with $(1 - 2t)^{-1}$, the mgf of χ_2^2, if and only if $\rho = 0$. We have shown that Y is distributed as χ_2^2 if and only if $\rho = 0$, that is if and only if X_1, X_2 are independent. ▲

We have not said anything yet about the difference of random variables which are individually distributed as Chi-squares.

> The difference of two Chi-square random variables may or may not be another Chi-square variable. See the Example 4.7.3.

Example 4.7.3 Suppose that we have X_1, X_2, X_3, X_4 which are iid $N(0,1)$. Let us denote $U_1 = X_1^2 + X_2^2 + X_3^2$ and $U_2 = X_3^2$. Since X_i^2's are iid $\chi_1^2, i = 1, 2, 3, 4$, by the reproductive property of Chi-square distributions we claim that U_1 is distributed as χ_3^2. Obviously, U_2 is distributed as χ_1^2. At the same time, we have $U_1 - U_2 = X_1^2 + X_2^2$ which is distributed as χ_2^2. Here, the difference of two Chi-square random variables is distributed as another Chi-square variable. Next, let us we denote $U_3 = X_3^2 + X_4^2$ which is distributed as χ_2^2. But, $U_1 - U_3 = X_1^2 + X_2^2 - X_4^2$ and this random variable cannot have a Chi-square distribution. We can give a compelling reason to support this conclusion. We note that $U_1 - U_3$ can take negative values with positive probability. In order to validate this claim, we use the pdf of $F_{2,1}$ and express $P\{X_1^2 + X_2^2 - X_4^2 < 0\}$ as

$$P\{X_1^2 + X_2^2 < X_4^2\} = P\{F_{2,1} < \tfrac{1}{2}\} = \int_0^{1/2}(1 + 2u)^{-3/2}du,$$

which simplifies to $1 - 2^{-1/2} \approx .29289$. ▲

> Let $X_1, ..., X_n$ be iid $N(\mu, \sigma^2)$ and S^2 be the sample variance. Then, $(n-1)S^2/\sigma^2$ is distributed as Chi-square. Is it possible that $(n-1)S^2/\sigma^2$ is a Chi-square random variable without the assumed normality of the X's? See the Exercise 4.7.4.

4.7.3   The Student's t Distribution

In the Definition 4.5.1, we introduced the Student's t distribution. In defining W, the Student's t variable, how crucial was it for the standard normal variable X and the Chi-square variable Y to be independent? The answer is: independence between the numerator and denominator was indeed very crucial. We elaborate this with the help of two examples.

Example 4.7.4 Suppose that Z stands for a standard normal variable and we let $X = Z, Y = Z^2$. Clearly, X is $N(0,1)$ and Y is χ_1^2, but they are *dependent* random variables. Along the line of construction of the random variable W, let us write $W = X/\sqrt{Y/\nu} = X/\mid X \mid$ and it is clear that W can take only two values, namely -1 and 1, each with probability $\frac{1}{2}$ since $P(X > 0) = P(X \leq 0) = \frac{1}{2}$. That is, W is a Bernoulli variable instead of being a t variable. Here, W has a discrete distribution! ▲

Example 4.7.5 Let X_1, X_2 be independent, X_1 being $N(0,1)$ and X_2 being χ_n^2. Define $X = X_1, Y = X_1^2 + X_2$, and obviously we have $X^2 < Y$ w.p. 1, so that X and Y are indeed *dependent* random variables. Also by choice, X is $N(0,1)$ and Y is χ_{n+1}^2. Now, consider the Definition 4.5.1 again and write

$$W = X \div \sqrt{(n+1)^{-1}Y} \qquad (4.7.4)$$

so that we have $\mid W \mid \leq (n+1)^{1/2}$. A random variable W with such a restricted domain space can not be distributed as the Student's t variable. The domain space of the t variable must be the real line, $\Re$. Note, however, that the random variable W in (4.7.4) has a continuous distribution unlike the scenario in the previous Example 4.7.4. ▲

4.7.4   The F Distribution

In the Definition 4.5.2, the construction of the F random variable was explained. The question is whether the independence of the two Chi-squares, namely X and Y, is essential in that definition. The answer is "yes" as the following example shows.

Example 4.7.6 Let U_1, U_2 be independent, U_1 be χ_m^2 and U_2 be χ_n^2. Define $X = U_1, Y = U_1 + U_2$, and we note that $Y > X$ w.p.1. Hence, obviously X, Y are *dependent* random variables. Also, X is χ_m^2 and Y is χ_{m+n}^2. Now, we look at the Definition 4.5.2 and express U as

$$m^{-1}U_1 \div (m+n)^{-1}(U_1 + U_2) = m^{-1}(m+n)\{U_1/(U_1 + U_2)\}, \quad (4.7.5)$$

so that we have $0 < U < (m+n)/m$. A random variable U with such a restricted domain space cannot have F distribution. The domain space of F variable must be $\Re^+$. Note, however, that the random variable U in (4.7.5) has a continuous distribution. ▲

4.8 Selected Review in Matrices and Vectors

We briefly summarize some useful notions involving matrices and vectors *made up of real numbers alone*. Suppose that $A_{m \times n}$ is such a matrix having m rows and n columns. Sometimes, we rewrite A as $(\mathbf{a}_1, ..., \mathbf{a}_n)$ where $\mathbf{a}_i$ stands for the ith column vector of $A, i = 1, ..., n$.

The *transpose* of A, denoted by A', stands for the matrix whose rows consist of the columns $\mathbf{a}_1, ..., \mathbf{a}_n$ of A. In other words, in the matrix A', the row vectors are respectively $\mathbf{a}_1', ..., \mathbf{a}_n'$. When $m = n$, we say that A is a *square* matrix.

> Each entry in a matrix is assumed to be a real number.

The *determinant* of a square matrix $A_{n \times n}$ is denoted by $det(A)$. If we have two square matrices $A_{n \times n}$ and $B_{n \times n}$, then one has

$$det(AB) = det(BA) = det(A)det(B). \qquad (4.8.1)$$

The *rank* of a matrix $A_{m \times n} = (\mathbf{a}_1, ..., \mathbf{a}_n)$, denoted by $R(A)$, stands for the maximum number of linearly independent vectors among $\mathbf{a}_1, ..., \mathbf{a}_n$. It can be shown that

$$R(A) = R(A') = R(AA') = R(A'A). \qquad (4.8.2)$$

A square matrix $A_{n \times n} = (\mathbf{a}_1, ..., \mathbf{a}_n)$ is called *non-singular* or *full rank* if and only if $R(A) = n$, that is all its column vectors $\mathbf{a}_1, ..., \mathbf{a}_n$ are linearly independent. In other words, $A_{n \times n}$ is non-singular or full rank if and only if the column vectors $\mathbf{a}_1, ..., \mathbf{a}_n$ form a *minimal generator* of $\Re^n$.

A square matrix $B_{n \times n}$ is called an *inverse* of $A_{n \times n}$ if and only if $AB = BA = I_{n \times n}$, the identity matrix. The inverse matrix of $A_{n \times n}$, if it exists, is unique and it is customarily denoted by A^{-1}. It may be worthwhile to recall the following result.

> For a matrix $A_{n \times n}$: A^{-1} exists $\Leftrightarrow R(A) = n \Leftrightarrow det(A) \neq 0$.

For a 2×2 matrix $A = \begin{pmatrix} a_{11} & a_{12} \\ a_{21} & a_{22} \end{pmatrix}$, one has $det(A) = a_{11}a_{22} - a_{12}a_{21}$.

Let us suppose that $a_{11}a_{22} \neq a_{12}a_{21}$, that is $det(A) \neq 0$. In this situation, the inverse matrix can be easily found. One has:

$$A^{-1} = \frac{1}{det(A)} \begin{pmatrix} a_{22} & -a_{12} \\ -a_{21} & a_{11} \end{pmatrix} \text{ whenever } a_{11}a_{22} \neq a_{12}a_{21}. \quad (4.8.4)$$

A square matrix $A_{n \times n}$ is called *orthogonal* if and only if A', the transpose of A, is the inverse of A. If $A_{n \times n}$ is orthogonal then it can be checked that $det(A) = \pm 1$.

A square matrix $A = (a_{ij})_{n \times n}$ is called *symmetric* if and only if $A' = A$, that is $a_{ij} = a_{ji}$, $1 \leq i \neq j \leq n$.

Let $A = (a_{ij}), i, j = 1, ..., n$, be a $n \times n$ matrix. Denote the $l \times l$ sub-matrix $B_l = (a_{pq}), p, q = 1, ..., l, l = 1, ..., n$. The lth *principal minor* is defined as the $det(B_l), l = 1, ..., n$.

For a symmetric matrix $A_{n \times n}$, an expression such as $x'Ax$ with $x \in \Re^n$ is customarily called a *quadratic form*.

A symmetric matrix $A_{n \times n}$ is called *positive semi definite* (p.s.d.) if (a) the quadratic form $x'Ax \geq 0$ for all $x \in \Re^k$, and (b) the quadratic form $x'Ax = 0$ for some non-zero $x \in \Re^n$.

A symmetric matrix $A_{n \times n}$ is called *positive definite* (p.d.) if (a) the quadratic form $x'Ax \geq 0$ for all $x \in \Re^n$, and (b) the quadratic form $x'Ax = 0$ if and only if $x = 0$.

> A symmetric matrix A is p.d. if and only if
> *all* principal minors of A are positive. $\qquad (4.8.5)$

A symmetric matrix $A_{n \times n}$ is called *negative definite* (n.d.) if (a) the quadratic form $x'Ax \leq 0$ for all $x \in \Re^n$, and (b) the quadratic form $x'Ax = 0$ if and only if $x = 0$. In other words, a symmetric matrix $A_{n \times n}$ is n.d. if and only if $-A$ is positive definite.

> A symmetric matrix A is n.d. if and only if all odd order
> principal minors are negative and all even order principal $\qquad (4.8.6)$
> minors are positive.

Let us now look at some *partitioned matrices* A, B where

$$A_{m \times n} = \begin{pmatrix} P_{u \times v} & Q_{u \times t} \\ R_{w \times v} & S_{w \times t} \end{pmatrix}, B_{n \times q} = \begin{pmatrix} E_{v \times c} & F_{v \times d} \\ G_{t \times c} & H_{t \times d} \end{pmatrix}, \quad (4.8.7)$$

where m, n, u, v, w, t, c, d are all positive integers such that $u + w = m, v + t = n$ and $c + d = q$. Then, one has

$$A' = \begin{pmatrix} P'_{v \times u} & R'_{v \times w} \\ Q'_{t \times u} & S'_{t \times w} \end{pmatrix}, AB = \begin{pmatrix} PE + QG & PF + QH \\ RE + SG & RF + SH \end{pmatrix}. \quad (4.8.8)$$

Again, let us consider a partitioned matrix A from (4.8.7) where $P_{u \times u}$ and $S_{w \times w}$ are square matrices where $u + w = n$. Then, one has

$$
\boxed{det(A) = det(P)\, det(S - RP^{-1}Q) \text{ provided that } P^{-1} \text{ exists.}}
\qquad (4.8.9)
$$

Let us reconsider the partitioned matrix $A_{n \times n}$ which was used in (4.8.9). Then, we can write

$$
\boxed{
\begin{aligned}
A^{-1} &= \begin{pmatrix} P^{-1} + FE^{-1}F' & -FE^{-1} \\ -E^{-1}F' & E^{-1} \end{pmatrix} \text{ where} \\[4pt]
E &= S - Q'P^{-1}Q, F = P^{-1}Q, \text{ under the assumption} \\
&\text{that all the } \textit{inverse matrices} \text{ involved } \textit{exist.}
\end{aligned}
}
\qquad (4.8.10)
$$

Next, we summarize an important tool which helps us to find the **maximum or minimum of a real valued function of two variables**. The result follows:

Suppose that $f(\mathbf{x})$ is a real valued function of a two-dimensional variable $\mathbf{x} = (x_1, x_2) \in (a_1, b_1) \times (a_2, b_2) \subseteq \Re^2$, having continuous second-order partial derivatives $\frac{\partial^2}{\partial x_1^2} f(\mathbf{x})$, $\frac{\partial^2}{\partial x_2^2} f(\mathbf{x})$ and $\frac{\partial^2}{\partial x_1 \partial x_2} f(\mathbf{x})$ $\left(\equiv \frac{\partial^2}{\partial x_2 \partial x_1} f(\mathbf{x}) \right)$, everywhere in an open rectangle $(a_1, b_1) \times (a_2, b_2)$. Suppose also that for some point $\boldsymbol{\xi} = (\xi_1, \xi_2) \in (a_1, b_1) \times (a_2, b_2)$, one has

$$
\frac{\partial}{\partial x_1} f(\mathbf{x}) \mid_{\mathbf{x}=\boldsymbol{\xi}} = \frac{\partial}{\partial x_2} f(\mathbf{x}) \mid_{\mathbf{x}=\boldsymbol{\xi}} = 0.
$$

Now, let us denote the matrix of the second-order partial derivatives

$$
H \equiv H(\mathbf{x}) = \begin{pmatrix} \frac{\partial^2}{\partial x_1^2} f(\mathbf{x}) & \frac{\partial^2}{\partial x_1 \partial x_2} f(\mathbf{x}) \\[6pt] \frac{\partial^2}{\partial x_1 \partial x_2} f(\mathbf{x}) & f(\mathbf{x}) \end{pmatrix}.
\qquad (4.8.11)
$$

Then,

(i) $f(\mathbf{x})$ has a minimum at $\mathbf{x} = \boldsymbol{\xi}$ if the matrix H evaluated at the point $\mathbf{x} = \boldsymbol{\xi}$ is positive definite (p.d.);

(ii) $f(\mathbf{x})$ has a maximum at $\mathbf{x} = \boldsymbol{\xi}$ if the matrix H evaluated at the point $\mathbf{x} = \boldsymbol{\xi}$ is negative definite (n.d.).

$$
\qquad (4.8.12)
$$

4.9   Exercises and Complements

4.1.1 (Example 4.1.1 Continued) Suppose that Z_1, Z_2 are iid standard normal variables. Evaluate $P(Z_1^2 < bZ_2^2)$ with some fixed but arbitrary $b > 0$. {*Hint*: Express the required probability as the double integral, $\int\int_{z_1^2/z_2^2 \le b} (2\pi)^{-1} exp\left\{-\frac{1}{2}(z_1^2 + z_2^2)\right\} dz_1 dz_2$. Use the substitutions from (4.1.1) to rewrite this as $(2\pi)^{-1} \int\int_{\tan^2(\theta) \le b} exp\{-r/2\} d\theta dr$ which will change to $(2\pi)^{-1} \int_{r=0}^{\infty} \int_{\theta=-\tan^{-1}(\sqrt{b})}^{\tan^{-1}(\sqrt{b})} exp\{-r/2\} d\theta dr$. This last integral can then be replaced by $\pi^{-1} \int_{r=0}^{\infty} \frac{1}{2} exp\{-r/2\} dr \int_{\theta=-\tan^{-1}(\sqrt{b})}^{\tan^{-1}(\sqrt{b})} d\theta = 2\pi^{-1} \tan^{-1}(\sqrt{b})$. Also, look at the Exercise 4.5.7.}

4.2.1 (Example 4.2.1 Continued) Consider the two random variables X_1, X_2 from the Example 4.2.1. Find the pmf's of the following random variables which are functions of X_1, X_2.

(i) $U = X_1 - X_2$;

(ii) $V = 2X_1 + \frac{1}{2}X_2$;

(iii) $W = X_1^2 + X_2$.

4.2.2 Let X_1, X_2 be iid Poisson$(\lambda), \lambda > 0$. Derive the distribution of the random variable $U = X_1 + X_2$. Then, use mathematical induction to show that $U = \Sigma_{i=1}^{n} X_i$ has the Poisson$(n\lambda)$ distribution if $X_1, ..., X_n$ are iid Poisson(λ). Next, evaluate $P(X_1 - X_2 = v)$ explicitly when $v = 0, 1, 2, 3$. Proceeding this way, is it possible to write down the pmf of the random variable $V = X_1 - X_2$? {*Hint*: In the first part, follow along the Examples 4.2.2-4.2.3.}

4.2.3 In a Bernoulli experiment, suppose that $X_1 =$ number of trials needed to observe the first success, and $X_2 =$ number of trials needed since the first success to observe the second success, where the trials are assumed independent having the common success probability $p, 0 < p < 1$. That is, X_1, X_2 are assumed iid having the Geometric(p) distribution defined earlier in (1.7.7). First, find the distribution of the random variable $U = X_1 + X_2$. Next, with k such iid random variables $X_1, ..., X_k$, derive the pmf of the random variable $U = \Sigma_{i=1}^{k} X_i$. The distribution of U is called *Negative Binomial* with parameters (k, p). A different parameterization was given in (1.7.9).

4.2.4 Let the pdf of X be $f(x) = \frac{1}{\beta} exp(-\frac{x}{\beta})I(x > 0)$ with $\beta > 0$. Find the pdf's of the following random variables which are functions of X.

(i) $U = X^2$;

(ii) $V = X^3$.

4.2.5 In general, suppose that $X_1, ..., X_n$ are iid continuous random variables with the common pdf and df respectively given by $f(x)$ and $F(x)$. Derive the joint pdf of the ith order statistic $Y_i = X_{n:i}$ and the jth order statistic $Y_j = X_{n:j}, 1 \leq i < j \leq n$ by solving the following parts in the order they are given. Define $U_1 = \Sigma_{i=1}^n I(X_i \leq u_1), U_2 = \Sigma_{i=1}^n I(u_1 < X_i \leq u_2), U_3 = n - U_1 - U_2$. Now, show that

(i) (U_1, U_2, U_3) is distributed as multinomial with $k = 3, p_1 = F(u_1), p_2 = F(u_2) - F(u_1)$, and $p_3 = 1 - p_1 - p_2 = 1 - F(u_2)$;

(ii) $P\{Y_i \leq u_1 \cap Y_j \leq u_2\} = P\{U_1 \geq i \cap (U_1 + U_2) \geq j\}$;

(iii) $P\{U_1 \geq i \cap (U_1 + U_2) \geq j\} = \Sigma_{k=i}^{j-1} \Sigma_{l=j-k}^{n-k} P\{U_1 = k \cap U_2 = l\} + P\{U \geq j\}$;

(iv)   the expression in part (iii) can be rewritten as
$\Sigma_{k=i}^{j-1} \Sigma_{l=j-k}^{n-k} \frac{n!}{k!l!(n-k-l)!} \{F(u_1)\}^k \{F(u_2) - F(u_1)\}^l \times \{1 - F(u_2)\}^{n-k-l}$;

(v)   the joint pdf of Y_i, Y_j, denoted by $f(y_i, y_j)$, can be directly obtained by evaluating $\frac{\partial^2}{\partial y_i \partial y_j} \{P\{Y_i \leq y_i \cap Y_j \leq y_j\}\}$ first, and then simplifying it.

4.2.6 Verify (4.2.7). {*Hint*: Use the Exercise 4.2.5.}

4.2.7 Suppose that $X_1, ..., X_n$ are iid uniform random variables on the interval $(0, \theta), \theta \in \Re$. Denote the ith order statistic $Y_i = X_{n:i}$ where $X_{n:1} \leq X_{n:2} \leq ... \leq X_{n:n}$. Define $U = Y_n - Y_1, V = \frac{1}{2}(Y_n + Y_1)$ where U and V are respectively referred to as the *range* and *midrange* for the data $X_1, ..., X_n$. This exercise shows how to derive the pdf of the range, U.

(i)   Show that the joint pdf of Y_1 and Y_n is given by $h(y_1, y_n) = n(n-1)\theta^{-n}(y_n - y_1)^{n-2} I(0 < y_1 < y_n < \theta)$. {*Hint*: Refer to the equation (4.2.7).};

(ii)   Transform (Y_1, Y_n) to (U, V) where $U = Y_n - Y_1, V = \frac{1}{2}(Y_n + Y_1)$. Show that the joint pdf of (U, V) is given by $g(u, v) = n(n-1)\theta^{-n}u^{n-2}$ when $0 < u < \theta, \frac{1}{2}u < v < \theta - \frac{1}{2}u$, and zero otherwise;

(iii)   Show that the pdf of U is given by $g_1(u) = n(n-1)\theta^{-n} \times (\theta - u)u^{n-2} I(0 < u < \theta)$. {*Hint*: $g_1(u) = \int_{\frac{1}{2}u}^{\theta - \frac{1}{2}u} g(u, v)dv$, for $0 < u < \theta$.};

(iv)   When $\theta = 1$, does the pdf of the *range*, U, derived in part (iii) correspond to that of one of the standard random variables from the Section 1.7?

4.2.8 Suppose that $X_1, ..., X_n$ are iid uniform random variables on the interval $(\theta, \theta + 1), \theta \in \Re$. Denote the ith order statistic $Y_i = X_{n:i}$ where

$X_{n:1} \leq X_{n:2} \leq \ldots \leq X_{n:n}$. Define $U = Y_n - Y_1, V = \frac{1}{2}(Y_n + Y_1)$ where recall that U and V are respectively referred to as the *range* and *midrange* for the data $X_1, \ldots, X_n$. This exercise shows how to derive the pdf of the *range*, U.

(i)   Show that the joint pdf of Y_1 and Y_n is given by $h(y_1, y_n) = n(n-1)(y_n - y_1)^{n-2}I(\theta < y_1 < y_n < \theta + 1)$. {*Hint:* Refer to the equation (4.2.7).};

(ii)   Transform (Y_1, Y_n) to (U, V) where $U = Y_n - Y_1, V = \frac{1}{2}(Y_n + Y_1)$. Show that the joint pdf of (U, V) is given by $g(u, v) = n(n-1)u^{n-2}$ when $0 < u < 1, \theta + \frac{1}{2}u < v$ $< \theta + 1 - \frac{1}{2}u$, and zero otherwise;

(iii)   Show that the pdf of U is given by $g_1(u) = n(n-1) \times$ $(1-u)u^{n-2}I(0 < u < 1)$. {*Hint:* $g_1(u) = \int_{\theta + \frac{1}{2}u}^{\theta + 1 - \frac{1}{2}u} g(u, v)dv$, for $0 < u < 1$.};

(iv)   Does the pdf of the *range*, U, derived in part (iii) correspond to that of one of the standard random variables from the Section 1.7?

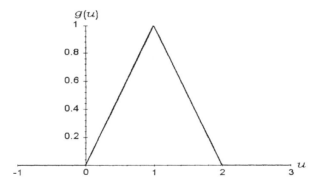

Figure 4.9.1. Triangular PDF $g(u)$ from the Exercise 4.2.9

4.2.9 (Triangular Distribution) Suppose that X_1 and X_2 are independent random variables, distributed uniformly on the interval $(0, 1)$ with the respective pdf's $f_1(x_1) = I(0 < x_1 < 1)$ and $f_2(x_2) = I(0 < x_2 < 1)$. Show that the pdf of $U = X_1 + X_2$ is given by

$$g(u) = \begin{cases} u & \text{if } 0 < u < 1 \\ 2 - u & \text{if } 1 < u < 2 \\ 0 & \text{elsewhere.} \end{cases}$$

The random variable U is said to follow the *triangular distribution*. Look at the plot of $g(u)$ in the Figure 4.9.1. {*Hint:* It is obvious that the pdf $g(u)$ of

U will be zero when $u \leq 0$ or $u \geq 2$, but it will be positive when $0 < u < 2$. Thus, using the convolution result, for $0 < u < 1$, we can write the df of U as $G(u) = \int_{x_1=0}^{u} \left[\int_{x_2=0}^{u-x_1} dx_2 \right] dx_1 = \int_{x_1=0}^{u} (u - x_1) dx_1 = \frac{1}{2}u^2$. Hence, for $0 < u < 1$ we have $g(u) = \frac{d}{du}G(u) = u$. On the other hand, for $1 \leq u < 2$, we can write the df of U as $G(u) = 1 - \int_{x_2=u-1}^{1} \left[\int_{x_1=u-x_2}^{1} dx_1 \right] dx_2 = 1 - \int_{x_2=u-1}^{1} (1-u+x_2) dx_2 = -\frac{1}{2}u^2 + 2u - 1$ so that $g(u) = \frac{d}{du}G(u) = 2-u$.}

4.2.10 Suppose that X_1 and X_2 are independent random variables, distributed uniformly on the intervals $(0,1)$ and $(0,2)$ with the respective pdf's $f_1(x_1) = I(0 < x_1 < 1)$ and $f_2(x_2) = \frac{1}{2}I(0 < x_2 < 2)$. Use the convolution result to derive the form of the pdf $g(u)$ of $U = X_1 + \frac{1}{2}X_2$ with $0 < u < 2$.

4.2.11 (Example 4.2.10 Continued) Let U, V be iid $N(\mu, \sigma^2)$. Find the pdf of $W = U + V$ by the convolution approach. Repeat the exercise for the random variable $T = U - V$.

4.2.12 (Example 4.2.11 Continued) Provide all the intermediate steps in the Example 4.2.11.

4.2.13 Prove a version of the Theorem 4.2.1 for the product, $V = X_1 X_2$, of two independent random variables X_1 and X_2 with their respective pdf's given by $f_1(x_1)$ and $f_2(x_2)$. Show that the pdf of V is given by $h(v) = \int_{x_1=-\infty}^{\infty} \frac{1}{x_1} f_2(\frac{v}{x_1}) f_1(x_1) dx_1$ for $v \in \Re$. Show that $h(v)$ can be equivalently written as $\int_{x_2=-\infty}^{\infty} \frac{1}{x_2} f_1(\frac{v}{x_2}) f_2(x_2) dx_2$. {*Caution*: Assume that V can not take the value zero.}

4.2.14 Suppose that X_1, X_2, X_3 are iid uniform random variables on the interval $(0,1)$.

    (*i*)   Find the joint pdf of $X_{3:1}, X_{3:2}, X_{3:3}$;

    (*ii*)   Derive the pdf of the median, $X_{3:2}$;

    (*iii*)   Derive the pdf of the range, $X_{3:3} - X_{3:1}$.

{*Hint*: Use the Exercise 4.2.7}

4.2.15 Suppose that $X_1, ..., X_n$ are iid $N(0,1)$. Let us denote $Y_i = 4\Phi(X_i)$ where $\Phi(u) = \{\sqrt{2\pi}\}^{-1} \int_{-\infty}^{u} e^{-x^2/2} dx, u \in \Re, i = 1, ..., n$.

    (*i*)   Find the pdf of the random variable $U = \Pi_{i=1}^{n} Y_i$;

    (*ii*)   Evaluate $E[U]$ and $V[U]$;

    (*iii*)   Find the marginal pdf's of $Y_{n:1}, Y_{n:n}$ and $Y_{n:n} - Y_{n:1}$.

4.2.16 Suppose that X_1 is uniform on the interval $(0,2)$, X_2 has its pdf $\frac{1}{2}xI(0 < x < 2)$, and X_3 has its pdf $\frac{3}{8}x^2 I(0 < x < 2)$. Suppose also that X_1, X_2, X_3 are independent. Derive the marginal pdf's of $X_{3:1}$ and $X_{3:3}$. {*Hint*: Follow along the Example 4.2.8.}

4.2.17 Suppose that Z is the standard normal random variable. Denote $X = \max\{|Z|, 1/|Z|\}$.

(i)   Find the expression of $P\{X \le x\}$ for $x \ge 1$. Observe that $P\{X \le x\} = 0$ for $x < 1$;

(ii) Use part (i) to show that the pdf of X is given by $f(x) = 2\{\phi(x) + x^{-2}\phi(x^{-1})\}I(x \ge 1)$.

{*Hint*: For $x \ge 1$, one can write $P\{X \le x\} = P\{x^{-1} \le |Z| \le x\} = 2P\{x^{-1} \le Z \le x\} = 2\{\Phi(x) - \Phi(x^{-1})\}$. This is part (i). Differentiating this expression with respect to x will lead to part (ii).}

4.2.18 (Exercise 4.2.9 Continued) Suppose that X_1, X_2 and X_3 are independent random variables, distributed uniformly on the interval $(0, 1)$ with the respective pdf's $f_1(x_1) = I(0 < x_1 < 1), f_2(x_2) = I(0 < x_2 < 1)$ and $f_3(x_3) = I(0 < x_3 < 1)$. Use the convolution result to derive the form of the pdf $h(v)$ of $V = X_1 + X_2 + X_3$ with $0 < v < 3$. {*Hint*: Note that $V = U + X_3$ where $U = X_1 + X_2$ and its pdf $g(u)$ was derived in the Exercise 4.2.9 for $0 < u < 2$. Now, use the convolution result again.}

4.2.19 (Example 4.2.9 Continued) Suppose that X_1, X_2 and X_3 are independent random variables, distributed exponentially with the respective pdf's $f_1(x_1) = e^{-x_1}I(x_1 \in \Re^+), f_2(x_2) = e^{-x_2}I(x_2 \in \Re^+)$ and $f_3(x_3) = e^{-x_3}I(x_3 \in \Re^+)$. Use the convolution result to show that the pdf $h(v)$ of $V = X_1 + X_2 + X_3$ with $v \in \Re^+$ matches with that of the Gamma$(3, 1)$ distribution. {*Hint*: Note that $V = U + X_3$ where $U = X_1 + X_2$ and its pdf $g(u)$ was derived in the Example 4.2.9 for $u \in \Re^+$. Now, use the convolution result again.}

4.2.20 Suppose that $X_1, ..., X_n$ are iid Uniform$(0, 1)$ random variables where $n = 2m + 1$ for some positive integer m. Consider the sample median, that is $U = X_{n:m+1}$ which is the order statistic in the middle.

(i)   Show that the pdf of U is given by $g(u) = cu^m(1 - u)^m \times I(0 < u < 1)$ where $c = (2m + 1)!/(m!)^2$. Is this one of the standard distributions listed in the Section 1.7?

(ii) Show that $E(U) = \frac{1}{2}$ and $V(U) = \frac{1}{4(n+2)}$.

4.3.1 Suppose that $X_1, ..., X_n$ are iid Poisson(λ). Use the mgf technique to show that $U = \Sigma_{i=1}^n X_i$ has the Poisson$(n\lambda)$ distribution. {*Hint*: Follow along the Examples 4.3.1-4.3.2.}

4.3.2 Suppose that $X_1, ..., X_k$ are iid Geometric(p), $0 < p < 1$. Find the distribution of $U = \Sigma_{i=1}^k X_i$. The distribution of U is called Negative Binomial with parameters (k, p). A different parameterization was given in (1.7.9).

4.3.3 Complete the arguments in the Example 4.3.5.

4.3.4 Consider the two related questions.
  (i)   Suppose that a random variable U has its pdf given by $g(u)$
        $= \frac{1}{2\sigma}e^{-|u|/\sigma}I(-\infty < u < \infty)$. First show that the mgf of U is
        given by $(1- \sigma^2 t^2)^{-1}$ for $|t| < \sigma^{-1}$, with $0 < \sigma < \infty$;
  (ii) Next, suppose that X_1, X_2 are iid with the common pdf $f(x)$
        $= \frac{1}{\beta}exp(-\frac{x}{\beta})I(x > 0)$, with $\beta > 0$. Using part (i), derive
        the pdf of $V = X_1 - X_2$.

4.3.5 Let Y_1, Y_2 be iid $N(0,4)$. Use the mgf technique to derive the distribution of $U = Y_1^2 + Y_2^2$. Then, obtain the exact algebraic expressions for
  (i) $P(U > 1)$;
  (ii) $P(1 < U < 2)$;
  (iii) $P(|U - 2| \geq 2.3)$.

4.3.6 Suppose that Z is the standard normal variable. Derive the expression of $E[exp(tZ^2)]$ directly by integration. Hence, obtain the distribution of Z^2.

4.3.7 (Exercise 4.3.6 Continued) Suppose that the random variable X is distributed as $N(\mu, \sigma^2), \mu \in \Re, \sigma \in \Re^+$. Obtain the expression of the mgf of the random variable X^2, that is $E[exp(tX^2)]$.

4.3.8 Prove the Theorem 4.3.2.

4.3.9 Suppose that $U_1, ..., U_n$ are iid random variables with the common pdf given by $g(u) = \frac{1}{2\sigma}e^{-|u|/\sigma}I(-\infty < u < \infty), \sigma \in \Re^+$. Then, use the part (ii) or (iii) from the Theorem 4.3.2 to obtain the distribution of the random variable $Y = \Sigma_{i=1}^n |U_i|$.

4.4.1 Let X be a random variable with the pdf

$$f(x) = \begin{cases} 2(1 - x) & \text{if } 0 < x < 1 \\ 0 & \text{elsewhere.} \end{cases}$$

Find the marginal pdf's of U, V, and W defined as follows:
  (i) $U = 2X - 1$;   (ii) $V = X^2$;   (iii) $W = \sqrt{X}$.

4.4.2 Suppose that a random variable X has the Raleigh density given by

$$f(x) = \begin{cases} \frac{2x}{\theta} exp(-x^2/\theta) & \text{if } x > 0 \\ 0 & \text{elsewhere.} \end{cases}$$

Derive the pdf of $U = X^2$.

4.4.3 (Example 4.4.3 Continued) Suppose that Z has the standard normal distribution. Let us denote $Y = |Z|$. Find the pdf of Y. {*Caution:* The transformation from Z to Y is *not* one-to-one. Follow along the Example 4.4.3.}

4.4.4 (Example 4.4.3 Continued) Suppose that Z has the standard normal distribution. Let us denote $Y = |Z|^3$. Find the pdf of Y. {*Caution*: The transformation from Z to Y is *not* one-to-one. Follow along the Example 4.4.3.}

4.4.5 (Example 4.4.4 Continued) Suppose that X has the pdf $f(x) = \frac{1}{2}exp\{-|x|\}I(x \in \Re)$. Obtain the pdf of $Y = |X|^3$. {*Caution*: The transformation from X to Y is *not* one-to-one. Follow along the Example 4.4.4.}

4.4.6 (Example 4.4.7 Continued) Verify all the details.

4.4.7 (Example 4.4.8 Continued) Verify all the details.

4.4.8 Suppose that $X_1, ..., X_n$ are iid Uniform$(0,1)$. Let us define $Y_i = X_{n:i}/X_{n:i-1}$, $i = 1, ..., n$ with $X_{n:0} = 1$. Find the joint distribution of $Y_2, ..., Y_n$. {*Hint*: First, consider taking *log* and then use the results from Examples 4.4.2 and 4.4.10.}

4.4.9 (Example 4.4.9 Continued) Suppose that $X_1, ..., X_n$ are iid $N(\mu, \sigma^2)$ with $n \geq 2$. Use the Helmert variables $Y_1, ..., Y_n$ from the Example 4.4.9 to show that $(\overline{X} - 1)^2$ and $S + S^{2/3}$ are independently distributed.

4.4.10 Suppose that (U, V) has the following joint pdf:

$$f(u,v) = \begin{cases} exp\{-\theta u - \theta^{-1}v\} & \text{if } 0 < u, v < \infty \\ 0 & \text{elsewhere,} \end{cases}$$

where $\theta > 0$. Define $X = UV$ and $Y = U/V$.

(i)   Find the joint pdf of (X, Y);
(ii)   Find the marginal pdf's of X, Y;
(iii) Find the conditional pdf of Y given $X = x$.

4.4.11 Let X_1, X_2 be iid having the Gamma(α, β) distribution with $\alpha > 0, \beta > 0$. Let us denote $U = X_1 + X_2, V = X_2/X_1$. Find the marginal distributions of U, V.

4.4.12 Let X_1, X_2 be iid $N(\mu, \sigma^2)$. Define $U = X_1 + X_2$, $V = X_1 - X_2$.

(i)   Find the joint pdf of (U, V);
(ii)   Evaluate the correlation coefficient between U and V;
(iii) Evaluate the following: $E\{(X_1 - X_2)^2 \mid X_1 + X_2 = x\}$; $E\{(X_1 + X_2)^2 \mid X_1 = X_2\}$; $V\{(X_1 - X_2)^2 \mid X_1 + X_2 = x\}$.

4.4.13 Suppose that X has the following pdf with $-\infty < \mu < \infty$, $0 < \sigma < \infty$:

$$f(x) = \begin{cases} \{\sigma\sqrt{2\pi}\}^{-1}x^{-1}exp\left\{-\frac{1}{2\sigma^2}(log(x) - \mu)^2\right\} & \text{if } x > 0 \\ 0 & \text{elsewhere.} \end{cases}$$

One will recall from (1.7.27) that this pdf is known as the lognormal density and the corresponding X is called a lognormal random variable. Suppose that X_1, X_2 are iid having the common pdf $f(x)$. Let r and s be arbitrary, but fixed real numbers. Then, find the pdf of the random variable $X_1^r X_2^s$. {*Hint*: Does taking *log* help?}

4.4.14 Suppose that X_1, X_2, X_3 are iid Gamma$(\alpha, \beta), \alpha > 0, \beta > 0$. Define $U_1 = X_1 + X_2 + X_3, U_2 = X_2/(X_1 + X_2)$, and $U_3 = X_3/(X_1 + X_2 + X_3)$. Solve the following problems.

  (i)   Show that one can express $x_1 = u_1(1 - u_2)(1 - u_3), x_2 = u_1 u_2(1 - u_3), x_3 = u_1 u_3$. Is this transformation one-to-one?

  (ii)   Determine the matrix J from (4.4.3) and show that $| \det(J) | = u_1^2(1 - u_3)$;

  (iii) Start out with the joint pdf of X_1, X_2, X_3. Use the transformation $(x_1, x_2, x_3) \to (u_1, u_2, u_3)$ to obtain the joint pdf of U_1, U_2, U_3;

  (iv)   Argue that U_1, U_2, U_3 are distributed independently. Show that (a) U_1 is distributed as Gamma$(3\alpha, \beta)$, (b) U_2 is distributed as Beta(α, α), and (c) U_3 is distributed as Beta$(\alpha, 2\alpha)$. In this part, recall the beta function and the Beta pdf from (1.6.25)-(1.6.26) and (1.7.35) respectively.

4.4.15 Suppose that X_1, X_2, X_3, X_4 are iid Gamma$(\alpha, \beta), \alpha > 0, \beta > 0$. Define $U_1 = X_1 + X_2 + X_3 + X_4, U_2 = X_2/(X_1 + X_2), U_3 = X_3/(X_1 + X_2 + X_3)$, and $U_4 = X_4/(X_1 + X_2 + X_3 + X_4)$. Solve the following problems.

  (i)   Show that one can express $x_1 = u_1(1 - u_2)(1 - u_3)(1 - u_4)$, $x_2 = u_1 u_2(1 - u_3)(1 - u_4), x_3 = u_1 u_3(1 - u_4), x_4 = u_1 u_4$. Is this transformation one-to-one?

  (ii)   Determine the matrix J from (4.4.3) and show that $| \det(J) | = u_1(1 - u_2)(1 - u_3)(1 - u_4)$;

  (iii) Start out with the joint pdf of X_1, X_2, X_3, X_4. Use the transformation $(x_1, x_2, x_3, x_4) \to (u_1, u_2, u_3, u_4)$ to obtain the joint pdf of U_1, U_2, U_3, U_4;

  (iv)   Argue that U_1, U_2, U_3, U_4 are distributed independently Show that (a) U_1 is distributed as Gamma$(4\alpha, \beta)$, (b) U_2 is distributed as Beta(α, α), (c) U_3 is distributed as Beta$(\alpha, 2\alpha)$, and (d) U_4 is Beta$(\alpha, 3\alpha)$. Recall the beta function and Beta pdf from (1.6.25)-(1.6.26) and (1.7.35) respectively.

4.4.16 (Example 4.4.13 Continued) Provide all the details in the Example 4.4.13.

4.4.17 (Example 4.4.14 Continued) Provide all the details in the Example 4.4.14.

4.4.18 (Example 4.4.16 Continued) Suppose that (X_1, X_2) is distributed as $N_2(0, 0, \sigma^2, \sigma^2, \rho)$ with $0 < \sigma < \infty, -1 < \rho < 1$. Define $U = X_1/X_2$ and $V = X_1$.

(i)   Find the joint pdf of U and V;

(ii)   In part (i), integrate out V and derive the marginal pdf of U;

(iii)   When $\rho = 0$, the pdf in part (ii) coincides with the standard Cauchy pdf, namely, $\pi^{-1}(1 + u^2)^{-1}$ for $-\infty < u < \infty$;

(iv)   When $\rho \neq 0$, the pdf in part (ii) coincides with the Cauchy pdf with appropriate location and scale depending on ρ.

4.4.19 Suppose that (X_1, X_2) has the following joint pdf:

$$f(x_1, x_2) = \begin{cases} 1 & \text{if } 0 < x_1 < 2, 0 < x_2 < 1, 2x_2 \leq x_1 \\ 0 & \text{elsewhere.} \end{cases}$$

Find the pdf of $U = X_1 - X_2$. {*Hint*: Use the transformation $u = x_1 - x_2$ and $v = x_1$.}

4.4.20 Suppose that (X_1, X_2) has the following joint pdf:

$$f(x_1, x_2) = \begin{cases} 4x_1 x_2 \exp\{-(x_1^2 + x_2^2)\} & \text{if } 0 < x_1, x_2 < \infty \\ 0 & \text{elsewhere.} \end{cases}$$

Show that the pdf of $U = \sqrt{X_1^2 + X_2^2}$ is given by

$$h(u) = 2u^3 \exp\{-u^2\} I(0 < u < \infty).$$

4.4.21 Suppose that (X_1, X_2) has the following joint pdf:

$$f(x_1, x_2) = \begin{cases} x_1 + x_2 & \text{if } 0 < x_1, x_2 < 1 \\ 0 & \text{elsewhere.} \end{cases}$$

Let $U = X_1 + X_2$ and $V = X_1 X_2$.

(i)   Show that the pdf of U is $a(u) = \begin{cases} u^2 & \text{if } 0 < u \leq 1 \\ 2u - u^2 & \text{if } 1 \leq u < 2 \\ 0 & \text{elsewhere;} \end{cases}$

(ii)   Show that the pdf of V is $b(v) = \begin{cases} 2 - 2v & \text{if } 0 < v < 1 \\ 0 & \text{elsewhere.} \end{cases}$

4.4.22 Suppose that X_1, X_2 are iid random variables. We are told that $X_1 + X_2$ and $X_1 - X_2$ are independently distributed. Show that the common pdf of X_1, X_2 must be same as the normal density. {*Hint*: Can the Remark 4.4.4 be used to solve this problem?}

4.4.23 Suppose that X_1, X_2, X_3 are three iid random variables. We are also told that $X_1 + X_2 + X_3$ and $(X_1 - X_2, X_1 - X_3)$ are independently distributed. Show that the common pdf of X_1, X_2, X_3 must be the normal pdf. {*Hint*: Can the Remark 4.4.4 be used to solve this problem?}

4.5.1 In the Theorem 4.5.1, provide all the details in the derivation of the pdf of t_ν, the Student's t with ν degrees of freedom.

4.5.2 Consider the pdf $h(w)$ of t_ν, the Student's t with ν degrees of freedom, for $w \in \Re$. In the case $\nu = 1$, show that $h(w)$ reduces to $\frac{1}{\pi(1+w^2)}$, the Cauchy density.

4.5.3 Generalize the result given in the Example 4.5.2 in the case of the k-sample problem with $k(\geq 3)$.

4.5.4 With $n(\geq 4)$ iid random variables from $N(0, \sigma^2)$, consider the Helmert variables $Y_1, ..., Y_4$ from Example 4.4.9. Hence, obtain the distribution of $W = Y_1/\sqrt{Y_2^2 + Y_3^2 + Y_4^2}$ and that of W^2?

4.5.5 Suppose that $X_1, ..., X_n$ are iid random variables having a common pdf given by $f(x) = \sigma^{-1}exp\{-(x - \mu)/\sigma\}I(x > \mu)$ for $-\infty < \mu < \infty, 0 < \sigma < \infty$. Here, μ is the location parameter and σ is the scale parameter. Consider (4.4.18) and let $U = n(X_{n:1} - \mu)/\sigma$, $V = 2T/\sigma$.

(i)   Find the pdf of U and $E(U^r)$ for all fixed numbers $r > 0$;
(ii)   Evaluate $E(V^r)$ for all fixed real numbers r;
(iii) What is the distribution of $W = \sigma U \div [T/(n - 1)]$?

{*Caution*: In parts (i)-(ii), watch for the condition on n as needed.}

4.5.6 Suppose that $X_1, ..., X_n$ are iid $N(\mu, \sigma^2)$. Recall the Helmert variables $Y_1, ..., Y_n$ from the Example 4.4.9.

(i)   Find the pdf of Y_i/Y_j for $i \neq j = 2, ..., n$;
(ii)   Find the pdf of $Y_i/|Y_j|$ for $i \neq j = 2, ..., n$;
(iii) Find the pdf of Y_i^2/Y_j^2 for $i \neq j = 2, ..., n$;
(iv)   Find the pdf of Y_i^2/T where $T = \sum\limits_{2 \leq k \leq n, k \neq i} Y_k^2$ for $i = 2, ..., n$;
(v)   Find the pdf of $(Y_i^2 + Y_j^2)/U$ where $U = \sum\limits_{2 \leq k \leq n, k \neq i, k \neq j} Y_k^2$
       for $i \neq j = 2, ..., n$.

{*Hint*: Recall the distributional properties of the Helmert variables.}

4.5.7 (Exercise 4.1.1 Continued) Let Z_1, Z_2 be iid standard normal. From the Exercise 4.5.2, note that the random variable Z_1/Z_2, denoted by X, has the Cauchy pdf given in (1.7.31). Use this sampling distribution to evaluate $P(Z_1^2 < bZ_2^2)$ where $b(> 0)$ is fixed but arbitrary. {*Hint*: Express the required probability as $P(-\sqrt{b} < X < \sqrt{b})$ and then integrate the Cauchy pdf within the appropriate interval. Does this match with the answer given for the Exercise 4.1.1?}

4.5.8 (Exercise 4.5.7 Continued) Let Z_1, Z_2 be iid standard normal. Evaluate $P(Z_1^{-1}(Z_1 + Z_2) \leq c)$ where c is a fixed but arbitrary real number.

4.5.9 (Exercise 4.5.8 Continued) Let Z_1, Z_2 be iid standard normal. Evaluate $P\left(Z_1^2(Z_1^2 + Z_2^2)^{-1} \leq c\right)$ where $0 < c < 1$ is a fixed but arbitrary number.

4.5.10 Verify the expression of the variance of W, which is distributed as t_ν, given in (4.5.4).

4.5.11 Verify the expression of the mean and variance of U, which is distributed as F_{ν_1, ν_2}, given in (4.5.11)-(4.5.12).

4.6.1 (Example 4.6.1 Continued) Use transformations directly to show that $(X_1, \overline{X})$ is distributed as $N_2(\mu, \mu, \sigma^2, \frac{1}{n}\sigma^2, \frac{1}{\sqrt{n}})$ when $X_1, ..., X_n$ are iid $N(\mu, \sigma^2)$.

4.6.2 (Example 4.6.2 Continued) Use transformations directly to find the bivariate normal distribution of $(aX_1 + bX_2, \overline{X})$ when $X_1, ..., X_n$ are iid $N(\mu, \sigma^2)$. Here, a and b are fixed non-zero real numbers.

4.6.3 Show that the bivariate normal density from (3.6.1) can be expressed in the form given by (4.6.1).

4.6.4 Verify the properties given in (4.6.2)-(4.6.3) for the multivariate normal distribution.

4.6.5 Let $X_1, ..., X_n$ be iid $N(\mu, \sigma^2)$. Then, find the joint distributions of

   (i) $Y_2, ..., Y_n$ where $Y_i = X_i - X_1, i = 2, ..., n$;
   (ii) $U_2, ..., U_n$ where $U_i = X_i - X_{i-1}, i = 2, ..., n$.

{*Hint*: Use the Definition 4.6.1 for the multivariate normality.}

4.6.6 (Exercise 4.6.5 Continued) Solve the Exercise 4.6.5 by using direct transformation techniques.

4.6.7 Suppose that $X_1, ..., X_n$ are iid $N(\mu, \sigma^2), n \geq 2$. Use the variables $U_1, ..., U_n$ where $U_1 = \Sigma_{i=1}^n X_i, U_j = X_1 - X_j$ for $j = 2, ..., n$.
   (i)   Show that $\mathbf{U} = (U_1, ..., U_n)$ has the n-dimensional normal distribution;
   (ii)   Show that U_1 and $(U_2, ..., U_n)$ are independent;
   (iii)   Use parts (i)-(ii) to derive the distribution of $\overline{X}$;
   (iv)   Express the sample variance S^2 as a function of $U_2, ..., U_n$ alone. Hence, show that $\overline{X}$ and S^2 are independently distributed.

{*Hint*: Use the Definition 4.6.1 for the multivariate normality to solve part (i). Also, observe that $\binom{n}{2}S^2$ can be rewritten as $\underset{1 \leq i < j \leq n}{\Sigma\Sigma} \frac{(X_i - X_j)^2}{2}$.}

4.6.8 Suppose that $(X_1, Y_1), ..., (X_n, Y_n)$ are iid $N_2(\mu_1, \mu_2, \sigma_1^2, \sigma_2^2, \rho)$ with $-\infty < \mu_1, \mu_2 < \infty, 0 < \sigma_1^2, \sigma_2^2 < \infty$ and $-1 < \rho < 1, n \geq 2$. Then, by appealing to the Definition 4.6.1 of a multivariate normal distribution, show that

$$(\overline{X}, \overline{Y}) \text{ is distributed as } N_2(\mu_1, \mu_2, n^{-1}\sigma_1^2, n^{-1}\sigma_2^2, \rho^*) \text{ with some } \rho^*.$$

{*Hint*: Look at (4.6.6).}

4.6.9 Suppose that $(X_1, Y_1), ..., (X_n, Y_n)$ are iid $N_2(\mu_1, \mu_2, \sigma_1^2, \sigma_2^2, \rho)$ with $-\infty < \mu_1, \mu_2 < \infty, 0 < \sigma_1^2, \sigma_2^2 < \infty$ and $-1 < \rho < 1, n \geq 3$. Consider the Pearson correlation coefficient r defined by (4.6.7). The pdf of r when $\rho = 0$ is given by

$$f(r) = c \, (1 - r^2)^{\frac{1}{2}(n-4)} \, for \, -1 < r < 1,$$

where $c = \Gamma\left(\frac{1}{2}(n-1)\right) \left\{ \sqrt{\pi} \, \Gamma\left(\frac{1}{2}(n-2)\right) \right\}^{-1}$. Use transformation techniques to derive the following result:

$$r(n-2)^{1/2}(1 - r^2)^{-1/2} \text{ has the Student's } t_\nu \text{ distribution}$$
$$\text{with } \nu = (n-2) \text{ degrees of freedom when } \rho = 0.$$

4.6.10 Suppose that $\mathbf{X}_1, ..., \mathbf{X}_n$ are iid $N_2(\boldsymbol{\mu}, \boldsymbol{\Sigma})$ where $\boldsymbol{\mu} \in \Re^2$ and $\boldsymbol{\Sigma}$ is a 2×2 p.d. matrix, $n \geq 2$. Let us denote $\overline{\mathbf{X}} = n^{-1}\Sigma_{i=1}^n \mathbf{X}_i$ and $\mathbf{W} = \Sigma_{i=1}^n (\mathbf{X}_i - \overline{\mathbf{X}})(\mathbf{X}_i - \overline{\mathbf{X}})'$. Prove the following sampling distributions:

(i) $\overline{\mathbf{X}}$ is distributed as $N_2(\boldsymbol{\mu}, n^{-1}\boldsymbol{\Sigma})$;

(ii) $n(\overline{\mathbf{X}} - \boldsymbol{\mu})' \boldsymbol{\Sigma}^{-1}(\overline{\mathbf{X}} - \boldsymbol{\mu})$ is distributed as χ_2^2.

{*Hint*: Apply the Definition 4.6.1 in part (i). To prove part (ii), without any loss of generality, assume that $n = 1$ and $\boldsymbol{\mu} = \mathbf{0}$. Now, show that $n(\overline{\mathbf{x}} - \boldsymbol{\mu})' \boldsymbol{\Sigma}^{-1}(\overline{\mathbf{x}} - \boldsymbol{\mu})$ can be written as $\frac{1}{2(1-\rho^2)} \left[\frac{x_1^2}{\sigma_1^2} - \frac{2\rho x_1 x_2}{\sigma_1 \sigma_2} + \frac{x_2^2}{\sigma_2^2} \right]$ which is further split up as $\left[\frac{1}{4(1+\rho)} \left\{ \frac{x_1}{\sigma_1} + \frac{x_2}{\sigma_2} \right\}^2 + \frac{1}{4(1-\rho)} \left\{ \frac{x_1}{\sigma_1} - \frac{x_2}{\sigma_2} \right\}^2 \right]$. At this stage, can the reproductive property of Chi-squares be applied?}

4.6.11 Suppose that the random vector $\mathbf{X}$, where $\mathbf{X}' = (X_1, ..., X_p)$, has a p-dimensional normal distribution with the mean vector $\mathbf{0}$ and the $p \times p$ p.d. dispersion matrix $\boldsymbol{\Sigma}$, denoted by $N_p(\mathbf{0}, \boldsymbol{\Sigma})$. We assume that each diagonal entry in $\boldsymbol{\Sigma}$ is 1. In other words, the $(i, j)^{th}$ entry in the matrix $\boldsymbol{\Sigma}$ corresponds to the population correlation coefficient ρ_{ij} between the variables $X_i, X_j, 1 \leq i \neq j \leq p$. Suppose that Y is a positive real valued random variable distributed as χ_ν^2. It is also assumed that Y and $(X_1, ..., X_p)$ are independent. Let us denote p new random variables $T_i = X_i \div (Y/\nu)^{1/2}, i = 1, ..., p$. Jointly, however, $(T_1, ..., T_p)$ is said to have the p-dimensional t distribution which depends on the correlation matrix $\boldsymbol{\Sigma}$, denoted by $Mt_p(\nu, \boldsymbol{\Sigma})$. Assume that $\boldsymbol{\Sigma}^{-1}$ exists. Then, show that the *joint* *pdf* of $T_1, ..., T_p$, distributed as $Mt_p(\nu, \boldsymbol{\Sigma})$, is given by

$$
\begin{aligned}
&f(t_1, ..., t_p) \\
&= \frac{\Gamma((\nu + p)/2)}{(\pi\nu)^{p/2}\Gamma(\nu/2)(det(\boldsymbol{\Sigma}))^{1/2}} \left(1 + \nu^{-1}\mathbf{t}'\boldsymbol{\Sigma}^{-1}\mathbf{t}\right)^{-(\nu+p)/2}, \quad \mathbf{t} \in \Re^p,
\end{aligned}
$$

with $\mathbf{t}' = (t_1, ..., t_p)$. {*Hint*: Start with the joint pdf of $X_1, ..., X_p$ and Y. From this, find the joint pdf of $T_1, ..., T_p$ and Y by transforming the variables $(x_1, ..., x_p, y)$ to $(t_1, ..., t_p, y)$. Then, integrate this latter joint pdf with respect to y.}

4.6.12 (Exercise 4.6.5 Continued) Let $X_1, ..., X_n$ be iid $N(\mu, \sigma^2)$. Recall that $Y_i = X_i - X_1, i = 2, ..., n$. Suppose that U is distributed as $\sigma^2\chi_\nu^2$ and also U is independent of the X's. Find the joint pdf of Y_i/U for $i = 2, ..., n$.

4.6.13 Consider the independent random variables $X_0, X_1, ..., X_p$ where X_i is distributed as $\chi_{\nu_i}^2, i = 0, 1, ..., p$. We denote the new random variables $U_i = (X_i/\nu_i) \div (X_0/\nu_0), i = 1, ..., p$. Jointly, $(U_1, ..., U_p)$ is said to have the p-dimensional F distribution, denoted by $MF_p(\nu_0, \nu_1, ..., \nu_p)$. Using transformation techniques, show that the *joint pdf* of $U_1, ..., U_p$, distributed as $MF_p(\nu_0, \nu_1, ..., \nu_p)$, is given by

$$
f(u_1, ..., u_p) = \frac{\Gamma(\nu/2)\Pi_{i=0}^p \nu_i^{\nu_i/2}}{\Pi_{i=0}^p \Gamma(\nu_i/2)} \frac{\Pi_{i=1}^p u_i^{(\nu_i - 2)/2}}{(\nu_0 + \Sigma_{i=1}^p \nu_i u_i)^{\nu/2}}, \quad \mathbf{u} \in \Re^{+p},
$$

with $\mathbf{u}' = (u_1, ..., u_p)$ and $\nu = \Sigma_{i=0}^p \nu_i$. {*Hint*: Start with the joint pdf of $X_0, X_1, ..., X_p$. From this, find the joint pdf of $U_1, ..., U_p$ and X_0 by transforming $(x_1, ..., x_p, x_0)$ to $(u_1, ..., u_p, x_0)$. Then, integrate this latter pdf with respect to x_0.}

4.6.14 Suppose that (U_1, U_2) is distributed as $N_2(5, 15, 8, 8, \rho)$ for some $\rho \in (-1, 1)$. Let $X_1 = U_1 + U_2$ and $X_2 = U_1 - U_2$. Show that X_1 and X_2 are independently distributed. {*Hint*: Is (X_1, X_2) jointly distributed as a bivariate normal random vector?}

4.7.1 (Examples 3.6.3 and 4.7.1 Continued) Suppose that a pair of random variables (X_1, X_2) has the joint pdf given by

$$
\begin{aligned}
f(x_1, x_2) &= \alpha\ f(x_1, x_2; 0, 0, 1, 1, \rho) \\
&+ (1 - \alpha)\ f(x_1, x_2; 0, 0, 1, 1, -\rho),
\end{aligned}
$$

with $0 < \alpha < 1$, where $f(x_1, x_2; \mu_1, \mu_2, \sigma_1^2, \sigma_2^2, \rho)$ stands for the bivariate normal pdf defined in (3.6.1) with means μ_1, μ_2, variances σ_1^2, σ_2^2, and the correlation coefficient ρ with $0 < \rho < 1$. Show that $X_1 + X_2$ is not normally distributed. {*Hint:* Let $u = x_1 + x_2, v = x_1 - x_2$, and find the joint pdf of the transformed random variable (U, V). Integrate this joint pdf with respect to v and show that the marginal pdf of U does not match with the pdf of a normal random variable.}

4.7.2 (Example 3.6.3 and Exercise 4.7.1 Continued) Suppose that a pair of random variables (X_1, X_2) has the joint pdf given by

$$
\begin{aligned}
f(x_1, x_2) \quad &= \alpha \ f(x_1, x_2; \mu, \theta, \sigma^2, \tau^2, \rho) \\
&+ (1 - \alpha) \ f(x_1, x_2; \mu, \theta, \sigma^2, \tau^2, -\rho),
\end{aligned}
$$

with $0 < \alpha < 1$, where $f(x_1, x_2; \mu, \theta, \sigma^2, \tau^2, \rho)$ stands for the bivariate normal pdf defined in (3.6.1) with $\mu_1 = \mu, \mu_2 = \theta, \sigma_1^2 = \sigma^2, \sigma_2^2 = \tau^2$ and $0 < \rho < 1$. Show that marginally X_1 is $N(\mu, \sigma^2)$, X_2 is $N(\theta, \tau^2)$, but $X_1 + X_2$ is not normally distributed.

4.7.3 (Example 4.7.2 Continued) Provide the missing details in the Example 4.7.2.

4.7.4 If $X_1, ..., X_n$ were iid $N(\mu, \sigma^2)$ and $S^2 = (n - 1)^{-1} \Sigma_{i=1}^n (X_i - \overline{X})^2$ was the sample variance, then we could claim that $(n - 1)S^2/\sigma^2$ was distributed as a Chi-square random variable with $(n-1)$ degrees of freedom for $n \geq 2$. Does such a conclusion necessarily need the assumption of the normality or the iid nature of the X sequence? In general there can be different kinds of scenarios on hand. The following problems indicate some of the possibilities. Think through each case carefully and summarize the story each situation unfolds.

(i)   Consider $Y_0, Y_1, ..., Y_n$ which are iid $N(0, 1)$, and define then $X_i = Y_i + Y_0$, $i = 1, ..., n$. Obviously, $\overline{X} = \overline{Y} + Y_0$ where $\overline{X}$ and $\overline{Y}$ are respectively the sample means of $X_1, ..., X_n$ and $Y_1,$ $..., Y_n$. Now, $(n - 1)S^2 = \Sigma_{i=1}^n (X_i - \overline{X})^2 = \Sigma_{i=1}^n (Y_i - \overline{Y})^2$, which would be distributed as a Chi-square random variable with $(n - 1)$ degrees of freedom. Note that $X_1, ..., X_n$ happen to be normal variables, but these are dependent;

(ii) In part (i), observe that Y_0 can be allowed to have any arbitrary distribution while on the other hand Y_0 and $(Y_1, ...$ $, Y_n)$ need not be independent either. Yet, the conclusion that $(n - 1)S^2$ is distributed as a Chi-square random variable with $(n - 1)$ degrees of freedom holds. Y_0 can even be a discrete random variable!

5

Concepts of Stochastic Convergence

5.1 Introduction

The concept of the convergence of a sequence of real numbers $\{a_n; n \geq 1\}$ to another real number a as $n \to \infty$ is well understood. Now, we are about to define the notions of "convergence" of a sequence of real valued random variables $\{U_n; n \geq 1\}$ to some constant u or a random variable U, as $n \to \infty$. But, before we do so, let us adopt a minor change of notation. In the previous chapters, we wrote $\overline{X}, S^2$ and so on, because the sample size n was held fixed. Instead, we will prefer writing $\overline{X}_n, S_n^2$ and so on, in order to make the dependence on the sample size n more explicit.

First, let us see why we need to explore the concepts of stochastic convergence. We may, for example, look at $\overline{X}_n$ or $X_{n:n}$ to gather information about the average or the record value (e.g. the average or record rainfall) in a population. In a situation like this, we are looking at a sequence of random variables $\{U_n; n \geq 1\}$ where $U_n = \overline{X}_n$ or $X_{n:n}$. Different probability calculations would then involve n and the distribution of U_n. Now, what can we say about the sampling distribution of U_n? It turns out that an exact answer for a simple-minded question like this is nearly impossible to give under full generality. When the population was normal or gamma, for example, we provided some exact answers throughout Chapter 4 when $U_n = \overline{X}_n$. On the other hand, the distribution of $X_{n:n}$ is fairly intractable for random samples from a normal or gamma population. If the population distribution happened to be uniform, we found the exact pdf of $X_{n:n}$ in Chapter 4, but in this case the exact pdf of $\overline{X}_n$ becomes too complicated even when n is three or four! In situations like these, we may want to examine the behavior of the random variables $\overline{X}_n$ or $X_{n:n}$, for example, when n becomes large so that we may come up with useful approximations.

But then having to study the sequence $\{\overline{X}_n; n \geq 1\}$ or $\{X_{n:n}; n \geq 1\}$, for example, is not the same as studying an infinite sequence of real numbers! The sequences $\{\overline{X}_n; n \geq 1\}$ or $\{X_{n:n}; n \geq 1\}$, and in general $\{U_n; n \geq 1\}$, are stochastic in nature.

This chapter introduces two fundamental concepts of convergence for a sequence of real valued random variables $\{U_n; n \geq 1\}$. First, in the Section

5.2, we discuss the notion of the *convergence in probability* (denoted by $\overset{P}{\to}$) and the *weak law of large numbers* (WLLN). We start with a weak version of the WLLN, referred to as the *Weak* WLLN, followed by a stronger version, referred to as Khinchine's WLLN. In the Section 5.3, we discuss the notion of the *convergence in distribution or law* (denoted by $\overset{\mathcal{L}}{\to}$). We provide Slutsky's Theorem in Section 5.3 which sets some of the ground rules for manipulations involving these modes of convergence. The *central limit theorem* (CLT) is discussed in Section 5.3 for both the sample mean $\overline{X}_n$ and sample variance S_n^2. This chapter ends (Section 5.4) with some interesting large-sample properties of the Chi-square, t, and F distributions.

5.2   Convergence in Probability

Definition 5.2.1 *Consider a sequence of real valued random variables $\{U_n; n \geq 1\}$. Then, U_n is said to converge in probability to a real number u as $n \to \infty$, denoted by $U_n \overset{P}{\to} u$, if and only if the following condition holds:*

$$P\{|U_n - u| > \varepsilon\} \to 0 \text{ as } n \to \infty, \text{ for every fixed } \varepsilon(> 0). \quad (5.2.1)$$

In other words, $U_n \overset{P}{\to} u$ means this: the probability that U_n will stay away from u, even by a small margin ε, can be made arbitrarily small for large enough n, that is for all $n \geq n_0$ for some $n_0 \equiv n_0(\varepsilon)$. The readers are familiar with the notion of convergence of a sequence of real numbers $\{a_n; n \geq 1\}$ to another real number a, as $n \to \infty$. In (5.2.1), for some fixed $\varepsilon \ (> 0)$, let us denote $p_n(\varepsilon) = P\{|U_n - u| > \varepsilon\}$ which turns out to be a non-negative real number. In order for U_n to converge to u in probability, all we ask is that $p_n(\varepsilon) \to 0$ as $n \to \infty$, for all fixed $\varepsilon(> 0)$.

Next we state another definition followed by some important results. Then, we give a couple of examples.

Definition 5.2.2 *A sequence of real valued random variables $\{U_n; n \geq 1\}$ is said to converge to another real valued random variable U in probability as $n \to \infty$ if and only if $U_n - U \overset{P}{\to} 0$ as $n \to \infty$.*

Now, we move to prove what is known as the *weak law of large numbers* (WLLN). We note that different versions of WLLN are available. Let us begin with one of the simplest versions of the WLLN. Later, we introduce a stronger version of the same result. To set these two weak laws of large numbers apart, the first one is referred to as the *Weak* WLLN.

Theorem 5.2.1 (Weak WLLN) *Let $X_1, ..., X_n$ be iid real valued random variables with $E(X_1) = \mu$ and $V(X_1) = \sigma^2$, $-\infty < \mu < \infty$, $0 <$*

$\sigma < \infty$. *Write $\overline{X}_n (= n^{-1}\Sigma_{i=1}^n X_i)$ for the sample mean. Then, $\overline{X}_n \xrightarrow{P} \mu$ as $n \to \infty$.*

Proof Consider arbitrary but otherwise fixed $\varepsilon(> 0)$ and use Tchebysheff's inequality (Theorem 3.9.3) to obtain

$$
\begin{aligned}
P\left\{ |\,\overline{X}_n - \mu\,| \geq \varepsilon \right\} &= P\left\{ (\overline{X}_n - \mu)^2 \geq \varepsilon^2 \right\} \\
&\leq \varepsilon^{-2} E\{(\overline{X}_n - \mu)^2\} = \sigma^2/(n\varepsilon^2).
\end{aligned}
\tag{5.2.2}
$$

Thus, we have $0 \leq P\left\{ |\,\overline{X}_n - \mu\,| \geq \varepsilon \right\} \leq \sigma^2/(n\varepsilon^2) \to 0$ as $n \to \infty$. Hence, $P\left\{ |\,\overline{X}_n - \mu\,| \geq \varepsilon \right\} \to 0$ as $n \to \infty$ and by the Definition 5.2.1 one claims that $\overline{X}_n \xrightarrow{P} \mu$ as $n \to \infty$. ∎

Intuitively speaking, the Weak WLLN helps us to conclude that the sample mean $\overline{X}_n$ of iid random variables may be expected to hang around the population average μ with high probability, if the sample size n is large enough and $0 < \sigma < \infty$.

Example 5.2.1 Let $X_1, ..., X_n$ be iid $N(\mu, \sigma^2)$, $-\infty < \mu < \infty$, $0 < \sigma < \infty$, and write $\overline{X}_n = n^{-1}\Sigma_{i=1}^n X_i$, $S_n^2 = (n-1)^{-1}\Sigma_{i=1}^n(X_i - \overline{X}_n)^2$ for $n \geq 2$. Then, we show that $S_n^2 \xrightarrow{P} \sigma^2$ as $n \to \infty$. From (4.4.9) recall that we can express S_n^2 as $(n-1)^{-1}\Sigma_{i=2}^n Y_i^2$ where $Y_2, ..., Y_n$ are the Helmert variables. These Helmert variables are iid $N(0, \sigma^2)$. Observe that $V(Y_2^2) = E(Y_2^4) - E^2(Y_2^2) = 3\sigma^4 - (\sigma^2)^2 = 2\sigma^4$, which is finite. In other words, the sample variance S_n^2 has the representation of a sample mean of iid random variables with a finite variance. Thus, the Weak WLLN immediately implies that $S_n^2 \xrightarrow{P} E(Y_2^2) = \sigma^2$ as $n \to \infty$. ▲

Under very mild additional conditions, one can conclude that $S_n^2 \xrightarrow{P} \sigma^2$ as $n \to \infty$, without the assumption of normality of the X's. Refer to Example 5.2.11.

Example 5.2.2 Let $X_1, ..., X_n$ be iid Bernoulli(p), $0 < p < 1$. We know that $E(X_1) = p$ and $V(X_1) = p(1-p)$, and thus by the Weak WLLN, we conclude that $\overline{X}_n \xrightarrow{P} E(X_1) = p$ as $n \to \infty$. ▲

In the following, we state a generalized version of the Weak WLLN. In some problems, this result could come in handy.

Theorem 5.2.2 *Let $\{T_n;\ n \geq 1\}$ be a sequence of real valued random variables such that with some $r(> 0)$ and $a \in \Re$, one can claim that $\xi_{r,n} = E\{|\,T_n - a\,|^r\} \to 0$ as $n \to \infty$. Then, $T_n \xrightarrow{P} a$ as $n \to \infty$.*

Proof For arbitrary but otherwise fixed ε (> 0), we apply the Markov inequality (Theorem 3.9.1) and write

$$P\{|\,T_n - a\,|\geq \varepsilon\,\} = P\{|\,T_n - a\,|^r \geq \varepsilon^r\,\} \leq \xi_{r,n}\varepsilon^{-r} \to 0 \text{ as } n \to \infty.$$

Thus, we have $0 \leq P\{|\,T_n - a\,|\geq \varepsilon\} \leq \xi_{r,n}\varepsilon^{-r} \to 0$ as $n \to \infty$. That is, $P\{|\,T_n - a\,|\geq \varepsilon\} \leq \xi_{r,n}\varepsilon^{-r} \to 0$ as $n \to \infty$. ∎

> In Example 5.2.3, the Weak WLLN does not apply,
> but Theorem 5.2.2 does.

Example 5.2.3 Let $X_1, ..., X_n$ be identically distributed with $-\infty < \mu < \infty$, $0 < \sigma < \infty$, but instead of assuming independence among these X's, we only assume that $Cov(X_i, X_j) = 0$, for $i \neq j = 1, ..., n, n \geq 1$. Again, we have $V(\overline{X}_n) = n^{-1}\sigma^2$ which converges to zero as $n \to \infty$. Thus, by the Theorem 5.2.2, $\overline{X}_n \xrightarrow{P} E(X_1) = \mu$ as $n \to \infty$. ▲

> Example 5.2.3 shows that the conclusion from the Weak WLLN
> holds under less restrictive assumption of uncorrelation among
> the X's rather than the independence among those X's.

Example 5.2.4 Here we directly apply the Theorem 5.2.2. Let $X_1, ..., X_n$ be iid Uniform$(0, \theta)$ with $\theta > 0$. From Example 4.2.7, recall that $T_n = X_{n:n}$, the largest order statistic, has the pdf given by

$$g(t) = nt^{n-1}\theta^{-n}I(0 < t < \theta). \tag{5.2.3}$$

Thus, $E(T_n) = \int_0^\theta tg(t)dt = n\theta^{-n}\int_0^\theta t^n dt = n(n+1)^{-1}\theta$, and similarly we get $E(T_n^2) = n(n+2)^{-1}\theta^2$. That is, $E\{(T_n - \theta)^2\} = 2(n+1)^{-1}(n+2)^{-1}\theta^2$ which converges to zero as $n \to \infty$. Now, we apply the Theorem 5.2.2 with $a = \theta$ and $r = 2$ to conclude that $X_{n:n} \xrightarrow{P} \theta$ as $n \to \infty$. ▲

We can, however, come to the same conclusion without referring to the Theorem 5.2.2. Look at the next example.

Example 5.2.5 (Example 5.2.4 Continued) Let $X_1, ..., X_n$ be iid uniform on the interval $(0, \theta)$ with $\theta > 0$. Let us directly apply the Definition 5.2.1 to show that $T_n = X_{n:n} \xrightarrow{P} \theta$ as $n \to \infty$. Recall the pdf of T_n from (5.2.3). Now, for arbitrary but otherwise fixed $\varepsilon > 0$, one has

$$P\{|\,T_n - \theta\,|\geq \varepsilon\,\} = P\{\theta - T_n \geq \varepsilon\,\} = \begin{cases} 0 & \text{if } \varepsilon \geq \theta \\ p_n & \text{if } \varepsilon < \theta, \end{cases} \tag{5.2.4}$$

where p_n is positive. We simply need to evaluate the $\lim p_n$ as $n \to \infty$. Now, with $0 < \varepsilon < \theta$, we have $p_n = P(T_n \leq \theta - \varepsilon) = n\theta^{-n}\int_0^{\theta-\varepsilon} t^{n-1}dt =$

$(1 - \frac{\varepsilon}{\theta})^n \to 0$ as $n \to \infty$, since $0 < 1 - \varepsilon\theta^{-1} < 1$. Hence, by the Definition 5.2.1, we can claim that $T_n \xrightarrow{P} \theta$ as $n \to \infty$. ▲

> Weak WLLN (Theorem 5.2.1) can be strengthened considerably. One claims: $\overline{X}_n \xrightarrow{P} E(X_1) = \mu$ as $n \to \infty$ so long as the X_i's are iid with finite μ. Assuming the finiteness of σ^2 is not essential.

In what follows, we state a stronger version of the weak law of large numbers. Its proof is involved and hence it is omitted. Some references are given in the Exercise 5.2.2.

Theorem 5.2.3 (Khinchine's WLLN) *Let* $X_1, ..., X_n$ *be iid real valued random variables with* $E(X_1) = \mu, -\infty < \mu < \infty$. *Then,* $\overline{X}_n \xrightarrow{P} \mu$ *as* $n \to \infty$.

Example 5.2.6 In order to appreciate the importance of Khinchine's WLLN (Theorem 5.2.3), let us consider a sequence of iid random variables $\{X_n; n \geq 1\}$ where the distribution of X_1 is given as follows:

$$
\begin{array}{llllllll}
X_1 \text{ values:} & 1 & 2 & 3 & . & . & . & i & . & . & . \\
\text{Probabilities:} & \frac{c}{1^3} & \frac{c}{2^3} & \frac{c}{3^3} & . & . & . & \frac{c}{i^3} & . & . & .
\end{array}
\tag{5.2.5}
$$

with $c^{-1} = \Sigma_{i=1}^{\infty} \frac{1}{i^3}$ so that $0 < c < \infty$. This is indeed a probability distribution. Review the Examples 2.3.1-2.3.2 as needed. We have $E(X_1) = c\Sigma_{i=1}^{\infty} \frac{1}{i^2}$ which is finite. However, $E(X_1^2) = c\Sigma_{i=1}^{\infty} \frac{1}{i}$, but this infinite series is not finite. Thus, we have a situation where μ is finite but σ^2 is not finite for the sequence of the iid X's. Now, in view of Khinchine's WLLN, we conclude that $\overline{X}_n \xrightarrow{P} E(X_1) = \mu$ as $n \to \infty$. From the Weak WLLN, we would not be able to reach this conclusion. ▲

We now move to discuss other important aspects. It may be that a sequence of random variables U_n converges to u in probability as $n \to \infty$, but then one should not take it for granted that $E(U_n) \to u$ as $n \to \infty$. On the other hand, one may be able to verify that $E(U_n) \to u$ as $n \to \infty$ in a problem, but then one should not jump to conclude that $U_n \xrightarrow{P} u$ as $n \to \infty$. To drive the point home, we first look at the following sequence of random variables:

$$
\begin{array}{lll}
U_n \text{ values:} & 1 & n \\
\text{Probabilities:} & 1 - \frac{1}{n} & \frac{1}{n}
\end{array}
\tag{5.2.6}
$$

It is simple enough to see that $E(U_n) = 2 - \frac{1}{n} \to 2$ as $n \to \infty$, but $U_n \xrightarrow{P} 1$ as $n \to \infty$. Next, look at a sequence of random variables V_n:

$$
\begin{array}{lll}
V_n \text{ values:} & 1 & n^2 \\
\text{Probabilities:} & 1 - \frac{1}{n} & \frac{1}{n}
\end{array}
\tag{5.2.7}
$$

Here again, it easily follows that $V_n \xrightarrow{P} 1$ as $n \to \infty$, but $E(V_n) = 1 - \frac{1}{n} + n \to \infty$ as $n \to \infty$. The relationship between the two notions involving the convergence in probability (that is, $U_n \xrightarrow{P} u$ as $n \to \infty$) and the *convergence in mean* (that is, $E(U_n) \to u$ as $n \to \infty$) is complicated. The details are out of scope for this book. Among other sources, one may refer to Sen and Singer (1993, Chapter 2) and Serfling (1980).

$\{U_n; n \geq 1\}$ may converge to u in probability, but this fact by itself may not imply that $E(U_n) \to u$. Also, a fact that $E(U_n) \to u$ alone may not imply that $U_n \xrightarrow{P} u$. Carefully study the random variables given in (5.2.6) and (5.2.7).

The Exercise 5.2.22 gives some sufficient conditions under which one claims: $U_n \xrightarrow{P} u$ as $n \to \infty \Rightarrow E[g(U_n)] \to g(u)$ as $n \to \infty$. The Exercise 5.2.23 gives some applications.

In textbooks at this level, the next theorem's proof is normally left out as an exercise. The proof is not hard, but if one is mindful to look into all the cases and sub-cases, it is not that simple either. We include the proof with the hope that the readers will think through the bits and pieces analytically, thereby checking each step with the care it deserves.

Theorem 5.2.4 *Suppose that we have two sequences of real valued random variables* $\{U_n, V_n; n \geq 1\}$ *such that* $U_n \xrightarrow{P} u$, $V_n \xrightarrow{P} v$ *as* $n \to \infty$. *Then, we have:*

(i) $U_n \pm V_n \xrightarrow{P} u \pm v$ *as* $n \to \infty$;

(ii) $U_n V_n \xrightarrow{P} uv$ *as* $n \to \infty$;

(iii) $U_n/V_n \xrightarrow{P} u/v$ *as* $n \to \infty$ *if* $P(V_n = 0) = 0$ *for all* $n \geq 1$ *and* $v \neq 0$.

Proof (i) We start with an arbitrary but otherwise fixed $\varepsilon(> 0)$ and write

$$P\{| (U_n + V_n) - (u + v) | \geq \varepsilon\}$$

$$\leq P\{| U_n - u | + | V_n - v | \geq \varepsilon\},$$

by the triangular inequality (1.6.30)   (5.2.8)

$$\leq P\{| U_n - u | \geq \tfrac{1}{2}\varepsilon \ \cup \ | V_n - v | \geq \tfrac{1}{2}\varepsilon\}$$

$$\leq P\{| U_n - u | \geq \tfrac{1}{2}\varepsilon\} + P\{| V_n - v | \geq \tfrac{1}{2}\varepsilon\}.$$

The second inequality follows because $[| U_n - u | < \tfrac{1}{2}\varepsilon \cap | V_n - v | < \tfrac{1}{2}\varepsilon]$ implies that $[| U_n - u | + | V_n - v | < \varepsilon]$, so that $[| U_n - u | + | V_n - v | \geq \varepsilon]$

implies $[\mid U_n - u \mid \geq \frac{1}{2}\varepsilon \cup \mid V_n - v \mid \geq \frac{1}{2}\varepsilon]$. The third inequality follows because $P(A \cup B) \leq P(A) + P(B)$ for any two events A and B. From the fact that $U_n \overset{P}{\to} u$, $V_n \overset{P}{\to} v$ as $n \to \infty$, both the probabilities in the last step in (5.2.8) converge to zero as $n \to \infty$, and hence the lhs in (5.2.8), which is non-negative, converges to zero as $n \to \infty$, for all fixed but arbitrary $\varepsilon(> 0)$. Thus, $U_n + V_n \overset{P}{\to} u + v$ as $n \to \infty$. The case of $U_n - V_n$ can be tackled similarly. It is left as the Exercise 5.2.6. This completes the proof of part (i). ◆

(ii) We start with arbitrary but otherwise fixed $\varepsilon(> 0)$ and write

$$P\{\mid U_n V_n - uv \mid \geq \varepsilon\}$$

$$= \begin{cases} P\{\mid U_n V_n \mid \geq \varepsilon\} & \text{if } v = 0 \\ P\{\mid U_n(V_n - v) + vU_n \mid \geq \varepsilon\} & \text{if } u = 0,\ v \neq 0 \\ P\{\mid U_n(V_n - v) + v(U_n - u) \mid \geq \varepsilon\} & \text{if } u \neq 0,\ v \neq 0 \end{cases} \tag{5.2.9}$$

Case 1: $u = 0, v = 0$

Observe that $[\mid U_n \mid < \sqrt{\varepsilon} \cap \mid V_n \mid < \sqrt{\varepsilon}]$ implies that $[\mid U_n V_n \mid < \varepsilon]$, that is $[\mid U_n V_n \mid \geq \varepsilon]$ implies $[\mid U_n \mid \geq \sqrt{\varepsilon} \cup \mid V_n \mid \geq \sqrt{\varepsilon}]$. Hence, we have $P\{\mid U_n V_n \mid \geq \varepsilon\} \leq P\{\mid U_n \mid \geq \sqrt{\varepsilon} \cup \mid V_n \mid \geq \sqrt{\varepsilon}\} \leq P\{\mid U_n \mid \geq \sqrt{\varepsilon}\} + P\{\mid V_n \mid \geq \sqrt{\varepsilon}\}$, and both these probabilities converge to zero as $n \to \infty$, and thus $P\{\mid U_n V_n \mid \geq \varepsilon\} \to 0$ as $n \to \infty$.

Case 2: $u \neq 0, v = 0$

Observe by triangular inequality that $\mid U_n \mid = \mid (U_n - u) + u \mid \leq \mid U_n - u \mid + \mid u \mid$ and hence $P\{\mid U_n \mid \geq 2 \mid u \mid\} \leq P\{\mid U_n - u \mid + \mid u \mid \geq 2 \mid u \mid\} = P\{\mid U_n - u \mid \geq \mid u \mid\}$ which converges to zero as $n \to \infty$, since $U_n \overset{P}{\to} u$ as $n \to \infty$. Thus, we claim that $P\{\mid U_n \mid \geq 2 \mid u \mid\} \to 0$ as $n \to \infty$. Let us now write

$$P\{\mid U_n V_n \mid \geq \varepsilon\}$$
$$\leq P\{\mid U_n V_n \mid \geq \varepsilon \cap \mid U_n \mid \geq 2 \mid u \mid\} + P\{\mid U_n V_n \mid \geq \varepsilon \cap \mid U_n \mid < 2 \mid u \mid\}$$
$$\leq P\{\mid U_n \mid \geq 2 \mid u \mid\} + P\{2 \mid u \mid \mid V_n \mid \geq \varepsilon \cap \mid U_n \mid < 2 \mid u \mid\}$$
$$\leq P\{\mid U_n \mid \geq 2 \mid u \mid\} + P\{\mid V_n \mid \geq \frac{1}{2}\varepsilon \mid u \mid^{-1}\}.$$

But, we had verified earlier that the first term in the last step converges to zero as $n \to \infty$, while the second term also converges to zero as $n \to \infty$ simply because $V_n \overset{P}{\to} 0$ as $n \to \infty$. Thus the lhs converges to zero as $n \to \infty$.

Case 3: $u = 0, v \neq 0$

The proof in this case is similar to the one in the previous case and hence it is left out as the Exercise 5.2.6.

Case 4: $u \neq 0, v \neq 0$

From (5.2.9), using similar arguments as before, we observe that

$$
\begin{aligned}
P\{ &| U_n(V_n - v) + v(U_n - u) | \geq \varepsilon \} \\
&\leq P\{| U_n \| V_n - v | \geq \tfrac{1}{2}\varepsilon \} + P\{| v \| U_n - u | \geq \tfrac{1}{2}\varepsilon \} \\
&\leq P\{| U_n \| V_n - v | \geq \tfrac{1}{2}\varepsilon \cap | U_n | \leq 2 | u | \} \\
&\quad + P\{| U_n \| V_n - v | \geq \tfrac{1}{2}\varepsilon \cap | U_n | > 2 | u | \} \\
&\quad + P\{| U_n - u | \geq \tfrac{1}{2}\varepsilon | v |^{-1} \} \\
&\leq P\{| V_n - v | \geq \tfrac{1}{4}\varepsilon | u |^{-1} \} + P\{| U_n | > 2 | u | \} \\
&\quad + P\{| U_n - u | \geq \tfrac{1}{2}\varepsilon | v |^{-1} \}.
\end{aligned}
\tag{5.2.10}
$$

In the last step in (5.2.10), the first and third terms certainly converge to zero since $U_n \overset{P}{\to} u \neq 0$, $V_n \overset{P}{\to} v \neq 0$ as $n \to \infty$. For the middle term in the last step of (5.2.10), let us again use the triangular inequality and proceed as follows:

$$
P\{| U_n - u | > | u | \} \geq P\{| U_n | - | u | > | u | \} = P\{| U_n | > 2 | u | \}
$$

which implies that $P\{| U_n | > 2 | u | \} \to 0$ as $n \to \infty$, since $U_n \overset{P}{\to} u \neq 0$ as $n \to \infty$. Hence the lhs in (5.2.10), converges to zero as $n \to \infty$. This completes the proof of part (ii). ◆

(iii) In view of part (ii), we simply proceed to prove that $V_n^{-1} \overset{P}{\to} v^{-1}$ as $n \to \infty$. Note that $| V_n - v | \geq | v | - | V_n |$ which implies that

$$
P\{| V_n - v | < \tfrac{1}{2} | v | \} \leq P\{| v | - | V_n | < \tfrac{1}{2} | v | \} \leq P\{| V_n | > \tfrac{1}{2} | v | \}.
\tag{5.2.11}
$$

Observe that the lhs of (5.2.11) converges to unity as $n \to \infty$ because $V_n \overset{P}{\to} v$ as $n \to \infty$. This implies

$$
\lim_{n \to \infty} P\{| V_n | > \tfrac{1}{2} | v | \} = 1.
\tag{5.2.12}
$$

Let us denote $W_n = | V_n - v | \{ | V_n \| v | \}^{-1}$. Then, for any arbitrary but otherwise fixed $\varepsilon(> 0)$, we can write

$$
\begin{aligned}
P\{ &| V_n^{-1} - v^{-1} | > \varepsilon \} \\
&= P\{ W_n > \varepsilon \cap | V_n | > \tfrac{1}{2} | v | \} \\
&\quad + P\{ W_n > \varepsilon \cap | V_n | \leq \tfrac{1}{2} | v | \} \\
&= P\{| V_n - v | > \tfrac{1}{2}\varepsilon | v |^2 \} + P\{| V_n | \leq \tfrac{1}{2} | v | \},
\end{aligned}
$$

which $\to 0$ as $n \to \infty$. The proof of part (iii) is now complete. ∎

Convergence in probability property is closed under the operations: addition, subtraction, multiplication and division. **Caution:** Division by 0 or ∞ is not allowed.

Example 5.2.7 Suppose that $X_1, ..., X_n$ are iid $N(\mu, \sigma^2), -\infty < \mu < \infty, 0 < \sigma^2 < \infty, n \geq 2$. Let us consider the sample mean $\overline{X}_n$ and the sample variance S_n^2. We know that $\overline{X}_n \overset{P}{\to} \mu$, $S_n^2 \overset{P}{\to} \sigma^2$ as $n \to \infty$. Thus by the Theorem 5.2.4, part (i), we conclude that $\overline{X}_n \pm S_n^2 \overset{P}{\to} \mu \pm \sigma^2$ as $n \to \infty$, and $\overline{X}_n / S_n^2 \overset{P}{\to} \mu/\sigma^2$ as $n \to \infty$. ▲

Let us again apply the Theorem 5.2.4. Suppose that $U_n \overset{P}{\to} u$ as $n \to \infty$. Then, by the Theorem 5.2.4, part (i), we can obviously claim, for example, that $U_n - u \overset{P}{\to} 0$, $U_n + u \overset{P}{\to} 2u$ as $n \to \infty$. Then, one may write

$$U_n^2 - u^2 = (U_n - u)(U_n + u) \overset{P}{\to} 0 \text{ as } n \to \infty,$$

in view of the Theorem 5.2.4, part (ii). That is, one can conclude: $U_n^2 \overset{P}{\to} u^2$ as $n \to \infty$. On the other hand, one could alternatively think of U_n^2 as $U_n V_n$ where $V_n = U_n$ and directly apply the Theorem 5.2.4, part (ii) also. The following theorem gives a more general result.

Theorem 5.2.5 *Suppose that we have a sequence of real valued random variables $\{U_n; n \geq 1\}$ and that $U_n \overset{P}{\to} u$ as $n \to \infty$. Let $g(.)$ be a real valued continuous function. Then, $g(U_n) \overset{P}{\to} g(u)$ as $n \to \infty$.*

Proof A function $g(x)$ is continuous at $x = u$ provided the following holds. Given arbitrary but otherwise fixed $\varepsilon(> 0)$, there exists some positive number $\delta \equiv \delta(\varepsilon)$ for which

$$|g(x) - g(u)| \geq \varepsilon \Rightarrow |x - u| \geq \delta. \tag{5.2.13}$$

Hence for large enough $n(\geq n_0 \equiv n_0(\varepsilon))$, we can write

$$0 \leq P\{|g(U_n) - g(u)| \geq \varepsilon\} \leq P\{|U_n - u| \geq \delta\}, \tag{5.2.14}$$

and the upper bound $\to 0$ as $n \to \infty$. Now, the proof is complete. ∎

Example 5.2.8 (Example 5.2.4 Continued) Let $X_1, ..., X_n$ be iid uniform on the interval $(0, \theta)$ with $\theta > 0$. Recall from the Example 5.2.4 that $T_n = X_{n:n}$, the largest order statistic, converges in probability to θ. In view of the Theorem 5.2.5, obviously $T_n^2 = X_{n:n}^2 \overset{P}{\to} \theta^2$ as $n \to \infty$, by considering the continuous function $g(x) = x^2$ for $x > 0$. Similarly, for example, $T_n^{\frac{1}{3}} \overset{P}{\to} \theta^{\frac{1}{3}}$ as $n \to \infty$, by considering the continuous function $g(x) = x^{\frac{1}{3}}$ for $x > 0$. By the same token, we can also claim, for example, that $\sin(X_{n:n}) \overset{P}{\to} \sin(\theta)$ as $n \to \infty$. ▲

Example 5.2.9 Let $X_1, ..., X_n$ be iid Poisson(λ) with $\lambda > 0$ and look at $\overline{X}_n$, the sample mean. By the Weak WLLN, we immediately claim that $\overline{X}_n \overset{P}{\to} \lambda$ as $n \to \infty$. By combining the results from the two Theorems 5.2.4 and 5.2.5, it also immediately follows, for example, that

$$\overline{X}_n^3 \left\{ 3\sqrt{\overline{X}_n} - \overline{X}_n + 5 \right\} \overset{P}{\to} \lambda^3 \left\{ 3\sqrt{\lambda} - \lambda + 5 \right\} \text{ as } n \to \infty;$$

$$3\overline{X}_n + n^{-1} \overset{P}{\to} 3\lambda \text{ as } n \to \infty.$$

But then can one also claim that

$$(\overline{X}_n)^{-3} \overset{P}{\to} \lambda^{-3} \text{ as } n \to \infty? \tag{5.2.15}$$

Before jumping into a conclusion, one should take into consideration the fact that $P\{\overline{X}_n = 0\} = P\{\cap_{i=1}^n X_i = 0\} = e^{-n\lambda}$, which is positive for every fixed $n \geq 1$ whatever be $\lambda(> 0)$. One can certainly conclude that

$$(\overline{X}_n + n^{-\gamma})^{-3} \overset{P}{\to} \lambda^{-3} \text{ as } n \to \infty \text{ with arbitrary} \atop \text{but fixed } \gamma(> 0). \tag{5.2.16}$$

We ask the reader to sort out the subtle difference between the question raised in (5.2.15) and the statement made in (5.2.16). ▲

Under fair bit of generality, we may claim that $\overline{X}_n \overset{P}{\to} \mu$ as $n \to \infty$. But, inspite of the result in the Theorem 5.2.4, part (iii), we may not be able to claim that $\{\overline{X}_n\}^{-\alpha} \overset{P}{\to} \mu^{-\alpha}$ as $n \to \infty$ when $\alpha > 0$. The reason is that μ may be zero or $P(\overline{X}_n = 0)$ may be positive for all n. See the Example 5.2.9.

Example 5.2.10 Let $X_1, ..., X_n$ be iid random variables with $E(X_1) = \mu$ and $V(X_1) = \sigma^2$, $-\infty < \mu < \infty$, $0 < \sigma < \infty$. Define $U_n = n^\gamma(\overline{X}_n - \mu)^2$ with $0 < \gamma < 1$. Note that $U_n > 0$ w.p. 1. Now, for any arbitrary $\varepsilon(> 0)$, by the Markov inequality, we observe that

$$P(U_n > \varepsilon) \leq \varepsilon^{-1} E(U_n) = \varepsilon^{-1} n^\gamma (\sigma^2 n^{-1}) \to 0 \text{ as } n \to \infty.$$

Hence, by the Definition 5.2.1, $U_n \overset{P}{\to} 0$ as $n \to \infty$. ▲

Remark 5.2.1 Under the setup of the Example 5.2.10, using the Theorems 5.2.4-5.2.5, it is obvious that $V_n = (\overline{X}_n - \mu)^2 \xrightarrow{P} 0$ as $n \to \infty$. But note that the term $n^\gamma \to \infty$ as $n \to \infty$ and the term V_n is inflated by this growth factor. It is noteworthy in the Example 5.2.10 that the inflated random variable $n^\gamma V_n \xrightarrow{P} 0$ as $n \to \infty$ if $0 < \gamma < 1$ is held fixed.

Example 5.2.11 Let $X_1, ..., X_n$ be iid random variables with $E(X_1) = \mu$ and $V(X_1) = \sigma^2$, $-\infty < \mu < \infty$, $0 < \sigma < \infty$, $n \geq 2$. As usual, write $\overline{X}_n, S_n^2$ for the sample mean and variance respectively. It is easy to see that

$$E\{(n-1)S_n^2\} = E\{\Sigma_{i=1}^n X_i^2\} - nE(\overline{X}_n^2) = n(\sigma^2 + \mu^2) - n(\tfrac{\sigma^2}{n} + \mu^2),$$

which reduces to $(n-1)\sigma^2$ so that $E\{S_n^2\} = \sigma^2$, without assuming any special distribution. But, the calculation of $V(S_n^2)$ is not so simple when the distribution of the X's is unspecified. In the Section 2.3, we had defined the central moments. In particular, we denoted the third and fourth central moments by $\mu_3 = E\{(X - \mu)^3\}$, $\mu_4 = E\{(X - \mu)^4\}$ respectively. Assume that $0 < \mu_4 < \infty$ and $\mu_4 > \sigma^4$. Since $V(S_n^2)$ does not depend on the specific value of μ, without any loss of generality, let us *pretend* from this point onward that $\mu = 0$. Look at the Exercise 5.2.17. Now,

$$\begin{aligned}
(n-1)S_n^2 &= \Sigma_{i=1}^n X_i^2 - \tfrac{1}{n}\left(\Sigma_{i=1}^n X_i\right)^2 \\
&= \Sigma_{i=1}^n X_i^2 - \tfrac{1}{n}\left\{\Sigma_{i=1}^n X_i^2 + 2\sum_{1 \leq i < j \leq n}\sum X_i X_j\right\} \qquad (5.2.17) \\
&= \tfrac{n-1}{n}\Sigma_{i=1}^n X_i^2 - \tfrac{2}{n}\sum_{1 \leq i < j \leq n}\sum X_i X_j,
\end{aligned}$$

which implies that

$$\begin{aligned}
(n-1)^2 V(S_n^2) &= \tfrac{(n-1)^2}{n^2}V\left(\Sigma_{i=1}^n X_i^2\right) + \tfrac{4}{n^2}V\left(\sum_{1 \leq i < j \leq n}\sum X_i X_j\right) \\
&\quad - \tfrac{4(n-1)}{n^2}Cov\left\{\Sigma_{i=1}^n X_i^2, \sum_{1 \leq i < j \leq n}\sum X_i X_j\right\}.
\end{aligned}$$
$$(5.2.18)$$

But, the X's are iid and hence

$$\begin{aligned}
V\left(\Sigma_{i=1}^n X_i^2\right) &= \Sigma_{i=1}^n V\left(X_i^2\right) = nV(X_1^2) \\
&= n\left\{E(X_1^4) - E^2(X_1^2)\right\} = n(\mu_4 - \sigma^4),
\end{aligned} \qquad (5.2.19)$$

and also,

$$\begin{aligned}
V\left(\sum_{1 \leq i < j \leq n}\sum X_i X_j\right) &= V\{X_1 X_2 + X_1 X_3 + ... + X_1 X_n \\
&\quad + X_2 X_3 + ... + X_2 X_n + ... + X_{n-1}X_n\}.
\end{aligned} \qquad (5.2.20)$$

The variance of each term in (5.2.20) would be equal to $E(X_1^2 X_2^2) - E^2(X_1 X_2) = E(X_1^2)E(X_2^2) - E^2(X_1)E^2(X_2) = \sigma^4$, since (i) we pretend that $\mu = 0$, and (ii) the X's are assumed independent. Also, the covariances such as

$$Cov(X_1 X_2, X_1 X_3) = E(X_1^2 X_2 X_3) - E(X_1 X_2)E(X_1 X_3) = 0,$$

and hence from (5.2.20) we get

$$V\left\{ \sum_{1 \le i < j \le n} X_i X_j \right\} = \sigma^4 \{ 1 + 2 + ... + (n-1) \} = \tfrac{1}{2} n(n-1)\sigma^4. \quad (5.2.21)$$

See (1.6.11) for the sum of successive positive integers. Next, by taking each covariance term into consideration, similarly one has

$$Cov\left(\Sigma_{i=1}^n X_i^2, \sum_{1 \le i < j \le n} X_i X_j \right) = nCov\left(X_1^2, \sum_{1 \le i < j \le n} X_i X_j \right) = 0. \quad (5.2.22)$$

Now combining (5.2.17)-(5.2.22), we obtain

$$(n-1)^2 V(S_n^2) = \left(\tfrac{n-1}{n} \right)^2 n(\mu_4 - \sigma^4) + \tfrac{4}{n^2} \tfrac{n(n-1)}{2} \sigma^4,$$

so that one writes

$$V(S_n^2) = \tfrac{1}{n}(\mu_4 - \sigma^4) + \tfrac{2}{n(n-1)} \sigma^4 = \tfrac{1}{n} \left\{ \mu_4 - \tfrac{n-3}{n-1} \sigma^4 \right\}. \quad (5.2.23)$$

Now, obviously $V(S_n^2) \to 0$ as $n \to \infty$. Using the Theorem 5.2.2 with $r = 2$ and $a = \sigma^2$, it follows that $S_n^2 \xrightarrow{P} \sigma^2$ as $n \to \infty$. By applying the Theorem 5.2.5, we can also conclude, for example, that $S_n \xrightarrow{P} \sigma$ as $n \to \infty$ or $S_n^{-1} \xrightarrow{P} \sigma^{-1}$ as $n \to \infty$. ▲

Remark 5.2.2 If the X's considered in the Example 5.2.11 were assumed iid $N(\mu, \sigma^2)$, then we would have had $\mu_4 = 3\sigma^4$ so that (5.2.23) would reduce to

$$V(S_n^2) = 2(n-1)^{-1}\sigma^4. \quad (5.2.24)$$

Note that (5.2.24) is also directly verified by observing that S_n^2 is distributed as $\sigma^2(n-1)^{-1}\chi_{n-1}^2$ so that one can write $V\left\{ \sigma^2(n-1)^{-1}\chi_{n-1}^2 \right\} = \sigma^4(n-1)^{-2}V\left(\chi_{n-1}^2 \right) = \sigma^4(n-1)^{-2}2(n-1) = 2(n-1)^{-1}\sigma^4$.

Example 5.2.12 (Example 5.2.11 Continued) Let us continue working under the non-normal case and define $T_n = n^\gamma (S_n^2 - \sigma^2)^2$ where $0 < \gamma < 1$. Now, by the Markov inequality and (5.2.23), with $k_n = \mu_4 - (n-3)(n-1)^{-1}\sigma^4$ and for any fixed $\varepsilon(> 0)$, we can write:

$$P\{T_n > \varepsilon\} \le \varepsilon^{-1} E(T_n) = k_n \varepsilon^{-1} n^{\gamma-1} \to 0 \text{ as } n \to \infty.$$

Hence, by the Definition 5.2.1, $T_n \xrightarrow{P} 0$ as $n \to \infty$, whenever μ_4 is finite and $\mu_4 > \sigma^4$. ▲

5.3   Convergence in Distribution

Definition 5.3.1 *Consider a sequence of real valued random variables $\{U_n;$ $n \geq 1\}$ and another real valued random variable U with the respective distribution functions $F_n(u) = P(U_n \leq u)$, $F(u) = P(U \leq u)$, $u \in \Re$. Then, U_n is said to converge in distribution to U as $n \to \infty$, denoted by $U_n \overset{\mathcal{L}}{\to} U$, if and only if $F_n(u) \to F(u)$ pointwise at all continuity points u of $F(.)$. The distribution of U is referred to as the limiting or asymptotic (as $n \to \infty$) distribution of U_n.*

In other words, $U_n \overset{\mathcal{L}}{\to} U$ means this: with every fixed u, once we compute $P(U_n \leq u)$, it turns out to be a real number, say, a_n. For the convergence in distribution of U_n to U, all we ask for is that the sequence of non-negative real numbers $a_n \to a$ as $n \to \infty$, where $a = P(U \leq u)$, u being any continuity point of $F(.)$.

It is known that the totality of all the discontinuity points of any df F can be at most countably infinite. Review Theorem 1.6.1 and the examples from (1.6.5)-(1.6.9).

Example 5.3.1 Let $X_1, ..., X_n$ be iid Uniform$(0, \theta)$ with $\theta > 0$. From Example 4.2.7, recall that the largest order statistic $T_n = X_{n:n}$ has the pdf given by $g(t) = nt^{n-1}\theta^{-n}I(0 < t < \theta)$. The df of T_n is given by

$$G_n(t) = \begin{cases} 0 & \text{if} \quad t \leq 0 \\ (t/\theta)^n & \text{if} \quad 0 < t < \theta \\ 1 & \text{if} \quad t \geq \theta. \end{cases} \qquad (5.3.1)$$

Now let $U_n = n(\theta - T_n)/\theta$. Obviously, $P(U_n > 0) = 1$ and from (5.3.1) we can write:

$$F_n(u) = P(U_n \leq u) = \begin{cases} 0 & \text{if} \quad u \leq 0 \\ 1 - \left(1 - \frac{u}{n}\right)^n & \text{if} \quad u > 0, \end{cases} \qquad (5.3.2)$$

so that

$$\lim_{n \to \infty} F_n(u) = \begin{cases} 0 & \text{if} \quad u \leq 0 \\ 1 - e^{-u} & \text{if} \quad u > 0, \end{cases} \qquad (5.3.3)$$

since $\lim_{n \to \infty} \left(1 - \frac{u}{n}\right)^n = e^{-u}$. See (1.6.13) for some of the results on limits. Now consider a random variable U with its pdf given by $f(u) = e^{-u}I(u > 0)$, so that U is the standard exponential random variable, while its distribution function is given by

$$F(u) = \begin{cases} 0 & \text{if} \quad u \leq 0 \\ 1 - e^{-u} & \text{if} \quad u > 0. \end{cases} \qquad (5.3.4)$$

It is clear that all points $u \in (-\infty, 0) \cup (0, \infty)$ are continuity points of the function $F(u)$, and for all such u, $\lim_{n \to \infty} F_n(u) = F(u)$. Hence, we would say that $U_n \overset{\mathcal{L}}{\to} U$ as $n \to \infty$. Hence, asymptotically U_n has the standard exponential distribution, that is the Gamma$(1, 1)$ distribution. ▲

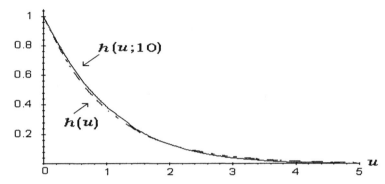

Figure 5.3.1. PDF $h(u; n)$ of U_n When $n = 10$ and PDF
$h(u)$ from the Example 5.3.2

Example 5.3.2 (Example 5.3.1 Continued) Using the expression of the df of U_n from (5.3.2), one can immediately check that the pdf of U_n is given by $h(u; n) = \left(1 - \frac{u}{n}\right)^{n-1} I(u > 0)$ for $n = 1, 2, \ldots$. The pdf of the limiting distribution is given by $h(u) = e^{-u} I(u > 0)$. How quickly does the distribution of U_n converge to the distribution of U? When we compare the plots of $h(u; 10)$ and $h(u)$ in the Figure 5.3.1, we hardly notice any difference between them. What it implies, from a practical point of view, is this: Even if the sample size n is as small as ten, the distribution of U_n is *approximated* remarkably well by the limiting distribution. ▲

In general, it may be hard to proceed along the lines of our Example 5.3.1. There may be two concerns. First, we must have the explicit expression of the df of U_n and second, we must be able to examine this df's asymptotic behavior as $n \to \infty$. Between these two concerns, the former is likely to create more headache. So we are literally forced to pursue an indirect approach involving the moment generating functions.

Let us suppose that the mgf's of the real valued random variables U_n and U are both finite and let these be respectively denoted by $M_n(t) \equiv M_{U_n}(t)$, $M(t) \equiv M_U(t)$ for $| t | < h$ with some $h(> 0)$.

Theorem 5.3.1 *Suppose that* $M_n(t) \to M(t)$ *for* $| t | < h$ *as* $n \to \infty$. *Then,*

$$U_n \overset{\mathcal{L}}{\to} U \text{ as } n \to \infty.$$

A proof of this result is beyond the scope of this book. One may refer to Serfling (1980) or Sen and Singer (1993).

The point is that once $M(t)$ is found, it must correspond to a unique random variable, say, U. In the Section 2.4, we had talked about identifying a distribution uniquely (Theorem 2.4.1) with the help of a finite mgf. We can use the same result here too.

Also, one may recall that the mgf of the sum of n independent random variables is same as the product of the n individual mgf's (Theorem 4.3.1). This result was successfully exploited earlier in the Section 4.3. In the following example and elsewhere, we will repeatedly exploit the form of the mgf of a linear function of iid random variables all over again.

Let $X_1, ..., X_n$ be independent real valued random variables. Let X_i have its finite mgf $M_{X_i}(t) = E(e^{tX_i})$ for $i = 1, ..., n$. Then, the mgf of $W = \Sigma_{i=1}^n a_i X_i$ is given by $\Pi_{i=1}^n M_{X_i}(ta_i)$ where $a_1, ..., a_n$ are any arbitrary but otherwise fixed real numbers.

Example 5.3.3 Let $X_1, ..., X_n$ be iid Bernoulli(p) with $p = \frac{1}{2}$, and we denote $U_n = 2\sqrt{n}(\overline{X}_n - .5)$. Now, the mgf $M_n(t)$ of U_n is given by

$$
\begin{aligned}
E\{exp(tU_n)\} &= exp\left(-t\sqrt{n}\right) E\left\{exp\left(\frac{2t}{\sqrt{n}}\Sigma_{i=1}^n X_i\right)\right\} \\
&= exp\left(-t\sqrt{n}\right) \frac{1}{2^n}\left\{1 + exp(2t/\sqrt{n})\right\}^n,
\end{aligned}
$$

since $E\left(exp(tX_1)\right) = \frac{1}{2}(1 + e^t)$. In other words, one has

$$
M_n(t) = \frac{1}{2^n}\left\{exp(-t/\sqrt{n}) + exp(t/\sqrt{n})\right\}^n = \left\{1 + \frac{1}{2}n^{-1}t^2 + R_n\right\}^n,
$$
(5.3.5)

since $e^x = 1 + \frac{x}{1!} + \frac{x^2}{2!} + \frac{x^3}{3!} + ...$ and the remainder term R_n is of the order $O(n^{-2})$ so that $\lim_{n\to\infty} n^2 R_n$ is finite. In (5.3.5), the expression in the last step converges to $exp(\frac{1}{2}t^2)$ as $n \to \infty$, because $\left(1 + \frac{a}{n}\right)^n \to e^a$ as $n \to \infty$. But recall from (2.3.16) that $M(t) = e^{\frac{1}{2}t^2}$ is the mgf of a standard normal variable, and thus we claim that $2\sqrt{n}(\overline{X}_n - .5) \xrightarrow{\mathcal{L}} N(0, 1)$ as $n \to \infty$. ▲

The Example 5.3.4 uses the mgf technique to show that for large n, the distribution of $(\chi_n^2 - n)/\sqrt{2n}$ approaches the standard normal distribution. The Section 5.4 gives another justification for the same result.

Example 5.3.4 Suppose that X_n is distributed as the Chi-square with n degrees of freedom. Denote $U_n = \frac{1}{\sqrt{2n}}(X_n - n), n \geq 1$. The question is whether $U_n \overset{\mathcal{L}}{\to} U$, some appropriate random variable, as $n \to \infty$. For $t < 1/\sqrt{2}$, let us start with the mgf $M_n(t)$ and write

$$
\begin{aligned}
E\{exp(tU_n)\} &= exp\left\{-t\sqrt{\tfrac{n}{2}}\right\} E\left\{exp\left(\tfrac{t}{\sqrt{2n}}X_n\right)\right\} \\
&= exp\left\{-t\sqrt{\tfrac{n}{2}}\right\}\left(1 - \tfrac{2t}{\sqrt{2n}}\right)^{-n/2}.
\end{aligned}
$$

Here, we use the form of the mgf of χ_n^2 from (2.3.28), namely, $E\{exp(sX_n)\} = (1 - 2s)^{-n/2}$ for $s < \tfrac{1}{2}$. Hence, with $s = \tfrac{t}{\sqrt{2n}}$, we obtain

$$
M_n(t) = \left\{\left(1 - t\sqrt{\tfrac{2}{n}}\right)^{-\sqrt{n/2}} e^{-t}\right\}^{\sqrt{n/2}}.
$$

Thus, for large n, we can write

$$
\begin{aligned}
log(M_n(t)) &= \sqrt{\tfrac{n}{2}}\left\{-\sqrt{\tfrac{n}{2}}log\left[1 - t\sqrt{\tfrac{2}{n}}\right] - t\right\} \\
&= \sqrt{\tfrac{n}{2}}\left\{-\sqrt{\tfrac{n}{2}}\left[-t\sqrt{\tfrac{2}{n}} - t^2\tfrac{1}{n} + O(n^{-3/2})\right] - t\right\} \\
&= \tfrac{1}{2}t^2 + O(\tfrac{1}{n^{1/2}}).
\end{aligned}
$$

Here, we used the series expansion of $log(1-x)$ from (1.6.15). We have thus proved that $M_n(t) \to exp\left(\tfrac{1}{2}t^2\right)$ as $n \to \infty$. Hence, by the Theorem 5.3.1, we conclude that $\tfrac{1}{\sqrt{2n}}(X_n - n) \overset{\mathcal{L}}{\to} N(0,1)$ as $n \to \infty$, since $M(t) = exp\left(\tfrac{1}{2}t^2\right)$ happens to be the mgf of the standard normal distribution. ▲

5.3.1 Combination of the Modes of Convergence

In this section, we summarize some important results without giving their proofs. One would find the proofs in Sen and Singer (1993, Chapter 2). One may also refer to Serfling (1980) and other sources.

   Theorem 5.3.2 *Suppose that a sequence of real valued random variables U_n converges in probability to another real valued random variable U as $n \to \infty$. Then, $U_n \overset{\mathcal{L}}{\to} U$ as $n \to \infty$.*

The converse of Theorem 5.3.2 is not necessarily true. That is, it is possible to have $U_n \overset{\mathcal{L}}{\to} U$, but $U_n \overset{P}{\nrightarrow} U$ as $n \to \infty$. Refer to Exercise 5.3.14.

Theorem 5.3.3 (Slutsky's Theorem) *Let us consider two sequences of real valued random variables* $\{U_n, V_n; n \geq 1\}$, *another real valued random variable* U, *and a fixed real number* v. *Suppose that* $U_n \xrightarrow{\mathcal{L}} U$, $V_n \xrightarrow{P} v$ *as* $n \to \infty$. *Then, we have as* $n \to \infty$:

(i) $U_n \pm V_n \xrightarrow{\mathcal{L}} U \pm v$;

(ii) $U_n V_n \xrightarrow{\mathcal{L}} vU$;

(iii) $U_n V_n^{-1} \xrightarrow{\mathcal{L}} Uv^{-1}$ *provided that* $P\{V_n = 0\} = 0$ *for all* n *and* $v \neq 0$.

Example 5.3.5 (Examples 5.3.1-5.3.2 Continued) Recall $T_n = X_{n:n}$ in the Uniform$(0, \theta)$ situation and let us define $V_n = T_n^2 \{T_n + \frac{1}{n}\}^{-1}$. We already know that $T_n \xrightarrow{P} \theta$ as $n \to \infty$. By using the Theorem 5.2.4 we can conclude that $V_n \xrightarrow{P} \theta$ as $n \to \infty$. We had proved that $U_n = n(\theta - T_n)/\theta \xrightarrow{\mathcal{L}} U$ as $n \to \infty$ where U has the standard exponential distribution. Hence, by Slutsky's Theorem, we can immediately claim that $G_n = n(\theta - T_n)/T_n \xrightarrow{\mathcal{L}} U$ as $n \to \infty$. We can also claim that $H_n = n(\theta - T_n)/V_n \xrightarrow{\mathcal{L}} U$ as $n \to \infty$. It is not easy to obtain the df of H_n and then proceed directly with the Definition 5.3.1 to show that $H_n \xrightarrow{\mathcal{L}} U$ as $n \to \infty$. ▲

> In the Exercise 5.3.2, we had suggested a direct approach using the Definition 5.3.1 to show that $G_n = n(\theta - T_n)/T_n \xrightarrow{\mathcal{L}} U$.

5.3.2   The Central Limit Theorems

First we discuss the central limit theorems for the *standardized* sample mean and the sample variance.

> Let $X_1, ..., X_n$ be iid random samples from a population with mean μ and variance σ^2, $n \geq 2$. Consider the sample mean $\overline{X}_n$ and the sample variance S_n^2. In the literature, $\sqrt{n}(\overline{X}_n - \mu)/\sigma$ is called the *standardized* version of the *sample mean* when σ is known, and $\sqrt{n}(\overline{X}_n - \mu)/S_n$ is called the *standardized* version of the *sample mean* when σ is unknown.

The following theorem, known as the *central limit theorem* (CLT), provides under generality the asymptotic distribution for the standardized version of the sample mean $\overline{X}_n$ of iid real valued random variables having a positive and finite variance.

Theorem 5.3.4 (Central Limit Theorem) *Let $X_1, ..., X_n$ be iid real valued random variables having the common mean μ and variance σ^2, $-\infty < \mu < \infty$ and $0 < \sigma < \infty$. Then, as $n \to \infty$, we have*

$$\frac{\sqrt{n}(\overline{X}_n - \mu)}{\sigma} \xrightarrow{\pounds} N(0,1).$$

A careful treatment of the proof of the CLT under such generality requires working knowledge of the characteristic functions and hence it is out of scope of this book. The reader may look at Sen and Singer (1993, pp. 107-108) for a proof of this result and other versions of the CLT.

If one however assumes additionally that the mgf of X_1 exists, then a proof of the CLT can be constructed fairly easily, essentially along the lines of the Examples 5.3.3-5.3.4. The details are left out as the Exercise 5.3.3.

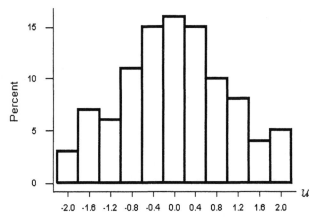

Figure 5.3.2. Histogram of 100 Values of $u = \sqrt{n}(\overline{x}_n - 10)/2$ from the $N(10, 4)$ Population When $n = 10$

In case the common pdf happened to be normal with mean μ and variance σ^2, we had shown by means of the Helmert transformation in Chapter 4 that $\sqrt{n}(\overline{X}_n - \mu)/\sigma$ would be distributed *exactly* as the standard normal variable *whatever be the sample size* n. Using the MINITAB Release 12.1, we have drawn random samples from the $N(\mu = 10, \sigma^2 = 4)$ population with $n = 10$ and replicated the process. Having first fixed n, in the ith replication we drew n random samples $x_{1i}, ..., x_{ni}$ which led to the value of the sample mean, $\overline{x}_{ni} = n^{-1}\Sigma_{j=1}^n x_{ji}$ for $i = 1, ..., 100$. We then obtained the histogram of the 100 randomly observed values of the standardized sample mean, namely, $u_i = \sqrt{n}(\overline{x}_{ni} - 10)/2, i = 1, ..., 100$. The Figure 5.3.2 shows the histogram which gives an *impression* of a standard normal pdf.

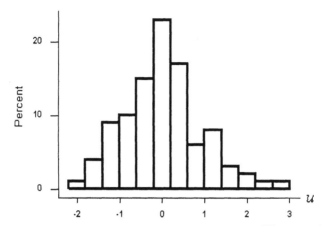

Figure 5.3.3. Histogram of 100 Values of $u = \sqrt{n}(\overline{x}_n - 10)/2$
from the Gamma(25, .4) Population When $n = 10$

The CLT on the other hand talks about the asymptotic distribution of $\sqrt{n}(\overline{X}_n - \mu)/\sigma$, whatever be the distribution of the parent population with finite $\sigma(> 0)$. So, we ran the following experiments with MINITAB Release 12.1.

We considered the Gamma($\alpha = 25, \beta = .4$) population which has its mean $\mu = \alpha\beta = 10$ and variance $\sigma^2 = \alpha\beta^2 = 4$. Then, having first fixed $n = 10$, in the ith replication we drew n random samples $x_{1i}, ..., x_{ni}$ which led to the value of the sample mean, $\overline{x}_{ni} = n^{-1}\Sigma_{j=1}^{n}x_{ji}$ for $i = 1, ..., 100$. We have plotted the histogram of the 100 randomly observed values of the standardized sample mean, namely, $u_i = \sqrt{n}(\overline{x}_{ni} - 10)/2, i = 1, ..., 100$. We give this plot in the Figure 5.3.3.

The CLT claims that the histogram in the Figure 5.3.3 should approximately resemble the pdf of a standard normal variable for large n. We note that this histogram, though a little skewed to the right, creates an *impression* of a standard normal pdf when the sample size is 10.

Next, we considered the Uniform($10 - a, 10 + a$) population with $a = \sqrt{12}$ which has its mean $\mu = 10$ and variance $\sigma^2 = 4$. This uniform distribution is symmetric about μ. Then, having first fixed $n = 10$, in the ith replication we drew n random samples $x_{1i}, ..., x_{ni}$ which led to the value of the sample mean, $\overline{x}_{ni} = n^{-1}\Sigma_{j=1}^{n}x_{ji}$ for $i = 1, ..., 100$. We have then plotted the histogram of the 100 randomly observed values of the standardized sample mean, namely, $u_i = \sqrt{n}(\overline{x}_{ni} - 10)/2, i = 1, ..., 100$. We give this plot in the

Figure 5.3.4. We add that the earlier comments made in the context of the Figure 5.3.3 remain valid when the Figure 5.3.4 is inspected.

In general, with a larger (than 10) sample size n, one will notice more clearly the symmetry in the resulting histograms around zero, whatever be the parent population pmf or pdf with finite $\sigma(> 0)$.

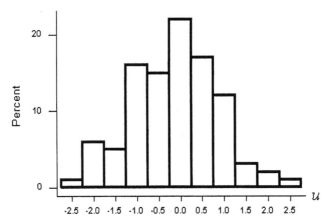

Figure 5.3.4. Histogram of 100 Random $u = \sqrt{n}(\overline{x}_n - 10)/2$ Values from the Uniform$(10 - a, 10 + a)$ Population When $n = 10, a = \sqrt{12}$

The three histograms presented in the Figures 5.3.2-5.3.4 are comparable to each other in the sense that the three parent populations under consideration have the same mean, $\mu = 10$ and variance, $\sigma^2 = 4$. The reader should pursue more explorations like these as exercises.

Example 5.3.6 Let $X_1, ..., X_n$ be iid $N(\mu, \sigma^2), -\infty < \mu < \infty, 0 < \sigma < \infty$. For $n \geq 2$, let $S_n^2 = (n-1)^{-1}\Sigma_{i=1}^n (X_i - \overline{X}_n)^2$ be the sample variance. From (4.4.9), recall that $S_n^2 = (n-1)^{-1}\Sigma_{i=2}^n Y_i^2$ where these Y_i's are the Helmert variables distributed as iid $N(0, \sigma^2)$. This implies that S_n^2 is indeed a sample mean of $(n-1)$ iid random variables. The CLT then immediately leads us to conclude that $U_n = \sqrt{n-1}\left(S_n^2 - E(Y_1^2)\right) \overset{\mathcal{L}}{\to} N\left(0, V(Y_1^2)\right)$ as $n \to \infty$. But, one has $E(Y_1^2) = \sigma^2$ and $V(Y_1^2) = E(Y_1^4) - E^2(Y_1^2) = 3\sigma^4 - \sigma^4 = 2\sigma^4$. That is in this case, we have $U_n = \sqrt{n-1}\left(S_n^2 - \sigma^2\right) \overset{\mathcal{L}}{\to} N\left(0, 2\sigma^4\right)$ as $n \to \infty$. Let us now denote $V_n = \{n/(n-1)\}^{1/2}$ and view it as a sequence of degenerate random variables to claim: $V_n \overset{P}{\to} 1$ as $n \to \infty$. Next, we apply Slutsky's Theorem to conclude that $\sqrt{n}\left(S_n^2 - \sigma^2\right) = U_n V_n \overset{\mathcal{L}}{\to} N\left(0, 2\sigma^4\right)$ as $n \to \infty$. ▲

Example 5.3.7 First note that we can express

$$\sqrt{n}\left(\overline{X}_n^2 - \mu^2\right) = \{\sqrt{n}(\overline{X}_n - \mu)/\sigma\}\{\sigma(\overline{X}_n + \mu)\},$$

and hence by Slutsky's Theorem, under the same conditions as the CLT, we can conclude that $\sqrt{n}\left(\overline{X}_n^2 - \mu^2\right) \overset{\mathcal{L}}{\to} N(0, 4\mu^2\sigma^2)$ as $n \to \infty$ if $\mu \neq 0$. This is so because $\sqrt{n}(\overline{X}_n - \mu)/\sigma \overset{\mathcal{L}}{\to} N(0,1)$ and $\sigma(\overline{X}_n + \mu) \overset{P}{\to} 2\mu\sigma$ as $n \to \infty$. Similarly, we can easily handle the convergence in distribution for the sequence of random variables (i) $U_n = \sqrt{n}\left(\overline{X}_n^3 - \mu^3\right)$ if $\mu \neq 0$ or (ii) $U_n = \sqrt{n}\left(\sqrt{\overline{X}_n} - \sqrt{\mu}\right)$ if $\mu > 0$ and $\overline{X}_n$ is positive w.p.1 for all n. These are left out as the Exercise 5.3.4. ▲

The following result is a very useful follow-up on the CLT. Suppose that one is able to use the CLT to verify that $\sqrt{n}(T_n - \theta)$ converges in distribution to $N(0, \sigma^2)$ as $n \to \infty$, with some appropriate θ and $\sigma(> 0)$, where $\{T_n; n \geq 1\}$ is a sequence of real valued random variables. But then, does $\sqrt{n}\{g(T_n) - g(\theta)\}$ converge to an appropriate normal variable for reasonable $g(.)$ functions? Review the Example 5.3.7. The following theorem, a nice blend of the CLT (Theorem 5.3.4) and Slutsky's Theorem, answers this question affirmatively.

Theorem 5.3.5 (Mann-Wald Theorem) *Suppose that* $\{T_n; n \geq 1\}$ *is a sequence of real valued random variables such that* $\sqrt{n}(T_n - \theta) \overset{\mathcal{L}}{\to} N(0, \sigma^2)$ *as* $n \to \infty$ *where* $\sigma^2(> 0)$ *may also depend on* θ. *Let* $g(.)$ *be a continuous real valued function such that* $\frac{d}{d\theta}g(\theta)$, *denoted by* $g'(\theta)$, *is finite and non-zero. Then, we have:*

$$\sqrt{n}\{g(T_n) - g(\theta)\} \overset{\mathcal{L}}{\to} N\left(0, \{\sigma g'(\theta)\}^2\right) \text{ as } n \to \infty.$$

Proof We proceed along the proof given in Sen and Singer (1993, pp. 231-232). Observe that

$$\sqrt{n}\{g(T_n) - g(\theta)\} = U_n V_n, \tag{5.3.6}$$

where we denote

$$U_n = \sqrt{n}(T_n - \theta), \quad V_n = \{g(T_n) - g(\theta)\}/(T_n - \theta).$$

Next, note that

$$(T_n - \theta) = \tfrac{1}{\sqrt{n}}U_n,$$

and since $U_n \overset{\mathcal{L}}{\to} N(0, \sigma^2)$ as $n \to \infty$, by Slutsky's Theorem, part (ii), we conclude that $T_n - \theta \overset{P}{\to} 0$ as $n \to \infty$. Thus, $V_n \overset{P}{\to} g'(\theta)$ as $n \to \infty$, by the definition of $g'(\theta)$ itself. Recall that for some fixed x, one has: $g'(x) = \lim_{y \to x}$ $[\{g(y) - g(x)\}/(y - x)]$. We again apply Slutsky's Theorem, part (ii), in conjunction with (5.3.6) to complete the proof. ∎

Remark 5.3.1 Suppose that the X's are as in the CLT. Then we immediately conclude that $\sqrt{n}(\overline{X}_n^2 - \mu^2) \xrightarrow{\pounds} N(0, 4\mu^2\sigma^2)$ as $n \to \infty$, once we plug in $g(x) = x^2, x \in \Re$ and apply the Mann-Wald Theorem. We had shown this result in a different way in the Example 5.3.7. Now, we can also claim that $\sqrt{n}\left(\overline{X}_n^q - \mu^q\right) \xrightarrow{\pounds} N(0, \{q\mu^{q-1}\sigma\}^2)$ as $n \to \infty$, by plugging in $g(x) = x^q$, $x \in \Re$, for any fixed non-zero real number q ($\neq 1$). We leave verification of this result as an exercise. Of course, we tacitly assume that q is chosen in such a way that both $\overline{X}_n^q$ and μ^q remain well-defined. Note that for *negative* values of q, we need to assume that the probability of $\overline{X}_n$ being exactly zero is in fact zero and that μ is also non-zero.

Example 5.3.8 Let $X_1, ..., X_n$ be iid Poisson(λ) with $\lambda > 0$. Observe that in this case $\mu = \sigma^2 = \lambda$ and then, by the CLT, $\sqrt{n}(\overline{X}_n - \lambda)/\sqrt{\lambda} \xrightarrow{\pounds} N(0,1)$ as $n \to \infty$. Then, by using the Remark 5.3.1 and Mann-Wald Theorem, we immediately see, for example, that $\sqrt{n}(\overline{X}_n^3 - \lambda^3) \xrightarrow{\pounds} N(0, 9\lambda^5)$ as $n \to \infty$. ▲

Now we move towards an appropriate CLT for the *standardized sample variance*. This result again utilizes a nice blend of the CLT (Theorem 5.3.4) and Slutsky's Theorem.

Theorem 5.3.6 (Central Limit Theorem for the Sample Variance) *Let* $X_1, ..., X_n$ *be iid random variables with mean* μ, *variance* $\sigma^2 (> 0)$, $\mu_4 = E\left\{(X_1 - \mu)^4\right\}$, *and we assume that* $0 < \mu_4 < \infty$ *as well as* $\mu_4 > \sigma^4$. *Denote* $\overline{X}_n = n^{-1}\Sigma_{i=1}^n X_i$ *and* $S_n^2 = (n-1)^{-1}\Sigma_{i=1}^n(X_i - \overline{X}_n)^2$ *for* $n \geq 2$. *Then, we have:*

$$\sqrt{n}\left(S_n^2 - \sigma^2\right) \xrightarrow{\pounds} N(0, \mu_4 - \sigma^4) \text{ as } n \to \infty.$$

Proof Let us first work with $W_n = (n-1)n^{-1}S_n^2$. We denote $Y_i = (X_i - \mu)^2$, $i = 1, ..., n$, $\overline{Y}_n = n^{-1}\Sigma_{i=1}^n Y_i$, and write

$$W_n = \tfrac{1}{n}\Sigma_{i=1}^n(X_i - \overline{X}_n)^2 = \tfrac{1}{n}\Sigma_{i=1}^n Y_i - \left\{\overline{X}_n - \mu\right\}^2. \tag{5.3.7}$$

From (5.3.7), observe that

$$\sqrt{n}(W_n - \sigma^2) = \sqrt{n}\left(\overline{Y}_n - \sigma^2\right) - \sqrt{n}\left\{\overline{X}_n - \mu\right\}^2 = U_n + V_n, \text{ say.} \tag{5.3.8}$$

Notice that Y's are iid with mean $= E(Y_1) = \sigma^2$ and variance $= E(Y_1^2) - E^2(Y_1) = \mu_4 - \sigma^4$ which is assumed finite and positive. Hence, by the CLT, we have $U_n \overset{\mathcal{L}}{\to} N(0, \mu_4 - \sigma^4)$ as $n \to \infty$. Also, from the Example 5.2.10 it follows that $V_n \overset{P}{\to} 0$ as $n \to \infty$. Thus by Slutsky's Theorem, part (i), we conclude that $\sqrt{n}(W_n - \sigma^2) \overset{\mathcal{L}}{\to} N(0, \mu_4 - \sigma^4)$ as $n \to \infty$. Next, we write

$$\sqrt{n}\left(S_n^2 - \sigma^2\right) = \sqrt{n}\left(\tfrac{n}{n-1}W_n - \sigma^2\right) = \sqrt{n}\left(W_n - \sigma^2\right) - \tfrac{\sqrt{n}}{n-1}W_n,$$
(5.3.9)

and reapply Slutsky's Theorem. The result then follows since $\sqrt{n}(W_n - \sigma^2)$ $\overset{\mathcal{L}}{\to} N(0, \mu_4 - \sigma^4)$ and $W_n \overset{P}{\to} \sigma^2$ as $n \to \infty$, so that $\sqrt{n}(n-1)^{-1}W_n \overset{P}{\to} 0$ as $n \to \infty$. ∎

Remark 5.3.2 One can obtain the result mentioned in the Example 5.3.6 from this theorem by noting that $\mu_4 = 3\sigma^4$ in the normal case.

Example 5.3.9 Under the setup of the Theorem 5.3.6, using Example 5.2.11 and Slutsky's Theorem, we can immediately conclude the following result when the X's are iid but non-normal: $\sqrt{n}\left(S_n^2 - \sigma^2\right) S_n^{-2} \overset{\mathcal{L}}{\to}$ $N(0, \mu_4\sigma^{-4} - 1)$ as $n \to \infty$. In the literature, $\mu_4\sigma^{-4}$ is traditionally denoted by β_2, which is customarily referred to as a measure of the *kurtosis* in the parent population. ▲

The following result is along the lines of the Theorem 5.2.5. It shows that the property of the convergence in distribution is preserved under a continuous transformation. We state it without giving its proof.

Theorem 5.3.7 *Suppose that we have a sequence of real valued random variables $\{U_n; n \geq 1\}$ and another real valued random variable U. Suppose also that $U_n \overset{\mathcal{L}}{\to} U$ as $n \to \infty$. Let $g(.)$ be a real valued continuous function. Then, $g(U_n) \overset{\mathcal{L}}{\to} g(U)$ as $n \to \infty$.*

Example 5.3.10 Let $X_1, ..., X_n$ be iid real valued random variables having the common mean μ and variance σ^2, $-\infty < \mu < \infty$ and $0 < \sigma < \infty$. From the CLT we know that $\sqrt{n}(\overline{X}_n - \mu)/\sigma \overset{\mathcal{L}}{\to} N(0, 1)$ as $n \to \infty$. Thus, using the Theorem 5.3.7 with $g(x) = x^2, x \in \Re$, we can immediately conclude that $n(\overline{X}_n - \mu)^2/\sigma^2 \overset{\mathcal{L}}{\to} \chi_1^2$ as $n \to \infty$, since the square of a standard normal variable has a Chi-square distribution with one degree of freedom. ▲

Example 5.3.11 (Example 5.3.10 Continued) Let $X_1, ..., X_n$ be iid real valued random variables having the common mean μ and variance σ^2, $-\infty < \mu < \infty$ and $0 < \sigma < \infty$. From the CLT we know that $\sqrt{n}(\overline{X}_n - \mu)/\sigma \overset{\mathcal{L}}{\to}$ $N(0, 1)$ as $n \to \infty$. Thus, using the Theorem 5.3.7 with $g(x) = |x|, x \in \Re$, we can immediately conclude that $\sqrt{n}\left|(\overline{X}_n - \mu)/\sigma\right| \xrightarrow{\mathcal{L}} |Z|$ as $n \to \infty$ where Z is the standard normal variable. ▲

> For the approximation of the Binomial(n, p) distribution with the $N(np, np(1 - p))$ distribution, for large n and fixed $0 < p < 1$, refer to the Exercise 5.3.18.

5.4   Convergence of Chi-square, t, and F Distributions

In this section, we discuss some asymptotic properties of the Chi-square, t, and F distributions. One will notice that most of the probabilistic and statistical tools introduced thus far come together in these topics.

5.4.1   The Chi-square Distribution

First, consider a random variable U_ν which is distributed as χ^2_ν. Now, $E(\nu^{-1}U_\nu) = 1$, $V(\nu^{-1}U_\nu) = \nu^{-2}V(U_\nu) = 2\nu^{-1} \to 0$ as $\nu \to \infty$, and hence by applying the Weak WLLN, we can claim immediately that

$$\nu^{-1}\chi^2_\nu \xrightarrow{P} 1 \text{ as } \nu \to \infty. \tag{5.4.1}$$

Let $X_1, ..., X_\nu, ...$ be iid χ^2_1 so that we may view χ^2_ν as $\Sigma^\nu_{i=1}X_i$, and hence we can write $\nu^{-1}\chi^2_\nu = \nu^{-1}\Sigma^\nu_{i=1}X_i = \overline{X}_\nu$. Now, we can apply the CLT (Theorem 5.3.4) to claim that $\sqrt{\nu}(\overline{X}_\nu - \mu)/\sigma \xrightarrow{\mathcal{L}} N(0, 1)$ as $\nu \to \infty$ with $\mu = E(X_1) = 1$, $\sigma^2 = V(X_1) = 2$, and hence, $(2\nu)^{-1/2}(\chi^2_\nu - \nu) \xrightarrow{\mathcal{L}} N(0, 1)$ as $\nu \to \infty$. In the Example 5.3.4, we arrived at the same conclusion with a fairly different approach. For practical purposes, we would say:

> The distribution of $(\sqrt{2\nu})^{-1}(\chi^2_\nu - \nu)$ is approximated by the $N(0, 1)$ distribution when ν is large.

$$(5.4.2)$$

5.4.2   The Student's t Distribution

The Student's t random variable was described in the Definition 4.5.1. Let $W_\nu = X \div \left(Y_\nu\nu^{-1}\right)^{1/2}$ where X is the standard normal variable and Y_ν is χ^2_ν, while X and Y_ν are assumed independent. In other words, W_ν has the

Student's t distribution with ν degrees of freedom. Now, since $Y_\nu \nu^{-1} \overset{P}{\to} 1$ according to (5.4.1), so does $\sqrt{Y_\nu \nu^{-1}} \overset{P}{\to} 1$ as $\nu \to \infty$ by the Theorem 5.2.5 with $g(x) = x^{1/2}, x > 0$. Hence, applying Slutsky's Theorem, we conclude that $W_\nu \overset{\mathcal{L}}{\to} N(0,1)$ as $\nu \to \infty$. For practical purposes, we would say:

> The t distribution with ν degrees of freedom is approximated by the standard normal distribution when ν is large.   (5.4.3)

5.4.3 The F Distribution

The F_{ν_1,ν_2} random variable was introduced by the Definition 4.5.2 as $\nu_1^{-1}X_{\nu_1} \div \nu_2^{-1}Y_{\nu_2}$ where X_{ν_1}, Y_{ν_2} are independent, X_{ν_1} is $\chi^2_{\nu_1}$ and Y_{ν_2} is $\chi^2_{\nu_2}$. Now, if ν_1 is held fixed, but ν_2 is allowed to go to infinity, then, $\nu_2^{-1}Y_{\nu_2} \overset{P}{\to} 1$. Next, we apply Slutsky's Theorem. Hence, as $\nu_2 \to \infty$, we conclude that $F_{\nu_1,\nu_2} \overset{\mathcal{L}}{\to} \nu_1^{-1}X_{\nu_1}$, that is, for practical purposes, we would say:

> $F_{\nu_1,\nu_2} \overset{\mathcal{L}}{\to} \chi^2_{\nu_1} \div \nu_1$ as $\nu_2 \to \infty$, when ν_1 is held fixed;
> $F_{\nu_1,\nu_2} \overset{\mathcal{L}}{\to} \nu_2 \div \chi^2_{\nu_2}$ as $\nu_1 \to \infty$, when ν_2 is held fixed.   (5.4.4)

5.4.4 Convergence of the PDF and Percentage Points

From (1.7.30), let us recall the pdf of the Student's t random variable with ν degrees of freedom, denoted by t_ν. The pdf of t_ν, indexed by ν, is given by

$$f_\nu(x) = a\left(1 + \frac{1}{\nu}x^2\right)^{-(\nu+1)/2} \quad \text{for } -\infty < x < \infty, \quad (5.4.5)$$

with $a \equiv a(\nu) = \{\sqrt{\nu\pi}\}^{-1}\Gamma(\frac{1}{2}(\nu+1))\{\Gamma(\frac{1}{2}\nu)\}^{-1}, \nu = 1, 2, 3, \dots$. We had seen in (5.4.3) that $t_\nu \overset{\mathcal{L}}{\to} Z$, the standard normal variable, as $\nu \to \infty$. But, is it true that $f_\nu(x) \to \phi(x)$, the pdf of the standard normal variable, for each fixed $x \in \Re$, as $\nu \to \infty$? The answer follows.

Using the limiting value of the ratio of gamma functions from (1.6.24), we obtain

$$\left(\tfrac{1}{2}\nu\right)^{-1/2} \frac{\Gamma\left(\frac{1}{2}(\nu+1)\right)}{\Gamma\left(\frac{1}{2}\nu\right)} \to 1 \text{ as } \nu \to \infty. \quad (5.4.6)$$

Next, we rewrite

$$f_\nu(x)$$

$$= \frac{1}{\sqrt{\nu\pi}\left(\frac{1}{2}\nu\right)^{-1/2}} \left(\tfrac{1}{2}\nu\right)^{-1/2} \frac{\Gamma\left(\frac{1}{2}(\nu+1)\right)}{\Gamma\left(\frac{1}{2}\nu\right)} \times$$

$$\left(1+\frac{1}{\nu}x^2\right)^{-\nu/2}\left(1+\frac{1}{\nu}x^2\right)^{-1/2}$$

$$= \frac{1}{\sqrt{2\pi}}\left\{\left(\tfrac{1}{2}\nu\right)^{-1/2}\frac{\Gamma\left(\frac{1}{2}(\nu+1)\right)}{\Gamma\left(\frac{1}{2}\nu\right)}\right\} \times$$

$$\left(1+\frac{1}{\nu}x^2\right)^{-\nu/2}\left(1+\frac{1}{\nu}x^2\right)^{-1/2}.$$

Hence, one obtains

$$f_\nu(x) \to \frac{1}{\sqrt{2\pi}}exp\{-\tfrac{1}{2}x^2\} \text{ as } \nu \to \infty,$$

using (5.4.6) and the facts that $\lim_{\nu\to\infty} (1 + \frac{1}{\nu}x^2)^{-\frac{1}{2}\nu} = e^{-\frac{1}{2}x^2}$ and $\lim_{\nu\to\infty}$ $(1+\frac{1}{\nu}x^2)^{-\frac{1}{2}} = 1$. Refer to the Section 1.6 as needed. Thus, we conclude:

$$\boxed{\begin{array}{c}\text{The pdf of } t_\nu \text{ at } x \text{ converges to } \phi(x), \text{ the pdf of a standard}\\ \text{normal variable, as } \nu \to \infty, \text{ for each } x \in \Re.\end{array}} \qquad (5.4.7)$$

With $0 < \alpha < 1$, let us define the upper $100\alpha\%$ point $t_{\nu,\alpha}$ for the Student's t_ν distribution through the following relationship:

$$P\{t_\nu > t_{\nu,\alpha}\} = \alpha, \qquad (5.4.8)$$

that is, the area under the pdf of t_ν on the rhs of the point $t_{\nu,\alpha}$ is α. Similarly let z_α stand for the upper $100\alpha\%$ point for the standard normal distribution. These percentage points have been tabulated in the Appendix. It is known that for any *fixed* $0 < \alpha < 1$, one has

$$\lim_{\nu\to\infty} t_{\nu,\alpha} = z_\alpha \text{ where } P\{Z > z_\alpha\} = P\{t_\nu > t_{\nu,\alpha}\} = \alpha. \qquad (5.4.9)$$

But what is interesting is that this *convergence is also monotone*. For any fixed $0 < \alpha < 1$, it is known that

$$t_{\nu,\alpha} > t_{\nu+1,\alpha}, \nu = 1, 2, 3, \dots . \qquad (5.4.10)$$

One should refer to Ghosh (1973) and DasGupta and Perlman (1974) for many related details.

For large ν, one can expand $t_{\nu,\alpha}$ in terms of z_α and ν. E. A. Cornish and R. A. Fisher systematically developed techniques for deriving expansions of the percentile of a distribution in terms of that of its limiting form. These expansions are customarily referred to as the *Cornish-Fisher expansions*. The Cornish-Fisher expansion of $t_{\nu,\alpha}$ in terms of z_α and ν is given below: For large values of ν,

$$t_{\nu,\alpha} = z_\alpha + \tfrac{1}{4}z_\alpha(z_\alpha^2+1)\nu^{-1} + \tfrac{1}{96}z_\alpha(5z_\alpha^4 + 16z_\alpha^2 + 3)\nu^{-2} + O(\nu^{-3}). \quad (5.4.11)$$

The reader may refer to Johnson and Kotz (1970, p. 102) for related details.

At this point, the reader may ask the following question: For some fixed $0 < \alpha < 1$, how good is the expression in the rhs of (5.4.11) as an approximation for $t_{\nu,\alpha}$ when ν is small or moderate? Let us write

$$t_{\nu,\alpha,\text{approx}} = z_\alpha + \tfrac{1}{4}z_\alpha(z_\alpha^2+1)\nu^{-1} + \tfrac{1}{96}z_\alpha(5z_\alpha^4 + 16z_\alpha^2 + 3)\nu^{-2}. \quad (5.4.12)$$

The Table 5.4.1 provides the values of z_α, correct up to four decimal places, and those of $t_{\nu,\alpha}$, correct up to three decimal places. We have computed $t_{\nu,\alpha,\text{approx}}$, correct up to five decimal places, for $\alpha = .05, .10$ and $\nu = 10, 20, 30$. The values of z_α and $t_{\nu,\alpha}$ are respectively obtained from Lindley and Scott (1995) and the Student's t Table 14.3.3. The entries in the Table 5.4.1 show clearly that the approximation (5.4.12) works well.

Table 5.4.1. Comparing $t_{\nu,\alpha}$ with $t_{\nu,\alpha,\text{approx}}$

	$\alpha = .10$	$z_\alpha = 1.2816$	$\alpha = .05$	$z_\alpha = 1.6449$
ν	$t_{\nu,\alpha}$	$t_{\nu,\alpha,\text{approx}}$	$t_{\nu,\alpha}$	$t_{\nu,\alpha,\text{approx}}$
10	1.3722	1.37197	1.8125	1.81149
20	1.3253	1.32536	1.7247	1.72465
30	1.3104	1.31046	1.6973	1.69727

Next, recall from (1.7.34) that for the F random variable with ν_1, ν_2 degrees of freedom, written as F_{ν_1,ν_2}, the pdf is given by,

$$f_{\nu_1,\nu_2}(x) = bx^{\frac{1}{2}\nu_1 - 1}\left(1 + \frac{\nu_1}{\nu_2}x\right)^{-\frac{1}{2}(\nu_1+\nu_2)} \quad \text{for } 0 < x < \infty, \quad (5.4.13)$$

indexed by ν_1, ν_2, with $b \equiv b(\nu_1, \nu_2) = (\nu_1/\nu_2)^{\frac{1}{2}\nu_1}\Gamma((\nu_1+\nu_2)/2)\{\Gamma(\nu_1/2) \times \Gamma(\nu_2/2)\}^{-1}, \nu_1, \nu_2 = 1, 2, 3, \dots$. We had seen earlier in (5.4.4) that $F_{\nu_1,\nu_2} \xrightarrow{\mathcal{L}} \frac{1}{\nu_1}\chi^2_{\nu_1}$, when ν_1 is held fixed, but $\nu_2 \to \infty$. But, is it true that $f_{\nu_1,\nu_2}(x) \to$

$g(x)$ where $g(x)$ is the pdf of $\frac{1}{\nu_1}\chi^2_{\nu_1}$ at x, for each fixed $x \in \Re^+$, as $\nu_2 \to \infty$? The answer follows.

Using the limiting value of the ratio of gamma functions from (1.6.24), we again obtain

$$\left(\tfrac{1}{2}\nu_2\right)^{-\nu_1/2} \frac{\Gamma\left(\tfrac{1}{2}(\nu_1 + \nu_2)\right)}{\Gamma\left(\tfrac{1}{2}\nu_2\right)} \to 1 \text{ as } \nu_2 \to \infty. \tag{5.4.14}$$

Next, we rewrite

$$f_{\nu_1,\nu_2}(x) = \frac{\nu_1^{\nu_1/2}}{2^{\nu_1/2}\Gamma\left(\frac{\nu_1}{2}\right)} \frac{\Gamma\left(\tfrac{1}{2}(\nu_1 + \nu_2)\right)}{\left(\frac{\nu_2}{2}\right)^{\nu_1/2}\Gamma\left(\frac{\nu_2}{2}\right)} x^{\frac{1}{2}\nu_1-1} \times$$

$$\left(1 + \frac{\nu_1}{\nu_2}x\right)^{-\frac{1}{2}\nu_2} \left(1 + \frac{\nu_1}{\nu_2}x\right)^{-\frac{1}{2}\nu_1},$$

and obtain

$$f_{\nu_1,\nu_2}(x) \to \{2^{\nu_1/2}\Gamma(\nu_1/2)\}^{-1}\nu_1^{\nu_1/2} x^{\frac{1}{2}\nu_1-1} e^{-\frac{1}{2}\nu_1 x} \text{ as } \nu_2 \to \infty,$$

using (5.4.14) as well as the facts that $\lim_{\nu_2\to\infty} (1 + \frac{\nu_1}{\nu_2}x)^{-\frac{1}{2}\nu_2} = e^{-\frac{1}{2}\nu_1 x}$ and $\lim_{\nu_2\to\infty} (1 + \frac{\nu_1}{\nu_2}x)^{-\frac{1}{2}\nu_1} = 1$. Refer to the Section 1.6 as needed. We conclude:

$$\boxed{\begin{array}{c} f_{\nu_1,\nu_2}(x) \text{ converges to } f^*(x) = \dfrac{\nu_1^{\nu_1/2}}{2^{\nu_1/2}\Gamma\left(\frac{1}{2}\nu_1\right)} x^{\frac{1}{2}\nu_1-1} e^{-\frac{1}{2}\nu_1 x}, \\[2mm] \text{the pdf of the random variable } \frac{1}{\nu_1}\chi^2_{\nu_1}, \text{ at the point } x, \text{ when } \nu_1 \text{ is} \\ \text{held fixed, as } \nu_2 \to \infty, \text{ for each } x \in \Re^+. \end{array}}$$

$$\tag{5.4.15}$$

Let $F_{\nu_1,\nu_2,\alpha}$ be the upper $100\alpha\%$ point of the F_{ν_1,ν_2} distribution for any fixed $0 < \alpha < 1$. Unlike the percentiles $t_{\nu,\alpha}$ for the t_ν distribution, the corresponding F percentiles $F_{\nu_1,\nu_2,\alpha}$ do not satisfy a clear-cut monotonicity property. The scenario here is complicated and one may consult the two articles of Ghosh (1973) and DasGupta and Perlman (1974) cited earlier for details. The percentile $F_{\nu_1,\nu_2,\alpha}$ does have the *Cornish-Fisher expansion* in terms of the upper $100\alpha\%$ point $\chi^2_{\nu_1,\alpha}$ from the $\chi^2_{\nu_1}$ distribution depending upon ν_1, ν_2, α for large ν_2 when ν_1 is kept fixed. The explicit details are due to Scheffé and Tukey (1944). Refer to Johnson and Kotz (1970, p. 84) for specific details in this regard.

In the same vein, when $\nu_1 = 2$, but $\nu_2 \to \infty$, we can easily study the behavior of $F_{2,\nu_2,\alpha}$ in terms of $\chi^2_{2,\alpha}$ for any fixed $0 < \alpha < 1$. We rely upon very simple and familiar tools for this purpose. In this special case, we provide the interesting details so that we can reinforce the techniques developed in this textbook.

Let us denote c for the upper $100\alpha\%$ point of the distribution of $V = \frac{1}{2}\chi^2_2$, that is

$$\alpha = P\{V > c\} = P\{\chi^2_2 > 2c\} = \int_{2c}^{\infty} \frac{1}{2}e^{-\frac{1}{2}x}dx = e^{-c}, \qquad (5.4.16)$$

and hence we have $c \equiv c(\alpha) = -log(\alpha)$. Suppose that X and Y are respectively distributed as χ^2_2 and $\chi^2_{\nu_2}$, and suppose also that they are independent. Let $h(y)$ be the pdf of Y at $y(> 0)$. Let us simply write d instead of $F_{2,\nu_2,\alpha}$ so that

$$\alpha = P\{F_{2,\nu_2} > d\} = P\left\{X > \frac{2d}{\nu_2}Y\right\}$$

$$= \int_0^{\infty} P\left\{X > \frac{2d}{\nu_2}y \mid Y = y\right\} h(y)dy.$$

Since X and Y are independent, one obtains

$$\alpha = \int_0^{\infty} P\left\{X > \frac{2d}{\nu_2}y\right\} h(y)dy = \int_0^{\infty} exp\left\{-\frac{d}{\nu_2}y\right\} h(y)dy$$

$$= \left\{1 + \frac{2d}{\nu_2}\right\}^{-\frac{1}{2}\nu_2},$$

writing the mgf of $\chi^2_{\nu_2}$ from (2.3.28). Thus, one has the following exact relationship between d and α:

$$d \equiv d(\nu_2, \alpha) = \frac{1}{2}\nu_2\left\{\alpha^{-2/\nu_2} - 1\right\}. \qquad (5.4.17)$$

One can show that d is a strictly decreasing function of ν_2. We leave this out as an exercise. Next, let us explore how we can expand d as a function of ν_2. We use a simple trick. Observe that

$$d(\nu_2, \alpha)$$

$$= \frac{1}{2}\nu_2\left\{e^{-2log(\alpha)/\nu_2} - 1\right\}$$

$$= \frac{1}{2}\nu_2\left\{1 - \frac{2log(\alpha)}{\nu_2} + \frac{2(log(\alpha))^2}{\nu_2^2} - \frac{4(log(\alpha))^3}{3\nu_2^3} + O(\nu_2^{-4}) - 1\right\},$$

using the expansion of e^x. Thus, one gets

$$d(\nu_2, \alpha) = -log(\alpha) + \frac{(log(\alpha))^2}{\nu_2} - \frac{2(log(\alpha))^3}{3\nu_2^2} + O\left(\nu_2^{-3}\right), \qquad (5.4.18)$$

which is the *Cornish-Fisher expansion* of $F_{2,\nu_2,\alpha}$ for large ν_2.

One may ask the following question: How good is the expression in (5.4.18) as an approximation for $F_{2,\nu_2,\alpha}$ when ν_2 is small or moderate? Let us write

$$F_{2,\nu_2,\alpha,\text{approx}} = -log(\alpha) + (log(\alpha))^2\,\nu_2^{-1} - \tfrac{2}{3}(log(\alpha))^3\,\nu_2^{-2}. \qquad (5.4.19)$$

The Table 5.4.2 provides the values of $log(\alpha)$, correct up to five decimal places, and those of $F_{2,\nu_2,\alpha}$, correct up to three decimal places. We have computed $F_{2,\nu_2,\alpha,\text{approx}}$, correct up to five decimal places, for $\alpha = .05, .10$ and $\nu_2 = 10, 20, 30$. The values of $F_{2,\nu_2,\alpha}$ are obtained from Lindley and Scott (1995) and the F Table 14.3.4. The entries in the Table 5.4.2 show clearly that the approximation (5.4.19) works well.

Table 5.4.2. Comparing $F_{2,\nu_2,\alpha}$ with $F_{2,\nu_2,\alpha,\text{approx}}$

ν_2	$\alpha = .10$ $log(\alpha) = -2.30259$		$\alpha = .05$ $log(\alpha) = -2.99573$	
	$F_{2,\nu_2,\alpha}$	$F_{2,\nu_2,\alpha,\text{approx}}$	$F_{2,\nu_2,\alpha}$	$F_{2,\nu_2,\alpha,\text{approx}}$
10	2.9245	2.91417	4.1028	4.07240
20	2.5893	2.58803	3.4928	3.48926
30	2.4890	2.48836	3.3160	3.31479

5.5   Exercises and Complements

5.2.1 Let $X_1, ..., X_n$ be iid having the following common discrete probability mass function:

$$X \text{ values:} \quad -2 \quad 0 \quad 1 \quad 3 \quad 4$$
$$\text{Probabilities:} \quad .2 \quad .05 \quad .1 \quad .15 \quad .5$$

Is there some real number a such that $\overline{X}_n \xrightarrow{P} a$ as $n \to \infty$? Is there some positive real number b such that $S_n^2 \xrightarrow{P} b$ as $n \to \infty$? {*Hint*: Find the values of μ, σ and then try to use the Weak WLLN and Example 5.2.11.}

5.2.2 Prove Khinchine's WLLN (Theorem 5.2.3). {*Hint*: For specific ideas, see Feller (1968, pp. 246-248) or Rao (1973, p.113).}

5.2.3 Let $X_1, ..., X_n$ be iid $N(\mu, 1), -\infty < \mu < \infty$. Consider the following cases separately:

(i) $T_n = exp(\overline{X}_n)$;   (ii) $T_n = log(|\overline{X}_n|)$;

(iii) $T_n = exp(\overline{X}_n^2 + 2\overline{X}_n)$;   (iv) $T_n = |\overline{X}_n|^3 \, exp(\overline{X}_n)$.

Is there some real number a in each case such that $T_n \overset{P}{\to} a$ as $n \to \infty$? {*Hint*: Can one use the Theorems 5.2.4-5.2.5 together?}

5.2.4 Let $X_1, ..., X_n$ be iid having the common pdf

$$f(x) = \begin{cases} 2/x^3 & \text{if } 1 < x < \infty \\ 0 & \text{elsewhere,} \end{cases}$$

Is there a real number a such that $\overline{X}_n \overset{P}{\to} a$ as $n \to \infty$? {*Hint*: Is the Weak WLLN applicable here?}

5.2.5 (Example 5.2.6 Continued) Suppose that $X_1, ..., X_n$ are iid having the following common distribution: $P(X_1 = i) = c/i^p, i = 1, 2, 3, ...$ and $2 < p < 3$. Here, $c \equiv c(p)(> 0)$ is such that $\Sigma_{i=1}^{\infty} P(X_1 = i) = 1$. Is there a real number $a \equiv a(p)$ such that $\overline{X}_n \overset{P}{\to} a$ as $n \to \infty$, for all fixed $2 < p < 3$?

5.2.6 In the Theorem 5.2.4, part (i), construct a proof of the result that $U_n - V_n \overset{P}{\to} u - v$ as $n \to \infty$. Also, prove part (ii) when $u = 0, v \neq 0$.

5.2.7 (Example 5.2.5 Continued) Let $X_1, ..., X_n$ be iid Uniform$(0, \theta)$ with $\theta > 0$. Consider $X_{n:n}$, the largest order statistic. Find the range of $\gamma(> 0)$ such that $n^\gamma \{X_{n:n} - \theta\} \overset{P}{\to} 0$ as $n \to \infty$. Does $X_{n:n} - X_{n-1:n-1} \overset{P}{\to} 0$ as $n \to \infty$? {*Hint*: In the first part, follow the approach used in the Example 5.2.10 and later combine with the Theorem 5.2.4.}

5.2.8 Obtain the expression of $V(S_n^2)$ using (5.2.23) when the X's are iid with

(i) Bernoulli$(p), 0 < p < 1$;   (ii) Poisson$(\lambda), \lambda > 0$;

(iii) Geometric$(p), 0 < p < 1$;   (iv) $N(\mu, \mu^2), \mu > 0$.

In part (ii), show that $V[S^2] > V[\overline{X}]$ for all fixed $\lambda(> 0)$.

5.2.9 (Example 4.4.12 Continued) Suppose that $X_1, ..., X_n$ are iid random variables having the negative exponential distribution with the common pdf given by $f(x) = \sigma^{-1} exp\{-(x - \mu)/\sigma\} I(x > \mu)$ with $-\infty < \mu < \infty, 0 < \sigma < \infty$. Let $X_{n:1}$ be the smallest order statistic and $T_n = \Sigma_{i=1}^{n}(X_i - X_{n:1})$. Show that

(i) $X_{n:1} \overset{P}{\to} \mu$ as $n \to \infty$;

(ii) $V_n = (n - 1)^{-1} T_n \overset{P}{\to} \sigma$ as $n \to \infty$;

(iii) $X_{n:1}/V_n \overset{P}{\to} \mu/\sigma$ as $n \to \infty$.

5.2.10 Consider a sequence of real valued random variables $\{T_n; n > k\}$, and suppose that $n^{-1}T_n \overset{P}{\to} \theta \, (\neq 0)$ as $n \to \infty$. Is it true that for some real number a, the random variable $\left(\frac{n-k}{n}\right)^{T_n} \overset{P}{\to} a$ as $n \to \infty$? {*Hint*: Can one apply Slutsky's Theorem after taking the natural logarithm?}

5.2.11 (Exercise 5.2.10 Continued) Let $X_1, ..., X_n$ be iid Poisson(λ) with $\lambda > 0$, and let $T_n = \Sigma_{i=1}^n X_i$ for $n > k$. Show that $\left(\frac{n-k}{n}\right)^{\Sigma_{i=1}^n X_i} \overset{P}{\to} e^{-k\lambda}$ as $n \to \infty$. Also, find the number $c(> 0)$ such that $\overline{X}_n/S_n^2 \overset{P}{\to} c$ as $n \to \infty$.

5.2.12 Consider a sequence of real valued random variables $\{T_n; n \geq 1\}$, and suppose that $T_n \overset{P}{\to} a \, (> 0)$ as $n \to \infty$. Let us define $X_n = I(T_n > \frac{1}{2}a)$, $n \geq 1$. Does $X_n \overset{P}{\to} 0$ or 1 as $n \to \infty$? Suppose that $Y_n = I(T_n > \frac{3}{2}a)$, $n \geq 1$. Does $Y_n \overset{P}{\to} 0$ or 1 as $n \to \infty$? {*Hint*: Try and apply the Definition 5.2.1 directly. Can one use the Markov inequality here?}

5.2.13 *(i)* Consider the two sequences of random variables $\{U_n; n \geq 1\}$ and $\{V_n; n \geq 1\}$ respectively defined in (5.2.6)-(5.2.7). Verify that $U_n \overset{P}{\to} 1$, $V_n \overset{P}{\to} 1$ as $n \to \infty$. Using Theorem 5.2.4, it will immediately follow that $U_n V_n^{-1} \overset{P}{\to} 1$ and $U_n V_n \overset{P}{\to} 1$ as $n \to \infty$.

(ii) Additionally suppose that U_n and V_n are independent for all $n \geq 1$. In this situation, first obtain the probability distributions of $U_n V_n^{-1}$ and $U_n V_n$. Hence, show directly, that is without appealing to Slutsky's Theorem, that $U_n V_n^{-1} \overset{P}{\to} 1$ and $U_n V_n \overset{P}{\to} 1$ as $n \to \infty$.

5.2.14 Suppose that $(X_i, Y_i), i = 1, ..., 2n$, are iid $N_2(0, 0, 1, 1, \rho)$ with $-1 < \rho < 1$. Recall the bivariate normal distribution from the Section 3.6. Let us denote

$$U_i = \begin{cases} 0 & \text{if } X_{2i-1}Y_{2i-1} + X_{2i}Y_{2i} \leq 0 \\ 1 & \text{if } X_{2i-1}Y_{2i-1} + X_{2i}Y_{2i} > 0 \end{cases}$$

with $i = 1, 2, ..., n$. Consider now the sample mean $\overline{U}_n = n^{-1}\Sigma_{i=1}^n U_i$, and denote $T_n = 2\overline{U}_n - 1$.

(i)   Show that the U_i's are iid Bernoulli with $p = \frac{1}{2}(1 + \rho)$;

(ii) Show that $T_n \overset{P}{\to} \rho$ as $n \to \infty$.

{*Hint*: Note that $p = P(U_1 = 1) = P(X_1Y_1 + X_2Y_2 > 0)$. But, one can write, for example, $X_1Y_1 = \frac{1}{4}\{(X_1+Y_1)^2 - (X_1-Y_1)^2\}$, so that $p = P\{(X_1 + Y_1)^2 - (X_1 - Y_1)^2 + (X_2 + Y_2)^2 - (X_2 - Y_2)^2 > 0\} = P\{(X_1 + Y_1)^2 + (X_2 + Y_2)^2 > (X_1 - Y_1)^2 + (X_2 - Y_2)^2\}$. Verify that $U = (X_1 + Y_1)^2 + (X_2 + Y_2)^2$ is independent of $V = (X_1 - Y_1)^2 + (X_2 - Y_2)^2$. Find the distributions of U and V. Then, rewrite p as the probability of an appropriate event defined in terms of the $F_{2,2}$ random variable and ρ. Thus, p can be evaluated with the help of the integration of the pdf of $F_{2,2}$.}

5.2.15 Suppose that $X_1, ..., X_n$ are iid having the common $N(0,1)$ distribution, and let us denote $T_n = n^{-1}\Sigma_{i=1}^n \mid X_i \mid, U_n = \sqrt{n^{-1}\Sigma_{i=1}^n \mid X_i \mid}$. Determine the positive real numbers a and b such that $T_n \overset{P}{\to} a, U_n \overset{P}{\to} b$ as $n \to \infty$. {*Hint*: We write $E[\mid X_1 \mid] = (\sqrt{2\pi})^{-1} \int_{-\infty}^{\infty} \mid x \mid exp\{-\frac{1}{2}x^2\}dx = (\sqrt{2\pi})^{-1} \int_0^{\infty} x exp\{-\frac{1}{2}x^2\}dx - (\sqrt{2\pi})^{-1} \int_{-\infty}^0 u exp\{-\frac{1}{2}u^2\}du$, with $u = -x$, so that $E[\mid X_1 \mid] = 2(\sqrt{2\pi})^{-1} \int_0^{\infty} u exp\{-\frac{1}{2}u^2\}du$, which can be rewritten as $2(\sqrt{2\pi})^{-1} \int_0^{\infty} exp\{-v\}dv = \sqrt{2/\pi}$ by substituting $v = \frac{1}{2}u^2$. Can one of the weak laws of large numbers be applied now?}

5.2.16 Suppose that $X_1, ..., X_n$ are iid with the pdf $f(x) = \frac{1}{8}e^{-|x|/4}I(x \in \Re)$ and $n \geq 3$. Denote $T_n = n^{-1}\Sigma_{i=1}^n X_i, U_n = (n-2)^{-1}\Sigma_{i=1}^n |X_i|$ and $V_n = \sqrt{(n-1)^{-1}\Sigma_{i=1}^n X_i^2}$. Determine the real numbers a, b and c such that $T_n \overset{P}{\to} a, U_n \overset{P}{\to} b, V_n \overset{P}{\to} c$ as $n \to \infty$.

5.2.17 (Example 5.2.11 Continued) Let $X_1, ..., X_n$ be iid random variables with $E(X_1) = \mu$ and $V(X_1) = \sigma^2$, $-\infty < \mu < \infty$, $0 < \sigma < \infty, n \geq 2$. Let $S_n^2 = (n-1)^{-1}\Sigma_{i=1}^n (X_i - \overline{X}_n)^2$ be the sample variance. Suppose that $Y_i = aX_i + b$ where $a(> 0)$ and b are fixed numbers, $i = 1, ..., n$. Show that the new sample variance based on the Y_i's is given by $a^2 S_n^2$.

5.2.18 Suppose that $(X_1, Y_1), ..., (X_n, Y_n)$ are iid $N_2(\mu_1, \mu_2, \sigma_1^2, \sigma_2^2, \rho)$ where $-\infty < \mu_1, \mu_2 < \infty, 0 < \sigma_1^2, \sigma_2^2 < \infty$ and $-1 < \rho < 1$. Let us denote Pearson's sample correlation coefficient defined in (4.6.7) by r_n instead of r. Show that $r_n \overset{P}{\to} \rho$ as $n \to \infty$. {*Hint*: Use Theorem 5.2.5.}

5.2.19 (Exercise 5.2.11 Continued) Denote $T_n = I(\overline{X}_n > \frac{1}{2}\lambda)$, $n \geq 1$. Does $T_n \overset{P}{\to} 0$ or 1 as $n \to \infty$? Prove your claim.

5.2.20 Let $X_1, ..., X_n, ...$ be iid random variables with the finite variance. Let us denote

$$T_n = \frac{6}{n(n+1)(2n+1)}\Sigma_{i=1}^n i^2 X_i, n = 1, 2, ...$$

Show that $T_n \overset{P}{\to} E[X_1]$ as $n \to \infty$. {*Hint*: Verify that $E[T_n] = E[X_1]$ for all n. In this problem, one will need expressions for both $\Sigma_{i=1}^n i^2$ and $\Sigma_{i=1}^n i^4$. Review (1.6.11) as needed.}

5.2.21 Let $X_1, ..., X_n, ...$ be iid random variables where

$$P\{X_1 = 2^{r-2log(r)}\} = 1/2^r \text{ for } r = 1, 2, ...$$

Does $\overline{X}_n \overset{P}{\to} c$ as $n \to \infty$ for some appropriate c? {*Hint:* Can one of the weak laws of large numbers be applied here? Is it true that $a^{log(r)} = r^{log(a)}$ for any two positive numbers a and r?}

5.2.22 (i) (**Monotone Convergence Theorem**) Consider $\{U_n; n \geq 1\}$, a sequence of real valued random variables. Suppose that $U_n \overset{P}{\to} u$ as $n \to \infty$. Let $g(x), x \in \Re$, be an *increasing* real valued function. Then, $E[g(U_n)] \to g(u)$ as $n \to \infty$.

(ii) (**Dominated Convergence Theorem**) Consider $\{U_n; n \geq 1\}$, a sequence of real valued random variables. Suppose that $U_n \overset{P}{\to} u$ as $n \to \infty$. Let $g(x), x \in \Re$, be a real valued function such that $|g(U_n)| \leq W$ and $E[W]$ is finite. Then, $E[g(U_n)] \to g(u)$ as $n \to \infty$.

{*Note:* Proofs of these two results are beyond the scope of this book. The part (i) is called the Monotone Convergence Theorem, a special form of which was stated earlier in Exercise 2.2.24. The part (ii) is called the Dominated Convergence Theorem. The usefulness of these two results is emphasized in Exercise 5.2.23.}

5.2.23 Let $X_1, ..., X_n$ be iid non-negative random variables with $\mu \in \Re^+, \sigma \in \Re^+$. We denote $U_n = \overline{X}_n, T_n = U_n^2(1 + U_n^2)^{-1}, n \geq 1$.

(i)   Does $E\{e^{U_n}\}$ converge to some real number as $n \to \infty$?

(ii)   Does $E\{e^{-2U_n}\}$ converge to some real number as $n \to \infty$?

(iii)   Does $E\{\sin(U_n)\}$ converge to some real number as $n \to \infty$?

(iv)   Does $E\{\Phi(U_n)\}$ converge to some real number as $n \to \infty$

where $\Phi(x) = (\sqrt{2\pi})^{-1} \int_{-\infty}^{x} exp\{-y^2/2\} dy, x \in \Re^+$?

(v)   Does $E(T_n)$ converge to some real number as $n \to \infty$?

{*Hints:* By the WLLN, we first claim that $U_n \overset{P}{\to} \mu$ as $n \to \infty$. Let $g(x) = e^x, x \in \Re^+$ be our *increasing* real valued function. Then, by the Exercise 5.2.22, part (i), we conclude that $E[g(U_n)] = E\{e^{U_n}\} \to e^\mu$ as $n \to \infty$. For the second part, note that $g(x) = e^{-2x}, x \in \Re^+$ is bounded between two fixed numbers, zero and one. For the third part, note that $g(x) = \sin(x), x \in \Re^+$ is bounded between ± 1. Thus, by the Exercise 5.2.22, part (ii), we conclude that $E[g(U_n)]$ will converge to $e^{-2\mu}$ (or $\sin(\mu)$) as $n \to \infty$ in part (ii) (or (iii)) respectively. How about parts (iv)-(v)?}

5.2.24 Let $X_1, ..., X_n$ be iid Poisson$(\lambda), 0 < \lambda < \infty$. We denote $U_n = \overline{X}_n, n \geq 1$. From Exercise 5.2.23, it follows that $E[e^{U_n}] \to e^\lambda$ and $E[e^{-2U_n}] \to e^{-2\lambda}$ as $n \to \infty$. Verify these two limiting results directly by using the mgf of a Poisson random variable. {*Hint:* Observe that $e^{t/n} = 1 + \frac{t}{n} + \frac{1}{2!}(\frac{t}{n})^2 + ...$. Then, note that $E[e^{tU_n}] = [exp\{-\lambda + \lambda e^{t/n}\}]^n = exp\{-n\lambda +$

$n\lambda e^{t/n}\} \to e^{t\lambda}$ as $n \to \infty.$ }

5.3.1 (Exercise 5.2.1 Continued) Find a and $b(> 0)$ such that we can claim: $\sqrt{n}(\overline{X}_n - a)/b \xrightarrow{\mathcal{L}} N(0,1)$ as $n \to \infty$.

5.3.2 (Example 5.3.5 Continued) Let $X_1, ..., X_n$ be iid Uniform$(0, \theta)$ with $\theta > 0$ and suppose that $T_n = X_{n:n}$, the largest order statistic. First, find the df of the random variable $G_n = n(\theta - T_n)/T_n$, and then use the Definition 5.3.1 to show directly that the limiting distribution of G_n is the standard exponential.

5.3.3 Prove CLT using the mgf technique assuming that the X_i's are iid having a finite mgf for $\mid t \mid < h$ with some $h > 0$. {*Hint*: Follow the derivations in Examples 5.3.3-5.3.4 closely.}

5.3.4 (Example 5.3.7 Continued) Under the same conditions as the CLT, along the lines of Example 5.3.7, derive the asymptotic distribution of

(i) $\sqrt{n}(\overline{X}_n^3 - \mu^3)$ as $n \to \infty$ if $\mu \neq 0$;

(ii) $\sqrt{n}(\sqrt{\overline{X}_n} - \sqrt{\mu})$ as $n \to \infty$ if $\mu > 0$ and $\overline{X}_n$ is positive w.p.1.
In this problem, avoid using the Mann-Wald Theorem.

5.3.5 (Exercise 5.3.2 Continued) Let $X_1, ..., X_n$ be iid Uniform$(0, \theta)$ with $\theta > 0$ and suppose that $T_n = X_{n:n}$, the largest order statistic. Does $n\{\theta - X_{n:n}\}/X_{n:n}^r$ converge to some appropriate random variable in distribution as $n \to \infty$? {*Hint*: Can Slutsky's Theorem be used here?}

5.3.6 (Exercise 5.2.3 Continued) Let $X_1, ..., X_n$ be iid $N(\mu, 1), -\infty < \mu < \infty$. Consider the following cases separately:

(i) $T_n = exp(\overline{X}_n)$;   (ii) $T_n = log(\mid \overline{X}_n \mid)$;

(iii) $T_n = exp(\overline{X}_n^2 + 2\overline{X}_n)$;   (iv) $T_n = \mid \overline{X}_n \mid^3 exp(\overline{X}_n)$.

Find suitable $a_n, b_n(> 0)$ associated with each T_n such that $(T_n - a_n)/b_n \xrightarrow{\mathcal{L}} N(0,1)$ as $n \to \infty$. {*Hint*: Is the Mann-Wald Theorem helpful here?}

5.3.7 (Example 5.3.8 Continued) Find the number $\eta(> 0)$ such that $\sqrt{n}(S_n - \sigma) \xrightarrow{\mathcal{L}} N(0, \eta^2)$ as $n \to \infty$. Here, S_n^2 is the sample variance obtained from iid random variables $X_1, ..., X_n, n \geq 2$. Solve this problem separately when the X_i's are (i) normal and (ii) non-normal, under appropriate moment assumptions. {*Hint*: Is the Mann-Wald Theorem helpful here?}

5.3.8 (Exercise 5.3.7 Continued) Find the number $\eta(> 0)$ such that $\sqrt{n}(S_n^u - \sigma^u) \xrightarrow{\mathcal{L}} N(0, \eta^2)$ as $n \to \infty$, where $u(\neq 0)$ is some fixed real number. *Hence,* that is without referring to Slutsky's Theorem, find the number $\xi(> 0)$ such that $\sqrt{n}(S_n^u - \sigma^u)/S_n^u \xrightarrow{\mathcal{L}} N(0, \xi^2)$ as $n \to \infty$ where $u(\neq 0)$ is some fixed real number. Solve this problem separately when the X's are (i) normal and (ii) non-normal, under appropriate moment assumptions.

5.3.9 (Exercise 5.2.9 Continued) Suppose that $X_1, ..., X_n$ are iid random variables having the negative exponential distribution with the common pdf given by

$$f(x) \quad = \sigma^{-1} exp\{-(x - \mu)/\sigma\}I(x > \mu)$$
$$\text{for } -\infty < \mu < \infty, 0 < \sigma < \infty.$$

Let $X_{n:1}$ be the smallest order statistic and $T_n = \Sigma_{i=1}^n (X_i - X_{n:1})$. Show that $n\{X_{n:1} - \mu\}/V_n \overset{\pounds}{\to} Y$, the standard exponential random variable, as $n \to \infty$.

5.3.10 Let $X_1, ..., X_n$ be iid with the lognormal distribution having the following pdf with $-\infty < \mu < \infty$, $0 < \sigma < \infty$:

$$f(x) = \begin{cases} \{\sigma\sqrt{2\pi}\}^{-1}x^{-1}exp \left\{-\frac{1}{2\sigma^2}(log(x) - \mu)^2\right\} & \text{if } x > 0 \\ 0 & \text{elsewhere} \end{cases}$$

Denote $T_n = (\Pi_{i=1}^n X_i)^{1/n}$, the geometric mean of $X_1, ..., X_n$.

(i)   Find the real number $c(> 0)$ such that $T_n \overset{P}{\to} c$ as $n \to \infty$;

(ii) Find the positive real numbers a_n, b_n such that $(T_n - a_n)/b_n \overset{\pounds}{\to} N(0, 1)$ as $n \to \infty$.

5.3.11 Let $X_1, ..., X_n$ be iid with the uniform distribution on the interval $(0, \theta), \theta > 0$. Denote $T_n = (\Pi_{i=1}^n X_i)^{1/n}$, the geometric mean of $X_1, ..., X_n$.

(i)   Find $c(> 0)$ such that $T_n \overset{P}{\to} c$ as $n \to \infty$;

(ii) Find $a_n, b_n(> 0)$ such that $(T_n - a_n)/b_n \overset{\pounds}{\to} N(0, 1)$ as $n \to \infty$.

5.3.12 (Exercise 5.2.15 Continued) Suppose that $X_1, ..., X_n$ are iid having the common $N(0, 1)$ distribution. We denote $T_n = n^{-1}\Sigma_{i=1}^n \mid X_i \mid$ and $U_n = \sqrt{n^{-1}\Sigma_{i=1}^n \mid X_i \mid}$. Find suitable numbers $c_n, d_n(> 0), c_n', d_n'(> 0)$ such that $(T_n - c_n)/d_n \overset{\pounds}{\to} N(0, 1)$ and $(U_n - c_n')/d_n' \overset{\pounds}{\to} N(0, 1)$ as $n \to \infty$.

5.3.13 (Exercise 5.2.16 Continued) Suppose that $X_1, ..., X_n$ are iid with the pdf $f(x) = \frac{1}{8}e^{-|x|/4}I(x \in \Re)$ and $n \geq 3$. Denote $T_n = n^{-1}\Sigma_{i=1}^n X_i, U_n = (n - 2)^{-1}\Sigma_{i=1}^n |X_i|$ and $V_n = \sqrt{(n - 1)^{-1}\Sigma_{i=1}^n X_i^2}$. Find suitable numbers $c_n, d_n(> 0), c_n', d_n'(> 0), c_n'', d_n''(> 0)$ such that $(T_n - c_n)/d_n \overset{\pounds}{\to} N(0, 1), (U_n - c_n')/d_n' \overset{\pounds}{\to} N(0, 1)$ and $(V_n - c_n'')/d_n'' \overset{\pounds}{\to} N(0, 1)$ as $n \to \infty$.

5.3.14 This exercise shows that the converse of the Theorem 5.3.2 is not necessarily true. Let $\{U_n, U; n \geq 1\}$ be a sequence of random variables such that U_n and U are independent, U_n is $N(0, 1 + \frac{1}{n})$, and U is $N(0,1)$, for each $n \geq 1$. Show that

(i) $U_n \overset{\mathcal{L}}{\to} U$ as $n \to \infty$. {*Hint:* For $x \in \Re$, check that $P(U_n \leq x)$
   $= \Phi\left(x/\sqrt{1+\frac{1}{n}}\right) \to \Phi(x)$ as $n \to \infty$.};

(ii) $U_n \overset{P}{\nrightarrow} U$ as $n \to \infty$. {*Hint:* Since U_n and U are independent,
   $U_n - U$ is distributed as $N(0, 2 + \frac{1}{n})$. Hence, $U_n - U \overset{P}{\nrightarrow} 0$
   as $n \to \infty$.}.

5.3.15 Let $X_1, ..., X_n$ be iid Uniform$(0,1)$. Let us denote a sequence of random variables, $U_n = \sqrt{n}(\overline{X}_n - \frac{1}{2}), n \geq 1$.

(i) Show that the mgf, $M_{X_1}(t) = \frac{1}{t}(e^t - 1), t \neq 0$;

(ii) Show that the mgf, $M_{U_n}(t) = \{\frac{1}{2}(e^a - e^{-a})/a\}^n$ where a
   $= \frac{1}{2}\frac{t}{\sqrt{n}}$;

(iii) Show that $\lim_{n \to \infty} M_{U_n}(t) = \lim_{n \to \infty} \left[1 + \frac{t^2}{24n} + O(n^{-2})\right]^n$
   $= exp\{t^2/24\}$;

(iv) Use part (iii) to argue that $U_n \overset{\mathcal{L}}{\to} N(0, \frac{1}{12})$ as $n \to \infty$.

5.3.16 Let $X_1, ..., X_k$ be iid $N(0,1)$ where $k = 2^n, n \geq 1$. We denote

$$U_n = \frac{X_1}{X_2} + \frac{X_3}{X_4} + ... + \frac{X_{k-1}}{X_k} \text{ with } k = 2^n$$
$$V_n = X_1^2 + X_2^2 + ... + X_n^2$$

Find the limiting (as $n \to \infty$) distribution of $W_n = U_n/V_n$. {*Hint:* Let $Y_j = \frac{X_{2j-1}}{X_{2j}}, j = 1, ..., n$. Each random variable Y_j has the Cauchy pdf $f(y) = \pi^{-1}(1+y^2)^{-1}, -\infty < y < \infty$. Also, $\frac{1}{n}U_n$ has the same Cauchy pdf. Is it possible to conclude that $W_n \overset{\mathcal{L}}{\to} W$ as $n \to \infty$ where W has the same Cauchy pdf? Would Slutsky's Theorem suffice?}

5.3.17 Let $X_1, ..., X_n$ be iid $N(0,1), n \geq 1$. We denote:

$$U_n = \frac{\sqrt{n}\{X_1 + X_2 + ... + X_n\}}{X_1^2 + X_2^2 + ... + X_n^2}$$

Show that $U_n \overset{\mathcal{L}}{\to} N(0,1)$ as $n \to \infty$. {*Hint:* Use both CLT and Slutsky's Theorem.}

5.3.18 (Normal Approximation to the Binomial Distribution)
Suppose that $X_1, ..., X_n$ are iid Bernoulli$(p), 0 < p < 1, n \geq 1$. We know

that $U_n = \Sigma_{i=1}^n X_i$ is distributed as Binomial$(n, p), n \geq 1$. Apply the CLT to show that

$$\frac{(U_n - np)}{\sqrt{np(1-p)}} \xrightarrow{\mathcal{L}} N(0,1) \text{ as } n \to \infty.$$

In other words, for practical problems, the Binomial(n, p) distribution can be approximated by the $N(np, np(1-p))$ distribution, for large n and fixed $0 < p < 1$.

5.3.19 (Exercise 5.3.18 Continued) Let $X_1, ..., X_n$ be iid Bernoulli(p) with $0 < p < 1, n \geq 1$. Let us denote $V_n = n^{-1}\Sigma_{i=1}^n X_i$. Show that

$$\frac{\sqrt{n}\left(V_n^{-1} - p^{-1}\right)}{\sqrt{(1-p)p^{-3}}} \xrightarrow{\mathcal{L}} N(0,1) \text{ as } n \to \infty.$$

{*Hint*: In view of the Exercise 5.3.18, would the Mann-Wald Theorem help?}

5.3.20 (Exercises 5.3.18-5.3.19 Continued) Suppose that $X_1, ..., X_n$ are iid Bernoulli$(p), 0 < p < 1, n \geq 1$. Let us denote $V_n = n^{-1}\Sigma_{i=1}^n X_i$ and $W_n = V_n(1 - V_n)$. Show that

$$\frac{\sqrt{n}\left(W_n - p(1-p)\right)}{|1 - 2p|\sqrt{(1-p)p}} \xrightarrow{\mathcal{L}} N(0,1) \text{ as } n \to \infty.$$

when $p \neq \frac{1}{2}$. {*Hint*: In view of the Exercise 5.3.18, would the Mann-Wald Theorem help?}

5.3.21 Suppose that $X_1, ..., X_n$ are iid with the common pdf

$$f(x) = cexp\{3x - 4x^2\} \text{ for } x \in \Re,$$

where $c(> 0)$ is an appropriate constant.

(*i*)   Find k such that $\overline{X}_n \xrightarrow{P} k$ as $n \to \infty$;

(*ii*)   Find $a_n, b_n(> 0)$ such that $(\overline{X}_n - a_n)/b_n \xrightarrow{\mathcal{L}} N(0,1)$ as $n \to \infty$.

5.4.1 (Example 4.5.2 Continued) Suppose that the random variables $X_{i1}, ..., X_{in_i}$ are iid $N(\mu_i, \sigma^2), i = 1, 2$, and that the X_{1j}'s are independent of the X_{2j}'s. With $n_i \geq 2, \mathbf{n} = (n_1, n_2)$, let us denote

$$\overline{X}_{in_i} = n_i^{-1}\Sigma_{j=1}^{n_i} X_{ij}, \ S_{in_i}^2 = (n_i - 1)^{-1}\Sigma_{j=1}^{n_i}(X_{ij} - \overline{X}_{in_i})^2$$

$$S_{Pn}^2 = (n_1 + n_2 - 2)^{-1}\left\{(n_1 - 1)S_{1n_1}^2 + (n_2 - 1)S_{2n_2}^2\right\}$$

for $i = 1, 2$. Consider the two-sample t random variable $t_\nu = \{n_1^{-1} + n_2^{-1}\}^{-\frac{1}{2}}[(\overline{X}_{1n_1} - \overline{X}_{2n_2}) - (\mu_1 - \mu_2)]S_{Pn}^{-1}$ which has the Student's t distribution with $\nu \equiv \nu(\mathbf{n}) = (n_1 + n_2 - 2)$ degrees of freedom. Does t_ν converge to some appropriate random variable in distribution as both n_1 and $n_2 \to \infty$ in such a way that $n_1/n_2 \to k(>0)$?

5.4.2 Let $X_1, ..., X_n$ be iid $N(\mu_1, \sigma^2)$, $Y_1, ..., Y_n$ be iid $N(\mu_2, 3\sigma^2)$ where $-\infty < \mu_1, \mu_2 < \infty, 0 < \sigma < \infty$. Also suppose that the X's and Y's are independent. Denote for $n \ge 2$, $\overline{X}_n = n^{-1}\Sigma_{i=1}^n X_i, \overline{Y}_n = n^{-1}\Sigma_{i=1}^n Y_i, S_{1n}^2 = (n-1)^{-1}\Sigma_{i=1}^n(X_i - \overline{X}_n)^2, S_{2n}^2 = (n-1)^{-1}\Sigma_{i=1}^n(Y_i - \overline{Y}_n)^2$, and $T_n = S_{1n}^2 + \frac{1}{3}S_{2n}^2$.

- (i) Show that $V_n = \frac{1}{2\sigma}\sqrt{n}(\overline{X}_n - \overline{Y}_n - \mu_1 + \mu_2)$ is distributed as $N(0, 1)$;
- (ii) Show that $(n-1)T_n/\sigma^2$ is distributed as $\chi^2_{2(n-1)}$;
- (iii) Are V_n, T_n independent?
- (iv) Show that $U_n = \sqrt{n}(\overline{X}_n - \overline{Y}_n - \mu_1 + \mu_2)/\sqrt{2T_n}$ is distributed as the Student's $t_{2(n-1)}$;
- (v) Show that $T_n \xrightarrow{P} 2\sigma^2$ as $n \to \infty$;
- (vi) Show that $U_n \xrightarrow{\mathcal{L}} N(0, 1)$ as $n \to \infty$.

5.4.3 Suppose that $X_1, ..., X_m$ are iid exponential with mean $\beta(>0)$, $Y_1, ..., Y_n$ are iid exponential with mean $\eta(>0)$, and that the X's are independent of the Y's. Define $T_{m,n} = \overline{X}_m/\overline{Y}_n$. Then,

- (i) Show that $T_{m,n}$ is distributed as $\frac{\beta}{\eta}F_{2m,2n}$;
- (ii) Determine the asymptotic distribution of $T_{m,n}$ as $n \to \infty$, when m is kept fixed;
- (iii) Determine the asymptotic distribution of $T_{m,n}$ as $m \to \infty$, when n is kept fixed.

5.4.4 Verify the limiting ratio of the gamma functions in (5.4.6).

5.4.5 Verify the limiting ratio of the gamma functions in (5.4.14).

5.4.6 Show that the percentile point $d \equiv d(\nu_2, \alpha)$ given by (5.4.17) is a strictly decreasing function of ν_2 whatever be fixed $\alpha \in (0, 1)$. {*Hint:* Take the derivative of d with respect to ν_2 and show that this derivative is negative. This approach is not entirely fair because ν_2 is after all a discrete variable $\in \{1, 2, 3, ...\}$. A rigorous approach should investigate the behavior of $d(\nu_2 + 1, \alpha) - d(\nu_2, \alpha)$ for $\nu_2 \in \{1, 2, 3, ...\}$. The "derivative approach" however should drive the point home.}

5.4.7 Consider the random variables $T_1, ..., T_p$ defined in (4.6.12) whose joint distribution was the multivariate t, denoted by $Mt_p(\nu, \Sigma)$. Derive the limiting distribution of $(T_1, ..., T_p)$ as $\nu \to \infty$. Show that the pdf of the $Mt_p(\nu, \Sigma)$ distribution given by (4.6.13) converges to the pdf of the corresponding limiting random variable $\mathbf{U}$ as $\nu \to \infty$. Identify this random variable $\mathbf{U}$ by name. {*Hint:* In the second part, use techniques similar to those used in Section 5.4.4.}

5.4.8 Consider the random variables $U_1, ..., U_p$ defined in (4.6.15) whose joint distribution was the multivariate F, denoted by $MF_p(\nu_0, \nu_1, ..., \nu_p)$. Derive the limiting distribution of $(U_1, ..., U_p)$ as $\nu_0 \to \infty$ but $\nu_1, ..., \nu_p$ are held fixed. Show that the pdf of the $MF_p(\nu_0, \nu_1, ..., \nu_p)$ distribution given by (4.6.16) converges to the pdf of the corresponding limiting random variable $\mathbf{W}$ as $\nu_0 \to \infty$ but $\nu_1, ..., \nu_p$ are held fixed. Identify this random variable $\mathbf{W}$ by name. {*Hint*: In the second part, use techniques similar to those used in Section 5.4.4.}

5.4.9 (Exercise 5.4.1 Continued) Suppose that the random variables $X_{i1}, ..., X_{in}$ are iid $N(\mu_i, \sigma^2), i = 1, 2$, and that the X_{1j}'s are independent of the X_{2j}'s. With $n \geq 2$, let us denote

$$\overline{X}_{in} = n^{-1}\Sigma_{j=1}^n X_{ij}, \ S_{in}^2 = (n-1)^{-1}\Sigma_{j=1}^n (X_{ij} - \overline{X}_{in})^2,$$

$$S_{Pn}^2 = \tfrac{1}{2}\left\{S_{1n}^2 + S_{2n}^2\right\},$$

for $i = 1, 2$. Consider the random variable $T_{in} = \sqrt{n}(\overline{X}_{in} - \mu_i)/S_{Pn}, i = 1, 2$. Show that (T_{1n}, T_{2n}) has an appropriate bivariate t distribution, for all fixed $n \geq 2$. Find the limiting distribution of (T_{1n}, T_{2n}) as $n \to \infty$.

5.4.10 (Exercise 5.4.9 Continued) Suppose that the random variables $X_{i1}, ..., X_{in}$ are iid $N(\mu_i, \sigma^2), i = 1, ..., 4$, and that the X_{ij}'s are independent of the X_{lj}'s for all $i \neq l = 1, ..., 4$. With $n \geq 2$, let us denote

$$\overline{X}_{in} = n^{-1}\Sigma_{j=1}^n X_{ij}, \ S_{in}^2 = (n-1)^{-1}\Sigma_{j=1}^n (X_{ij} - \overline{X}_{in})^2,$$

$$S_{Pn}^2 = \tfrac{1}{4}\left\{S_{1n}^2 + ... + S_{4n}^2\right\},$$

for $i = 1, ..., 4$. Consider the random variable $T_{in} = \sqrt{n}(\overline{X}_{in} - \mu_i)/S_{Pn}, i = 1, ..., 4$. Show that $(T_{1n}, ..., T_{4n})$ has an appropriate four-dimensional t distribution, for all fixed $n \geq 2$. Find the limiting distribution of $(T_{1n}, ..., T_{4n})$ as $n \to \infty$.

5.4.11 (Exercise 5.4.10 Continued) Suppose that the random variables $X_{i1}, ..., X_{in_i}$ are iid $N(\mu_i, \sigma^2), i = 1, ..., 4$, and that the X_{ij}'s are independent of the X_{lj}'s for all $i \neq l = 1, ..., 4$. With $n_i \geq 2$, let us denote

$$\overline{X}_{in_i} = n_i^{-1}\Sigma_{j=1}^{n_i} X_{ij}, \ S_{in_i}^2 = (n_i-1)^{-1}\Sigma_{j=1}^{n_i} (X_{ij} - \overline{X}_{in_i})^2,$$

for $i = 1, ..., 4$. Suppose that $n_1 = n_2 = n_3 = k$ and $n_4 = n$. Consider the random variable $T_{in} = S_{ik}^2/S_{4n}^2, i = 1, 2, 3$. Show that (T_{1n}, T_{2n}, T_{3n}) has an appropriate three-dimensional F distribution, for all fixed $n \geq 2$. Find the limiting distribution of (T_{1n}, T_{2n}, T_{3n}) as $n \to \infty$.

6

Sufficiency, Completeness, and Ancillarity

6.1 Introduction

Sir Ronald Aylmer Fisher published several path-breaking articles in the 1920's which laid the foundation of statistical inference. Many fundamental concepts and principles of statistical inference originated in the works of Fisher. The most exciting thing about these concepts is that these are still alive, well, and indispensable. Perhaps the deepest of all statistical concepts and principles is what is known as *sufficiency*. The concept of sufficiency originated from Fisher (1920) and later it blossomed further, again in the hands of Fisher (1922). First we introduce the notion of sufficiency which helps in summarizing data without any loss of *information*.

Consider a scenario like this. From past experience, suppose that a market analyst postulates the monthly income per household in a small town to be *normally distributed* with the unknown population mean μ and population standard deviation $\sigma = \$800$. In order to guess the unknown μ, twenty one households are randomly selected, independently of each other, from the population, leading to the observations $X_1 = x_1, X_2 = x_2, ..., X_{21} = x_{21}$. At this point, the market analyst may be debating between the appropriateness of using $\overline{x}$, the observed value of the sample mean $\overline{X}$, as the guess or using $x_{21:11}$, the observed value of the sample median $X_{21:11}$ instead. Now, the question is this: which guess should the market analyst use in this situation? Since the income distribution is assumed normal, $\overline{X}$ should be used because $\overline{X}$ is sufficient for μ as we will see later. On the other hand, the sample median, $X_{21:11}$ is not sufficient for μ. Once we develop the idea of sufficiency in Section 6.2, it will be clear that the summary obtained via $\overline{X}$ preserves all the information contained in the whole data $\mathbf{X} = (X_1, ..., X_{21})$, whereas in the alternative summary obtained via $X_{21:11}$, some information from the data $\mathbf{X}$ will be lost. The common phrases such as the *estimator, statistic, information,* and *sufficiency* would all be defined shortly.

Section 6.2 includes two ways to find *sufficient statistics* in a statistical model. The first method involves the direct calculation of the conditional distribution of the data given the value of a particular statistic, while the second approach consists of the classical *Neyman factorization* of a *likelihood function*. We include specific examples to highlight some approaches to verify whether a statistic is or is not sufficient.

In Section 6.3, the notion of *minimal sufficiency* is introduced and a fundamental result due to Lehmann and Scheffé (1950) is discussed. This result helps us in locating, in some sense, the best sufficient statistic, if it exists. We had seen in Section 3.8 that many standard statistical models such as the binomial, Poisson, normal, gamma and several others belong to an exponential family. It is often a simple matter to locate the minimal sufficient statistic and its distribution in an exponential family. One gets a glimpse of this in the Theorems 6.3.3-6.3.4.

The Section 6.4 provides the idea of quantifying *information* in both one- and two-parameter situations, but we do so in a fairly elementary fashion. By means of examples, we show that the information contained in the whole data is indeed preserved by the sufficient statistics. In the one-parameter case, we compare the information content in a non-sufficient statistic with that in the data and find the extent of the lost information if a non-sufficient statistic is used as a summary.

The topic of *ancillarity* is discussed in Section 6.5, again moving deeper into the concepts and highlighting the fact that ancillary statistics can be useful in making statistical inferences. We include the *location, scale,* and *location-scale families* of distributions in Section 6.5.1. The Section 6.6 introduces the concept of *completeness* and discusses some of the roles complete sufficient statistics play within the realm of statistical inference. Section 6.6.2 highlights Basu's Theorem from Basu (1955a).

6.2 Sufficiency

Suppose that we start a statistical investigation with observable iid random variables $X_1, ..., X_n$, having a common pmf or pdf $f(x), x \in \mathcal{X}$, the domain space for x. Here, n is the sample size which is assumed *known*. Practically speaking, we like to think that we are going to observe $X_1, ..., X_n$ from a *population* whose distribution is approximated well by $f(x)$. In the example discussed in the introduction, the market analyst is interested in the income distribution of households per month which is denoted by $f(x)$, with some appropriate space $\mathcal{X}$ for x. The income distribution may be indexed by some *parameter* (or *parameter vector*) $\theta($ or $\boldsymbol{\theta})$ which captures important features of the distribution. A practical significance of indexing with the parameter θ(or $\boldsymbol{\theta}$) is that once we know the value of θ(or $\boldsymbol{\theta}$), the population distribution $f(x)$ would then be completely specified.

For simplicity, however, let us assume first that θ is a single parameter and denote the population pmf or pdf by $f(x;\theta)$ so that the dependence of the features of the underlying distribution on the parameter θ becomes explicit. In classical statistics, we assume that this parameter θ is *fixed but otherwise unknown* while all possible values of $\theta \in \Theta$, called the *parameter space*, $\Theta \subseteq \Re$, the real line. For example, in a problem, we may postulate that the X's are distributed as $N(\mu,\sigma^2)$ where μ is the unknown parameter, $-\infty < \mu < \infty$, but $\sigma(> 0)$ is known. In this case we may denote the pdf by $f(x;\theta)$ with $\theta = \mu$ while the parameter space $\Theta = \Re$. But if both the parameters μ and σ^2 are unknown, the population density would be denoted by $f(x;\boldsymbol{\theta})$ where the parameter vector is $\boldsymbol{\theta} = (\mu,\sigma^2) \in \Theta = \Re \times \Re^+$. This is the idea behind *indexing* a population distribution by the unknown parameters in general.

> From the context, it should become clear whether the unknown parameter is real valued (θ) or vector valued $(\boldsymbol{\theta})$.

Consider again the observable real valued iid random variables $X_1, ..., X_n$ from a population with the common pmf or pdf $f(x;\theta)$ where $\theta(\in \Theta)$ is the unknown parameter. Our quest for gaining information about the unknown parameter θ can safely be characterized as the core of *statistical inference*. The data, of course, has all the *information* about θ even though we have not yet specified how to quantify this "information." In Section 6.4, we address this. A data can be large or small, and it may be nice or cumbersome, but it is ultimately incumbent upon the experimenter to summarize this data so that all interesting features are captured by its summary. That is, the summary should preferably have the exact same "information" about the unknown parameter θ as does the original data. If one can prepare such a summary, then this would be as good as the whole data as far as the information content regarding the unknown parameter θ is concerned. We would call such a summary *sufficient* for θ and make this basic idea more formal as we move along.

Definition 6.2.1 *Any observable real or vector valued function* $T \equiv T(X_1, ..., X_n)$, *of the random variables* $X_1, ..., X_n$ *is called a statistic.*

Some examples of statistics are $\overline{X}, X_1(X_2 - X_{n:n}), \Sigma_{i=1}^n X_i, S^2$ and so on. As long as the numerical evaluation of T, having observed a specific data $X_1 = x_1, ..., X_n = x_n$, does not depend on any unknown quantities, we will call T a statistic. Supposing that $X_1, ..., X_n$ are iid $N(\mu,\sigma^2)$ where μ is unknown, but σ is known, $T = \overline{X}$ is a statistic because the value of T associated with any observed data $x_1, ..., x_n$ can be explicitly calculated. In the same example, however, the standardized form of $\overline{X}$, namely $\sqrt{n}(\overline{X} -$

$\mu)/\sigma$, is not a statistic because it involves the unknown parameter μ, and hence its value associated with any observed data $x_1, ..., x_n$ can not be calculated.

Definition 6.2.2 *A real valued statistic T is called sufficient (for the unknown parameter θ) if and only if the conditional distribution of the random sample $\mathbf{X} = (X_1, ..., X_n)$ given $T = t$ does not involve θ, for all $t \in \mathcal{T}$, the domain space for T.*

In other words, *given* the value t of a *sufficient statistic T, conditionally* there is no more "information" left in the original data regarding the unknown parameter θ. Put another way, we may think of $\mathbf{X}$ trying to tell us a story about θ, but once a sufficient summary T becomes available, the original story then becomes redundant. Observe that the whole data $\mathbf{X}$ is always sufficient for θ in this sense. But, we are aiming at a "shorter" summary statistic which has the same amount of information available in $\mathbf{X}$. Thus, once we find a sufficient statistic T, we will focus only on the summary statistic T. Before we give other details, we define the concept of *joint sufficiency* of a vector valued statistic $\mathbf{T}$ for an unknown parameter θ.

Definition 6.2.3 *A vector valued statistic $\mathbf{T} \equiv (T_1, ..., T_k)$ where $T_i \equiv T_i(X_1, ..., X_n), i = 1, ..., k$, is called jointly sufficient (for the unknown parameter θ) if and only if the conditional distribution of $\mathbf{X} = (X_1, ..., X_n)$ given $\mathbf{T} = \mathbf{t}$ does not involve θ, for all $\mathbf{t} \in \mathcal{T} \subseteq \Re^k$.*

The Section 6.2.1 shows how the conditional distribution of $\mathbf{X}$ given $T = t$ can be evaluated. The Section 6.2.2 provides the celebrated Neyman factorization which plays a fundamental role in locating sufficient statistics.

6.2.1 The Conditional Distribution Approach

With the help of examples, we show how the Definition 6.2.2 can be applied to find sufficient statistics for an unknown parameter θ.

Example 6.2.1 Suppose that $X_1, ..., X_n$ are iid Bernoulli(p), where p is the unknown parameter, $0 < p < 1$. Here, $\mathcal{X} = \{0, 1\}$, $\theta = p$, and $\Theta = (0, 1)$. Let us consider the specific statistic $T = \Sigma_{i=1}^n X_i$. Its values are denoted by $t \in \mathcal{T} = \{0, 1, 2, ..., n\}$. We verify that T is sufficient for p by showing that the conditional distribution of $(X_1, ..., X_n)$ given $T = t$ does not involve p, whatever be $t \in \mathcal{T}$. From the Examples 4.2.2-4.2.3, recall that T has the Binomial(n, p) distribution. Now, we obviously have:

$$P(X_1 = x_1 \cap ... \cap X_n = x_n \mid T = t) = 0 \quad \text{if } \Sigma_{i=1}^n x_i \neq t.$$

But, when $\Sigma_{i=1}^n x_i = t$, since $A = \{\cap_{i=1}^n X_i = x_i\}$ is a subset of $B = \{T = t\}$, we can write

$$
\begin{aligned}
P\left\{[\cap_{i=1}^{n} X_i = x_i] \mid T = t\right\} \\
= P\left\{[\cap_{i=1}^{n} X_i = x_i] \cap T = t\right\} / P(T = t) \\
= P\left\{[\cap_{i=1}^{n} X_i = x_i]\right\} / P(T = t) \\
= \Pi_{i=1}^{n} P\left\{X_i = x_i\right\} / P(T = t),
\end{aligned}
$$

since the X's are independent.

Thus, one has

$$
\begin{aligned}
P\left\{[\cap_{i=1}^{n} X_i = x_i] \mid T = t\right\} \\
= \left\{p^{\Sigma_{i=1}^{n} x_i}(1-p)^{n-\Sigma_{i=1}^{n} x_i}\right\} / \left\{\binom{n}{t} p^t (1-p)^{n-t}\right\} \\
= 1/\binom{n}{t},
\end{aligned}
\tag{6.2.1}
$$

which is free from p. In other words, $\Sigma_{i=1}^{n} X_i$ is a sufficient statistic for the unknown parameter p. ▲

Example 6.2.2 Suppose that $X_1, ..., X_n$ are iid Poisson(λ) where λ is the unknown parameter, $0 < \lambda < \infty$. Here, $\mathcal{X} = \{0, 1, 2, ...\}$, $\theta = \lambda$, and $\Theta = (0, \infty)$. Let us consider the specific statistic $T = \Sigma_{i=1}^{n} X_i$. Its values are denoted by $t \in \mathcal{T} = \{0, 1, 2, ...\}$. We verify that T is sufficient for λ by showing that the conditional distribution of $(X_1, ..., X_n)$ given $T = t$ does not involve λ, whatever be $t \in \mathcal{T}$. From the Exercise 4.2.2 recall that T has the Poisson($n\lambda$) distribution. Now, we obviously have:

$$
P(X_1 = x_1 \cap ... \cap X_n = x_n \mid T = t) = 0 \quad \text{if } \Sigma_{i=1}^{n} x_i \neq t.
$$

But, when $\Sigma_{i=1}^{n} x_i = t$, since $A = \{\cap_{i=1}^{n} X_i = x_i\}$ is a subset of $B = \{T = t\}$, we can write

$$
\begin{aligned}
P\left\{[\cap_{i=1}^{n} X_i = x_i] \mid T = t\right\} \\
= P\left\{[\cap_{i=1}^{n} X_i = x_i] \cap T = t\right\} / P(T = t) \\
= P\left\{[\cap_{i=1}^{n} X_i = x_i]\right\} / P(T = t).
\end{aligned}
$$

Hence, one gets

$$
\begin{aligned}
P\left\{[\cap_{i=1}^{n} X_i = x_i] \mid T = t\right\} \\
= \Pi_{i=1}^{n} \left\{e^{-\lambda} \lambda^{x_i} / x_i!\right\} / \left\{e^{-n\lambda} (n\lambda)^t / t!\right\}, \\
\text{since the } X\text{'s are independent} \\
= \left\{e^{-n\lambda} \lambda^{\Sigma_{i=1}^{n} x_i} / \Pi_{i=1}^{n} (x_i!)\right\} / \left\{e^{-n\lambda} (n\lambda)^t / t!\right\} \\
= \left\{t! / \Pi_{i=1}^{n} (x_i!)\right\} n^{-t},
\end{aligned}
\tag{6.2.2}
$$

which is free from λ. In other words, $\Sigma_{i=1}^{n} X_i$ is a sufficient statistic for the unknown parameter λ. ▲

Example 6.2.3 Suppose that X has the *Laplace* or *double exponential* pdf given by $f(x; \theta) = \frac{1}{2\theta} e^{-|x|/\theta} I(-\infty < x < \infty)$ where $\theta \, (> 0)$ is an unknown parameter. Let us consider the statistic $T = | X |$. The difference between X and T is that T provides the magnitude of X, but not its sign. Conditionally given $T = t \, (> 0)$, X can take one of the two possible values, namely t or $-t$, each with probability $\frac{1}{2}$. In other words, the conditional distribution of X given $T = t$ does not depend on the unknown parameter θ. Hence, $| X |$ is a sufficient statistic for θ. ▲

Example 6.2.4 Suppose that X_1, X_2 are iid $N(\theta, 1)$ where θ is unknown, $-\infty < \theta < \infty$. Here, $\mathcal{X} = \Re$ and $\Theta = \Re$. Let us consider the specific statistic $T = X_1 + X_2$. Its values are denoted by $t \in \mathcal{T} = \Re$. Observe that the conditional pdf of (X_1, X_2) at (x_1, x_2) would be zero if $x_1 + x_2 \neq t$ when $T = t$ has been observed. So we may work with data points (x_1, x_2) such that $x_1 + x_2 = t$ once $T = t$ is observed. Given $T = t$, only one of the two X's is a free-standing variable and so we will have a valid pdf of one of the X's. Let us verify that T is sufficient for θ by showing that the conditional distribution of X_1 given $T = t$ does not involve θ. Now following the Definition 4.6.1 of multivariate normality and the Example 4.6.1, we can claim that the joint distribution of (X_1, T) is $N_2(\theta, 2\theta, 1, 2, 1/\sqrt{2})$, and hence the conditional distribution of X_1 given $T = t$ is normal with its mean $= \theta + \frac{1}{\sqrt{2}} \frac{1}{\sqrt{2}}(t - 2\theta) = \frac{1}{2}t$ and conditional variance $= 1 - (1/\sqrt{2})^2 = \frac{1}{2}$, for all $t \in \Re$. Refer to the Theorem 3.6.1 for the expressions of the conditional mean and variance. This conditional distribution is clearly free from θ. In other words, T is a sufficient statistic for θ. ▲

How can we show that a statistic is *not sufficient* for θ?
Discussions follow.

If T is not sufficient for θ, then it follows from the Definition 6.2.2 that the conditional pmf or pdf of $X_1, ..., X_n$ given $T = t$ must depend on the unknown parameter θ, for some possible $x_1, ..., x_n$ and t.

In a discrete case, suppose that for some chosen data $x_1, ..., x_n$, the conditional probability $P\{X_1 = x_1, ..., X_n = x_n \mid T = t\}$, involves the parameter θ. Then, T can not be sufficient for θ.
Look at Examples 6.2.5 and 6.2.7.

Example 6.2.5 (Example 6.2.1 Continued) Suppose that X_1, X_2, X_3 are iid Bernoulli(p) where p is unknown, $0 < p < 1$. Here, $\mathcal{X} = \{0, 1\}$, $\theta = p$, and $\Theta = (0, 1)$. We had verified that the statistic $T = \Sigma_{i=1}^3 X_i$ was sufficient for θ. Let us consider another statistic $U = X_1 X_2 + X_3$. The question is whether U is a sufficient statistic for p. Observe that

$$
\begin{aligned}
P(U = 0) \\
= P\{X_1 X_2 = 0 \cap X_3 = 0\} \\
= P\{[X_1 = 0 \cap X_2 = 0 \cap X_3 = 0] \\
\cup [X_1 = 0 \cap X_2 = 1 \cap X_3 = 0] \\
\cup [X_1 = 1 \cap X_2 = 0 \cap X_3 = 0]\} \\
= (1 - p)^3 + 2p(1 - p)^2 \\
= (1 - p)^2 (1 + p).
\end{aligned}
\tag{6.2.3}
$$

Now, since $\{X_1 = 1 \cap X_2 = 0 \cap X_3 = 0\}$ is a subset of $\{U = 0\}$, we have

$$
\begin{aligned}
P\{[X_1 = 1 \cap X_2 = 0 \cap X_3 = 0] \mid U = 0\} \\
= P\{X_1 = 1 \cap X_2 = 0 \cap X_3 = 0\} / P(U = 0) \\
= p(1 - p)^2 / \{(1 - p)^2 (1 + p)\} = p/(1 + p).
\end{aligned}
$$

This conditional probability depends on the true value of p and so we claim that the statistic U is *not* sufficient for p. That is, after the completion of the n trials of the Bernoulli experiment, if one is merely told the observed value of the statistic U, then some information about the unknown parameter p would be lost. ▲

In the continuous case, we work with the same basic idea. If for some data $x_1, ..., x_n$, the conditional pdf given $T = t$, $f_{\mathbf{X} | T = t}(x_1, ..., x_n)$, involves the parameter θ, then the statistic T can not be sufficient for θ. Look at the Example 6.2.6.

Example 6.2.6 (Example 6.2.4 Continued) Suppose that X_1, X_2 are iid $N(\theta, 1)$ where θ is unknown, $-\infty < \theta < \infty$. Here, $\mathcal{X} = \Re$ and $\Theta = \Re$. Let us consider a statistic, for example, $T = X_1 + 2X_2$ while its values are denoted by $t \in \mathcal{T} = \Re$. Let us verify that T is *not* sufficient for θ by showing that the conditional distribution of X_1 given $T = t$ involves θ. Now, following the Definition 4.6.1 and the Example 4.6.1, we can claim that the joint distribution of (X_1, T) is $N_2(\theta, 3\theta, 1, 5, 1/\sqrt{5})$, and hence the conditional distribution of X_1 given $T = t$ is normal with its mean $= \theta + \frac{1}{\sqrt{5}}$ $\frac{1}{\sqrt{5}}(t - 3\theta) = \frac{1}{5}(t + 2\theta)$ and variance $= 1 - (1/\sqrt{5})^2 = \frac{4}{5}$, for $t \in \Re$. Refer to the Theorem 3.6.1 as needed. Since this conditional distribution depends on the unknown parameter θ, we conclude that T is *not* sufficient for θ. That is, merely knowing the value of T after the experiment, some information about the unknown parameter θ would be lost. ▲

Example 6.2.7 Suppose that X has the exponential pdf given by $f(x) = \lambda e^{-\lambda x} I(x > 0)$ where $\lambda(> 0)$ is the unknown parameter. Instead of the original data, suppose that we are only told whether $X \leq 2$ or $X > 2$, that is we merely observe the value of the statistic $T \equiv I(X > 2)$. Is the statistic T sufficient for λ? In order to check, let us proceed as follows: Note that we can express $P\{X > 3 \mid T = 1\}$ as

$$P\{X > 3 \cap X > 2\}/P\{X > 2\} = P\{X > 3\}/P\{X > 2\} = e^{-\lambda},$$

which depends on λ and hence T is *not* a sufficient statistic for λ. ▲

The methods we pursued in the Examples 6.2.1-6.2.7 closely followed the definition of a sufficient statistic. But, such direct approaches to verify the sufficiency or non-sufficiency of a statistic may become quite cumbersome. More importantly, in the cited examples we had started with specific statistics which we could eventually prove to be either sufficient or non-sufficient by evaluating appropriate conditional probabilities. But, what is one supposed to do in situations where a suitable candidate for a sufficient statistic can not be guessed readily? A more versatile technique follows.

6.2.2   The Neyman Factorization Theorem

Suppose that we have at our disposal, observable real valued iid random variables $X_1, ..., X_n$ from a population with the common pmf or pdf $f(x; \theta)$. Here, the unknown parameter is θ which belongs to the parameter space Θ.

Definition 6.2.4 *Consider the (observable) real valued iid random variables $X_1, ..., X_n$ from a population with the common pmf or pdf $f(x; \theta)$, where the unknown parameter $\theta \in \Theta$. Once we have observed $X_i = x_i$, $i = 1, ..., n$, the likelihood function is given by*

$$L(\theta) = \Pi_{i=1}^{n} f(x_i; \theta), \theta \in \Theta. \tag{6.2.4}$$

In the discrete case, $L(\theta)$ stands for $P_\theta\{X_1 = x_1 \cap ... \cap X_n = x_n\}$, that is the probability of the data on hand when θ obtains. In the continuous case, $L(\theta)$ stands for the joint pdf at the observed data point $(x_1, ..., x_n)$ when θ obtains.

It is not essential however for the X's to be real valued or that they be iid. But, in many examples, they will be so. If the X's happen to be vector valued or if they are not iid, then the corresponding joint pmf or pdf of

$X_i = x_i$, $i = 1, ..., n$, would stand for the corresponding likelihood function $L(\theta)$. We will give several examples of $L(\theta)$ shortly.

One should note that once the data $\{x_i; i = 1, ..., n\}$ has been observed, there are no random quantities in (6.2.4), and so the likelihood $L(.)$ is simply treated as a function of the unknown parameter θ alone.

> The sample size n is assumed known and fixed before the data collection begins.

One should note that θ can be real or vector valued in this general discussion, however, let us pretend for the time being that θ is a real valued parameter. Fisher (1922) discovered the fundamental idea of factorization. Neyman (1935a) rediscovered a refined approach to factorize the likelihood function in order to find sufficient statistics for θ. Halmos and Savage (1949) and Bahadur (1954) gave more involved measure-theoretic treatments.

Theorem 6.2.1 (Neyman Factorization Theorem) *Consider the likelihood function $L(\theta)$ from (6.2.4). A real valued statistic $T = T(X_1, ..., X_n)$ is sufficient for the unknown parameter θ if and only if the following factorization holds:*

$$L(\theta) = g\left(T(x_1, ..., x_n); \theta\right) h(x_1, ..., x_n), \text{for all } x_1, ..., x_n \in \mathcal{X}, \quad (6.2.5)$$

where the two functions $g(.; \theta)$ and $h(.)$ are both nonnegative, $h(x_1, ..., x_n)$ is free from θ, and $g\left(T(x_1, ..., x_n); \theta\right)$ depends on $x_1, ..., x_n$ only through the observed value $T(x_1, ..., x_n)$ of the statistic T.

Proof For simplicity, we will provide a proof only in the *discrete* case. Let us write $\mathbf{X} = (X_1, ..., X_n)$ and $\mathbf{x} = (x_1, ..., x_n)$. Let the two sets A and B respectively denote the events $\mathbf{X} = \mathbf{x}$ and $T(\mathbf{X}) = T(\mathbf{x})$, and observe that $A \subseteq B$.

Only if part: Suppose that T is sufficient for θ. Now, we write

$$\begin{aligned} L(\theta) \quad &= P_\theta\{\mathbf{X} = \mathbf{x}\} \\ &= P_\theta\{\mathbf{X} = \mathbf{x} \cap T(\mathbf{X}) = T(\mathbf{x})\}, \text{ since } A \subseteq B \quad (6.2.6) \\ &= P_\theta\{T(\mathbf{X}) = T(\mathbf{x})\}P_\theta\{\mathbf{X} = \mathbf{x} \,|T(\mathbf{X}) = T(\mathbf{x})\}. \end{aligned}$$

Comparing (6.2.5)-(6.2.6), let us denote $g\left(T(x_1, ..., x_n); \theta\right) = P_\theta\{T(\mathbf{X}) = T(\mathbf{x})\}$ and $h(x_1, ..., x_n) = P_\theta\{\mathbf{X} = \mathbf{x} \,|T(\mathbf{X}) = T(\mathbf{x})\}$. But, we have assumed that T is sufficient for θ and hence by the Definition 6.2.2 of sufficiency, the conditional probability $P_\theta\{\mathbf{X} = \mathbf{x} \,|T(\mathbf{X}) = T(\mathbf{x})\}$ *cannot* depend on the parameter θ. Thus, the function $h(x_1, ..., x_n)$ so defined may depend only on $x_1, ..., x_n$. The factorization given in (6.2.5) thus holds. The "only if" part is now complete. ♦

If part: Suppose that the factorization in (6.2.5) holds. Let us denote the pmf of T by $p(t; \theta)$. Observe that the pmf of T is given by $p(t; \theta) = P_\theta\{T(\mathbf{X}) = t\} = \sum_{\mathbf{y}: T(\mathbf{y})=t} \{\Pi_{i=1}^n f(y_i; \theta)\} = \sum_{\mathbf{y}: T(\mathbf{y})=t} L(\theta)$. It is easy to see that

$$P_\theta\{\mathbf{X} = \mathbf{x} \mid T(\mathbf{X}) = t\} = 0 \quad \text{if } T(\mathbf{x}) \neq t. \tag{6.2.7}$$

For all $\mathbf{x} \in \mathcal{X}$ such that $T(\mathbf{x}) = t$ and $p(t; \theta) \neq 0$, we can express $P_\theta\{\mathbf{X} = \mathbf{x} \mid T(\mathbf{X}) = t\}$ as

$$
\begin{aligned}
L(\theta)/p(t; \theta) &= g\left(T(\mathbf{x}); \theta\right) h(\mathbf{x})/p(t; \theta) \\
&= g\left(t; \theta\right) h(\mathbf{x}) / \left\{ \sum_{\mathbf{y}: T(\mathbf{y})=t} L(\theta) \right\} \\
&= g\left(t; \theta\right) h(\mathbf{x}) / \left\{ \sum_{\mathbf{y}: T(\mathbf{y})=t} g(T(\mathbf{y}); \theta) h(\mathbf{y}) \right\},
\end{aligned}
$$

because of factorization in (6.2.5). Hence, one gets

$$
\begin{aligned}
P_\theta\{\mathbf{X} = \mathbf{x} \mid T(\mathbf{X}) = t\} &= g\left(t; \theta\right) h(\mathbf{x}) / \{g(t; \theta)\} \left\{ \sum_{\mathbf{y}: T(\mathbf{y})=t} h(\mathbf{y}) \right\} \\
&= h(\mathbf{x}) / \left\{ \sum_{\mathbf{y}: T(\mathbf{y})=t} h(\mathbf{y}) \right\} = q(\mathbf{x}), \quad \text{say,}
\end{aligned}
$$
$$\tag{6.2.8}$$

where $q(\mathbf{x})$ does not depend upon θ. Combining (6.2.7)-(6.2.8), the proof of the "if part" is now complete. ∎

In the statement of the Theorem 6.2.1, notice that we *do not* demand that $g\left(T(x_1, ..., x_n); \theta\right)$ must be the pmf or the pdf of $T(X_1, ..., X_n)$. It is essential, however, that the function $h(x_1, ..., x_n)$ must be entirely free from θ.

It should be noted that the splitting of $L(\theta)$ may not be unique, that is there may be more than one way to determine the function $h(.)$ so that (6.2.5) holds. Also, there can be different versions of the sufficient statistics.

Remark 6.2.1 We mentioned earlier that in the Theorem 6.2.1, it was not essential that the random variables $X_1, ..., X_n$ and the unknown parameter θ be all real valued. Suppose that $\mathbf{X}_1, ..., \mathbf{X}_n$ are iid p-dimensional random variables with the common pmf or pdf $f(\mathbf{x}; \boldsymbol{\theta})$ where the unknown parameter $\boldsymbol{\theta}$ is vector valued, $\boldsymbol{\theta} \in \Theta \subseteq \Re^q$. The Neyman Factorization Theorem can be stated under this generality. Let us consider the likelihood function, $L(\boldsymbol{\theta}) = \Pi_{i=1}^n f(\mathbf{x}_i; \boldsymbol{\theta})$.

A vector valued statistic $\mathbf{T} = (T_1, ..., T_k)$ is *jointly sufficient* for $\boldsymbol{\theta} = (\theta_1, ..., \theta_q)$ if and only if $L(\boldsymbol{\theta}) = g(\mathbf{T}; \boldsymbol{\theta})h(\mathbf{x}_1, ..., \mathbf{x}_n)$ for all $\mathbf{x}_1, ..., \mathbf{x}_n \in \mathcal{X} \subseteq \Re^p$, with both $g(.; \boldsymbol{\theta}), h(.)$ non-negative, $g(.; \boldsymbol{\theta})$ depending upon $\boldsymbol{\theta}$ only through $\mathbf{T}$, but $h(.)$ is free from $\boldsymbol{\theta}$.

$$(6.2.9)$$

Example 6.2.8 (Example 6.2.1 Continued) Suppose that $X_1, ..., X_n$ are iid Bernoulli(p) where p is unknown, $0 < p < 1$. Here, $\mathcal{X} = \{0, 1\}$, $\theta = p$, and $\Theta = (0, 1)$. Then,

$$L(p) = \Pi_{i=1}^n p^{x_i}(1-p)^{1-x_i} = p^{\Sigma_{i=1}^n x_i}(1-p)^{n-\Sigma_{i=1}^n x_i}, \qquad (6.2.10)$$

which looks like the factorization provided in (6.2.5) where $g\left(\Sigma_{i=1}^n x_i; p\right) = p^{\Sigma_{i=1}^n x_i}(1-p)^{n-\Sigma_{i=1}^n x_i}$ and $h(x_1, ..., x_n) = 1$ for all $x_1, ..., x_n \in \{0, 1\}$. Hence, the statistic $T = T(X_1, ..., X_n) = \Sigma_{i=1}^n X_i$ is sufficient for p. From (6.2.10), we could instead view $L(\theta) = g(x_1, ..., x_n; p)h(x_1, ..., x_n)$ with, say, $g(x_1, ..., x_n; p) = \Pi_{i=1}^n p^{x_i}(1-p)^{n-x_i}$ and $h(x_1, ..., x_n) = 1$. That is, one could claim that $\mathbf{X} = (X_1, ..., X_n)$ was sufficient too for p. But, $\Sigma_{i=1}^n X_i$ provides a significantly reduced summary compared with $\mathbf{X}$, the whole data. We will have more to say on this in the Section 6.3. ▲

Example 6.2.9 (Example 6.2.2 Continued) Suppose that $X_1, ..., X_n$ are iid Poisson(λ) where λ is unknown, $0 < \lambda < \infty$. Here, $\mathcal{X} = \{0, 1, 2, ...\}$, $\theta = \lambda$, and $\Theta = (0, \infty)$. Then,

$$L(\lambda) = \Pi_{i=1}^n \left\{ e^{-\lambda}\lambda^{x_i}/x_i! \right\} = e^{-n\lambda}\lambda^{\Sigma_{i=1}^n x_i} \left(\Pi_{i=1}^n x_i!\right)^{-1}, \qquad (6.2.11)$$

which looks like the factorization provided in (6.2.5) with $g\left(\Sigma_{i=1}^n x_i; \lambda\right) = e^{-n\lambda}\lambda^{\Sigma_{i=1}^n x_i}$ and $h(x_1, ..., x_n) = \left(\Pi_{i=1}^n x_i!\right)^{-1}$ for all $x_1, ..., x_n \in \{0, 1, 2, ...\}$. Hence, the statistic $T = T(X_1, ..., X_n) = \Sigma_{i=1}^n X_i$ is sufficient for λ. Again, from (6.2.11) one can say that the whole data $\mathbf{X}$ is sufficient too, but $\Sigma_{i=1}^n X_i$ provides a significantly reduced summary compared with $\mathbf{X}$. ▲

Example 6.2.10 Suppose that $X_1, ..., X_n$ are iid $N(\mu, \sigma^2)$ where μ and σ are both assumed unknown, $-\infty < \mu < \infty$, $0 < \sigma < \infty$. Here, we may denote $\boldsymbol{\theta} = (\mu, \sigma)$ so that $\mathcal{X} = \Re$ and $\Theta = \Re \times \Re^+$. We wish to find *jointly sufficient* statistics for $\boldsymbol{\theta}$. Now, we have

$$L(\boldsymbol{\theta}) = \{\sigma\sqrt{2\pi}\}^{-n}exp\left\{-\tfrac{1}{2}\left(\Sigma_{i=1}^n x_i^2 - 2\mu\Sigma_{i=1}^n x_i + n\mu^2\right)/\sigma^2\right\}, \quad (6.2.12)$$

which looks like the factorization provided in (6.2.9) where one writes $g(\Sigma_{i=1}^n x_i, \Sigma_{i=1}^n x_i^2; \boldsymbol{\theta}) = \sigma^{-n}exp\left\{-\tfrac{1}{2}\left(\Sigma_{i=1}^n x_i^2 - 2\mu\Sigma_{i=1}^n x_i + n\mu^2\right)/\sigma^2\right\}$ and

$h(x_1, ..., x_n) = \{\sqrt{2\pi}\}^{-n}$ for all $(x_1, ..., x_n) \in \Re^n$. In other words, $\mathbf{T} = \mathbf{T}(X_1, ..., X_n) = \left(\Sigma_{i=1}^n X_i, \Sigma_{i=1}^n X_i^2\right)$ is *jointly* sufficient for (μ, σ^2). ▲

> If $\mathbf{T}$ is a sufficient statistic for θ, then any statistic $\mathbf{T}'$ which
> is a *one-to-one* function of $\mathbf{T}$ is also sufficient for θ.

Example 6.2.11 (Example 6.2.10 Continued) We have $\overline{X} = n^{-1}\Sigma_{i=1}^n X_i$, $S^2 = (n-1)^{-1}\{\Sigma_{i=1}^n X_i^2 - n^{-1}(\Sigma_{i=1}^n X_i)^2\}$, and so it is clear that the transformation from $\mathbf{T} = \left(\Sigma_{i=1}^n X_i, \Sigma_{i=1}^n X_i^2\right)$ to $\mathbf{T}' = \left(\overline{X}, S^2\right)$ is one-to-one. Hence, in the Example 6.2.10, we can also claim that $(\overline{X}, S^2)$ is jointly sufficient for $\theta = (\mu, \sigma^2)$. ▲

> Let us emphasize another point. Let $\mathbf{T}$ be a sufficient statistic for
> θ. Consider another statistic $\mathbf{T}'$, an arbitrary function of $\mathbf{T}$. Then,
> the statistic $\mathbf{T}'$ itself is not necessarily sufficient for θ.
> Look at the earlier Example 6.2.7.

An arbitrary function of a sufficient statistic $\mathbf{T}$ need not be sufficient for θ. Suppose that X is distributed as $N(\theta, 1)$ where $-\infty < \theta < \infty$ is the unknown parameter. Obviously, $T = X$ is sufficient for θ. One should check that the statistic $T' = |X|$, a function of T, is *not* sufficient for θ.

Remark 6.2.2 In a two-parameter situation, suppose that the Neyman factorization (6.2.9) leads to a statistic $\mathbf{T} = (T_1, T_2)$ which is *jointly* sufficient for $\theta = (\theta_1, \theta_2)$. But, the joint sufficiency of $\mathbf{T}$ should not be misunderstood to imply that T_1 is sufficient for θ_1 or T_2 is sufficient for θ_2.

> From the joint sufficiency of the statistic $\mathbf{T} = (T_1, ..., T_p)$ for
> $\theta = (\theta_1, ..., \theta_p)$, one should not be tempted to claim that the
> statistic T_i is componentwise sufficient for $\theta_i, i = 1, ..., p$.
> Look at the Example 6.2.12. In some cases, the statistic $\mathbf{T}$
> and θ may not even have the same number of components!

Example 6.2.12 (Example 6.2.11 Continued) In the $N(\mu, \sigma^2)$ case when both the parameters are unknown, recall from the Example 6.2.11 that $(\overline{X}, S^2)$ is jointly sufficient for (μ, σ^2). This is very different from trying to answer a question like this: Is $\overline{X}$ sufficient for μ or is S^2 sufficient for σ^2? We can legitimately talk only about the *joint* sufficiency of the statistic $(\overline{X}, S^2)$ for $\theta = (\mu, \sigma^2)$.

To appreciate this fine line, let us think through the example again and **pretend** for a moment that one could claim componentwise sufficiency. But, since $(\overline{X}, S^2)$ is jointly sufficient for (μ, σ^2), we can certainly claim that $(S^2, \overline{X})$ is also jointly sufficient for $\boldsymbol{\theta} = (\mu, \sigma^2)$. Now, how many readers would be willing to push forward the idea that componentwise, S^2 is sufficient for μ or $\overline{X}$ is sufficient for σ^2! Let us denote $U = \overline{X}$ and let $g(u; n)$ be the pdf of U when the sample size is n.

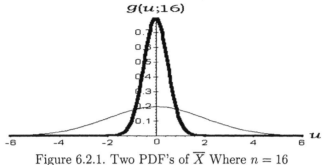

Figure 6.2.1. Two PDF's of $\overline{X}$ Where $n = 16$

In the Figure 6.2.1, the two pdf's of $\overline{X}$ when $\mu = 0$ and $n = 16$, for example, are certainly very different from one another. Relatively speaking, the darker pdf gives the impression that σ is "small" whereas the lighter pdf gives the impression that σ is "large". There should be no doubt that $\overline{X}$ provides *some information* about σ^2. In fact $\overline{X}$ has some information about *both* μ and σ^2, whereas S^2 has information about σ^2 alone. ▲

Example 6.2.13 Suppose that $X_1, ..., X_n$ are iid Uniform$(0, \theta)$, where $\theta \ (> 0)$ is unknown. Here, $\mathcal{X} = (0, \theta)$ and $\Theta = \Re^+$. We wish to find a sufficient statistic for θ. Now,

$$
\begin{aligned}
L(\theta) &= \Pi_{i=1}^n \{\theta^{-1} I(0 < x_i < \theta)\} \\
&= \theta^{-n} I(0 < x_{n:n} < \theta) I(0 < x_{n:1} < x_{n:n}),
\end{aligned}
\tag{6.2.13}
$$

where $x_{n:1}$ and $x_{n:n}$ are respectively the observed smallest and largest order statistics. The last step in (6.2.13) looks exactly like the Neyman factorization provided in (6.2.5) where $g(x_{n:n}; \theta) = \theta^{-n} I(0 < x_{n:n} < \theta)$ and $h(x_1, ..., x_n) = I(0 < x_{n:1} < x_{n:n})$ for all $x_1, ..., x_n \in (0, \theta)$. Hence, the statistic $T = T(X_1, ..., X_n) = X_{n:n}$ is sufficient for θ. ▲

> It is not crucial that the X's be iid for the Neyman factorization of the likelihood function to lead to a *(jointly) sufficient* statistic $(\mathbf{T})T$. Look at the Example 6.2.14.

In many examples and exercises the X's are often assumed iid. But, if the X's are not iid, then all we have to do is to carefully write down the likelihood function $L(\theta)$ as the corresponding joint pmf or pdf of the random variables $X_1, ..., X_n$. The Neyman factorization would then hold.

Example 6.2.14 Suppose that X_1, X_2 are independent random variables, with their respective pdf's $f(x_1; \theta) = \theta e^{-\theta x_1} I(0 < x_1 < \infty)$ and $g(x_2; \theta) = 2\theta e^{-2\theta x_2} I(0 < x_2 < \infty)$, where $\theta > 0$ is an unknown parameter. For $0 < x_1, x_2 < \infty$, the likelihood function is given by the joint pdf, namely

$$L(\theta) = f(x_1; \theta)g(x_2; \theta) = 2\theta^2 e^{-\theta(x_1 + 2x_2)}. \qquad (6.2.14)$$

From (6.2.14) it is clear that the Neyman factorization holds and hence the statistic $T = X_1 + 2X_2$ is sufficient for θ. Here, X_1, X_2 are not identically distributed, and yet the factorization theorem has been fruitful. ▲

The following result shows a simple way to find sufficient statistics when the pmf or the pdf belongs to the *exponential family*. Refer back to the Section 3.8 in this context. The proof follows easily from the factorization (6.2.9) and so we leave it out as the Exercise 6.2.15.

Theorem 6.2.2 (Sufficiency in the Exponential Family) *Suppose that $X_1, ..., X_n$ are iid with the common pmf or the pdf belonging to the k-parameter exponential family defined by (3.8.4), namely*

$$f(x; \boldsymbol{\theta}) = a(\boldsymbol{\theta})g(x)exp\{\Sigma_{i=1}^{k} b_i(\boldsymbol{\theta})R_i(x)\}$$

with appropriate forms for $g(x) \geq 0$, $a(\boldsymbol{\theta}) \geq 0$, $b_i(\boldsymbol{\theta})$ and $R_i(x)$, $i = 1, ..., k$. Suppose that the regulatory conditions stated in (3.8.5) hold. Denote the statistic $T_j = \Sigma_{i=1}^{n} R_j(X_i), j = 1, ..., k$. Then, the statistic $\mathbf{T} = (T_1, ..., T_k)$ is jointly sufficient for $\boldsymbol{\theta}$.

The sufficient statistics derived earlier in the Examples 6.2.8-6.2.11 can also be found directly by using the Theorem 6.2.2. We leave these as the Exercise 6.2.14.

6.3 Minimal Sufficiency

We noted earlier that the whole data $\mathbf{X}$ must always be sufficient for the unknown parameter $\boldsymbol{\theta}$. But, we aim at reducing the data by means of summary statistics in lieu of considering $\mathbf{X}$ itself. From the series of examples 6.2.8-6.2.14, we found that the Neyman factorization provided sufficient statistics which were substantially "reduced" compared with **X**. As a principle, we should use the "shortest sufficient" summary in lieu of handling the original data. Two pertinent questions may arise: What is a natural way to define the "shortest sufficient" summary statistic? The next question is how to get hold of such a "shortest sufficient" summary, if indeed there is one?

Lehmann and Scheffé (1950) developed a precise mathematical formulation of the concept known as *minimal sufficiency* and they gave a technique that helps to locate minimal sufficient statistics. Lehmann and Scheffé (1955, 1956) included important followups.

Definition 6.3.1 *A statistic* **T** *is called minimal sufficient for the unknown parameter* θ *or simply minimal sufficient if and only if*

(i)   **T** *is sufficient for* θ, *and*

(ii)   **T** *is minimal or "shortest" in the sense that* **T** *is a function of any other sufficient statistic.*

Let us think about this concept for a moment. We do want to summarize the whole data **X** by reducing it to some appropriate statistic such as $\overline{X}$, the median (M), or the histogram, and so on. Suppose that in a particular situation, the summary statistic $\mathbf{T} = (\overline{X}, M)$ turns out to be *minimal sufficient* for θ. Can we reduce this summary any further? Of course, we can. We may simply look at, for example, $T_1 = \overline{X}$ or $T_2 = M$ or $T_3 = \frac{1}{2}(\overline{X} + M)$. Can T_1, T_2, or T_3 individually be sufficient for θ? The answer is no, none of these could be sufficient for θ. Because if, for example T_1 by itself was sufficient for θ, then $\mathbf{T} = (\overline{X}, M)$ would have to be a function of T_1 in view of the requirement in part (ii) of the Definition 6.3.1. But, $\mathbf{T} = (\overline{X}, M)$ can not be a function of T_1 because we can not uniquely specify the value of **T** from our knowledge of the value of T_1 alone. A minimal sufficient summary **T** can not be reduced any further to another sufficient summary statistic. In this sense, a minimal sufficient statistic **T** may be looked upon as the best sufficient statistic.

In the Definition 6.3.1, the part (i) is often verified via Neyman factorization, but the verification of part (ii) gets more involved. In the next subsection, we state a theorem due to Lehmann and Scheffé (1950) which provides a direct approach to find minimal sufficient statistics for θ.

6.3.1   The Lehmann-Scheffé Approach

The following theorem was proved in Lehmann and Scheffé (1950). This result is an essential tool to locate a minimal sufficient statistic when it exists. Its proof, however, requires some understanding of the correspondence between a statistic and so called *partitions* it induces on the sample space.

Let us look at the original data $\mathbf{X} = (X_1, ..., X_n)$ where $\mathbf{x} = (x_1, ..., x_n) \in \mathcal{X}^n$. Consider a statistic $\mathbf{T} \equiv \mathbf{T}(X_1, ..., X_n)$, that is $\mathbf{T}$ is a mapping from $\mathcal{X}^n$ onto some space $\mathcal{T}$, say. For $\mathbf{t} \in \mathcal{T}$, let $\mathcal{X}_{\mathbf{t}} = \{\mathbf{x} : \mathbf{x} \in \mathcal{X}^n$ such that $\mathbf{T}(\mathbf{x}) = \mathbf{t}\}$ which are disjoint subsets of $\mathcal{X}^n$ and also $\mathcal{X}^n = \cup_{\mathbf{t} \in \mathcal{T}} \mathcal{X}_{\mathbf{t}}$. In other words, the collection of subsets $\{\mathcal{X}_{\mathbf{t}} : \mathbf{t} \in \mathcal{T}\}$ forms a *partition* of the space $\mathcal{X}^n$. Often, $\{\mathcal{X}_{\mathbf{t}} : \mathbf{t} \in \mathcal{T}\}$ is also called the partition of $\mathcal{X}^n$ *induced* by the statistic $\mathbf{T}$.

Theorem 6.3.1 (Minimal Sufficient Statistics) *Let us consider the function* $h(\mathbf{x}, \mathbf{y}; \boldsymbol{\theta}) = \Pi_{i=1}^n f(x_i; \boldsymbol{\theta}) / \Pi_{i=1}^n f(y_i; \boldsymbol{\theta})$, *the ratio of the likelihood functions from (6.2.9) at* $\mathbf{x}$ *and* $\mathbf{y}$, *where* $\boldsymbol{\theta}$ *is the unknown parameter and* $\mathbf{x}, \mathbf{y} \in \mathcal{X}^n$. *Suppose that we have a statistic* $\mathbf{T} \equiv \mathbf{T}(X_1, ..., X_n) = (T_1, ..., T_k)$ *such that the following conditions hold:*

> *Having any two arbitrary but otherwise fixed data points* $\mathbf{x} = (x_1, ..., x_n)$, $\mathbf{y} = (y_1, ..., y_n)$ *from*
> $\mathcal{X}^n$, *the expression* $h(\mathbf{x}, \mathbf{y}; \boldsymbol{\theta})$ *does not involve the*   (6.3.1)
> *unknown parameter* $\boldsymbol{\theta}$ *if and only if* $\mathbf{T}(\mathbf{x}) = \mathbf{T}(\mathbf{y})$,
> *that is* $T_i(\mathbf{x}) = T_i(\mathbf{y}), i = 1, ..., k$.

Then, the statistic $\mathbf{T}$ *is minimal sufficient for the parameter* $\boldsymbol{\theta}$.

Proof We first show that $\mathbf{T}$ is a sufficient statistic for $\boldsymbol{\theta}$ and then we verify that $\mathbf{T}$ is also minimal. For simplicity, let us assume that $f(\mathbf{x}; \boldsymbol{\theta})$ is positive for all $\mathbf{x} \in \mathcal{X}^n$ and $\boldsymbol{\theta}$.

Sufficiency part: Start with $\{\mathcal{X}_{\mathbf{t}} : \mathbf{t} \in \mathcal{T}\}$ which is the partition of $\mathcal{X}^n$ induced by the statistic $\mathbf{T}$. In the subset $\mathcal{X}_{\mathbf{t}}$, let us select and fix an element $\mathbf{x}_{\mathbf{t}}$. If we look at an arbitrary element $\mathbf{x} \in \mathcal{X}^n$, then this element $\mathbf{x}$ belongs to $\mathcal{X}_{\mathbf{t}}$ for some unique $\mathbf{t}$ so that both $\mathbf{x}$ and $\mathbf{x}_{\mathbf{t}}$ belong to the same set $\mathcal{X}_{\mathbf{t}}$. In other words, one has $\mathbf{T}(\mathbf{x}) = \mathbf{T}(\mathbf{x}_{\mathbf{t}})$. Thus, by invoking the "if part" of the statement in (6.3.1), we can claim that $h(\mathbf{x}, \mathbf{x}_{\mathbf{t}}; \boldsymbol{\theta})$ is free from $\boldsymbol{\theta}$. Let us then denote $h(\mathbf{x}) \equiv h(\mathbf{x}, \mathbf{x}_{\mathbf{t}}; \boldsymbol{\theta}), \mathbf{x} \in \mathcal{X}^n$. Hence, we write

$$\Pi_{i=1}^n f(x_i; \boldsymbol{\theta}) = \Pi_{i=1}^n f(x_{\mathbf{t}i}; \boldsymbol{\theta}) h(\mathbf{x}) = g(\mathbf{T}(\mathbf{x}); \boldsymbol{\theta}) h(\mathbf{x}).$$

where $\mathbf{x}_{\mathbf{t}} = (x_{\mathbf{t}1}, ..., x_{\mathbf{t}n})$. In view of the Neyman Factorization Theorem, the statistic $\mathbf{T}(\mathbf{x})$ is thus sufficient for $\boldsymbol{\theta}$. ◆

Minimal part: Suppose $\mathbf{U} = \mathbf{U}(\mathbf{X})$ is another sufficient statistic for $\boldsymbol{\theta}$. Then, by the Neyman Factorization Theorem, we can write

$$\Pi_{i=1}^n f(x_i; \boldsymbol{\theta}) = g_0(\mathbf{U}(\mathbf{x}); \boldsymbol{\theta}) h_0(\mathbf{x})$$

for some appropriate $g_0(.; \boldsymbol{\theta})$ and $h_0(.)$. Here, $h_0(.)$ does not depend upon $\boldsymbol{\theta}$. Now, for any two sample points $\mathbf{x} = (x_1, ..., x_n)$, $\mathbf{y} = (y_1, ..., y_n)$ from $\mathcal{X}^n$ such that $\mathbf{U}(\mathbf{x}) = \mathbf{U}(\mathbf{y})$, we obtain

$$h(\mathbf{x}, \mathbf{y}; \boldsymbol{\theta})$$
$$= \Pi_{i=1}^n f(x_i; \boldsymbol{\theta}) / \Pi_{i=1}^n f(y_i; \boldsymbol{\theta})$$
$$= \{g_0(\mathbf{U}(\mathbf{x}); \boldsymbol{\theta}) h_0(\mathbf{x})\} / \{g_0(\mathbf{U}(\mathbf{y}); \boldsymbol{\theta}) h_0(\mathbf{y})\}$$
$$= h_0(\mathbf{x}) / h_0(\mathbf{y}), \text{ since } g_0(\mathbf{U}(\mathbf{x}); \boldsymbol{\theta}) = g_0(\mathbf{U}(\mathbf{y}); \boldsymbol{\theta}).$$

Thus, $h(\mathbf{x}, \mathbf{y}; \boldsymbol{\theta})$ is free from $\boldsymbol{\theta}$. Now, by invoking the "only if" part of the statement in (6.3.1), we claim that $\mathbf{T}(\mathbf{x}) = \mathbf{T}(\mathbf{y})$. That is, $\mathbf{T}$ is a function of $\mathbf{U}$. Now, the proof is complete. ∎

Example 6.3.1 (Example 6.2.8 Continued) Suppose that $X_1, ..., X_n$ are iid Bernoulli(p), where p is unknown, $0 < p < 1$. Here, $\mathcal{X} = \{0, 1\}$, $\theta = p$, and $\Theta = (0, 1)$. Then, for two arbitrary data points $\mathbf{x} = (x_1, ..., x_n)$ and $\mathbf{y} = (y_1, ..., y_n)$, both from $\mathcal{X}$, we have:

$$\{\Pi_{i=1}^n f(x_i; \theta)\} / \{\Pi_{i=1}^n f(y_i; \theta)\}$$
$$= \Pi_{i=1}^n p^{x_i}(1-p)^{n-x_i} / \Pi_{i=1}^n p^{y_i}(1-p)^{n-y_i} \qquad (6.3.2)$$
$$= \left(p(1-p)^{-1}\right)^{\{\Sigma_{i=1}^n x_i - \Sigma_{i=1}^n y_i\}}.$$

From (6.3.2), it is clear that $\left(p(1-p)^{-1}\right)^{\{\Sigma_{i=1}^n x_i - \Sigma_{i=1}^n y_i\}}$ would become free from the unknown parameter p if and only if $\Sigma_{i=1}^n x_i - \Sigma_{i=1}^n y_i = 0$, that is, if and only if $\Sigma_{i=1}^n x_i = \Sigma_{i=1}^n y_i$. Hence, by the theorem of Lehmann-Scheffé, we claim that $T = \Sigma_{i=1}^n X_i$ is minimal sufficient for p. ▲

We had shown non-sufficiency of a statistic U in the Example 6.2.5. One can arrive at the same conclusion by contrasting U with a minimal sufficient statistic. Look at the Examples 6.3.2 and 6.3.5.

Example 6.3.2 (Example 6.2.5 Continued) Suppose that X_1, X_2, X_3 are iid Bernoulli(p), where p is unknown, $0 < p < 1$. Here, $\mathcal{X} = \{0, 1\}$, $\theta = p$, and $\Theta = (0, 1)$. We know that the statistic $T = \Sigma_{i=1}^3 X_i$ is minimal sufficient for p. Let $U = X_1 X_2 + X_3$, as in the Example 6.2.5, and the question is whether U is a sufficient statistic for p. Here, we prove again that the statistic U can not be sufficient for p. Assume that U is sufficient for p, and then $\Sigma_{i=1}^3 X_i$ must be a function of U, by the Definition 6.3.1 of minimal sufficiency. That is, knowing an observed value of U, we must be able to come up with a unique observed value of T. Now, the event $\{U = 0\}$

consists of the union of possibilities such as $\{X_1 = 0 \cap X_2 = 0 \cap X_3 = 0\}$, $\{X_1 = 0 \cap X_2 = 1 \cap X_3 = 0\}$, and $\{X_1 = 1 \cap X_2 = 0 \cap X_3 = 0\}$. Hence, if the event $\{U = 0\}$ is observed, we know then that either $T = 0$ or $T = 1$ must be observed. But, the point is that we cannot be sure about a unique observed value of T. Thus, T can not be a function of U and so there is a contradiction. Thus, U can not be sufficient for p. ▲

Example 6.3.3 (Example 6.2.10 Continued) Suppose that $X_1, ..., X_n$ are iid $N(\mu, \sigma^2)$, where $\boldsymbol{\theta} = (\mu, \sigma)$ and both μ, σ are unknown, $-\infty < \mu < \infty$, $0 < \sigma < \infty$. Here, $\mathcal{X} = \Re$ and $\Theta = \Re \times \Re^+$. We wish to find a minimal sufficient statistic for $\boldsymbol{\theta}$. Now, for two arbitrary data points $\mathbf{x} = (x_1, ..., x_n)$ and $\mathbf{y} = (y_1, ..., y_n)$, both from $\mathcal{X}$, we have:

$$
\begin{aligned}
&\{\Pi_{i=1}^n f(x_i; \boldsymbol{\theta})\}/\{\Pi_{i=1}^n f(y_i; \boldsymbol{\theta})\} \\
&= \Pi_{i=1}^n \left[\{\sigma\sqrt{2\pi}\}^{-1} exp\left\{-\tfrac{1}{2}(x_i - \mu)^2/\sigma^2\right\}\right] \times \\
&\qquad \left[\{\sigma\sqrt{2\pi}\}^{-1} exp\left\{-\tfrac{1}{2}(y_i - \mu)^2/\sigma^2\right\}\right]^{-1} \qquad (6.3.3) \\
&= exp\left\{\mu\left(\Sigma_{i=1}^n x_i - \Sigma_{i=1}^n y_i\right)\right\}/\sigma^2 \times \\
&\qquad exp\left\{-\tfrac{1}{2}\left(\Sigma_{i=1}^n x_i^2 - \Sigma_{i=1}^n y_i^2\right)/\sigma^2\right\}.
\end{aligned}
$$

From (6.3.3), it becomes clear that the last expression would not involve the unknown parameter $\boldsymbol{\theta} = (\mu, \sigma)$ if and only if $\Sigma_{i=1}^n x_i - \Sigma_{i=1}^n y_i = 0$ as well as $\Sigma_{i=1}^n x_i^2 - \Sigma_{i=1}^n y_i^2 = 0$, that is, if and only if $\Sigma_{i=1}^n x_i = \Sigma_{i=1}^n y_i$ and $\Sigma_{i=1}^n x_i^2 = \Sigma_{i=1}^n y_i^2$. Hence, by the theorem of Lehmann-Scheffé, we claim that $\mathbf{T} = (\Sigma_{i=1}^n X_i, \Sigma_{i=1}^n X_i^2)$ is minimal sufficient for (μ, σ). ▲

Theorem 6.3.2 *Any statistic which is a one-to-one function of a minimal sufficient statistic is itself minimal sufficient.*

Proof Suppose that a statistic $\mathbf{S}$ is minimal sufficient for $\boldsymbol{\theta}$. Let us consider another statistic $\mathbf{T} = \mathbf{h}(\mathbf{S})$ where $\mathbf{h}(.)$ is one-to-one. In Section 6.2.2, we mentioned that a one-to-one function of a (jointly) sufficient statistic is (jointly) sufficient and so $\mathbf{T}$ is sufficient. Let $\mathbf{U}$ be any other sufficient statistic for $\boldsymbol{\theta}$. Since, $\mathbf{S}$ is minimal sufficient, we must have $\mathbf{S} = \mathbf{g}(\mathbf{U})$ for some $\mathbf{g}(.)$. Then, we obviously have $\mathbf{T} = \mathbf{h}(\mathbf{S}) = \mathbf{h}(\mathbf{g}(\mathbf{U})) = \mathbf{h} \circ \mathbf{g}(\mathbf{U})$ which verifies the minimality of the statistic $\mathbf{T}$. ∎

Example 6.3.4 (Example 6.3.3 Continued) Suppose that $X_1, ..., X_n$ are iid $N(\mu, \sigma^2)$, where $\boldsymbol{\theta} = (\mu, \sigma)$ and both μ, σ are unknown, $-\infty < \mu < \infty$, $0 < \sigma < \infty$. We know that $\mathbf{T} = (\Sigma_{i=1}^n X_i, \Sigma_{i=1}^n X_i^2)$ is minimal sufficient for (μ, σ). Now, $(\overline{X}, S^2)$ being a one-to-one function of $\mathbf{T}$, we can claim that $(\overline{X}, S^2)$ is also minimal sufficient. ▲

Example 6.3.5 (Example 6.3.3 Continued) Suppose that X_1, X_2, X_3 are iid $N(\mu, \sigma^2)$ where μ is unknown, but σ is assumed known, $-\infty < \mu < \infty$,

$0 < \sigma < \infty$. Here, we write $\theta = \mu$, $\mathcal{X} = \Re$ and $\Theta = \Re$. It is easy to verify that the statistic $T = \Sigma_{i=1}^{3} X_i$ is minimal sufficient for θ. Now consider the statistic $U = X_1 X_2 + X_3$ and suppose that the question is whether U is sufficient for θ. Assume that U is sufficient for θ. But, then $\Sigma_{i=1}^{3} X_i$ must be a function of U, by the Definition 6.3.1 of minimal sufficiency. That is, knowing an observed value of U, we must be able to come up with a unique value of T. One can proceed in the spirit of the earlier Example 6.3.2 and easily arrive at a contradiction. So, U cannot be sufficient for θ. ▲

Example 6.3.6 (Example 6.2.13 Continued) Suppose that $X_1, ..., X_n$ are iid Uniform$(0, \theta)$, where θ (> 0) is unknown. Here, $\mathcal{X} = (0, \theta)$ and $\Theta = \Re^+$. We wish to find a minimal sufficient statistic for θ. For two arbitrary data points $\mathbf{x} = (x_1, ..., x_n)$ and $\mathbf{y} = (y_1, ..., y_n)$, both from $\mathcal{X}$, we have:

$$\{\Pi_{i=1}^{n} f(x_i; \theta)\} / \{\Pi_{i=1}^{n} f(y_i; \theta)\}$$
$$= \left\{\theta^{-n} I\left(0 < x_{n:n} < \theta\right) I(0 < x_{n:1} < x_{n:n})\right\} / \{\theta^{-n} \times \quad (6.3.4)$$
$$I(0 < y_{n:n} < \theta) I(0 < y_{n:1} < y_{n:n})\}.$$

Let us denote $a(\theta) = I\left(0 < x_{n:n} < \theta\right) / I(0 < y_{n:n} < \theta)$. Now, the question is this: Does the term $a(\theta)$ become free from θ if and only if $x_{n:n} = y_{n:n}$?

If we assume that $x_{n:n} = y_{n:n}$, then certainly $a(\theta)$ does not involve θ. It remains to show that when $a(\theta)$ does not involve θ, then we must have $x_{n:n} = y_{n:n}$. Let us assume that $x_{n:n} \neq y_{n:n}$, and then show that $a(\theta)$ must depend on the value of θ. Now, suppose that $x_{n:n} = 2, y_{n:n} = .5$. Then, $a(\theta) = 0, 1$ or $0/0$ when $\theta = 1, 3$ or $.1$ respectively. Clearly, $a(\theta)$ will depend upon the value of θ whenever $x_{n:n} \neq y_{n:n}$. We assert that the term $a(\theta)$ becomes free from θ if and only if $x_{n:n} = y_{n:n}$. Hence, by the theorem of Lehmann-Scheffé, we claim that $T = X_{n:n}$, the largest order statistic, is minimal sufficient for θ. ▲

Remark 6.3.1 Let $X_1, ..., X_n$ be iid with the pmf or pdf $f(x; \theta)$, $x \in \mathcal{X} \subseteq \Re$, $\theta = (\theta_1, ..., \theta_k) \in \Theta \subseteq \Re^k$. Suppose that a statistic $\mathbf{T} = \mathbf{T}(\mathbf{X}) = (T_1(\mathbf{X}), ..., T_r(\mathbf{X}))$ is minimal sufficient for $\boldsymbol{\theta}$. In general can we claim that $r = k$? The answer is no, we can not necessarily say that $r = k$. Suppose that $f(x; \theta)$ corresponds to the pdf of the $N(\theta, \theta)$ random variable with the unknown parameter $\theta > 0$ so that $k = 1$. The reader should verify that $\mathbf{T}(\mathbf{X}) = (\Sigma_{i=1}^{n} X_i, \Sigma_{i=1}^{n} X_i^2)$ is a minimal sufficient for θ so that $r = 2$. Here, we have $r > k$. On the other hand, suppose that X_1 is $N(\mu, \sigma^2)$ where μ and σ^2 are both unknown parameters. In this case one has $\boldsymbol{\theta} = (\mu, \sigma^2)$ so that $k = 2$. But, $T = X_1$ is minimal sufficient so that $r = 1$. Here, we have $r < k$. In many situations, of course, one would find that $r = k$. But, the point is that there is no guarantee that r would necessarily be same as k.

The following theorem provides a useful tool for finding minimal sufficient statistics within a rich class of statistical models, namely the *exponential family*. It is a hard result to prove. One may refer to Lehmann (1983, pp. 43-44) or Lehmann and Casella (1998) for some of the details.

Theorem 6.3.3 (Minimal Sufficiency in the Exponential Family) *Suppose that $X_1, ..., X_n$ are iid with the common pmf or the pdf belonging to the k-parameter exponential family defined by (3.8.4), namely*

$$f(x; \boldsymbol{\theta}) = a(\boldsymbol{\theta})g(x)exp\{\Sigma_{i=1}^{k}b_i(\boldsymbol{\theta})R_i(x)\} \qquad (6.3.5)$$

with some appropriate forms for $g(x) \geq 0$, $a(\boldsymbol{\theta}) \geq 0, b_i(\boldsymbol{\theta})$ and $R_i(x)$, $i = 1, ..., k$. Suppose that the regulatory conditions stated in (3.8.5) hold. Denote the statistic $T_j = \Sigma_{i=1}^{n}R_j(X_i), j = 1, ..., k$. Then, the statistic $\mathbf{T} = (T_1, ..., T_k)$ is (jointly) minimal sufficient for $\boldsymbol{\theta}$.

The following result provides the nature of the distribution itself of a minimal sufficient statistic when the common pmf or pdf comes from an exponential family. Its proof is beyond the scope of this book. One may refer to the Theorem 4.3 of Lehmann (1983) and Lemma 8 in Lehmann (1986). One may also review Barankin and Maitra (1963), Brown (1964), Hipp (1974), Barndorff-Nielsen (1978), and Lehmann and Casella (1998) to gain broader perspectives.

Theorem 6.3.4 (Distribution of a Minimal Sufficient Statistic in the Exponential Family) *Under the conditions of the Theorem 6.3.3, the pmf or the pdf of the minimal sufficient statistic $(T_1, ..., T_k)$ belongs to the k-parameter exponential family.*

> In each example, the data $\mathbf{X}$ was reduced enormously by the minimal sufficient summary. There are situations where no significant data reduction may be possible.
> See the Exercise 6.3.19 for specific examples.

6.4 Information

Earlier we have remarked that we wish to work with a sufficient or minimal sufficient statistic $\mathbf{T}$ because such a statistic will reduce the data *and* preserve all the "information" about $\boldsymbol{\theta}$ contained in the original data. Here, $\boldsymbol{\theta}$ may be real or vector valued. But, how much *information* do we have in the original data which we are trying to preserve? Now our major concern is to quantify the *information content* within some data. In order to keep the deliberations simple, we discuss the one-parameter and two-parameter situations separately. The notion of the information about an unknown parameter θ contained in the data was introduced by F. Y. Edgeworth in a series of papers, published in the *J. Roy. Statist. Soc.*, during 1908-1909. Fisher (1922) articulated the systematic development of this concept. The reader is referred to Efron's (1998, p.101) recent commentaries on (Fisher) information.

6.4.1 One-parameter Situation

Suppose that X is an observable real valued random variable with the pmf or pdf $f(x;\theta)$ where the unknown parameter $\theta \in \Theta$, an open subinterval of $\Re$, while the $\mathcal{X}$ space is *assumed* not to depend upon θ. We *assume* throughout that the partial derivative $\frac{\partial}{\partial\theta}f(x;\theta)$ is finite for all $x \in \mathcal{X}$, $\theta \in \Theta$. We also *assume* that we can interchange the derivative (with respect to θ) and the integral (with respect to x).

Definition 6.4.1 *The Fisher information or simply the information about θ, contained in the data, is given by*

$$\mathcal{I}_X(\theta) = E_\theta\left[\left\{\tfrac{\partial}{\partial\theta}logf(X;\theta)\right\}^2\right]. \qquad (6.4.1)$$

Example 6.4.1 Let X be Poisson(λ), $\lambda > 0$. Now,

$$logf(x;\lambda) = -\lambda + xlog(\lambda) - log(x!),$$

which implies that $\frac{\partial}{\partial\lambda}logf(x;\lambda) = -1 + x\lambda^{-1}$. Thus, we have

$$\mathcal{I}_X(\lambda) = E_\lambda\left[\left\{\tfrac{\partial}{\partial\lambda}logf(X;\lambda)\right\}^2\right] = E_\lambda\left[(X-\lambda)^2/\lambda^2\right] = \lambda^{-1}, \qquad (6.4.2)$$

since $E_\lambda[(X-\lambda)^2] = V(X) = \lambda$. That is, as we contemplate having larger and larger values of λ, the variability built in X increases, and hence it seems natural that the information about the unknown parameter λ contained in the data X will go down further and further. ▲

Example 6.4.2 Let X be $N(\mu,\sigma^2)$ where $\mu \in (-\infty,\infty)$ is the unknown parameter. Here, $\sigma \in (0,\infty)$ is assumed known. Now,

$$logf(x;\mu) = -\tfrac{1}{2}\left\{(x-\mu)^2/\sigma^2\right\} - log(\sigma\sqrt{2\pi}),$$

which implies that $\frac{\partial}{\partial\mu}logf(x;\mu) = (x-\mu)/\sigma^2$. Thus we have

$$\mathcal{I}_X(\mu) = E_\mu\left[\left\{\tfrac{\partial}{\partial\mu}logf(X;\mu)\right\}^2\right] = E_\mu\left[(X-\mu)^2/\sigma^4\right] = \sigma^{-2}, \qquad (6.4.3)$$

since $E_\mu[(X - \mu)^2] = V(X) = \sigma^2$. That is, as we contemplate having larger and larger values of σ, the variability built in X increases, and hence it seems natural that the information about the unknown parameter μ contained in the data X will go down further and further. ▲

The following result quantifies the information about the unknown parameter θ contained in a random sample $X_1, ..., X_n$ of size n.

Theorem 6.4.1 *Suppose that* $X_1, ..., X_n$ *are iid with the common pmf or pdf given by* $f(x; \theta)$. *We denote* $E_\theta\left[\left\{\frac{\partial}{\partial\theta}logf(X_1; \theta)\right\}^2\right] = \mathcal{I}_{X_1}(\theta)$, *the information contained in the observation* X_1. *Then, the information* $\mathcal{I}_{\mathbf{X}}(\theta)$, *contained in the random sample* $\mathbf{X} = (X_1, ..., X_n)$, *is given by*

$$\mathcal{I}_{\mathbf{X}}(\theta) = n\mathcal{I}_{X_1}(\theta) \text{ for all } \theta \in \Theta. \tag{6.4.4}$$

Proof Denote the observed data $\mathbf{x} = (x_1, ..., x_n)$ and rewrite the likelihood function from (6.2.4) as

$$L(\theta; \mathbf{x}) = \Pi_{i=1}^n f(x_i; \theta). \tag{6.4.5}$$

Hence we have $logL(\theta; \mathbf{x}) = \Sigma_{i=1}^n logf(x_i; \theta)$. Now, utilizing (6.4.1), one can write down the information contained in the data $\mathbf{X}$ as follows:

$$
\begin{aligned}
\mathcal{I}_{\mathbf{X}}(\theta) &= E_\theta\left[\left\{\frac{\partial}{\partial\theta}logL(\theta; \mathbf{X})\right\}^2\right] \\
&= E_\theta\left[\left\{\Sigma_{i=1}^n\frac{\partial}{\partial\theta}logf(X_i; \theta)\right\}^2\right] \\
&= E_\theta\left[\Sigma_{i=1}^n\left\{\frac{\partial}{\partial\theta}logf(X_i; \theta)\right\}^2\right] \\
&\quad + E_\theta\left[\underset{1\le i\ne j\le n}{\Sigma\Sigma}\left\{\frac{\partial}{\partial\theta}logf(X_i; \theta)\right\}\left\{\frac{\partial}{\partial\theta}logf(X_j; \theta)\right\}\right].
\end{aligned}
\tag{6.4.6}
$$

Since the X's are iid, we have $E_\theta\left[\left\{\frac{\partial}{\partial\theta}logf(X_i; \theta)\right\}^2\right] = \mathcal{I}_{X_1}(\theta)$ for each $i = 1, ..., n$, and hence the first term included in the end of (6.4.6) amounts to $n\mathcal{I}_{X_1}(\theta)$. Next, the second term in the end of (6.4.6) can be expressed as

$$
\begin{aligned}
&E_\theta\left[\underset{1\le i\ne j\le n}{\Sigma\Sigma}\left\{\frac{\partial}{\partial\theta}logf(X_i; \theta)\right\}\left\{\frac{\partial}{\partial\theta}logf(X_j; \theta)\right\}\right] \\
&= \underset{1\le i\ne j\le n}{\Sigma\Sigma}E_\theta\left[\left\{\frac{\partial}{\partial\theta}logf(X_i; \theta)\right\}\left\{\frac{\partial}{\partial\theta}logf(X_j; \theta)\right\}\right] \\
&= \underset{1\le i\ne j\le n}{\Sigma\Sigma}E_\theta\left[\left\{\frac{\partial}{\partial\theta}logf(X_i; \theta)\right\}\right]E_\theta\left[\left\{\frac{\partial}{\partial\theta}logf(X_j; \theta)\right\}\right]
\end{aligned}
\tag{6.4.7}
$$

since X_i and X_j are independent, $1 \le i \ne j \le n$

$$= n(n-1)E_\theta^2\left[\left\{\frac{\partial}{\partial\theta}logf(X_1; \theta)\right\}\right],$$

since the X's are identically distributed.

Now, let us write

$$\int_{\mathcal{X}} f(x;\theta)\,dx = 1$$

so that $0 = \frac{d}{d\theta}(1) = \frac{d}{d\theta}\int_{\mathcal{X}} f(x;\theta)\,dx = \int_{\mathcal{X}} \frac{\partial}{\partial\theta}\left[e^{log f(x;\theta)}\right]dx$. Hence one obtains

$$0 = \int_{\mathcal{X}} f(x;\theta)\frac{\partial}{\partial\theta}[log f(x;\theta)]\,dx = E_\theta\left[\left\{\frac{\partial}{\partial\theta}log f(X_1;\theta)\right\}\right]. \qquad (6.4.8)$$

Next, combining (6.4.6)-(6.4.8), we conclude that $\mathcal{I}_{\mathbf{X}}(\theta) = n\mathcal{I}_{X_1}(\theta)$. ∎

Suppose that we have collected random samples $X_1, ..., X_n$ from a population and we have evaluated the information $\mathcal{I}_{\mathbf{X}}(\theta)$ contained in the *data* $\mathbf{X} = (X_1, ..., X_n)$. Next, suppose also that we have a *statistic* $T = T(\mathbf{X})$ in mind for which we have evaluated the information $\mathcal{I}_T(\theta)$ contained in T. If it turns out that $\mathcal{I}_T(\theta) = \mathcal{I}_{\mathbf{X}}(\theta)$, can we then claim that the statistic T is indeed sufficient for θ? The answer is yes, we certainly can. We state the following result without supplying its proof. One may refer to Rao (1973, result (iii), p. 330) for details. In a recent exchange of personal communications, C. R. Rao has provided a simple way to look at the Theorem 6.4.2. In the Exercise 6.4.15, we have given an outline of Rao's elegant proof of this result. In the Examples 6.4.3-6.4.4, we find opportunities to apply this theorem.

Theorem 6.4.2 *Suppose that* $\mathbf{X}$ *is the whole data and* $T = T(\mathbf{X})$ *is some statistic. Then,* $\mathcal{I}_{\mathbf{X}}(\theta) \geq \mathcal{I}_T(\theta)$ *for all* $\theta \in \Theta$. *The two information measures match with each other for all* θ *if and only if* T *is a sufficient statistic for* θ.

Example 6.4.3 (Example 6.4.1 Continued) Suppose that $X_1, ..., X_n$ are iid Poisson(λ), where λ (> 0) is the unknown parameter. We already know that $T = \Sigma_{i=1}^{n}X_i$ is a minimal sufficient statistic for λ and T is distributed as Poisson($n\lambda$). But, let us now pursue T from the information point of view. One can start with the pmf $g(t;\lambda)$ of T and verify that

$$\mathcal{I}_T(\lambda) = n\lambda^{-1}$$

as follows: Let us write $log\{g(t;\lambda)\} = -n\lambda + t log(n\lambda) - log(t!)$ which implies that $\frac{\partial}{\partial\lambda}log\{g(t;\lambda)\} = -n + t\lambda^{-1}$. So, $\mathcal{I}_T(\lambda) = E_\lambda\left[\left\{\frac{\partial}{\partial\lambda}log\{g(T;\lambda)\}\right\}^2\right] = E_\lambda\left[(T - n\lambda)^2/\lambda^2\right] = n\lambda^{-1}$ since $E_\lambda[(T - n\lambda)^2] = V(T) = n\lambda$.

On the other hand, from (6.4.4) and Example 6.4.1, we can write $\mathcal{I}_{\mathbf{X}}(\lambda) = n\mathcal{I}_{X_1}(\lambda) = n\lambda^{-1}$. That is, T preserves the available information from the whole data $\mathbf{X}$. The Theorem 6.4.2 implies that the statistic T is indeed sufficient for λ. ▲

Example 6.4.4 (Example 6.4.2 Continued) Let $X_1, ..., X_n$ be iid $N(\mu, \sigma^2)$ where $\mu \in (-\infty, \infty)$ is the unknown parameter. Here $\sigma \in (0, \infty)$ is assumed known. We had shown earlier that the statistic $T = \overline{X}$ was sufficient for μ. Let us now pursue T from the information point of view. The statistic T is distributed as $N(\mu, n^{-1}\sigma^2)$ so that one can start with the pdf $g(t; \mu)$ of T and verify that

$$\mathcal{I}_T(\mu) = n\sigma^{-2}$$

as follows: Let us write $log\{g(t; \mu)\} = -\frac{1}{2}\left\{n(t - \mu)^2/\sigma^2\right\} - log(\sigma\sqrt{2n^{-1}\pi})$, which implies that $\frac{\partial}{\partial \mu}log\{g(t; \mu)\} = n(t - \mu)/\sigma^2$. Hence, we have $\mathcal{I}_T(\mu) = E_\mu\left[\left\{\frac{\partial}{\partial \mu}log\{g(T; \mu)\}\right\}^2\right] = E_\mu\left[n^2(T - \mu)^2/\sigma^4\right] = n\sigma^{-2}$ since $E_\mu[(T - \mu)^2] = V(\overline{X}) = n^{-1}\sigma^2$. From the Example 6.4.2, however, we know that the information contained in one single observation is $\mathcal{I}_{X_1}(\mu) = \sigma^{-2}$ and thus in view of (6.4.4), we have $\mathcal{I}_{\mathbf{X}}(\mu) = n\mathcal{I}_{X_1}(\mu) = n\sigma^{-2}$. That is, T preserves the available information from the whole data $\mathbf{X}$. Now, the Theorem 6.4.2 would imply that the statistic T is indeed sufficient for λ. ▲

Remark 6.4.1 Suppose that the pmf or pdf $f(x; \theta)$ is such that $\frac{\partial^2}{\partial \theta^2}f(x; \theta)$ is finite for all $x \in \mathcal{X}$ and $E_\theta\left[\frac{\partial^2}{\partial \theta^2}f(X; \theta)\right]$ is finite for all $\theta \in \Theta$. Then the *Fisher information* defined earlier can be alternatively evaluated using the following expression:

$$\mathcal{I}_X(\theta) = -E_\theta\left[\frac{\partial^2}{\partial \theta^2}log f(X; \theta)\right]. \qquad (6.4.9)$$

We leave its verification as an exercise.

Example 6.4.5 (Example 6.4.1 Continued) Use (6.4.9) and observe that $\frac{\partial^2}{\partial \lambda^2}log f(x; \lambda) = -x\lambda^{-2}$ so that $\mathcal{I}_X(\lambda) = -E_\lambda\left[\frac{\partial^2}{\partial \lambda^2}log f(X; \lambda)\right] = -E_\lambda[-X\lambda^{-2}] = \lambda^{-1}$. ▲

Example 6.4.6 (Example 6.4.2 Continued) Use (6.4.9) and observe that $\frac{\partial^2}{\partial \mu^2}log f(x; \mu) = -\sigma^{-2}$ so that $\mathcal{I}_X(\lambda) = -E_\lambda\left[\frac{\partial^2}{\partial \mu^2}log f(X; \mu)\right] = -E_\lambda[-\sigma^{-2}] = \sigma^{-2}$. ▲

> In the Exercise 6.4.16, we pursue an idea like this: Suppose that a statistic T is not sufficient for θ. Can we say something about how non-sufficient T is for θ?

6.4.2 Multi-parameter Situation

When the unknown parameter $\boldsymbol{\theta}$ is multidimensional, the definition of the Fisher information gets more involved. To keep the presentation simple, we only discuss the case of a two-dimensional parameter. Suppose that X is an observable real valued random variable with the pmf or pdf $f(x; \boldsymbol{\theta})$ where the parameter $\boldsymbol{\theta} = (\theta_1, \theta_2) \in \Theta$, an open rectangle $\subseteq \Re^2$, and the $\mathcal{X}$ space does not depend upon $\boldsymbol{\theta}$. We *assume* throughout that $\frac{\partial}{\partial \theta_i} f(x; \boldsymbol{\theta})$ exists, $i = 1, 2$, for all $x \in \mathcal{X}$, $\boldsymbol{\theta} \in \Theta$, and that we can also interchange the partial derivative (with respect to θ_1, θ_2) and the integral (with respect to x).

Definition 6.4.2 *Let us extend the earlier notation as follows. Denote* $I_{ij}(\boldsymbol{\theta}) = E_{\boldsymbol{\theta}} \left[\left\{ \frac{\partial}{\partial \theta_i} log f(X; \boldsymbol{\theta}) \right\} \left\{ \frac{\partial}{\partial \theta_j} log f(X; \boldsymbol{\theta}) \right\} \right]$, *for* $i, j = 1, 2$. *The Fisher information matrix or simply the information matrix about* $\boldsymbol{\theta}$ *is given by*

$$\mathcal{I}_X(\boldsymbol{\theta}) = \begin{pmatrix} I_{11}(\boldsymbol{\theta}) & I_{12}(\boldsymbol{\theta}) \\ I_{21}(\boldsymbol{\theta}) & I_{22}(\boldsymbol{\theta}) \end{pmatrix}. \tag{6.4.10}$$

Remark 6.4.2 In situations where $\frac{\partial^2}{\partial \theta_i \partial \theta_j} f(x; \boldsymbol{\theta})$ exists for all $x \in \mathcal{X}$, for all $i, j = 1, 2$, and for all $\boldsymbol{\theta} \in \Theta$, we can alternatively write

$$I_{ij}(\boldsymbol{\theta}) = -E_{\boldsymbol{\theta}} \left[\frac{\partial^2}{\partial \theta_i \partial \theta_j} log f(X; \boldsymbol{\theta}) \right] \text{ for } i, j = 1, 2, \tag{6.4.11}$$

and express the information matrix $\mathcal{I}_X(\boldsymbol{\theta})$ accordingly. We have left this as an exercise.

> Theorem 6.4.1 holds if $\boldsymbol{\theta}$ is multidimensional. If $\boldsymbol{\theta}$ is two-dimensional and $X_1, ..., X_n$ are iid with the pmf or pdf $f(x; \boldsymbol{\theta})$, then under mild regularity conditions (stated before) the Definition 6.4.2), we would have:
> $$\mathcal{I}_{\mathbf{X}}(\boldsymbol{\theta}) = n\mathcal{I}_{X_1}(\boldsymbol{\theta}).$$
  $\qquad (6.4.12)$

Having a statistic $T = T(X_1, ..., X_n)$, however, the associated information matrix about $\boldsymbol{\theta}$ will simply be calculated as $\mathcal{I}_T(\boldsymbol{\theta})$ where one would replace the original pmf or pdf $f(x; \boldsymbol{\theta})$ by that of T, namely $g(t; \boldsymbol{\theta})$, $t \in \mathcal{T}$. When we compare two statistics T_1 and T_2 in terms of their information content about a single unknown parameter θ, we simply look at the two one-dimensional quantities $\mathcal{I}_{T_1}(\theta)$ and $\mathcal{I}_{T_2}(\theta)$, and compare these two numbers. The statistic associated with the larger information content would be more appealing. But, when $\boldsymbol{\theta}$ is two-dimensional, in order to compare the two statistics T_1 and T_2, we have to consider their individual two-dimensional *information matrices* $\mathcal{I}_{T_1}(\boldsymbol{\theta})$ and $\mathcal{I}_{T_2}(\boldsymbol{\theta})$. It would be tempting to say that T_1 *is more informative* about $\boldsymbol{\theta}$ than T_2 provided that

the matrix $\mathcal{I}_{T_1}(\theta) - \mathcal{I}_{T_2}(\theta)$ is positive semi definite. $\qquad (6.4.13)$

We add that a version of the Theorem 6.4.2 holds in the multiparameter case as well. One may refer to Section 5a.3 of Rao (1973).

Example 6.4.7 Let $X_1, ..., X_n$ be iid $N(\mu, \sigma^2)$ where $\mu \in (-\infty, \infty)$ and $\sigma^2 \in (0, \infty)$ are both unknown parameters. Denote $\boldsymbol{\theta} = (\mu, \sigma^2), \mathbf{X} = (X_1, ..., X_n)$. Let us evaluate the information matrix for $\mathbf{X}$. First, a single observation X_1 has its pdf

$$f(x; \boldsymbol{\theta}) = \{\sigma\sqrt{2\pi}\}^{-1}exp\{-\tfrac{1}{2\sigma^2}(x - \mu)^2\} \quad \text{for } x \in (-\infty, \infty)$$

so that one has

$$\tfrac{\partial}{\partial\mu}logf(x; \boldsymbol{\theta}) = \sigma^{-2}(x - \mu), \quad \tfrac{\partial}{\partial\sigma^2}logf(x; \boldsymbol{\theta}) = \tfrac{1}{2\sigma^2}[\sigma^{-2}(x - \mu)^2 - 1].$$

Hence we obtain

$$
\begin{aligned}
I_{11}(\boldsymbol{\theta}) &= E_{\boldsymbol{\theta}}\left[\left(\tfrac{\partial}{\partial\mu}logf(X_1; \boldsymbol{\theta})\right)^2\right] \\
&= E_{\boldsymbol{\theta}}\left(\sigma^{-4}(X_1 - \mu)^2\right) = \sigma^{-2}, \\
I_{22}(\boldsymbol{\theta}) &= E_{\boldsymbol{\theta}}\left[\left(\tfrac{\partial}{\partial\sigma^2}logf(X_1; \boldsymbol{\theta})\right)^2\right] \\
&= \tfrac{1}{4\sigma^4}E_{\boldsymbol{\theta}}\left([\sigma^{-2}(X_1 - \mu)^2 - 1]^2\right) = \tfrac{1}{2}\sigma^{-4},
\end{aligned}
\tag{6.4.14}
$$

since $\sigma^{-2}(X_1 - \mu)^2$ is χ_1^2 so that $E[\sigma^{-2}(X_1 - \mu)^2] = 1, V[\sigma^{-2}(X_1 - \mu)^2] = 2$. Next, we have

$$
\begin{aligned}
I_{12}(\boldsymbol{\theta}) = I_{21}(\boldsymbol{\theta}) &= E_{\boldsymbol{\theta}}\left[\left\{\tfrac{\partial}{\partial\mu}logf(X_1; \boldsymbol{\theta})\right\}\left\{\tfrac{\partial}{\partial\sigma^2}logf(X_1; \boldsymbol{\theta})\right\}\right] \\
&= \tfrac{1}{2\sigma^4}E_{\boldsymbol{\theta}}\left[\sigma^{-3}(X_1 - \mu)^3 - (X_1 - \mu)\right] \\
&= 0, \text{ since } X_1 \text{ is symmetric about } \mu,
\end{aligned}
\tag{6.4.15}
$$

so that combining (6.4.14)-(6.4.15) with (6.4.10), we obtain the following information matrix for one single observation X_1:

$$\mathcal{I}_{X_1}(\boldsymbol{\theta}) = \begin{pmatrix} I_{11}(\boldsymbol{\theta}) & I_{12}(\boldsymbol{\theta}) \\ I_{21}(\boldsymbol{\theta}) & I_{22}(\boldsymbol{\theta}) \end{pmatrix} = \begin{pmatrix} \sigma^{-2} & 0 \\ 0 & \tfrac{1}{2}\sigma^{-4} \end{pmatrix}. \tag{6.4.16}$$

Utilizing (6.4.12), we obtain the information matrix,

$$\mathcal{I}_{\mathbf{X}}(\boldsymbol{\theta}) = n\mathcal{I}_{X_1}(\boldsymbol{\theta}) = \begin{pmatrix} n\sigma^{-2} & 0 \\ 0 & \tfrac{1}{2}n\sigma^{-4} \end{pmatrix}, \tag{6.4.17}$$

for the whole data $\mathbf{X}$. ▲

Example 6.4.8 (Example 6.4.7 Continued) Let $\overline{X} = n^{-1}\Sigma_{i=1}^n X_i$, the sample mean. We are aware that $\overline{X}$ is distributed as $N(\mu, n^{-1}\sigma^2)$ and its pdf is given by

$$g(x; \boldsymbol{\theta}) = \sqrt{n}\{\sigma\sqrt{2\pi}\}^{-1} \, exp\left\{-\tfrac{n}{2\sigma^2}(x - \mu)^2\right\} \quad \text{for } x \in (-\infty, \infty)$$

so that one has

$$\tfrac{\partial}{\partial\mu}log\{g(x; \boldsymbol{\theta})\} = n\sigma^{-2}(x - \mu),$$
$$\tfrac{\partial}{\partial\sigma^2}log\{g(x; \boldsymbol{\theta})\} = \tfrac{1}{2\sigma^2}[n\sigma^{-2}(x - \mu)^2 - 1].$$

Hence we have

$$
\begin{aligned}
I_{11}(\boldsymbol{\theta}) &= E_{\boldsymbol{\theta}}\left[\left(\tfrac{\partial}{\partial\mu}log\{g(\overline{X}; \boldsymbol{\theta})\}\right)^2\right] \\
&= E_{\boldsymbol{\theta}}\left[n^2\sigma^{-4}(\overline{X} - \mu)^2\right] = n\sigma^{-2},
\end{aligned}
\tag{6.4.18}
$$

$$
\begin{aligned}
I_{22}(\boldsymbol{\theta}) &= E_{\boldsymbol{\theta}}\left[\left(\tfrac{\partial}{\partial\sigma^2}log\{g(\overline{X}; \boldsymbol{\theta})\}\right)^2\right] \\
&= \tfrac{1}{4\sigma^4}E_{\boldsymbol{\theta}}\left([n\sigma^{-2}(\overline{X} - \mu)^2 - 1]^2\right) = \tfrac{1}{2}\sigma^{-4}.
\end{aligned}
$$

We can again show that $I_{12}(\boldsymbol{\theta}) = I_{21}(\boldsymbol{\theta}) = 0$ corresponding to $\overline{X}$. Utilizing (6.4.18), we obtain the following information matrix corresponding to the statistic $\overline{X}$:

$$\mathcal{I}_{\overline{X}}(\boldsymbol{\theta}) = \begin{pmatrix} n\sigma^{-2} & 0 \\ 0 & \tfrac{1}{2}\sigma^{-4} \end{pmatrix}. \tag{6.4.19}$$

Comparing (6.4.17) and (6.4.19), we observe that

$$\mathcal{I}_{\mathbf{X}}(\boldsymbol{\theta}) - \mathcal{I}_{\overline{X}}(\boldsymbol{\theta}) = \begin{pmatrix} 0 & 0 \\ 0 & \tfrac{1}{2}(n - 1)\sigma^{-4} \end{pmatrix}, \tag{6.4.20}$$

which is a positive semi definite matrix. That is, if we summarize the whole data $\mathbf{X}$ only through $\overline{X}$, then there is some loss of information. In other words, $\overline{X}$ does not preserve all the information contained in the data $\mathbf{X}$ when μ and σ^2 are both assumed unknown. ▲

Example 6.4.9 (Example 6.4.7 Continued) Suppose that we consider the sample variance, $S^2 = (n - 1)^{-1}\Sigma_{i=1}^n(X_i - \overline{X})^2$. We are aware that $Y = (n - 1)S^2/\sigma^2$ is distributed as χ_{n-1}^2 for $n \geq 2$, and so with $c = \{2^{(n-1)/2}\Gamma(\tfrac{1}{2}(n - 1))\}^{-1}$, the pdf of Y is given by

$$c \, exp\{-\tfrac{1}{2}y\}y^{(n-3)/2} \quad \text{for } 0 < y < \infty.$$

Hence with $d = (n - 1)^{(n-1)/2}c$, the pdf of S^2 is given by

$$h(x; \boldsymbol{\theta}) = d\sigma^{-(n-1)} \, exp\left\{-\tfrac{1}{2}(n - 1)x/\sigma^2\right\} x^{(n-3)/2} \quad \text{for } x \in (0, \infty)$$

so that one has

$$\frac{\partial}{\partial \mu} log\{h(x; \boldsymbol{\theta})\} = 0,$$
$$\frac{\partial}{\partial \sigma^2} log\{h(x; \boldsymbol{\theta})\} = \frac{1}{2\sigma^2}[(n-1)\sigma^{-2}S^2 - (n-1)].$$

Hence we obtain

$$I_{11}(\boldsymbol{\theta}) = 0, \quad I_{22}(\boldsymbol{\theta}) = \tfrac{1}{2}(n-1)\sigma^{-4}. \tag{6.4.21}$$

Obviously, $I_{12}(\boldsymbol{\theta}) = I_{21}(\boldsymbol{\theta}) = 0$ corresponding to S^2. Utilizing (6.4.21), we obtain the information matrix corresponding to the statistic S^2, namely,

$$\mathcal{I}_{S^2}(\boldsymbol{\theta}) = \begin{pmatrix} 0 & 0 \\ 0 & \tfrac{1}{2}(n-1)\sigma^{-4} \end{pmatrix}. \tag{6.4.22}$$

Comparing (6.4.17) and (6.4.22), we observe that

$$\mathcal{I}_{\mathbf{X}}(\boldsymbol{\theta}) - \mathcal{I}_{S^2}(\boldsymbol{\theta}) = \begin{pmatrix} n\sigma^{-2} & 0 \\ 0 & \tfrac{1}{2}\sigma^{-4} \end{pmatrix}, \tag{6.4.23}$$

which is a positive semi definite matrix. That is, if we summarize the whole data $\mathbf{X}$ only through S^2, then there is certainly some loss of information when μ and σ^2 are both assumed unknown. ▲

Example 6.4.10 (Examples 6.4.8-6.4.9 Continued) Individually, whether we consider the statistic $\overline{X}$ or S^2, both lose some information in comparison with $\mathcal{I}_{\mathbf{X}}(\boldsymbol{\theta})$, the information contained in the whole data $\mathbf{X}$. This is clear from (6.4.20) and (6.4.23). But recall that $\overline{X}$ and S^2 are *independently* distributed, and hence we note that

$$\mathcal{I}_{\overline{X},S^2}(\boldsymbol{\theta}) = \mathcal{I}_{\overline{X}}(\boldsymbol{\theta}) + \mathcal{I}_{S^2}(\boldsymbol{\theta}) = \mathcal{I}_{\mathbf{X}}(\boldsymbol{\theta}).$$

That is, the lost information when we consider only $\overline{X}$ or S^2 is picked up by the other statistic. ▲

In the Example 6.4.10, we tacitly used a particular result which is fairly easy to prove. For the record, we merely state this result while its proof is left as the Exercise 6.4.11.

Theorem 6.4.3 *Suppose that* $X_1, ..., X_n$ *are iid with the common pmf or pdf given by* $f(x; \boldsymbol{\theta})$. *We denote the whole data* $\mathbf{X} = (X_1, ..., X_n)$. *Suppose that we have two statistics* $T_1 = T_1(\mathbf{X}), T_2 = T_2(\mathbf{X})$ *at our disposal and* T_1, T_2 *are distributed independently. Then, the information matrix* $\mathcal{I}_{\mathbf{T}}(\boldsymbol{\theta})$ *is given by*

$$\mathcal{I}_{\mathbf{T}}(\boldsymbol{\theta}) = \mathcal{I}_{T_1}(\boldsymbol{\theta}) + \mathcal{I}_{T_2}(\boldsymbol{\theta}).$$

Let us now go back for a moment to (6.4.10) for the definition of the information matrix $\mathcal{I}_{\mathbf{X}}(\boldsymbol{\theta})$. Now suppose that $\mathbf{Y} = \mathbf{h}(\mathbf{X})$ where the function $\mathbf{h}(.) : \mathcal{X} \rightarrow \mathcal{Y}$ is one-to-one. It should be intuitive enough to guess that $\mathcal{I}_{\mathbf{X}}(\boldsymbol{\theta}) = \mathcal{I}_{\mathbf{Y}}(\boldsymbol{\theta})$. For the record, we now state this result formally.

Theorem 6.4.4 *Let* $\mathbf{X}$ *be an observable random variable with its pmf or pdf* $f(\mathbf{x}; \boldsymbol{\theta})$ *and the information matrix* $\mathcal{I}_{\mathbf{X}}(\boldsymbol{\theta})$. *Suppose that* $\mathbf{Y} = \mathbf{h}(\mathbf{X})$ *where the function* $\mathbf{h}(.) : \mathcal{X} \rightarrow \mathcal{Y}$ *is one-to-one. Then, the information matrix about the unknown parameter* $\boldsymbol{\theta}$ *contained in* $\mathbf{Y}$ *is same as that in* $\mathbf{X}$, *that is*

$$\mathcal{I}_{\mathbf{X}}(\boldsymbol{\theta}) = \mathcal{I}_{\mathbf{Y}}(\boldsymbol{\theta}).$$

Proof In order to keep the deliberations simple, we consider only a real valued continuous random variable X and a real valued unknown parameter θ. Recall that we can write $\mathcal{I}_X(\theta) = E_\theta \left[\{ \frac{\partial}{\partial\theta} log f(X; \theta) \}^2 \right]$. Note that $x = h^{-1}(y)$ is well-defined since $h(.)$ is assumed one-to-one. Now, using the transformation techniques from (4.4.1), observe that the pdf of Y can be expressed as

$$g(y; \theta) = f\left(h^{-1}(y); \theta\right) \mid \frac{d}{dy} h^{-1}(y) \mid \text{ for } y \in \mathcal{Y}.$$

Thus, one immediately writes

$$\begin{aligned}
\mathcal{I}_Y(\theta) &= E_Y \left[\{ \tfrac{\partial}{\partial\theta} log\{g(Y; \theta)\} \}^2 \right] \\
&= E_Y \left[\{ \tfrac{\partial}{\partial\theta} log\{f\left(h^{-1}(Y); \theta\right)\} \}^2 \right] \\
&= E_X \left[\{ \tfrac{\partial}{\partial\theta} log\{f\left(X; \theta\right)\} \}^2 \right] \\
&= \mathcal{I}_X(\theta).
\end{aligned}$$

The vector valued case and the case of discrete $\mathbf{X}$ can be disposed off with minor modifications. These are left out as Exercise 6.4.12. ∎

6.5 Ancillarity

The concept called *ancillarity* of a statistic is perhaps the furthest away from sufficiency. A sufficient statistic $\mathbf{T}$ preserves all the information about θ contained in the data $\mathbf{X}$. To contrast, an ancillary statistic $\mathbf{T}$ *by itself* provides *no information* about the unknown parameter θ. We are not implying that an ancillary statistic is necessarily bad or useless. Individually, an ancillary statistic would not provide any information about θ, but such statistics can play useful roles in statistical methodologies. In the mid 1920's, R. A. Fisher introduced this concept and he frequently revisited it in his writings. This concept evolved from Fisher (1925a) and later it blossomed into the vast area of *conditional inferences*. In his 1956 book, Fisher emphasized many positive aspects of ancillarity in analyzing real data. Some of these ideas will be explored in this and the next section. For fuller discussions of *conditional inference* one may look at Basu (1964), Hinkley (1980a) and Ghosh (1988). The interesting article of Reid (1995) also provides an assessment of conditional inference procedures.

Consider the real valued observable random variables $X_1, ..., X_n$ from some population having the common pmf or pdf $f(x; \boldsymbol{\theta})$, where the unknown parameter vector $\boldsymbol{\theta}$ belongs to the parameter space $\Theta \subseteq \Re^p$. Let us continue writing $\mathbf{X} = (X_1, ..., X_n)$ for the data, and denote a vector valued statistic by $\mathbf{T} = \mathbf{T}(\mathbf{X})$.

Definition 6.5.1 *A statistic* $\mathbf{T}$ *is called ancillary for* $\boldsymbol{\theta}$ *or simply ancillary provided that the pmf or the pdf of* $\mathbf{T}$, *denoted by* $g(\mathbf{t})$ *for* $\mathbf{t} \in \mathcal{T}$, *does not involve the unknown parameter* $\boldsymbol{\theta} \in \Theta$.

Example 6.5.1 Suppose that $X_1, ..., X_n$ are iid $N(\theta, 1)$ where θ is the unknown parameter, $-\infty < \theta < \infty, n \geq 3$. A statistic $T_1 = X_1 - X_2$ is distributed as $N(0, 2)$ whatever be the value of the unknown parameter θ. Hence T_1 is ancillary for θ. Another statistic $T_2 = X_1 + ... + X_{n-1} - (n-1)X_n$ is distributed as $N(0, n(n-1))$ whatever be the value of the unknown parameter θ. Hence T_2 is also ancillary for θ. The sample variance S^2 is distributed as $(n-1)^{-1}\chi^2_{n-1}$ whatever be the value of the unknown parameter θ and hence S^2 is ancillary too for θ. ▲

Example 6.5.2 Suppose that $X_1, ..., X_n$ are iid $N(\mu, \sigma^2)$, $\boldsymbol{\theta} = (\mu, \sigma^2)$, $-\infty < \mu < \infty, 0 < \sigma^2 < \infty, n \geq 2$. Here both the parameters μ and σ are assumed unknown. Let us reconsider the statistics T_1 or T_2 defined in the Example 6.5.1. Now, T_1, T_2 are respectively distributed as $N(0, 2\sigma^2)$ and $N(0, n(n-1)\sigma^2)$ respectively, and both these distributions clearly depend upon $\boldsymbol{\theta}$. Thus, T_1 or T_2 is no longer ancillary for $\boldsymbol{\theta}$ in this situation. The sample variance S^2 is distributed as $(n-1)^{-1}\sigma^2\chi^2_{n-1}$ and hence S^2 is not ancillary either for $\boldsymbol{\theta}$. But, consider another statistic $T_3 = (X_1 - X_2)/S$ where S^2 is the sample variance. Denoting $Y_i = (X_i - \mu)/\sigma$, observe that in terms of the Y's, we can equivalently rewrite T_3^2 as

$$U_3 = \{Y_1 - Y_2\}^2 / \left\{ (n-1)^{-1}\Sigma_{i=1}^n (Y_i - \overline{Y})^2 \right\}. \qquad (6.5.1)$$

Since $Y_1, ..., Y_n$ are iid $N(0, 1)$, starting with the likelihood function of $\mathbf{Y} = (Y_1, ..., Y_n)$, and then by using transformations, one can find the pdf of U_3. We leave it as an exercise. But, since the likelihood function of $\mathbf{Y}$ would not involve θ to begin with, the pdf of U_3 would not involve θ. In other words, T_3 is an ancillary *statistic* here. Note that we do not need the explicit pdf of U_3 to conclude this. ▲

In Example 6.5.2, note that $T_4 = \sqrt{n}(\overline{X} - \mu)/S$ has a Student's t distribution with $(n-1)$ degrees of freedom which is free from θ. But, we do not talk about its ancillarity or non-ancillarity since T_4 is not a statistic. T_3, however, was a statistic. The expression used in (6.5.1) was merely a device to argue that the distribution of the *statistic* T_3 in the Example 6.5.2 was free from θ.

Example 6.5.3 Suppose that $X_1, ..., X_n$ are iid with the common pdf $f(x; \lambda) = \lambda e^{-\lambda x} I(x > 0)$ where $\lambda(> 0)$ is the unknown parameter with $n \geq 2$. Let us write $S^2 = (n-1)^{-1} \Sigma_{i=1}^{n}(X_i - \overline{X})^2$ and denote $U = \Sigma_{i=1}^{n} X_i$, $T_1 = X_1/X_n, T_2 = X_2/U, T_3 = (X_1 + X_3)/S$. Define $Y_i = \lambda X_i$ for $i = 1, ..., n$ and it is obvious that the joint distribution of $\mathbf{Y} = (Y_1, ..., Y_n)$ does not involve the unknown parameter λ. Next, one can rewrite the statistic T_1 as Y_1/Y_n and its pdf cannot involve λ. So, T_1 is ancillary. Also, the statistic T_2 can be rewritten as $Y_2/\{\Sigma_{i=1}^{n} Y_i\}$ and its pdf cannot involve λ. So, T_2 is ancillary. Similarly one can argue that T_3 is also ancillary. The details are left out as Exercise 6.5.2. ▲

Example 6.5.4 (Example 6.5.1 Continued) Suppose that X_1, X_2, X_3 are iid $N(\theta, 1)$ where θ is the unknown parameter, $-\infty < \theta < \infty$. Denote $T_1 = X_1 - X_2, T_2 = X_1 + X_2 - 2X_3$, and consider the two dimensional statistic $\mathbf{T} = (T_1, T_2)$. Note that any linear function of T_1, T_2 is also a linear function of X_1, X_2, X_3, and hence it is distributed as a univariate normal random variable. Then, by the Definition 4.6.1 of the multivariate normality, it follows that the statistic $\mathbf{T}$ is distributed as a bivariate normal variable. More specifically, one can check that $\mathbf{T}$ is distributed as $N_2(0, 0, 2, 6, 0)$ which is free from θ. In other words, $\mathbf{T}$ is an ancillary statistic for θ. ▲

Example 6.5.5 (Example 6.5.2 Continued) Suppose that $X_1, ..., X_n$ are iid $N(\mu, \sigma^2)$, $\theta = (\mu, \sigma^2)$, $-\infty < \mu < \infty, 0 < \sigma^2 < \infty, n \geq 4$. Here, both the parameters μ and σ are assumed unknown. Let S^2 be the sample variance and $T_1 = (X_1 - X_3)/S, T_2 = (X_1 + X_3 - 2X_2)/S, T_3 = (X_1 - X_3 + 2X_2 - 2X_4)/S$, and denote the statistic $\mathbf{T} = (T_1, T_2, T_3)$. Follow the technique used in the Example 6.5.2 to show that $\mathbf{T}$ is ancillary for θ. ▲

We remarked earlier that a statistic which is ancillary for the unknown parameter θ can play useful roles in the process of inference making. The following examples would clarify this point.

> It is possible to have statistics T_1, T_2 such that (i) T_2 has some information about θ, but it is not sufficient for θ, (ii) T_1 is ancillary for θ, and yet (iii) (T_1, T_2) is jointly sufficient for θ. Look at the Example 6.5.6.   (6.5.2)

Example 6.5.6 (Example 6.5.1 Continued) Suppose that X_1, X_2 are iid $N(\theta, 1)$ where θ is the unknown parameter, $-\infty < \theta < \infty$. The statistic $T_1 = X_1 - X_2$ is ancillary for θ. Consider another statistic $T_2 = X_1$. One would recall from Example 6.4.4 that $\mathcal{I}_{T_2}(\theta) = 1$ whereas $\mathcal{I}_{\mathbf{X}}(\theta) = 2$, and so the statistic T_2 can *not* be sufficient for θ. Here, while T_2 is not sufficient for θ, it has some information about θ, but T_1 itself has no information about θ. Now, if we are told the observed value of the statistic $\mathbf{T} = (T_1, T_2)$, then we can reconstruct the original data $\mathbf{X} = (X_1, X_2)$ *uniquely*. That is, the statistic $\mathbf{T} = (T_1, T_2)$ turns out to be *jointly sufficient* for the unknown parameter θ. ▲

> It is possible to have statistics T_1, T_2 such that (i) T_1 has no information about θ, (ii) T_2 has no information about θ, and yet (iii) (T_1, T_2) is jointly minimal sufficient for θ. Look at the Example 6.5.7.   (6.5.3)

Example 6.5.7 This example is due to D. Basu. Suppose that (X, Y) is a bivariate normal variable distributed as $N_2(0, 0, 1, 1, \rho)$, introduced in Section 3.6, where the unknown parameter is the correlation coefficient $\rho \in (-1, 1)$. Now consider the two statistics $T_1 = X$ and $T_2 = Y$. Since T_1 and T_2 have individually both standard normal distributions, it follows that T_1 is ancillary for ρ and so is T_2. But, note that the statistic $\mathbf{T} = (T_1, T_2)$ is minimal sufficient for the unknown parameter ρ. What is remarkable is that the statistic T_1 has no information about ρ and the statistic T_2 has no information about ρ, but the statistic (T_1, T_2) *jointly* has all the information about ρ. ▲

The situation described in (6.5.2) has been highlighted in the Example 6.5.6 where we note that (T_1, T_2) is sufficient for θ, but (T_1, T_2) is *not* minimal sufficient for θ. Instead, $2T_2 - T_1$ is minimal sufficient in the Example 6.5.6. The situation described in (6.5.3) has been highlighted in the Example 6.5.7 where we note especially that (T_1, T_2) is minimal sufficient for θ. In other words, there are remarkable differences between the situations described by these two Examples.

> Let us now calculate the information $\mathcal{I}_{X,Y}(\rho)$ where ρ is the correlation coefficient in a bivariate normal population.

With $c(\rho) = \{2\pi\sqrt{1 - \rho^2}\}^{-1}$, from (3.6.2), one can write down the joint pdf of (X, Y):

$$f(x, y; \rho) = c(\rho)\ exp\left[-\tfrac{1}{2}\left(x^2 - 2\rho xy + y^2\right)/(1 - \rho^2)\right]. \qquad (6.5.4)$$

In order to derive the expression for the Fisher information $\mathcal{I}_{X,Y}(\rho)$, one may proceed with the natural logarithm of $f(x, y; \rho)$. Next, differentiate the natural logarithm with respect to ρ, followed by squaring it, and then evaluating the expectation of that expression after replacing (x, y) with (X, Y). This direct approach becomes quite involved and it is left as an exercise. Let us, however, adopt a different approach in the following example.

Example 6.5.8 (Example 6.5.7 Continued) Suppose that (X, Y) is distributed as $N_2(0, 0, 1, 1, \rho)$, where the unknown parameter is the correlation coefficient $\rho \in (-1, 1)$. Define $U = X - Y$, $V = X + Y$, and notice that (U, V) can be uniquely obtained from (X, Y) and vice versa. In other words, $\mathcal{I}_{U,V}(\rho)$ and $\mathcal{I}_{X,Y}(\rho)$ should be exactly same in view of the Theorem 6.4.4. But, observe that (U, V) is distributed as $N_2(0, 0, 2(1 - \rho), 2(1 + \rho), 0)$, that is U and V are independent random variables. Refer back to the Section 3.7 as needed. Hence, using the Theorem 6.4.3, we immediately conclude that $\mathcal{I}_{U,V}(\rho) = \mathcal{I}_U(\rho) + \mathcal{I}_V(\rho)$. So, we can write:

$$\mathcal{I}_{X,Y}(\rho) = \mathcal{I}_{U,V}(\rho) = \mathcal{I}_U(\rho) + \mathcal{I}_V(\rho). \qquad (6.5.5)$$

That is, it will suffice to evaluate $\mathcal{I}_U(\rho)$ and $\mathcal{I}_V(\rho)$ separately. It is clear that U is $N(0, 2(1 - \rho))$ while V is $N(0, 2(1 + \rho))$. The pdf of U is given by

$$g(u; \rho) = \{4\pi(1 - \rho)\}^{-1/2}exp[-u^2/\{4(1 - \rho)\}], \quad -\infty < u < \infty,$$

so that one has

$$\tfrac{\partial}{\partial \rho}log\{g(u; \rho)\} = \tfrac{1}{2}(1 - \rho)^{-1} - \tfrac{1}{4}u^2(1 - \rho)^{-2}.$$

Hence, we write

$$\begin{aligned}
\mathcal{I}_U&(\rho)\\
&= E_\rho\left[\tfrac{1}{4}(1 - \rho)^{-2} - \tfrac{1}{4}U^2(1 - \rho)^{-3} + \tfrac{1}{16}U^4(1 - \rho)^{-4}\right]\\
&= \tfrac{1}{4}(1 - \rho)^{-2} - \tfrac{1}{4}2(1 - \rho)(1 - \rho)^{-3} + \tfrac{12}{16}(1 - \rho)^2(1 - \rho)^{-4}\\
&= \tfrac{1}{2}(1 - \rho)^{-2},
\end{aligned}$$

since $E_\rho[U^2] = 2(1 - \rho)$, $E_\rho[U^4] = 3\left\{\sqrt{2(1 - \rho)}\right\}^4$. Similarly, one would verify that $\mathcal{I}_V(\rho) = \tfrac{1}{2}(1 + \rho)^{-2}$. Then, from (6.5.5), we can obviously write:

$$\mathcal{I}_{X,Y}(\rho) = \frac{1}{2}(1 - \rho)^{-2} + \frac{1}{2}(1 + \rho)^{-2} = (1 + \rho^2)(1 - \rho^2)^{-2}. \qquad (6.5.6)$$

A direct verification of (6.5.6) using the expression of $f(x, y; \rho)$ is left as the Exercise 6.5.17. ▲

Example 6.5.9 Let $X_1, X_2, ...$ be a sequence of iid Bernoulli(p), $0 < p < 1$. One may think of $X_i = 1$ or 0 according as the ith toss of a coin results in a head (H) or tail (T) where $P(H) = 1 - P(T) = p$ in each independent toss, $i = 1, ..., n$. Once the coin tossing experiment is over, suppose that we are only told how many times the particular coin was tossed and nothing else. That is, we are told what n is, but nothing else about the random samples $X_1, ..., X_n$. In this situation, knowing n alone, can we expect to gain any knowledge about the unknown parameter p? Of course, the answer should be "no," which amounts to saying that the sample size n is indeed ancillary for p. ▲

6.5.1 The Location, Scale, and Location-Scale Families

In Chapter 3, we discussed a very special class of distributions, namely the exponential family. Now, we briefly introduce a few other special ones which are frequently encountered in statistics. Let us start with a pdf $g(x), x \in \Re$ and construct the following families of distributions defined through the $g(.)$ function:

$$
\begin{aligned}
&(i) \quad \mathcal{F}_1 = \{f(x; \theta) = g(x - \theta) : \theta \in \Re, x \in \Re\}; \\
&(ii) \quad \mathcal{F}_2 = \{f(x; \delta) = \delta^{-1} g(x/\delta) : \delta \in \Re^+, x \in \Re\}; \\
&(iii) \quad \mathcal{F}_3 = \{f(x; \theta, \delta) = \delta^{-1} g((x - \theta)/\delta) : \theta \in \Re, \\
&\qquad\qquad \delta \in \Re^+, x \in \Re\}.
\end{aligned}
\tag{6.5.7}
$$

The understanding is that θ, δ may belong to some appropriate subsets of $\Re, \Re^+$ respectively. The reader should check that the corresponding members $f(.)$ from the families $\mathcal{F}_1, \mathcal{F}_2, \mathcal{F}_3$ are indeed pdf's themselves.

> The distributions defined via parts (i), (ii), and (iii) in (6.5.7) are respectively said to belong to the *location, scale,* and *location-scale family.*

We often say that the families $\mathcal{F}_1, \mathcal{F}_2, \mathcal{F}_3$ are respectively *indexed* by (i) the parameter θ, (ii) the parameter δ, and (iii) the parameters θ and δ. In part (i) θ is called a *location parameter*, (ii) δ is called a *scale parameter*, and (iii) θ, δ are respectively called the *location* and *scale parameters*.

For example, the collection of $N(\mu, 1)$ distributions, with $\mu \in \Re$, forms a *location family*. To see this, let $g(x) = \phi(x) = \left(\sqrt{2\pi}\right)^{-1} exp\{-x^2/2\}, x \in \Re$

and then write $\mathcal{F}_1 = \{f(x; \mu) = \phi(x - \mu) : \mu \in \Re, x \in \Re\}$. Next, the collection of $N(0, \sigma^2)$ distributions, with $\sigma \in \Re^+$, forms a *scale family*. To see this, let us simply write $\mathcal{F}_2 = \{f(x; \sigma) = \sigma^{-1}\phi(x/\sigma) : \sigma \in \Re^+, x \in \Re\}$. The collection of $N(\mu, \sigma^2)$ distributions, with $\mu \in \Re, \sigma \in \Re^+$, forms a *location-scale family*. We simply write $\mathcal{F}_3 = \{f(x; \mu, \sigma) = \sigma^{-1}\phi((x - \mu)/\sigma) : \mu \in \Re, \sigma \in \Re^+, x \in \Re\}$. We have left other examples as exercises.

In a location family, the role of the location parameter θ is felt in the "movement" of the pdf along the x-axis as different values of θ are contemplated. The $N(0, 1)$ distribution has its center of symmetry at the point $x = 0$, but the $N(\mu, 1)$ distribution's center of symmetry moves along the x-axis, to the right or left of $x = 0$, depending upon whether μ is positive or negative, without changing anything with regard to the shape of the probability density curve. In this sense, the mean μ of the normal distribution serves as a *location parameter*. For example, in a large factory, we may look at the monthly wage of each employee and postulate that the distribution of wage as $N(\mu, \sigma^2)$ where $\sigma = \$100$. After the negotiation of a new contract, suppose that each employee receives $50 monthly raise. Then the distribution moves to the right with its new center of symmetry at $\mu + 50$. The intrinsic shape of the distribution can not change in a situation like this.

In a scale family, the role of the scale parameter δ is felt in "squeezing or expanding" the pdf along the x-axis as different values of δ are contemplated. The $N(0, 1)$ distribution has its center of symmetry at the point $x = 0$. The $N(0, \delta^2)$ distribution's center of symmetry stays put at the point $x = 0$, but depending on whether δ is larger or smaller than one, the shape of the density curve will become more flat or more peaked, compared with the standard normal, around the center. In this sense, the variance δ^2 of the normal distribution serves as a *scale parameter*. Suppose that we record the heights (in inches) of individuals and we postulate the distribution of these heights as $N(70, \sigma^2)$. If heights are measured in centimeters instead, then the distribution would *appear* more spread out around the new center. One needs to keep in mind that recording the heights in centimeters would amount to multiplying each original observation X measured in inches by 2.54.

In a location-scale family, one would notice movement of the distribution along the x-axis as well as the squeezing or expansion effect in the shape. We may be looking at a data on the weekly maximum temperature in a city recorded over a period, in Fahrenheit (F) or Celsius (C). If one postulates a normal distribution for the temperatures, changing the unit of measurements from Fahrenheit to Celsius would amount to shifts in both the origin and scale. One merely needs to recall the relationship $\frac{1}{5}C =$

$\frac{1}{9}(F - 32)$ between the two units, Fahrenheit and Celsius.

At this point, one may ask the following question. What is the relevance of such special families of distributions in the context of ancillarity? It may help if one goes back to the Examples 6.5.1-6.5.5 and thinks through the process of how we had formed some of the ancillary statistics. Suppose that $X_1, ..., X_n$ are iid random variables having the common pdf $f(x)$, *indexed* by some appropriate parameter(s). Then, we can conclude the following.

Let the common pdf of $X_1, ..., X_n$ belong to the *location* family $\mathcal{F}_1$ from the equation (6.5.7). Then the statistic $U = (X_1 - X_n, X_2 - X_n, ..., X_{n-1} - X_n)$ is ancillary.

(6.5.8)

Let the common pdf of $X_1, ..., X_n$ belong to the *scale* family $\mathcal{F}_2$ from the equation (6.5.7). Then, the statistic

$$V = \left(\frac{X_1}{X_n}, \frac{X_2}{X_n}, ..., \frac{X_{n-1}}{X_n} \right) \text{ is ancillary.}$$

(6.5.9)

Let the common pdf of $X_1, ..., X_n$ belong to the *location-scale* family $\mathcal{F}_3$ from the equation (6.5.7). Then, the statistic

$$W = \left(\frac{X_1 - X_n}{S}, \frac{X_2 - X_n}{S}, ..., \frac{X_{n-1} - X_n}{S} \right) \text{ is ancillary}$$

where S^2 is the sample variance.

(6.5.10)

One should not, however, get the impression that (6.5.8)-(6.5.10) list the unique or in some sense the "best" ancillary statistics. These summary statements and ancillary statistics should be viewed as building blocks to arrive at many forms of ancillary statistics.

6.5.2   Its Role in the Recovery of Information

In Examples 6.5.6-6.5.7, we had seen how ancillary statistics could play significant roles in conjunction with non-sufficient statistics. Suppose that T_1 is a non-sufficient statistic for θ and T_2 is ancillary for θ. In other words, in terms of the information content, $\mathcal{I}_{T_1}(\theta) < \mathcal{I}_{\mathbf{X}}(\theta)$ where $\mathbf{X}$ is the whole data and $\mathcal{I}_{T_2}(\theta) = 0$ for all $\theta \in \Theta$. Can we recover all the information contained in $\mathbf{X}$ by reporting T_1 while conditioning on the observed value of T_2? The answer is: we can do so and it is a fairly simple process. Such a process of conditioning has far reaching implications as emphasized by

Fisher (1934,1956) in the famous "Nile" example. One may also refer to Basu (1964), Hinkley (1980a), Ghosh (1988) and Reid (1995) for fuller discussions of *conditional inference*.

The approach goes through the following steps. One first finds the conditional pdf of T_1 at the point $T_1 = u$ given that $T_2 = v$, denoted by $g_{T_1 \mid v}(u; \theta)$. Using this conditional pdf, one obtains the information content, namely $\mathcal{I}_{T_1 \mid v}(\theta)$ following the Definition 6.4.1. In other words,

$$\mathcal{I}_{T_1 \mid v}(\theta) = E_\theta \left[\left\{ \tfrac{\partial}{\partial \theta} log\{g_{T_1 \mid v}(T_1; \theta)\} \right\}^2 \right]. \qquad (6.5.11)$$

In general, the expression of $\mathcal{I}_{T_1 \mid v}(\theta)$ would depend on v, the value of the ancillary statistic T_2. Next, one averages $\mathcal{I}_{T_1 \mid v}(\theta)$ over all possible values v, that is to take $E_{T_2}[\mathcal{I}_{T_1 \mid T_2}(\theta)]$. Once this last bit of averaging is done, it will coincide with the information content in the joint statistic (T_1, T_2), that is

$$\mathcal{I}_{T_1, T_2}(\theta) = E_{T_2} \left[\mathcal{I}_{T_1 \mid T_2}(\theta) \right]. \qquad (6.5.12)$$

This analysis provides a way to recover, in the sense of (6.5.12), the lost information due to reporting T_1 alone via conditioning on the ancillary statistic T_2. Few examples follow.

Example 6.5.10 (Example 6.5.1 Continued) Let X_1, X_2 be iid $N(\theta, 1)$ where $\theta \in (-\infty, \infty)$ is an unknown parameter. We know that $\overline{X}$ is sufficient for θ. Now, $\overline{X}$ is distributed as $N(\theta, \tfrac{1}{2})$ so that we can immediately write $\mathcal{I}_{\overline{X}}(\theta) = 2$. Now, $T_1 = X_1$ is not sufficient for θ since $\mathcal{I}_{X_1}(\theta) = 1 < \mathcal{I}_{\overline{X}}(\theta)$. That is, if we report only X_1 after the data (X_1, X_2) has been collected, there will be some loss of information. Next, consider an ancillary statistic, $T_2 = X_1 - X_2$ and now the joint distribution of (T_1, T_2) is $N_2(\theta, 0, 1, 2, \rho = \tfrac{1}{\sqrt{2}})$. Hence, using Theorem 3.6.1, we find that the conditional distribution of T_1 given $T_2 = v$ is $N(\theta + \tfrac{1}{2}v, \tfrac{1}{2})$, $v \in (-\infty, \infty)$. Thus, we first have $\mathcal{I}_{T_1 \mid v}(\theta) = E_{T_1 \mid v} \left[4(T_1 - \theta - \tfrac{1}{2}v)^2 \right] = 2$ and since this expression does not involve v, we then have $E_{T_2} \left[\mathcal{I}_{T_1 \mid T_2}(\theta) \right] = 2$ which equals $\mathcal{I}_{\overline{X}}(\theta)$. In other words, by conditioning on the ancillary statistic T_2, we have recovered the full information which is $\mathcal{I}_{\overline{X}}(\theta)$. ▲

Example 6.5.11 (Example 6.5.7 Continued) Suppose that (X, Y) is distributed as $N_2(0, 0, 1, 1, \rho)$ where the unknown parameter is the correlation coefficient $\rho \in (-1, 1)$. Now consider the two statistics X and Y. Individually, both $T_1 = X$ and $T_2 = Y$ are ancillary for ρ. Again, we utilize (6.5.11)-(6.5.12). Using the Theorem 3.6.1, we note that the conditional distribution of X given $Y = y$ is $N(\rho y, 1 - \rho^2)$ for $y \in (-\infty, \infty)$. That is, with $x \in (-\infty, \infty)$, we can write

$$f_{X \mid Y=y}(x; \rho) = \left\{ \sqrt{2\pi} \sqrt{1 - \rho^2} \right\}^{-1} exp \left[-\tfrac{1}{2}(x - \rho y)^2 / (1 - \rho^2) \right]$$

so that we have

$$\frac{\partial}{\partial \rho} log f_{X| Y=y}(x; \rho) = \frac{\rho}{1 - \rho^2} - \left[\frac{\rho(x - \rho y)^2}{(1 - \rho^2)^2} - \frac{y(x - \rho y)}{(1 - \rho^2)} \right].$$

In other words, the information about ρ contained in the conditional distribution of $T_1 \mid T_2 = v, v \in \Re$, is given by

$$E_{T_1|T_2=v} \left[\left\{ \frac{\partial}{\partial \rho} log f_{T_1| T_2=v}(T_1; \rho) \right\}^2 \right]$$

$$= E_{T_1|T_2=v} \left[\frac{\rho^2}{(1-\rho^2)^2} - \frac{2\rho}{1-\rho^2} \left\{ \frac{\rho(T_1-\rho v)^2}{(1-\rho^2)^2} - \frac{v(T_1-\rho v)}{(1-\rho^2)} \right\} \right]$$

$$\quad + E_{T_1|T_2=v} \left[\frac{v^2(T_1-\rho v)^2}{(1-\rho^2)^2} + \frac{\rho^2(T_1-\rho v)^4}{(1-\rho^2)^4} - \frac{2\rho v(T_1-\rho v)^3}{(1-\rho^2)^3} \right] \quad (6.5.13)$$

$$= \frac{\rho^2}{(1-\rho^2)^2} - \frac{2\rho}{1-\rho^2} \left\{ \frac{\rho(1-\rho^2)}{(1-\rho^2)^2} \right\} + \frac{y^2(1-\rho^2)}{(1-\rho^2)^2} + \frac{\rho^2 3(1-\rho^2)^2}{(1-\rho^2)^4}$$

$$= \frac{2\rho^2}{(1-\rho^2)^2} + \frac{v^2}{(1-\rho^2)},$$

which depends on the value v unlike what we had in the Example 6.5.10. Then, the information contained in (X, Y) will be given by

$$\mathcal{I}_{X,Y}(\rho) \quad = E_{T_2} \left[E_{T_1| T_2=v} \left(\left\{ \frac{\partial}{\partial \rho} log f_{T_1| T_2=v}(T_1; \rho) \right\}^2 \right) \right]$$

$$= E_{T_2} \left[\frac{2\rho^2}{(1-\rho^2)^2} + \frac{T_2^2}{(1-\rho^2)} \right] \quad (6.5.14)$$

$$= \frac{2\rho^2}{(1-\rho^2)^2} + \frac{1}{(1-\rho^2)} = \frac{1+\rho^2}{(1-\rho^2)^2} .$$

In other words, even though the statistic X tells us nothing about ρ, by averaging the conditional (on the statistic Y) information in X, we have recovered the full information about ρ contained in the whole data (X, Y). Refer to the Example 6.5.8. ▲

6.6   Completeness

Consider a real valued random variable X whose pmf or pdf is given by $f(x; \theta)$ for $x \in \mathcal{X}$ and $\theta \in \Theta$. Let $T = T(X)$ be a statistic and suppose that its pmf or pdf is denoted by $g(t; \theta)$ for $t \in \mathcal{T}$ and $\theta \in \Theta$.

Definition 6.6.1 *The collection of pmf's or pdf's denoted by $\{g(t; \theta)$: $\theta \in \Theta\}$ is called the family of distributions induced by the statistic T.*

Definition 6.6.2 *The family of distributions $\{g(t; \theta)$: $\theta \in \Theta\}$, induced by a statistic T, is called complete if and only if the following condition* *holds. Consider any real valued function $h(t)$ defined for $t \in \mathcal{T}$, having finite expectation, such that*

$$E_\theta\left[h(T)\right] = 0 \ \text{for all } \theta \in \Theta \ \text{implies } h(t) \equiv 0 \ w.p.1.$$

In other words,

$$\int_{t \in \mathcal{T}} h(t)g(t;\theta)dt = 0 \ \text{for all } \theta \in \Theta \ \text{implies } h(t) \equiv 0 \ w.p.1. \quad (6.6.1)$$

Definition 6.6.3 *A statistic T is said to be complete if and only if the family of distributions $\{g(t;\theta): \theta \in \Theta\}$ induced by T is complete.*

In these definitions, observe that neither the statistic T nor the parameter θ has to be necessarily real valued. If we have, for example, a vector valued statistic $\mathbf{T} = (T_1, T_2, ..., T_k)$ and $\boldsymbol{\theta}$ is also vector valued, then the requirement in (6.6.1) would obviously be replaced by the following:

$$\int \cdots \int_{\mathbf{t} \in \mathcal{T}} h(\mathbf{t})g(\mathbf{t};\boldsymbol{\theta})\Pi_{i=1}^k dt_i = 0 \ \text{for all } \boldsymbol{\theta} \in \Theta \ \text{implies}$$
$$h(\mathbf{t}) \equiv 0 \ w.p.1. \quad (6.6.2)$$

Here, $h(\mathbf{t})$ is a real valued function of $\mathbf{t} = (t_1, ..., t_k) \in \mathcal{T}$ and $g(\mathbf{t};\boldsymbol{\theta})$ is the pmf or the pdf of the statistic $\mathbf{T}$ at the point $\mathbf{t}$.

The concept of *completeness* was introduced by Lehmann and Scheffé (1950) and further explored measure-theoretically by Bahadur (1957). Next we give two examples. More examples will be given in the subsection 6.6.1.

Example 6.6.1 Suppose that a *statistic T* is distributed as Bernoulli(p), $0 < p < 1$. Let us examine if T is a complete statistic according to the Definition 6.6.3. The pmf induced by T is given by $g(t;p) = p^t(1-p)^{1-t}$, $t = 0, 1$. Consider any real valued function $h(t)$ such that $E_p[h(T)] = 0$ for all $0 < p < 1$. Now, let us focus on the possible values of $h(t)$ when $t = 0, 1$, and write

$$E_p[h(T)] = (1-p)h(0) + ph(1) = p\{h(1) - h(0)\} + h(0) = 0,$$
$$\text{for all } p \in (0,1). \quad (6.6.3)$$

Observe that the middle step in (6.6.3) is linear in p and so it *may* be zero for exactly one value of p between zero and unity. But we are demanding that $p\{h(1) - h(0)\} + h(0)$ must be zero for infinitely many values of p in $(0,1)$. Hence, this expression must be identically equal to zero which means that the constant term as well as the coefficient of p must individually be both zero. That is, we must have $h(0) = 0$ and $h(1) - h(0) = 0$ so that $h(1) = 0$. In other words, we have $h(t) \equiv 0$ for $t = 0, 1$. Thus, T is a complete statistic. ▲

Example 6.6.2 Suppose $g(t; 0, \sigma) = \{\sigma\sqrt{2\pi}\}^{-1}exp\{-\frac{1}{2}t^2/\sigma^2\}$, $-\infty < t < \infty, \sigma \in \Re^+$. Is the family of distributions $\{g(t; 0, \sigma): \sigma > 0\}$ complete? The answer is "no," it is not complete. In order to prove this claim, consider the function $h(t) = t$ and then observe that $E_\sigma[h(T)] = E_\sigma[T] = 0$ for all $0 < \sigma < \infty$, but $h(t)$ is not identically zero for all $t \in \Re$. Hence, the claim is true. ▲

6.6.1 Complete Sufficient Statistics

The completeness property of a statistic T looks like a mathematical concept. From the statistical point of view, however, this concept can lead to important results when a *complete statistic* T also happens to be *sufficient*.

Definition 6.6.4 *A statistic* **T** *is called complete sufficient for an unknown parameter* $\boldsymbol{\theta}$ *if and only if* (i) **T** *is sufficient for* $\boldsymbol{\theta}$ *and* (ii) **T** *is complete.*

In Section 6.6.3, we present a very useful theorem, known as Basu's Theorem, which needs a fusion of both concepts: sufficiency and completeness. Theorem 6.6.2 and (6.6.8) would help one to claim the minimality of a complete sufficient statistic. Other important applications which exploit the existence of a complete sufficient statistic will be mentioned in later chapters.

Example 6.6.3 Let $X_1, ..., X_n$ be iid Bernoulli(p), $0 < p < 1$. We already know that $T = \Sigma_{i=1}^n X_i$ is a sufficient statistic for p. Let us verify that T is complete too. Since T is distributed as Binomial(n, p), the pmf induced by T is given by $g(t; p) = \binom{n}{t}p^t(1 - p)^{n-t}$, $t \in \mathcal{T} = \{0, 1, ..., n\}$. Consider any real valued function $h(t)$ such that $E_p[h(T)] = 0$ for all $0 < p < 1$. Now, let us focus on the possible values of $h(t)$ for $t = 0, 1, ..., n$. With $\gamma = p(1 - p)^{-1}$, we can write

$$E_p[h(T)] = \Sigma_{t=0}^n h(t)\binom{n}{t}p^t(1 - p)^{n-t} = (1 - p)^n\Sigma_{t=0}^n\binom{n}{t}h(t)\gamma^t. \quad (6.6.4)$$

From (6.6.4) we observe that $E_p[h(T)]$ has been expressed as a polynomial of the nth degree in terms of the variable $\gamma \in (0, \infty)$. A nth degree polynomials in γ may be equal to zero for at most n values of $\gamma \in (0, \infty)$. If we are forced to assume that

$$E_p[h(T)] = 0 \text{ for all } \gamma \in (0, \infty),$$

then we conclude that $\Sigma_{t=0}^n\binom{n}{t}h(t)\gamma^t \equiv 0$, that is $\binom{n}{t}h(t) \equiv 0$ for all $t = 0, 1, ..., n$. In other words, $h(t) \equiv 0$ for all $t = 0, 1, ..., n$, proving the completeness of the sufficient statistic T. ▲

Example 6.6.4 Let $X_1, ..., X_n$ be iid Poisson(λ) where $\lambda(> 0)$ is the unknown parameter. We know that $T = \Sigma_{i=1}^n X_i$ is a sufficient statistic for λ. Let us verify that T is also complete. Since T is distributed as Poisson($n\lambda$), the pmf induced by T is given by $g(t; \lambda) = e^{-n\lambda}(n\lambda)^t/t!$, $t \in \mathcal{T} = \{0, 1, 2, ...\}$. Consider any real valued function $h(t)$ such that $E_\lambda[h(T)] = 0$ for all $0 < \lambda < \infty$. Now, let us focus on the possible values of $h(t)$ for $t = 0, 1, 2, ...$. With $k(t) = h(t)\, n^t/t!$, we can write

$$E_\lambda[h(T)] = \Sigma_{t=0}^\infty h(t) e^{-n\lambda}(n\lambda)^t/t! = e^{-n\lambda}\Sigma_{t=0}^\infty k(t)\lambda^t. \tag{6.6.5}$$

From (6.6.5) we observe that $E_\lambda[h(T)]$ has been expressed as an infinite power series in the variable λ belonging to $(0, \infty)$. The collection of such infinite power series forms a vector space generated by the set $\mathcal{G} = \{1, \lambda, \lambda^2, \lambda^3, ..., \lambda^t, ...\}$. Also, $\mathcal{G}$ happens to be the *smallest generator* in the sense that if any element is dropped from $\mathcal{G}$, then $\mathcal{G}$ will be unable to generate all infinite power series in λ. In other words, the vectors belonging to $\mathcal{G}$ are *linearly independent*. So, if we are forced to assume that

$$E_\lambda[h(T)] = 0 \text{ for all } \lambda \in (0, \infty),$$

we will then conclude that $\Sigma_{t=0}^\infty k(t)\lambda^t \equiv 0$, that is $k(t) \equiv 0$ for all $t = 0, 1, 2, ...$. Hence, we conclude that $h(t) \equiv 0$ for all $t = 0, 1, 2, ...$, proving the completeness of the sufficient statistic T. ▲

> In general, a sufficient or minimal sufficient statistic **T** for an unknown parameter $\boldsymbol{\theta}$ is not necessarily complete. Look at the Examples 6.6.5-6.6.6.

Example 6.6.5 Let $X_1, ..., X_n$ be iid Normal(θ, θ) where $\theta(> 0)$ is the unknown parameter. One should verify that $\mathbf{T} = (\overline{X}, S^2)$ is a minimal sufficient statistic for θ, but note that $E_\theta(\overline{X}) = \theta$ and $E_\theta(S^2) = \theta$ for all $\theta > 0$. That is, for all $\theta > 0$, we have $E_\theta(\overline{X} - S^2) = 0$. Consider $h(\mathbf{T}) = \overline{X} - S^2$ and then we have $E_\theta(h(\mathbf{T})) \equiv 0$ for all $\theta > 0$, but $h(\mathbf{t}) = \overline{x} - s^2$ is not identically zero for all $\mathbf{t} \in \Re \times \Re^+$. Hence, the minimal sufficient statistic **T** can not be complete. ▲

Example 6.6.6 Let $X_1, ..., X_n$ be iid with the common pdf given by $\theta^{-1}exp\{-(x - \theta)/\theta\}I(x > \theta)$ where $\theta(> 0)$ is the unknown parameter. We note that the likelihood function $L(\theta)$ is given by

$$\theta^{-n}exp\{-\Sigma_{i=1}^n(x_i - \theta)/\theta\}I(x_{n:1} > \theta)I(x_{n:1} < x_{n:n} < \infty) \tag{6.6.6}$$

and so it follows that $\mathbf{U} = (X_{n:1}, \Sigma_{i=1}^n X_i)$ is a minimal sufficient statistic for θ. But, the statistic $\mathbf{T} = (X_{n:1}, \Sigma_{i=1}^n(X_i - X_{n:1}))$ is a one-to-one function of

U, and hence by the Theorem 6.3.2, **T** itself is a minimal sufficient statistic for θ. But recall that $E_\theta(X_{n:1}) = (1 + n^{-1})\theta$ and $E_\theta[(n-1)^{-1}\Sigma_{i=1}^n(X_i - X_{n:1})] = \theta$ for all $\theta > 0$. Let us denote $h(\mathbf{t}) = (1 + n^{-1})^{-1}x_{n:1} - (n-1)^{-1}\Sigma_{i=1}^n(x_i - x_{n:1})$ for all $\mathbf{t} \in \mathcal{T} = (\theta, \infty) \times \Re^+$, so that $E_\theta[h(\mathbf{T})] = E_\theta[(1+n^{-1})^{-1}X_{n:1} - (n-1)^{-1}\Sigma_{i=1}^n(X_i - X_{n:1})] \equiv 0$ for all $\theta > 0$. But obviously $h(\mathbf{t})$ is not identically zero for all $\mathbf{t} \in \mathcal{T} = (\theta, \infty) \times \Re^+$. Hence, the minimal sufficient statistic **T** can not be complete. ▲

In the Examples 6.6.5-6.6.6, one can easily find other nontrivial real valued functions $h(\mathbf{t})$ such that $E_\theta[h(\mathbf{T})] \equiv 0$ for all θ, but $h(\mathbf{t})$ is not identically zero. We leave these as exercises.

Theorem 6.6.1 *Suppose that a statistic* $\mathbf{T} = (T_1, ..., T_k)$ *is complete. Let* $\mathbf{U} = (U_1, ..., U_k)$ *be another statistic with* $\mathbf{U} = f(\mathbf{T})$ *where* $g : \mathcal{T} \to \mathcal{U}$ *is one-to-one. Then,* **U** *is complete.*

Proof For simplicity, let us pretend that T, U have continuous distributions and that T, U and θ are all real valued. Let the pdf of T be denoted by $g(t; \theta)$. Since $f(.)$ is one-to-one, we can write:

$$E_\theta[a(U)] = \int_{\mathcal{U}} a(u)g(f^{-1}(u); \theta) \mid \frac{d}{du}f^{-1}(u) \mid du = \int_{\mathcal{T}} a(f(t))g(t; \theta)dt.$$
(6.6.7)

Now, assume that U is not complete. Then, there is a function $a(U)$ such that $E_\theta[a(U)] \equiv 0$ but $P_\theta\{u : a(u) \neq 0\} > 0$ for all $\theta \in \Theta$. From (6.6.7) we can then claim that $E_\theta[h(T)] \equiv 0$ for all $\theta \in \Theta$ where $h \equiv a \circ f$ stands for the composition mapping. But, this function $h(.)$ is non-zero with positive probability which contradicts the assumed completeness property of T. ∎

Are complete sufficient statistics also minimal sufficient? The answer is "yes." From the Examples 6.6.5-6.6.6, it should be clear, however, that the converse is not necessarily true. Refer to Lehmann and Scheffé (1950) and Bahadur (1957).   (6.6.8)

Now, we state a remarkably general result (Theorem 6.6.2) in the case of the exponential family of distributions. One may refer to Lehmann (1986, pp. 142-143) for a proof of this result.

Theorem 6.6.2 (Completeness of the Minimal Sufficient Statistic in the Exponential Family) *Suppose that* $X_1, ..., X_n$ *are iid with the common pmf or the pdf belonging to the k-parameter exponential family defined by*

$$f(x; \boldsymbol{\theta}) = q(\boldsymbol{\theta})p(x)exp\{\Sigma_{i=1}^k\theta_iR_i(x)\}$$
(6.6.9)

with some appropriate forms for $p(x) \geq 0$, $q(\boldsymbol{\theta}) \geq 0$, θ_i *and* $R_i(x)$, $i = 1, ..., k$. *Suppose that the regulatory conditions stated in (3.8.5) hold. Denote the statistic* $T_j = \Sigma_{i=1}^n R_j(X_i), j = 1, ..., k$. *Then, the (jointly) minimal sufficient statistic* $\mathbf{T} = (T_1, ..., T_k)$ *for* $\boldsymbol{\theta}$ *is complete.*

Theorem 6.6.2 will sometimes provide an easy route to claim the completeness of a sufficient statistic for $\boldsymbol{\theta}$ as long as $f(x; \boldsymbol{\theta})$ is given by (6.6.9) and the parameter space $\Theta(\subseteq \Re^k)$ includes a k-dimensional rectangle.

Example 6.6.7 Let $X_1, ..., X_n$ be iid $N(\mu, \sigma^2)$ with $(\mu, \sigma) \in \Theta = \Re \times \Re^+$ where μ and σ are both assumed unknown. The pdf of $N(\mu, \sigma^2)$ has the form given in (6.6.9) where $k = 2, \boldsymbol{\theta} = (\theta_1, \theta_2)$ with $\theta_1 = \mu/\sigma^2, \theta_2 = 1/\sigma^2$, $R_1(x) = x, R_2(x) = x^2$, so that the minimal sufficient statistic is $\mathbf{T} = \mathbf{T}(\mathbf{X}) = (T_1(\mathbf{X}), T_2(\mathbf{X}))$ where $T_1(\mathbf{X}) = \Sigma_{i=1}^n X_i$ and $T_2(\mathbf{X}) = \Sigma_{i=1}^n X_i^2$. In view of the Theorem 6.6.2, the statistic $\mathbf{T}(\mathbf{X})$ is complete. ▲

Example 6.6.8 Let $X_1, ..., X_n$ be iid Gamma(α, β) so that $f(x; \alpha, \beta) = \{\beta^\alpha \Gamma(\alpha)\}^{-1} exp(-x/\beta) x^{\alpha-1}$ with $(\alpha, \beta) \in \Re^+ \times \Re^+, x \in \Re^+$ where the parameters α and β are both assumed unknown. This pdf has the form given in (6.6.9) where $k = 2, \boldsymbol{\theta} = (\theta_1, \theta_2)$ with $\theta_1 = 1/\beta, \theta_2 = \alpha, R_1(x) = -x, R_2(x) = log(x)$, so that the sufficient statistic is $\mathbf{T}(\mathbf{X}) = (T_1(\mathbf{X}), T_2(\mathbf{X}))$ where $T_1(\mathbf{X}) = \Sigma_{i=1}^n X_i$ and $T_2(\mathbf{X}) = \Sigma_{i=1}^n log(X_i) = log\{\Pi_{i=1}^n X_i\}$. In view of the Theorem 6.6.2, the statistic $\mathbf{T}$ is complete. ▲

Example 6.6.9 In the Example 6.6.7, if μ is assumed unknown, but σ is known, then $T(\mathbf{X}) = \Sigma_{i=1}^n X_i$ or equivalently $\overline{X}$ is a complete sufficient statistic for μ. In that same example, if μ is known, but σ is unknown, then $T(\mathbf{X}) = \Sigma_{i=1}^n (X_i - \mu)^2$ is a complete sufficient statistic for σ. In the Example 6.6.8, if α is unknown, but β is known, then $T(\mathbf{X}) = \Pi_{i=1}^n X_i$ is a complete sufficient statistic for α. In this case if α is known, but β is unknown, then $T(\mathbf{X}) = \Sigma_{i=1}^n X_i$ or equivalently $\overline{X}$ is a complete sufficient statistic for β. These results follow immediately from the Theorem 6.6.2. ▲

Example 6.6.10 (Example 6.6.4 Continued) Suppose that we have $X_1, ..., X_n$ iid Poisson(λ) where $\lambda(> 0)$ is the unknown parameter. The common pmf $f(x; \lambda) = e^{-\lambda} \lambda^x / x!$ with $\mathcal{X} = \{0, 1, 2, ...\}, \theta = log(\lambda)$ has the same representation given in (6.6.9) where $k = 1, p(x) = (x!)^{-1}$, $q(\theta) = exp\{e^{-\theta}\}$ and $R(x) = x$. Hence, in view of the Theorem 6.6.2, the sufficient statistic $T = \Sigma_{i=1}^n X_i$ is then complete. ▲

Theorem 6.6.2 covers a lot of ground by helping to prove the completeness property of sufficient statistics. But it fails to reach out to non-exponential family members. The case in point will become clear from the Example 6.6.11.

Example 6.6.11 (Example 6.2.13 Continued) Suppose that $X_1, ..., X_n$ are iid Uniform$(0, \theta)$, $\theta(> 0)$ being the unknown parameter. We know that $T(\mathbf{X}) = X_{n:n}$ is a minimal sufficient statistic for θ and the pdf of T is given by $f(t; \theta) = nt^{n-1}\theta^{-n}I(0 < t < \theta)$ which does not belong to the exponential family defined by (6.6.9) with $k = 1$. But, we show directly that T is complete. Let $h(t)$, $0 < t < \theta$, be any arbitrary real valued function such that $E_\theta[h(T)] = 0$ for all $\theta > 0$ and we can write

$$0 = \tfrac{d}{d\theta}E_\theta[h(T)] = \tfrac{d}{d\theta}\int_0^\theta h(t)nt^{n-1}\theta^{-n}dt = n\theta^{-1}h(\theta),$$

which proves that $h(\theta) \equiv 0$ for all $\theta > 0$. We have now shown that the minimal sufficient statistic T is complete. See (1.6.16)-(1.6.17) for the rules on differentiating an integral. ▲

6.6.2 Basu's Theorem

Suppose that $\mathbf{X} = (X_1, ..., X_n)$ has the likelihood function L which depends on some unknown parameter θ and the observed value $\mathbf{x}$. It is not essential to assume that $X_1, ..., X_n$ are iid in the present setup. Consider now two *statistics* $\mathbf{U} = \mathbf{U}(\mathbf{X})$ and $\mathbf{W} = \mathbf{W}(\mathbf{X})$. In general, showing that the two statistics $\mathbf{U}$ and $\mathbf{W}$ are independent is a fairly tedious process. Usually, one first finds the joint pmf or pdf of $(\mathbf{U}, \mathbf{W})$ and then shows that it can be factored into the two marginal pmf's or pdf's of $\mathbf{U}$ and $\mathbf{W}$.

The following theorem, known as *Basu's Theorem*, provides a scenario under which we can prove *independence* of two appropriate statistics painlessly. Basu (1955a) came up with this elegant result which we state here under full generality.

Theorem 6.6.3 (Basu's Theorem) *Suppose that we have two vector valued statistics,* $\mathbf{U} = \mathbf{U}(\mathbf{X})$ *which is complete sufficient for* $\boldsymbol{\theta}$ *and* $\mathbf{W} = \mathbf{W}(\mathbf{X})$ *which is ancillary for* $\boldsymbol{\theta}$. *Then,* $\mathbf{U}$ *and* $\mathbf{W}$ *are independently distributed.*

Proof For simplicity, we supply a proof only in the discrete case. The proof in the continuous situation is similar. Suppose that the domain spaces for $\mathbf{U}$ and $\mathbf{W}$ are respectively denoted by $\mathcal{U}$ and $\mathcal{W}$.

In order to prove that $\mathbf{U}$ and $\mathbf{W}$ are independently distributed, we need to show that

$$P_\theta\{\mathbf{W} = \mathbf{w} \mid \mathbf{U} = \mathbf{u}\} = P_\theta(\mathbf{W} = \mathbf{w}) \text{ for all } \mathbf{w} \in \mathcal{W},$$
$$\mathbf{u} \in \mathcal{U}, \text{ and } \boldsymbol{\theta} \in \Theta. \tag{6.6.10}$$

Now, for $\mathbf{w} \in \mathcal{W}$, let us denote $P_\theta(\mathbf{W} = \mathbf{w}) = h(\mathbf{w})$. Obviously $h(\mathbf{w})$ is free from $\boldsymbol{\theta}$ since $\mathbf{W}$'s distribution does not involve the parameter $\boldsymbol{\theta}$.

But, observe that $P_\theta\{\mathbf{W} = \mathbf{w} \mid \mathbf{U} = \mathbf{u}\}$ must be free from θ since $\mathbf{U}$ is a sufficient statistic for θ. So, let us write $g(\mathbf{u}) = P_\theta\{\mathbf{W} = \mathbf{w} \mid \mathbf{U} = \mathbf{u}\}$. Now, $E_\theta[g(\mathbf{U})] = P_\theta(\mathbf{W} = \mathbf{w})$ which we had denoted earlier by $h(\mathbf{w})$, and this is free from θ. Hence, we have verified that $\{g(\mathbf{U}) - h(\mathbf{w})\}$ is a genuine *statistic*.

Now, we note that $E_\theta[g(\mathbf{U}) - h(\mathbf{w})] \equiv 0$ for all $\theta \in \Theta$ and use the fact that the statistic $\mathbf{U}$ is complete too. Thus, by the Definition 6.6.3 we must have $g(\mathbf{u}) - h(\mathbf{w}) \equiv 0$ w.p.1, that is, $g(\mathbf{u}) \equiv h(\mathbf{w})$ for all $\mathbf{w} \in \mathcal{W}$, $\mathbf{u} \in \mathcal{U}$. In other words, we have shown the validity of (6.6.10). ∎

Example 6.6.12 Let $X_1, ..., X_n$ be iid $N(\mu, \sigma^2)$ with $n \geq 2$, $(\mu, \sigma) \in \Re \times \Re^+$ where μ is unknown, but σ is known. Let $U = \overline{X}$ which is complete sufficient statistic for μ. Observe that $W = S^2$ is an ancillary statistic for μ. The fact that W is ancillary can be claimed from its explicit distribution or by appealing to the location family of distributions. Now, by Basu's Theorem, the two *statistics* $\overline{X}$ and S^2 are then independently distributed. The Theorem 4.4.2 showed this via Helmert transformation. ▲

Example 6.6.13 (Example 6.6.12 Continued) Let $V = X_{n:n} - X_{n:1}$, the sample range. Then, $\overline{X}$ and S/V are independently distributed. Also, $\overline{X}$ and $(X_{n:n} - \overline{X})$ are independent. In the same spirit, $\overline{X}$ and $(X_{n:1} - \overline{X})^2$ are independent. Is $\overline{X}$ independent of $\mid X_{n:n} - \overline{X} \mid /S$? The ancillarity of the corresponding statistics can be verified by appealing to the location family of distributions. We leave out the details as exercises. ▲

Example 6.6.14 Let $X_1, ..., X_n$ be iid $N(\mu, \sigma^2)$ with $n \geq 2$, $(\mu, \sigma) \in \Re \times \Re^+$ where σ is unknown, but μ is known. Let $U^2 = \Sigma_{i=1}^n (X_i - \mu)^2$ which is a complete sufficient statistic for σ^2. Observe that $W = (X_1 - \overline{X})/U$ is ancillary for σ. Immediately we can claim that $\Sigma_{i=1}^n (X_i - \mu)^2$ and W are independently distributed. Also, $\Sigma_{i=1}^n (X_i - \mu)^2$ and $(X_{n:1} - \overline{X})/S$ are independent where S^2 is the customary sample variance. Here, ancillarity of the corresponding statistics can be verified by appealing to the scale family of distributions. We leave out the details as exercises. ▲

In the $N(\mu, \sigma^2)$ case with μ, σ both unknown, Basu's Theorem can be used to show that $\overline{X}$ and S^2 are independent. See the Example 6.6.15. Compare its elegance with the brute-force approach given in the Example 4.4.9 via Helmert transformations.

Example 6.6.15 Suppose that $X_1, ..., X_n$ are iid $N(\mu, \sigma^2)$ with $n \geq 2$, $(\mu, \sigma) \in \Re \times \Re^+$ where μ and σ are both assumed unknown. Let us see how we can apply Basu's Theorem to show that $\overline{X}$ and S^2 are independently distributed. This clever application was originated by D. Basu. Let us have any two sets $A \subseteq \Re, B \subseteq \Re^+$ and we wish to verify that

$$P_{\mu,\sigma}\left\{\overline{X} \in A \cap S^2 \in B\right\} = P_{\mu,\sigma}\left\{\overline{X} \in A\right\} P_{\mu,\sigma}\left\{S^2 \in B\right\},$$
$$\text{for all fixed values of } (\mu, \sigma) \in \Re \times \Re^+. \tag{6.6.11}$$

We now work with a *fixed* but otherwise arbitrary value $\sigma = \sigma_0(> 0)$. In this situation, we may *pretend* that μ is really the only unknown parameter so that we are thrown back to the setup considered in the Example 6.6.12, and hence claim that $\overline{X}$ is complete sufficient for μ and S^2 is ancillary for μ. Thus, having *fixed* $\sigma = \sigma_0$, using Basu's Theorem we claim that $\overline{X}$ and S^2 will be independently distributed. That is, for *all* $\mu \in \Re$ and *fixed* $\sigma_0(> 0)$, we have so far shown that

$$P_{\mu,\sigma_0}\left\{\overline{X} \in A \cap S^2 \in B\right\} = P_{\mu,\sigma_0}\left\{\overline{X} \in A\right\} P_{\mu,\sigma_0}\left\{S^2 \in B\right\}. \tag{6.6.12}$$

But, then (6.6.12) holds for *any fixed value* $\sigma_0 \in \Re^+$. That is, we can claim the validity of (6.6.12) for all $(\mu, \sigma_0) \in \Re \times \Re^+$. There is no difference between what we have shown and what we started out to prove in (6.6.11). Hence, (6.6.11) holds. ▲

From the Example 6.6.15, the reader may think that we used the sufficiency property of $\overline{X}$ and ancillarity property of S^2. But if μ, σ are both unknown, then certainly $\overline{X}$ is *not* sufficient and S^2 is *not* ancillary. So, one may think that the previous proof must be wrong. But, note that we used the following two *facts* only: when $\sigma = \sigma_0$ is *fixed* but arbitrary, $\overline{X}$ *is sufficient* and S^2 *is ancillary.*

Example 6.6.16 (Example 6.6.15 Continued) In the Example 6.6.15, the statistic $\mathbf{U} = (\overline{X}, S^2)$ is complete sufficient for $\boldsymbol{\theta} = (\mu, \sigma^2)$ while $W = (X_1 - X_2)/S$ is a statistic whose distribution does not depend upon $\boldsymbol{\theta}$. Here, we may use the characteristics of a location-scale family. To check directly that W has a distribution which is free from $\boldsymbol{\theta}$, one may pursue as follows: Let $Y_i = (X_i - \mu)/\sigma$ which are iid standard normal, $i = 1, ..., n$, and then note that the statistic W can also be expressed as $(Y_1 - Y_2)/S^*$ where $S^{*2} = (n-1)^{-1}\Sigma_{i=1}^n(Y_i - \overline{Y})^2$ with $\overline{Y} = n^{-1}\Sigma_{i=1}^n Y_i$. The statistics $\mathbf{U}$ and W are independent by virtue of Basu's Theorem. In this deliberation, the ancillary statistic W can be vector valued too. For example, with $n \geq 3$, suppose that we define $\mathbf{W}^* = (\{X_1 - X_2\}/S, \{X_2 - X_3\}/|X_1 + X_2 - 2X_3|)$. As before, we can rewrite $\mathbf{W}^*$ as $(\{Y_1 - Y_2\}/S^*, \{Y_2 - Y_3\}/|Y_1 + Y_2 - 2Y_3|)$ where we recall that the Y's are iid standard normal, and hence $\mathbf{W}^*$ is ancillary for $\boldsymbol{\theta}$. Hence, $\mathbf{U}$ and $\mathbf{W}^*$ are independent by virtue of Basu's Theorem. ▲

In the Examples 6.6.12-6.6.15, the arguments revolved around sufficiency and completeness to come face to face with the result that $\overline{X}$ and S^2 are independent. A reader may get the wrong impression that the completeness property is essential to claim that $\overline{X}$ and S^2 are independent. But, note that $(\overline{X}, S^2)$ is *not* a complete statistic when we have random samples from $N(\theta, \theta)$ or $N(\theta, \theta^2)$ population, with $\theta > 0$. Yet it is true that $\overline{X}$ and S^2 are independent in such situations. Refer to the Example 4.4.9.

Example 6.6.17 (Example 6.6.15 Continued) Suppose that $X_1, ..., X_n$ are iid $N(\mu, \sigma^2)$ with $(\mu, \sigma) \in \Re \times \Re^+$ where μ and σ are both assumed unknown. By Basu's Theorem, one can immediately claim that the statistic $(\overline{X}, S^2)$ and $(X_1 - \overline{X})/(X_{n:n} - X_{n:1})$ are independent. ▲

Example 6.6.18 (Example 6.6.11 Continued) Suppose that $X_1, ..., X_n$ are iid Uniform$(0, \theta)$ with $n \geq 2, \theta(> 0)$ being the unknown parameter. We know that $U = X_{n:n}$ is a complete sufficient statistic for θ. Let $W = X_{n:1}/X_{n:n}$ which is ancillary for θ. Hence, by Basu's Theorem, $X_{n:n}$ and $X_{n:1}/X_{n:n}$ are independently distributed. Also, $X_{n:n}$ and $\overline{X}/S$ are independent, since $\overline{X}/S$ is ancillary for θ where S^2 stands for the sample variance. Using a similar argument, one can also claim that $(X_{n:n} - X_{n:1})/S$ and $X_{n:n}$ are independent. One may look at the scale family of distributions to verify the ancillarity property of the appropriate statistics. ▲

Remark 6.6.1 We add that a kind of the converse of Basu's Theorem was proved later in Basu (1958). Further details are omitted for brevity.

6.7 Exercises and Complements

6.2.1 Suppose that X_1, X_2 are iid Geometric(p), that is the common pmf is given by $f(x; p) = p(1 - p)^x$, $x = 0, 1, 2, ...$ where $0 < p < 1$ is the unknown parameter. By means of the conditional distribution approach, show that $X_1 + X_2$ is sufficient for p.

6.2.2 Suppose that $X_1, ..., X_m$ are iid Bernoulli(p), $Y_1, ..., Y_n$ are iid Bernoulli(q), and that the X's are independent of the Y's where $0 < p < 1$ is the unknown parameter with $q = 1 - p$. By means of the conditional distribution approach, show that $\Sigma_{i=1}^m X_i - \Sigma_{j=1}^n Y_j$ is sufficient for p. {*Hint*: Instead of looking at the data $(X_1, ..., X_m, Y_1, ..., Y_n)$, can one justify looking at $(X_1, ..., X_m, 1 - Y_1, ..., 1 - Y_n)$?}

6.2.3 Suppose that X is distributed as $N(0, \sigma^2)$ where $0 < \sigma < \infty$ is the unknown parameter. Along the lines of the Example 6.2.3, by means of the conditional distribution approach, show that the statistic $| X |$ is sufficient for σ^2.

6.2.4 Suppose that X is $N(\theta, 1)$ where $-\infty < \theta < \infty$ is the unknown parameter. By means of the conditional distribution approach, show that $| X |$ can not be sufficient for θ.

6.2.5 (Exercise 6.2.2 Continued) Suppose that $m = 2, n = 3$. By means of the conditional distribution approach, show that $X_1 + Y_1 Y_2$ can not be sufficient for p.

6.2.6 Suppose that $X_1, ..., X_m$ are distributed as iid Poisson(λ), $Y_1, ..., Y_n$ are iid Poisson(2λ), and that the X's are independent of the Y's where $0 < \lambda < \infty$ is the unknown parameter. By means of the conditional distribution approach, show that $\Sigma_{i=1}^m X_i + \Sigma_{j=1}^n Y_j$ is sufficient for λ.

6.2.7 (Exercise 6.2.6 Continued) Suppose that $m = 4, n = 5$. Show that $X_1 + Y_1$ can not be sufficient for λ.

6.2.8 (Exercise 6.2.6 Continued) Suppose that $m = 4, n = 0$. Show that $X_1 - X_2$ can not be sufficient for λ.

6.2.9 (Example 6.2.5 Continued) Let $X_1, ..., X_4$ be iid Bernoulli(p) where $0 < p < 1$ is the unknown parameter. Consider the statistic $U = X_1(X_3 + X_4) + X_2$. By means of the conditional distribution approach, show that the statistic U is not sufficient for p.

6.2.10 Let $X_1, ..., X_n$ be iid $N(\mu, \sigma^2)$ where $-\infty < \mu < \infty, 0 < \sigma < \infty$. Use the Neyman Factorization Theorem to show that

(i) $\overline{X} = n^{-1}\Sigma_{i=1}^n X_i$ is sufficient for μ if σ is known;

(ii) $n^{-1}\Sigma_{i=1}^n(X_i - \mu)^2$ is sufficient for σ if μ is known.

6.2.11 Let $X_1, ..., X_n$ be iid having the common pdf $\sigma^{-1}exp\{-(x - \mu)/\sigma\}I(x > \mu)$ where $-\infty < \mu < \infty, 0 < \sigma < \infty$. Use the Neyman Factorization Theorem to show that

(i) $X_{n:1}$, the smallest order statistic, is sufficient for μ if σ is known;

(ii) $n^{-1}\Sigma_{i=1}^n(X_i - \mu)$ is sufficient for σ if μ is known;

(iii) $(X_{n:1}, \Sigma_{i=1}^n(X_i - X_{n:1}))$ is jointly sufficient for (μ, σ) if both parameters are unknown.

6.2.12 Let $X_1, ..., X_n$ be iid having the Beta(α, β) distribution with the parameters α and β, so that the common pdf is given by

$$f(x; \alpha, \beta) = \{b(\alpha, \beta)\}^{-1}x^{\alpha-1}(1 - x)^{\beta-1}, \ 0 < x < 1, \alpha > 0, \beta > 0,$$

where $b(\alpha, \beta) = \Gamma(\alpha)\Gamma(\beta)\{\Gamma(\alpha + \beta)\}^{-1}$. Refer to (1.7.35) as needed. Use the Neyman Factorization Theorem to show that

(i) $\Pi_{i=1}^n X_i$ is sufficient for α if β is known;

(ii) $\Pi_{i=1}^n (1 - X_i)$ is sufficient for β if α is known;

(iii) $(\Pi_{i=1}^n X_i, \Pi_{i=1}^n (1 - X_i))$ is jointly sufficient for (α, β) if both parameters are unknown.

6.2.13 Suppose that $X_1, ..., X_n$ are iid with the Uniform distribution on the interval $(\theta - \frac{1}{2}, \theta + \frac{1}{2})$, that is the common pdf is given by $f(x; \theta) = I(\theta - \frac{1}{2} < x < \theta + \frac{1}{2})$ where $\theta (> 0)$ is the unknown parameter. Show that $(X_{n:1}, X_{n:n})$ is jointly sufficient for θ.

6.2.14 Solve the Examples 6.2.8-6.2.11 by applying the Theorem 6.2.2 based on the exponential family.

6.2.15 Prove the Theorem 6.2.2 using the Neyman Fctorization Theorem.

6.2.16 Suppose that $X_1, ..., X_n$ are iid with the Rayleigh distribution, that is the common pdf is

$$f(x; \theta) = 2\theta^{-1} x \exp(-x^2/\theta) I(x > 0),$$

where $\theta (> 0)$ is the unknown parameter. Show that $\Sigma_{i=1}^n X_i^2$ is sufficient for θ.

6.2.17 Suppose that $X_1, ..., X_n$ are iid with the Weibull distribution, that is the common pdf is

$$f(x; \alpha) = \alpha^{-1} \beta x^{\beta-1} \exp(-x^\beta/\alpha) I(x > 0),$$

where $\alpha (> 0)$ is the unknown parameter, but $\beta (> 0)$ is assumed known. Show that $\Sigma_{i=1}^n X_i^\beta$ is sufficient for α.

6.3.1 (Exercise 6.2.10 Continued) Let $X_1, ..., X_n$ be iid $N(\mu, \sigma^2)$ where $-\infty < \mu < \infty, 0 < \sigma < \infty$. Show that

(i) $\overline{X} = n^{-1} \Sigma_{i=1}^n X_i$ is minimal sufficient for μ if σ is known;

(ii) $n^{-1} \Sigma_{i=1}^n (X_i - \mu)^2$ is minimal sufficient for σ if μ is known.

6.3.2 (Exercise 6.2.11 Continued) Let $X_1, ..., X_n$ be iid having the common pdf $\sigma^{-1} \exp\{-(x - \mu)/\sigma\} I(x > \mu)$ where $-\infty < \mu < \infty, 0 < \sigma < \infty$. Show that

(i) $X_{n:1}$, the smallest order statistic, is minimal sufficient for μ if σ is known;

(ii) $n^{-1} \Sigma_{i=1}^n (X_i - \mu)$ is minimal sufficient for σ if μ is known;

(iii) $(X_{n:1}, \Sigma_{i=1}^n (X_i - X_{n:1}))$ is minimal sufficient for (μ, σ) if both parameters are unknown.

6.3.3 Solve the Exercises 6.2.5-6.2.8 along the lines of the Examples 6.3.2 and 6.3.5.

6.3.4 Show that the pmf or pdf corresponding to the distributions such as Binomial(n, p), Poisson(λ), Gamma(α, β), $N(\mu, \sigma^2)$, Beta(α, β) belong to the exponential family defined in (6.3.5) when

    (i) $0 < p < 1$ is unknown;
    (ii) $\lambda \in \Re^+$ is unknown;
    (iii) $\mu \in \Re$ is unknown but $\sigma \in \Re^+$ is known;
    (iv) $\sigma \in \Re^+$ is unknown but $\mu \in \Re$ is known;
    (v) $\mu \in \Re$ and $\sigma \in \Re^+$ are both unknown;
    (vi) $\alpha \in \Re^+$ is known but $\beta \in \Re^+$ is unknown;
    (vii) $\alpha \in \Re^+, \beta \in \Re^+$ are both unknown.

In each case, obtain the minimal sufficient statistic(s) for the associated unknown parameter(s).

6.3.5 (Exercise 6.2.1 Continued) Let $X_1, ..., X_n$ be Geometric(p), that is the common pmf is given by $f(x; p) = p(1 - p)^x$, $x = 0, 1, 2, ...$ where $0 < p < 1$ is the unknown parameter. Show that this pmf belongs to the exponential family defined in (6.3.5). Hence, show that $\Sigma_{i=1}^n X_i$ is minimal sufficient for p.

6.3.6. Show that the common pdf given in the Exercises 6.2.16-6.2.17 respectively belongs to the exponential family.

6.3.7 Suppose that $X_1, ..., X_n$ are iid with the Uniform distribution on the interval $(\theta - \frac{1}{2}, \theta + \frac{1}{2})$, that is the common pdf is given by $f(x; \theta) = I(\theta - \frac{1}{2} < x < \theta + \frac{1}{2})$ where $\theta (> 0)$ is the unknown parameter. Show that $(X_{n:1}, X_{n:n})$ is jointly minimal sufficient for θ.

6.3.8 Let $X_1, ..., X_n$ be iid $N(\theta, \theta^2)$ where $0 < \theta < \infty$ is the unknown parameter. Derive the minimal sufficient statistic for θ. Does the common pdf belong to the exponential family (6.3.5)? {*Hint*: If it does belong to the exponential family, it should be a 2-parameter family, but it is not so. Does the parameter space include a 2-dimensional rectangle?}

6.3.9 Let $X_1, ..., X_n$ be iid $N(\theta, \theta)$ where $0 < \theta < \infty$ is the unknown parameter. Derive the minimal sufficient statistic for θ. Does the common pdf belong to the exponential family (6.3.5)? {*Hint*: If it does belong to the exponential family, it should be a 2-parameter family, but it is not so. Does the parameter space include a 2-dimensional rectangle?}

6.3.10 Let $X_1, ..., X_n$ be iid having a negative exponential distribution with the common pdf $\theta^{-1} exp\{-(x - \theta)/\theta\} I(x > \theta)$ where $0 < \theta < \infty$ is the unknown parameter. Derive the minimal sufficient statistics for θ. Does the common pdf belong to the exponential family (6.3.5)? {*Hint*: If it does belong to the exponential family, it should be a 2-parameter family, but it is not so. Does the parameter space include a 2-dimensional rectangle?}

6.3.11 Let $X_1, ..., X_n$ be iid having a negative exponential distribution with the common pdf $\theta^{-2} exp\{-(x-\theta)/\theta^2\}I(x > \theta)$ where $0 < \theta < \infty$ is the unknown parameter. Derive the minimal sufficient statistics for θ. Does the common pdf belong to the exponential family (6.3.5)? {*Hint*: If it does belong to the exponential family, it should be a 2-parameter family, but it is not so. Does the parameter space include a 2-dimensional rectangle?}

6.3.12 Let $X_1, ..., X_n$ be iid having the common Uniform distribution on the interval $(-\theta, \theta)$ where $0 < \theta < \infty$ is the unknown parameter. Derive the minimal sufficient statistic for θ.

6.3.13 Let $X_1, ..., X_m$ be iid $N(\mu_1, \sigma^2)$, $Y_1, ..., Y_n$ be iid $N(\mu_2, \sigma^2)$, and also let the X's be independent of the Y's where $-\infty < \mu_1, \mu_2 < \infty$, $0 < \sigma < \infty$ are the unknown parameters. Derive the minimal sufficient statistics for (μ_1, μ_2, σ^2).

6.3.14 (Exercise 6.3.13 Continued) Let $X_1, ..., X_m$ be iid $N(\mu_1, \sigma^2)$, $Y_1, ..., Y_n$ be iid $N(\mu_2, k\sigma^2)$, and also let the X's be independent of the Y's where $-\infty < \mu_1, \mu_2 < \infty$, $0 < \sigma < \infty$ are the unknown parameters. Assume that the number k (> 0) is known. Derive the minimal sufficient statistic for (μ_1, μ_2, σ^2).

6.3.15 Let $X_1, ..., X_m$ be iid Gamma(α, β), $Y_1, ..., Y_n$ be iid Gamma$(\alpha, k\beta)$, and also let the X's be independent of the Y's with $0 < \alpha, \beta < \infty$ where β is the only unknown parameter. Assume that the number k (> 0) is known. Derive the minimal sufficient statistic for β.

6.3.16 (Exercise 6.2.12 Continued) Let $X_1, ..., X_n$ be iid having a Beta distribution with its parameters $\alpha = \beta = \theta$ where θ (> 0) is unknown. Derive the minimal sufficient statistic for θ.

6.3.17 Let $X_1, ..., X_m$ be iid $N(\mu, \sigma^2)$, $Y_1, ..., Y_n$ be iid $N(0, \sigma^2)$, and also let the X's be independent of the Y's where $-\infty < \mu < \infty$, $0 < \sigma < \infty$ are the unknown parameters. Derive the minimal sufficient statistics for (μ, σ^2).

6.3.18 Let $X_1, ..., X_m$ be iid $N(\mu, \sigma^2)$, $Y_1, ..., Y_n$ be iid $N(0, k\sigma^2)$, and also let the X's be independent of the Y's where $-\infty < \mu < \infty$, $0 < \sigma < \infty$ are the unknown parameters. Assume that the number k (> 0) is known. Derive the minimal sufficient statistics for (μ, σ^2).

6.3.19 Suppose that $X_1, ..., X_n$ are iid with the common pdf given by one of the following:

(i) $f(x;\theta) = exp\{-(x-\theta)\}/[1 + exp\{-(x-\theta)\}]^2, x \in \Re, \theta \in \Re$,
which is called the *logistic* distribution;

(ii) $f(x;\theta) = \frac{1}{\pi}\{1 + (x-\theta)^2\}^{-1}, x \in \Re, \theta \in \Re$;

(iii) $f(x;\theta) = \frac{1}{2}exp\{-|x-\theta|\}, x \in \Re, \theta \in \Re$.

In each case, show that the order statistics $(X_{n:1}, ..., X_{n:n})$ is minimal sufficient for the *unknown location parameter* θ. That is, we do not achieve any significant reduction of the original data $\mathbf{X} = (X_1, ..., X_n)$. {*Hint*: Part (i) is proved in Lehmann (1983, p. 43). Parts (ii) and (iii) can be handled similarly.}

6.4.1 Let $X_1, ..., X_n$ be iid Bernoulli(p) where $0 < p < 1$ is the unknown parameter. Evaluate $\mathcal{I}_{\mathbf{X}}(p)$, the information content in the whole data $\mathbf{X} = (X_1, ..., X_n)$. Compare $\mathcal{I}_{\mathbf{X}}(p)$ with $\mathcal{I}_{\overline{X}}(p)$. Can the Theorem 6.4.2 be used here to claim that $\overline{X}$ is sufficient for p?

6.4.2 (Exercise 6.4.1 Continued) Let $X_1, ..., X_n$ be iid Bernoulli(p) where $0 < p < 1$ is the unknown parameter, $n \geq 3$. Let $T = X_1 + X_2, U = X_1 + X_2 + 2X_3$. Compare $\mathcal{I}_{\mathbf{X}}(p)$ with $\mathcal{I}_T(p)$ and $\mathcal{I}_U(p)$. Can T be sufficient for p? Can U be sufficient for p? {*Hint*: Try to exploit the Theorem 6.4.2.}

6.4.3 Verify the results given in equations (6.4.9) and (6.4.11).

6.4.4 (Exercise 6.2.1 Continued) Let X_1, X_2 be iid Geometric(p) where $0 < p < 1$ is the unknown parameter. Let $\mathbf{X} = (X_1, X_2)$, and $T = X_1 + X_2$ which is sufficient for p. Evaluate $\mathcal{I}_{\mathbf{X}}(p)$ and $\mathcal{I}_T(p)$, and then compare these two information contents.

6.4.5 (Exercise 6.2.10 Continued) In a $N(\mu, \sigma^2)$ distribution where $-\infty < \mu < \infty$, $0 < \sigma < \infty$, suppose that only μ is known. Show that $\mathcal{I}_{U^2}(\sigma^2) > \mathcal{I}_{S^2}(\sigma^2)$ where $U^2 = n^{-1}\Sigma_{i=1}^n(X_i - \mu)^2$ and $S^2 = (n-1)^{-1}\Sigma_{i=1}^n(X_i - \overline{X})^2, n \geq 2$. Would it then be fair to say that there is no point in using the statistic S^2 while making inferences about σ^2 when μ is assumed known?

6.4.6 (Exercise 6.2.11 Continued) In the two-parameter negative exponential distribution, if only μ is known, show that $\mathcal{I}_V(\sigma) > \mathcal{I}_T(\sigma)$ where $V = n^{-1}\Sigma_{i=1}^n(X_i - \mu)$ and $T = (n-1)^{-1}\Sigma_{i=1}^n(X_i - X_{n:1}), n \geq 2$. Would it then be fair to say that there is no point in using the statistic T while making inferences about σ when μ is assumed known?

6.4.7 (Exercise 6.3.17 Continued) Suppose that we have $X_1, ..., X_m$ are iid $N(\mu, \sigma^2)$, $Y_1, ..., Y_n$ are iid $N(0, \sigma^2)$, the X's are independent of the Y's where $-\infty < \mu < \infty$, $0 < \sigma < \infty$ are the unknown parameters. Suppose that $\mathbf{T} = \mathbf{T}(\mathbf{X}, \mathbf{Y})$ is the minimal sufficient statistic for $\boldsymbol{\theta} = (\mu, \sigma^2)$. Evaluate the expressions of the information matrices $\mathcal{I}_{\mathbf{X},\mathbf{Y}}(\boldsymbol{\theta})$ and $\mathcal{I}_{\mathbf{T}}(\boldsymbol{\theta})$, and then compare these two information contents.

6.4.8 (Example 6.2.5 Continued) Consider the statistics $X_1 X_2$ and U.

Find the pmf of $X_1 X_2$ and then using the fact that $X_1 X_2$ is independent of X_3, derive the pmf of U. Hence, find the expressions of $\mathcal{I}_{\mathbf{X}}(p), \mathcal{I}_{(X_1 X_2, X_3)}(p)$ and $\mathcal{I}_U(p)$ where $\mathbf{X} = (X_1, X_2, X_3)$. Show that $\mathcal{I}_U(p) < \mathcal{I}_{(X_1 X_2, X_3)}(p) < \mathcal{I}_{\mathbf{X}}(p)$. This shows that U or $(X_1 X_2, X_3)$ can not be sufficient for p.

6.4.9 (Exercise 6.2.16 Continued) Suppose that $X_1, ..., X_n$ are iid with the Rayleigh distribution, that is the common pdf is

$$f(x; \theta) = 2\theta^{-1} x \exp(-x^2/\theta) I(x > 0),$$

where $\theta(> 0)$ is the unknown parameter. Denote the statistic $T = \Sigma_{i=1}^n X_i^2$. Evaluate $\mathcal{I}_{\mathbf{X}}(\theta)$ and $\mathcal{I}_T(\theta)$. Are these two information contents the same? What conclusions can one draw from this comparison?

6.4.10 (Exercise 6.2.17 Continued) Suppose that $X_1, ..., X_n$ are iid with the Weibull distribution, that is the common pdf is

$$f(x; \theta) = \alpha^{-1} \beta x^{\beta-1} \exp(-x^\beta/\alpha) I(x > 0),$$

where $\alpha(> 0)$ is the unknown parameter, but $\beta(> 0)$ is assumed known. Denote the statistic $T = \Sigma_{i=1}^n X_i^\beta$. Evaluate $\mathcal{I}_{\mathbf{X}}(\theta)$ and $\mathcal{I}_T(\theta)$. Are these two information contents the same? What conclusions can one draw from this comparison?

6.4.11 Prove the Theorem 6.4.3. {*Hint*: One may proceed along the lines of the proof given in the case of the Theorem 6.4.1.}

6.4.12 Prove the Theorem 6.4.4 when

(i) $\mathbf{X}, \boldsymbol{\theta}$ are both vector valued, but $\mathbf{X}$ has a continuous distribution;

(ii) X, θ are both real valued, but X has a discrete distribution.

6.4.13 (Exercise 6.3.8 Continued) Let $X_1, ..., X_n$ be iid $N(\theta, \theta^2)$ where $0 < \theta < \infty$ is the unknown parameter. Evaluate $\mathcal{I}_{\mathbf{X}}(\theta)$ and $\mathcal{I}_{\overline{X}}(\theta)$, and compare these two quantities. Is it possible to claim that $\overline{X}$ is sufficient for θ?

6.4.14 (Exercise 6.3.9 Continued) Let $X_1, ..., X_n$ be iid $N(\theta, \theta)$ where $0 < \theta < \infty$ is the unknown parameter. Evaluate $\mathcal{I}_{\mathbf{X}}(\theta)$ and $\mathcal{I}_{\overline{X}}(\theta)$, and compare these two quantities. Is it possible to claim that $\overline{X}$ is sufficient for θ?

6.4.15 (Proof of the Theorem 6.4.2) Let θ be a real valued parameter. Suppose that $\mathbf{X}$ is the whole data and $T = T(\mathbf{X})$ is some statistic. Then, show that $\mathcal{I}_{\mathbf{X}}(\theta) \geq \mathcal{I}_T(\theta)$ for all $\theta \in \Theta$. Also verify that the two information measures match with each other for all θ if and only if T is a sufficient statistic for θ. {*Hint*: This interesting idea was included in a recent personal communication from C. R. Rao. Note that the likelihood function $f(\mathbf{x}; \theta)$ can be written as $g(t; \theta)h(\mathbf{x} \mid T = t; \theta)$ where $g(t; \theta)$ is the pdf or pmf of T and $h(\mathbf{x} \mid T = t; \theta)$ is the conditional pdf or pmf of $\mathbf{X}$ given that $T = t$. From this equality, first derive the identity: $\mathcal{I}_{\mathbf{X}}(\theta) = \mathcal{I}_T(\theta) + \mathcal{I}_{\mathbf{X}|T}(\theta)$ for all $\theta \in \Theta$. This implies that $\mathcal{I}_{\mathbf{X}}(\theta) \geq \mathcal{I}_T(\theta)$ for all $\theta \in \Theta$. One will have $\mathcal{I}_{\mathbf{X}}(\theta) = \mathcal{I}_T(\theta)$ if and only if $\mathcal{I}_{\mathbf{X}|T}(\theta) = 0$, that is $h(\mathbf{x} \mid T = t; \theta)$ must be free from θ. It then follows from the definition of sufficiency that T is sufficient for θ.}

6.4.16 Suppose that X_1, X_2 are iid $N(\theta, 1)$ where the unknown parameter $\theta \in \Re$. We know that $T = X_1 + X_2$ is sufficient for θ and also that $\mathcal{I}_T(\theta) = 2$. Next, consider a statistic $U_p = X_1 + pX_2$ where p is a known positive number.

(i)   Show that $\mathcal{I}_{U_p}(\theta) = (1 + p)^2/(1 + p^2)$;

(ii)   Show that $1 < \mathcal{I}_{U_p}(\theta) \leq 2$ for all $p(> 0)$;

(iii) Show that $\mathcal{I}_{U_p}(\theta) = 2$ if and only if $p = 1$;

(iv)   An expression such as $\mathcal{I}_{U_p}(\theta)/\mathcal{I}_T(\theta)$ may be used to quantify the extent of *non-sufficiency* or the fraction of the lost information due to using U_p instead of utilizing the sufficient statistic T. Analytically, study the behavior of $\mathcal{I}_{U_p}(\theta)/\mathcal{I}_T(\theta)$ as a function of $p(> 0)$;

(v)   Evaluate $\mathcal{I}_{U_p}(\theta)/\mathcal{I}_T(\theta)$ for $p = .90, .95, .99, 1.01, 1.05, 1.1$. Is U_p too *non-sufficient* for θ in practice when $p = .90, .95, .99, 1.01$?

(vi)   In practice, if an experimenter is willing to accept at the most 1% lost information compared with the full information $\mathcal{I}_T(\theta)$, find the range of values of p for which the non-sufficient statistic U_p will attain the goal.

{*Note*: This exercise exploits much of what has been learned conceptually from the Theorem 6.4.2.}

6.5.1 (Example 6.5.2 Continued) Find the pdf of U_3 defined in (6.5.1). Hence, show that T_3 is ancillary for θ.

6.5.2 (Example 6.5.3 Continued) Show that T_3 is ancillary for λ.

6.5.3 (Example 6.5.4 Continued) Show that $\mathbf{T} = (T_1, T_2)$ is distributed as $N_2(0, 0, 2, 6, 0)$. Are T_1, T_2 independent?

6.5.4 (Example 6.5.5 Continued) Show that $\mathbf{T}$ is ancillary for $\boldsymbol{\theta}$.

6.5.5 (Exercise 6.3.7 Continued) Let $X_1, ..., X_n$ be iid having the common Uniform distribution on the interval $(\theta - \frac{1}{2}, \theta + \frac{1}{2})$ where $-\infty < \theta < \infty$ is the unknown parameter. Show that $X_{n:n} - X_{n:1}$ is an ancillary statistic for θ.

6.5.6 (Exercise 6.3.12 Continued) Let $X_1, ..., X_n$ be iid having the common Uniform distribution on the interval $(-\theta, \theta)$ where $0 < \theta < \infty$ is the unknown parameter. Show that $X_{n:n}/X_{n:1}$ is an ancillary statistic for θ. Also show that $X_{n:n}/(X_{n:n} - X_{n:1})$ is an ancillary statistic for θ.

6.5.7 (Curved Exponential Family) Suppose that (X, Y) has a particular curved exponential family of distributions with the joint pdf given by

$$f(x, y; \theta) = \begin{cases} exp\{-\theta x - \theta^{-1}y\} & \text{if } 0 < x, y < \infty \\ 0 & \text{elsewhere,} \end{cases}$$

where $\theta(> 0)$ is the unknown parameter. This distribution was discussed by Fisher (1934,1956) in the context of the famous "Nile" example. Denote $U = XY, V = X/Y$. Show that U is ancillary for θ, but (U, V) is minimal sufficient for θ, whereas V by itself is not sufficient for θ. {*Hint*: The answers to the Exercise 4.4.10 would help in this problem.}

6.5.8 Show that the families $\mathcal{F}_1, \mathcal{F}_2, \mathcal{F}_3$ defined via (6.5.7) include only genuine pdf's.

6.5.9 (Exercise 6.3.8 Continued) Let $X_1, ..., X_n$ be iid $N(\theta, \theta^2)$ where $0 < \theta < \infty$ is the unknown parameter. Construct several ancillary statistics for θ. Does the common pdf belong to one of the special families defined via (6.5.7)?

6.5.10 (Exercise 6.3.9 Continued) Let $X_1, ..., X_n$ be iid $N(\theta, \theta)$ where $0 < \theta < \infty$ is the unknown parameter. Construct several ancillary statistics for θ. Does the common pdf belong to one of the special families defined via (6.5.7)?

6.5.11 (Exercise 6.3.10 Continued) Let $X_1, ..., X_n$ be iid having the common pdf $\theta^{-1}exp\{-(x - \theta)/\theta\}I(x > \theta)$ where $0 < \theta < \infty$ is the unknown parameter. Construct several ancillary statistics for θ. Does the common pdf belong to one of the special families defined via (6.5.7)?

6.5.12 (Exercise 6.3.11 Continued) Let $X_1, ..., X_n$ be iid having the common pdf $\theta^{-2}exp\{-(x - \theta)/\theta^2\}I(x > \theta)$ where $0 < \theta < \infty$ is the unknown parameter. Construct several ancillary statistics for θ. Does the common pdf belong to one of the special families defined via (6.5.7)?

6.5.13 Use the Definition 6.4.1 of the information and the pdf from (6.5.4) to derive directly the expression for $\mathcal{I}_{X,Y}(\rho)$. {*Hint*: Take the *log* of the pdf from (6.5.4). Then take the partial derivative of this with respect to ρ. Next, square this partial derivative and take its expectation.}

6.5.14 Suppose that (X, Y) is distributed as $N_2(0, 0, \sigma^2, \sigma^2, \rho)$ where $-1 < \rho < 1$ is the unknown parameter while $0 < \sigma \, (\neq 1) < \infty$ is assumed known. Evaluate the expression for the information matrix $\mathcal{I}_{X,Y}(\rho)$ along the lines of the Example 6.5.8. Next, along the lines of the Example 6.5.11, recover the lost information in X by means of conditioning on the ancillary statistic Y.

6.5.15 (Example 6.5.8 Continued) Suppose that (X, Y) is distributed as $N_2(0, 0, \sigma^2, \sigma^2, \rho)$ where $\boldsymbol{\theta} = (\sigma^2, \rho)$ with the unknown parameters σ^2, ρ where $0 < \sigma < \infty, -1 < \rho < 1$. Evaluate the expression for the information matrix $\mathcal{I}_{X,Y}(\boldsymbol{\theta})$. {*Hint*: Does working with $U = X + Y, V = X - Y$ help in the derivation?}

6.5.16 Suppose that $\mathbf{X}' = (X_1, ..., X_p)$ where $\mathbf{X}$ is distributed as multivariate normal $N_p\left(0, \sigma^2[(1 - \rho)I_{p \times p} + \rho \mathbf{11}']\right)$ with $\mathbf{1}' = (1, 1, ...1)$, $\sigma \in \Re^+$ and $-(p - 1)^{-1} < \rho < 1$. We assume that $\boldsymbol{\theta} = (\sigma^2, \rho)$ where σ^2, ρ are the unknown parameters, $0 < \sigma < \infty, -1 < \rho < 1$. Evaluate the expression for the information matrix $\mathcal{I}_{\mathbf{X}}(\boldsymbol{\theta})$. {*Hint*: Try the Helmert transformation from the Example 2.4.9 on $\mathbf{X}$ to generate p *independent* normal variables each with zero mean, and variances depending on both σ^2 and ρ, while $(p - 1)$ of these variances are all equal but different from the pth one. Is it then possible to use the Theorem 6.4.3?}

6.5.17 (Example 6.5.8 Continued) Suppose that (X, Y) is distributed as $N_2(0, 0, 1, 1, \rho)$ where ρ is the unknown parameter, $-1 < \rho < 1$. Start with the pdf of (X, Y) and then directly apply the equivalent formula from the equation (6.4.9) for the evaluation of the expression of $\mathcal{I}_{X,Y}(\rho)$.

6.6.1 Suppose that X_1, X_2 are iid Poisson(λ) where $\lambda(> 0)$ is the unknown parameter. Consider the family of distributions induced by the statistic $\mathbf{T} = (X_1, X_2)$. Is this family, indexed by λ, complete?

6.6.2 (Exercise 6.3.5 Continued) Let $X_1, ..., X_n$ be iid Geometric(p), that is the common pmf is given by $f(x; p) = p(1 - p)^x$, $x = 0, 1, 2, ...$ where $0 < p < 1$ is the unknown parameter. Is the statistic $\sum_{i=1}^{n} X_i$ complete sufficient for p? {*Hint*: Is it possible to use the Theorem 6.6.2 here?}

6.6.3 Let $X_1, ..., X_n$ be iid $N(\theta, \theta^2)$ where $0 < \theta < \infty$ is the unknown parameter. Is the minimal sufficient statistic complete? Show that $\overline{X}$ and S^2 are independent.{*Hint*: Can the Example 4.4.9 be used here to solve the second part?}

6.6.4 Let $X_1, ..., X_n$ be iid $N(\theta, \theta)$ where $0 < \theta < \infty$ is the unknown parameter. Is the minimal sufficient statistic complete? Show that $\overline{X}$ and S^2 are independent.{*Hint*: Can the Example 4.4.9 be used here to solve the second part?}

6.6.5 Let $X_1, ..., X_n$ be iid having a negative exponential distribution with the common pdf $\theta^{-1} exp\{-(x - \theta)/\theta\}I(x > \theta)$ where $0 < \theta < \infty$ is the unknown parameter. Is the minimal sufficient statistic complete? Show that $X_{n:1}$ and $\Sigma_{i=1}^n(X_i - X_{n:1})$ are independent.{*Hint*: Can the Example 4.4.12 be used here to solve the second part?}

6.6.6 Let $X_1, ..., X_n$ be iid having a negative exponential distribution with the common pdf $\theta^{-2}exp\{-(x - \theta)/\theta^2\}I(x > \theta)$ where $0 < \theta < \infty$ is the unknown parameter. Is the minimal sufficient statistic complete? Show that $X_{n:1}$ and $\Sigma_{i=1}^n(X_i - X_{n:1})$ are independent.{*Hint*: Can the Example 4.4.12 be used here to solve the second part?}

6.6.7 Let $X_1, ..., X_n$ be iid having the common Uniform distribution on the interval $\left(\theta - \frac{1}{2}, \theta + \frac{1}{2}\right)$where $-\infty < \theta < \infty$ is the unknown parameter. Is the minimal sufficient statistic $\mathbf{T} = (X_{n:1}, X_{n:n})$ complete?

6.6.8 Let $X_1, ..., X_n$ be iid having the common Uniform distribution on the interval $(-\theta, \theta)$ where $0 < \theta < \infty$ is the unknown parameter. Find the minimal sufficient statistic T for θ. Is the statistic T complete?{*Hint*: Use indicator functions appropriately so that the problem reduces to the common pdf's of the random variables $|X_i|, i = 1, ..., n.$}

6.6.9 Let $X_1, ..., X_m$ be iid $N(\mu_1, \sigma^2)$, $Y_1, ..., Y_n$ be iid $N(\mu_2, \sigma^2)$, the X's are independent of the Y's where $-\infty < \mu_1, \mu_2 < \infty, 0 < \sigma < \infty$ are all unknown parameters. Is the minimal sufficient statistic for (μ_1, μ_2, σ^2) complete?{*Hint*: Is it possible to use the Theorem 6.6.2 here?}

6.6.10 (Exercise 6.3.13 Continued) Let $X_1, ..., X_m$ be iid $N(\mu_1, \sigma^2)$, $Y_1, ..., Y_n$ be iid $N(\mu_2, k\sigma^2)$, the X's be independent of the Y's where $-\infty < \mu_1, \mu_2 < \infty, 0 < \sigma < \infty$ are all unknown parameters, but k (> 0) is known. Is the minimal sufficient statistic for (μ_1, μ_2, σ^2) complete? {*Hint*: Is it possible to use the Theorem 6.6.2 here?}

6.6.11 (Exercise 6.3.15 Continued) Let $X_1, ..., X_m$ be iid Gamma(α, β), $Y_1, ..., Y_n$ be iid Gamma$(\alpha, k\beta)$, the X's be independent of the Y's, with $0 < \alpha, \beta < \infty$ where β is the *only* unknown parameter. Assume that the number k (> 0) is known. Is the minimal sufficient statistic for β complete?{*Hint*: Is it possible to use the Theorem 6.6.2 here?}

6.6.12 (Exercise 6.3.16 Continued) Let $X_1, ..., X_n$ be iid having a Beta distribution with its parameters $\alpha = \beta = \theta$ where θ (> 0) is unknown. Is the minimal sufficient statistic for θ complete? {*Hint*: Is it possible to use the Theorem 6.6.2 here?}

6.6.13 (Exercise 6.6.9 Continued) Let the X's be independent of the Y's, $X_1, ..., X_m$ be iid $N(\mu_1, \sigma^2)$, $Y_1, ..., Y_n$ be iid $N(\mu_2, \sigma^2)$ where $-\infty < \mu_1, \mu_2 < \infty, 0 < \sigma < \infty$ are all unknown parameters. Use Basu's Theorem along the lines of the Example 6.6.15 to show that $(\overline{X}, \overline{Y})$ is distributed independently of $T = \Sigma_{i=1}^{m}(X_i - \overline{X})^2 + \Sigma_{i=1}^{n}(Y_i - \overline{Y})^2$. Is $(\overline{X} - \overline{Y})^3$ distributed independently of T? Is $\{(X_{n:n} - \overline{X}) - (Y_{n:n} - \overline{Y})\}^2/T$ distributed independently of T? Is $\{(X_{n:n} - \overline{X}) - (Y_{n:n} - \overline{Y})\}^2/(X_{n:n} - X_{n:1})^2$ distributed independently of T? {*Hint*: Can the characteristics of a location-scale family and (6.5.10) be used here?}

6.6.14 (Exercise 6.6.8 Continued) Let $X_1, ..., X_n$ be iid having the common Uniform distribution on the interval $(-\theta, \theta)$ where $0 < \theta < \infty$ is the unknown parameter. Let us denote $T = \max\limits_{1 \le i \le n} |X_i|$. Is T distributed independently of $|X_{n:1}|/X_{n:n}$? Is T distributed independently of $(X_1 - X_2)^2/\{X_{n:1}X_{n:n}\}$? {*Hint*: Can the characteristics of a scale family and (6.5.9) be used here?}

6.6.15 (Exercise 6.6.14 Continued) Let $X_1, ..., X_n$ be iid having the common Uniform distribution on the interval $(-\theta, \theta)$ where $0 < \theta < \infty$ is the unknown parameter. Let us denote $T = \max\limits_{1 \le i \le n} |X_i|$. Is T distributed independently of the two-dimensional statistic (U_1, U_2) where $U_1 = |X_{n:1}|/X_{n:n}, U_2 = (X_1 - X_2)^2/\{X_{n:1}X_{n:n}\}$? {*Hint*: Can the characteristics of a scale family and (6.5.9) be used here?}

6.6.16 (Exercise 6.6.13 Continued) Let $X_1, ..., X_m$ be iid $N(\mu_1, \sigma^2)$, $Y_1, ..., Y_n$ be iid $N(\mu_2, \sigma^2)$, the X's be independent of the Y's where $-\infty < \mu_1, \mu_2 < \infty, 0 < \sigma < \infty$ are all unknown parameters. Use Basu's Theorem to check whether the following two-dimensional statistics

$$\left(\{\overline{X} - \overline{Y}\}^3, \{\overline{X} - \overline{Y}\}^2/T\right), \text{ and}$$

$$\left(\{\overline{X} - Y_{n:n} - X_1 + Y_2\}^2/T, \ \{\overline{X} - \overline{Y} - X_2 + Y_{n:1}\}^3/|X_{n:n} - X_{n:1}|^3\right)$$

are distributed independently where $T = \Sigma_{i=1}^{m}(X_i - \overline{X})^2 + \Sigma_{i=1}^{n}(Y_i - \overline{Y})^2$. {*Hint*: Can the characteristics of a location-scale family and (6.5.10) be used here?}

6.6.17 Let $X_1, ..., X_n$ be iid having a negative exponential distribution with the common pdf $\sigma^{-1}exp\{-(x-\theta)/\sigma\}I(x > \theta)$ where θ and σ are both unknown parameters, $-\infty < \theta < \infty, 0 < \sigma < \infty$. Argue as in the Example 6.6.15 to show that $X_{n:1}$ and $\Sigma_{i=1}^{n}(X_i - X_{n:1})$ are independent.

6.6.18 (Exercise 6.3.2 Continued) Let $X_1, ..., X_n$ be iid having the common pdf $\sigma^{-1}exp\{-(x - \mu)/\sigma\}I(x > \mu)$ where $-\infty < \mu < \infty, 0 < \sigma < \infty$. Show that

(*i*) $X_{n:1}$, the smallest order statistic, is complete if μ is unknown but σ is known;

(*ii*) $n^{-1}\Sigma_{i=1}^{n}(X_i - \mu)$ is complete if σ is unknown but μ is known;

(*iii*) $(X_{n:1}, \Sigma_{i=1}^{n}(X_i - X_{n:1}))$ is complete if both μ, σ are unknown.

{*Hint*: Part (ii) follows from the Theorem 6.6.2. One may look at Lehmann (1983, pp. 47-48) for details.}

6.6.19 Let $X_1, ..., X_4$ be iid $N(\mu, \sigma^2)$ where $-\infty < \mu < \infty$, $0 < \sigma < \infty$ are both unknown parameters. Evaluate the expression for

$$Cov\left(\frac{X_1 - X_4}{|\overline{X} - X_3|}, \frac{|\overline{X}|}{S}\right),$$

where $\overline{X} = \frac{1}{4}\Sigma_{i=1}^4 X_i$ and $S^2 = \frac{1}{3}\Sigma_{i=1}^4 (X_i - \overline{X})^2$. {*Note*: Can Basu's Theorem be used here?}

7

Point Estimation

7.1 Introduction

We begin with iid *observable* random variables $X_1, ..., X_n$ having a common pmf or pdf $f(x)$, $x \in \mathcal{X}$, the domain space for x. Here, n is assumed known and it is referred to as the (fixed) sample size. We like to think of a *population* distribution which in practice may perhaps be approximated by $f(x)$. For example, in a population of two thousand juniors in a college campus, we may be interested in the average $GPA(= X)$ and its distribution which is denoted by $f(x)$ with an appropriate domain space $\mathcal{X} = (0, 4)$ for x. The population distribution of X may be characterized or indexed by some *parameter* $\boldsymbol{\theta}$, for example the median GPA of the population. The practical significance of $\boldsymbol{\theta}$ is that once we fix a value of $\boldsymbol{\theta}$, the population distribution $f(x)$ would then be completely specified. Thus, we denote the pmf or pdf by $f(x; \boldsymbol{\theta})$, instead of just $f(x)$, so that its dependence on a few specific population-features (that is, $\boldsymbol{\theta}$) is now made explicit. The idea of indexing a population distribution by the unknown parameter $\boldsymbol{\theta}$ was also discussed in Chapter 6.

We suppose that the parameter $\boldsymbol{\theta}$ is fixed but otherwise unknown and that the possible values of $\boldsymbol{\theta}$ belong to a *parameter space* $\Theta \subseteq \Re^k$. For example, we may be able to postulate that the X's are distributed as $N(\mu, \sigma^2)$ where μ is the only unknown parameter, $-\infty < \mu < \infty, 0 < \sigma < \infty$. In this case we may denote the pdf of X by $f(x; \mu)$ where $\theta = \mu \in \Theta = \Re, \mathcal{X} = \Re$. If μ and σ^2 are both unknown, then the population density would be denoted by $f(x; \boldsymbol{\theta})$ where $\boldsymbol{\theta} = (\mu, \sigma^2) \in \Theta = \Re \times \Re^+, \mathcal{X} = \Re$.

The Definition 7.2.1 gives a formal statement of what an estimator of a parameter is. In this chapter, we consider only *point estimation* problems. In Section 7.2, we first apply the *method of moments*, originated by Karl Pearson (1902), to find estimators of $\boldsymbol{\theta}$. This approach is ad hoc in nature, and hence we later introduce a more elaborate way of finding estimators by the *method of maximum likelihood*. The latter approach was pioneered by Fisher (1922,1925a,1934).

One may arrive at different choices of estimators for the unknown parameter $\boldsymbol{\theta}$ and hence some criteria to compare their performances are addressed in Section 7.3. One criterion which stands out more than anything else is called *unbiasedness* of an estimator and we follow this up with a notion of the *best estimator* among all unbiased estimators. Sections 7.4 and 7.5 include several fundamental results, for example, the Rao-Blackwell Theorem, Cramér-Rao inequality, and Lehmann-Scheffé Theorems. These machineries are useful in finding the best unbiased estimator of θ in different situations. The Section 7.6 addresses a situation which arises when the *Rao-Blackwellization* technique is used but the minimal sufficient statistic is not complete. In Section 7.7, an attractive large sample criterion called *consistency*, proposed by Fisher (1922), is discussed.

7.2   Finding Estimators

Consider iid and *observable* real valued random variables $X_1, ..., X_n$ from a population with the common pmf or pdf $f(x; \theta)$ where the unknown parameter $\theta \in \Theta \subseteq \Re^k$. It is not essential for the X's to be real valued or iid. But, in many examples they will be so and hence we assume that the X's are real valued and iid unless specified otherwise. As before, we denote $\mathbf{X} = (X_1, ..., X_n)$.

Definition 7.2.1 *An estimator or a point estimator of the unknown parameter θ is merely a function $\mathbf{T} \equiv \mathbf{T}(X_1, ..., X_n)$ which is allowed to depend only on the observable random variables $X_1, ..., X_n$. That is, once a particular data $\mathbf{X} = \mathbf{x}$ has been observed, the numerical value of $\mathbf{T}(\mathbf{x})$ must be computable. We distinguish between $T(\mathbf{X})$ and $T(\mathbf{x})$ by referring to them as an estimator and an estimate of θ respectively.*

An arbitrary estimator T of a real valued parameter θ, for example, can be practically any function which depends on the observable random variables alone. In some problem, we may think of $X_1, X_1^2 - \overline{X}, X_{n:n} - S, \overline{X}, S^2$ and so on as competing estimators. At this point, the only restriction we have to watch for is that T must be computable in order to qualify to be called an estimator. In the following sections, two different methods are provided for locating competing estimators of θ.

7.2.1   The Method of Moments

During the late nineteenth and early twentieth centuries, Karl Pearson was the key figure in the major methodological developments in statistics. During his long career, Karl Pearson pioneered on many fronts. He originated innovative ideas of curve fitting to observational data and did fundamental research with correlations and causation in a series of multivariate data from anthropometry and astronomy . The method of moments was introduced by Pearson (1902).

The methodology is very simple. Suppose that $\theta = (\theta_1, ..., \theta_k)$. Derive the first k theoretical moments of the distribution $f(x; \theta)$ and pretend that they are equal to the corresponding sample moments, thereby obtaining k equations in k unknown parameters $\theta_1, ..., \theta_k$. Next, simultaneously solve these k equations for $\theta_1, ..., \theta_k$. The solutions are then the estimators of $\theta_1, ..., \theta_k$. Refer back to the Section 2.3 as needed. To be more specific, we proceed as follows: We write

$$\eta_1 \equiv \eta_1(\theta_1, ..., \theta_k) = \int_{\mathcal{X}} x f(x; \theta) dx \overset{\text{Let}}{=} n^{-1} \Sigma_{i=1}^n X_i$$

$$\eta_2 \equiv \eta_2(\theta_1, ..., \theta_k) = \int_{\mathcal{X}} x^2 f(x; \theta) dx \overset{\text{Let}}{=} n^{-1} \Sigma_{i=1}^n X_i^2$$

$$\vdots \qquad \qquad \qquad \vdots \qquad \qquad (7.2.1)$$

$$\eta_k \equiv \eta_k(\theta_1, ..., \theta_k) = \int_{\mathcal{X}} x^k f(x; \theta) dx \overset{\text{Let}}{=} n^{-1} \Sigma_{i=1}^n X_i^k$$

Now, having observed the data $\mathbf{X} = \mathbf{x}$, the expressions given in the rhs of (7.2.1) can all be evaluated, and hence we will have k separate equations in k unknown parameters $\theta_1, ..., \theta_k$. These equations are solved simultaneously. The following examples would clarify the technique.

Often $\widehat{\theta}$ is written for an estimator of the unknown parameter θ.

Example 7.2.1 (Example 6.2.8 Continued) Suppose that $X_1, ..., X_n$ are iid Bernoulli(p) where p is unknown, $0 < p < 1$. Here $\mathcal{X} = \{0, 1\}$, $\theta = p$ and $\Theta = (0, 1)$. Observe that $\eta_1 \equiv \eta_1(\theta) = E_p[X_1] = p$, and let us pretend that $p = n^{-1} \Sigma_{i=1}^n X_i = \overline{X}$, the sample mean. Hence, $T = \overline{X}$ would be the estimator of p obtained by the method of moments. We write $\widehat{p} = \overline{X}$ which happens to be sufficient for the parameter p too. ▲

Example 7.2.2 (Example 6.2.10 Continued) Suppose that $X_1, ..., X_n$ are iid $N(\mu, \sigma^2)$ where μ, σ^2 are both unknown with $n \geq 2$, $\theta = (\mu, \sigma^2)$, $\theta_1 = \mu$, $\theta_2 = \sigma^2$, $-\infty < \mu < \infty$, $0 < \sigma < \infty$. Here $\mathcal{X} = \Re$ and $\Theta = \Re \times \Re^+$. Observe that $\eta_1 \equiv \eta_1(\theta_1, \theta_2) = E_\theta[X_1] = \mu$ and $\eta_2 \equiv \eta_2(\theta_1, \theta_2) = E_\theta[X_1^2] = \mu^2 + \sigma^2$, so that (7.2.1) would lead to the two equations,

$$\mu = n^{-1} \Sigma_{i=1}^n X_i \text{ and } \mu^2 + \sigma^2 = n^{-1} \Sigma_{i=1}^n X_i^2.$$

After solving these two equations simultaneously for μ and σ^2, we obtain the estimators $\widehat{\mu} = \overline{X}$ and $\widehat{\sigma}^2 = n^{-1} \Sigma_{i=1}^n X_i^2 - \overline{X}^2 = n^{-1} \Sigma_{i=1}^n (X_i - \overline{X})^2$. The estimator of σ^2 coincides with $(n-1)n^{-1}S^2$. Note, however, that $(\widehat{\mu}, \widehat{\sigma}^2)$ is sufficient for θ too. ▲

Example 7.2.3 Suppose that $X_1, ..., X_n$ are iid $N(0, \sigma^2)$ where $\theta = \sigma^2$ is unknown, $0 < \sigma < \infty$. Here $\mathcal{X} = \Re^+$ and $\Theta = \Re^+$. Observe that $\eta_1 \equiv \eta_1(\theta) = E_\theta[X_1] = 0$ for all θ. In this situation, it is clear that the equation given by the first moment in (7.2.1) does not lead to anything interesting and one may *arbitrarily* move to use the second moment. Note that $\eta_2 \equiv \eta_2(\theta) = E_\theta[X_1^2] = \sigma^2$ so that (7.2.1) will now lead to $\widehat{\sigma}^2 = n^{-1}\Sigma_{i=1}^n X_i^2$. After such *ad hoc* adjustment, the method of moment estimator turns out to be sufficient for σ^2. ▲

> The method of moments is an ad hoc way to find estimators. Also, this method may not lead to estimators which are functions of minimal sufficient statistics for θ. Look at Examples 7.2.4-7.2.5.

If any of the theoretical moments $\eta_1, \eta_2, ..., \eta_k$ is zero, then one will continue to work with the first k non-zero theoretical moments so that (7.2.1) may lead to sensible solutions. Of course, there is a lot of arbitrariness in this approach.

Example 7.2.4 Suppose that $X_1, ..., X_n$ are iid Poisson(λ) where $\theta = \lambda$ is unknown, $0 < \lambda < \infty$. Here $\mathcal{X} = \{0, 1, 2, ...\}$ and $\Theta = \Re^+$. Now, $\eta_1 \equiv \eta_1(\theta) = E_\theta[X_1] = \lambda$ and $\eta_2 \equiv \eta_2(\theta) = E_\theta[X_1^2] = \lambda + \lambda^2$. Suppose that instead of starting with η_1 in (7.2.1), we start with η_2 and equate this with $n^{-1}\Sigma_{i=1}^n X_i^2$. This then provides the equation

$$\lambda + \lambda^2 = n^{-1}\Sigma_{i=1}^n X_i^2,$$

which leads to the estimator $\widehat{\lambda} = \frac{1}{2}[\{4n^{-1}\Sigma_{i=1}^n X_i^2 + 1\}^{1/2} - 1]$. However, if we had started with η_1, we would have ended up with the estimator $\widehat{\lambda} = \overline{X}$, a minimal sufficient statistic for λ. The first estimator is *not* sufficient for λ. From this example, one can feel the sense of arbitrariness built within this methodology. ▲

Example 7.2.5 Suppose that $X_1, ..., X_n$ are iid Uniform$(0, \theta)$ where $\theta(> 0)$ is the unknown parameter. Here $\eta_1 \equiv \eta_1(\theta) = E_\theta[X_1] = \frac{1}{2}\theta$ so that by equating this with $\overline{X}$ we obtain $\widehat{\theta} = 2\overline{X}$ which is *not* sufficient for θ. Recall that $X_{n:n}$ is a minimal sufficient statistic for θ. ▲

7.2.2   The Method of Maximum Likelihood

The method of moments appeared quite simple but it was ad hoc and arbitrary in its approach. In the Example 7.2.3 we saw that we could not equate

η_1 and $\overline{X}$ to come up with an estimator of σ^2. From the Examples 7.2.4-7.2.5, it is clear that the method of moments may lead to estimators which will depend upon non-sufficient statistics. Next, on top of this, if we face situations where theoretical moments are infinite, we can not hope to apply this method. R. A. Fisher certainly realized the pitfalls of this methodology and started criticizing Karl Pearson's way of finding estimators early on. Fisher (1912) was critical of Pearson's approach of curve fitting and wrote on page 54 that "The method of moments ... though its arbitrary nature is apparent" and went on to formulate the *method of maximum likelihood* in the same paper. Fisher's preliminary ideas took concrete shapes in a path-breaking article appearing in 1922 and followed by more elaborate discussions laid out in Fisher (1925a,1934).

Consider $X_1, ..., X_n$ which are iid with the common pmf or pdf $f(x; \boldsymbol{\theta})$ where $x \in \mathcal{X} \subseteq \Re$ and $\boldsymbol{\theta} = (\theta_1, ..., \theta_k) \in \Theta \subseteq \Re^k$. Here $\theta_1, ..., \theta_k$ are all assumed unknown and thus $\boldsymbol{\theta}$ is an unknown vector valued parameter. Recall the notion of a *likelihood function* defined in (6.2.4). Having observed the data $\mathbf{X} = \mathbf{x}$, we write down the likelihood function

$$L(\boldsymbol{\theta}) = \Pi_{i=1}^n f(x_i; \boldsymbol{\theta}). \tag{7.2.2}$$

Note that the observed data $\mathbf{x} = (x_1, ..., x_n)$ is arbitrary but otherwise held fixed.

> Throughout this chapter and the ones that follow, we essentially pay attention to the likelihood function when it is positive.

Definition 7.2.2 *The maximum likelihood estimate of $\boldsymbol{\theta}$ is the value $\widehat{\boldsymbol{\theta}} \equiv \widehat{\boldsymbol{\theta}}(\mathbf{x})$ for which $L(\widehat{\boldsymbol{\theta}}) = \underset{\boldsymbol{\theta} \in \Theta}{Sup}\, L(\boldsymbol{\theta})$. The maximum likelihood estimator (MLE) of $\boldsymbol{\theta}$ is denoted by $\widehat{\boldsymbol{\theta}}(\mathbf{X})$. If we write $\widehat{\boldsymbol{\theta}}$, the context will dictate whether it is referring to an estimate or an estimator of $\boldsymbol{\theta}$.*

When the X's are discrete, $L(\boldsymbol{\theta})$ stands for $P_{\boldsymbol{\theta}}\{\mathbf{X} = \mathbf{x}\}$, that is the probability of observing the type of data on hand when $\boldsymbol{\theta}$ is the true value of the unknown parameter. The MLE is interpreted as the value of $\boldsymbol{\theta}$ which maximizes the chance of observing the particular data we already have on hand. Instead when the X's are continuous, a similar interpretation is given by replacing the probability statement with an analogous statement using the joint pdf of $\mathbf{X}$.

As far as the definition of MLE goes, there is no hard and fast dictum regarding any specific mathematical method to follow in order to locate where $L(\boldsymbol{\theta})$ attains its supremum. If $L(\boldsymbol{\theta})$ is a twice differentiable function of $\boldsymbol{\theta}$, then we may apply the standard techniques from differential calculus to find $\widehat{\theta}$. See the Section 1.6 for some review. Sometimes we take the natural logarithm of $L(\theta)$ first, and then maximize the logarithm instead to obtain $\widehat{\theta}$.

There are situations where one finds a unique solution $\widehat{\theta}$ or situations where we find more than one solution $\widehat{\theta}$ which globally maximize $L(\theta)$. In some situations there may not be any solution $\widehat{\theta}$ which will globally maximize $L(\theta)$. But, from our discourse it will become clear that quite often a unique MLE $\widehat{\theta}$ of θ will exist and we would be able to find it explicitly. In the sequel, we temporarily write c throughout to denote a *generic* constant which does not depend upon the unknown parameter θ.

Example 7.2.6 Suppose that $X_1, ..., X_n$ are iid $N(\mu, \sigma^2)$ where μ is unknown but σ^2 is known. Here we have $-\infty < \mu < \infty$, $0 < \sigma < \infty$ and $\mathcal{X} = \Re$. The likelihood function is given by

$$L(\mu) = \Pi_{i=1}^{n}\{\sigma\sqrt{2\pi}\}^{-n}exp\{-\tfrac{1}{2\sigma^2}\Sigma_{i=1}^{n}(x_i - \mu)^2\}, \qquad (7.2.3)$$

which is to be maximized with respect to μ. This is *equivalent* to maximizing $logL(\mu)$ with respect to μ. Now, we have

$$logL(\mu) = c - (2\sigma^2)^{-1}\Sigma_{i=1}^{n}(x_i - \mu)^2,$$

and hence, $\frac{d}{d\mu}logL(\mu) = -\tfrac{1}{2\sigma^2}\frac{d}{d\mu}\{\Sigma_{i=1}^{n}(x_i - \mu)^2\} = -\tfrac{1}{2\sigma^2}\{-2\Sigma_{i=1}^{n}(x_i - \mu)\} = \Sigma_{i=1}^{n}(x_i-\mu)/\sigma^2$. Next, equate $\frac{d}{d\mu}logL(\mu)$ to zero and solve for μ. But, $\frac{d}{d\mu}logL(\mu) = 0$ implies that $\mu = \overline{x}$, and so we would say that $\widehat{\mu} = \overline{X}$. At this step, our only concern should be to decide whether $\mu = \overline{x}$ really maximizes $logL(\mu)$. Towards that end, observe that $\frac{d^2}{d\mu^2}logL(\mu)\mid_{\mu=\overline{x}} = -n/\sigma^2$ which is negative, and this shows that $L(\mu)$ is globally maximized at $\mu = \overline{x}$. Thus the MLE for μ is $\overline{X}$, the sample mean.

Suppose that we had observed the following set of data from a normal population: $x_1 = 11.4058, x_2 = 9.7311, x_3 = 8.2280, x_4 = 8.5678$ and $x_5 = 8.6006$ with $n = 5$ and $\sigma = 1$.

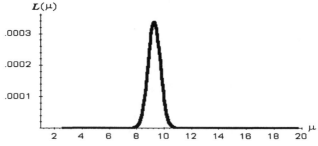

Figure 7.2.1. Likelihood Function $L(\mu)$ When the Mean μ
Varies from $2.5 - 19.7$.

In the Figure 7.2.1, the variable μ runs from 2.5 through 19.7. It becomes *practically* clear from the Figure 7.2.1 that $L(\mu)$ attains its maximum at only one point which is around 9.3. Using MAPLE, we found that the likelihood function $L(\mu)$ was maximized at $\mu = 9.3067$. For the observed data, the sample mean happens to be $\overline{x} = \frac{46.5333}{5} = 9.30666$. In the end, we may add that the observed data was generated from a normal population with $\mu = 10$ and $\sigma = 1$ using MINITAB Release 12.1. ▲

Example 7.2.7 Suppose that $X_1, ..., X_n$ are iid $N(\mu, \sigma^2)$ where μ and σ^2 are both unknown, $\boldsymbol{\theta} = (\mu, \sigma^2)$, $-\infty < \mu < \infty$, $0 < \sigma < \infty$, $n \geq 2$. Here we have $\mathcal{X} = \Re$ and $\Theta = \Re \times \Re^+$. We wish to find the MLE for $\boldsymbol{\theta}$. In this problem, the likelihood function is given by

$$L(\mu, \sigma^2) = \Pi_{i=1}^n \{2\pi\}^{-n/2} \{\sigma^2\}^{-n/2} exp\{-\tfrac{1}{2\sigma^2}\Sigma_{i=1}^n (x_i - \mu)^2\}, \qquad (7.2.4)$$

which is to be maximized with respect to both μ and σ^2. This is *equivalent* to maximizing $logL(\mu, \sigma^2)$ with respect to both μ and σ^2. Now, one has

$$logL(\mu, \sigma^2) = c - \tfrac{n}{2}log(\sigma^2) - \tfrac{1}{2\sigma^2}\Sigma_{i=1}^n (x_i - \mu)^2,$$

which leads to

$$\tfrac{\partial}{\partial\mu}logL(\mu, \sigma^2) = \tfrac{1}{\sigma^2}\Sigma_{i=1}^n(x_i - \mu),$$

$$\tfrac{\partial}{\partial\sigma^2}logL(\mu, \sigma^2) = -\tfrac{n}{2\sigma^2} + \tfrac{1}{2\sigma^4}\Sigma_{i=1}^n(x_i - \mu)^2.$$

Then, we equate both these partial derivatives to zero and solve the resulting equations simultaneously for μ and σ^2. But, $\tfrac{\partial}{\partial\mu}logL(\mu, \sigma^2) = 0$ and $\tfrac{\partial}{\partial\sigma^2}logL(\mu, \sigma^2) = 0$ imply that $\Sigma_{i=1}^n(x_i - \mu) = 0$ so that $\mu = \overline{x}$, as well as $-\tfrac{n}{2}\sigma^{-2} + \tfrac{1}{2}\sigma^{-4}\Sigma_{i=1}^n(x_i - \overline{x})^2 = 0$ thereby leading to $\sigma^2 = n^{-1}\Sigma_{i=1}^n(x_i - \overline{x})^2 = u$, say. Next, the only concern is whether $L(\mu, \sigma^2)$ given by (7.2.4) is globally maximized at $(\mu, \sigma^2) = (\overline{x}, u)$. We need to obtain the matrix H of the second-order partial derivatives of $logL(\mu, \sigma^2)$ and show that H evaluated at $(\mu, \sigma^2) = (\overline{x}, u)$ is negative definite (n.d.). See (4.8.12) for some review. Now, we have

$$H = \begin{pmatrix} \tfrac{\partial^2}{\partial\mu^2}logL(\mu, \sigma^2) & \tfrac{\partial^2}{\partial\sigma^2\partial\mu}logL(\mu, \sigma^2) \\ \tfrac{\partial^2}{\partial\mu\partial\sigma^2}logL(\mu, \sigma^2) & \tfrac{\partial^2}{\partial(\sigma^2)^2}logL(\mu, \sigma^2) \end{pmatrix}$$

$$= \begin{pmatrix} -n\sigma^{-2} & -\Sigma_{i=1}^n(x_i - \mu)\sigma^{-4} \\ -\Sigma_{i=1}^n(x_i - \mu)\sigma^{-4} & \tfrac{n}{2}\sigma^{-4} - \Sigma_{i=1}^n(x_i - \mu)^2\sigma^{-6} \end{pmatrix},$$

which evaluated at $(\mu, \sigma^2) = (\overline{x}, u)$ reduces to the matrix

$$G = \begin{pmatrix} -nu^{-1} & 0 \\ 0 & -\tfrac{1}{2}nu^{-2} \end{pmatrix}.$$

The matrix G would be n.d. if and only if its odd order principal minors are negative and all even order principal minors are positive. Refer to (4.8.6) as needed. In this case, the first diagonal is $-nu^{-1}$ which is negative and $det(G) = \frac{1}{2}n^2u^{-3}$ which is positive. In other words, G is a n.d. matrix. Thus, $L(\mu, \sigma^2)$ is globally maximized at $(\mu, \sigma^2) = (\overline{x}, u)$. That is, the MLE of μ and σ^2 are respectively $\widehat{\mu} = \overline{X}$ and $\widehat{\sigma}^2 = n^{-1}\Sigma_{i=1}^n(X_i - \overline{X})^2$. ▲

Next we include few examples to highlight the point that $L(\theta)$ may *not* be a differentiable function of θ where $L(\theta)$ attains the global maximum. In such situations, the process of finding the MLE turns out little different on a case by case basis.

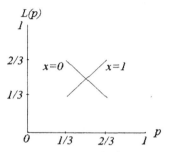

Figure 7.2.2. $L(p)$ When $\Theta = [\frac{1}{3}, \frac{2}{3}]$ and $x = 0$ or $x = 1$

Example 7.2.8 Suppose that we have a single observation X which is distributed as Bernoulli(p) where $0 \leq p \leq 1$ is the unknown parameter. Here, we have

$$L(p) = \begin{cases} 1 - p & \text{if } x = 0 \\ p & \text{if } x = 1. \end{cases}$$

Whether we observe $x = 0$ or 1, the resulting likelihood function $L(p)$ is not differentiable at the *end points*. But, by simply drawing a picture of $L(p)$ one can verify that (i) when $x = 0$ then $L(p)$ is maximized if p is the smallest, that is if $p = 0$, and (ii) when $x = 1$ then $L(p)$ is maximized if p is the largest, that is if $p = 1$. Hence the MLE of p is $\widehat{p} = X$ when $\Theta = [0, 1]$.

But, if the parameter space happens to be $\Theta = [\frac{1}{3}, \frac{2}{3}]$ instead, then what will be the MLE of p? Again, $L(p)$ is maximized at the end points where $L(p)$ is not differentiable. By examining the simple picture of $L(p)$ in the Figure 7.2.2, it becomes clear in this situation that (i) when $x = 0$, $L(p)$ is maximized if p is the smallest, that is if $p = \frac{1}{3}$, and (ii) when $x = 1$, $L(p)$ is maximized if p is the largest, that is if $p = \frac{2}{3}$. Hence the MLE of p is $\widehat{p} = \frac{1}{3}(X + 1)$ if the parameter space happens to be $\Theta = [\frac{1}{3}, \frac{2}{3}]$. ▲

Example 7.2.9 Suppose that $X_1, ..., X_n$ are iid Uniform($0, \theta$) where $0 < \theta < \infty$ is the unknown parameter. Here $\mathcal{X} = (0, \theta)$ and $\Theta = \Re^+$. We wish to find the MLE for θ. The likelihood function is

$$L(\theta) = \theta^{-n} I(0 < x_{n:n} < \theta) I(0 < x_{n:1} < x_{n:n}),$$

which is maximized at an end point. By drawing a simple picture of $L(\theta)$ it should be apparent that $L(\theta)$ is maximized when $\theta = x_{n:n}$. That is the MLE of θ is $\widehat{\theta} = X_{n:n}$, the largest order statistic. ▲

Example 7.2.10 Suppose that $X_1, ..., X_n$ are iid Poisson(λ) where $0 < \lambda < \infty$ is the unknown parameter. Here $\mathcal{X} = \{0, 1, 2, ...\}$ and $\Theta = \Re^+$. We wish to find the MLE for θ. The likelihood function is

$$L(\lambda) = \Pi_{i=1}^{n}\{e^{-\lambda}\lambda^{x_i}(x_i!)^{-1}\} = [\Pi_{i=1}^{n}(x_i!)^{-1}]e^{-n\lambda}\lambda^{\Sigma_{i=1}^{n}x_i} \qquad (7.2.5)$$

so that one has

$$logL(\lambda) = c - n\lambda + \{\Sigma_{i=1}^{n}x_i\}log(\lambda).$$

Now $L(\lambda)$ is to be maximized with respect to λ and that is equivalent to maximizing $logL(\lambda)$ with respect to λ. We have $\frac{d}{d\lambda}logL(\lambda) = -n + \{\Sigma_{i=1}^{n}x_i\}\lambda^{-1}$ which when equated to zero provides the solution $\lambda = \overline{x}$. If $\overline{x} > 0$, then $\frac{d}{d\lambda}logL(\lambda) = 0$ when $\lambda = \overline{x}$. In this situation it is easy to verify that $\frac{d^2}{d\lambda^2}logL(\lambda)\mid_{\lambda=\overline{x}}$ is negative. Hence the MLE is $\widehat{\lambda} = \overline{X}$ whenever $\overline{X} > 0$. But there is a fine point here. If $\overline{x} = 0$, which is equivalent to saying that $x_1 = ... = x_n = 0$, the likelihood function in (7.2.5) does not have a global maximum. In this case $L(\lambda) = e^{-n\lambda}$ which becomes larger as $\lambda(> 0)$ is allowed to become smaller. In other words if $\overline{x} = 0$, an MLE for λ can not be found. Observe, however, that $P_\lambda\{\overline{X} = 0\} = P_\lambda\{X_i = 0$ for all $i = 1, ..., n\} = exp(-n\lambda)$ which will be negligible for "large" values of $n\lambda$.

Table 7.2.1. Values of $P_\lambda\{\overline{X} = 0\}$

$n\lambda :$	3	4	5	6	7
$P_\lambda\{\overline{X} = 0\} :$	.0498	.0183	.0067	.0025	.0009

By looking at these entries, one may form some subjective opinion about what values of $n\lambda$ should perhaps be considered "large" in a situation like this. ▲

In the Binomial(n, p) situation where $0 < p < 1$ is the unknown parameter, the problem of deriving the MLE of p would hit a snag similar to what we found in the Example 7.2.10 when $\overline{X}$ is 0 or 1. Otherwise the MLE of p would be $\widehat{p} = \overline{X}$. If the parameter space is replaced by $0 \leq p \leq 1$, then of course the MLE of p would be $\widehat{p} = \overline{X}$. We leave this as the Exercise 7.2.8.

Remark 7.2.1 In the Examples 7.2.6-7.2.10, the MLE of the unknown parameter θ turned out to be either a (minimal) sufficient statistic or its function. This observation should not surprise anyone. With the help of Neyman factorization one can rewrite $L(\theta)$ as $g(\mathbf{T}; \theta)h(\mathbf{x})$ where $\mathbf{T}$ is a (minimal) sufficient statistic for θ. Now, any maximizing technique for the likelihood function with respect to θ would reduce to the problem of maximizing $g(\mathbf{T}; \theta)$ with respect to θ. Thus the MLE of θ in general will be a function of the associated (minimal) sufficient statistic $\mathbf{T}$. The method of moment estimator of θ may lack having this attractive feature. Refer to the Examples 7.2.4-7.2.5.

The MLE $\widehat{\theta}$ has a very attractive property referred to as its *invariance property*: That is, if $\widehat{\theta}$ is the MLE of θ, then $g(\widehat{\theta})$ is the MLE of $g(\theta)$.

The MLE can be quite versatile. It has a remarkable feature which is known as its *invariance property*. This result will be useful to derive the MLE of parametric functions of θ without a whole lot of effort as long as one already has the MLE $\widehat{\theta}$. We state this interesting result as a theorem without giving its proof. It was proved by Zehna (1966).

Theorem 7.2.1 (Invariance Property of MLE) *Consider the likelihood function $L(\theta)$ defined in (7.2.2). Suppose that the MLE of θ ($\in \Theta \subseteq \Re^k$) exists and it is denoted by $\widehat{\theta}$. Let $g(.)$ be a function, not necessarily one-to-one, from $\Re^k$ to a subset of $\Re^m$. Then, the MLE of the parametric function $g(\theta)$ is given by $g(\widehat{\theta})$.*

Example 7.2.11 (Example 7.2.7 Continued) Suppose that $X_1, ..., X_n$ are iid $N(\mu, \sigma^2)$ with $\theta = (\mu, \sigma^2)$ where μ, σ^2 are both unknown, $-\infty < \mu < \infty$, $0 < \sigma < \infty, n \geq 2$. Here $\mathcal{X} = \Re$ and $\Theta = \Re \times \Re^+$. We recall that the MLE of μ and σ^2 are respectively $\widehat{\mu} = \overline{X}$ and $\widehat{\sigma}^2 = U = n^{-1}\Sigma_{i=1}^n (X_i - \overline{X})^2$. Then, in view of the MLE's invariance property, the MLE of (i) σ would be $\sqrt{U}$, (ii) $\mu + \sigma$ would be $\overline{X} + \sqrt{U}$, (iii) μ^2/σ^2 would be $\overline{X}^2/U$. ▲

Note that the function $g(.)$ in the Theorem 7.2.1 is not necessarily one-to-one for the *invariance property* of the MLE to hold.

Example 7.2.12 (Example 7.2.9 Continued) Suppose that $X_1, ..., X_n$ are iid Uniform$(0, \theta)$ where $0 < \theta < \infty$ is the unknown parameter. Here $\mathcal{X} = (0, \theta)$ and $\Theta = \Re^+$. We recall that the MLE for θ is $X_{n:n}$. In view of the MLE's invariance property, one can verify that the MLE of (i) θ^2 would be $X_{n:n}^2$, (ii) $1/\theta$ would be $1/X_{n:n}$, (iii) $(\theta - 1)\sqrt{\theta + 1}$ would be $(X_{n:n} - 1)\sqrt{X_{n:n} + 1}$. ▲

7.3 Criteria to Compare Estimators

Let us assume that a population's pmf or pdf depends on some unknown vector valued parameter $\theta \in \Theta(\subseteq \Re^k)$. In many instances, after observing the random variables $X_1, ..., X_n$, we will often come up with several competing estimators for $\tau(\theta)$, a real valued parametric function of interest. How should we then proceed to compare the performances of rival estimators and then ultimately decide which one is perhaps the "best"? The first idea is introduced in Section 7.3.1 and the second one is formalized in Sections 7.3.2-7.3.3.

7.3.1 Unbiasedness, Variance, and Mean Squared Error

In order to set the stage, right away we start with two simple definitions. These are followed by some examples as usual. Recall that $\tau(\theta)$ is a real valued parametric function of θ.

Definition 7.3.1 *A real valued statistic* $T \equiv T(X_1, ..., X_n)$ *is called an unbiased estimator of* $\tau(\theta)$ *if and only if* $E_\theta(T) = \tau(\theta)$ *for all* $\theta \in \Theta$. *A statistic* $T \equiv T(X_1, ..., X_n)$ *is called a biased estimator of* $\tau(\theta)$ *if and only if T is not unbiased for* $\tau(\theta)$.

Definition 7.3.2 *For a real valued estimator T of* $\tau(\theta)$, *the amount of bias or simply the bias is given by*

$$B_\theta(T) = E_\theta(T) - \tau(\theta), \ \theta \in \Theta.$$

Gauss (1821) introduced originally the concept of an unbiased estimator in the context of his theory of least squares. Intuitively speaking, an unbiased estimator of $\tau(\theta)$ hits its target $\tau(\theta)$ on the average and the corresponding bias is then exactly zero for all $\theta \in \Theta$. In statistical analysis, the unbiasedness property of an estimator is considered very attractive. The class of unbiased estimators can be fairly rich. Thus, when comparing rival estimators, *initially* we restrict ourselves to consider the unbiased ones only. Then we choose the estimator from this bunch which appears to be the "best" according to an appropriate criteria.

In order to clarify the ideas, let us consider a specific population or universe which is described as $N(\mu, \sigma^2)$ where $\mu \in \Re$ is unknown but $\sigma \in \Re^+$ is assumed known. Here $\mathcal{X} = \Re$ and $\Theta = \Re$. The problem is one of estimating the population mean μ. From this universe, we observe the iid random variables $X_1, ..., X_4$. Let us consider several rival estimators of μ

defined as follows:

$$T_1 = X_1 + X_4 \qquad T_2 = \tfrac{1}{2}(X_1 + X_3) \qquad T_3 = \overline{X}$$
$$T_4 = \tfrac{1}{3}(X_1 + X_3) \quad T_5 = X_1 + T_2 - X_4 \quad T_6 = \tfrac{1}{10}\Sigma_{i=1}^4 i X_i \qquad (7.3.1)$$

Based on $X_1, ..., X_4$, one can certainly form many other rival estimators for μ. Observe that $E_\mu(T_1) = 2\mu$, $E_\mu(T_2) = \mu$, $E_\mu(T_3) = \mu$, $E_\mu(T_4) = \tfrac{2}{3}\mu$, $E_\mu(T_5) = \mu$ and $E_\mu(T_6) = \mu$. Thus, T_1 and T_4 are both biased estimators of μ, but T_2, T_3, T_5 and T_6 are unbiased estimators of μ. If we wish to estimate μ *unbiasedly*, then among T_1 through T_6, we should only include T_2, T_3, T_5, T_6 for further considerations.

Definition 7.3.3 *Suppose that the real valued statistic $T \equiv T(X_1, ..., X_n)$ is an estimator of $\tau(\theta)$. Then, the mean squared error (MSE) of the estimator T, sometimes denoted by MSE_T, is given by $E_\theta[(T - \tau(\theta))^2]$. If T is an unbiased estimator of $\tau(\theta)$, then the MSE is the variance of T, denoted by $V_\theta(T)$.*

Note that we have independence between the X's. Thus, utilizing the Theorem 3.4.3 in the case of the example we worked with little earlier, we obtain

$$V_\mu(T_2) = \tfrac{1}{4}V_\mu(X_1 + X_3) = \tfrac{1}{4}[V_\mu(X_1) + V_\mu(X_3)] = \tfrac{1}{2}\sigma^2,$$
$$V_\mu(T_3) = V_\mu(\overline{X}) = \tfrac{1}{4}\sigma^2,$$
$$V_\mu(T_5) = V_\mu(\tfrac{3}{2}X_1) + V_\mu(\tfrac{1}{2}X_3) + V_\mu(X_4) = [\tfrac{9}{4} + \tfrac{1}{4} + 1]\sigma^2 = \tfrac{7}{2}\sigma^2,$$
$$V_\mu(T_6) = \tfrac{1}{100}\Sigma_{i=1}^4 V_\mu(iX_i) = \tfrac{1}{100}\sigma^2\Sigma_{i=1}^4 i^2 = \tfrac{3}{10}\sigma^2.$$
$$(7.3.2)$$

In other words, these are the mean squared errors associated with the unbiased estimators T_2, T_3, T_5 and T_6.

How would one evaluate the MSE of the biased estimators T_1 and T_4? The following result will help in this regard.

Theorem 7.3.1 *If a statistic T is used to estimate a real valued parametric function $\tau(\theta)$, then MSE_T, the MSE associated with T, is given by*

$$E_\theta[(T - \tau(\theta))^2] = V_\theta(T) + [E_\theta(T) - \tau(\theta)]^2,$$

which amounts to saying that the MSE is same as the variance plus the square of the bias.

Proof Let us write $E_\theta(T) = \xi(\theta)$. Then, we have

$$E_\theta[(T - \tau(\theta))^2]$$
$$= E_\theta\left[(\{T - \xi(\theta)\} + \{\xi(\theta) - \tau(\theta)\})^2\right]$$
$$= E_\theta\left[\{T - \xi(\theta)\}^2\right] + E_\theta\left[\{\xi(\theta) - \tau(\theta)\}^2\right]$$
$$+ 2E_\theta\left[\{T - \xi(\theta)\}\{\xi(\theta) - \tau(\theta)\}\right]$$
$$= V_\theta(T) + B_\theta^2(T) + 2E_\theta\left[\{T - \xi(\theta)\}\{\xi(\theta) - \tau(\theta)\}\right].$$

Now, since $\xi(\theta) - \tau(\theta)$ is a fixed real number, we can write $E_\theta[\{T - \xi(\theta)\}\{\xi(\theta) - \tau(\theta)\}] = \{\xi(\theta) - \tau(\theta)\} E_\theta[\{T - \xi(\theta)\}] = 0$. Hence the result follows. ∎

Now, we can evaluate the MSE_{T_1} as $V_\mu(T_1) + [E_\mu(T_1) - \mu]^2 = 2\sigma^2 + (2\mu - \mu)^2 = \mu^2 + 2\sigma^2$. The evaluation of MSE_{T_4} is left as the Exercise 7.3.6.

It is possible sometimes to have T_1 and T_2 which are respectively biased and unbiased estimators of $\tau(\theta)$, but $\text{MSE}_{T_1} < \text{MSE}_{T_2}$. In other words, *intuitively* a biased estimator may be preferable if its average squared error is smaller. Look at the Example 7.3.1.

Example 7.3.1 Let $X_1, ..., X_n$ be iid $N(\mu, \sigma^2)$ where μ, σ^2 are both unknown, $\theta = (\mu, \sigma^2)$, $-\infty < \mu < \infty$, $0 < \sigma < \infty$, $n \geq 2$. Here $\mathcal{X} = \Re$ and $\Theta = \Re \times \Re^+$. Our goal is the estimation of a parametric function $\tau(\theta) = \sigma^2$, the population variance. Consider the customary sample variance $S^2 = (n-1)^{-1}\Sigma_{i=1}^n(X_i - \overline{X})^2$. We know that S^2 is unbiased for σ^2. One will also recall that $(n-1)S^2/\sigma^2$ is distributed as χ_{n-1}^2 and hence $V_\theta(S^2) = 2\sigma^4(n-1)^{-1}$. Next, consider another estimator for σ^2, namely $T = (n+1)^{-1}\Sigma_{i=1}^n(X_i - \overline{X})^2$ which can be rewritten as $(n-1)(n+1)^{-1}S^2$. Thus, $E_\theta(T) = (n-1)(n+1)^{-1}\sigma^2 \neq \sigma^2$ and so T is a *biased estimator* of σ^2. Next, we evaluate

$$V_\theta(T) = (n+1)^{-2}\sigma^4 V_\theta[(n-1)S^2/\sigma^2] = 2(n-1)(n+1)^{-2}\sigma^4.$$

Then we apply the Theorem 7.3.1 to express MSE_T as

$$2(n-1)(n+1)^{-2}\sigma^4 + \sigma^4\left[(n-1)(n+1)^{-1} - 1\right]^2 = 2(n+1)^{-1}\sigma^4,$$

which is smaller than $V_\theta(S^2)$. That is, S^2 is unbiased for σ^2 and T is biased for σ^2 but MSE_T is smaller than MSE_{S^2} for all θ. Refer to the Exercise 7.3.2 to see how one comes up with an estimator such as T for σ^2. ▲

In the context of the example we had been discussing earlier in this section, suppose that we consider two other estimators $T_7 = \frac{1}{2}X_1$ and

$T_8 \equiv 0$ for μ. Obviously both are biased estimators of μ. In view of the Theorem 7.3.1, the MSE$_{T_7}$ would be $\frac{1}{4}(\sigma^2 + \mu^2)$ whereas the MSE$_{T_8}$ would simply be μ^2. Between these two estimators T_7 and T_8, as far as the *smaller MSE criterion* goes, T_7 will be deemed better if and only if $\sigma^2 < 3\mu^2$.

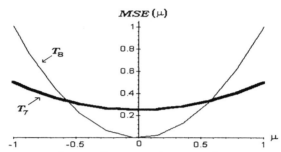

Figure 7.3.1. Curves Corresponding to MSE$_{T_7}$ and MSE$_{T_8}$
When $\sigma = 1$

In the Figure 7.3.1, we have plotted both MSE$_{T_7}$ (thick curve) and MSE$_{T_8}$ (thin curve) assuming that $\sigma = 1$. In this case, we claim that T_7 is better than T_8 (that is, MSE$_{T_7}$ < MSE$_{T_8}$) if and only if $|\mu| > 1/\sqrt{3} \approx .58$. This conclusion is validated by the Figure 7.3.1. This means that between the two estimators T_7 and T_8, we should prefer T_7 if and only if $|\mu| > 1/\sqrt{3}$. But, we do not know μ to begin with! In other words, it will be impossible to choose between T_7 and T_8 in practice. These two estimators are *not comparable*. The reader should try and find other estimators in this case which are not comparable among themselves.

> Sometimes estimators' MSE's may not be comparable to each other. Look at the Figure 7.3.1.

In the next section, we consider unbiased estimators of $\tau(\theta)$ and compare the performances among those unbiased estimators only.

7.3.2   Best Unbiased and Linear Unbiased Estimators

Let us discuss how to define the phrase "best" unbiased estimator or the "best" linear unbiased estimator of a real valued parametric function $\tau(\theta)$ as long as there is at least one unbiased estimator of $\tau(\theta)$. In the previous section, examples were given where we could find competing unbiased estimators of $\tau(\theta)$.

There are situations where a parametric function of interest has
no unbiased estimator. In such situations, naturally we *do not*
proceed to seek out the best unbiased estimator.
Look at the Example 7.3.2.

Example 7.3.2 Suppose that $X_1, ..., X_n$ are iid Bernoulli(p) where $0 <
p < 1$ is unknown and $\theta = p$. We now wish to estimate the parametric
function $\tau(\theta) = 1/p$. Of course $T = \Sigma_{i=1}^n X_i$ is sufficient for p and we also
know that T is distributed as Binomial(n, p) for all fixed $0 < p < 1$. In this
case, there is *no unbiased* estimator of $\tau(\theta)$. In order to prove this result,
all we need to show is that no function of T can be unbiased for $\tau(\theta)$. So
let us *assume* that $h(T)$ is an unbiased estimator of $\tau(\theta)$ and write, for all
$0 < p < 1$,

$$p^{-1} = E_p[h(T)] = \Sigma_{t=0}^n h(t) \binom{n}{t} p^t (1-p)^{n-t},$$

which can be rewritten as

$$\Sigma_{t=0}^n h(t) \binom{n}{t} p^{t+1} (1-p)^{n-t} - 1 = 0. \tag{7.3.3}$$

But the lhs in (7.3.3) is a polynomial of degree $n+1$ in the variable p and it
must be zero for all $p \in (0,1)$. That is the lhs in (7.3.3) must be identically
equal to zero and hence we must have $-1 \equiv 0$ which is a contradiction.
That is, there is no unbiased estimator of $1/p$ based on T. In this situation,
there is no point in trying to find the "best" unbiased estimator of $1/p$.
Look at a closely related Exercise 7.3.7. ▲

In the Example 7.3.2, we could not find any unbiased estimator
of p^{-1}. But, one can *appropriately* modify the sampling
scheme itself and then find an unbiased estimator of p^{-1}.
Look at the Example 7.3.3.

Example 7.3.3 (Example 7.3.2 Continued) In many areas, for example, ecology, entomology, genetics, forestry and wild life management, the
problem of estimating p^{-1} is very important. In the previous example, the
sample size n was held fixed and out of the n independent runs of the
Bernoulli experiment, we counted the number (T) of observed successes.
If the data was collected in that fashion, then we had shown the nonexis-
tence of any unbiased estimator of p^{-1}. But we do not suggest that p^{-1}
can not be estimated unbiasedly whatever be the process of data collection.

One may run the experiment a little differently as follows. Let $N =$ the number of runs needed to observe the first success so that the pmf of the random variable N is given by $P(N = y) = p(1 - p)^{y-1}$, $y = 1, 2, 3, ...$ which means that N is a Geometric(p) random variable. One can verify that $E_p[N] = p^{-1}$ for all $p \in (0, 1)$. That is the sample size N is an unbiased estimator of p^{-1}. This method of data collection, known as the *inverse binomial sampling*, is widely used in applications mentioned before. In his landmark paper, Petersen (1896) gave the foundation of *capture-recapture sampling*. The 1956 paper of the famous geneticist, J. B. S. Haldane, is cited frequently. Look at the closely related Exercise 7.3.8. ▲

> How should we go about comparing performances of
> any two unbiased estimators of $\tau(\boldsymbol{\theta})$?

This is done by comparing the variances of the rival estimators. Since the rival estimators are assumed unbiased for $\tau(\boldsymbol{\theta})$, it is clear that a smaller variance will indicate a smaller average (squared) error. So, if T_1, T_2 are two unbiased estimators of $\tau(\boldsymbol{\theta})$, then T_1 is *preferable* to (or *better* than) T_2 if $V_\theta(T_1) \leq V_\theta(T_2)$ for all $\boldsymbol{\theta} \in \Theta$ but $V_\theta(T_1) < V_\theta(T_2)$ for some $\boldsymbol{\theta} \in \Theta$. Now, in the class of unbiased estimators of $\tau(\boldsymbol{\theta})$, the one having the smallest variance is called the *best unbiased estimator* of $\tau(\boldsymbol{\theta})$. A formal definition is given shortly.

Using such a general principle, by looking at (7.3.2), it becomes apparent that among the unbiased estimators T_2, T_3, T_5 and T_6 for the unknown mean μ, the estimator T_3 is the best one to use because it has the smallest variance.

Definition 7.3.4 *Assume that there is at least one unbiased estimator of the unknown real valued parametric function $\tau(\boldsymbol{\theta})$. Consider the class $\mathcal{C}$ of all unbiased estimators of $\tau(\boldsymbol{\theta})$. An estimator $T \in \mathcal{C}$ is called the best unbiased estimator or the uniformly minimum variance unbiased estimator (UMVUE) of $\tau(\boldsymbol{\theta})$ if and only if for all estimators $T^* \in \mathcal{C}$, we have*

$$V_\theta(T) \leq V_\theta(T^*) \text{ for all } \boldsymbol{\theta} \in \Theta. \tag{7.3.4}$$

In Section 7.4 we will introduce several approaches to locate the UMVUE. But first let us focus on a smaller subset of $\mathcal{C}$ for simplicity. Suppose that we have located estimators $T_1, ..., T_k$ which are all *unbiased* for $\tau(\boldsymbol{\theta})$ such that $V_\theta(T_i) = \delta^2$ and the T_i's are pairwise uncorrelated, $0 < \delta < \infty, i = 1, ..., k$. Denote a new subclass of estimators

$$\mathcal{D} = \left\{ T : T = \Sigma_{i=1}^k \alpha_i T_i \text{ where } \alpha_i \in \Re, i = 1, ..., k, \Sigma_{i=1}^k \alpha_i = 1 \right\}. \tag{7.3.5}$$

For any $T \in \mathcal{D}$ we have: $E_\theta(T) = E_\theta(\Sigma_{i=1}^k \alpha_i T_i) = \Sigma_{i=1}^k \alpha_i E_\theta(T_i) = \tau(\boldsymbol{\theta})$ for all $\boldsymbol{\theta}$ which shows that any estimator T chosen from $\mathcal{D}$ is an unbiased estimator of $\tau(\boldsymbol{\theta})$. Hence, it is obvious that $\mathcal{D} \subseteq \mathcal{C}$. Now we wish to address the following question.

> What is the best estimator of $\tau(\boldsymbol{\theta})$ within the smaller class $\mathcal{D}$?

That is, which estimator from class $\mathcal{D}$ has the smallest variance? From the following theorem one will see that the answer is indeed very simple.

Theorem 7.3.2 *Within the class of estimators $\mathcal{D}$, the one which has the smallest variance corresponds to $\alpha_i = k^{-1}, i = 1, ..., k$. That is, the best unbiased estimator of $\tau(\boldsymbol{\theta})$ within the class $\mathcal{D}$ turns out to be $\overline{T} = k^{-1}\Sigma_{i=1}^k T_i$ which is referred to as the best linear (in $T_1, ..., T_k$) unbiased estimator (BLUE) of $\tau(\boldsymbol{\theta})$.*

Proof Since the T_i's are pairwise uncorrelated, for any typical estimator T from the class $\mathcal{D}$, we have

$$V_\theta(T) = V_\theta(\Sigma_{i=1}^k \alpha_i T_i) = \Sigma_{i=1}^k \alpha_i^2 V_\theta(T_i) = \delta^2 \Sigma_{i=1}^k \alpha_i^2. \qquad (7.3.6)$$

From (7.3.6) it is now clear that we need to

minimize $\Sigma_{i=1}^k \alpha_i^2$ subjected to the restriction that $\Sigma_{i=1}^k \alpha_i = 1$.

But, observe that

$$\Sigma_{i=1}^k \alpha_i^2 = \Sigma_{i=1}^k \left(\alpha_i - k^{-1}\right)^2 + k^{-1}. \qquad (7.3.7)$$

Hence, $\Sigma_{i=1}^k \alpha_i^2 \geq k^{-1}$ for all choices of $\alpha_i, i = 1, ..., k$ such that $\Sigma_{i=1}^k \alpha_i = 1$. But, from (7.3.7) we see that $\Sigma_{i=1}^k \alpha_i^2 = k^{-1}$, the smallest possible value, if and only if $\Sigma_{i=1}^k \left(\alpha_i - k^{-1}\right)^2 = 0$. That is, $V_\theta(T)$ would be minimized if and only if $\alpha_i = k^{-1}, i = 1, ..., k$. ∎

Example 7.3.4 Suppose that $X_1, ..., X_n$ are iid $N(\mu, \sigma^2)$ where μ, σ are both unknown, $-\infty < \mu < \infty, 0 < \sigma < \infty$. Among all linear (in $X_1, ..., X_n$) unbiased estimators of μ, the BLUE turns out to be $\overline{X}$, the sample mean. This follows immediately from the Theorem 7.3.2. ▲

> In the Example 7.3.4, $\overline{X}$ turned out to be the *best* among *linear unbiased* estimators ($\in \mathcal{D}$) of μ. But, $\overline{X}$ must be compared with *all unbiased* estimators of μ before we can hope to find the UMVUE of μ. That is, the estimator $\overline{X}$ has to be compared with nonlinear unbiased estimators of μ, for example, $X_1 + X_2^2 - X_3^2$, $X_3 + |X_2| - |X_3|$, $\frac{1}{2}(X_1 + X_3) + |X_2 - X_4|^{3/2} - |X_3 - X_4|^{3/2}$, ... and others. $(7.3.8)$

7.4   Improved Unbiased Estimators via Sufficiency

It appears that one of the first examples of UMVUE was found by Aitken and Silverstone (1942). The UMVUE's as suitable and unique functions of sufficient statistics were investigated by Halmos (1946), Kolmogorov (1950a), and more generally by Rao (1947).

In a problem suppose that we can get hold of at least one unbiased estimator for a *real valued* parametric function $\tau(\boldsymbol{\theta})$. Now the question is this: If we start with some unbiased estimator T for $\tau(\boldsymbol{\theta})$, however trivial this estimator may appear to be, can we improve upon T? That is, can we revise the initial unbiased estimator in order to come up with another unbiased estimator T' of $\tau(\boldsymbol{\theta})$ such that $V_{\theta}(T') < V_{\theta}(T)$ for all $\boldsymbol{\theta} \in \Theta$?

Let us illustrate. Suppose that $X_1, ..., X_n$ are iid Bernoulli(p) where $0 < p < 1$ is unknown. We wish to estimate the parameter p unbiasedly. Consider $T = X_1$ and obviously T is an unbiased estimator of p. But note that T can take one of two possible values 0 or 1. Such an estimator, even though unbiased, may appear naive and useless. T may be unreasonable as an estimator of the parameter p which lies between zero and one. But, this criticism against T should not be too bothersome because T is not the estimator which would be recommended for use in practice. The question is, even if we start with T, can we hope to improve upon this initial estimator in the sense of *reducing* the variance? For all practical purposes, the answer is in the affirmative. The general machinery comes next.

7.4.1   The Rao-Blackwell Theorem

The technique to improve upon an initial unbiased estimator of $\tau(\boldsymbol{\theta})$ is customarily referred to as the *Rao-Blackwellization* in the statistical literature. C. R. Rao and D. Blackwell independently published fundamental papers respectively in 1945 and 1947 which included this path-breaking idea. Neither Rao nor Blackwell knew about the other's paper for quite some time because of disruptions due to the war. We first state and prove this fundamental result.

Theorem 7.4.1 (Rao-Blackwell Theorem) *Let T be an unbiased estimator of a real valued parametric function $\tau(\boldsymbol{\theta})$ where the unknown parameter $\boldsymbol{\theta} \in \Theta \subset \Re^k$. Suppose that $\mathbf{U}$ is a jointly sufficient statistic for $\boldsymbol{\theta}$. Define $g(\mathbf{u}) = E_{\theta}[T \mid \mathbf{U} = \mathbf{u}]$, for $\mathbf{u}$ belonging to $\mathcal{U}$, the domain space of $\mathbf{U}$. Then, the following results hold:*

(i)   *Define $W = g(\mathbf{U})$. Then, W is an unbiased estimator of $\tau(\boldsymbol{\theta})$;*

(ii) $V_{\theta}[W] \leq V_{\theta}[T]$ *for all $\boldsymbol{\theta} \in \Theta$, with the equality holding if and only if T is the same as W w. p. 1.*

Proof (i) Since **U** is sufficient for $\boldsymbol{\theta}$, the conditional distribution of T given $\mathbf{U} = \mathbf{u}$ can not depend upon the unknown parameter $\boldsymbol{\theta}$, and this remains true for all $\mathbf{u} \in \mathcal{U}$. This clearly follows from the Definition 6.2.3 of sufficiency. Hence, $g(\mathbf{u})$ is a function of $\mathbf{u}$ and it is free from $\boldsymbol{\theta}$ for all $\mathbf{u} \in \mathcal{U}$. In other words, $W = g(\mathbf{U})$ is indeed a real valued statistic and so we can call it an estimator. Using the Theorem 3.3.1, part (i), we can write $E[X] = E_Y[E(X \mid Y)]$ where X and Y are any two random variables with finite expectations. Hence we have for all $\boldsymbol{\theta} \in \Theta$,

$$\tau(\boldsymbol{\theta}) = E_{\boldsymbol{\theta}}[T] = E_{\mathbf{U}}[E\{T \mid \mathbf{U}\}] = E_{\boldsymbol{\theta}}[g(\mathbf{U})] = E_{\boldsymbol{\theta}}[W], \qquad (7.4.1)$$

which shows that W is an unbiased estimator of $\tau(\boldsymbol{\theta})$. ♦

(ii) Let us now proceed as follows for all $\boldsymbol{\theta} \in \Theta$:

$$\begin{aligned} V_{\boldsymbol{\theta}}[T] &= E_{\boldsymbol{\theta}}[\{T - \tau(\boldsymbol{\theta})\}^2] \\ &= E_{\boldsymbol{\theta}}[\{(T - W) + (W - \tau(\boldsymbol{\theta}))\}^2] \\ &= E_{\boldsymbol{\theta}}[\{W - \tau(\boldsymbol{\theta})\}^2] + E_{\boldsymbol{\theta}}[\{T - W\}^2] \qquad (7.4.2) \\ &\quad + 2E_{\boldsymbol{\theta}}[(T - W)(W - \tau(\boldsymbol{\theta}))] \\ &= V_{\boldsymbol{\theta}}[W] + E_{\boldsymbol{\theta}}[\{T - W\}^2], \end{aligned}$$

since we have

$$\begin{aligned} E_{\boldsymbol{\theta}}[(T - W)&(W - \tau(\boldsymbol{\theta}))] \\ &= E_{\mathbf{U}}\left[E[\{T - g(\mathbf{U})\}\{g(\mathbf{U}) - \tau(\boldsymbol{\theta})\} \mid \mathbf{U}]\right] \\ &= E_{\mathbf{U}}\left[\{g(\mathbf{U}) - \tau(\boldsymbol{\theta})\}E\{T - g(\mathbf{U}) \mid \mathbf{U}\}\right] \\ &= E_{\mathbf{U}}[\{g(\mathbf{U}) - \tau(\boldsymbol{\theta})\}\{g(\mathbf{U}) - g(\mathbf{U})\}] = 0. \end{aligned}$$

Now, from (7.4.2), the first conclusion in part (ii) is obvious since $\{T - W\}^2$ is non-negative w.p.1 and thus $E_{\boldsymbol{\theta}}[\{T - W\}^2] \geq 0$ for all $\boldsymbol{\theta} \in \Theta$. For the second conclusion in part (ii), notice again from (7.4.2) that $V_{\boldsymbol{\theta}}[W] = V_{\boldsymbol{\theta}}[T]$ for all $\boldsymbol{\theta} \in \Theta$ if and only if $E_{\boldsymbol{\theta}}[\{T - W\}^2] = 0$ for all $\boldsymbol{\theta} \in \Theta$, that is if and only if T is the same as W w.p.1. The proof is complete. ∎

One attractive feature of the Rao-Blackwell Theorem is that there is no need to guess the functional form of the final unbiased estimator of $\tau(\boldsymbol{\theta})$. Sometimes guessing the form of the final unbiased estimator of $\tau(\boldsymbol{\theta})$ may be hard to do particularly when estimating some unusual parametric function. One will see such illustrations in the Examples 7.4.5 and 7.4.7.

Example 7.4.1 Suppose that $X_1, ..., X_n$ are iid Bernoulli(p) where $0 < p < 1$ is unknown. We wish to estimate $\tau(p) = p$ unbiasedly. Consider $T = X_1$ which is an unbiased estimator of p. We were discussing this example just before we introduced the Rao-Blackwell Theorem. The possible values of T are 0 or 1. Of course, $U = \Sigma_{i=1}^{n} X_i$ is sufficient for p. The domain space for U is $\mathcal{U} = \{0, 1, 2, ..., n\}$. Let us write for $u \in \mathcal{U}$,

$$
\begin{aligned}
E_p[T \mid U = u] \\
= [1 \times P_p\{T = 1 \mid U = u\}] + [0 \times P_p\{T = 0 \mid U = u\}] \\
= P_p\{X_1 = 1 \mid U = u\} \\
= P_p\{X_1 = 1 \cap U = u\}/P_p\{U = u\} \\
= P_p\{X_1 = 1 \cap \Sigma_{i=2}^{n} X_i = u - 1\}/P_p\{\Sigma_{i=1}^{n} X_i = u\}.
\end{aligned}
\tag{7.4.3}
$$

Next, observe that $\Sigma_{i=1}^{n} X_i$ is Binomial(n, p) and $\Sigma_{i=2}^{n} X_i$ is Binomial$(n - 1, p)$. Also X_1 and $\Sigma_{i=2}^{n} X_i$ are independently distributed. Thus, we can immediately rewrite (7.4.3) as

$$
\begin{aligned}
E_p[X_1 \mid U = u] \\
= P_p\{X_1 = 1\} P_p\{\Sigma_{i=2}^{n} X_i = u - 1\}/P_p\{\Sigma_{i=1}^{n} X_i = u\} \\
= p\binom{n-1}{u-1} p^{u-1} (1 - p)^{n-u}/\{\binom{n}{u} p^u (1 - p)^{n-u}\} \\
= \binom{n-1}{u-1}/\binom{n}{u} = u/n = \overline{x}.
\end{aligned}
$$

That is, the Rao-Blackwellized version of the initial unbiased estimator X_1 turns out to be $\overline{X}$, the sample mean, even though T was indeed a very naive and practically useless initial estimator of p. Now note that $V_p[T] = p(1-p)$ and $V_p[W] = p(1 - p)/n$ so that $V_p[W] < V_p[T]$ if $n \geq 2$. When $n = 1$, the sufficient statistic is X_1 and so if one starts with $T = X_1$, then the final estimator obtained through Rao-Blackwellization would remain X_1. That is, when $n = 1$, we will not see any improvement over T through the Rao-Blackwellization technique. ▲

Start with an unbiased estimator T of a parametric function $\tau(\theta)$. The process of conditioning T given a sufficient (for θ) statistic $\mathbf{U}$ is referred to as *Rao-Blackwellization*. The *refined* estimator W is often called the Rao-Blackwellized version of T. This technique is remarkable because one always comes up with an *improved* unbiased estimator W for $\tau(\theta)$ except in situations where the initial estimator T itself is already a function of the sufficient statistic $\mathbf{U}$.

Example 7.4.2 (Example 7.4.1 Continued) Suppose that $X_1, ..., X_n$ are iid Bernoulli(p) where $0 < p < 1$ is unknown, with $n \geq 2$. Again, we wish to estimate $\tau(p) = p$ unbiasedly. Now consider a different initial estimator $T = \frac{1}{2}(X_1 + X_2)$ and obviously T is an unbiased estimator of p. The possible values of T are $0, \frac{1}{2}$ or 1 and again $U = \Sigma_{i=1}^{n} X_i$ is a sufficient statistic for p. Then, $E_p[T \mid U = u] = \frac{1}{2}\{E_p[X_1 \mid U = u] + E_p[X_2 \mid U = u]\} = \frac{1}{2}\{\overline{x} + \overline{x}\} = \overline{x}$. In other words, if one starts with $\frac{1}{2}(X_1 + X_2)$ as the initial unbiased estimator of p, then after going through the process of Rao-Blackwellization, one again ends up with $\overline{X}$ as the refined unbiased estimator of p. Observe that $V_p[T] = p(1-p)/2$ and $V_p[\overline{X}] = p(1-p)/n$ so that $V_p[\overline{X}] < V_p[X_1]$ if $n \geq 3$. When $n = 2$, the sufficient statistic is T, and so if one happens to start with T, then the final estimator obtained through the process of Rao-Blackwellization would remain T. In other words, when $n = 2$, we will not see any improvement over T through the Rao-Blackwellization technique. ▲

Example 7.4.3 (Example 7.4.1 Continued) Suppose that $X_1, ..., X_n$ are iid Bernoulli(p) where $0 < p < 1$ is unknown, with $n \geq 2$. We wish to estimate $\tau(p) = p(1-p)$ unbiasedly. Consider $T = X_1(1 - X_2)$ which is an unbiased estimator of $\tau(p)$. The possible values of T are 0 or 1. Again, $U = \Sigma_{i=1}^{n} X_i$ is the sufficient statistic for p with the domain space $\mathcal{U} = \{0, 1, 2, ..., n\}$. Let us denote $h_p(u) = P_p\{\Sigma_{i=1}^{n} X_i = u\}$ and then write for $u \in \mathcal{U}$:

$$E_p[X_1(1 - X_2) \mid U = u] = P_p\{X_1(1 - X_2) = 1 \mid U = u\},$$

since $X_1(1 - X_2)$ takes the value 0 or 1 only. Thus, we express $E_p[X_1(1 - X_2) \mid U = u]$ as

$$
\begin{aligned}
&P_p\{X_1 = 1 \cap X_2 = 0 \cap U = u\}/P_p\{U = u\} \\
&= P_p\{X_1 = 1 \cap X_2 = 0 \cap \Sigma_{i=3}^{n} X_i = u - 1\}/h_p(u).
\end{aligned}
\tag{7.4.4}
$$

Next observe that $\Sigma_{i=1}^{n} X_i$ is Binomial(n, p), $\Sigma_{i=3}^{n} X_i$ is Binomial$(n - 2, p)$, and also that $X_1, X_2, \Sigma_{i=3}^{n} X_i$ are independently distributed. Thus, we can rewrite (7.4.4) as

$$
\begin{aligned}
&E_p[X_1(1 - X_2) \mid U = u] \\
&= P_p\{X_1 = 1\}P_p\{X_2 = 0\}P_p\{\Sigma_{i=3}^{n} X_i = u - 1\}/h_p(u) \\
&= p(1 - p)\binom{n-2}{u-1}p^{u-1}(1 - p)^{n-u-1}/\{\binom{n}{u}p^u(1 - p)^{n-u}\} \\
&= \binom{n-2}{u-1}/\binom{n}{u} = u(n - u)/\{n(n - 1)\},
\end{aligned}
$$

which is the same as $n(n - 1)^{-1}\overline{x}(1 - \overline{x})$. That is, the Rao-Blackwellized version of the initial unbiased estimator $X_1(1 - X_2)$ turns out to be $n(n - 1)^{-1}\overline{X}(1 - \overline{X})$. For the Bernoulli random samples, since the X's are either 0 or 1, observe that the sample variance in this situation turns out to be $S^2 = (n - 1)^{-1}\{\Sigma_{i=1}^{n} X_i^2 - n\overline{X}^2\} = (n - 1)^{-1}\{\Sigma_{i=1}^{n} X_i - n\overline{X}^2\} = (n - 1)^{-1}\{n\overline{X} - n\overline{X}^2\} = n(n - 1)^{-1}\overline{X}(1 - \overline{X})$ which coincides with the Rao-Blackwellized version. ▲

Example 7.4.4 Suppose that $X_1, ..., X_n$ are iid Poisson(λ) where $0 < \lambda < \infty$ is unknown. We wish to estimate $\tau(\lambda) = \lambda$ unbiasedly. Consider $T = X_1$ which is an unbiased estimator of λ and $U = \Sigma_{i=1}^n X_i$ is a sufficient statistic for λ. The domain space for U is $\mathcal{U} = \{0, 1, 2, ..\}$. Now, for $u \in \mathcal{U}$, conditionally given $U = u$, the statistic T can take one of the possible values from the set $\mathcal{T} = \{0, 1, 2, ..., u\}$. Thus, for $u \in \mathcal{U}$, we can write

$$E_\lambda[T \mid U = u] = \Sigma_{t=0}^u t P_\lambda\{T = t \mid U = u\}. \tag{7.4.5}$$

Now, we have to find the expression for $P_\lambda\{T = t \mid U = u\}$ for all fixed $u \in \mathcal{U}$ and $t \in \mathcal{T}$. Let us write

$$\begin{aligned} P_\lambda&\{T = t \mid U = u\} \\ &= P_\lambda\{T = t \cap U = u\}/P_\lambda\{U = u\} \\ &= P_\lambda\{X_1 = t \cap \Sigma_{i=2}^n X_i = u - t\}/P_\lambda\{\Sigma_{i=1}^n X_i = u\}. \end{aligned} \tag{7.4.6}$$

But observe that $\Sigma_{i=1}^n X_i$ is Poisson($n\lambda$), $\Sigma_{i=2}^n X_i$ is Poisson($(n-1)\lambda$), and also that $X_1, \Sigma_{i=2}^n X_i$ are independently distributed. Thus, from (7.4.6) we can express $P_\lambda\{T = t \mid U = u\}$ as

$$\begin{aligned} P_\lambda&\{X_1 = t\} P_\lambda\{\Sigma_{i=2}^n X_i = u - t\}/P_\lambda\{\Sigma_{i=1}^n X_i = u\} \\ &= \frac{e^{-\lambda}\lambda^t}{t!} \frac{e^{-(n-1)\lambda}\{(n-1)\lambda\}^{u-t}}{(u-t)!} \frac{u!}{e^{-n\lambda}\{n\lambda\}^u} \\ &= \binom{u}{t}\left(\frac{1}{n}\right)^t \left(1 - \frac{1}{n}\right)^{u-t}. \end{aligned} \tag{7.4.7}$$

That is, the conditional distribution of T given $U = u$ is Binomial($u, \frac{1}{n}$). Hence, combining (7.4.5) and (7.4.7), we note that $E_\lambda[T \mid U = u] = u/n = \bar{x}$. In other words, the Rao-Blackwellized version of T is $\overline{X}$, the sample mean. ▲

> We started with trivial unbiased estimators in the Examples 7.4.1 -7.4.4. In these examples, perhaps one could intuitively guess the the improved unbiased estimator. The Rao-Blackwell Theorem did not lead to any surprises here. The Examples 7.4.5 and 7.4.7 are however, different because we are forced to begin with naive unbiased estimators. Look at the Exercises 7.4.6 and 7.4.8 for more of the same.

Example 7.4.5 (Example 7.4.4 Continued) Suppose that $X_1, ..., X_n$ are iid Poisson(λ) where $0 < \lambda < \infty$ is unknown and $n \geq 2$. We wish to estimate $\tau(\lambda) = e^{-\lambda}$ unbiasedly. Consider $T = I(X_1 = 0)$ which is an unbiased estimator of $\tau(\lambda)$ since $E_\lambda[T] = P_\lambda\{X_1 = 0\} = e^{-\lambda}$. Consider

$U = \Sigma_{i=1}^n X_i$, the sufficient statistic for λ. The domain space for U is $\mathcal{U} = \{0, 1, 2, ..\}$. Now, for all $u \in \mathcal{U}$, conditionally given $U = u$, the statistic T can take one of the possible values 0 or 1. Thus, for $u \in \mathcal{U}$, we can write

$$
\begin{aligned}
E_\lambda[T \mid U = u] \\
&= P_\lambda\{T = 1 \mid U = u\} \\
&= P_\lambda\{X_1 = 0 \cap \Sigma_{i=1}^n X_i = u\}/P_\lambda\{\Sigma_{i=1}^n X_i = u\} \\
&= P_\lambda\{X_1 = 0 \cap \Sigma_{i=2}^n X_i = u\}/P_\lambda\{\Sigma_{i=1}^n X_i = u\}.
\end{aligned}
\tag{7.4.8}
$$

Note that $\Sigma_{i=1}^n X_i$ is Poisson($n\lambda$), $\Sigma_{i=2}^n X_i$ is Poisson($(n-1)\lambda$) whereas X_1 and $\Sigma_{i=2}^n X_i$ are independently distributed. Thus, from (7.4.8) we rewrite $E_\lambda[T \mid U = u]$ as

$$
\begin{aligned}
P_\lambda\{X_1 = 0\}P_\lambda\{\Sigma_{i=2}^n X_i = u\}/P_\lambda\{\Sigma_{i=1}^n X_i = u\} \\
= e^{-\lambda}\frac{e^{-(n-1)\lambda}\{(n-1)\lambda\}^u}{u!}\frac{u!}{e^{-n\lambda}\{n\lambda\}^u} = \left(\frac{n-1}{n}\right)^u,
\end{aligned}
\tag{7.4.9}
$$

and hence the Rao-Blackwellized version of the estimator T is $W = (1 - n^{-1})^{\Sigma_{i=1}^n X_i}$. Now we know the expression for W and so we can directly evaluate $E_\lambda[W]$. We use the form of the mgf of a Poisson random variable, namely $E_\lambda\left[e^{sU}\right] = exp\{n\lambda(e^s - 1)\}$, and then replace s with $log(1 - n^{-1})$ to write

$$
E_\lambda[W] = exp\{\lambda(e^s - 1)\} = exp\{n\lambda(1 - n^{-1} - 1)\} = e^{-\lambda}.
$$

In other words, W is an unbiased estimator of $e^{-\lambda}$, but this should not be surprising. The part (i) of the Rao-Blackwell Theorem leads to the same conclusion. Was there any way to guess the form of the estimator W before actually going through Rao-Blackwellization? How should one estimate $e^{-2\lambda}$ or $e^{-3\lambda}$? One should attack these problems via Rao-Blackwellization or mgf as we just did. We leave these and other related problems as Exercise 7.4.3. ▲

Example 7.4.6 Suppose that $X_1, ..., X_n$ are iid $N(\mu, \sigma^2)$ where μ is unknown but σ^2 is known with $-\infty < \mu < \infty$, $0 < \sigma < \infty$ and $\mathcal{X} = \Re$. We wish to estimate $\tau(\mu) = \mu$ unbiasedly. Consider $T = X_1$ which is an unbiased estimator of μ. Consider $U = n^{-1}\Sigma_{i=1}^n X_i$, the sufficient statistic for μ. The domain space for U is $\mathcal{U} = \Re$. Now, for $u \in \mathcal{U}$, conditionally given $U = u$, the distribution of the statistic T is $N\left(u, \sigma^2(1 - n^{-1})\right)$. Refer to the Section 3.6 on the bivariate normal distribution as needed. Now, $E_\mu[T \mid U = u] = u = \overline{x}$. That is, the Rao-Blackwellized version of the initial unbiased estimator T turns out to be $\overline{X}$. ▲

Example 7.4.7 (Example 7.4.6 Continued) Suppose that $X_1, ..., X_n$ are iid $N(\mu, \sigma^2)$ where μ is unknown but σ^2 is known with $-\infty < \mu < \infty$, $0 < \sigma < \infty$ and $\mathcal{X} = \Re$. We wish to estimate the parametric function $\tau(\mu) = P_\mu\{X_1 > a\}$ unbiasedly where a is a fixed and *known* real number. Consider $T = I(X_1 > a)$ and obviously T is an unbiased estimator of $\tau(\mu)$. Consider $U = n^{-1}\Sigma_{i=1}^n X_i$, the sufficient statistic for μ. The domain space for U is $\mathcal{U} = \Re$. For $u \in \mathcal{U}$, conditionally given $U = u$, the distribution of the statistic T is $N\left(u, \sigma^2(1 - n^{-1})\right)$. In other words, we have

$$E_\mu[T \mid U = u] = P_\mu\{X_1 > a \mid U = u\}$$
$$= 1 - \Phi\left([a - u]/\{\sigma^2[1 - n^{-1}]\}^{1/2}\right).$$

That is, the Rao-Blackwellized version of the initial unbiased estimator T is $W = 1 - \Phi\left([a - \overline{X}]/\{\sigma^2[1 - n^{-1}]\}^{1/2}\right)$. Was there a way to guess the form of the improved estimator W before we used Rao-Blackwellization? The Exercise 7.4.4 poses a similar problem for estimating the two-sided probability. A much harder problem is to estimate the same parametric function $\tau(\mu)$ when both μ and σ are unknown. Kolmogorov (1950a) gave a very elegant solution for this latter problem. The Exercise 7.4.6 shows the important steps. ▲

Example 7.4.8 (Example 7.4.6 Continued) Suppose that $X_1, ..., X_n$ are iid $N(\mu, \sigma^2)$ where μ is unknown but σ^2 is known with $-\infty < \mu < \infty$, $0 < \sigma < \infty$ and $\mathcal{X} = \Re$. We wish to estimate $\tau(\mu) = \mu^2$ unbiasedly. Consider $T = X_1^2 - \sigma^2$ which is an unbiased estimator of $\tau(\mu)$. Consider again $U = n^{-1}\Sigma_{i=1}^n X_i$, a sufficient statistic for μ. The domain space for U is $\mathcal{U} = \Re$. As before, for $u \in \mathcal{U}$, conditionally given $U = u$, the distribution of the statistic T is $N\left(u, \sigma^2(1 - n^{-1})\right)$. In other words, we can express $E_\mu[T \mid U = u]$ as

$$E_\mu\{X_1^2 - \sigma^2 \mid U = u\} = \sigma^2(1 - n^{-1}) + u^2 - \sigma^2 = u^2 - n^{-1}\sigma^2.$$

That is, the Rao-Blackwellized version of the initial unbiased estimator T of μ^2 is $W = \overline{X}^2 - n^{-1}\sigma^2$. What initial estimator T would one start with if we wished to estimate $\tau(\mu) = \mu^3$ unbiasedly? Try to answer this question first before looking at the Exercise 7.4.9. ▲

Remark 7.4.1 In the Example 7.4.8, we found $W = \overline{X}^2 - n^{-1}\sigma^2$ as the final unbiased estimator of μ^2. Even though W is unbiased for μ^2, one may feel somewhat uneasy to use this estimator in practice. It is true that $P_\mu\{W < 0\}$ is positive whatever be n, μ and σ. The parametric function μ^2 is non-negative, but the final unbiased estimator W can be negative with positive probability! From the Figure 7.4.1 one can see the behavior of $P_\mu\{W < 0\}$ as a function of n when $\mu = .1$ but $\sigma = 1, 2, 3$. The unbiasedness criterion at times may create an awkward situation like this. In this particular case, however, the MLE of μ^2 is $\overline{X}^2$, by virtue of the invariance property (Theorem 7.2.1). The MLE is a biased estimator but it is always non-negative.

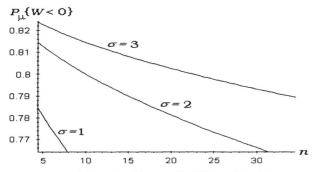

Figure 7.4.1. Plot of n Versus $P_\mu\{W < 0\}$ When $\mu = .1$

In the preceding Examples 7.4.1-7.4.8, we got started with a naive unbiased estimator for the parametric functions of interest. But once the initial unbiased estimator was refined through Rao-Blackwellization, the *improved* unbiased estimator appeared quickly. Now, the question is whether such a refined estimator is in fact the UMVUE of the associated parametric function. We have indeed found the unique UMVUE in the given examples, but we can not validate this claim by relying upon the Rao-Blackwell Theorem alone. We give extra machineries next to pinpoint the UMVUE. In Section 7.6.1, we briefly discuss a scenario where the refinement via Rao-Blackwellization does not lead to the UMVUE.

7.5 Uniformly Minimum Variance Unbiased Estimator

This section provides specific tools to derive the uniformly minimum variance unbiased estimator (UMVUE) of parametric functions when there is a UMVUE. The first approach relies upon the celebrated *Cramér-Rao inequality*. C. R. Rao and H. Cramér independently discovered, under mild regularity conditions, a lower bound for the variance of an unbiased estimator of a real valued parametric function $\tau(\theta)$ where $\theta \in \Theta(\subseteq \Re)$ in their classic papers, Rao (1945) and Cramér (1946b). Neither of them was aware about the other's results for quite some time due to the war. We will first prove this famous inequality.

Next, we discuss a fundamental result (Theorem 7.5.2) due to Lehmann and Scheffé (1950) which leaps out of the Rao-Blackwell Theorem and it helps us go very far in our search for the UMVUE. The Lehmann-Scheffé (1950,1955,1956) series of papers is invaluable in this regard.

7.5.1 The Cramér-Rao Inequality and UMVUE

Lehmann (1983) referred to this inequality as the "information inequality," a name which was suggested by Savage (1954). Lehmann (1983, p. 145, Section 9) wrote, "The first version of the information inequality appears to have been given by Fréchet (1943). Early extensions and rediscoveries are due to Darmois (1945), Rao (1945), and Cramér (1946b)." We will, however, continue to refer to this inequality by its commonly used name, the *Cramér-Rao inequality,* for the ease of (i) locating cross-references and (ii) going for a literature search among the available books and other sources.

This bound for the variance, customarily called the *Cramér-Rao lower bound (*CRLB*),* for unbiased estimators of $\tau(\theta)$ is appreciated in many problems where one can (i) derive the expression of the CRLB, and (ii) easily locate an unbiased estimator of $\tau(\theta)$ whose variance happens to co-incide with the CRLB. In situations like these, one has then precisely found the UMVUE for $\tau(\theta)$.

Consider iid real valued and *observable* random variables $X_1, ..., X_n$ from a population with the common pmf or pdf $f(x; \theta)$ where the unknown parameter $\theta \in \Theta \subseteq \Re$ and $x \in \mathcal{X} \subseteq \Re$. Recall that we denote $\mathbf{X} = (X_1, ..., X_n)$. Let us *pretend* that we are working with the pdf and hence the expectations of functions of random variables would be written as appropriate multiple integrals. In the case of discrete random variables, one would replace the integrals by the appropriate finite or infinite sums, as the case may be.

Standing Assumptions: Let us assume that the *support* $\mathcal{X}$ does not involve θ and the first partial derivative of $f(x; \theta)$ with respect to θ and the integrals with respect to $\mathbf{x} = (x_1, ..., x_n)$ are interchangeable.

Theorem 7.5.1 (Cramér-Rao Inequality) *Suppose that $T = T(\mathbf{X})$ is an unbiased estimator of a real valued parametric function $\tau(\theta)$, that is $E_\theta(T) = \tau(\theta)$ for all $\theta \in \Theta$. Assume also that $\frac{d}{d\theta}\tau(\theta)$, denoted by $\tau'(\theta)$, exists and it is finite for all $\theta \in \Theta$. Then, for all $\theta \in \Theta$, under the standing assumptions we have:*

$$V_\theta(T) \geq \frac{\{\tau'(\theta)\}^2}{nE_\theta\left[\left\{\frac{\partial}{\partial\theta}[\log f(X_1; \theta)]\right\}^2\right]} \qquad (7.5.1)$$

The expression on the rhs of the inequality in (7.5.1) is called the Cramér-Rao lower bound (CRLB).

Proof Without any loss of generality, let us assume that $0 < V_\theta(T) < \infty$. By the definition of expectation, we can write

$$\tau(\theta) = \int_\mathcal{X} \cdots \int_\mathcal{X} T(x_1, ..., x_n) \Pi_{i=1}^n f(x_i; \theta) \Pi_{i=1}^n dx_i, \qquad (7.5.2)$$

which implies that

$$\begin{aligned}
\tau'(\theta) &= \tfrac{d}{d\theta} \left[\int_\mathcal{X} \cdots \int_\mathcal{X} T(x_1, ..., x_n) \Pi_{i=1}^n f(x_i; \theta) \Pi_{i=1}^n dx_i \right] \\
&= \int_\mathcal{X} \cdots \int_\mathcal{X} T(x_1, ..., x_n) \left[\tfrac{\partial}{\partial\theta} \Pi_{i=1}^n f(x_i; \theta) \right] \Pi_{i=1}^n dx_i.
\end{aligned} \qquad (7.5.3)$$

Now, observe that

$$\Pi_{i=1}^n f(x_i; \theta) = exp\{\Sigma_{i=1}^n log f(x_i; \theta)\} \text{ for } x \in \mathcal{X}, \qquad (7.5.4)$$

so that by using the chain rule of differentiation, we get

$$\begin{aligned}
\tfrac{\partial}{\partial\theta} &\left[\Pi_{i=1}^n f(x_i; \theta) \right] \\
&= exp\{\Sigma_{i=1}^n log f(x_i; \theta)\} \tfrac{\partial}{\partial\theta} \{\Sigma_{i=1}^n log f(x_i; \theta)\} \qquad (7.5.5) \\
&= \{\Sigma_{i=1}^n \tfrac{\partial}{\partial\theta} [log \ f(x_i; \theta)]\} \Pi_{i=1}^n f(x_i; \theta).
\end{aligned}$$

Next, let us denote $Y = \Sigma_{i=1}^n \tfrac{\partial}{\partial\theta}[log f(X_i; \theta)]$. Note that Y is not an observable random variable because it involves the unknown parameter θ. We now combine (7.5.4) and (7.5.5) to rewrite

$$\begin{aligned}
\tau'(\theta) &\\
&= \int_\mathcal{X} \cdots \int_\mathcal{X} T(x_1, ..., x_n)\{\Sigma_{i=1}^n \tfrac{\partial}{\partial\theta}[log f(x_i; \theta)]\} \Pi_{i=1}^n f(x_i; \theta) \Pi_{i=1}^n dx_i \\
&= E_\theta\{TY\}.
\end{aligned} \qquad (7.5.6)$$

Also, one obviously has $\int_\mathcal{X} f(x; \theta) dx = 1$ so that one writes

$$0 = \tfrac{d}{d\theta} \int_\mathcal{X} f(x; \theta) dx = \int_\mathcal{X} \left[\tfrac{\partial}{\partial\theta} f(x; \theta) \right] dx = \int_\mathcal{X} \left[\tfrac{\partial}{\partial\theta} log f(x; \theta) \right] f(x; \theta) dx. \qquad (7.5 .7)$$

Hence, we have

$$E_\theta[Y] = E_\theta \left\{ \Sigma_{i=1}^n \tfrac{\partial}{\partial\theta}[log f(X_i; \theta)] \right\} = \Sigma_{i=1}^n E_\theta \left\{ \tfrac{\partial}{\partial\theta}[log f(X_i; \theta)] \right\} = 0$$

for all $\theta \in \Theta$, since the X's have identical distributions. Thus, (7.5.6) leads us to conclude that

$$\tau'(\theta) = E_\theta\{TY\} = Cov_\theta(T, Y),$$

which can be rewritten as

$$\{\tau'(\theta)\}^2 = Cov_\theta^2(T, Y) \leq V_\theta(T) V_\theta(Y), \qquad (7.5.8)$$

by virtue of the Cauchy-Schwarz inequality or the covariance inequality (Theorems 3.9.5-3.9.6). Recall that Y is the sum of n iid random variables and thus in view of (7.5.7), we obtain

$$V_\theta(Y) = nV_\theta\left\{\tfrac{\partial}{\partial\theta}[log\,f(X_1;\theta)]\right\} = nE_\theta\left[\{\tfrac{\partial}{\partial\theta}[log\,f(X_1;\theta)]\}^2\right]. \qquad (7.5.9)$$

Now the inequality (7.5.1) follows by combining (7.5.8) and (7.5.9). ∎

Remark 7.5.1 It is easy to see that the CRLB given by the rhs of the inequality in (7.5.1) would be attained by the variance of the estimator T for all $\theta \in \Theta$ if and only if we can conclude the strict equality in (7.5.8), that is if and only if the statistic T and the random variable Y are linearly related w.p.1. That is, the CRLB will be attained by the variance of T if and only if

$$T - a(\theta) = b(\theta)Y \text{ w.p.1, for all } \theta \in \Theta, \qquad (7.5.10)$$

with some fixed real valued functions $a(.)$ and $b(.)$.

Remark 7.5.2 By combining the CRLB from (7.5.1) and the expression for the information $\mathcal{I}_X(\theta)$, defined in (6.4.1), we can immediately restate the Cramér-Rao inequality under the same standard assumptions as follows:

$$V_\theta(T) \geq \frac{\{\tau'(\theta)\}^2}{n\mathcal{I}_{X_1}(\theta)} \qquad (7.5.11)$$

where $\mathcal{I}_{X_1}(\theta)$ is the information about the unknown parameter θ in one single observation X_1.

> We will interchangeably use the form of the Cramér-Rao inequality given by (7.5.1) or (7.5.11). The CRLB would then correspond to the expressions found on the rhs of either of these two equations.

Since the denominator in the CRLB involves the *information*, we would rely heavily upon some of the worked out examples from Section 6.4.

Example 7.5.1 Let $X_1, ..., X_n$ be iid Poisson(λ) where $\lambda(> 0)$ is the unknown parameter. Let us consider $\tau(\lambda) = \lambda$ so that $\tau'(\lambda) = 1$. Now, $\overline{X}$ is an unbiased estimator of λ and $V_\lambda(\overline{X}) = n^{-1}\lambda$. Recall from the Example 7.4.4 that we could not claim that $\overline{X}$ was the UMVUE of λ. Can we now claim that $\overline{X}$ is the UMVUE of λ? In (6.4.2) we find $\mathcal{I}_{X_1}(\lambda) = \lambda^{-1}$ and hence from the rhs of (7.5.11) we see that the CRLB $= 1/(n\lambda^{-1}) = \lambda/n$

which coincides with the expression for $V_\lambda(\overline{X})$. That is, the estimator $\overline{X}$ attains the smallest possible variance among all unbiased estimators of λ. Then, $\overline{X}$ must be the UMVUE of λ. ▲

Example 7.5.2 Let $X_1, ..., X_n$ be iid Bernoulli(p) where $0 < p < 1$ is the unknown parameter. Let us consider $\tau(p) = p$ so that $\tau'(p) = 1$. Now, $\overline{X}$ is an unbiased estimator of p and $V_p(\overline{X}) = n^{-1}p(1-p)$. Recall from the Example 7.4.1 that we could not claim that $\overline{X}$ was the UMVUE of p. Can we now claim that $\overline{X}$ is the UMVUE of p? Let us first derive the expression for $\mathcal{I}_{X_1}(p)$. Observe that

$$log f(x; p) = x log(p) + (1 - x) log(1 - p)$$

so that $\frac{\partial}{\partial p} log f(x; p) = xp^{-1} - (1 - x)(1 - p)^{-1} = (x - p)\{p(1 - p)\}^{-1}$. Hence, one evaluates $\mathcal{I}_{X_1}(p)$ as

$$E_p\left[\left\{\tfrac{\partial}{\partial p}[log f(x_1; p)]\right\}^2\right] = \{p(1 - p)\}^{-2}V_p(X_1) = \{p(1 - p)\}^{-1}.$$

$$(7.5.12)$$

Thus from (7.5.11), we have the CRLB $= 1/(n\{p(1 - p)\}^{-1}) = p(1 - p)/n$ which coincides with the expression for $V_p(\overline{X})$. That is, $\overline{X}$ attains the smallest possible variance among all unbiased estimators of p. Then, $\overline{X}$ must be the UMVUE of p. ▲

Example 7.5.3 Suppose that $X_1, ..., X_n$ are iid $N(\mu, \sigma^2)$ where μ is unknown but σ^2 is known. Here we have $-\infty < \mu < \infty$, $0 < \sigma < \infty$ and $\mathcal{X} = \Re$. We wish to estimate $\tau(\mu) = \mu$ unbiasedly. Consider $\overline{X}$ which is obviously an unbiased estimator of μ. Is $\overline{X}$ the UMVUE of μ? Example 7.4.6 was not decisive in this regard. In (6.4.3) we find $\mathcal{I}_{X_1}(\mu) = \sigma^{-2}$ so that from (7.5.11) we have the CRLB $= 1/(n\,\sigma^{-2}) = \sigma^2/n$ which coincides with the expression for $V_\mu(\overline{X})$. Then, $\overline{X}$ must be the UMVUE of μ. ▲

In these examples, we thought of a "natural" unbiased estimator of the parametric function of interest and this estimator's variance happened to coincide with the CRLB. So, in the end we could claim that the estimator we started with was in fact the UMVUE of $\tau(\theta)$. The reader has also surely noted that these UMVUE's agreed with the Rao-Blackwellized versions of some of the naive initial unbiased estimators.

> We are interested in unbiased estimators of $\tau(\theta)$. We found the Rao-Blackwellized version W of an initial unbiased estimator T. But, the variance of this improved estimator W may not attain the CRLB. Look at the following example.

Example 7.5.4 (Example 7.4.5 Continued) Suppose that $X_1, ..., X_n$ are iid Poisson(λ) where $0 < \lambda < \infty$ is unknown with $n \geq 2$. We wish to estimate $\tau(\lambda) = e^{-\lambda}$ unbiasedly. In the Example 7.4.5, we had started with $T = I(X_1 = 0)$ but its Rao-Blackwellized version was in fact $W = (1 - n^{-1})^{\Sigma_{i=1}^n X_i}$. Does $V_\lambda(W)$ attain the CRLB? Recall that $\Sigma_{i=1}^n X_i$ is Poisson$(n\lambda)$ so that its mgf is given by

$$E_\lambda[e^{s\Sigma_{i=1}^n X_i}] = exp\{n\lambda(e^s - 1)\}. \tag{7.5.13}$$

Let us use (7.5.13) with $s = 2log(1 - n^{-1})$ to claim that $E_\lambda[W^2] = exp\{n\lambda(e^s - 1)\} = exp\{n\lambda[(\frac{n-1}{n})^2 - 1]\} = exp\{-2\lambda + n^{-1}\lambda\}$. Hence, we obtain

$$V_\lambda[W] = exp\{-2\lambda + n^{-1}\lambda\} - exp(-2\lambda) = e^{-2\lambda}(e^{\lambda/n} - 1). \tag{7.5.14}$$

Now, we have $\tau'(\lambda) = -e^{-\lambda}$ and, from (6.4.2), $\mathcal{I}_{X_1}(\lambda) = \lambda^{-1}$. Utilizing (7.5.11) we obtain the CRLB $= \{\tau'(\lambda)\}^2/(n\lambda^{-1}) = n^{-1}\lambda e^{-2\lambda}$. Now, for $x > 0$, observe that $e^x > 1 + x$. Hence, from (7.5.14) we obtain

$$V_\lambda[W] = e^{-2\lambda}(e^{\lambda/n} - 1) > e^{-2\lambda}(1 + \tfrac{\lambda}{n} - 1) = e^{-2\lambda}\tfrac{\lambda}{n} = CRLB.$$

In other words, $V_\lambda[W]$ *does not* attain the CRLB. ▲

Question remains whether the estimator W in the Example 7.5.4 is the UMVUE of $e^{-\lambda}$. It is clear that the CRLB alone *may not* point toward the UMVUE. Example 7.5.5 is also similar.

Example 7.5.5 (Example 7.4.8 Continued) Suppose that $X_1, ..., X_n$ are iid $N(\mu, \sigma^2)$ where μ is unknown but σ^2 is known with $-\infty < \mu < \infty$, $0 < \sigma < \infty$ and $\mathcal{X} = \Re$. We wish to estimate $\tau(\mu) = \mu^2$ unbiasedly. In the Example 7.4.8, we found the Rao-Blackwellized unbiased estimator $W = \overline{X}^2 - n^{-1}\sigma^2$ of $\tau(\mu)$. Let us first obtain the expression of the variance of the estimator W as follows:

$$V_\mu[W] = E_\mu[\overline{X}^4] - E_\mu^2[\overline{X}^2] = E_\mu[\overline{X}^4] - \{\mu^2 + n^{-1}\sigma^2\}^2. \tag{7.5.15}$$

The first term in (7.5.15) is evaluated next. Recall that $E_\mu[(\overline{X} - \mu)] = 0$, $E_\mu[(\overline{X} - \mu)^2] = n^{-1}\sigma^2$, $E_\mu[(\overline{X} - \mu)^3] = 0$, and $E_\mu[(\overline{X} - \mu)^4] = 3n^{-2}\sigma^4$ since $\overline{X}$ has $N(\mu, n^{-1}\sigma^2)$ distribution. Thus, we have

$$\begin{aligned} E_\mu[\overline{X}^4] &= E_\mu[\{(\overline{X} - \mu) + \mu\}^4] \\ &= E_\mu[(\overline{X} - \mu)^4] + 6\mu^2 E_\mu[(\overline{X} - \mu)^2] + \mu^4 \\ &= 3n^{-2}\sigma^4 + 6n^{-1}\mu^2\sigma^2 + \mu^4, \end{aligned}$$

and in view of (7.5.15) we obtain

$$V_\mu[W] = 2n^{-2}\sigma^4 + 4n^{-1}\mu^2\sigma^2. \tag{7.5.16}$$

From (7.5.11), we have the CRLB $= \{\tau'(\mu)\}^2/(n\sigma^{-2}) = 4n^{-1}\mu^2\sigma^2$ and comparing this with (7.5.16) it is clear that $V_\mu[W] > $ CRLB for all μ. That is, the CRLB *is not* attained. ▲

7.5.2 The Lehmann-Scheffé Theorems and UMVUE

In situations similar to those encountered in the Examples 7.5.4-7.5.5, it is clear that neither the Rao-Blackwell Theorem nor the Cramér-Rao inequality may help in deciding whether an unbiased estimator W is the UMVUE of $\tau(\theta)$. An alternative approach is provided in this section.

If the Rao-Blackwellized version in the end always comes up with the same refined estimator regardless of which unbiased estimator of $\tau(\theta)$ one initially starts with, then of course one has found the UMVUE for $\tau(\theta)$. Lehmann and Scheffé's (1950) notion of a complete statistic, introduced in Section 6.6, plays a major role in this area. We first prove the following result from Lehmann and Scheffé (1950).

Theorem 7.5.2 (Lehmann-Scheffé Theorem I) *Suppose that T is an unbiased estimator of the real valued parametric function $\tau(\boldsymbol{\theta})$ where the unknown parameter $\boldsymbol{\theta} \in \Theta \subseteq \Re^k$. Suppose that $\mathbf{U}$ is a complete (jointly) sufficient statistic for $\boldsymbol{\theta}$. Define $g(\mathbf{u}) = E_{\boldsymbol{\theta}}[T \mid \mathbf{U} = \mathbf{u}]$, for $\mathbf{u}$ belonging to $\mathcal{U}$, the domain space of $\mathbf{U}$. Then, the statistic $W = g(\mathbf{U})$ is the unique (w.p.1) UMVUE of $\tau(\boldsymbol{\theta})$.*

Proof The Rao-Blackwell Theorem assures us that in order to search for the best unbiased estimator of $\tau(\boldsymbol{\theta})$, we need only to focus on unbiased estimators which are functions of $\mathbf{U}$ alone. We already know that W is a function of $\mathbf{U}$ and it is an unbiased estimator of $\tau(\boldsymbol{\theta})$. Suppose that there is another unbiased estimator W^* of $\tau(\boldsymbol{\theta})$ where W^* is also a function of $\mathbf{U}$. Define $h(\mathbf{U}) = W - W^*$ and then we have

$$E_{\boldsymbol{\theta}}[h(\mathbf{U})] = E_{\boldsymbol{\theta}}[W - W^*] = \tau(\boldsymbol{\theta}) - \tau(\boldsymbol{\theta}) \equiv 0 \text{ for all } \boldsymbol{\theta} \in \Theta. \tag{7.5.17}$$

Now, we use the Definition 6.6.2 of the completeness of a statistic. Since $\mathbf{U}$ is a complete statistic, from (7.5.17) it follows that $h(\mathbf{U}) \equiv 0$ w.p.1, that is we must have $W = W^*$ w.p.1. The result then follows. ■

In our quest for finding the UMVUE of $\tau(\boldsymbol{\theta})$, we need not always have to go through the conditioning with respect to the complete sufficient statistic $\mathbf{U}$. In some problems, the following alternate and yet equivalent result may be more directly applicable. We state the result without proving it. Its proof can be easily constructed from the preceding proof of Theorem 7.5.2.

Theorem 7.5.3 (Lehmann-Scheffé Theorem II) *Suppose that* **U** *is a complete sufficient statistic for* $\boldsymbol{\theta}$ *where the unknown parameter* $\boldsymbol{\theta} \in \Theta \subseteq \mathfrak{R}^k$. *Suppose that a statistic* $W = g(\mathbf{U})$ *is an unbiased estimator of the real valued parametric function* $\tau(\boldsymbol{\theta})$. *Then,* W *is the unique (w.p.1) UMVUE of* $\tau(\boldsymbol{\theta})$.

> In the Examples 7.5.6-7.5.8, neither the Rao-Blackwell Theorem nor the Cramér-Rao inequality helps in identifying the UMVUE. But, the Lehmann-Scheffé approach is right on the money.

Example 7.5.6 (Example 7.5.4 Continued) Suppose that $X_1, ..., X_n$ are iid Poisson(λ) where $0 < \lambda < \infty$ is unknown with $n \geq 2$. We wish to estimate $\tau(\lambda) = e^{-\lambda}$ unbiasedly. The Rao-Blackwellized *unbiased* estimator of $\tau(\lambda)$ was $W = (1 - n^{-1})^{\Sigma_{i=1}^n X_i}$ and its variance was strictly larger than the CRLB. But, W depends only on the complete sufficient statistic $U = \Sigma_{i=1}^n X_i$. Hence, in view of the Lehmann-Scheffé Theorems, W is the unique (w.p.1) UMVUE for $\tau(\lambda)$. ▲

Example 7.5.7 (Example 7.5.5 Continued) Suppose that $X_1, ..., X_n$ are iid $N(\mu, \sigma^2)$ where μ is unknown but σ^2 is known with $-\infty < \mu < \infty$, $0 < \sigma < \infty$ and $\mathcal{X} = \mathfrak{R}$. We wish to estimate $\tau(\mu) = \mu^2$ unbiasedly. We found the Rao-Blackwellized *unbiased* estimator $W = \overline{X}^2 - n^{-1}\sigma^2$ of $\tau(\mu)$ and but its variance was strictly larger than the CRLB. Now, W depends only on the complete sufficient statistic $U = \Sigma_{i=1}^n X_i$. Hence, in view of the Lehmann-Scheffé Theorems, W is the unique (w.p.1) UMVUE for $\tau(\mu)$. ▲

Example 7.5.8 (Example 7.4.3 Continued) Suppose that $X_1, ..., X_n$ are iid Bernoulli(p) where $0 < p < 1$ is unknown with $n \geq 2$. We wish to estimate $\tau(p) = p(1 - p)$ unbiasedly. Recall that the Rao-Blackwellized version of the *unbiased* estimator turned out to be $W = \frac{n}{n-1}\overline{X}(1 - \overline{X})$. But, W depends only on the complete sufficient statistic $\Sigma_{i=1}^n X_i$. Hence, in view of the Lehmann-Scheffé Theorems, W is the unique (w.p.1) UMVUE for $\tau(p)$. ▲

Let us add that the Rao-Blackwellized estimator in the Example 7.4.7 is also the UMVUE of the associated parametric function. The verification is left as Exercise 7.5.2.

> Next we give examples where the Cramér-Rao inequality is not applicable but we can conclude the UMVUE property of the natural unbiased estimator via the Lehmann-Scheffé Theorems.

Example 7.5.9 Let $X_1, ..., X_n$ be iid Uniform$(0, \theta)$ where $\theta(> 0)$ is the unknown parameter. Now, $U = X_{n:n}$, the largest order statistic, is complete sufficient for θ. We wish to estimate $\tau(\theta) = \theta$ unbiasedly. Since U has its pdf given by $nu^{n-1}\theta^{-n}I(0 < u < \theta)$ we have $E_\theta[X_{n:n}] = n(n+1)^{-1}\theta$. Hence, $E_\theta[(n+1)n^{-1}X_{n:n}] = \theta$ so that $W = (n+1)n^{-1}X_{n:n}$ is an unbiased estimator of θ. Thus, by the Lehmann-Scheffé Theorems, $(n+1)n^{-1}X_{n:n}$ is the UMVUE for θ. Here, the domain space $\mathcal{X}$ for the X variable depends on θ itself. Hence, an approach through the Cramér-Rao inequality is not feasible. ▲

Example 7.5.10 Suppose that $X_1, ..., X_n$ are iid $N(\mu, \sigma^2)$ where μ and σ are both unknown, $-\infty < \mu < \infty$, $0 < \sigma < \infty$ and $\mathcal{X} = \Re$. Let us write $\boldsymbol{\theta} = (\mu, \sigma^2) \in \Theta = \Re \times \Re^+$ and we wish to estimate $\tau(\boldsymbol{\theta}) = \mu$ unbiasedly. Recall from the Example 6.6.7 that $\mathbf{U} = (\overline{X}, S^2)$ is a complete sufficient statistic for $\boldsymbol{\theta}$. Obviously, $\overline{X}$ is an unbiased estimator of $\tau(\boldsymbol{\theta})$ whereas $\overline{X}$ is a function of $\mathbf{U}$ only. Thus, by the Lehmann-Scheffé Theorems, $\overline{X}$ is the UMVUE for μ. Similarly, one can show that S^2 is the UMVUE for σ^2. The situation here involves two unknown parameters μ and σ. Hence, an approach through the Cramér-Rao inequality *as stated* will not be appropriate. ▲

Example 7.5.11 (Example 7.5.10 Continued) Suppose that $X_1, ..., X_n$ are iid $N(\mu, \sigma^2)$ where μ and σ are both unknown, $-\infty < \mu < \infty$, $0 < \sigma < \infty$ and $\mathcal{X} = \Re$. Let us write $\boldsymbol{\theta} = (\mu, \sigma^2) \in \Theta = \Re \times \Re^+$ and we wish to estimate $\tau(\boldsymbol{\theta}) = \mu + \sigma$ unbiasedly. Now, $\mathbf{U} = (\overline{X}, S^2)$ is a complete sufficient statistic for $\boldsymbol{\theta}$. From (2.3.26) for the moments of a gamma variable, it follows that $E_\theta[S] = a_n\sigma$ where $a_n = \sqrt{2}\Gamma(\frac{n}{2})\{\Gamma(\frac{n-1}{2})\sqrt{n-1}\}^{-1}$. Thus, $W = \overline{X} + a_n^{-1}S$ is an unbiased estimator of $\tau(\boldsymbol{\theta})$ and W depends only on $\mathbf{U}$. Hence, by the Lehmann-Scheffé Theorems, $\overline{X} + a_n^{-1}S$ is the UMVUE for $\mu + \sigma$. Similarly one can derive the UMVUE for the parametric function $\mu\sigma^k$ where k is any real number. For $k < 0$, however, one should be particularly cautious about the minimum required n. We leave this out as Exercise 7.5.7. ▲

Example 7.5.12 (Example 4.4.12 Continued) Suppose that $X_1, ..., X_n$ are iid with the common negative exponential pdf $f(x; \boldsymbol{\theta}) = \sigma^{-1}exp\{-(x-\mu)/\sigma\}I(x > \mu)$ where μ and σ are both unknown, $\boldsymbol{\theta} = (\mu, \sigma) \in \Theta = \Re \times \Re^+, n \geq 2$. Here, we wish to estimate $\tau(\boldsymbol{\theta}) = \mu + \sigma$, that is the mean of the population, unbiasedly. Recall from Exercises 6.3.2 and 6.6.18 that $\mathbf{U} = (X_{n:1}, \Sigma_{i=1}^n(X_i - X_{n:1}))$ is a complete sufficient statistic for $\boldsymbol{\theta}$. Now, $\overline{X}$ is an unbiased estimator of $\tau(\boldsymbol{\theta})$ and note that we can rewrite $\overline{X} = X_{n:1} + n^{-1}\Sigma_{i=1}^n(X_i - X_{n:1})$ which shows that $\overline{X}$ depends only on $\mathbf{U}$. Thus, by the Lehmann-Scheffé Theorems, $\overline{X}$ is the UMVUE for $\mu + \sigma$. Exploiting (2.3.26), one can derive the UMVUE for the parametric function $\mu\sigma^k$ where k is any real number. For $k < 0$, however, one should be particularly cautious about the minimum required n. We leave this out as Exercise 7.5.8. ▲

7.5.3 A Generalization of the Cramér-Rao Inequality

In the statement of the Cramér-Rao inequality, Theorem 7.5.1, the random variables $X_1, ..., X_n$ do not necessarily have to be iid. The inequality holds with minor modifications even if the X's are not identically distributed but they are independent.

Let us suppose that we have iid real valued *observable* random variables $X_{j1}, ..., X_{jn_j}$ from a population with the common pmf or pdf $f_j(x; \theta)$ where X_j's and X_l's are independent for all $j \neq l = 1, ..., k$, $x \in \mathcal{X} \subseteq \Re$. Here, we have the unknown parameter $\theta \in \Theta \subseteq \Re$. We denote $\mathbf{X} = (X_{11}, ..., X_{1n_1}, ..., X_{k1}, ..., X_{kn_k})$ and $\mathbf{x} = (x_{11}, ..., x_{1n_1}, ..., x_{k1}, ..., x_{kn_k})$ with $n = \Sigma_{i=1}^k n_i, \mathbf{x} \in \mathcal{X}^n$. Let us *pretend* that we are working with the pdf's and hence the expectations of functions of the random variables would be written as appropriate multiple integrals. In the case of discrete random variables, one will replace the integrals by the corresponding finite or infinite sums, as the case may be. Let us denote the information content of X_{j1} by $\mathcal{I}_{X_j}(\theta)$ which is calculated as $E_\theta \left[\left(\frac{\partial}{\partial \theta}[log f_j(X; \theta)] \right)^2 \right]$. We now state a generalized version of the Theorem 7.5.1. Its proof is left as Exercise 7.5.16.

Standing Assumptions: Let us assume that the *support* $\mathcal{X}$ does not involve θ and the first partial derivative of $f_j(x; \theta), j = 1, ..., k$ with respect to θ and the integrals with respect to $\mathbf{x}$ are interchangeable.

Theorem 7.5.4 *Suppose that $T = T(\mathbf{X})$ is an unbiased estimator of a real valued parametric function $\tau(\theta)$, that is $E_\theta(T) = \tau(\theta)$ for all $\theta \in \Theta$. Assume also that $\frac{d}{d\theta}\tau(\theta)$, denoted by $\tau'(\theta)$, exists and is finite for all $\theta \in \Theta$. Then, for all $\theta \in \Theta$, under the standing assumptions we have:*

$$V_\theta(T) \geq \frac{\{\tau'(\theta)\}^2}{\Sigma_{j=1}^k n_j \mathcal{I}_{X_j}(\theta)} \tag{7.5.18}$$

The expression on the rhs of the inequality in (7.5.18) is called the Cramér-Rao lower bound (CRLB) as before.

Example 7.5.13 Let $X_{11}, ..., X_{1n_1}$ be iid Exponential(θ) so that their common pdf is given by $f_1(x; \theta) = \theta^{-1} e^{-x/\theta}$ with the unknown parameter $\theta \in \Theta = (0, \infty)$ and $\mathcal{X} = (0, \infty)$. Also, suppose that $X_{21}, ..., X_{2n_2}$ are iid Gamma(α, θ), that is, their common pdf is given by $f_2(x; \theta) = \{\theta^\alpha \Gamma(\alpha)\}^{-1} e^{-x/\theta} x^{\alpha-1}$ with the same unknown parameter θ and $\mathcal{X} = (0, \infty)$, but $\alpha(> 0)$ is assumed known. Let us assume that the X_1's are independent of the X_2's. In an experiment on reliability and survival analyses, one may have a situation like this where a combination of two or more statistical models, depending on the same unknown parameter θ, may be appropriate. By direct calculations we obtain $\mathcal{I}_{X_1}(\theta) = \theta^{-2}$ and $\mathcal{I}_{X_2}(\theta) = \alpha \theta^{-2}$. That is, the CRLB from (7.5.18) for the variance of unbiased estimators of $\tau(\theta)$ would be $\theta^2(n_1 + n_2\alpha)^{-1}$. Obviously, $U = \Sigma_{i=1}^{n_1}X_{1i} + \Sigma_{j=1}^{n_2}X_{2j}$ is sufficient for θ and $E_\theta\{U\} = (n_1 + n_2\alpha)\theta$ so that $(n_1 + n_2\alpha)^{-1}U$ is an unbiased estimator of θ. But, note that $V_\theta\{(n_1 + n_2\alpha)^{-1}U\} = (n_1 + n_2\alpha)^{-2}(n_1\theta^2 + n_2\alpha\theta^2) = \theta^2(n_1 + n_2\alpha)^{-1}$, the same as the CRLB. Hence, we conclude that $(n_1 + n_2\alpha)^{-1}\{\Sigma_{i=1}^{n_1}X_{1i} + \Sigma_{j=1}^{n_2}X_{2j}\}$ is the UMVUE of θ. ▲

Example 7.5.14 Suppose that $X_{11}, ..., X_{1n_1}$ are iid $N(\mu, 1)$ with the unknown parameter $\mu \in (-\infty, \infty)$ and $\mathcal{X} = (-\infty, \infty)$. Also, suppose that $X_{21}, ..., X_{2n_2}$ are iid $N(2\mu, \sigma^2)$ involving the same unknown parameter μ and $\mathcal{X} = (-\infty, \infty)$, but $\sigma(> 0)$ is assumed known. Let us assume that the X_1's are independent of the X_2's. By direct calculations we have $\mathcal{I}_{X_1}(\mu) = 1$ and $\mathcal{I}_{X_2}(\mu) = 4\sigma^{-2}$. That is, the CRLB from (7.5.18) for the variance of unbiased estimators of $\tau(\mu) = \mu$ is given by $(n_1 + 4n_2\sigma^{-2})^{-1}$. Obviously, $U = \Sigma_{i=1}^{n_1}X_{1i} + 2\sigma^{-2}\Sigma_{j=1}^{n_2}X_{2j}$ is sufficient for μ and $E_\mu\{U\} = (n_1 + 4n_2\sigma^{-2})\mu$ so that $(n_1 + 4n_2\sigma^{-2})^{-1}U$ is an unbiased estimator of μ. But, note that $V_\mu\{(n_1 + 4n_2\sigma^{-2})^{-1}U\} = (n_1 + 4n_2\sigma^{-2})^{-2}(n_1 + 4n_2\sigma^{-2}) = (n_1 + 4n_2\sigma^{-2})^{-1}$, the same as the CRLB. Hence, we conclude that $(n_1 + 4n_2\sigma^{-2})^{-1}\{\Sigma_{i=1}^{n_1}X_{1i} + 2\sigma^{-2}\Sigma_{j=1}^{n_2}X_{2j}\}$ is the UMVUE of μ. ▲

7.5.4   Evaluation of Conditional Expectations

Thus far we aimed at finding the UMVUE of a real valued parametric function $\tau(\boldsymbol{\theta})$. With some practice, in many problems one will remember the well-known UMVUE for some of the parametric functions. Frequently, such UMVUE's will depend on complete sufficient statistics. Combining these information we may easily find the expressions of special types of conditional expectations.

Suppose that T and $\mathbf{U}$ are statistics and we wish to find the expression for $E_\theta[T \mid \mathbf{U} = \mathbf{u}]$. The general approach is involved. One will first derive the conditional distribution of T given $\mathbf{U}$ and then directly evaluate the integral $\int_T tf(t \mid \mathbf{U} = \mathbf{u}; \theta)dt$. In some situations we can avoid this cumbersome process. The following result may be viewed as a restatement of the Lehmann-Scheffé Theorems but it seems that it has not been included elsewhere in its present form.

Theorem 7.5.5 *Suppose that $X_1, ..., X_n$ are iid real valued random variables with the common pmf or pdf given by $f(x; \theta)$ with $x \in \mathcal{X} \subseteq \Re, \theta \in \Theta \subseteq \Re^k$. Consider two statistics T and $\mathbf{U}$ where T is real valued but $\mathbf{U}$ may be vector valued such that $E_\theta[T] = \tau(\boldsymbol{\theta})$, a real valued parametric function. Suppose that $\mathbf{U}$ is complete sufficient for $\boldsymbol{\theta}$ and a known real valued* *statistic* $g(\mathbf{U})$ *is the UMVUE for* $\tau(\boldsymbol{\theta})$. *Then, we have:*

$$E_{\boldsymbol{\theta}}[T \mid \mathbf{U} = \mathbf{u}] = g(\mathbf{u}) \text{ for all } \boldsymbol{\theta} \in \Theta, \mathbf{u} \in \mathcal{U}.$$

Proof Since the statistic $\mathbf{U}$ is assumed both complete and sufficient, by virtue of the Lehmann-Scheffé Theorems it follows that $E_{\boldsymbol{\theta}}[T \mid \mathbf{U}]$ must be the best unbiased estimator of $\tau(\boldsymbol{\theta})$. But, $g(\mathbf{U})$ is the unique UMVUE of $\tau(\boldsymbol{\theta})$ and hence the result follows immediately. ∎

Example 7.5.15 Suppose that $X_1, ..., X_n$ are iid $N(\mu, \sigma^2)$ where μ is unknown but σ^2 is known with $n \geq 4, -\infty < \mu < \infty, 0 < \sigma < \infty$ and $\mathcal{X} = \Re$. Let $T = X_1^2 + 2X_3 - X_4^2$ which is unbiased for $\tau(\mu) = 2\mu$ and $U = \overline{X}$. But, U is complete sufficient for μ and hence $g(U) = 2U$ is the unique UMVUE for $\tau(\mu)$. Thus, in view of the Theorem 7.5.5 we can immediately write $E_{\mu}[T \mid \overline{X}] = 2\overline{X}$. Instead if we had $T = (X_1 + X_2)^2$, then $E_{\mu}[T] = 4\mu^2 + 2\sigma^2 = \tau(\mu)$. But, the unique UMVUE of μ^2 was earlier (Example 7.5.7) found to be $\overline{X}^2 - n^{-1}\sigma^2$. Hence in view of the Theorem 7.5.5, we can immediately write $E_{\mu}[T \mid \overline{X}] = 4\{\overline{X}^2 - n^{-1}\sigma^2\} + 2\sigma^2$. ▲

> Suppose that $X_1, ..., X_n$ are random samples from Poisson(λ), $0 < \lambda < \infty$. Among other things, the following example shows easily that $V_{\lambda}(S^2) > V_{\lambda}(\overline{X})$ for all λ.

Example 7.5.16 Suppose that $X_1, ..., X_n$ are iid Poisson(λ) with $0 < \lambda < \infty$ unknown and $n \geq 2$. Let us denote $T = S^2$, the sample variance, and $U = \overline{X}$. Obviously, $E_{\lambda}[T] = \lambda = \tau(\lambda)$. But, U is complete sufficient and so U is the unique UMVUE for $\tau(\lambda)$. Thus, in view of the Theorem 7.5.5, we can write $E_{\lambda}[T \mid \overline{X}] = \overline{X}$.

Now, we use the Theorem 3.3.1, part (ii) to rewrite $V_{\lambda}(S^2)$ as

$$V_{\lambda}[E(S^2 \mid \overline{X})] + E_{\lambda}[V(S^2 \mid \overline{X})] = V_{\lambda}[\overline{X}] + E_{\lambda}[V(S^2 \mid \overline{X})],$$

which exceeds $V_{\lambda}[\overline{X}]$, for all λ, because $V(S^2 \mid \overline{X})$ is a positive random variable whose expectation is positive. For the direct calculation of $V_{\lambda}(S^2)$, however, refer to the expression given by (5.2.23). ▲

Example 7.5.17 Let $X_1, ..., X_n$ be iid Uniform($0, \theta$) where $\theta(> 0)$ is the unknown parameter with $n \geq 2$. Then, $U = X_{n:n}$ is complete sufficient for θ. Consider $T = X_1^2 + X_2$. Obviously, $E_{\theta}[T] = \frac{1}{2}\theta + \frac{1}{3}\theta^2 = \tau(\theta)$. Now, since U has its pdf given by $nu^{n-1}\theta^{-n}I(0 < u < \theta)$, we have $E_{\theta}[X_{n:n}] = n(n+1)^{-1}\theta$ and $E_{\theta}[X_{n:n}^2] = n(n+2)^{-1}\theta^2$. Hence, the unique UMVUE of $\tau(\theta)$ is given by $\frac{1}{2}(n+1)n^{-1}X_{n:n} + \frac{1}{3}(n+2)n^{-1}X_{n:n}^2$. Thus, in view of the Theorem 7.5.5, we can write $E_\theta[T \mid X_{n:n}] = \frac{1}{2}(n+1)n^{-1}X_{n:n} + \frac{1}{3}(n+2)n^{-1}X_{n:n}^2$. If one first obtains the conditional pdf of T given $X_{n:n}$ and evaluates $E_\theta[T \mid X_{n:n}]$ directly, then one will realize what a clever approach the present one has been. ▲

Example 7.5.18 (Example 7.5.17 Continued) Let $X_1, ..., X_n$ be iid random variables distributed uniformly on $(0, \theta)$ where $\theta(> 0)$ is the unknown parameter with $n \geq 2$. Again $U = X_{n:n}$ is complete sufficient for θ. Consider $T = \overline{X}^2$. Obviously, $E_\theta[T] = \frac{1}{4}\theta^2(1 + \frac{1}{3n}) = \tau(\theta)$. One should verify that $E_\theta[T \mid X_{n:n}] = \frac{1}{4}(1 + \frac{1}{3n})(1 + \frac{2}{n})X_{n:n}^2$. We leave it out as Exercise 7.5.21. Again, if one first obtains the conditional pdf of T given $X_{n:n}$ and evaluates $E_\theta[T \mid X_{n:n}]$ directly, then one will realize what a clever approach the present one has been. ▲

7.6   Unbiased Estimation Under Incompleteness

Suppose that it we start with two statistics T and T', both estimating the same real valued parametric function $\tau(\boldsymbol{\theta})$ unbiasedly. Then, we individually condition T and T' given the sufficient statistic $\mathbf{U}$ and come up with the refined Rao-Blackwellized estimators $W = E_\theta[T \mid \mathbf{U}]$ and $W' = E_\theta[T' \mid \mathbf{U}]$ respectively. It is clear that (i) both W, W' would be unbiased estimators of $\tau(\boldsymbol{\theta})$, (ii) W will have a smaller variance than that of T , and (iii) W' will have a smaller variance than that of T'.

> An important question to ask here is this: how will the two variances $V_\theta(W)$ and $V_\theta(W')$ stack up against each other?

If the statistic $\mathbf{U}$ happens to be complete, then the Lehmann-Scheffé Theorems will settle this question because W and W' will be identical estimators w.p.1. But if $\mathbf{U}$ is *not* complete, then it may be a different story. We highlight some possibilities by means of examples.

7.6.1   Does the Rao-Blackwell Theorem Lead to UMVUE?

Suppose that $X_1, ..., X_n$ are iid $N(\theta, \theta^2)$ having the unknown parameter $\theta \in \Theta = (0, \infty), \mathcal{X} = (-\infty, \infty), n \geq 2$. Of course, $\mathbf{U} = (\overline{X}, S^2)$ is sufficient for θ but $\mathbf{U}$ is not complete since one can check that $E_\theta\{n(n+1)^{-1}\overline{X}^2 - S^2\} \equiv 0$ for all $\theta \in \Theta$ and yet $n(n+1)^{-1}\overline{X}^2 - S^2$ is not identically zero w.p.1. Now, let us work with $T = \overline{X}$, $T' = a_n^{-1}S$ where $a_n = \sqrt{2}\Gamma(\frac{n}{2})\{\Gamma(\frac{n-1}{2})\sqrt{n-1}\}^{-1}$. From (2.3.26) it follows that $E_\theta[S] = a_n\theta$ and hence T' is unbiased for

θ while obviously so is T itself. Now, $\overline{X}, S^2$ are independently distributed and hence we claim the following:

$$E_\theta[\overline{X} \mid \mathbf{U}] = \overline{X}, \quad E_\theta[S \mid \mathbf{U}] = S.$$

Thus, the Rao-Blackwellized versions of T and T' are respectively $\overline{X}(= W)$ and $a_n^{-1}S(= W')$. Clearly W and W' are both unbiased estimators for the unknown parameter θ. But observe that $V_\theta(W) = n^{-1}\theta^2$ while $V_\theta(W') = (a_n^{-2} - 1)\theta^2$. Next, note that $\Gamma(x + 1) = x\Gamma(x)$ for $x > 0, \Gamma(\frac{1}{2}) = \sqrt{\pi}$ and evaluate $V_\theta(W')$. Look at the Table 7.6.1.

Table 7.6.1. Comparing the Two Variances $V_\theta(W)$ and $V_\theta(W')$

n	$\theta^{-2}V_\theta(W)$	Exact $\theta^{-2}V_\theta(W')$	Approx. $\theta^{-2}V_\theta(W')$
2	.5000	$(\pi/2) - 1$	.5708
3	.3333	$(4/\pi) - 1$	.2732
4	.2500	$(3\pi/8) - 1$	.1781
5	.2000	$\{32/(9\pi)\} - 1$	.1318

Upon inspecting the entries in the Table 7.6.1, it becomes apparent that W is better than W' in the case $n = 2$. But when $n = 3, 4, 5$, we find that W' is better than W.

For large n, however, $V_\theta(W')$ can be approximated. Using (1.6.24) to approximate the ratios of gamma functions, we obtain

$$x^{b-a}\Gamma(x + a)\{\Gamma(x + b)\}^{-1}$$
$$= 1 + \tfrac{1}{2}(a - b)(a + b - 1)x^{-1} + O(x^{-2}) \text{ as } x \to \infty. \tag{7.6.1}$$

Now, we apply (7.6.1) with $x = \frac{1}{2}n$, $a = -\frac{1}{2}$ and $b = 0$ to write

$$\begin{aligned} a_n^{-2} &= \tfrac{1}{2}(n - 1)\tfrac{2}{n}\left\{1 + \tfrac{3}{4n} + O(n^{-2})\right\}^2 \\ &= \left(1 - \tfrac{1}{n}\right)\left\{1 + \tfrac{3}{2n} + O(n^{-2})\right\} \\ &= 1 + \tfrac{1}{2n} + O(n^{-2}) \text{ as } n \to \infty. \end{aligned} \tag{7.6.2}$$

In other words, from (7.6.2) it follows that for large n, we have

$$V_\theta(W') = (a_n^{-2} - 1)\theta^2 \sim \tfrac{1}{2n}\theta^2. \tag{7.6.3}$$

Using MAPLE, the computed values of $\theta^{-2}V_\theta(W')$ came out to be $2.6653\times 10^{-2}, 1.7387\times 10^{-2}, 1.2902\times 10^{-2}, 1.0256\times 12^{-2}, 5.0632\times 10^3$ and 2.5157×10^{-3} respectively when $n = 20, 30, 40, 50, 100$ and 200. The approximation given by (7.6.3) seems to work well for $n \geq 40$.

Khan (1968) considered this estimation problem but he first focussed on the status of the MLE. Our goal is slightly different. Utilizing the two initial unbiased estimators T, T' we focussed on the Rao-Blackwellized versions and compared their variances. This was not emphasized in Khan (1968). We find that W' can be better than W even if n is small (≥ 3) but for large n, the estimator W' can have approximately fifty percent less variation than the estimator W.

Khan (1968) proceeded to derive another unbiased estimator for θ which performed better than both W and W'. The basic idea was simple. Let us look at the following class of unbiased estimators of θ:

$$\{T^*: T^* \equiv T^*(\alpha) = \alpha W + (1 - \alpha)W' \text{ with } 0 \leq \alpha \leq 1\}. \tag{7.6.4}$$

For each $\alpha \in [0, 1]$, $T^*(\alpha)$ is unbiased for θ and hence the one having the smallest variance should be more attractive than either W or W'. Since $\overline{X}$ and S^2 are independent, we have

$$V_\theta(T^*) = \alpha^2 V_\theta(W) + (1 - \alpha)^2 V_\theta(W').$$

Now, $V_\theta(T^*)$ can be minimized directly with respect to the choice of α. The optimal choice of α is given by

$$\alpha^* = V_\theta(W')\{V_\theta(W) + V_\theta(W')\}^{-1} = n(a_n^{-2} - 1)\{1 + n(a_n^{-2} - 1)\}^{-1}. \tag{7.6.5}$$

In view of (7.6.3), α^* reduces to $\frac{1}{3}$ for large n. With α^* determined by (7.6.5), the corresponding unbiased estimator $T^*(\alpha^*)$ would have its variance smaller than that of both W and W'. One will find interesting decision theoretic considerations in Gleser and Healy (1976). Lehmann (1983, page 89) mentions an unpublished thesis of Unni (1978) which showed that a UMVUE of θ did not exist in the present situation.

> In the preceding discussion, W and W' were the Rao-Blackwellized versions of two unbiased estimators T and T'. We found an unbiased estimator T^* of θ such that $V_\theta(T^*) < V_\theta(W)$ and $V_\theta(T^*) < V_\theta(W')$ for $\theta > 0$. (7.6.6)

Example 7.6.1 Let us suppose that one has $X_1, ..., X_n$ iid with the common pdf $f(x; \theta) = \theta^{-1} exp\{-(x-\theta)/\theta\} I(x > \theta)$ with $\theta \in \Theta = \Re^+$ where θ is the unknown parameter. Here, we wish to estimate θ unbiasedly. Recall from the Exercise 6.6.5 that the statistic $\mathbf{U} = (X_{n:1}, \Sigma_{i=1}^n (X_i - X_{n:1}))$ is not complete but it is sufficient for θ. One should check into the ramifications of the preceding discussions in the context of the present problem. We leave this out as Exercise 7.6.1. ▲

7.7 Consistent Estimators

The *consistency* is a large sample property of an estimator. R. A. Fisher had introduced the concept of *consistency* in his 1922 paper. Let us suppose that we have $\{T_n \equiv T_n(X_1, ..., X_n); n \geq 1\}$, a sequence of estimators for some unknown real valued parametric function $\tau(\boldsymbol{\theta})$ where $\boldsymbol{\theta} \in \Theta \subseteq \Re^k$. We emphasize the dependence of the estimators on the sample size n by indexing with subscript n. It may help to recall the concept of the *convergence in probability*, denoted by $\xrightarrow{P}$, which was laid out by the Definition 5.2.1.

Definition 7.7.1 *Consider $\{T_n \equiv T_n(X_1, ..., X_n); n \geq 1\}$, a sequence of estimators for some unknown real valued parametric function $\tau(\boldsymbol{\theta})$ where $\boldsymbol{\theta} \in \Theta \subseteq \Re^k$. Then, T_n is said to be consistent for $\tau(\boldsymbol{\theta})$ if and only if $T_n \xrightarrow{P} \tau(\boldsymbol{\theta})$ as $n \to \infty$. Also, T_n is called inconsistent for $\tau(\boldsymbol{\theta})$ if T_n is not consistent for $\tau(\boldsymbol{\theta})$.*

> An estimator T_n may be biased for $\tau(\boldsymbol{\theta})$ and yet T_n may be consistent for $\tau(\boldsymbol{\theta})$.

Estimators found by the method of moments are often smooth functions of averages of powers of the X's and so they are consistent for the associated parametric functions. One should verify that all the estimators derived in the Examples 7.2.1-7.2.5 are indeed consistent for the parameters of interest. The MLE's derived in the Examples 7.2.6-7.2.7 and also in the Examples 7.2.9-7.2.12 are all consistent.

Let us particularly draw attention to the Examples 7.2.5 and 7.2.9. In the Uniform$(0, \theta)$ case, the method of moment estimator (Example 7.2.5) for θ turned out to be $T_n = 2\overline{X}$ and by the Weak WLLN (Theorem 5.2.1) it follows that $T_n \xrightarrow{P} \theta$ as $n \to \infty$. That is, T_n is a consistent estimator of θ even though T_n is not exclusively a function of the minimal sufficient statistic $X_{n:n}$. The MLE (Example 7.2.9) of θ is $X_{n:n}$ and recall from Example 5.2.5 that $X_{n:n} \xrightarrow{P} \theta$ as $n \to \infty$. Thus, $X_{n:n}$ is a consistent estimator for θ. Note that $X_{n:n}$ is not unbiased for θ.

All the UMVUE's derived in Section 7.5 are consistent for the parameters of interest. Their verifications are left as exercises. Let us, however, look at the Example 7.5.4 where the unbiased estimator $T_n = (1 - n^{-1})^{\Sigma_{i=1}^n X_i}$ was found for the parametric function $e^{-\lambda}$ in the case of the Poisson(λ) distribution. Let us rewrite

$$log\{T_n\} = n\{log(1 - n^{-1})\}\overline{X} = \overline{X}\{log(1 - n^{-1})^n\} = U_n log(V_n),$$

where $U_n = \overline{X} \xrightarrow{P} \lambda$, $V_n = (1-n^{-1})^n \to e^{-1}$ as $n \to \infty$. In this deliberation, it will help to think of V_n as a sequence of degenerate random variables. Then, one can repeatedly apply the Theorems 5.2.4-5.2.5 and conclude that $log\{T_n\} \xrightarrow{P} \lambda \, log(e^{-1}) = -\lambda$ and hence $T_n \xrightarrow{P} e^{-\lambda}$ as $n \to \infty$. That is, $(1 - n^{-1})^{\Sigma_{i=1}^{n}X_i}$ is a consistent estimator of $e^{-\lambda}$.

One should also verify that the estimators W, W', and $T^*(\alpha^*)$ defined via (7.6.4)-(7.6.5) are also consistent for the unknown parameter θ in the $N(\theta, \theta^2)$ population. Details are left out as Exercise 7.7.5.

Having found a number of estimators in so many statistical models which are consistent for the parameters of interest, we wish to emphasize again that *consistency* is after all a *large sample* property. In statistics, one invariably works with data of fixed size n and hence one should not rely upon the consistency property of an estimator *alone* to claim any superior performance on behalf of the estimator under consideration.

D. Basu, among others, emphasized the limited usefulness of the concept of consistency of an estimator. Basu gave the following forceful example. Let $X_1, ..., X_n$ be iid $N(\theta, 1)$ where $\theta \in (-\infty, \infty)$ is the unknown parameter. Define a sequence of estimators of θ as follows:

$$T_n = \begin{cases} 0 & \text{if} \quad n \le k \\ n^{-1}\Sigma_{i=1}^{n}X_i & \text{if} \quad n \ge k+1, \end{cases} \qquad (7.7.1)$$

where k is fixed but presumably very large. Since $n^{-1}\Sigma_{i=1}^{n}X_i$ is consistent for θ, it follows from the definition of the estimator T_n that it would be consistent for θ too. But, Basu would argue that in real life how many times does one encounter a sample of size n larger than a million! So, he would focus on the estimator T_n when $k = 10^6$. Now, if an experimenter is committed to use the estimator T_n from (7.7.1), then for all practical purposes, regardless of what the observed data dictates, one will end up guessing that θ is zero and yet such an estimator would be consistent for θ! This construction of T_n unfortunately created an impression that there was something inherently wrong with the concept of consistency. Can a reader think of any practical scenario where T_n defined by (7.7.1) will be used to estimate θ?

Basu wanted to emphasize that the consistency of an estimator is not a useful property in the practice of statistics. But, on the other hand, the given T_n will almost certainly never be used in practice, and so it remains vague what this example actually conveys. What this example might have demonstrated was this: The consistency of an estimator may not be a useful property *in itself* in the practice of statistics. This is exactly what we had said before we gave Basu's example! In other words, Basu's example has not been nearly as damaging to the concept of consistency as some non-Fisherians may have liked to believe.

It has now become a common place for some authors of Bayesian papers to claim that a proposed "Bayes estimator" or "Bayes decision" also satisfies the "consistency" property in some sense. In the Bayesian literature, some authors refer to this as the Bayesian-Frequentist compromise. Will it then be fair to say that the (Fisherian) consistency property has not been such a bad concept after all? The "history" does have an uncanny ability to correct its own course from time to time! At the very least, it certainly feels that way.

A simple truth is that R. A. Fisher never suggested that one should choose an estimator because of its consistency property alone. The fact is that in many "standard" problems, some of the usual estimators such as the MLE, UMVUE or estimators obtained via method of moments are frequently consistent for the parameter(s) of interest.

The common sense dictates that the consistency property has to take a back seat when considered in conjunction with the sufficiency property. The estimator T_n defined in (7.7.1) is *not* sufficient for θ for any reasonable fixed-sample-size n when $k = 10^6$. On the other hand, the sample mean $U_n = n^{-1}\Sigma_{i=1}^n X_i$ is sufficient for θ and it also happens to be *consistent* for θ.

We should add that, in general, an MLE $\widehat{\theta}$ may not be consistent for θ. Some examples of inconsistent MLE's were constructed by Neyman and Scott (1948) and Basu (1955b). In Chapter 12, under mild regularity conditions, we quote a result in (12.2.3) showing that an MLE $\widehat{\theta}$ is indeed consistent for the parameter θ it is supposed to estimate in the first place. This result, in conjunction with the invariance property (Theorem 7.2.1), make the MLE's very appealing in practice.

7.8   Exercises and Complements

7.2.1 (Example 7.2.3 Continued) Suppose that $X_1, ..., X_n$ are iid $N(0, \sigma^2)$ where $\theta = \sigma^2$ is unknown, $0 < \sigma < \infty$. Here we have $\mathcal{X} = \Theta = \Re$ and $\eta_1 \equiv \eta_1(\theta) = E_\theta[X_1] = 0$, $\eta_3 \equiv \eta_3(\theta) = E_\theta[X_1^3] = 0$ for all θ. It is clear that the equations given by the first and third moments in (7.2.1) do not lead to anything interesting. In the Example 7.2.3 we had used the expression of $\eta_2(\theta)$. Now we may *arbitrarily* move to the fourth moment and write $\eta_4 \equiv \eta_4(\theta) = E_\theta[X_1^4] = 3\sigma^4$. Using η_4, find an appropriate estimator of σ^2. How is this estimator different from the one obtained in the Example 7.2.3?

7.2.2 (Exercise 6.3.5 Continued) Suppose that $X_1, ..., X_n$ are iid distributed as Geometric(p) random variables with the common pmf given by $f(x; p) = p(1 - p)^x$, $x = 0, 1, 2, ...$, and $0 < p < 1$ is the unknown parameter. Find an estimator of p by the method of moments.

7.2.3 Suppose that $X_1, ..., X_n$ are iid distributed as Gamma(α, β) random variables where α and β are both unknown parameters, $0 < \alpha, \beta < \infty$. Derive estimators for α and β by the method of moments.

7.2.4 Suppose that $X_1, ..., X_n$ are iid whose common pdf is given by

$$f(x) = \begin{cases} (\theta + 1)x^\theta & \text{if } 0 < x < 1 \\ 0 & \text{elsewhere,} \end{cases}$$

where $\theta(> 0)$ is the unknown parameter. Derive an estimator for θ by the method of moments.

7.2.5 Suppose that $X_1, ..., X_n$ are iid distributed as Beta(θ, θ) random variables where $\theta(> 0)$ is the unknown parameter. Derive an estimator for θ by the method of moments.

7.2.6 (Exercise 7.2.2 Continued) Suppose that $X_1, ..., X_n$ are iid distributed as Geometric(p) random variables with the common pmf given by $f(x; p) = p(1 - p)^x$, $x = 0, 1, 2, ...$, and $0 < p < 1$ is the unknown parameter. Find the MLE of p. Is the MLE sufficient for p?

7.2.7 Suppose that $X_1, ..., X_n$ are iid Bernoulli(p) random variables where p is the unknown parameter, $0 \le p \le 1$.

(i)   Show that $\overline{X}$ is the MLE of p. Is the MLE sufficient for p?

(ii)   Derive the MLE for p^2;

(iiii) Derive the MLE for p/q where $q = 1 - p$;

(iv)   Derive the MLE for $p(1 - p)$.

{*Hint*: In parts (ii)-(iv), use the invariance property of the MLE from Theorem 7.2.1.}

7.2.8 (Exercise 7.2.7 Continued) Suppose that we have iid Bernoulli(p) random variables $X_1, ..., X_n$ where p is the unknown parameter, $0 < p < 1$. Show that $\overline{X}$ is the MLE of p when $\overline{X}$ is not zero or one. In the light of the Example 7.2.10, discuss the situation one faces when $\overline{X}$ is zero or one.

7.2.9 Suppose that $X_1, ..., X_n$ are iid $N(\mu, \sigma^2)$ where μ is known but σ^2 is unknown, $\theta = \sigma^2$, $-\infty < \mu < \infty$, $0 < \sigma < \infty, n \ge 2$.

(i)   Show that the MLE of σ^2 is $n^{-1}\Sigma_{i=1}^{n}(X_i - \mu)^2$;

(ii)   Is the MLE in part (i) sufficient for σ^2?

(iii) Derive the MLE for $1/\sigma$;

(iv)   Derive the MLE for $(\sigma + \sigma^{-1})^{1/2}$.

{*Hint*: In parts (iii)-(iv), use the invariance property of the MLE from Theorem 7.2.1.}

7.2.10 Suppose that $X_1, ..., X_n$ are iid with the Rayleigh distribution, that is the common pdf is

$$f(x; \theta) = 2\theta^{-1}x\,exp(-x^2/\theta)I(x > 0),$$

where $\theta(> 0)$ is the unknown parameter. Find the MLE for θ. Is the MLE sufficient for θ?

7.2.11 Suppose that $X_1, ..., X_n$ are iid with the Weibull distribution, that is the common pdf is

$$f(x; \alpha) = \alpha^{-1}\beta x^{\beta-1}\,exp(-x^\beta/\alpha)I(x > 0)$$

where $\alpha(> 0)$ is the unknown parameter, but $\beta(> 0)$ is assumed known. Find the MLE for α. Is the MLE sufficient for α?

7.2.12 (Exercise 6.2.11 Continued) Let $X_1, ..., X_n$ be iid having the common pdf $\sigma^{-1}exp\{-(x - \mu)/\sigma\}I(x > \mu)$ where μ and σ are both unknown, $-\infty < \mu < \infty, 0 < \sigma < \infty, n \geq 2$. Show that the MLE for μ and σ are respectively $X_{n:1}$, the smallest order statistic, and $n^{-1}\Sigma_{i=1}^n(X_i - X_{n:1})$. Then, derive the MLE's for $\mu/\sigma, \mu/\sigma^2$ and $\mu + \sigma$.

7.2.13 (Exercise 7.2.12 Continued) Let $X_1, ..., X_n$ be iid having the common pdf $\sigma^{-1}exp\{-(x - \mu)/\sigma\}I(x > \mu)$ where μ is known but σ is unknown, $-\infty < \mu < \infty, 0 < \sigma < \infty$. Show that the MLE for σ is given by $n^{-1}\Sigma_{i=1}^n(X_i - \mu)$.

7.2.14 (Exercise 6.3.12 Continued) Let $X_1, ..., X_n$ be iid having the common Uniform distribution on the interval $(-\theta, \theta)$ where $0 < \theta < \infty$ is the unknown parameter. Derive the MLE for θ. Is the MLE sufficient for θ? Also, derive the MLE's for θ^2 and θ^{-2}.

7.2.15 (Exercise 6.3.13 Continued) Let $X_1, ..., X_m$ be iid $N(\mu_1, \sigma^2)$, $Y_1, ..., Y_n$ be iid $N(\mu_2, \sigma^2)$, and also let the X's be independent of the Y's where $-\infty < \mu_1, \mu_2 < \infty, 0 < \sigma < \infty$ are the unknown parameters. Derive the MLE for (μ_1, μ_2, σ^2). Is the MLE sufficient for (μ_1, μ_2, σ^2)? Also, derive the MLE for $(\mu_1 - \mu_2)/\sigma$.

7.2.16 (Exercise 6.3.14 Continued) Let $X_1, ..., X_m$ be iid $N(\mu_1, \sigma^2)$, $Y_1, ..., Y_n$ be iid $N(\mu_2, k\sigma^2)$, and also let the X's be independent of the Y's where $-\infty < \mu_1, \mu_2 < \infty, 0 < \sigma < \infty$ are the unknown parameters. Assume that the number k (> 0) is known. Derive the MLE for (μ_1, μ_2, σ^2). Is the MLE sufficient for (μ_1, μ_2, σ^2)? Also, derive the MLE for $(\mu_1 - \mu_2)/\sigma$.

7.2.17 (Exercise 7.2.4 Continued) Suppose that $X_1, ..., X_n$ are iid whose common pdf is given by

$$f(x; \theta) = \begin{cases} (\theta + 1)x^\theta & \text{if } 0 < x < 1 \\ 0 & \text{elsewhere}, \end{cases}$$

where $\theta (> 0)$ is the unknown parameter. Derive the MLE for θ. Compare this MLE with the method of moments estimator obtained earlier. Is the MLE sufficient for θ?

7.2.18 (Exercise 7.2.12 Continued) Let $X_1, ..., X_n$ be iid having the common pdf $\sigma^{-1}exp\{-(x - \mu)/\sigma\}I(x > \mu)$ where μ and σ are both unknown, $-\infty < \mu < \infty, 0 < \sigma < \infty, n \geq 2$. Derive the method of moment estimators for μ and σ.

7.2.19 Suppose that $Y_1, ..., Y_n$ are independent random variables where Y_i is distributed as $N(\beta_0 + \beta_1 x_i, \sigma^2)$ with unknown parameters β_0, β_1. Here, $\sigma (> 0)$ is assumed known and x_i's are fixed real numbers with $\boldsymbol{\theta} = (\beta_0, \beta_1) \in \Re^2, i = 1, ..., n, n \geq 2$. Denote $a = \Sigma_{i=1}^{n}(x_i - \overline{x})^2$ with $\overline{x} = \frac{1}{n}\Sigma_{i=1}^{n}x_i$ and assume that $a > 0$. Suppose that the MLE's for β_0 and β_1 are respectively denoted by $\widehat{\beta}_0, \widehat{\beta}_1$. Now, consider the following *linear regression* problems.

(i)   Write down the likelihood function of $Y_1, ..., Y_n$;

(ii)   Show that $\widehat{\beta}_1 = \frac{1}{a}\Sigma_{i=1}^{n}(x_i - \overline{x})Y_i$ and $\widehat{\beta}_1$ is normally distributed with mean β_1 and variance σ^2/a;

(iii)   Show that $\widehat{\beta}_0 = \overline{Y} - \widehat{\beta}_1\overline{x}$ and $\widehat{\beta}_0$ is normally distributed with mean β_0 and variance $\sigma^2[\frac{1}{n} + \frac{\overline{x}^2}{a}]$.

7.2.20 Suppose that $X_1, ..., X_n$ are iid with the common pdf $f(x; \theta) = \theta^{-1}x\,exp\{-x^2/(2\theta)\}I(x > 0)$ where $0 < \theta < \infty$ is the unknown parameter. Estimate θ by the method of moments separately using the first and second population moments respectively. Between these two method of moments estimators of the parameter θ, which one should one prefer and why?

7.3.1 Suppose that $X_1, ..., X_4$ are iid $N(0, \sigma^2)$ where $0 < \sigma < \infty$ is the unknown parameter. Consider the following estimators:

$$T_1 = X_1^2 - X_2 + X_4, T_2 = \frac{1}{3}(X_1^2 + X_2^2 + X_4^2), T_3 = \frac{1}{4}\Sigma_{i=1}^{4}X_i^2,$$

$$T_4 = \frac{1}{3}\Sigma_{i=1}^{4}(X_i - \overline{X})^2, T_5 = \frac{1}{2}|X_1 - X_2|$$

(i)   Is T_i unbiased for σ^2, $i = 1, ..., 4$?

(ii)   Among the estimators T_1, T_2, T_3, T_4 for σ^2, which one has the smallest MSE?

(iii)   Is T_5 unbiased for σ? If not, find a suitable multiple of T_5 which is unbiased for σ. Evaluate the MSE of T_5.

7.3.2 (Example 7.3.1 Continued) Let $X_1, ..., X_n$ be iid $N(\mu, \sigma^2)$ where μ, σ are both unknown with $-\infty < \mu < \infty, 0 < \sigma < \infty, n \geq 2$. Denote $U = \Sigma_{i=1}^{n}(X_i - \overline{X})^2$. Let $V = cU$ be an estimator of σ^2 where $c(> 0)$ is a constant.

(i)   Find the MSE of V. Then, minimize this MSE with respect to c. Call this latter estimator W which has the smallest MSE among the estimators of σ^2 which are multiples of U;

(ii)   Show that estimator W coincides with $(n+1)^{-1}\Sigma_{i=1}^{n}(X_i - \overline{X})^2$ which was used in the Example 7.3.1.

7.3.3 (Exercise 7.2.13 Continued) Let $X_1, ..., X_n$ be iid having the common pdf $\sigma^{-1}exp\{-(x - \mu)/\sigma\}I(x > \mu)$ where μ, σ are both unknown, $-\infty < \mu < \infty, 0 < \sigma < \infty, n \geq 2$. Denote $U = \Sigma_{i=1}^{n}(X_i - X_{n:1})$. Let $V = cU$ be an estimator of σ where $c(> 0)$ is a constant.

(i)   Find the MSE of V. Then, minimize this MSE with respect to c. Call this latter estimator W which has the smallest MSE among the estimators of σ which are the multiples of U;

(ii)   How do the two estimators W and $(n-1)^{-1}\Sigma_{i=1}^{n}(X_i - X_{n:1})$ compare relative to their respective bias and MSE?

7.3.4 Suppose that $X_1, ..., X_n$ are iid from the following respective populations. Find the expressions for the BLUE of θ, the parameter of interest in each case.

(i)   The population is Poisson(λ) where $\theta = \lambda \in \Re^+$;

(ii)   The population is Binomial(n, p) where $\theta = p \in (0, 1)$;

(iii) The population has the pdf $f(x) = \frac{1}{2\sigma}exp(-|x|/\sigma), x \in \Re$ where $\theta = \sigma \in \Re^+$.

7.3.5 (Exercise 7.3.1 Continued) Suppose that $X_1, ..., X_n$ are iid $N(0, \sigma^2)$ where $0 < \sigma < \infty$ is the unknown parameter. Consider estimating σ unbiasedly by linear functions of $|X_i|, i = 1, ..., n$. Within this class of estimators, find the expression of the BLUE of σ. Next, evaluate the variance of the BLUE of σ.

7.3.6 Look at the estimator T_4 defined in (7.3.1). Evaluate its MSE.

7.3.7 (Example 7.3.2 Continued) Suppose that we have iid Bernoulli(p) random variables $X_1, ..., X_n$ where $0 < p < 1$ is an unknown parameter. Show that there is no unbiased estimator for the parametric function (i) $\tau(p) = p^{-1}(1 - p)^{-1}$, (ii) $\tau(p) = p/(1 - p)$.

7.3.8 (Example 7.3.3 Continued) Suppose that we have iid Bernoulli(p) random variables $X_1, X_2, ...$ where $0 < p < 1$ is an unknown parameter. Consider the parametric function $\tau(p) = p^{-2}$ and the observable random variable N defined in the Example 7.3.3. Use the expressions for the mean and variance of the Geometric distribution to find an estimator T involving N so that T is unbiased for $\tau(p)$.

7.4.1 Suppose that $X_1, ..., X_n$ are iid Bernoulli(p) where $0 < p < 1$ is an unknown parameter with $n \geq 2$. Consider the parametric function

$\tau(p) = p^2$. Start with the estimator $T = X_1 X_2$ which is unbiased for $\tau(p)$ and then derive the Rao-Blackwellized version of T. {*Hint*: Proceed along the lines of the Examples 7.4.1 and 7.4.3.}

7.4.2 (Example 7.4.3 Continued) Suppose that we have iid Bernoulli(p) random variables $X_1, ..., X_n$ where $0 < p < 1$ is an unknown parameter with $n \geq 3$. Consider the parametric function $\tau(p) = p^2(1 - p)$. Start with the estimator $T = X_1 X_2 (1 - X_3)$ which is unbiased for $\tau(p)$ and then derive the Rao-Blackwellized version of T. {*Hint*: Proceed along the lines of the Example 7.4.3.}

7.4.3 (Example 7.4.5 Continued) Suppose that we have iid Poisson(λ) random variables $X_1, ..., X_n$ where $0 < \lambda < \infty$ is an unknown parameter with $n \geq 4$. Consider the parametric function $\tau(\lambda)$ and the initial estimator T defined in each part and then derive the corresponding Rao-Blackwellized version W of T to estimate $\tau(\lambda)$ unbiasedly. Consider

(i) $\tau(\lambda) = \lambda e^{-\lambda}$ and start with $T = I(X_1 = 1)$. Verify first that T is unbiased for $\tau(\lambda)$. Derive W;

(ii) $\tau(\lambda) = e^{-2\lambda}$ and start with $T = I(X_1 = 0 \cap X_2 = 0)$. Verify first that T is unbiased for $\tau(\lambda)$. Derive W;

(iii) $\tau(\lambda) = e^{-3\lambda}$ and start with $T = I(X_1 = 0 \cap X_2 = 0 \cap X_3 = 0)$. Verify first that T is unbiased for $\tau(\lambda)$. Derive W;

(iv) $\tau(\lambda) = \lambda^2 e^{-\lambda}$ and start with $T = 2I(X_1 = 2)$. Verify first that T is unbiased for $\tau(\lambda)$. Derive W.

7.4.4 (Example 7.4.7 Continued) Suppose that $X_1, ..., X_n$ are iid $N(\mu, \sigma^2)$ where μ is unknown but σ is assumed known with $\mu \in \Re, \sigma \in \Re^+$. Let $\tau(\mu) = P_\mu\{|X_1| \leq a\}$ where a is some known positive real number. Find an initial unbiased estimator T for $\tau(\mu)$. Next, derive the Rao-Blackwellized version of T to estimate $\tau(\mu)$ unbiasedly.

7.4.5 (Exercise 7.4.4 Continued) Suppose that $X_1, ..., X_n$ are iid $N(\mu, \sigma^2)$ where μ is unknown but σ is assumed known with $\mu \in \Re, \sigma \in \Re^+, n \geq 3$. Let $\tau(\mu) = P_\mu\{X_1 + X_2 \leq a\}$ where a is some known real number. Find an initial unbiased estimator T for $\tau(\mu)$. Next, derive the Rao-Blackwellized version of T to estimate $\tau(\mu)$ unbiasedly.

7.4.6 (Example 7.4.7 Continued) Suppose that $X_1, ..., X_n$ are iid $N(\mu, \sigma^2)$ where μ, σ are both unknown with $\mu \in \Re, \sigma \in \Re^+, n \geq 2$. Let $\tau(\mu) = P_\mu\{X_1 > a\}$ where a is some known real number. Start with the initial unbiased estimator $T = I(X_1 > a)$ for $\tau(\mu)$. Next, derive the Rao-Blackwellized version W of T to estimate $\tau(\mu)$ unbiasedly. {*Hint*: Kolmogorov (1950a) first found the form of the final unbiased estimator W.

This elegant proof is due to Kolmogorov. Observe that one has the sufficient statistic $\mathbf{U} = (\overline{X}, S^*)$ for $\boldsymbol{\theta} = (\mu, \sigma)$ where $S^{*2} = \Sigma_{i=1}^{n}(X_i - \overline{X})^2$. Let $g(\mathbf{u}) = E_{\boldsymbol{\theta}}\{T \mid \overline{X}, S^*\} = P_{\boldsymbol{\theta}}\{X_1 > a \mid \overline{X}, S^*\} = P_{\boldsymbol{\theta}}\{Y > y_0 \mid \overline{X}, S^*\}$ with $Y = \sqrt{n}(X_1 - \overline{X})/\{(n-1)S^{*2}\}^{1/2}$ and $y_0 = \sqrt{n}(a - \overline{X})/\{(n-1)S^{*2}\}^{1/2}$. Next, verify that Y is distributed independently of $\mathbf{U}$, and the pdf of Y is given by $k(1-y^2)^{(n-4)/2}$ for $-1 < y < 1$ and some known positive constant k. Hence, $W = P_{\boldsymbol{\theta}}\{Y > y_0 \mid \overline{X}, S^*\} = k\int_{y_0}^{1} k(1 - y^2)^{(n-4)/2}dy.$ }

7.4.7 (Exercise 7.4.6 Continued) Suppose that $X_1, ..., X_n$ are iid $N(\mu, \sigma^2)$ where μ, σ are both unknown with $\mu \in \Re, \sigma \in \Re^+, n \geq 2$. Let $\tau(\mu) = P_\mu\{X_1 > a\}$ where a is some known real number. Consider the Rao-Blackwellized version W from the Exercise 7.4.6 which estimates $\tau(\mu)$ unbiasedly. Derive the form of W in its simplest form when $n = 4, 6$ and 8.

7.4.8 Suppose that $X_1, ..., X_n, X_{n+1}$ are iid Bernoulli(p) where $0 < p < 1$ is an unknown parameter. Denote the parametric function $\tau(p) = P_p\{\Sigma_{i=1}^{n}X_i > X_{n+1}\}$. Now, consider the initial unbiased estimator $T = I\left(\Sigma_{i=1}^{n}X_i > X_{n+1}\right)$ for $\tau(p)$. The problem is to find the Rao-Blackwellized version W of T. Here, it will be hard to guess the form of the final estimator. Now, proceed along the following steps.

(i)   Note that $U = \Sigma_{i=1}^{n+1}X_i$ is sufficient for p and that T is an unbiased estimator of $\tau(p)$;

(ii)   Observe that $W = E_p\{T \mid U = u\} = P_p\{T = 1 \mid U = u\}$, and find the expression for W; {*Hint*: $g(u) = P\{\Sigma_{i=1}^{n}X_i > X_{n+1} \mid U = u\} = P\{\Sigma_{i=1}^{n}X_i > X_{n+1} \cap U = u\}[P\{U = u\}]^{-1} = P\{u - X_{n+1} > X_{n+1} \cap U = u\}/P\{U = u\} = P[X_{n+1} < \frac{1}{2}u \cap \Sigma_{i=1}^{n+1}X_i = u]/P[\Sigma_{i=1}^{n+1}X_i = u].$};

(iii) Directly evaluate the ratio of the two probabilities in part (ii), and show that $g(u) = 0, n/(n+1), (n-1)/(n+1),$
$\{\binom{n}{3} + \binom{n}{2}\}/\binom{n+1}{3}, ..., \{\binom{n}{n} + \binom{n}{n-1}\}/\binom{n+1}{n}, \binom{n}{n}\}/\binom{n+1}{n+1}$
respectively when $u = 0, 1, 2, 3, ..., n, n + 1$;

(iv)   Using the explicit distribution of W given in part (iii), show that $E_p[W] = 1 - np^2q - q^n$ where $q = 1 - p$;

(v)   It is possible that so far the reader did not feel any urge to think about the explicit form of the parametric function $\tau(p)$. In part (iv), one first gets a glimpse of the expression of $\tau(p)$. Is the expression of $\tau(p)$ correct? In order to check, the reader is now asked to find the expression of $\tau(p)$ directly from its definition. {*Hint*: $\tau(p) = P_p\{\Sigma_{i=1}^{n}X_i > X_{n+1}\}$ $= P_p\{\Sigma_{i=1}^{n}X_i > X_{n+1} \cap X_{n+1} = 0\} + P_p\{\Sigma_{i=1}^{n}X_i > X_{n+1}$ $\cap X_{n+1} = 1\} = P_p\{\Sigma_{i=1}^{n}X_i > 0 \cap X_{n+1} = 0\} + P_p\{\Sigma_{i=1}^{n}X_i$

$> 1 \cap X_{n+1} = 1\} = (1 - q^n)q + (1 - q^n - npq^{n-1})p =$
$1 - np^2q - q^n$. Even if one had first found the expression
of $\tau(p)$, could one intuitively guess the form of W found in
part (ii)?}

7.4.9 (Example 7.4.8 Continued) Suppose that $X_1, ..., X_n$ are iid $N(\mu, \sigma^2)$
where μ is unknown but σ is assumed known with $\mu \in \Re, \sigma \in \Re^+, n \geq 2$. Let
$\tau(\mu) = \mu^3$. Find an initial unbiased estimator T for $\tau(\mu)$. Next, derive the
Rao-Blackwellized version of T to estimate $\tau(\mu)$ unbiasedly. {*Hint*: Start
with the fact that $E_\mu[(X_1 - \mu)^3] = 0$ and hence show that $T = X_1^3 - 3\sigma^2 X_1$
is an unbiased estimator for $\tau(\mu)$. Next, work with the third moment of the
conditional distribution of X_1 given that $\overline{X} = \overline{x}$.}

7.5.1 Prove Theorem 7.5.3 exploiting the arguments used in the proof
of Theorem 7.5.2.

7.5.2 (Example 7.4.7 Continued) Show that the Rao-Blackwellized estimator W is the UMVUE of $\tau(\mu)$.

7.5.3 (Exercise 6.2.16 Continued) Suppose that $X_1, ..., X_n$ are iid with
the Rayleigh distribution, that is the common pdf is

$$f(x; \theta) = 2\theta^{-1} x \exp(-x^2/\theta) I(x > 0),$$

where $\theta(> 0)$ is the unknown parameter. Find the UMVUE for (i) θ, (ii) θ^2
and (iii) θ^{-1}. {*Hint*: Start with $U = \Sigma_{i=1}^n X_i^2$ which is sufficient for θ. Show
that the distribution of U/θ is a multiple of χ_{2n}^2. Hence, derive unbiased
estimators for θ, θ^2 and θ^{-1} which depend only on U. Can the completeness
of U be justified with the help of the Theorem 6.6.2?}

7.5.4 (Exercise 7.5.3 Continued) Find the CRLB for the variance of
unbiased estimators of (i) θ, (ii) θ^2 and (iii) θ^{-1}. Is the CRLB attained by
the variance of the respective UMVUE obtained in the Exercise 7.5.3?

7.5.5 (Exercise 6.2.17 Continued) Suppose that $X_1, ..., X_n$ are iid with
the Weibull distribution, that is the common pdf is

$$f(x; \theta) = \alpha^{-1} \beta x^{\beta-1} \exp(-x^\beta/\alpha) I(x > 0),$$

where $\alpha(> 0)$ is the unknown parameter, but $\beta(> 0)$ is assumed known.
Find the UMVUE for (i) α, (ii) α^2 and (iii) α^{-1}. {*Hint*: Start with $U = \Sigma_{i=1}^n X_i^\beta$ which is sufficient for θ. Show that the distribution of U/α is a
multiple of χ_{2n}^2. Hence, derive unbiased estimators for α, α^2 and α^{-1} which
depend only on U. Can the completeness of U be justified with the help of
the Theorem 6.6.2?}.

7.5.6 (Exercise 7.5.5 Continued) Find the CRLB for the variance of
unbiased estimators of (i) α, (ii) α^2 and (iii) α^{-1}. Is the CRLB attained
by the variance of the respective UMVUE obtained in the Exercise 7.5.5?

7.5.7 (Example 7.5.11 Continued) Let $X_1, ..., X_n$ be iid $N(\mu, \sigma^2)$ where μ, σ are both unknown with $\mu \in \Re, \sigma \in \Re^+, n \geq 2$. Let $\theta = (\mu, \sigma)$ and $\tau(\theta) = \mu \sigma^k$ where k is a known and fixed real number. Derive the UMVUE for $\tau(\theta)$. Pay particular attention to any required minimum sample size n which may be needed.{*Hint*: Use (2.3.26) and the independence between $\overline{X}$ and S to first derive the expectation of $\overline{X}S^k$ where S^2 is the sample variance. Then make some final adjustments.}

7.5.8 (Example 7.5.12 Continued) Let $X_1, ..., X_n$ be iid having the common pdf $\sigma^{-1}exp\{-(x - \mu)/\sigma\}I(x > \mu)$ where μ, σ are both unknown with $-\infty < \mu < \infty, 0 < \sigma < \infty, n \geq 2$. Let $\theta = (\mu, \sigma)$ and $\tau(\theta) = \mu \sigma^k$ where k is a known and fixed real number. Derive the UMVUE for $\tau(\theta)$. Pay particular attention to any required minimum sample size n which may be needed. {*Hint*: Use (2.3.26) and the independence between $X_{n:1}$ and $Y = \Sigma_{i=1}^n(X_i - X_{n:1})$ to first derive the expectation of $\overline{X}Y^k$. Then make some final adjustments.}

7.5.9 Suppose that $X_1, ..., X_n$ are iid Uniform$(-\theta, \theta)$ where θ is the unknown parameter, $\theta \in \Re^+$. Let $\tau(\theta) = \theta^k$ where k is a known and fixed positive real number. Derive the UMVUE for $\tau(\theta)$. {*Hint*: Verify that $U = |X_{n:n}|$ is complete sufficient for θ. Find the pdf of U/θ to first derive the expectation of U^k. Then make some final adjustments.}

7.5.10 In each Example 7.4.1-7.4.8, argue that the Rao-Blackwellized estimator W is indeed the UMVUE for the respective parametric function. In the single parameter problems, verify whether the variance of the UMVUE attains the corresponding CRLB.

7.5.11 Suppose that $X_1, ..., X_n, X_{n+1}$ are iid $N(\mu, \sigma^2)$ where μ is unknown but σ is assumed known with $\mu \in \Re, \sigma \in \Re^+, n \geq 2$. Consider the parametric function $\tau(\mu) = P_\mu\{\Sigma_{i=1}^n X_i > X_{n+1}\}$. The problem is to find the UMVUE for $\tau(\mu)$. Start with $T = I(\Sigma_{i=1}^n X_i > X_{n+1})$ which is an unbiased estimator for $\tau(\mu)$. Now, proceed along the following steps.

(i)   Note that $U = \Sigma_{i=1}^{n+1} X_i$ is complete sufficient for μ and that T is an unbiased estimator of $\tau(\mu)$;

(ii)   Observe that $W = E_\mu\{T \mid U = u\} = P_\mu\{T = 1 \mid U = u\}$, and find the expression for W. {*Hint*: Write down explicitly the bivariate normal distribution of $T_1 = \Sigma_{i=1}^n X_i - X_{n+1}$ and U. Then, find the conditional probability, $P_\mu\{T_1 > 0 \mid U = u\}$ utilizing the Theorem 3.6.1. Next, argue that $T_1 > 0$ holds if and only if $T = 1$.}.

7.5.12 Suppose that $X_1, ..., X_n$ are iid Bernoulli(p) where $0 < p < 1$ is an unknown parameter. Consider the parametric function $\tau(p) = p + qe^2$ with $q = 1 - p$.

(i)   Find a suitable unbiased estimator T for $\tau(p)$;

(ii)   Since the complete sufficient statistic is $U = \Sigma_{i=1}^n X_i$, use the Lehmann-Scheffé theorems and evaluate the conditional expectation, $E_p[T \mid U = u]$;

(iii) Hence, derive the UMVUE for $\tau(p)$.

{*Hint*: Try and use the mgf of the X's appropriately.}

7.5.13 Suppose that $X_1, ..., X_n$ are iid Bernoulli(p) where $0 < p < 1$ is an unknown parameter. Consider the parametric function $\tau(p) = (p + qe^3)^2$ with $q = 1 - p$.

(i)   Find a suitable unbiased estimator T for $\tau(p)$;

(ii)   Since the complete sufficient statistic is $U = \Sigma_{i=1}^n X_i$, use the Lehmann-Scheffé theorems and evaluate the conditional expectation, $E_p[T \mid U = u]$;

(iii) Hence, derive the UMVUE for $\tau(p)$.

{*Hint*: Try and use the mgf of the X's appropriately.}

7.5.14 Suppose that $X_1, ..., X_n$ are iid $N(\mu, \sigma^2)$ where μ is unknown but σ is assumed known with $\mu \in \Re, \sigma \in \Re^+$. Consider the parametric function $\tau(\mu) = e^{2\mu}$. Derive the UMVUE for $\tau(\mu)$. Is the CRLB attained in this problem?

7.5.15 Suppose that $X_1, ..., X_n$ are iid $N(\mu, \sigma^2)$ where μ and σ are both assumed unknown with $\mu \in \Re, \sigma \in \Re^+, \boldsymbol{\theta} = (\mu, \sigma), n \geq 2$. Consider the parametric function $\tau(\boldsymbol{\theta}) = e^{2(\mu + \sigma^2)}$. Derive the UMVUE for $\tau(\boldsymbol{\theta})$.

7.5.16 Prove the Theorem 7.5.4 by appropriately modifying the lines of proof given for the Theorem 7.5.1.

7.5.17 Let $X_1, ..., X_m$ be iid $N(0, \sigma^2)$, $Y_1, ..., Y_n$ be iid $N(2, 3\sigma^2)$ where σ is unknown with $\sigma \in \Re^+$. Assume also that the X's are independent of the Y's. Derive the UMVUE for σ^2 and check whether the CRLB given by the Theorem 7.5.4 is attained in this case.

7.5.18 Suppose that $X_1, ..., X_m$ are iid Gamma($2, \beta$), $Y_1, ..., Y_n$ are iid Gamma($4, 3\beta$) where β is unknown with $\beta \in \Re^+$. Assume also that the X's are independent of the Y's. Derive the UMVUE for β and check whether the CRLB given by the Theorem 7.5.4 is attained in this case.

7.5.19 Suppose that X is distributed as $N(0, \sigma^2)$, Y has its pdf given by $g(y; \sigma^2) = (2\sigma^2)^{-1} exp\{-|y|/\sigma^2\}I(y \in \Re)$ where σ is unknown with $\sigma \in \Re^+$. Assume also that the X is independent of the Y. Derive the UMVUE for σ^2 and check whether the CRLB given by the Theorem 7.5.4 is attained in this case.

7.5.20 Let $X_1, ..., X_n$ be iid having the common pdf $\sigma^{-1} exp\{-(x - \mu)/\sigma\}I(x > \mu)$ where μ is known but σ is unknown with $-\infty < \mu <$

$\infty, 0 < \sigma < \infty, n \geq 5$. Let $U = X_{n:1}$, the smallest order statistic, which is complete sufficient for μ.

(i)   Find the conditional expectation, $E_\mu[\overline{X} \mid U]$;

(ii)   Find the conditional expectation, $E_\mu[X_1 \mid U]$;

(iii)   Find the conditional expectation, $E_\mu\{X_1 + e^{|X_1 - X_3|} - e^{|X_2 - X_4|} \mid U\}$.

{*Hint*: Try and use the Theorem 7.5.5 appropriately.}

7.5.21 (Example 7.5.18 Continued) Let $X_1, ..., X_n$ be iid random variables distributed uniformly on $(0, \theta)$ where $\theta(> 0)$ is the unknown parameter with $n \geq 2$. Show that $E_\theta[\overline{X}^2 \mid X_{n:n}] = \frac{1}{4}(1 + \frac{1}{3n})(1 + \frac{2}{n})X_{n:n}^2$. {*Hint*: Try and use the Theorem 7.5.5 appropriately.}

7.6.1 (Example 7.6.1 Continued) Suppose that $X_1, ..., X_n$ are iid with the common pdf $f(x; \theta) = \theta^{-1} exp\{-(x - \theta)/\theta\}I(x > \theta)$ with $\theta \in \Theta = \Re^+$ where θ is the unknown parameter and $n \geq 2$. Recall from Exercise 6.3.10 that $\mathbf{U} = (X_{n:1}, \Sigma_{i=1}^n(X_i - X_{n:1}))$ is minimal sufficient for θ but U is not complete. Here, we wish to estimate θ unbiasedly.

(i)   Show that $T = n(n+1)^{-1}X_{n:1}$ is an unbiased estimator for θ;

(ii)   Show that $T^{'} = (n-1)^{-1}\Sigma_{i=1}^n(X_i - X_{n:1})$ is also an unbiased estimator for θ;

(iii)   Along the lines of our discussions in the Section 7.6.1, derive the Rao-Blackwellized versions W and $W^{'}$ of the estimators T and $T^{'}$ respectively;

(iv)   Compare the variances of W and $W^{'}$. Any comments?

(v)   Define a class of unbiased estimators of θ which are convex combinations of W and $W^{'}$ along the lines of (7.6.4). Within this class find the estimator T^* whose variance is the smallest;

(vi)   Summarize your comments along the lines of the remarks made in (7.6.6).

7.7.1 Show that the estimators derived in the Examples 7.2.1-7.2.5 are consistent for the parametric function being estimated.

7.7.2 Show that the MLE's derived in the Examples 7.2.6-7.2.7 are consistent for the parametric function being estimated.

7.7.3 Show that the MLE's derived in the Examples 7.2.9-7.2.12 are consistent for the parametric function being estimated.

7.7.4 Show that the UMVUE's derived in the Section 7.5 are consistent for parametric function being estimated.

7.7.5 Show that the estimators $W, W^{'}$ and $T^*(\alpha^*)$ defined in (7.6.4)-(7.6.5) are consistent for the parameter θ being estimated.

7.7.6 (Exercises 7.2.10 and 7.5.4) Suppose that $X_1, ..., X_n$ are iid with the Rayleigh distribution, that is the common pdf is

$$f(x; \theta) = 2\theta^{-1}x\,exp(-x^2/\theta)I(x > 0),$$

where $\theta(> 0)$ is the unknown parameter. Show that the MLE's and the UMVUE's for θ, θ^2 and θ^{-1} are all consistent.

7.7.7 (Exercises 7.2.11 and 7.5.6) Suppose that $X_1, ..., X_n$ are iid with the Weibull distribution, that is the common pdf is

$$f(x; \theta) = \alpha^{-1}\beta x^{\beta-1}\,exp(-x^\beta/\alpha)I(x > 0),$$

where $\alpha(> 0)$ is the unknown parameter but $\beta(> 0)$ is assumed known. Show that the MLE's and the UMVUE's for α, α^2 and α^{-1} are all consistent.

7.7.8 (Exercises 7.2.4 and 7.2.17 Continued) Suppose that $X_1, ..., X_n$ are iid whose common pdf is given by

$$f(x) = \begin{cases} (\theta + 1)x^\theta & \text{if } 0 < x < 1 \\ 0 & \text{elsewhere,} \end{cases}$$

where $\theta(> 0)$ is the unknown parameter. Show that the method of moment estimator and the MLE for θ are both consistent.

7.7.9 Suppose that $X_1, ..., X_n$ are iid whose common pdf is given by

$$f(x) = \begin{cases} \{\sigma\sqrt{2\pi}\}^{-1}x^{-1}exp\{-\frac{1}{2\sigma^2}(log(x) - \mu)^2\} & \text{if } x > 0 \\ 0 & \text{elsewhere,} \end{cases}$$

where μ, σ are both assumed unknown with $-\infty < \mu < \infty, 0 < \sigma < \infty, \boldsymbol{\theta} = (\mu, \sigma)$. One will recall from (1.7.27) that this pdf is known as the lognormal density.

(i)   Evaluate the expression for $E[X_1^k]$, denoted by the parametric function $\tau(\boldsymbol{\theta})$, for any fixed $k(> 0)$;

(ii)   Derive the MLE, denoted by T_n, for $\tau(\boldsymbol{\theta})$;

(iii)   Show that T_n is consistent for $\tau(\boldsymbol{\theta})$.

7.7.10 Suppose that $X_1, ..., X_n$ are iid $N(\mu, \sigma^2)$ where μ and σ are both assumed unknown with $\mu \in \Re, \sigma \in \Re^+, \boldsymbol{\theta} = (\mu, \sigma), n \geq 2$. First find the UMVUE $T \equiv T_n$ for the parametric function $\tau(\mu) = \mu + \mu^2$. Show that T_n is consistent for $\tau(\mu)$.

7.7.11 (Exercise 7.5.4) Suppose that $X_1, ..., X_n$ are iid with the Rayleigh distribution, that is the common pdf is

$$f(x; \theta) = 2\theta^{-1}x\,exp(-x^2/\theta)I(x > 0),$$

where $\theta(> 0)$ is the unknown parameter. Let $T_n(c) = c\Sigma_{i=1}^n X_i^2$ with $c > 0$ and consider estimating θ with $T_n(c)$. First find the MSE of $T_n(c)$ with $c(> 0)$ fixed. Then, minimize the MSE with respect to c. Denote the optimal choice for c by $c^* \equiv c_n^*$. Show that the minimum MSE estimator $T_n(c^*)$ is consistent for θ.

7.7.12 (Example 7.4.6 Continued) Let $X_1, ..., X_n$ be iid $N(\mu, \sigma^2)$ where μ, σ are both unknown with $\mu \in \Re, \sigma \in \Re^+, n \geq 2$. Let $\tau(\mu) = P_\mu\{X_1 < a\}$ where a is some known real number. First find the UMVUE $T \equiv T_n$ for $\tau(\mu)$. Show that T_n is consistent for $\tau(\mu)$.

7.7.13 (Example 7.4.6 Continued) Let $X_1, ..., X_n$ be iid $N(\mu, \sigma^2)$ where μ, σ are both unknown with $\mu \in \Re, \sigma \in \Re^+, n \geq 2$. Let $\tau(\mu) = P_\mu\{|X_1| < a\}$ where a is some known positive number. First find the UMVUE $T \equiv T_n$ for $\tau(\mu)$. Show that T_n is consistent for $\tau(\mu)$.

7.7.14 Suppose that $X_1, ..., X_n$ are iid Uniform$(0, \theta)$ where $\theta(> 0)$ is the unknown parameter. Let $T_n(c) = cX_{n:n}$ with $c > 0$ and consider estimating θ with $T_n(c)$. First find the MSE of $T_n(c)$ with $c(> 0)$ fixed. Then, minimize the MSE with respect to c. Denote the optimal choice for c by $c^* \equiv c_n^*$. Show that the minimum MSE estimator $T_n(c^*)$ is consistent for θ.

7.7.15 (Example 7.7.13 Continued) Let $X_1, ..., X_n$ be iid $N(\mu, \sigma^2)$ where μ, σ are both unknown with $\mu \in \Re, \sigma \in \Re^+, n \geq 2$. Let $\tau(\mu) = P_\mu\{|X_1| < a\}$ where a is some known positive number. Obtain the expression for $\tau(\mu)$ and thereby propose a consistent estimator U_n for $\tau(\mu)$. But, U_n must be different from the UMVUE T_n proposed earlier in the Exercise 7.7.13. {*Note:* A consistent estimator does not have to be unbiased.}

8

Tests of Hypotheses

8.1 Introduction

Suppose that a population pmf or pdf is given by $f(x; \theta)$ where $x \in \mathcal{X} \subseteq \Re$ and θ is an unknown parameter which belongs to a parameter space $\Theta \subseteq \Re$.

Definition 8.1.1 *A hypothesis is a statement about the unknown parameter θ.*

In a problem the parameter θ may refer to the population mean and the experimenter may hypothesize that $\theta \geq 100$ and like to examine the plausibility of such a hypothesis after gathering the sample evidence. But, a formulation of testing the plausibility or otherwise of a single hypothesis leads to some conceptual difficulties. Jerzy Neyman and Egon S. Pearson discovered fundamental approaches to test statistical hypotheses. The Neyman-Pearson collaboration first emerged (1928a,b) with formulations and constructions of tests through comparisons of likelihood functions. These blossomed into two landmark papers of Neyman and Pearson (1933a,b).

Neyman and Pearson formulated the problem of testing of hypotheses as follows. Suppose one is contemplating to choose between two plausible hypotheses

$$H_0 : \theta \in \Theta_0 \text{ versus } H_1 : \theta \in \Theta_1$$

where $\Theta_0 \subset \Theta, \Theta_1 \subset \Theta$ and $\Theta_0 \cap \Theta_1 = \varphi$, the empty set. Based on the evidence collected from random samples $X_1, ..., X_n$, obtained from the relevant population under consideration, the statistical problem is to select one hypothesis which seems more reasonable in comparison with the other. The experimenter may, for example, decide to opt for H_0 compared with H_1 based on the sample evidence. But we should not interpret this decision to indicate that H_0 is thus proven to be true. An experimenter's final decision to opt for the hypothesis H_0 (or H_1) will simply indicate which hypothesis appears more favorable based on the collected sample evidence. In reality it is possible, however, that neither H_0 nor H_1 is actually true. The basic question is this: Given the sample evidence, if one must decide in favor of H_0 or H_1, which hypothesis is it going to be? This chapter will build the methods of such decision-making.

We customarily refer to H_0 and H_1 respectively as the *null* and *alternative hypothesis*. The null hypothesis H_0 or the alternative hypothesis H_1 is called *simple* provided that Θ_0, Θ_1 are singleton subsets of Θ. That is, a hypothesis such as $H_0 : \theta = \theta_0$ or $H_1 : \theta = \theta_1$, with θ_0, θ_1 known, would be called a *simple hypothesis*. A hypothesis which is not simple is called *composite*. A hypothesis such as $H_0 : \theta > \theta_0$, $H_0 : \theta < \theta_0$, $H_1 : \theta \geq \theta_1$, $H_1 : \theta \leq \theta_1$, for example, are referred to as *one-sided* composite hypotheses. A hypothesis such as $H_1 : \theta \neq \theta_0$ or $H_1 : \theta \leq \theta_0 \cup \theta \geq \theta_1$ with $\theta_0 < \theta_1$, for example, is referred to as a *two-sided* composite hypothesis. In Section 8.2, we first formulate the two types of errors in the decision making and focus on the fundamental idea of a test. The Section 8.3 develops the concept of the *most powerful* (MP) test for choosing between a simple null versus simple alternative hypotheses. In Section 8.4, the idea of a *uniformly most powerful* (UMP) test is pursued when H_0 is simple but H_1 is one-sided. The Section 8.5 gives examples of two-sided alternative hypotheses testing situations and examines the possibilities of finding the UMP test. Section 8.5 touches upon the ideas of *unbiased* and *uniformly most powerful unbiased* (UMPU) tests.

8.2   Error Probabilities and the Power Function

It is emphasized again that H_0 and H_1 are two statements regarding the unknown parameter θ. Then we gather around the random samples $X_1, ..., X_n$ from the appropriate population and learn about the unknown value of θ. Intuitively speaking, the experimenter would favor H_0 or H_1 if the estimator $\widehat{\theta} \equiv \widehat{\theta}(X_1, ..., X_n)$ seems more likely to be observed under H_0 or H_1 respectively. But, at the time of decision making, the experimenter may make mistakes by favoring the wrong hypothesis simply because the decision is based on the evidence gathered from a random sample.

To understand the kinds of errors one can commit by choosing one hypothesis over the other, the following table may be helpful. An examination

Table 8.2.1. Type I and Type II Errors

Test Result or Decision	Nature's Choice	
	H_0 True	H_1 True
Accept H_0	No Error	Type II Error
Accept H_1	Type I Error	No Error

of the entries reveals that the experimenter may commit one of the two possible errors. One ends up rejecting the null hypothesis H_0 while H_0 is actually true or ends up accepting the null hypothesis H_0 while H_1 is true. Table 8.2.1 clearly suggests that the other two possible decisions are correct ones under given circumstances.

Once and for all, let us add that we freely interchange statements, for example, "Reject H_i" and "Accept H_j" for $i \neq j \in \{0, 1\}$.

Next, let us explain what we mean by a *test* of H_0 versus H_1. Having observed the data $\mathbf{X} = (X_1, ..., X_n)$, a test will guide us *unambiguously* to reject H_0 or accept H_0. This is accomplished by *partitioning* the sample space $\Re^n$ into two parts $\mathcal{R}$ and $\mathcal{R}^c$ corresponding to the respective final decision: "reject H_0" and "accept H_0".

$\mathcal{R}$ is constructed so that we reject H_0 whenever $\mathbf{x} \in \mathcal{R}$. The subset $\mathcal{R}$ is called the *rejection region* or the *critical region*.

Example 8.2.1 Let us walk through a simple example. Consider a population with the pdf $N(\theta, 1)$ where $\theta \in \Re$ is unknown. An experimenter postulates two possible hypotheses $H_0 : \theta = 5.5$ and $H_1 : \theta = 8$. A random sample $\mathbf{X} = (X_1, ..., X_9)$ is collected and denote $\overline{X} = \frac{1}{9} \Sigma_{i=1}^{9} X_i$. Some examples of "tests" are given below:

$$
\begin{array}{lll}
\text{Test \#1:} & \text{Reject } H_0 \text{ if and only if } X_1 > 7; & \\
\text{Test \#2:} & \text{Reject } H_0 \text{ if and only if } \frac{1}{2}(X_1 + X_2) > 7; & \\
\text{Test \#3:} & \text{Reject } H_0 \text{ if and only if } \overline{X} > 6; & (8.2.1) \\
\text{Test \#4:} & \text{Reject } H_0 \text{ if and only if } \overline{X} > 7.5. &
\end{array}
$$

We may summarize these tests in a different fashion: Let us rewrite

$$
\begin{array}{lll}
\text{Test \#1:} & \mathcal{R}_1 = \{\mathbf{x} = (x_1, ..., x_9) : x_1 > 7\}; & \\
\text{Test \#2:} & \mathcal{R}_2 = \{\mathbf{x} = (x_1, ..., x_9) : \frac{1}{2}(x_1 + x_2) > 7\}; & \\
\text{Test \#3:} & \mathcal{R}_3 = \{\mathbf{x} = (x_1, ..., x_9) : \overline{x} > 6\}; & (8.2.2) \\
\text{Test \#4:} & \mathcal{R}_4 = \{\mathbf{x} = (x_1, ..., x_9) : \overline{x} > 7.5\}. &
\end{array}
$$

Here, $\mathcal{R}_i$ is that part of the sample space $\Re^9$ where H_0 is rejected by means of the Test $\#i, i = 1, ..., 4$. ▲

Whenever H_0, H_1 are both simple hypotheses, we respectively write α and β for the Type I and II error probabilities:

$$
\begin{aligned}
\alpha = P\{\text{Type I error}\} \quad &= P\{\text{Rejecting } H_0 \text{ when } H_0 \text{ is true}\} \\
&= P\{\mathbf{X} \in \mathcal{R} \text{ when } H_0 \text{ is true}\}; \\
\beta = P\{\text{Type II error}\} \quad &= P\{\text{Accepting } H_0 \text{ when } H_1 \text{ is true}\} \\
&= P\{\mathbf{X} \in \mathcal{R}^c \text{ when } H_0 \text{ is true}\},
\end{aligned}
$$

$(8.2.3)$

where $\mathcal{R}$ denotes a generic rejection region for H_0.

Example 8.2.2 (Example 8.2.1 Continued) In the case of the Test #1, writing Z for a standard normal variable, we have:

$$\alpha = P\{X_1 > 7 \text{ when } \theta = 5.5\} = P\{Z > 1.5\} = .06681,$$
$$\beta = P\{X_1 \le 7 \text{ when } \theta = 8\} = P\{Z \le -1\} = .15866.$$

Proceeding similarly, we used MAPLE to prepare the following table for the values of α and β associated with the Tests #1-4 given by (8.2.1).

Table 8.2.2. Values of α and β for Tests #1-4 from (8.2.1)

Test #1 $\mathcal{R}_1$	Test #2 $\mathcal{R}_2$	Test #3 $\mathcal{R}_3$	Test #4 $\mathcal{R}_4$
$\alpha = .06681$ $\beta = .15866$	$\alpha = .01696$ $\beta = .07865$	$\alpha = .06681$ $\beta = .00000$	$\alpha = .00000$ $\beta = .06681$

Upon inspecting the entries in the Table 8.2.2, we can immediately conclude a few things. Between the Tests #1 and #2, we feel that the Test #2 appears better because both its error probabilities are smaller than the ones associated with the Test #1. Comparing the Tests #1 and #3 we can similarly say that Test #3 performs much better. In other words, while comparing Tests #1-3, we feel that the Test #1 certainly should not be in the running, but no clear-cut choice between Tests #2 and #3 emerges from this. One of these has a smaller value of α but has a larger value of β. If we must pick between the Tests #2-3, then we have to take into consideration the consequences of committing either error in practice. It is clear that an experimenter may not be able to accomplish this by looking at the values of α and β alone. Tests #3-4 point out a slightly different story. By down-sizing the rejection region $\mathcal{R}$ for the Test #4 in comparison with that of Test #3, we are able to make the α value for Test #4 practically zero, but this happens at the expense of a sharp rise in the value of β. ▲

From Table 8.2.2, we observe some special features which also hold in general. We summarize these as follows:

> All tests may not be comparable among themselves such as tests #2-3. By suitably adjusting the rejection region $\mathcal{R}$, we can make α (or β) as small as we would like, but then β (or α) will be on the rise as the sample size n is kept fixed.

So, then how should one proceed to define a test for H_0 versus H_1 which can be called the "best"? We discuss the Neyman-Pearson formulation of the testing problem in its generality in the next subsection.

Let us, however, discuss few other notions for the moment. Sometimes instead of focussing on the Type II error probability β, one considers what is known as the *power* of a test.

Definition 8.2.1 *The power or the power function of a test, denoted by $Q(\theta)$, is the probability of rejecting the null hypothesis H_0 when $\theta \in \Theta$ is the true parameter value. The power function is then given by*

$$Q(\theta) = P_\theta(\mathcal{R}) \text{ for all } \theta \in \Theta. \tag{8.2.4}$$

In a simple null versus simple alternative testing situation such as the one we had been discussing earlier, one obviously has $Q(\theta_0) = \alpha$ and $Q(\theta_1) = 1 - \beta$.

Example 8.2.3 (Example 8.2.1 Continued) One can verify that for the Test #4, the power function is given by $Q(\theta) = 1 - \Phi(22.5 - 3\theta)$ for all $\theta \in \Re$. ▲

A test is frequently laid out as in (8.2.1) or (8.2.2). There is yet another way to identify a test by what is called a *critical function* or *test function*. Recall that proposing a test is equivalent to partitioning the sample space $\mathcal{X}^n$ into two parts, critical region $\mathcal{R}$ and its complement $\mathcal{R}^c$.

Definition 8.2.2 *A function $\psi(.) : \mathcal{X}^n \to [0, 1]$ is called a critical function or test function where $\psi(\mathbf{x})$ stands for the probability with which the null hypothesis H_0 is rejected when the data $\mathbf{X} = \mathbf{x}$ has been observed, $\mathbf{x} \in \mathcal{X}^n$.*

In general, we can rewrite the power function defined in (8.2.4) as follows:

$$Q(\theta) = E_\theta\{\psi(\mathbf{X})\} \text{ for } \theta \in \Theta. \tag{8.2.5}$$

8.2.1 The Concept of a Best Test

In (8.2.3) we pointed out how Type I and II error probabilities are evaluated when testing between a simple null and simple alternative hypotheses. When the null and alternative hypotheses are composite, we refocus on the definition of the two associated error probabilities. Let us consider testing $H_0 : \theta \in \Theta_0$ versus $H_1 : \theta \in \Theta_1$ where $\Theta_0 \subset \Theta, \Theta_1 \subset \Theta$ and $\Theta_0 \cap \Theta_1 = \varphi$, the empty set.

Definition 8.2.3 *We start with a fixed number $\alpha \in (0, 1)$. A test for $H_0 : \theta \in \Theta_0$ versus $H_1 : \theta \in \Theta_1$ with its power function $Q(\theta)$ defined in (8.2.2) is called size α or level α according as $Sup_{\theta \in \Theta_0} Q(\theta) = \alpha$ or $\leq \alpha$ respectively.*

In defining a size α or level α test, what we are doing is quite simple. For every θ in the null space Θ_0, we look at the associated Type I error probability which coincides with $Q(\theta)$, and then consider the largest among all Type I error probabilities. $Sup_{\theta \in \Theta_0} Q(\theta)$ may be viewed as the worst possible Type I error probability. A test is called size α if $Sup_{\theta \in \Theta_0} Q(\theta) = \alpha$ whereas it is called level α if $Sup_{\theta \in \Theta_0} Q(\theta) \leq \alpha$. We may equivalently restrict our attention to a class of *test functions* such that $Sup_{\theta \in \Theta_0} E_\theta \{\psi(\mathbf{X})\} = \alpha$ or $Sup_{\theta \in \Theta_0} E_\theta \{\psi(\mathbf{X})\} \leq \alpha$. It should be obvious that a size α test is also a level α test.

It is important to be able to compare all tests for H_0 versus H_1 such that each has some common basic property to begin with. In Chapter 7, for example, we wanted to find the best estimator among all *unbiased* estimators of θ. The common basic property of each estimator was its unbiasedness. Now, in defining the "best" test for H_0 versus H_1, we compare only *the level α tests* among themselves.

Definition 8.2.4 *Consider the collection C of all level α tests for H_0 : $\theta \in \Theta_0$ versus $H_1 : \theta \in \Theta_1$ where $\Theta_0 \cap \Theta_1 = \varphi, \Theta_0 \cup \Theta_1 \subseteq \Theta$. A test belonging to C with its power function $Q(\theta)$ is called the best or the uniformly most powerful (UMP) level α test if and only if $Q(\theta) \geq Q^*(\theta)$ for all $\theta \in \Theta_1$ where $Q^*(\theta)$ is the power function of any other test belonging to C. If the alternative hypothesis is simple, that is if Θ_1 is a singleton set, then the best test is called the most powerful (MP) level α test.*

Among all level α tests our goal is to find the one whose power hits the maximum at *each point $\theta \in \Theta_1$*. Such a test would be UMP level α. In the simple null versus simple alternative case first, Neyman and Pearson (1933a,b) gave an explicit method to determine the MP level α test. This approach is described in the next section. We end this section with an example.

Example 8.2.4 (Example 8.2.1 Continued) Consider a population with the pdf $N(\theta, 1)$ where $\theta \in \Re$ is unknown. Again an experimenter postulates two hypotheses, $H_0 : \theta = 5.5$ and $H_1 : \theta = 8$. A random sample $\mathbf{X} = (X_1, ..., X_9)$ is collected. Let us look at the following tests:

Test #1: Reject H_0 if and only if $X_1 > 7.1449$;

Test #2: Reject H_0 if and only if $\frac{1}{4}(X_1 + ... + X_4) > 6.32245$;

Test #3: Reject H_0 if and only if $\overline{X} > 6.0483$.

Let us write $Q_i(\theta)$ for the power function of Test #$i, i = 1, 2, 3$. Using MAPLE, we verified that $Q_1(5.5) = Q_2(5.5) = Q_3(5.5) = .049995$. In other words, these are all level α tests where $\alpha = .049995$. In the Figure 8.2.1, we have plotted these three power functions. It is clear from this plot that among the three tests under investigation, the Test #3 has the maximum power at each point $\theta > 5.5$.

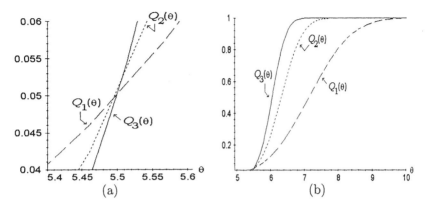

Figure 8.2.1. Power Functions for Tests #1, #2 and #3

We may add that the respective power at the point $\theta = 8$ is given by $Q_1(8) = .80375, Q_2(8) = .9996$, and $Q_3(8) = 1.0$. ▲

8.3   Simple Null Versus Simple Alternative Hypotheses

Here we elaborate the derivation of the MP level α test to choose between a simple null hypothesis H_0 and a simple alternative hypothesis H_1. We first prove the celebrated Neyman-Pearson Lemma which was originally formulated and proved by Neyman and Pearson (1933a).

8.3.1   Most Powerful Test via the Neyman-Pearson Lemma

Suppose that $X_1, ..., X_n$ are iid real valued random variables with the pmf or pdf $f(x; \theta)$, $\theta \in \Theta \subseteq \Re$. Let us continue to write $\mathbf{X} = (X_1, ..., X_n)$, $\mathbf{x} = (x_1, ..., x_n)$. The likelihood function is denoted by $L(\mathbf{x}; \theta)$ when θ is the true value. We wish to test

$$H_0 : \theta = \theta_0 \text{ versus } H_1 : \theta = \theta_1 \qquad (8.3.1)$$

where $\theta_0 \neq \theta_1$ but both $\theta_0, \theta_1 \in \Theta$, the parameter space. Under $H_i : \theta = \theta_i$, the data $\mathbf{x}$ has its likelihood function given by $L(\mathbf{x}; \theta_i), i = 0, 1$.

Let us focus our attention on comparing the powers associated with all level α tests where $0 < \alpha < 1$ is preassigned. Customarily the number α

is chosen small. In practice, one often chooses $\alpha = .10, .05$ or $.01$ unless otherwise stated. But the experimenter is free to choose any appropriate α value.

Note that $L(\mathbf{x}; \theta_i), i = 0, 1$, are two *completely specified* likelihood functions. Intuitively speaking, a test for H_0 versus H_1 comes down to the comparison of $L(\mathbf{x}; \theta_0)$ with $L(\mathbf{x}; \theta_1)$ and figure out which one is significantly larger. We favor the hypothesis associated with the significantly larger likelihood as the more plausible one. The following result gives a precise statement.

Theorem 8.3.1 (Neyman-Pearson Lemma) *Consider a test of H_0 versus H_1 stated in (8.3.1) with its rejection and acceptance regions for the null hypothesis H_0 defined as follows:*

$$\begin{aligned} \mathbf{x} \in \mathcal{R} \quad & \text{if } L(\mathbf{x}; \theta_1) > kL(\mathbf{x}; \theta_0) \\ \mathbf{x} \in \mathcal{R}^c \quad & \text{if } L(\mathbf{x}; \theta_1) < kL(\mathbf{x}; \theta_0) \end{aligned}$$

or equivalently, suppose that the test function has the form

$$\psi(\mathbf{x}) = \begin{cases} 1 & \text{if } L(\mathbf{x}; \theta_1) > kL(\mathbf{x}; \theta_0) \\ 0 & \text{if } L(\mathbf{x}; \theta_1) < kL(\mathbf{x}; \theta_0), \end{cases} \qquad (8.3.2)$$

where the constant $k (\geq 0)$ is so determined that

$$E_{\theta_0}\{\psi(\mathbf{X})\} = \alpha. \qquad (8.3.3)$$

Any test satisfying (8.3.2)-(8.3.3) is a MP level α test.

Proof We give a proof assuming that the X's are continuous random variables. The discrete case can be disposed off by replacing the integrals with the corresponding sums. First note that any test which satisfies (8.3.3) has size α and hence it is level α too.

We already have a level α test function $\psi(\mathbf{x})$ defined by (8.3.2)-(8.3.3). Let $\psi^*(\mathbf{x})$ be the test function of any other level α test. Suppose that $Q(\theta), Q^*(\theta)$ are respectively the power functions associated with the test functions ψ, ψ^*. Now, let us first verify that

$$\{\psi(\mathbf{x}) - \psi^*(\mathbf{x})\}\{L(\mathbf{x}; \theta_1) - kL(\mathbf{x}; \theta_0)\} \geq 0 \text{ for all } \mathbf{x} \in \mathcal{X}^n. \qquad (8.3.4)$$

Suppose that $\mathbf{x} \in \mathcal{X}^n$ is such that $\psi(\mathbf{x}) = 1$ which implies $L(\mathbf{x}; \theta_1) - kL(\mathbf{x}; \theta_0) > 0$, by the definition of ψ in (8.3.2). Also for such $\mathbf{x}$, one obviously has $\psi(\mathbf{x}) - \psi^*(\mathbf{x}) \geq 0$ since $\psi^*(\mathbf{x}) \in (0, 1)$. That is, if $\mathbf{x} \in \mathcal{X}^n$ is such that $\psi(\mathbf{x}) = 1$, we have verified (8.3.4). Next, suppose that $\mathbf{x} \in \mathcal{X}^n$ is such that $\psi(\mathbf{x}) = 0$ which implies $L(\mathbf{x}; \theta_1) - kL(\mathbf{x}; \theta_0) < 0$, by the definition of ψ in (8.3.2). Also for such $\mathbf{x}$ one obviously has $\psi(\mathbf{x}) - \psi^*(\mathbf{x}) \le 0$ since $\psi^*(\mathbf{x}) \in (0,1)$. Again (8.3.4) is validated.. Now, if $\mathbf{x} \in \mathcal{X}^n$ is such that $0 < \psi(\mathbf{x}) < 1$, then from (8.3.2) we must have $L(\mathbf{x}; \theta_1) - kL(\mathbf{x}; \theta_0) = 0$, and again (8.3.4) is validated. That is, (8.3.4) surely holds for all $\mathbf{x} \in \mathcal{X}^n$. Hence we have

$$
\begin{aligned}
0 \ \le &\int \ldots \int_{\mathcal{X}^n} \{\psi(\mathbf{x}) - \psi^*(\mathbf{x})\}\{L(\mathbf{x}; \theta_1) - kL(\mathbf{x}; \theta_0)\}\Pi_{i=1}^n dx_i \\
= &\int \ldots \int_{\mathcal{X}^n} \psi(\mathbf{x})\{L(\mathbf{x}; \theta_1) - kL(\mathbf{x}; \theta_0)\}\Pi_{i=1}^n dx_i \\
&- \int \ldots \int_{\mathcal{X}^n} \psi^*(\mathbf{x})\{L(\mathbf{x}; \theta_1) - kL(\mathbf{x}; \theta_0)\}\Pi_{i=1}^n dx_i \\
= &\{E_{\theta_1}[\psi(\mathbf{X})] - kE_{\theta_0}[\psi(\mathbf{X})]\} - \{E_{\theta_1}[\psi^*(\mathbf{X})] - kE_{\theta_0}[\psi^*(\mathbf{X})]\} \\
= &\{Q(\theta_1) - Q^*(\theta_1)\} - k\{Q(\theta_0) - Q^*(\theta_0)\}.
\end{aligned}
$$

$$(8.3.5)$$

Now recall that $Q(\theta_0)$ is the Type I error probability associated with the test ψ defined in (8.3.2) and thus $Q(\theta_0) = \alpha$ from (8.3.3). Also, $Q^*(\theta_0)$ is the similar entity associated with the test ψ^* which is assumed to have the level α, that is $Q^*(\theta_0) \le \alpha$. Thus, $Q(\theta_0) - Q^*(\theta_0) \ge 0$ and hence we can rewrite (8.3.5) as

$$Q(\theta_1) - Q^*(\theta_1) \ge k\{Q(\theta_0) - Q^*(\theta_0)\} \ge 0,$$

which shows that $Q(\theta_1) \ge Q^*(\theta_1)$. Hence, the test associated with ψ is at least as powerful as the one associated with ψ^*. But, ψ^* is any arbitrary level α test to begin with. The proof is now complete. ∎

Remark 8.3.1 Observe that the Neyman-Pearson Lemma rejects H_0 in favor of accepting H_1 provided that the ratio of two likelihoods under H_1 and H_0 is sufficiently large, that is if and only if $L(\mathbf{X}; \theta_1)/L(\mathbf{X}; \theta_0) > k$ for some suitable $k(\ge 0)$.

> *Convention*: The ratio $\frac{c}{0}$ is interpreted as infinity if $c > 0$
>
> and one if $c = 0$.

Remark 8.3.2 In the statement of the Neyman-Pearson Lemma, note that nothing has been said about the data points $\mathbf{x}$ which satisfy the equation $L(\mathbf{x}; \theta_1) = kL(\mathbf{x}; \theta_0)$. First, if the X's happen to be continuous random variables, then the set of such points $\mathbf{x}$ would have the probability zero. In other words, by not specifying exactly what to do when $L(\mathbf{x}; \theta_1) = kL(\mathbf{x}; \theta_0)$ in the continuous case amounts to nothing serious in practice. In a discrete case, however, one needs to *randomize* on the set of $\mathbf{x}$'s for which $L(\mathbf{x}; \theta_1) = kL(\mathbf{x}; \theta_0)$ holds so that the MP test has the size

α. See the Examples 8.3.6-8.3.7.

A final form of the MP or UMP level α test is written down in the *simplest implementable* form. That is, having fixed $\alpha \in (0,1)$, the cut-off point of the test defining the regions $\mathcal{R}, \mathcal{R}^c$ must explicitly be found analytically or from a standard statistical table.

Remark 8.3.3 We note that, for all practical purposes, the MP level α test given by the Neyman-Pearson Lemma is unique. In other words, if one finds another MP level α test with its test function ψ^* by some other method, then for all practical purposes, ψ^* and ψ will coincide on the two sets $\{\mathbf{x} \in \mathcal{X}^n : L(\mathbf{x}; \theta_1) > kL(\mathbf{x}; \theta_0)\}$ and $\{\mathbf{x} \in \mathcal{X}^n : L(\mathbf{x}; \theta_1) < kL(\mathbf{x}; \theta_0)\}$.

Convention: k is used as a *generic and nonstochastic* constant. k may not remain same from one step to another.

Example 8.3.1 Let $X_1, ..., X_n$ be iid $N(\mu, \sigma^2)$ with unknown $\mu \in \Re$, but assume that $\sigma \in \Re^+$ is known. With preassigned $\alpha \in (0,1)$ we wish to derive the MP level α test for $H_0 : \mu = \mu_0$ versus $H_1 : \mu = \mu_1$ where $\mu_1 > \mu_0$ and μ_0, μ_1 are two known real numbers. Both H_0, H_1 are simple hypotheses and the Neyman-Pearson Lemma applies. The likelihood function is given by

$$L(\mathbf{x}; \mu) = \{\sigma^2 2\pi\}^{-n/2} exp\{-(2\sigma^2)^{-1}\Sigma_{i=1}^n (x_i - \mu)^2\}, \ \mu \in \Re.$$

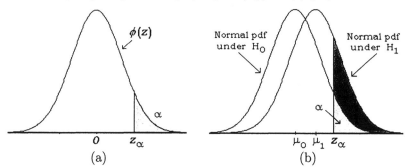

Figure 8.3.1. (a) Standard Normal PDF: Upper $100\alpha\%$ Point (b) Probability on the right of z_α Is Larger Under H_1 (darker plus lighter shaded areas) than Under H_0 (lighter shaded area)

The MP test will have the following form:

Reject H_0 if and only if $L(\mathbf{x}; \mu_1)/L(\mathbf{x}; \mu_0) > k$,

that is, we will reject the null hypothesis H_0 if and only if

$$exp\{\sigma^{-2}(\mu_1 - \mu_0)\Sigma_{i=1}^n x_i\} \text{ is large } (> k). \tag{8.3.6}$$

Now since $\mu_1 > \mu_0$, the condition in (8.3.6) can be rephrased as:

Reject H_0 if and only if $\Sigma_{i=1}^{n} X_i$, or equivalently the sample

mean $\overline{X}$, is large ($> k$).

(8.3.7)

Since $E_\mu[X] = \mu$, it does make sense to reject H_0 when $\overline{X}$ is large ($> k$) because the alternative hypothesis postulates a value μ_1 which is larger than μ_0. But the MP test given in (8.3.7) is not yet in the *implementable* form and the test must also be size α. Let us equivalently rewrite the same test as follows:

$$\text{Reject } H_0 \text{ if and only if } \sqrt{n}(\overline{X} - \mu_0)/\sigma > z_\alpha, \qquad (8.3.8)$$

where z_α is the upper $100\alpha\%$ point of the standard normal distribution. See the Figure 8.3.1. The form of the test given in (8.3.7) asks us to reject H_0 for *large enough* values of $\overline{X}$ while (8.3.8) equivalently asks us to reject H_0 for *large enough* values of $\sqrt{n}(\overline{X}-\mu_0)/\sigma$. Under H_0, observe that $\sqrt{n}(\overline{X}-\mu_0)/\sigma$ is a statistic, referred to as the *test statistic*, which is distributed as a standard normal random variable.

Here, the critical region $\mathcal{R} = \{\mathbf{x} = (x_1, ..., x_n) \in \Re^n : \sqrt{n}(\overline{x} - \mu_0)/\sigma > z_\alpha\}$. Now, we have:

Type I error probability

$$= P_{\mu_0}\left\{\sqrt{n}(\overline{X} - \mu_0)/\sigma > z_\alpha\right\} = \alpha, \text{ by the choice of } z_\alpha.$$

Thus, we have the MP level α test by the Neyman-Pearson Lemma. ▲

Example 8.3.2 (Example 8.3.1 Continued) Suppose that $X_1, ..., X_n$ are iid $N(\mu, \sigma^2)$ with unknown $\mu \in \Re$, but $\sigma \in \Re^+$ is known. With preassigned $\alpha \in (0, 1)$ we wish to derive the MP level α test for $H_0 : \mu = \mu_0$ versus $H_1 : \mu = \mu_1$ where $\mu_1 < \mu_0$ and μ_0, μ_1 are two real numbers.

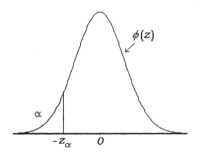

Figure 8.3.2. Standard Normal PDF: Lower $100\alpha\%$ Point

Since both H_0, H_1 are simple hypotheses, the Neyman-Pearson Lemma applies. The equation (8.3.6) will continue to hold but we now have $\mu_1 - \mu_0 < 0$. Thus the large values of $exp[\sigma^{-2}(\mu_1 - \mu_0)\Sigma_{i=1}^{n}x_i]$ will correspond to the small values of $\Sigma_{i=1}^{n}x_i$. In other words, the MP test would look like this:

Reject H_0 if and only if $\Sigma_{i=1}^{n}X_i$, or equivalently the sample mean $\overline{X}$, is small ($< k$).

This simplifies to the following form of the MP level α test:

$$\text{Reject } H_0 \text{ if and only if } \sqrt{n}(X - \mu_0)/\sigma < -z_\alpha. \qquad (8.3.9)$$

See the Figure 8.3.2. Since $E_\mu[X] = \mu$, it does make sense to reject H_0 when $\overline{X}$ is small because the alternative hypothesis postulates a value μ_1 which is smaller than μ_0. Under H_0, again observe that $\sqrt{n}(\overline{X} - \mu_0)/\sigma$ is a statistic, referred to as the *test statistic*, which has a standard normal distribution. Here, the critical region $\mathcal{R} = \{\mathbf{x} = (x_1, ..., x_n) \in \Re^n : \sqrt{n}(\overline{x} - \mu_0)/\sigma < -z_\alpha\}$. ▲

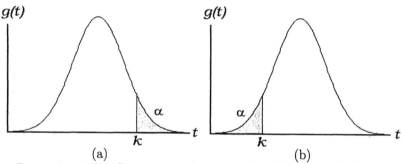

Figure 8.3.3. The Shaded Area is α, the Type I Error Probability:
(a) $\mathcal{R} \equiv \{t \in \mathcal{T} : t > k\}$ (b) $\mathcal{R} \equiv \{t \in \mathcal{T} : t < k\}$

Suppose that we wish to test $H_0 : \theta = \theta_0$ versus $H_1 : \theta = \theta_1$ at the level α where $\theta_1 > \theta_0$ and θ_0, θ_1 are two known real numbers. In a number of problems, we may discover that we reject H_0 when an appropriate *test statistic* T exceeds some number k. This was the situation in the Example 8.3.1 where we had $T = \sqrt{n}(\overline{X} - \mu_0)/\sigma$ and $k = z_\alpha$. Here, the alternative hypothesis was on the upper side (of μ_0) and the rejection region $\mathcal{R}$ (for H_0) fell on the upper side too.

Instead we may wish to test $H_0 : \theta = \theta_0$ versus $H_1 : \theta = \theta_1$ at the level α where $\theta_1 < \theta_0$ and θ_0, θ_1 are two known real numbers . In a number of problems, we may discover that we reject H_0 when an appropriate *test statistic* T falls under some number k. This was the situation in the Example 8.3.2 where we had $T = \sqrt{n}(\overline{X} - \mu_0)/\sigma$ and $k = -z_\alpha$. Here, the alternative hypothesis was on the lower side (of μ_0) and the rejection region $\mathcal{R}$ (for H_0) fell on the lower side too.

In general, the *cut-off point* k has to be determined from the distribution $g(t)$, that is the pmf or pdf of the test statistic T under H_0. The pmf or pdf of T under H_0 specifies what is called a *null distribution*. We have summarized the upper- and lower-sided critical regions in the Figure 8.3.3.

Example 8.3.3 Suppose that $X_1, ..., X_n$ are iid with the common pdf $b^{-1}exp(-x/b)$ for $x \in \Re^+$ with unknown $b \in \Re^+$. With preassigned $\alpha \in (0, 1)$, we wish to obtain the MP level α test for $H_0 : b = b_0$ versus $H_1 : b = b_1(> b_0)$ where b_0, b_1 are two positive numbers. Both H_0, H_1 are simple hypothesis and the Neyman-Pearson Lemma applies. The likelihood function is given by

$$L(\mathbf{x}; b) = b^{-n}exp\{-b^{-1}\Sigma_{i=1}^n x_i\}, \ b \in \Re^+.$$

The MP test will have the following form:

Reject H_0 if and only if $L(\mathbf{x}; b_1)/L(\mathbf{x}; b_0) > k$,

that is, we will reject the null hypothesis H_0 if and only if

$$exp\{(b_0^{-1} - b_1^{-1})\Sigma_{i=1}^n x_i\} \text{ is large } (> k). \tag{8.3.10}$$

Now since $b_1 > b_0$, the condition in (8.3.10) can be rephrased as:

Reject H_0 if and only if $\Sigma_{i=1}^n X_i$ is large $(> k)$. \hfill (8.3.11)

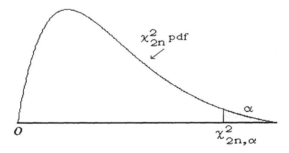

Figure 8.3.4. Chi-Square Upper $100\alpha\%$ Point with
Degrees of Freedom $2n$

But, the MP test described by (8.3.11) is not yet in the *implementable* form. Under H_0, observe that $b_0^{-1}X_i$'s are iid standard exponential random variables, so that $2b_0^{-1}X_i$'s are iid χ_2^2. Hence, $2b_0^{-1}\Sigma_{i=1}^n X_i$ is distributed as χ_{2n}^2 when H_0 is true. This test must also be size α. Let us equivalently rewrite the same test as follows:

$$\text{Reject } H_0 \text{ if and only if } 2b_0^{-1}\Sigma_{i=1}^n X_i > \chi_{2n,\alpha}^2, \qquad (8.3.12)$$

where $\chi_{2n,\alpha}^2$ is the upper $100\alpha\%$ point of the χ_{2n}^2 distribution. See, for example, the Figure 8.3.4. The test from (8.3.11) asks us to reject H_0 for *large enough* values of $\Sigma_{i=1}^n X_i$, whereas (8.3.12) asks us to reject H_0 for *large enough* values of $2b_0^{-1}\Sigma_{i=1}^n X_i$. We call $2b_0^{-1}\Sigma_{i=1}^n X_i$ the associated *test statistic*. The order of "largeness", that is the choice of k, depends on the normalized test statistic and its distribution under H_0. Under H_0, since $2b_0^{-1}\Sigma_{i=1}^n X_i$ is distributed as a χ_{2n}^2 random variable, we have:

The Type I error probability

$$= P_{b_0}\left\{2b_0^{-1}\Sigma_{i=1}^n X_i > \chi_{2n,\alpha}^2\right\} = \alpha, \text{ by the choice of } \chi_{2n,\alpha}^2.$$

Thus, we have the MP level α test. Here, the critical region $\mathcal{R} = \{\mathbf{x} = (x_1, ..., x_n) \in \Re^{+n} : 2b_0^{-1}\Sigma_{i=1}^n x_i > \chi_{2n,\alpha}^2\}$. ▲

Example 8.3.4 Suppose that $X_1, ..., X_n$ are iid having the Uniform$(0, \theta)$ distribution with the unknown parameter $\theta(> 0)$. With preassigned $\alpha \in (0, 1)$, we wish to obtain the MP level α test for $H_0 : \theta = \theta_0$ versus $H_1 : \theta = \theta_1(> \theta_0)$ where θ_0, θ_1 are two positive numbers. Both H_0, H_1 are simple hypothesis and the Neyman-Pearson Lemma applies. The likelihood function is given by

$$L(\mathbf{x}; \theta) = \theta^{-n}I(0 < x_{n:n} < \theta)I(0 < x_{1:n} < x_{n:n}), \quad \theta \in \Re^+.$$

In view of the Remark 8.3.1, the MP test will have the following form:

$$\text{Reject } H_0 \text{ if and only if } L(\mathbf{x}; \theta_1)/L(\mathbf{x}; \theta_0) > k,$$

that is, we will reject the null hypothesis H_0 if and only if

$$X_{n:n} \text{ is large } (> k). \qquad (8.3.13)$$

Note that the MP test given by (8.3.13) is *not* in the implementable form. Under H_0, the pdf of the statistic $T = X_{n:n}$ is given by nt^{n-1}/θ_0^n. Hence we can determine k as follows:

The Type I error probability

$$= P_{\theta_0}\{X_{n:n} > k\} = \int_k^{\theta_0} nt^{n-1}\theta_0^{-n}dt = (\theta_0^n - k^n)/\theta_0^n = \alpha,$$

provided that we choose $k = \theta_0(1 - \alpha)^{1/n}$ and implement test defined by (8.3.13). Here, the critical region $\mathcal{R} = \{\mathbf{x} = (x_1, ..., x_n) \in \Re^{+n} : x_{n:n} > \theta_0(1 - \alpha)^{1/n}\}$. ▲

> In the Neyman-Pearson Lemma, we assumed that θ was real valued. This assumption was not crucial. The unknown parameter can be vector valued too. It is crucial that the likelihood function involves no unknown component of θ under either hypothesis $H_i, i = 0, 1$ if θ is vector valued. Look at Example 8.3.5.

Example 8.3.5 Suppose that $X_1, ..., X_n$ are iid having the common pdf $b^{-\delta}[\Gamma(\delta)]^{-1} x^{\delta-1} exp(-x/b)$ with *two unknown parameters* $(\delta, b) \in \Re^+ \times \Re^+, x \in \Re^+$. With preassigned $\alpha \in (0, 1)$, we wish to obtain the MP level α test for $H_0 : (b = b_0, \delta = \delta^*)$ versus $H_1 : (b = b_1, \delta = \delta^*)$ where $b_1 > b_0$ are two positive numbers and δ^* is also a positive number. Both H_0 and H_1 are simple hypothesis and hence the Neyman-Pearson Lemma applies. The likelihood function is given by

$$L(\mathbf{x}; \delta, b) = b^{-n\delta}[\Gamma(\delta)]^{-n} exp\{-b^{-1}\Sigma_{i=1}^n x_i\} \{\Pi_{i=1}^n x_i\}^{\delta-1},$$
$$\text{for } (\delta, b) \in \Re^+ \times \Re^+.$$

The MP test will have the following form:

Reject H_0 if and only if $L(\mathbf{x}; \delta^*, b_1)/L(\mathbf{x}; \delta^*, b_0) > k$,

that is, we will reject the null hypothesis H_0 if and only if

$$exp\{(b_0^{-1} - b_1^{-1})\Sigma_{i=1}^n x_i\} \text{ is large } (> k). \tag{8.3.14}$$

This test must also have size α. Observe that, under H_0, the statistic $\Sigma_{i=1}^n X_i$ has the Gamma$(n\delta^*, b_0)$ distribution which is completely known for fixed values of n, δ^*, b_0 and α. Let us equivalently write the test as follows:

Reject H_0 if and only if $\Sigma_{i=1}^n X_i > g_{n,\delta^*,b_0,\alpha}$, \tag{8.3.15}

where $g_{n,\delta^*,b_0,\alpha}$ is the upper $100\alpha\%$ point of the Gamma$(n\delta^*, b_0)$ distribution. In the Table 8.3.1, $g_{n,\delta^*,b_0,\alpha}$ values are given for $\alpha = .01, .05, b_0 = 1$,

Table 8.3.1. Selected Values of $g_{n,\delta^*,b_0,\alpha}$ with $b_0 = 1$

	$\alpha = .05$			$\alpha = .01$		
	$n = 2$	$n = 5$	$n = 6$	$n = 2$	$n = 5$	$n = 6$
$\delta^* = 2$	1.3663	5.4254	6.9242	0.8234	4.1302	5.4282
$\delta^* = 3$	2.6130	9.2463	11.6340	1.7853	7.4767	9.6163

$n = 2, 5, 6$ and $\delta^* = 2, 3$. Express the critical region explicitly. ▲

In a discrete case, one applies the Neyman-Pearson Lemma, but employs *randomization*. Look at Examples 8.3.6-8.3.7.

In all the examples, we have so far dealt with continuous distributions only. The reader should recall the Remark 8.3.2. If the X's are discrete, then we carefully use randomization whenever $L(\mathbf{x}; \theta_1) = kL(\mathbf{x}; \theta_0)$. The next two examples emphasize this concept.

Example 8.3.6 Suppose that $X_1, ..., X_n$ are iid Bernoulli(p) where $p \in (0, 1)$ is the unknown parameter. With preassigned $\alpha \in (0, 1)$, we wish to derive the MP level α test for $H_0 : p = p_0$ versus $H_1 : p = p_1 (> p_0)$ where p_0, p_1 are two numbers from the interval $(0, 1)$. Both H_0 and H_1 are simple hypothesis and hence the Neyman-Pearson Lemma applies. Then, writing $T = \Sigma_{i=1}^n X_i$, $t = \Sigma_{i=1}^n x_i$, the likelihood function is given by

$$L(\mathbf{x}; p) = p^t (1-p)^{n-t}, \quad p \in (0, 1).$$

The MP test will have the following form:

Reject H_0 if and only if $L(\mathbf{x}; p_1)/L(\mathbf{x}; p_0) > k$,

that is, we will reject the null hypothesis H_0 if

$$\{[p_1(1-p_0)]/[p_0(1-p_1)]\}^t \{(1-p_1)/(1-p_0)\}^n \text{ is large } (> k). \quad (8.3.16)$$

Now since $p_1 > p_0$, we have $[p_1(1-p_0)]/[p_0(1-p_1)] > 1$. So, the "large values" of the lhs in (8.3.16) correspond to the "large values" of $\Sigma_{i=1}^n X_i$. Hence, the MP test defined in (8.3.16) can be rephrased as:

Reject H_0 if $\Sigma_{i=1}^n X_i$ is large $(> k)$.

We may then write down the corresponding test function as follows:

$$\psi(\mathbf{X}) = \begin{cases} 1 & \text{if} \quad \Sigma_{i=1}^n X_i > k \\ \gamma & \text{if} \quad \Sigma_{i=1}^n X_i = k \\ 0 & \text{if} \quad \Sigma_{i=1}^n X_i < k, \end{cases} \quad (8.3.17)$$

where a positive integer k and $\gamma \in (0, 1)$ are to be chosen in such a way that the test has the size α. Observe that $\Sigma_{i=1}^n X_i$ has the Binomial(n, p_0) distribution under H_0. First, we determine the *smallest integer* k such that $P_{p_0}\{\Sigma_{i=1}^n X_i > k\} < \alpha$ and let

$$\gamma = [\alpha - P_{p_0}\{\Sigma_{i=1}^n X_i > k\}]/P_{p_0}\{\Sigma_{i=1}^n X_i = k\}, \quad (8.3.18)$$

where one has

$$P_{p_0}\{\Sigma_{i=1}^n X_i = k\} = \binom{n}{k}p_0^k(1 - p_0)^{n-k},$$
$$P_{p_0}\{\Sigma_{i=1}^n X_i > k\}= \Sigma_{u=k+1}^n \binom{n}{u}p_0^u(1 - p_0)^{n-u}. \tag{8.3.19}$$

Now, with k and γ defined by (8.3.18), one can check that the Type I error probability is

$$\gamma P_{p_0}\{\Sigma_{i=1}^n X_i = k\} + P_{p_0}\{\Sigma_{i=1}^n X_i > k\} = \alpha.$$

Thus, we have the MP level α test. If $\Sigma_{i=1}^n X_i = k$, then one rejects H_0 with probability γ. For example, if $\gamma = .135$, then consider three-digit random numbers $000, 001, ..., 134, 135, ..., 999$ and look at a random number table to draw one three-digit number. If we come up with one of the numbers $000, 001, ...$ or 134, then and only then H_0 will be rejected. This is what is known as *randomization*. The following table provides the values of k and γ for some specific choices of n and α.

Table 8.3.2. Values of k and γ in the Bernoulli Case

$n = 10$	$\alpha = .10$		$n = 10$	$\alpha = .05$	
p_0	k	γ	p_0	k	γ
.2	4	.763	.2	4	.195
.4	6	.406	.6	8	.030

$n = 20$	$\alpha = .05$		$n = 25$	$\alpha = .10$	
p_0	k	γ	p_0	k	γ
.3	9	.031	.5	16	.757
.5	14	.792	.6	18	.330

The reader should verify some of the entries in this table by using the expressions given in (8.3.19). ▲

Example 8.3.7 Suppose that $X_1, ..., X_n$ are iid Poisson(λ) where $\lambda \in \Re^+$ is the unknown parameter. With preassigned $\alpha \in (0, 1)$, we wish to derive the MP level α test for $H_0 : \lambda = \lambda_0$ versus $H_1 : \lambda = \lambda_1(> \lambda_0)$ where λ_0, λ_1 are two positive numbers. Both H_0 and H_1 are simple hypothesis and the Neyman-Pearson Lemma applies Then, writing $T = \Sigma_{i=1}^n X_i$, $t = \Sigma_{i=1}^n x_i$, the likelihood function is given by

$$L(\mathbf{x}; \lambda) = e^{-n\lambda}(n\lambda)^t/t!, \ \lambda \in (0, \infty).$$

The MP test will have the following form:

$$\text{Reject } H_0 \text{ if } L(\mathbf{x}; \lambda_1)/L(\mathbf{x}; \lambda_0) > k,$$

that is, we will reject the null hypothesis H_0 if

$$[\lambda_1/\lambda_0]^t e^{-n(\lambda_1-\lambda_0)} \text{ is large } (> k). \tag{8.3.20}$$

Now since $\lambda_1 > \lambda_0$, the "large values" of the lhs in (8.3.20) correspond to the "large values" of $\Sigma_{i=1}^n X_i$. Hence, the MP test defined by (8.3.20) can be rephrased as:

Reject H_0 if $\Sigma_{i=1}^n X_i$ is large $(> k)$.

We then write down the test function as follows:

$$\psi(\mathbf{X}) = \begin{cases} 1 & \text{if} \quad \Sigma_{i=1}^n X_i > k \\ \gamma & \text{if} \quad \Sigma_{i=1}^n X_i = k \\ 0 & \text{if} \quad \Sigma_{i=1}^n X_i < k, \end{cases} \tag{8.3.21}$$

where a positive integer k and $\gamma \in (0,1)$ are to be chosen in such a way that the test has the size α. Observe that $\Sigma_{i=1}^n X_i$ has Poisson$(n\lambda_0)$ distribution under H_0. First, we determine the *smallest integer* value of k such that $P_{\lambda_0}\{\Sigma_{i=1}^n X_i > k\} < \alpha$ and let

$$\gamma = [\alpha - P_{\lambda_0}\{\Sigma_{i=1}^n X_i > k\}]/P_{\lambda_0}\{\Sigma_{i=1}^n X_i = k\}, \tag{8.3.22}$$

where

$$\begin{aligned} P_{\lambda_0}\{\Sigma_{i=1}^n X_i = k\} &= e^{-n\lambda_0}(n\lambda_0)^t/t!, \\ P_{\lambda_0}\{\Sigma_{i=1}^n X_i > k\} &= \Sigma_{u=k+1}^n e^{-n\lambda_0}(n\lambda_0)^u/u!. \end{aligned} \tag{8.3.23}$$

Now, with k and γ defined by (8.3.22), one can check that the Type I error probability is

$$\gamma P_{\lambda_0}\{\Sigma_{i=1}^n X_i = k\} + P_{\lambda_0}\{\Sigma_{i=1}^n X_i > k\} = \alpha.$$

Thus, we have the MP level α test. If $\Sigma_{i=1}^n X_i = k$, then one would employ appropriate *randomization* and reject H_0 with probability γ. The following table provides some values of k, γ for specific choices of n and α. The reader

Table 8.3.3. Values of k and γ in the Poisson Case

$n=10$ $\alpha=.10$			$n=10$ $\alpha=.05$		
λ_0	k	γ	λ_0	k	γ
.15	3	.274	.15	4	.668
.30	5	.160	.35	7	.604

$n=20$ $\alpha=.05$			$n=25$ $\alpha=.10$		
λ_0	k	γ	λ_0	k	γ
.40	13	.534	.28	10	.021
.50	15	.037	.40	14	.317

should verify some of the entries in this table by using the expressions given in (8.3.23). ▲

A MP level α test always depends on (jointly) sufficient statistics.

In each example, the reader has noticed that the MP level α test always depended only on the (jointly) sufficient statistics. This is not a coincidence. Suppose that $\mathbf{T} = \mathbf{T}(X_1, ..., X_n)$ is a (jointly) sufficient statistic for $\boldsymbol{\theta}$. By the Neyman factorization (Theorem 6.2.1), we can split the likelihood function as follows:

$$L(\mathbf{x}; \boldsymbol{\theta}) = g\left(\mathbf{T}(\mathbf{x}); \boldsymbol{\theta}\right)) h(\mathbf{x}) \text{ for all } \mathbf{x} \in \mathcal{X}, \qquad (8.3.24)$$

where $h(\mathbf{x})$ does not involve $\boldsymbol{\theta}$. Next, recall that the MP test rejects $H_0 : \boldsymbol{\theta} = \boldsymbol{\theta}_0$ in favor of accepting $H_1 : \boldsymbol{\theta} = \boldsymbol{\theta}_1$ for large values of $L(\mathbf{x}; \boldsymbol{\theta}_1)/L(\mathbf{x}; \boldsymbol{\theta}_0)$. But, in view of the factorization in (8.3.24), we can write

$$L(\mathbf{x}; \boldsymbol{\theta}_1)/L(\mathbf{x}; \boldsymbol{\theta}_0) = g\left(\mathbf{T}(\mathbf{x}); \boldsymbol{\theta}_1\right)) / g\left(\mathbf{T}(\mathbf{x}); \boldsymbol{\theta}_0\right)),$$

which implies that the MP test rejects $H_0 : \boldsymbol{\theta} = \boldsymbol{\theta}_0$ in favor of accepting $H_1 : \boldsymbol{\theta} = \boldsymbol{\theta}_1$, for large values of $g\left(\mathbf{T}(\mathbf{x}); \boldsymbol{\theta}_1\right)) / g\left(\mathbf{T}(\mathbf{x}); \boldsymbol{\theta}_0\right))$. Thus, it should not surprise anyone to see that the MP tests in all the examples ultimately depended on the (jointly) sufficient statistic $\mathbf{T}$.

8.3.2 Applications: No Parameters Are Involved

In the statement of the Neyman-Pearson Lemma, we assumed that θ was a real valued parameter and the common pmf or pdf was indexed by θ. We mentioned earlier that this assumption was not really crucial. It is essential to have a known and unique likelihood function under both H_0, H_1. In other words, as long as H_0, H_1 are both simple hypothesis, the Neyman-Pearson Lemma will provide the MP level α test explicitly Some examples follow.

Example 8.3.8 Suppose that X is an observable random variable with its pdf given by $f(x), x \in \Re$. Consider two functions defined as follows:

$$f_0(x) = \begin{cases} \frac{3}{64}x^2 & \text{if } 0 < x < 4 \\ 0 & \text{elsewhere,} \end{cases} \qquad f_1(x) = \begin{cases} \frac{3}{16}\sqrt{x} & \text{if } 0 < x < 4 \\ 0 & \text{elsewhere.} \end{cases}$$

We wish to determine the MP level α test for

$$H_0 : f(x) = f_0(x) \text{ versus } H_1 : f(x) = f_1(x) \qquad (8.3.25)$$

In view of the Remark 8.3.1, the MP test will have the following form:

$$\text{Reject } H_0 \text{ if and only if } f_1(x)/f_0(x) > k. \qquad (8.3.26)$$

But, $f_1(x)/f_0(x) = 4x^{-3/2}$ and hence the test given in (8.3.26) can be rewritten as:

$$\text{Reject } H_0 \text{ if and only if } X \text{ is small } (< k). \qquad (8.3.27)$$

This test must also have the size α, that is we require:

$$\alpha = P\{X < k \text{ when } f(x) = f_0(x)\} = \int_0^k \tfrac{3}{64}x^2 dx = \tfrac{1}{64}k^3,$$

so that $k = 4\alpha^{1/3}$. With $k = 4\alpha^{1/3}$, one would implement the MP level α test given by (8.3.27). The associated power calculation can be carried out as follows:

$$P\{X < k \text{ when } f(x) = f_1(x)\} = \int_0^k \tfrac{3}{16}\sqrt{x}dx = \alpha^{1/2}$$

when $k = 4\alpha^{1/3}$. When $\alpha = .05, .01$ and $.001$, the power is respectively $.223606, .1$ and $.031623$. ▲

Example 8.3.9 (Example 8.3.8 Continued) Suppose that X_1 and X_2 are observable random variables with the common pdf given by $f(x), x \in \Re$. Consider the two functions $f_0(x)$ and $f_1(x)$ defined in the Example 8.3.8. As in the earlier example, we wish to determine the MP level α test for

$$H_0 : f(x) = f_0(x) \text{ versus } H_1 : f(x) = f_1(x).$$

The reader should check that the MP test has the following form:

$$\text{Reject } H_0 \text{ if and only if } X_1 X_2 \text{ is small } (< k). \qquad (8.3.28)$$

Let us write $F_0(x)$ for the distribution function which corresponds to the pdf $f_0(x)$ so that

$$F_0(x) = \begin{cases} 0 & \text{if } x \leq 0 \\ \tfrac{x^3}{64} & \text{if } 0 < x < 4 \\ 1 & \text{if } x \geq 4. \end{cases}$$

Under H_0, it is known that $-2log\{F_0(X_i)\}, i = 1, 2$, is distributed as iid χ_2^2. One may go back to the Example 4.2.5 in this context. The test defined via (8.3.28) must also have size α, that is, we require:

$$\begin{aligned} \alpha &= P\{X_1 X_2 < k \text{ when } f(x) = f_0(x)\} \\ &= P\left\{\tfrac{X_1^3}{64}\tfrac{X_2^3}{64} < k \text{ when } f(x) = f_0(x)\right\} \\ &= P\left[\Sigma_{i=1}^2 \left(-2log\{F_0(X_i)\}\right) > k \text{ when } f(x) = f_0(x)\right] \\ &= P\left\{\chi_4^2 > k\right\}. \end{aligned}$$

Thus, one implements the MP level α test as follows:

Reject H_0 if and only if $log[2^{24}/(X_1 X_2)^6] > \chi^2_{4,\alpha}$

where $\chi^2_{4,\alpha}$ is the upper $100\alpha\%$ point of the χ^2_4 distribution. We leave out the power calculation as an exercise. ▲

Example 8.3.10 Suppose that X is an observable random variable with the pdf $f(x), x \in \Re$. We wish to test

$$H_0 : f(x) = \pi^{-1/2} exp\{-x^2\}, \; x \in \Re$$

versus

$$H_1 : f(x) = \pi^{-1}(1 + x^2)^{-1}, \; x \in \Re.$$

That is, to decide between the *simple null hypothesis* which specifies that X is distributed as $N(0, \frac{1}{2})$ versus the *simple alternative hypothesis* which specifies that X has the Cauchy distribution. In view of the Remark 8.3.1, the MP test will have the following form:

Reject H_0 if and only if $\pi^{-1/2}(1 + x^2)^{-1} exp(x^2) > k.$ \qquad (8.3.29)

Now, we want to determine the points x for which the function $(1 + x^2)^{-1} exp(x^2)$ becomes "large". Let us define a new function, $g(y) = (1 + y)^{-1} exp(y), 0 < y < \infty$. We claim that $g(y)$ is increasing in y. In order to verify this claim, it will be simpler to consider $h(y) = log(g(y)) = y - log(1 + y)$ instead. Note that $\frac{d}{dy} h(y) = y/(1 + y) > 0$ for $0 < y < \infty$. That is, $h(y)$ is increasing in y so that $g(y)$ is then increasing in y too, since the *log* operation is monotonically increasing. Hence, the "large" values of $(1 + x^2)^{-1} exp(x^2)$ would correspond to the "large" values of x^2 or equivalently, the "large" values of $|x|$. Thus, the MP test given by (8.3.29) will have the following simple form:

Reject H_0 if and only if $|X|$ is large $(> k)$. \qquad (8.3.30)

This test must also have the size α, that is

$$\alpha = P\{|X| > k \text{ when } X \text{ is } N(0, \tfrac{1}{2})\} = P\left\{|N(0,1)| > k\sqrt{2}\right\},$$

so that we have $k = \frac{1}{\sqrt{2}} z_{\alpha/2}$ where $z_{\alpha/2}$ stands for the upper $50\alpha\%$ point of the standard normal distribution. With this choice of k, one would implement the MP level α test given by (8.3.30). For example, if $\alpha = .05$, then $z_{\alpha/2} = 1.96$ so that $k \approx 1.385929$. The associated power calculation can be carried out as follows:

$$P\{|X| > k \text{ when } X \text{ is Cauchy}(0, 1)\}$$

$$= \frac{1}{\pi} \int_{-\infty}^{-k} \frac{dx}{(1+x^2)} + \frac{1}{\pi} \int_{k}^{\infty} \frac{dx}{(1+x^2)} = \frac{2}{\pi} \int_{k}^{\infty} \frac{dx}{(1+x^2)},$$

since Cauchy$(0,1)$ pdf is symmetric around $x = 0$. Hence, the test's power is given by $1 - \frac{2}{\pi}\arctan(k)$ with $k = \frac{1}{\sqrt{2}}z_{\alpha/2}$. When $\alpha = .05$, the power turns out to be $.39791$. ▲

Look at the Exercises 8.3.8-8.3.10 and the Exercise 8.3.20.

8.3.3 Applications: Observations Are Non-IID

In the formulation of the Neyman-Pearson Lemma, it is not crucial that the X's be identically distributed or they be independent. In the general situation, all one has to work with is the explicit forms of the *likelihood function* under *both simple hypotheses* H_0 and H_1. We give two examples.

Example 8.3.11 Suppose that X_1 and X_2 are independent random variables respectively distributed as $N(\mu, \sigma^2)$ and $N(\mu, 4\sigma^2)$ where $\mu \in \Re$ is unknown and $\sigma \in \Re^+$ is assumed known. Here, we have a situation where the X's are independent but they are not identically distributed. We wish to derive the MP level α test for $H_0 : \mu = \mu_0$ versus $H_1 : \mu = \mu_1(> \mu_0)$. The likelihood function is given by

$$L(\mathbf{x}; \mu) = (4\pi\sigma^2)^{-1}exp\{-\tfrac{1}{2\sigma^2}(x_1 - \mu)^2 - \tfrac{1}{8\sigma^2}(x_2 - \mu)^2\}, \qquad (8.3.31)$$

where $\mathbf{x} = (x_1, x_2)$. Thus, one has

$$L(\mathbf{x}; \mu_1)/L(\mathbf{x}; \mu_0) = exp\{\tfrac{1}{4\sigma^2}(4x_1 + x_2)(\mu_1 - \mu_0)\}exp\{\tfrac{5}{8\sigma^2}(\mu_1^2 - \mu_0^2)\}.$$

In view of the Neyman-Pearson Lemma, we would reject H_0 if and only if $L(\mathbf{x}; \mu_1)/L(\mathbf{x}; \mu_0)$ is "large", that is when $4X_1 + X_2 > k$. This will be the MP level α test when k is chosen so that $P_{\mu_0}\{4X_1 + X_2 > k\} = \alpha$. But, under H_0, the statistic $4X_1 + X_2$ is distributed as $N(5\mu_0, 20\sigma^2)$. So, we can rephrase the MP level α test as follows:

Reject H_0 if and only if $\{4X_1 + X_2 - 5\mu_0\}/(\sqrt{20}\sigma) > z_\alpha$. \qquad (8.3.32)

We leave out the power calculation as Exercise 8.3.15. ▲

Example 8.3.12 Let us denote $\mathbf{X}' = (X_1, X_2)$ where $\mathbf{X}$ is assumed to have the bivariate normal distribution, $N_2(\mu, \mu, 1, 1, \frac{1}{\sqrt{2}})$ with the unknown parameter $\mu \in \Re$. Refer to Section 3.6 for a review of the bivariate normal distribution. We wish to derive the MP level α test for $H_0 : \mu = \mu_0$ versus $H_1 : \mu = \mu_1(> \mu_0)$. The likelihood function is given by

$$L(\mathbf{x}; \mu) = \pi^{-1}exp\{-(x_1 - \mu)^2 + (x_1 - \mu)(x_2 - \mu) - (x_2 - \mu)^2\}. \qquad (8.3.33)$$

Thus, one has

$$L(\mathbf{x}; \mu_1)/L(\mathbf{x}; \mu_0) = exp\{(x_1 + x_2)(\mu_1 - \mu_0)\}exp\{(\mu_0^2 - \mu_1^2)\}.$$

In view of the Neyman-Pearson Lemma, we would reject H_0 if and only if $L(\mathbf{x}; \mu_1)/L(\mathbf{x}; \mu_0)$ is "large", that is when $X_1 + X_2 > k$. This will be the MP level α test when k is chosen so that $P_{\mu_0}\{X_1 + X_2 > k\} = \alpha$. But, under H_0, the statistic $X_1 + X_2$ is distributed as $N(2\mu_0, 2 + \sqrt{2})$. So, we can rephrase the MP level α test as follows:

$$\text{Reject } H_0 \text{ if and only if } \{X_1 + X_2 - 2\mu_0\}/(\sqrt{2 + \sqrt{2}}) > z_\alpha. \quad (8.3.34)$$

We leave out the power calculation as Exercise 8.3.17. ▲

> Look at the related Exercises 8.3.16, 8.3.18 and 8.3.21.

8.4   One-Sided Composite Alternative Hypothesis

So far we have developed the Neyman-Pearson methodology to test a simple null hypothesis versus a simple alternative hypothesis. Next suppose that we have a simple null hypothesis $H_0 : \theta = \theta_0$, but the alternative hypothesis is $H_1 : \theta > \theta_0$. Here, $H_1 : \theta > \theta_0$ represents an upper-sided composite hypothesis. It is conceivable that the alternative hypothesis in another situation may be $H_1 : \theta < \theta_0$ which is a lower-sided composite hypothesis.

We recall the concept of the UMP level α test of H_0 versus H_1, earlier laid out in the Definition 8.2.4. A test with its power function $Q(\theta)$ would be called UMP level α provided that (i) $Q(\theta_0) \le \alpha$ and (ii) $Q(\theta)$ is maximized at *every point* θ satisfying H_1.

In other words, a test would be UMP level α provided that it has level α and its power $Q(\theta)$ is at least as large as the power $Q^*(\theta)$ of any other level α test, at *every point* $\theta > \theta_0$ or $\theta < \theta_0$, as the case may be. The following subsections lay down a couple of different approaches to derive the UMP level α tests.

8.4.1   UMP Test via the Neyman-Pearson Lemma

This method is very simple. We wish to obtain the UMP level α test for $H_0 : \theta = \theta_0$ versus $H_1 : \theta > \theta_0$. We start by fixing an arbitrary value $\theta_1 \in \Theta$ such that $\theta_1 > \theta_0$. Now, we invoke the Neyman-Pearson Lemma to conclude that there exists a MP level α test for deciding between the choices of two simple hypotheses θ_0 and θ_1. If this particular test, so determined, happens to remain unaffected by the choice of the specific value θ_1, then by the Definition 8.2.4, we already have on hand the required UMP level

α test for H_0 versus H_1.

| How can one prove this claim? One may argue as follows. |

Suppose that $Q(\theta)$ is the power function of the MP level α test between θ_0 and some fixed $\theta_1(> \theta_0)$. Suppose that there exists a level α test with its power function $Q^*(\theta)$ for choosing between θ_0 and some fixed θ^* ($> \theta_0$) such that $Q^*(\theta^*) > Q(\theta^*)$. But, we are working under the assumption that the MP level α test between θ_0 and θ_1 is also the MP level α test between θ_0 and θ^*, that is $Q(\theta^*)$ is the maximum among all level α tests between θ_0 and θ^*. This leads to a contradiction.

The case of a lower-sided composite alternative is handled analogously. Some examples follow.

Example 8.4.1 (Example 8.3.1 Continued) Suppose that $X_1, ..., X_n$ are iid $N(\mu, \sigma^2)$ with unknown $\mu \in \Re$, but assume that $\sigma \in \Re^+$ is known. With preassigned $\alpha \in (0, 1)$, we wish to obtain the UMP level α test for $H_0 : \mu = \mu_0$ versus $H_1 : \mu > \mu_0$ where μ_0 is a real number. Now, fix a value $\mu_1(> \mu_0)$ and then from (8.3.8) recall that the MP level α test between μ_0 and arbitrarily chosen μ_1 will have the following form:

$$\text{Reject } H_0 \text{ if and only if } \sqrt{n}(\overline{X} - \mu_0)/\sigma > z_\alpha, \qquad (8.4.1)$$

where z_α is the upper $100\alpha\%$ point of the standard normal distribution. See the Figure 8.3.1. Obviously this test does not depend on the specific choice of $\mu_1(> \mu_0)$. Hence, the test given by (8.4.1) is UMP level α for testing $H_0 : \mu = \mu_0$ versus $H_1 : \mu > \mu_0$. ▲

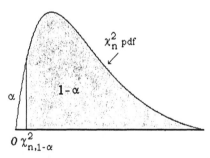

Figure 8.4.1. Chi-Square Lower $100\alpha\%$ or Upper $100(1 - \alpha)\%$
Point with Degrees of Freedom n

Example 8.4.2 Let $X_1, ..., X_n$ be iid $N(0, \sigma^2)$ with unknown $\sigma \in \Re^+$. With preassigned $\alpha \in (0, 1)$, we wish to obtain the UMP level α test for $H_0 : \sigma = \sigma_0$ versus $H_1 : \sigma < \sigma_0$ where σ_0 is a positive number. Now, fix a value $\sigma_1(< \sigma_0)$ and then verify that the MP level α test between σ_0 and arbitrarily chosen σ_1 will have the following form:

$$\text{Reject } H_0 \text{ if and only if } \Sigma_{i=1}^n X_i^2 < \sigma_0^2 \chi_{n,1-\alpha}^2, \qquad (8.4.2)$$

where $\chi_{n,1-\alpha}^2$ is the lower $100\alpha\%$ point of the χ_n^2 distribution. See, for example, the Figure 8.4.1. Obviously this test does not depend on the specific choice of $\sigma_1(< \sigma_0)$. Hence, the test given by (8.4.2) is UMP level α for testing $H_0 : \sigma = \sigma_0$ versus $H_1 : \sigma < \sigma_0$. ▲

Example 8.4.3 (Example 8.3.6 Continued) Suppose that $X_1, ..., X_n$ are iid Bernoulli(p) where $p \in (0,1)$ is the unknown parameter. With preassigned $\alpha \in (0,1)$, we wish to obtain the UMP level α test for $H_0 : p = p_0$ versus $H_1 : p > p_0$ where p_0 is a number from the interval $(0,1)$. Now, fix a value $p_1(> p_0)$ and then from (8.3.17) recall that the test function associated with the MP level α test between p_0 and arbitrarily chosen p_1 will have the following form:

$$\psi(\mathbf{X}) = \begin{cases} 1 & \text{if } \Sigma_{i=1}^n X_i > k \\ \gamma & \text{if } \Sigma_{i=1}^n X_i = k \\ 0 & \text{if } \Sigma_{i=1}^n X_i < k, \end{cases} \qquad (8.4.3)$$

where k, a positive integer, and $\gamma \in (0,1)$ are chosen so that the test described by (8.4.3) has size α. Obviously the test does not depend on the specific choice of $p_1(> p_0)$. Hence, the test given by (8.4.3) is UMP level α for testing $H_0 : p = p_0$ versus $H_1 : p > p_0$. ▲

One can similarly argue that the MP level α tests derived in the Examples 8.3.2-8.3.5 are also the UMP level α tests for the same simple null hypothesis versus the corresponding upper- or lower-sided composite alternative hypothesis. Verifications of such claims are left out as exercises.

The p-value of a test:

In tests of hypothesis, one often reports the final result of the test, that is whether one accepts H_0 or H_1 at a chosen level α. There is, however, another way to report the final result in practice using the notion called the *p-value*. We emphasize that the *p-value* is a fully data driven quantification of "error" and its interpretation is sometimes difficult.

Consider the Example 8.4.1. Suppose that we have the observed value $\overline{x}$ for the sample mean and we have $z_{calc} = \sqrt{n}(\overline{x} - \mu_0)/\sigma$, the calculated value of the *test statistic*, $\sqrt{n}(\overline{X} - \mu_0)/\sigma$.

Now, we may ask the following question:

> What is the probability of observing a data which is similar to the one on hand or *more extreme*, if H_0 happens to be true? This probability will be called the *p-value*.

Since the alternative hypothesis is $\mu > \mu_0$, here the phrase *more extreme* is interpreted as "$\overline{X} > \overline{x}$" and hence the *p-value* is given by

$$P_{\mu_0}\{\sqrt{n}(\overline{X} - \mu_0)/\sigma > z_{calc}\} = P\{Z > z_{calc}\},$$

which can be evaluated using the standard normal table.

If the alternative hypothesis is $\mu < \mu_0$, then the phrase *more extreme* will be interpreted as "$\overline{X} < \overline{x}$" and hence the *p-value* is going to be

$$P_{\mu_0}\{\sqrt{n}(\overline{X} - \mu_0)/\sigma < z_{calc}\} = P\{Z < z_{calc}\}.$$

This probability can again be evaluated using the standard normal table.

A test with a "small" *p-value* indicates that the null hypothesis is less plausible than the alternative hypothesis and in that case H_0 is rejected.

> A test would reject the null hypothesis H_0 at the chosen level α if and only if the associated *p-value* $< \alpha$.

Example 8.4.4 (Example 8.4.1 Continued) Suppose that $X_1, ..., X_{15}$ are iid $N(\mu, \sigma^2)$ with unknown $\mu \in \Re$ but $\sigma^2 = 9$. We wish to test $H_0 : \mu = 3.1$ versus $H_1 : \mu > 3.1$ at the 5% level. Now, suppose that the observed data gave $\overline{x} = 5.3$ so that $z_{calc} = \sqrt{15}(5.3 - 3.1)/3 \approx 2.8402$. Here, we have $z_\alpha = 1.645$ and hence according to the rule given by (8.4.1), we will reject H_0 at the 5% level since $z_{calc} > z_\alpha$.

On the other hand, the associated *p-value* is given by $P\{Z > 2.8402\} \approx 2.2543 \times 10^{-3}$. If we were told that the *p-value* was 2.2543×10^{-3} in the first place, then we would have immediately rejected H_0 at any level $\alpha > 2.2543 \times 10^{-3}$. ▲

8.4.2 *Monotone Likelihood Ratio Property*

We now define the monotone likelihood ratio (MLR) property for a family of pmf or pdf denoted by $f(x; \theta), \theta \in \Theta \subseteq \Re$. In the next subsection, we exploit this property to derive the UMP level α tests for one-sided null against one-sided alternative hypotheses in some situations. Suppose that $X_1, ..., X_n$ are iid with a common distribution $f(x; \theta)$ and as before let us continue to write $\mathbf{X} = (X_1, ..., X_n), \mathbf{x} = (x_1, ..., x_n)$.

Definition 8.4.1 *A family of distributions* $\{f(x; \theta): \theta \in \Theta\}$ *is said to have the monotone likelihood ratio (MLR) property in a real valued statistic* $T = T(\mathbf{X})$ *provided that the following holds: for all* $\{\theta^*, \theta\} \subset \Theta$ *and* $\mathbf{x} \in \mathcal{X}$, *we have*

$$\frac{L(\mathbf{x}; \theta^*)}{L(\mathbf{x}; \theta)} \text{ is non-decreasing in } T(\mathbf{x}) \text{ whenever } \theta^* > \theta. \qquad (8.4.4)$$

Remark 8.4.1 In this definition, all we need is that $L(\mathbf{x}; \theta^*)/L(\mathbf{x}; \theta)$ should be *non-increasing or non-decreasing in* $T(\mathbf{x})$ *whenever* $\theta^* > \theta$. If the likelihood ratio $L(\mathbf{x}; \theta^*)/L(\mathbf{x}; \theta)$ is non-increasing instead of being non-decreasing, then we shall see later that its primary effect would be felt in placing the rejection region $\mathcal{R}$ in the upper or lower tail of the distribution of the test statistic under H_0. One should realize that the statistic T would invariably be *sufficient* for the unknown parameter θ.

Example 8.4.5 Suppose that $X_1, ..., X_n$ are iid $N(\mu, \sigma^2)$ with unknown $\mu \in \Re$, but assume that $\sigma \in \Re^+$ is known. Consider arbitrary real numbers $\mu, \mu^*(> \mu)$, and then with $T(\mathbf{x}) = \Sigma_{i=1}^n x_i$, let us write

$$L(\mathbf{x}; \mu^*)/L(\mathbf{x}; \mu) = exp\{[(\mu^* - \mu)T(\mathbf{x})/\sigma^2] + [n(\mu^2 - \mu^{*2})/(2\sigma^2)]\},$$

which is increasing in T. Then, we have the MLR property in T. ▲

Example 8.4.6 Suppose that $X_1, ..., X_n$ are iid with the common pdf $f(x; b) = \frac{1}{b}e^{-x/b}I(x > 0)$ with unknown $b \in \Re^+$. Consider arbitrary real numbers $b, b^*(> b)$, and with $T(\mathbf{x}) = \Sigma_{i=1}^n x_i$, let us write

$$L(\mathbf{x}; b^*)/L(\mathbf{x}; b) = (b/b^*)^n exp\{(b^* - b)T(\mathbf{x})/(bb^*)\},$$

which is increasing in T. Then, we have the MLR property in T. ▲

Example 8.4.7 Suppose that $X_1, ..., X_n$ are iid Uniform$(0, \theta)$ with unknown $\theta \in \Re^+$. Consider arbitrary real numbers $\theta, \theta^*(> \theta)$, and with $T(\mathbf{x}) = x_{n:n}$, the largest order statistic, let us write

$$\frac{L(\mathbf{x}; \theta^*)}{L(\mathbf{x}; \theta)} = \left(\frac{\theta}{\theta^*}\right)^n \frac{I(0 < T(\mathbf{x}) < \theta^*)}{I(0 < T(\mathbf{x}) < \theta)},$$

which is non-decreasing in $T(\mathbf{x})$. Then, we have the MLR property in T. ▲

Next let us reconsider the real valued sufficient statistic $T = T(\mathbf{X})$ for the unknown parameter θ, and suppose that its family of pmf or pdf is given by $\{g(t; \theta): \theta \in \Theta \subseteq \Re\}$. Here the domain space for T is indicated by $t \in \mathcal{T} \subseteq \Re$. Suppose that $g(t; \theta)$ belongs to the *one-parameter exponential family*, given by the Definition 3.8.1. That is, we can express

$$g(t; \theta) = a(\theta)c(t)e^{tb(\theta)}, \theta \in \Theta, t \in \mathcal{T}, \tag{8.4.5}$$

where $b(\theta)$ is an increasing function of θ. Then, the family $\{g(t; \theta): \theta \in \Theta \subseteq \Re\}$ has the MLR property in T. Here, $a(\theta)$ and $b(\theta)$ can not involve t whereas $c(t)$ can not involve θ. In many distributions involving only a single real valued unknown parameter θ, including the binomial, Poisson, normal, and gamma, the family of pmf or pdf of the associated sufficient statistic T would be MLR in T. We leave out the verifications as exercises.

Some families of distributions may not enjoy the MLR property. The Exercise 8.4.6 gives an example.

8.4.3 UMP Test via MLR Property

The following useful result is due to Karlin and Rubin (1956). We simply state the result without giving its proof.

Theorem 8.4.1 (Karlin-Rubin Theorem) *Suppose that we wish to test $H_0 : \theta = \theta_0$ versus $H_1 : \theta > \theta_0$. Consider a real valued sufficient statistic $T = T(\mathbf{X})$ for $\theta \in \Theta(\subseteq \Re)$. Suppose that the family $\{g(t; \theta) : \theta \in \Theta\}$ of the pdf's induced by T has the MLR (non-decreasing) property. Then, the test function*

$$\psi(\mathbf{X}) = \begin{cases} 1 & \text{if } T(\mathbf{X}) > k \\ 0 & \text{if } T(\mathbf{X}) < k \end{cases} \qquad (8.4.6)$$

corresponds to the UMP level α test if k is so chosen that $E_{\theta_0}[\psi(\mathbf{X})] = \alpha$.

If the null hypothesis in the Karlin-Rubin Theorem is replaced by $H_0 : \theta \le \theta_0$, the test given by (8.4.6) continues to be UMP level α.

Remark 8.4.2 Suppose that the null hypothesis in Theorem 8.4.1 is replaced by $H_0 : \theta \le \theta_0$. Recall that the level of the test is defined by $\text{Sup}_{\theta \le \theta_0} P_\theta\{T(\mathbf{X}) > k\} = \alpha$. One may, however, note that the maximum Type I error probability is attained at the boundary point $\theta = \theta_0$ because of the MLR (non-decreasing) property in T. If the distribution of T happens to be discrete, then the theorem continues to hold but it will become necessary to *randomize* on the set $\{T = k\}$ as we did in the Examples 8.3.6-8.3.7. In the Karlin-Rubin Theorem, if the family $\{f(x; \theta) : \theta \in \Theta\}$ has instead the MLR *non-increasing property in* $T(\mathbf{X})$, then the UMP level α test function $\psi(\mathbf{X})$ would instead be 1 or 0 according as $T(\mathbf{X}) < k$ or $T(\mathbf{X}) > k$ respectively.

Example 8.4.8 (Example 8.4.1 Continued) Suppose that $X_1, ..., X_n$ are iid $N(\mu, \sigma^2)$ with unknown $\mu \in \Re$, but assume that $\sigma \in \Re^+$ is known. Let $T(\mathbf{X}) = \Sigma_{i=1}^n X_i$ which is sufficient for μ and its pdf has the MLR increasing property in T. One may use the representation in (8.4.5) and the remark about the one-parameter exponential family. Now, we wish to test the null hypothesis $H_0 : \mu \le \mu_0$ versus $H_1 : \mu > \mu_0$ with level α where μ_0 is a fixed real number. In view of the Karlin-Rubin Theorem and Remark 8.4.2, the UMP level α test will look like this:

$$\psi(\mathbf{x}) = \begin{cases} 1 & \text{if } T(\mathbf{x}) > k \\ 0 & \text{if } T(\mathbf{x}) \leq k, \end{cases}$$

or equivalently it can be written as

$$\psi(\mathbf{x}) = \begin{cases} 1 & \text{if } \sqrt{n}(\overline{X} - \mu_0)/\sigma > z_\alpha \\ 0 & \text{if } \sqrt{n}(\overline{X} - \mu_0)/\sigma \leq z_\alpha. \end{cases}$$

One will note that the Type I error probability at the boundary point $\mu = \mu_0$ in the null space is exactly α. One may also check directly that the same for any other $\mu < \mu_0$ is smaller than α as follows: Writing Z for a standard normal variable, for $\mu < \mu_0$, we get

$$\begin{aligned} P_\mu\{\text{Rejecting } H_0\} &= P_\mu\{\sqrt{n}(\overline{X} - \mu)/\sigma > z_\alpha + \sqrt{n}(\mu_0 - \mu)\sigma^{-1}\} \\ &= P\{Z > z_\alpha + \sqrt{n}(\mu_0 - \mu)\sigma^{-1}\}. \end{aligned}$$

Since $z_\alpha + \sqrt{n}(\mu_0 - \mu)\sigma^{-1} > z_\alpha$, we can now conclude that $P\{Z > z_\alpha + \sqrt{n}(\mu_0 - \mu)\sigma^{-1}\} < \alpha$. ▲

Example 8.4.9 Suppose that $X_1, ..., X_n$ are iid $N(\mu, \sigma^2)$ with known $\mu \in \Re$, but assume that $\sigma \in \Re^+$ is unknown. Let $T(\mathbf{x}) = \Sigma_{i=1}^n (x_i - \mu)^2$ and observe that $\sigma^{-2} T(\mathbf{X})$ is distributed as the χ_n^2 random variable. Thus, the pdf of T, which is a sufficient statistic for σ, has the MLR increasing property in T. One may use (8.4.5) and the remark about the one-parameter exponential family. We wish to test the null hypothesis $H_0 : \sigma \leq \sigma_0$ versus $H_1 : \sigma > \sigma_0$ with level α where σ_0 is a fixed positive real number. In view of the Karlin-Rubin Theorem and Remark 8.4.2, the UMP level α test will look like this:

$$\psi(\mathbf{x}) = \begin{cases} 1 & \text{if } T(\mathbf{x}) > k \\ 0 & \text{if } T(\mathbf{x}) \leq k, \end{cases}$$

or equivalently it can be written as

$$\psi(\mathbf{X}) = \begin{cases} 1 & \text{if } \Sigma_{i=1}^n (X_i - \mu)^2/\sigma_0^2 > \chi_{n,\alpha}^2 \\ 0 & \text{if } \Sigma_{i=1}^n (X_i - \mu)^2/\sigma_0^2 \leq \chi_{n,\alpha}^2, \end{cases}$$

where recall that $\chi_{n,\alpha}^2$ is the upper $100\alpha\%$ point of the χ_n^2 distribution. See, for example, the Figure 8.3.4. One will note that the Type I error probability at the boundary point $\sigma = \sigma_0$ in the null space is exactly α. One may also check directly that the same for any other $\sigma < \sigma_0$ is smaller than α as follows: For $\sigma < \sigma_0$, we get

$$\begin{aligned} P_\sigma\{\text{Rejecting } H_0\} &= P_\sigma\{\Sigma_{i=1}^n (X_i - \mu)^2/\sigma^2 > \chi_{n,\alpha}^2(\sigma_0/\sigma)^2\} \\ &= P\{\chi_n^2 > \chi_{n,\alpha}^2(\sigma_0/\sigma)^2\}. \end{aligned}$$

Since $\chi^2_{n,\alpha}(\sigma_0/\sigma)^2 > \chi^2_{n,\alpha}$, we claim that $P\{\chi^2_n > \chi^2_{n,\alpha}(\sigma_0/\sigma)^2\} < \alpha$. ▲

Example 8.4.10 Suppose that $X_1, ..., X_n$ are iid with the common pdf $f(x; \lambda) = \lambda e^{-\lambda x}I(x > 0)$ with unknown $\lambda \in \Re^+$. Let us consider $T(\mathbf{X}) = \Sigma_{i=1}^n X_i$ which is sufficient for λ and its pdf has the MLR *decreasing* property in T. We wish to test the null hypothesis $H_0 : \lambda \leq \lambda_0$ versus $H_1 : \lambda > \lambda_0$ with level α where λ_0 is a fixed positive real number. In view of the Karlin-Rubin Theorem and Remark 8.4.2, the UMP level α test will look like this:

$$\psi(\mathbf{x}) = \begin{cases} 1 & \text{if } T(\mathbf{x}) < k \\ 0 & \text{if } T(\mathbf{x}) \geq k, \end{cases}$$

or equivalently it can be written as

$$\psi(\mathbf{X}) = \begin{cases} 1 & \text{if } 2\Sigma_{i=1}^n X_i/\lambda_0 < \chi^2_{2n,1-\alpha} \\ 0 & \text{if } 2\Sigma_{i=1}^n X_i/\lambda_0 \geq \chi^2_{2n,1-\alpha}, \end{cases}$$

where recall that $\chi^2_{2n,1-\alpha}$ is the lower $100\alpha\%$ point of the χ^2_{2n} distribution. See, for example, the Figure 8.4.1. One will note that the Type I error probability at the boundary point $\lambda = \lambda_0$ in the null space is exactly α. One may also check directly, as in the two previous examples, that the same for any other $\lambda < \lambda_0$ is smaller than α. Note that for small values of $\Sigma_{i=1}^n X_i$, we are rejecting the null hypothesis $H_0 : \lambda \leq \lambda_0$ in favor of the alternative hypothesis $H_1 : \lambda > \lambda_0$. This is due to the fact that we have the MLR decreasing property in T instead of the increasing property. Also, observe that $E_\lambda(X) = \lambda^{-1}$ and thus "small" values of $\Sigma_{i=1}^n X_i$ would be associated with "large" values of λ which should lead to the rejection of H_0. Recall Remark 8.4.2 in this context. ▲

Example 8.4.11 Suppose that $X_1, ..., X_n$ are iid Uniform$(0, \theta)$ with unknown $\theta \in \Re^+$. Let us consider $T(\mathbf{X}) = X_{n:n}$ which is sufficient for θ and its pdf has the MLR increasing property in T. We wish to test the null hypothesis $H_0 : \theta \leq \theta_0$ versus $H_1 : \theta > \theta_0$ with level α where θ_0 is a fixed positive real number. In view of the Karlin-Rubin Theorem and Remark 8.4.2, the UMP level α test will look like this:

$$\psi(\mathbf{x}) = \begin{cases} 1 & \text{if } T(\mathbf{x}) > k \\ 0 & \text{if } T(\mathbf{x}) \leq k. \end{cases}$$

Let us now choose k so that the Type I error probability at the boundary point $\theta = \theta_0$ in the null space is exactly α. The pdf of T is given by $nt^{n-1}\theta^{-n}I(0 < t < \theta)$ when θ is the true value. So, we proceed as follows.

$$\alpha = P_{\theta_0}\{T > k\} = \int_k^{\theta_0} nt^{n-1}\theta_0^{-n}dt = 1 - [k/\theta_0]^n,$$

so that $k = (1 - \alpha)^{1/n}\theta_0$. One may also check directly that the Type I error probability for any other $\theta < \theta_0$ is smaller than α. ▲

> Sample size determination is important in practice when we wish to control both Type I and II error probabilities. See Exercise 8.4.25.

8.5 Simple Null Versus Two-Sided Alternative Hypotheses

Consider testing a simple null hypothesis $H_0 : \theta = \theta_0$ versus a two-sided alternative hypothesis $H_1 : \theta \neq \theta_0$ where θ_0 is a fixed value in the parameter space Θ. Will there exist a UMP level α test? The answer is "yes" in some situations and "no" in some others. We refrain from going into general discussions of what may or may not happen when the alternative hypothesis is two-sided.

In the case of the $N(\mu, \sigma^2)$ distribution with μ unknown but σ known, a UMP level α test fails to exist for deciding between a simple null hypothesis $H_0 : \mu = \mu_0$ and a two-sided alternative hypothesis $H_1 : \mu \neq \mu_0$. On the other hand, for the Uniform$(0, \theta)$ distribution with θ unknown, a UMP level α test exists for deciding between a simple null hypothesis $H_0 : \theta = \theta_0$ and a two-sided alternative hypothesis $H_1 : \theta \neq \theta_0$. In the next two subsections, we provide the details.

8.5.1 An Example Where UMP Test Does Not Exist

Suppose that $X_1, ..., X_n$ are iid $N(\mu, \sigma^2)$ with unknown $\mu \in \Re$ but known $\sigma \in \Re^+$. Consider the statistic $T(\mathbf{X}) = \Sigma_{i=1}^n X_i$ which is sufficient for μ. We wish to test the simple null hypothesis $H_0 : \mu = \mu_0$ against the two-sided alternative $H_1 : \mu \neq \mu_0$ with level α where μ_0 is a fixed real number. We wish to show that there exists no UMP level α test in this situation.

Suppose that there is a UMP level α test and let its test function be denoted by $\psi^*(\mathbf{X})$. Observe that $\psi^*(\mathbf{X})$ is then a UMP level α test for deciding between $H_0 : \mu = \mu_0$ versus $H_1' : \mu > \mu_0$. In the Example 8.4.1, however, the UMP level α for deciding between H_0 versus H_1' was written as

$$\psi_1(\mathbf{X}) = \begin{cases} 1 & \text{if } \sqrt{n}(\overline{X} - \mu_0)/\sigma > z_\alpha \\ 0 & \text{if } \sqrt{n}(\overline{X} - \mu_0)/\sigma \leq z_\alpha, \end{cases} \tag{8.5.1}$$

where z_α is the upper $100\alpha\%$ point of the standard normal distribution. See, for example, the Figure 8.3.1. The two test functions ψ^* and ψ_1 must coincide on the sets where ψ_1 is zero or one. One can similarly show that the test function corresponding to the UMP level α test for deciding between $H_0 : \mu = \mu_0$ versus $H_1'' : \mu < \mu_0$ can be written as

$$\psi_2(\mathbf{X}) = \begin{cases} 1 & \text{if } \sqrt{n}(\overline{X} - \mu_0)/\sigma < -z_\alpha \\ 0 & \text{if } \sqrt{n}(\overline{X} - \mu_0)/\sigma \geq -z_\alpha. \end{cases} \qquad (8.5.2)$$

But, ψ^* is also a UMP level α test for deciding between H_0 versus H_1''. Again, the two test functions ψ^* and ψ_2 must also then coincide whenever ψ_2 is zero or one. Now, we argue as follows.

Suppose that the observed data $\mathbf{x}$ is such that the test statistic's calculated value $\sqrt{n}(\overline{x} - \mu_0)/\sigma$ does not exceed $-z_\alpha$. Then, for such $\mathbf{x}$, we must have $\psi_1(\mathbf{x}) = 0, \psi_2(\mathbf{x}) = 1$. That is, on the part of the sample space where $\sqrt{n}(\overline{x} - \mu_0)/\sigma \leq -z_\alpha$, the test function $\psi^*(\mathbf{x})$ will fail to coincide with both $\psi_1(\mathbf{x}), \psi_2(\mathbf{x})$. So, we have arrived at a contradiction. In other words, there is no UMP level α test for deciding between $H_0 : \mu = \mu_0$ against $H_1 : \mu \neq \mu_0$.

8.5.2   An Example Where UMP Test Exists

Suppose that $X_1, ..., X_n$ are iid Uniform$(0, \theta)$ with unknown $\theta \in \Re^+$. Consider $T(\mathbf{X}) = X_{n:n}$, the largest order statistic, which is sufficient for θ. We wish to test the simple null hypothesis $H_0 : \theta = \theta_0$ against the two-sided alternative hypothesis $H_1 : \theta \neq \theta_0$ with level α where θ_0 is a fixed positive number. We will show that there exists a UMP level α test in this situation. This is one of many celebrated exercises from Lehmann (1986, p. 111).

As a follow-up of the earlier Example 8.4.11, let us first show that any test function $\psi^*(\mathbf{X})$ such that

$$E_{\theta_0}[\psi^*(\mathbf{X})] = \alpha, E_\theta[\psi^*(\mathbf{X})] \leq \alpha \text{ for } \theta \leq \theta_0, \text{ and}$$
$$\psi^*(\mathbf{X}) = 1 \text{ when } T(\mathbf{X}) > \theta_0 \qquad (8.5.3)$$

corresponds to a UMP level α test for deciding between $H_0' : \theta \leq \theta_0$ and $H_1' : \theta > \theta_0$.

Using the MLR property, from the Example 8.4.11, one may write down the test function of the UMP level α test for H_0' versus H_1' as follows:

$$\psi_0^*(\mathbf{X}) = \begin{cases} 1 & \text{if } T(\mathbf{X}) > \theta_0(1-\alpha)^{1/n} \\ 0 & \text{if } T(\mathbf{X}) \leq \theta_0(1-\alpha)^{1/n}. \end{cases} \qquad (8.5.4)$$

Now, for any $\theta > \theta_0$, the power function associated with the test function

ψ_0^* is given by

$$E_\theta[\psi_0^*(\mathbf{X})] = P_\theta\{T(\mathbf{X}) > \theta_0(1-\alpha)^{1/n}\} \quad = \int_{\theta_0(1-\alpha)^{1/n}}^\theta nt^{n-1}\theta^{-n}dt$$
$$= 1 - (1-\alpha)(\theta_0/\theta)^n.$$
$$(8.5.5)$$

Of course, the test function ψ^* corresponds to a level α test for H_0' versus H_1'. Let us now compute the power function associated with ψ^* and show that its power coincides with the expression given in (8.5.5). Let us write $g(t) = E[\psi^*(\mathbf{X}) \mid T(\mathbf{X}) = t]$ which can not depend upon θ because T is sufficient for θ and thus we have

$$\alpha = E_{\theta_0}[\psi^*(\mathbf{X})] = E_{\theta_0}[E\{\psi^*(\mathbf{X}) \mid T\}] = E_{\theta_0}[g(T)].$$

Now, $\psi^*(\mathbf{x}) = 1$ for any $t > \theta_0$ and hence $g(t) = 1$ if $t > \theta_0$. Thus, for any $\theta > \theta_0$, we can write

$$E_\theta[\psi^*(\mathbf{X})] = \int_0^{\theta_0} g(t)nt^{n-1}\theta^{-n}dt + \int_{\theta_0}^\theta g(t)nt^{n-1}\theta^{-n}dt$$
$$= (\theta_0/\theta)^n \int_0^{\theta_0} g(t)nt^{n-1}\theta_0^{-n}dt + \int_{\theta_0}^\theta nt^{n-1}\theta^{-n}dt,$$
$$\text{since } g(t) = 1 \text{ when } t > \theta_0 \qquad (8.5.6)$$
$$= (\theta_0/\theta)^n\alpha + \{1 - (\theta_0/\theta)^n\}$$
$$= 1 - (1-\alpha)(\theta_0/\theta)^n.$$

This last expression is the same as in (8.5.5). So, any test function ψ^* defined via (8.5.3) is UMP level α for testing H_0' versus H_1'. Now we are in position to state and prove the following result.

Theorem 8.5.1 *In the Uniform$(0, \theta)$ case, for testing the simple null hypothesis $H_0 : \theta = \theta_0$ against the two-sided alternative $H_1 : \theta \neq \theta_0$ where θ_0 is a fixed positive number, the test associated with*

$$\psi(\mathbf{X}) = \begin{cases} 1 & \text{if } T(\mathbf{X}) \geq \theta_0 \text{ or } T(\mathbf{X}) \leq \theta_0\alpha^{1/n} \\ 0 & \text{otherwise} \end{cases}$$

is UMP level α.

Proof First note that since $P_{\theta_0}\{T(\mathbf{X}) \geq \theta_0\} = 0$, one obviously has

$$E_{\theta_0}[\psi(\mathbf{X})] = P_{\theta_0}\{T(\mathbf{X}) \leq \theta_0\alpha^{1/n}\} = \int_0^{\theta_0\alpha^{1/n}} nt^{n-1}\theta_0^{-n}dt = \alpha.$$
$$(8.5.7)$$

Along the lines of the Example 8.4.11, we can easily show that a UMP level α test for deciding between $H_0 : \theta = \theta_0$ and $H_1'' : \theta < \theta_0$ would have its test function as follows:

$$\psi_0^{**}(\mathbf{X}) = \begin{cases} 1 & \text{if } T(\mathbf{X}) \leq \theta_0\alpha^{1/n} \\ 0 & \text{if } T(\mathbf{X}) > \theta_0\alpha^{1/n}. \end{cases}$$
$$(8.5.8)$$

One can use the test function $\psi(\mathbf{X})$ for testing H_0 versus $H_1^{''} : \theta < \theta_0$ too. In view of (8.5.7), we realize that $\psi(\mathbf{X})$ has level α and so for $\theta < \theta_0$ the two test functions $\psi(\mathbf{X}), \psi_0^{**}(\mathbf{X})$ must coincide. That is, $\psi(\mathbf{X})$ corresponds to a UMP level α test for H_0 versus $H_1^{''} : \theta < \theta_0$. Next, by comparing the form of $\psi(\mathbf{X})$ described in the statement of the theorem with $\psi^*(\mathbf{X})$ given in (8.5.3), we can claim that $\psi(\mathbf{X})$ also corresponds to a UMP level α test for H_0 versus $H_1^{''} : \theta > \theta_0$. Hence the result follows. ∎

8.5.3 Unbiased and UMP Unbiased Tests

It should be clear that one would prefer a UMP level α test to decide between two hypotheses, but the problem is that there may not exist such a test. Since, the whole class of level α tests may be very large in the first place, one may be tempted to look at a sub-class of competing tests and then derive the best test in that class. There are, however, different ways to consider smaller classes of tests. We restrict our attention to the *unbiased class of tests*.

Definition 8.5.1 *Consider testing $H_0 : \theta \in \Theta_0$ versus $H_1 : \theta \in \Theta_1$ where Θ_0, Θ_1 are subsets of Θ and Θ_0, Θ_1 are disjoint. A level α test is called unbiased if $Q(\theta)$, the power of the test, that is the probability of rejecting H_0, is at least α whenever $\theta \in \Theta_1$.*

Thus, a test with its test function $\psi(\mathbf{X})$ would be called unbiased level α provided the following two conditions hold:

$$
\begin{aligned}
&(i) \quad E_\theta\{\psi(\mathbf{X})\} \leq \alpha \text{ for all } \theta \in \Theta_0 \\
&(ii) \quad E_\theta\{\psi(\mathbf{X})\} \geq \alpha \text{ for all } \theta \in \Theta_1.
\end{aligned}
\qquad (8.5.9)
$$

The properties listed in (8.5.9) state that an unbiased test rejects the null hypothesis more frequently when the alternative hypothesis is true than in a situation when the null hypothesis is true. This is a fairly minimal demand on any reasonable test used in practice.

Next, one may set out to locate the UMP level α test within the class of level α unbiased tests. Such a test is called the *uniformly most powerful unbiased (UMPU)* level α test. An elaborate theory of unbiased tests was given by Neyman as well as Neyman and Pearson in a series of papers. The readers will find a wealth of information in Lehmann (1986).

When testing a simple null hypothesis $H_0 : \theta = \theta_0$ versus a simple alternative $H_1 : \theta = \theta_1$, the Neyman-Pearson Lemma came up with the MP level α test implemented by the test function $\psi(\mathbf{x})$ defined by (8.3.2). Now, consider another test function $\psi^*(\mathbf{x}) \equiv \alpha$, that is this randomized test rejects H_0 with probability α whatever be the observed data $\mathbf{x} \in \mathcal{X}^n$.

Note that ψ^* corresponds to a level α test and hence the power associated with ψ must be at least as large as the power associated with ψ^* since ψ is MP level α. Thus,

$$E_{\theta_1}\{\psi(\mathbf{X})\} \geq E_{\theta_1}\{\psi^*(\mathbf{X})\} = E_{\theta_1}\{\alpha\} = \alpha.$$

In other words, the MP level α test arising from the Neyman-Pearson Lemma is in fact unbiased.

Remark 8.5.1 In the Section 8.5.1, we had shown that there was no UMP level α test for testing a simple null against the two-sided alternative hypotheses about the unknown mean of a normal population with its variance known. But, there is a good test available for this problem. In Chapter 11, we will derive the *likelihood ratio test* for the same hypotheses testing problem. It would be shown that the level α likelihood ratio test would

$$\text{reject } H_0 \text{ if and only if } \sqrt{n}\,|\,\overline{X} - \mu_0\,|\,/\sigma > z_{\alpha/2}. \tag{8.5.10}$$

This coincides with the customary two-tailed test which is routinely used in practice. It will take quite some effort to verify that the test given by (8.5.10) is indeed the UMPU level α test. This derivation is out of scope at the level of this book. The readers nonetheless should be aware of this result.

8.6 Exercises and Complements

8.2.1 Suppose that X_1, X_2, X_3, X_4 are iid random variables from the $N(\theta, 4)$ population where $\theta(\in \Re)$ is the unknown parameter. We wish to test $H_0 : \theta = 2$ versus $H_1 : \theta = 5$. Consider the following tests:

Test #1: Reject H_0 if and only if $X_1 > 4.7$;

Test #2: Reject H_0 if and only if $\frac{1}{3}(X_1 + 2X_2) > 4.5$;

Test #3: Reject H_0 if and only if $\frac{1}{2}(X_1 + X_3) > 4.2$;

Test #4: Reject H_0 if and only if $\overline{X} > 4.1$.

Find Type I and Type II error probabilities for each test and compare the tests.

8.2.2 Suppose that X_1, X_2, X_3, X_4 are iid random variables from a population with the exponential distribution having unknown mean $\theta(\in \Re^+)$. We wish to test $H_0 : \theta = 4$ versus $H_1 : \theta = 8$. Consider the following tests:

Test #1: Reject H_0 if and only if $X_1 > 7$;

Test #2: Reject H_0 if and only if $\frac{1}{2}(X_1 + X_2) > 6.5$;

Test #3: Reject H_0 if and only if $\frac{1}{3}(X_1 + X_2 + X_3) > 6$;

Test #4: Reject H_0 if and only if $\overline{X} > 6$.

Find Type I and Type II error probabilities for each test and compare the tests.

8.2.3 Suppose that X_1, X_2, X_3, X_4 are iid random variables from the $N(\theta, 4)$ population where $\theta(\in \Re)$ is the unknown parameter. We wish to test $H_0 : \theta = 3$ versus $H_1 : \theta = 1$. Consider the following tests:

Test #1: Reject H_0 if and only if $X_3 < 2.7$;

Test #2: Reject H_0 if and only if $\frac{1}{3}(X_1 + 2X_2) < 2.3$;

Test #3: Reject H_0 if and only if $\frac{1}{2}(X_1 + X_3) < 2.3$;

Test #4: Reject H_0 if and only if $\overline{X} < 2$.

Find Type I and Type II error probabilities for each test and compare the tests.

8.2.4 Suppose that X_1, X_2, X_3, X_4 are iid random variables from a population with the exponential distribution having unknown mean $\theta(\in \Re^+)$. We wish to test $H_0 : \theta = 6$ versus $H_1 : \theta = 2$. Consider the following possible tests:

Test #1: Reject H_0 if and only if $X_1 < 4$;

Test #2: Reject H_0 if and only if $\frac{1}{2}(X_1 + X_2) < 3.5$;

Test #3: Reject H_0 if and only if $\frac{1}{3}(X_1 + X_2 + X_3) < 3.4$;

Test #4: Reject H_0 if and only if $\overline{X} < 2.8$.

Find Type I and Type II error probabilities for each test and compare the tests.

8.2.5 Suppose that X_1, X_2 are iid with the common pdf

$$f(x; \theta) = \begin{cases} \theta x^{\theta-1} & \text{if } 0 < x < 1 \\ 0 & \text{elsewhere} \end{cases}$$

where $\theta(> 0)$ is the unknown parameter. In order to test the null hypothesis $H_0 : \theta = 1$ against the alternative hypothesis $H_1 : \theta = 2$, we propose the critical region $\mathcal{R} = \{(x_1, x_2) \in \Re^2 : x_1 x_2 \geq \frac{3}{4}\}$.

    (i)   Show that the level $\alpha = \frac{1}{4} + \frac{3}{4} log(\frac{3}{4})$;

    (ii)   Show that power at $\theta = 2$ is $\frac{7}{16} + \frac{9}{8} log(\frac{3}{4})$.

{*Hints*: Observe that $\alpha = P_{\theta=1}(\mathcal{R}) = P_{\theta=1}\{X_1 X_2 \geq \frac{3}{4}\}$ which is written as $\int_{x_2=\frac{3}{4}}^{1} \left\{ \int_{x_1=\frac{3}{4}x_2^{-1}}^{1} dx_1 \right\} dx_2 = \int_{x_2=\frac{3}{4}}^{1} \{1 - \frac{3}{4}x_2^{-1}\} dx_2$. Similarly, power

$$= P_{\theta=2}(\mathcal{R}) = P_{\theta=2}\{X_1X_2 \geq \tfrac{3}{4}\} = 4\int_{x_2=\frac{3}{4}}^{1} x_2 \left\{\int_{x_1=\frac{3}{4}x_2^{-1}}^{1} x_1 dx_1\right\} dx_2 =$$
$$2\int_{x_2=\frac{3}{4}}^{1} x_2 \left\{1 - \tfrac{9}{16}x_2^{-2}\right\} dx_2.\}$$

8.2.6 Suppose that $X_1, ..., X_{10}$ are iid $N(\theta, 4)$ where $\theta(\in \mathfrak{R})$ is the unknown parameter. In order to test $H_0 : \theta = -1$ against $H_1 : \theta = 1$, we propose the critical region $\mathcal{R} = \{\mathbf{X} \in \mathfrak{R}^{10} : \Sigma_{i=1}^{10} iX_i > 0\}$. Find the level α and evaluate the power at $\theta = 1$.

8.3.1 Suppose that $X_1, ..., X_n$ are iid random variables from the $N(\mu, \sigma^2)$ population where μ is assumed known but σ is unknown, $\mu \in \mathfrak{R}, \sigma \in \mathfrak{R}^+$. We fix a number $\alpha \in (0, 1)$ and two positive numbers σ_0, σ_1.

(i)   Derive the MP level α test for $H_0 : \sigma = \sigma_0$ versus $H_1 : \sigma = \sigma_1$ $(> \sigma_0)$ in the simplest implementable form;

(ii) Derive the MP level α test for $H_0 : \sigma = \sigma_0$ versus $H_1 : \sigma = \sigma_1$ $(< \sigma_0)$ in the simplest implementable form;

In each part, draw the power function.

8.3.2 (Example 8.3.4 Continued) Suppose that $X_1, ..., X_n$ are iid having the Uniform$(0, \theta)$ distribution with unknown $\theta(> 0)$. With preassigned $\alpha \in (0, 1)$ and two positive numbers $\theta_1 < \theta_0$, derive the MP level α test for $H_0 : \theta = \theta_0$ versus $H_1 : \theta = \theta_1$ in the simplest implementable form. Perform the power calculations.

8.3.3 (Example 8.3.5 Continued) Suppose that $X_1, ..., X_n$ are iid with the common pdf $b^{-\delta}[\Gamma(\delta)]^{-1}x^{\delta-1}exp(-x/b)$, with *two unknown* parameters $(\delta, b) \in \mathfrak{R}^{+2}$. With preassigned $\alpha \in (0, 1)$, derive the MP level α test, in the simplest implementable form, for $H_0 : (b = b_0, \delta = \delta^*)$ versus $H_1 : (b = b_1, \delta = \delta^*)$ where $b_1 < b_0$ are two positive numbers and δ^* is a positive number.

8.3.4 Suppose that $X_1, ..., X_n$ are iid random variables from the $N(\mu, \sigma^2)$ population where μ is unknown but σ is assumed known, $\mu \in \mathfrak{R}, \sigma \in \mathfrak{R}^+$. In order to choose between the two hypotheses $H_0 : \mu = \mu_0$ versus $H_1 : \mu = \mu_1(> \mu_0)$, suppose that we reject H_0 if and only if $\overline{X} > c$ where c is a fixed number. Is there any $\alpha \in (0, 1)$ for which this particular test is MP level α?

8.3.5 (Exercise 8.3.2 Continued) Suppose that $X_1, ..., X_n$ are iid having the Uniform$(0, \theta)$ distribution with unknown $\theta(> 0)$. In order to choose between the two hypotheses $H_0 : \theta = \theta_0$ versus $H_1 : \theta = \theta_1$ where $\theta_1 < \theta_0$ are two positive numbers, suppose that we reject H_0 if and only if $X_{n:n} < c$ where c is a fixed positive number. Is there any $\alpha \in (0, 1)$ for which this test is MP level α?

8.3.6 (Example 8.3.6 Continued) Suppose that $X_1, ..., X_n$ are iid random variables having the Bernoulli(p) distribution where $p \in (0, 1)$ is the unknown parameter. With preassigned $\alpha \in (0,1)$, derive the randomized MP level α test for $H_0 : p = p_0$ versus $H_1 : p = p_1$ where $p_1 < p_0$ are two numbers from $(0,1)$. Explicitly find k and γ numerically when $n = 10, 15, 25, \alpha = .05, .10$ and $p_0 = .1, .5, .7$.

8.3.7 (Example 8.3.7 Continued) Suppose that $X_1, ..., X_n$ are iid random variables having the Poisson(λ) distribution where $\lambda \in \Re^+$ is the unknown parameter. With preassigned $\alpha \in (0,1)$, derive the randomized MP level α test for $H_0 : \lambda = \lambda_0$ versus $H_1 : \lambda = \lambda_1$ where $\lambda_1 < \lambda_0$ are two positive numbers. Explicitly find k and γ numerically when $n = 10, 15, 25, \alpha = .05, .10$ and $\lambda_0 = 1, 2$.

8.3.8 Suppose that X is an observable random variable with its pdf given by $f(x), x \in \Re$. Consider the two functions defined as follows:

$$f_0(x) = \begin{cases} 1 & \text{if } 0 < x < 1 \\ 0 & \text{elsewhere,} \end{cases} \qquad f_1(x) = \begin{cases} 3x^2 & \text{if } 0 < x < 1 \\ 0 & \text{elsewhere.} \end{cases}$$

Determine the MP level α test for

$$H_0 : f(x) = f_0(x) \text{ versus } H_1 : f(x) = f_1(x)$$

in the simplest implementable form. Perform the power calculations. {*Hint*: Follow the Example 8.3.8.}

8.3.9 Suppose that X_1, X_2, X_3 are observable iid random variables with the common pdf given by $f(x), x \in \Re$. Consider the two functions defined as follows:

$$f_0(x) = \begin{cases} 3x^2 & \text{if } 0 < x < 1 \\ 0 & \text{elsewhere,} \end{cases} \qquad f_1(x) = \begin{cases} 2x & \text{if } 0 < x < 1 \\ 0 & \text{elsewhere.} \end{cases}$$

Determine the MP level α test for

$$H_0 : f(x) = f_0(x) \text{ versus } H_1 : f(x) = f_1(x)$$

in the simplest implementable form. Perform the power calculations. {*Hint*: Follow the Example 8.3.9.}

8.3.10 Suppose that X is an observable random variable with its pdf given by $f(x), x \in \Re$. Consider the two functions defined as follows:

$$f_0(x) = \tfrac{1}{\pi}(1 + x^2)^{-1}, \quad f_1(x) = \tfrac{1}{2}exp\{-|x|\}.$$

Show that the MP level α test for

$$H_0 : f(x) = f_0(x) \text{ versus } H_1 : f(x) = f_1(x)$$

will reject H_0 if and only if $|X| < k$. Determine k as a function of α. Perform the power calculations. {*Hint*: Follow the Example 8.3.10.}

8.3.11 Suppose that $X_1, ..., X_n$ are iid having Uniform$(-\theta, \theta)$ distribution with the unknown parameter $\theta(> 0)$. In order to choose between the two hypotheses $H_0 : \theta = \theta_0$ versus $H_1 : \theta = \theta_1(> \theta_0)$ with two positive numbers θ_0, θ_1, suppose that we reject H_0 if and only if $|X_{n:n}| > c$ where c is a fixed positive number. Is there any $\alpha \in (0, 1)$ for which this test is MP level α? {*Hint*: Write the likelihood function and look at the minimal sufficient statistic for θ.}

8.3.12 Suppose that $X_1, ..., X_n$ are iid Geometric(p) where $p \in (0, 1)$ is the unknown parameter. With preassigned $\alpha \in (0, 1)$, derive the randomized MP level α test for $H_0 : p = p_0$ versus $H_1 : p = p_1(> p_0)$ where p_0, p_1 are two numbers from $(0, 1)$. Explicitly find k and γ numerically when $n = 4, 5, 6, \alpha = .05$ and $p_0 = .1, .5, .7$.

8.3.13 Let $X_1, ..., X_n$ be iid having the Rayleigh distribution with the common pdf $f(x; \theta) = 2\theta^{-1} x exp(-x^2/\theta) I(x > 0)$ where $\theta(> 0)$ is the unknown parameter. With preassigned $\alpha \in (0, 1)$, derive the MP level α test for $H_0 : \theta = \theta_0$ versus $H_1 : \theta = \theta_1(< \theta_0)$ where θ_0, θ_1 are two positive numbers, in the simplest implementable form.

8.3.14 Let $X_1, ..., X_n$ be iid having the Weibull distribution with the common pdf $f(x; a) = a^{-1} b x^{b-1} exp(-x^b/a) I(x > 0)$ where $a(> 0)$ is an unknown parameter but $b(> 0)$ is assumed known. With preassigned $\alpha \in (0, 1)$, derive the MP level α test for $H_0 : a = a_0$ versus $H_1 : a = a_1(> a_0)$ where a_0, a_1 are two positive numbers, in the simplest implementable form.

8.3.15 (Example 8.3.11 Continued) Evaluate the power of the MP level α test described by (8.3.32).

8.3.16 (Example 8.3.11 Continued) Suppose that X_1 and X_2 are independent random variables respectively distributed as $N(\mu, \sigma^2), N(3\mu, 2\sigma^2)$ where $\mu \in \Re$ is the unknown parameter and $\sigma \in \Re^+$ is assumed known. Derive the MP level α test for $H_0 : \mu = \mu_0$ versus $H_1 : \mu = \mu_1(< \mu_0)$. Evaluate the power of the test. {*Hint*: Write the likelihood function along the line of (8.3.31) and proceed accordingly.}

8.3.17 (Example 8.3.12 Continued) Evaluate the power of the MP level α test described by (8.3.34).

8.3.18 (Example 8.3.12 Continued) Let us denote $\mathbf{X}' = (X_1, X_2)$ where $\mathbf{X}$ is assumed to have the bivariate normal distribution, $N_2(\mu, \mu, 1, 1, \frac{1}{\sqrt{2}})$. Here, we consider $\mu(\in \Re)$ as the unknown parameter. Derive the MP level α test for $H_0 : \mu = \mu_0$ versus $H_1 : \mu = \mu_1(< \mu_0)$. Evaluate the power of the test. {*Hint*: Write the likelihood function along the line of (8.3.33) and proceed accordingly.}

8.3.19 Let $X_1, ..., X_n$ be iid positive random variables having the common lognormal pdf $f(x; \mu) = [x\sigma\sqrt{2\pi}]^{-1} \, exp\{-[log(x) - \mu]^2/(2\sigma^2)\} I(x > 0)$ with $-\infty < \mu < \infty, 0 < \sigma < \infty$. Here, μ is the only unknown parameter.

(i)   Find the minimal sufficient statistic for μ;

(ii)   Given $\alpha \in (0, 1)$, find the MP level α test for deciding between the null hypothesis $H_0 : \mu = \mu_0$ and the alternative hypothesis $H_1 : \mu = \mu_1 (> \mu_0)$ where μ_0, μ_1 are fixed real numbers.

8.3.20 Suppose that X is an observable random variable with its pdf given by $f(x), x \in \Re$. Consider the two functions defined as follows:

$$f_0(x) = \frac{1}{\sqrt{2\pi}} exp\{-x^2/2\}, \quad f_1(x) = \frac{2}{\Gamma(1/4)} exp\{-x^4\}.$$

Derive the MP level α test for

$$H_0 : f(x) = f_0(x) \text{ versus } H_1 : f(x) = f_1(x).$$

Perform the power calculations. {*Hint*: Follow the Example 8.3.10.}

8.3.21 (Exercise 8.3.18 Continued) Let us denote $\mathbf{X}' = (X_1, X_2)$ where $\mathbf{X}$ is assumed to have the bivariate normal distribution, $N_2(\mu, 2\mu, 1, 1, \frac{1}{\sqrt{2}})$. Here, we consider $\mu(\in \Re)$ as the unknown parameter. Derive the MP level α test for $H_0 : \mu = \mu_0$ versus $H_1 : \mu = \mu_1 (< \mu_0)$. Evaluate the power of the test. {*Hint*: Write down the likelihood function along the line of (8.3.33) and proceed accordingly.}

8.4.1 Let $X_1, ..., X_n$ be iid having the common Laplace pdf $f(x; b) = \frac{1}{2}b^{-1} exp(-|x - a|/b)I(x \in \Re)$ where $b(> 0)$ is an unknown parameter but $a(\in \Re)$ is assumed known. Show that the family of distributions has the MLR increasing property in T, the sufficient statistic for b.

8.4.2 Let $X_1, ..., X_n$ be iid having the Weibull distribution with the common pdf $f(x; a) = a^{-1}bx^{b-1} exp(-x^b/a)I(x > 0)$ where $a(> 0)$ is an unknown parameter but $b(> 0)$ is assumed known. Show that the family of distributions has the MLR increasing property in T, the sufficient statistic for a.

8.4.3 Let $X_1, ..., X_n$ be iid random variables with the Poisson(λ) distribution where $\lambda \in \Re^+$ is the unknown parameter. Show that the family of distributions has the MLR increasing property in T, the sufficient statistic for λ.

8.4.4 Suppose that $X_1, ..., X_n$ are iid having the Uniform$(-\theta, \theta)$ distribution with unknown $\theta(> 0)$. Show that the family of distributions has the MLR increasing property in $T = |X_{n:n}|$, the sufficient statistic for θ.

8.4.5 Let $X_1, ..., X_n$ be iid having the common negative exponential pdf $f(x; \mu) = \sigma^{-1} exp\{-(x - \mu)/\sigma\}I(x > \mu)$ where $\mu(\in \Re)$ is an unknown parameter but $\sigma(\in \Re^+)$ is assumed known. Show that the family of distributions has the MLR increasing property in $T = X_{n:1}$, the sufficient statistic for μ.

8.4.6 Let $X_1, ..., X_n$ be iid having the common Cauchy pdf $f(x; \theta) = \frac{1}{\pi}\{1 + (x - \theta)^2\}^{-1}I(x \in \Re)$ where $\theta(\in \Re)$ is an unknown parameter. Show that this family of distributions *does not* enjoy the MLR property. {*Hint*: Note that for any $\theta^* > \theta$, the likelihood ratio $L(\mathbf{x}; \theta^*)/L(\mathbf{x}; \theta) \to 1$ as x_i's converge to $\pm\infty$.}

8.4.7 Suppose that $X_1, ..., X_n$ are iid random variables from the $N(\mu, \sigma^2)$ population where μ is assumed known but σ is unknown, $\mu \in \Re, \sigma \in \Re^+$. We have fixed numbers $\alpha \in (0, 1)$ and $\sigma_0(> 0)$. Derive the UMP level α test for $H_0 : \sigma > \sigma_0$ versus $H_1 : \sigma \le \sigma_0$ in the simplest implementable form.

(i)   Use the Neyman-Pearson approach;

(ii)   Use the MLR approach.

8.4.8 Suppose that $X_1, ..., X_n$ are iid having the Uniform$(0, \theta)$ distribution with unknown $\theta(> 0)$. With preassigned $\alpha \in (0, 1)$, derive the UMP level α test for $H_0 : \theta > \theta_0$ versus $H_1 : \theta \le \theta_0$ where θ_0 is a positive number, in the simplest implementable form.

(i)   Use the Neyman-Pearson approach;

(ii)   Use the MLR approach.

8.4.9 Let $X_1, ..., X_n$ be iid with the common gamma pdf $f(x; \delta, b) = b^{-\delta}[\Gamma(\delta)]^{-1}x^{\delta-1}exp(-x/b)I(x > 0)$ with *two unknown* parameters $(\delta, b) \in \Re^+ \times \Re^+$. With preassigned $\alpha \in (0, 1)$, derive the UMP level α test, in the simplest implementable form, for $H_0 : (b < b_0, \delta = \delta^*)$ versus $H_1 : (b \ge b_0, \delta = \delta^*)$ where b_0 is a positive number and δ^* is also positive number.

(i)   Use the Neyman-Pearson approach;

(ii)   Use the MLR approach.

8.4.10 Suppose that $X_1, ..., X_n$ are iid random variables from a $N(\mu, \sigma^2)$ population where μ is unknown but σ is assumed known, $\mu \in \Re, \sigma \in \Re^+$. In order to choose between the two hypotheses $H_0 : \mu < \mu_0$ versus $H_1 : \mu \ge \mu_0$, suppose that we reject H_0 if and only if $\overline{X} > c$ where c is a fixed number. Is there any $\alpha \in (0, 1)$ for which this test is UMP level α?

8.4.11 Suppose that $X_1, ..., X_n$ are iid having the Uniform$(0, \theta)$ distribution with unknown $\theta(> 0)$. In order to choose between the two hypotheses $H_0 : \theta > \theta_0$ versus $H_1 : \theta \le \theta_0$ where θ_0 is a positive number, suppose that we reject H_0 if and only if $X_{n:n} < c$ where c is a fixed positive number. Is there any $\alpha \in (0, 1)$ for which this test is UMP level α?

8.4.12 Suppose that $X_1, ..., X_n$ are iid random variables having the Bernoulli(p) distribution where $p \in (0, 1)$ is the unknown parameter. With preassigned $\alpha \in (0, 1)$, derive the randomized UMP level α test for $H_0 : p = p_0$ versus $H_1 : p > p_0$ where p_0 is a number between 0 and 1.

8.4.13 Suppose that $X_1, ..., X_n$ are iid random variables having the Poisson(λ) distribution where $\lambda \in \Re^+$ is the unknown parameter. With preassigned $\alpha \in (0, 1)$, derive the randomized UMP level α test for $H_0 : \lambda = \lambda_0$ versus $H_1 : \lambda < \lambda_0$ where λ_0 is a positive number.

8.4.14 Suppose that $X_1, ..., X_n$ are iid having the Uniform($-\theta, \theta$) distribution with unknown $\theta (> 0)$. In order to choose between the two hypotheses $H_0 : \theta \leq \theta_0$ versus $H_1 : \theta > \theta_0$ where θ_0 is a positive number, suppose that we reject H_0 if and only if $|X_{n:n}| > c$ where c is a fixed known positive number. Is there any $\alpha \in (0, 1)$ for which this test is UMP level α?

8.4.15 Suppose that $X_1, ..., X_n$ are iid Geometric(p) where $p \in (0, 1)$ is the unknown parameter. With preassigned $\alpha \in (0, 1)$, derive the randomized UMP level α test for $H_0 : p \geq p_0$ versus $H_1 : p < p_0$ where p_0 is a number between 0 and 1.

8.4.16 Let $X_1, ..., X_n$ be iid having the Rayleigh distribution with the common pdf $f(x; \theta) = 2\theta^{-1} x \exp(-x^2/\theta) I(x > 0)$ where $\theta (> 0)$ is the unknown parameter. With preassigned $\alpha \in (0, 1)$, derive the UMP level α test for $H_0 : \theta \leq \theta_0$ versus $H_1 : \theta > \theta_0$ where θ_0 is a positive number, in the simplest implementable form.

(i)   Use the Neyman-Pearson approach;

(ii)   Use the MLR approach.

8.4.17 Let $X_1, ..., X_n$ be iid having the Weibull distribution with the common pdf $f(x; a) = a^{-1} b x^{b-1} \exp(-x^b/a) I(x > 0)$ where $a (> 0)$ is an unknown parameter but $b (> 0)$ is assumed known. With preassigned $\alpha \in (0, 1)$, derive the UMP level α test for $H_0 : a \leq a_0$ versus $H_1 : a > a_0$ where a_0 is a positive number, in the simplest implementable form.

(i)   Use the Neyman-Pearson approach;

(ii)   Use the MLR approach.

8.4.18 (Exercise 8.4.5 Continued) Let $X_1, ..., X_n$ be iid having the common negative exponential pdf $f(x; \mu) = \sigma^{-1} \exp\{-(x - \mu)/\sigma\} I(x > \mu)$ where $\mu (\in \Re)$ is an unknown parameter but $\sigma (\in \Re^+)$ is assumed known. With preassigned $\alpha \in (0, 1)$, derive the UMP level α test for $H_0 : \mu \leq \mu_0$ versus $H_1 : \mu > \mu_0$ where μ_0 is a real number, in the simplest implementable form.

8.4.19 (Exercise 8.4.1 Continued) Let $X_1, ..., X_n$ be iid having the common Laplace pdf $f(x; b) = \frac{1}{2} b^{-1} \exp(-|x - a|/b) I(x \in \Re)$ where $b (> 0)$ is an unknown parameter but $a (\in \Re)$ is assumed known. With preassigned

$\alpha \in (0,1)$, derive the UMP level α test for $H_0 : b \leq b_0$ versus $H_1 : b > b_0$ where b_0 is a positive real number, in the simplest implementable form.

8.4.20 (Exercise 8.3.19 Continued) Suppose that $X_1, ..., X_n$ are iid positive random variables having the common lognormal pdf

$$f(x; \mu) = [x\sigma\sqrt{2\pi}]^{-1} exp\left\{-[log(x) - \mu]^2/(2\sigma^2)\right\}$$

with $x > 0, -\infty < \mu < \infty, 0 < \sigma < \infty$. Here, μ is the only unknown parameter. Given the value $\alpha \in (0,1)$, derive the UMP level α test for deciding between the null hypothesis $H_0 : \mu = \mu_0$ and the alternative hypothesis $H_1 : \mu > \mu_0$ where μ_0 is a fixed real number. Describe the test in its simplest implementable form.

8.4.21 (Exercise 8.4.20 Continued) Let us denote the lognormal pdf

$$f(w; \mu) = [w\sigma\sqrt{2\pi}]^{-1} exp\left\{-[log(w) - \mu]^2/(2\sigma^2)\right\}$$

with $w > 0, -\infty < \mu < \infty, 0 < \sigma < \infty$. Suppose that $X_1, ..., X_m$ are iid positive random variables having the common pdf $f(x; \mu, 2)$, $Y_1, ..., Y_n$ are iid positive random variables having the common pdf $f(y; 2\mu, 3)$, and also that the X's and Y's are independent. Here, μ is the unknown parameter and $m \neq n$. Describe the tests in their simplest implementable forms.

(i)   Find the minimal sufficient statistic for μ;

(ii)   Given $\alpha \in (0,1)$, find the MP level α test to choose between the null hypothesis $H_0 : \mu = \mu_0$ and the alternative hypothesis $H_1 : \mu = \mu_1(> \mu_0)$ where μ_0, μ_1 are fixed real numbers;

(iii)   Given $\alpha \in (0,1)$, find the UMP level α test to choose between the null hypothesis $H_0 : \mu = 1$ and the alternative hypothesis $H_a : \mu > 1$.

{*Hints*: In part (i), use the Lehmann-Scheffé Theorems from Chapter 6 to claim that $T = \frac{1}{4}\Sigma_{i=1}^m log(X_i) + \frac{2}{9}\Sigma_{j=1}^n log(Y_j)$ is the minimal sufficient statistic. In parts (ii)-(iii), use the likelihood functions along the lines of the Examples 8.3.11-8.3.12 and show that one will reject H_0 if and only if $\{T - (\frac{1}{4}m + \frac{2}{9}n)\}/\sqrt{(\frac{1}{4}m + \frac{4}{9}n)} > z_\alpha$.}

8.4.22 Denote the gamma pdf $f(w; \mu) = [\mu^a \Gamma(a)]^{-1} exp\{-w/\mu\} w^{a-1}$ with $w > 0, 0 < \mu < \infty, 0 < a < \infty$. Suppose that $X_1, ..., X_m$ are iid positive random variables having the common pdf $f(x; \mu, 1)$, $Y_1, ..., Y_n$ are iid positive random variables having the common pdf $f(y; \mu, 3)$, and also that the X's and Y's are independent. Here, μ is the only unknown parameter and $m \neq n$.

(i)   Find the minimal sufficient statistic for μ;

(ii)   Given $\alpha \in (0, 1)$, find the MP level α test to choose between the null hypothesis $H_0 : \mu = \mu_0$ and the alternative hypothesis $H_1 : \mu = \mu_1(> \mu_0)$ where μ_0, μ_1 are fixed real numbers;

(iii) Given $\alpha \in (0, 1)$, find the UMP level α test to choose between the null hypothesis $H_0 : \mu = 1$ and the alternative hypothesis $H_a : \mu > 1$.

{*Hints*: In part (i), use the Lehmann-Scheffé Theorems from Chapter 6 to claim that $T = \Sigma_{i=1}^m X_i + \Sigma_{j=1}^n Y_j$ is the minimal sufficient statistic. In parts (ii)-(iii), use the likelihood functions along the lines of the Examples 8.3.11-8.3.12 and show that one will reject H_0 if and only if $T > k$.}

8.4.23 Let X_1, X_2 be iid with the common pdf $\theta^{-1}\exp(-x/\theta)I(x > 0)$ where $\theta > 0$. In order to test $H_0 : \theta = 2$ against $H_1 : \theta > 2$, we propose to use the critical region $\mathcal{R} = \{\mathbf{X} \in \Re^{+2} : X_1 + X_2 > 9.5\}$. Evaluate the level α and find the power function. Find the Type II error probabilities when $\theta = 4, 4.5$.

8.4.24 Let $X_1, ..., X_n$ be iid random variables having the common pdf $f(x; \theta) = \theta^{-1}x^{(1-\theta)/\theta}I(0 < x < 1)$ with $0 < \theta < \infty$. Here, θ is the unknown parameter. Given the value $\alpha \in (0, 1)$, derive the UMP level α test in its simplest implementable form, if it exists, for deciding between the null hypothesis $H_0 : \theta \leq \theta_0$ and the alternative hypothesis $H_1 : \theta > \theta_0$ where θ_0 is a fixed positive number.

8.4.25 (Sample Size Determination) Let $X_1, ..., X_n$ be iid $N(\mu, \sigma^2)$ where μ is assumed unknown but σ is known, $\mu \in \Re, \sigma \in \Re^+$. We have shown that given $\alpha \in (0, 1)$, the UMP level α test for $H_0 : \mu = \mu_0$ versus $H_1 : \mu > \mu_0$ rejects H_0 if and only if $\sqrt{n}(\overline{X} - \mu_0)/\sigma > z_\alpha$. The UMP test makes sure that it has the *minimum* possible Type II error probability at $\mu = \mu_1(> \mu_0)$ among all level α tests, but there is *no* guarantee that this minimum Type II error probability at $\mu = \mu_1$ will be "small" unless n is appropriately determined.

Suppose that we require the UMP test to have Type II error probability $\leq \beta \in (0, 1)$ for some specified value $\mu = \mu_1(> \mu_0)$. Show that the sample size n must be the smallest integer $\geq \{(z_\alpha + z_\beta)\sigma/(\mu_1 - \mu_0)\}^2$.

8.5.1 Let $X_1, ..., X_n$ be iid having the common negative exponential pdf $f(x; \theta) = b^{-1}exp\{-(x - \theta)/b\}I(x > \theta)$ where $b(> 0)$ is assumed known and $\theta(\in \Re)$ is the unknown parameter. With preassigned $\alpha \in (0, 1)$, derive the UMP level α test for $H_0 : \theta = \theta_0$ versus $H_1 : \theta \neq \theta_0$ where θ_0 is a fixed number, in the simplest implementable form. {*Hint*: Observe that $Y_i = exp\{-X_i/b\}, i = 1, ..., n$, are iid Uniform$(0, \delta)$ where the parameter $\delta = exp\{-\theta/b\}$. Now, exploit the UMP test from the Section 8.5.2.}

8.5.2 (Exercise 8.4.21 Continued) Denote the lognormal pdf $f(w; \mu) = [w\sigma\sqrt{2\pi}]^{-1} exp\left\{-[log(w) - \mu]^2/(2\sigma^2)\right\}$ with $w > 0, -\infty < \mu < \infty, 0 < \sigma < \infty$. Suppose that $X_1, ..., X_m$ are iid positive random variables having the common pdf $f(x; \mu, 2)$, $Y_1, ..., Y_n$ are iid positive random variables having the common pdf $f(y; 2\mu, 3)$, and also that the X's and Y's are independent. Here, μ is the only unknown parameter and $m \neq n$. Argue whether there does or does not exist a UMP level α test for deciding between the null hypothesis $H_0 : \mu = 1$ and the alternative hypothesis $H_a : \mu \neq 1$. {*Hint:* Try to exploit arguments similar to those used in the Section 8.5.1.}

8.5.3 Let X_1, X_2, X_3, X_4 be iid Uniform$(0, \theta)$ where $\theta > 0$. In order to test $H_0 : \theta = 1$ against $H_1 : \theta \neq 1$, suppose that we propose to use the critical region $\mathcal{R} = \{\mathbf{X} \in \Re^{+4} : X_{4:4} < \frac{1}{2}$ or $X_{4:4} > 1\}$. Evaluate the level α and the power function.

8.5.4 Let $X_1, ..., X_n$ be iid having the common exponential pdf $f(x; \theta) = \theta^{-1} exp\{-x/\theta\}I(x > 0)$ where $\theta(> 0)$ is assumed unknown. With preassigned $\alpha \in (0, 1)$, show that no UMP level α test for $H_0 : \theta = \theta_0$ versus $H_1 : \theta \neq \theta_0$ exists where θ_0 is a fixed positive number. {*Hint:* Try to exploit arguments similar to those used in the Section 8.5.1.}

8.5.5 Let $X_1, ..., X_n$ be iid $N(0, \sigma^2)$ with unknown $\sigma \in \Re^+$. With preassigned $\alpha \in (0, 1)$, we wish to test $H_0 : \sigma = \sigma_0$ versus $H_1 : \sigma \neq \sigma_0$ at the level α, where σ_0 is a fixed positive number. Show that no UMP level α test exists for testing H_0 versus H_1. {*Hint:* Try to exploit arguments similar to those used in the Section 8.5.1.}

9

Confidence Interval Estimation

9.1 Introduction

As the name *confidence interval* suggests, we will now explore methods to estimate an unknown parameter $\theta \in \Theta \subseteq \Re$ with the help of an interval. That is, we would construct two *statistics* $T_L(\mathbf{X}), T_U(\mathbf{X})$ based on the data $\mathbf{X}$ and propose the interval $(T_L(\mathbf{X}), T_U(\mathbf{X}))$ as the final estimator of θ. Sometimes the lower bound $T_L(\mathbf{X})$ may coincide with $-\infty$ and in that case the associated interval $(-\infty, T_U(\mathbf{X}))$ will be called an *upper confidence interval* for θ. On the other hand, sometimes the upper bound $T_U(\mathbf{X})$ may coincide with ∞ and in that case the associated interval $(T_L(\mathbf{X}), \infty)$ will be called a *lower confidence interval* for θ.

We may add that the concepts of the "fiducial distribution" and "fiducial intervals" originated with Fisher (1930) which led to persistent and substantial philosophical arguments. Among others, J. Neyman came down hard on Fisher on philosophical grounds and proceeded to give the foundation of the theory of confidence intervals. This culminated in Neyman's (1935b,1937) two path-breaking papers. After 1937, neither Neyman nor Fisher swayed from their respective philosophical stance. However, in the 1961 article, *Silver jubilee of my dispute with Fisher*, Neyman was a little kinder to Fisher in his exposition. It may not be entirely out of place to note that Fisher died on July 29, 1962. The articles of Buehler (1980), Lane (1980) and Wallace (1980) gave important perspectives of fiducial probability and distribution.

> We prefer to have both the lower and upper end points of the confidence interval J, namely $T_L(\mathbf{X})$ and $T_U(\mathbf{X})$, depend exclusively on the (minimal) sufficient statistics for θ.

Now, let us examine how we should measure the quality of a proposed confidence interval. We start with one fundamental concept defined as follows.

Definition 9.1.1 *The coverage probability associated with a confidence interval $J = (T_L(\mathbf{X}), T_U(\mathbf{X}))$ for the unknown parameter θ is measured by*

$$P_\theta \{\theta \in (T_L(\mathbf{X}), T_U(\mathbf{X}))\}. \tag{9.1.1}$$

The confidence coefficient corresponding to the confidence interval J is defined to be

$$Inf_{\theta \in \Theta} P_\theta \{\theta \in (T_L(\mathbf{X}), T_U(\mathbf{X}))\}. \tag{9.1.2}$$

However, the coverage probability $P_\theta \{\theta \in (T_L(\mathbf{X}), T_U(\mathbf{X}))\}$ will not involve the unknown parameter θ in many standard applications. In those situations, it will be easy to derive $Inf_{\theta \in \Theta} P_\theta \{\theta \in (T_L(\mathbf{X}), T_U(\mathbf{X}))\}$ because then it will coincide with the coverage probability itself. Thus, we will interchangeably use the two phrases, the *confidence coefficient* and the *coverage probability* when describing a confidence interval.

Customarily, we fix a small preassigned number $\alpha \in (0, 1)$ and require a confidence interval for θ with the confidence coefficient exactly $(1 - \alpha)$. We refer to such an interval as a $(1 - \alpha)$ or $100(1 - \alpha)\%$ *confidence interval estimator* for the unknown parameter θ.

Example 9.1.1 Suppose that X_1, X_2 are iid $N(\mu, 1)$ where $\mu(\in \Re)$ is the unknown parameter.

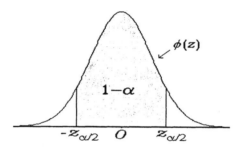

Figure 9.1.1. Standard Normal PDF: The Shaded Area Between $-z_{\alpha/2}$ and $z_{\alpha/2}$ with $z_{\alpha/2} = 1.96$ Is $1 - \alpha$ Where $\alpha = 0.05$

First consider $T_L(\mathbf{X}) = X_1 - 1.96, T_U(\mathbf{X}) = X_1 + 1.96$, leading to the confidence interval

$$J_1 = (X_1 - 1.96, X_1 + 1.96). \tag{9.1.3}$$

The associated coverage probability is given by

$$
\begin{aligned}
P_\mu \{\mu \in (X_1 - 1.96, X_1 + 1.96)\} \\
= P_\mu \{X_1 - 1.96 < \mu < X_1 + 1.96\} \\
= P\{|Z| < 1.96\}, \text{ where } Z = X_1 - \mu \text{ is distributed} \\
\text{as } N(0, 1) \text{ if } \mu \text{ is the true population mean,}
\end{aligned}
$$

which is .95 and it does not depend upon μ. So, the confidence coefficient associated with the interval J_1 is .95.

Next consider $T_L(\mathbf{X}) = \overline{X} - \frac{1.96}{\sqrt{2}}, T_U(\mathbf{X}) = \overline{X} + \frac{1.96}{\sqrt{2}}$, leading to the confidence interval $J_2 = \left(\overline{X} - \frac{1.96}{\sqrt{2}}, \overline{X} + \frac{1.96}{\sqrt{2}}\right)$. Let $Z = \sqrt{2}(\overline{X} - \mu)$ which is distributed as $N(0, 1)$ if μ is the true population mean. Now, the confidence coefficient is given by

$$P_\mu\left\{\mu \in \left(\overline{X} - \frac{1.96}{\sqrt{2}}, \overline{X} + \frac{1.96}{\sqrt{2}}\right)\right\} = P_\mu\left\{\overline{X} - \frac{1.96}{\sqrt{2}} < \mu < \overline{X} + \frac{1.96}{\sqrt{2}}\right\}$$
$$= P_\mu\left\{\left|\sqrt{2}(\overline{X} - \mu)\right| < 1.96\right\},$$

which is the same as $P\{|Z| < 1.96\} = .95$, whatever be μ. Between the two 95% confidence intervals J_1 and J_2 for μ, the interval J_2 appears superior because J_2 is shorter in length than J_1. Observe that the construction of J_2 is based on the sufficient statistic $\overline{X}$, the sample mean. ▲

In Section 9.2, we discuss some standard one-sample problems. The first approach discussed in Section 9.2.1 involves what is known as the *inversion* of a suitable test procedure. But, this approach becomes complicated particularly when two or more parameters are involved. A more flexible method is introduced in Section 9.2.2 by considering *pivotal* random variables. Next, we provide an interpretation of the confidence coefficient in Section 9.2.3. Then, in Section 9.2.4, we look into some notions of *accuracy measures* of confidence intervals. The Section 9.3 introduces a number of two-sample problems via pivotal approach. Simultaneous confidence regions are briefly addressed in Section 9.4.

It will become clear from this chapter that our focus lies in the methods of construction of *exact* confidence intervals. In discrete populations, for example binomial or Poisson, exact confidence interval procedures are often intractable. In such situations, one may derive useful approximate techniques assuming that the sample size is "large". These topics fall in the realm of large sample inferences and hence their developments are delegated to the Chapter 12.

9.2   One-Sample Problems

The first approach involves the inversion of a test procedure. Next, we provide a more flexible method using pivots which functionally depend only on the (minimal) sufficient statistics. Then, an interpretation is given for the confidence interval, followed by notions of accuracy measures associated with a confidence interval.

9.2.1 *Inversion of a Test Procedure*

In general, for testing a null hypothesis $H_0 : \theta = \theta_0$ against the alternative hypothesis $H_1 : \theta > \theta_0$ (or $H_1 : \theta < \theta_0$ or $H_1 : \theta \neq \theta_0$), we look at the subset of the sample space $\mathcal{R}^c$ which corresponds to the acceptance of H_0. In Chapter 8, we had called the subset $\mathcal{R}$ the *critical* or the *rejection region*. The subset $\mathcal{R}^c$ which corresponds to accepting H_0 may be referred to as the *acceptance region*. The construction of a confidence interval and its confidence coefficient are both closely tied in with the nature of the acceptance region $\mathcal{R}^c$ and the level of the test.

Example 9.2.1 (Example 8.4.1 Continued) Suppose that $X_1, ..., X_n$ are iid $N(\mu, \sigma^2)$ with the unknown parameter $\mu \in \Re$. We assume that $\sigma \in \Re^+$ is known. With preassigned $\alpha \in (0, 1)$, the UMP level α test for $H_0 : \mu = \mu_0$ versus $H_1 : \mu > \mu_0$ where μ_0 is a fixed real number, would be as follows:

$$\text{Reject } H_0 \text{ if and only if } \sqrt{n}(\overline{X} - \mu_0)/\sigma \geq z_\alpha \qquad (9.2.1)$$

where z_α is the upper $100\alpha\%$ point of the standard normal distribution. Refer to the Figure 9.2.1. The *acceptance region* (for H_0) then corresponds to

$$\mathcal{R}^c = \{\mathbf{x} \in \Re^n : \sqrt{n}(\overline{x} - \mu_0)/\sigma < z_\alpha\}. \qquad (9.2.2)$$

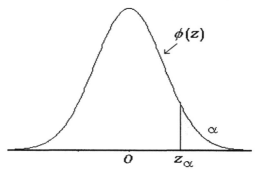

Figure 9.2.1. Standard Normal PDF: The Shaded Area
on the Right of z_α Is α

Since the test described by (9.2.2) has the level α, we can write

$$P_{\mu_0}\{\sqrt{n}(\overline{X} - \mu_0)/\sigma < z_\alpha\} \quad = 1 - P_{\mu_0}\{\sqrt{n}(\overline{X} - \mu_0)/\sigma \geq z_\alpha\}$$
$$= 1 - \text{Level of the Test}$$
$$= 1 - \alpha .$$

In other words, we can claim that

$$P_{\mu_0}\{\mu_0 > \overline{X} - z_\alpha n^{-1/2}\sigma\} = 1 - \alpha \text{ whatever be } \mu_0 \in \Re. \qquad (9.2.3)$$

Now, the equation (9.2.3) can be rewritten as

$$P_\mu\{\mu > \overline{X} - z_\alpha n^{-1/2}\sigma\} = 1 - \alpha \text{ whatever be } \mu \in \Re, \qquad (9.2.4)$$

and thus, we can claim that $(\overline{X} - z_\alpha n^{-1/2}\sigma, \infty)$ is a $100(1 - \alpha)\%$ lower confidence interval estimator for μ. ▲

The upper-sided α level test $\Rightarrow 100(1 - \alpha)\%$ lower confidence interval estimator $(T_L(\mathbf{X}), \infty)$ for θ.

The lower-sided α level test $\Rightarrow 100(1 - \alpha)\%$ upper confidence interval estimator $(-\infty, T_U(\mathbf{X}))$ for θ.

Example 9.2.2 Let $X_1, ..., X_n$ be iid with the common exponential pdf $\theta^{-1}exp\{-x/\theta\}I(x > 0)$ with the unknown parameter $\theta \in \Re^+$. With preassigned $\alpha \in (0, 1)$, the UMP level α test for $H_0 : \theta = \theta_0$ versus $H_1 : \theta > \theta_0$ where θ_0 is a fixed positive real number, would be as follows:

$$\text{Reject } H_0 \text{ if and only if } 2\Sigma_{i=1}^n X_i/\theta_0 \geq \chi^2_{2n,\alpha}, \qquad (9.2.5)$$

where $\chi^2_{2n,\alpha}$ is the upper $100\alpha\%$ point of the Chi-square distribution with $2n$ degrees of freedom. See the Figure 9.2.2.

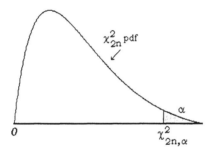

Figure 9.2.2. The Shaded Area on the Right of $\chi^2_{2n,\alpha}$ Is α

The *acceptance region* (for H_0) then corresponds to

$$\mathcal{R}^c = \{\mathbf{x} \in \Re^n : 2\Sigma_{i=1}^n X_i/\theta_0 < \chi^2_{2n,\alpha}\}. \qquad (9.2.6)$$

Since this test has level α, we can write

$$\begin{aligned} P_{\theta_0}\{2\Sigma_{i=1}^n X_i/\theta_0 < \chi^2_{2n,\alpha}\} \quad &= 1 - P_{\theta_0}\{2\Sigma_{i=1}^n X_i/\theta_0 \geq \chi^2_{2n,\alpha}\} \\ &= 1- \text{ Level of the Test} \\ &= 1 - \alpha. \end{aligned}$$

In other words, we can claim that

$$P_{\theta_0}\{\theta_0 > 2(\chi^2_{2n,\alpha})^{-1}\Sigma^n_{i=1}X_i\} = 1 - \alpha \text{ whatever be } \theta_0 \in \Re^+. \quad (9.2.7)$$

The equation (9.2.7) can be rewritten as

$$P_{\theta}\{\theta > 2(\chi^2_{2n,\alpha})^{-1}\Sigma^n_{i=1}X_i\} = 1 - \alpha \text{ whatever be } \theta \in \Re^+, \quad (9.2.8)$$

and thus we can claim that $\left(2(\chi^2_{2n,\alpha})^{-1}\Sigma^n_{i=1}X_i, \infty\right)$ is a $100(1-\alpha)\%$ lower confidence interval estimator for θ. ▲

Example 9.2.3 (Example 9.2.2 Continued) Let $X_1, ..., X_n$ be iid with the common exponential pdf $\theta^{-1}exp\{-x/\theta\}I(x > 0)$ with the unknown parameter $\theta \in \Re^+$. With preassigned $\alpha \in (0, 1)$, suppose that we invert the UMP level α test for $H_0 : \theta = \theta_0$ versus $H_1 : \theta < \theta_0$, where θ_0 is a positive real number.

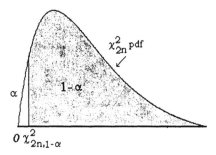

Figure 9.2.3. The Area on the Left (or Right) of $\chi^2_{2n,1-\alpha}$
Is α (or $1 - \alpha$)

Then, one arrives at $\left(0, 2(\chi^2_{2n,1-\alpha})^{-1}\Sigma^n_{i=1}X_i\right)$, a $100(1-\alpha)\%$ upper confidence interval estimator for θ, by inverting the UMP level α test. See the Figure 9.2.3. We leave out the details as an exercise. ▲

9.2.2   The Pivotal Approach

Let $X_1, ..., X_n$ be iid real valued random variables from a population with the pmf or pdf $f(x; \theta)$ for $x \in \mathcal{X}$ where $\theta(\in \Theta)$ is an unknown real valued parameter. Suppose that $T \equiv T(\mathbf{X})$ is a real valued (minimal) sufficient statistic for θ.

The family of pmf or pdf induced by the statistic T is denoted by $g(t; \theta)$ for $t \in \mathcal{T}$ and $\theta \in \Theta$. In many applications, $g(t; \theta)$ will belong to an appropriate location, scale, or location-scale family of distributions which were discussed in Section 6.5.1. We may expect the following results:

> *The Location Case*: With some $a(\theta)$, the distribution of
> $\{T - a(\theta)\}$ would not involve θ for any $\theta \in \Theta$.

> *The Scale Case*: With some $b(\theta)$, the distribution of
> $T/b(\theta)$ would not involve θ for any $\theta \in \Theta$.

> *The Location-Scale Case*: With some $a(\theta), b(\theta)$, the
> distribution of $\{T - a(\theta)\}/b(\theta)$ would not involve θ for any $\theta \in \Theta$.

Definition 9.2.1 *A pivot is a random variable* **U** *which functionally depends on both the (minimal) sufficient statistic* **T** *and* $\boldsymbol{\theta}$*, but the distribution of* **U** *does not involve* $\boldsymbol{\theta}$ *for any* $\boldsymbol{\theta} \in \Theta$.

In the location, scale, and location-scale situations, when U, T and θ are all real valued, the customary pivots are appropriate multiples of $\{T - a(\theta)\}, T/b(\theta)$ or $\{T - a(\theta)\}/b(\theta)$ respectively with suitable expressions of $a(\theta)$ and $b(\theta)$.

> We often demand that the distribution of the pivot must
> coincide with one of the standard distributions so that a
> standard statistical table can be utilized to determine the
> appropriate percentiles of the distribution.

Example 9.2.4 Let X be a random variable with its pdf $f(x; \theta) = \theta^{-1}\exp\{-x/\theta\}I(x > 0)$ where $\theta(> 0)$ is the unknown parameter. Given some $\alpha \in (0, 1)$, we wish to construct a $(1-\alpha)$ two-sided confidence interval for θ. The statistic X is minimal sufficient for θ and the pdf of X belongs to the scale family. The pdf of the pivot $U = X/\theta$ is given by $g(u) = e^{-u}I(u > 0)$. One can explicitly determine two positive numbers $a < b$ such that $P(U < a) = P(U > b) = \frac{1}{2}\alpha$ so that $P(a < U < b) = 1 - \alpha$. It can be easily checked that $a = -log(1 - \frac{1}{2}\alpha)$ and $b = -log(\frac{1}{2}\alpha)$.

> Since the distribution of U does not involve θ, we can
> determine both a and b depending exclusively upon α.

Now observe that

$$P(a < U < b) = 1 - \alpha \Rightarrow P_\theta\{\theta \in (b^{-1}X, a^{-1}X)\} = 1 - \alpha,$$

which shows that $J = (b^{-1}X, a^{-1}X)$ is a $(1 - \alpha)$ two-sided confidence interval estimator for θ. ▲

Example 9.2.5 Let $X_1, ..., X_n$ be iid Uniform$(0, \theta)$ where $\theta(> 0)$ is the unknown parameter. Given some $\alpha \in (0, 1)$, we wish to construct a $(1 - \alpha)$ two-sided confidence interval for θ. The statistic $T \equiv X_{n:n}$, the largest order statistic, is minimal sufficient for θ and the pdf of T belongs to the scale family. The pdf of the pivot $U = T/\theta$ is given by $g(u) = nu^{n-1}I(0 < u < 1)$. One can explicitly determine two numbers $0 < a < b < 1$ such that $P(U < a) = P(U > b) = \frac{1}{2}\alpha$ so that $P(a < U < b) = 1 - \alpha$. It can be easily checked that $a = (\frac{1}{2}\alpha)^{1/n}$ and $b = (1 - \frac{1}{2}\alpha)^{1/n}$. Observe that

$$P(a < U < b) = 1 - \alpha \Rightarrow P_\theta\{\theta \in (b^{-1}T, a^{-1}T)\} = 1 - \alpha,$$

which shows that $J = (b^{-1}X_{n:n}, a^{-1}X_{n:n})$ is a $(1 - \alpha)$ two-sided confidence interval estimator for θ. ▲

Example 9.2.6 Negative Exponential Location Parameter with Known Scale: Let $X_1, ..., X_n$ be iid with the common negative exponential pdf $f(x; \theta) = \sigma^{-1}exp\{-(x - \theta)/\sigma\}I(x > \theta)$. Here, $\theta \in \Re$ is the unknown parameter and we assume that $\sigma \in \Re^+$ is known. Given some $\alpha \in (0, 1)$, we wish to construct a $(1 - \alpha)$ two-sided confidence interval for θ. The statistic $T \equiv X_{n:1}$, the smallest order statistic, is minimal sufficient for θ and the pdf of T belongs to the location family. The pdf of the pivot $U = n(T - \theta)/\sigma$ is given by $g(u) = e^{-u}I(u > 0)$. One can explicitly determine a positive number b such that $P(U > b) = \alpha$ so that we will then have $P(0 < U < b) = 1 - \alpha$. It can be easily checked that $b = -log(\alpha)$. Observe that

$$P(0 < U < b) = 1 - \alpha \Rightarrow P_\theta\{\theta \in (T - bn^{-1}\sigma, T)\} = 1 - \alpha,$$

which shows that $J = (X_{n:1} - bn^{-1}\sigma, X_{n:1})$ is a $(1 - \alpha)$ two-sided confidence interval estimator for θ. ▲

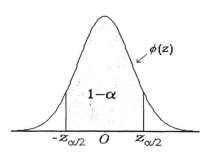

Figure 9.2.4. Standard Normal PDF: The Area on the
Right (or Left) of $z_{\alpha/2}$ (or $-z_{\alpha/2}$) Is $\alpha/2$

Example 9.2.7 Normal Mean with Known Variance: Let $X_1, ..., X_n$ be iid $N(\mu, \sigma^2)$ with the unknown parameter $\mu \in \Re$. We assume that

$\sigma \in \Re^{+}$ is known. Given some $\alpha \in (0,1)$, we wish to construct a $(1-\alpha)$ two-sided confidence interval for μ. The statistic $T \equiv \overline{X}$, the sample mean, is minimal sufficient for μ and T has the $N(\mu, \frac{1}{n}\sigma^2)$ distribution which belongs to the location family. The pivot $U = \sqrt{n}(T-\mu)/\sigma$ has the standard normal distribution. See the Figure 9.2.4. We have $P\{-z_{\alpha/2} < U < z_{\alpha/2}\} = 1 - \alpha$ which implies that

$$P_\mu \left\{ T - z_{\alpha/2} n^{-1/2}\sigma < \mu < T + z_{\alpha/2} n^{-1/2}\sigma \right\} = 1 - \alpha.$$

In other words,

$$\left(\overline{X} - z_{\alpha/2} n^{-1/2}\sigma, \overline{X} + z_{\alpha/2} n^{-1/2}\sigma \right) \qquad (9.2.9)$$

is a $(1 - \alpha)$ two-sided confidence interval estimator for μ. ▲

Example 9.2.8 Normal Mean with Unknown Variance: Suppose that $X_1, ..., X_n$ are iid $N(\mu, \sigma^2)$ with both unknown parameters $\mu \in \Re$ and $\sigma \in \Re^{+}, n \geq 2$. Given some $\alpha \in (0,1)$, we wish to construct a $(1 - \alpha)$ two-sided confidence interval for μ. Let $\overline{X}$ be the sample mean and $S^2 = (n-1)^{-1}\Sigma_{i=1}^{n}(X_i - \overline{X})^2$ be the sample variance. The statistic $T \equiv (\overline{X}, S)$ is minimal sufficient for (μ, σ). Here, the distributions of the X's belong to the location-scale family. The pivot $U = \sqrt{n}(\overline{X} - \mu)/S$ has the Student's t distribution with $(n - 1)$ degrees of freedom. So, we can say that $P\{-t_{n-1,\alpha/2} < U < t_{n-1,\alpha/2}\} = 1 - \alpha$ where $t_{n-1,\alpha/2}$ is the upper $100(1 - \frac{1}{2}\alpha)\%$ point of the Student's t distribution with $(n - 1)$ degrees of freedom. See the Figure 9.2.5.

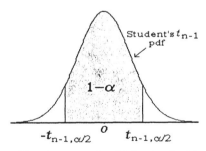

Figure 9.2.5. The Area on the Right (or Left) of $t_{n-1,\alpha/2}$ (or $-t_{n-1,\alpha/2}$) Is $\alpha/2$

Thus, we claim that

$$P_{\mu,\sigma} \left\{ \overline{X} - t_{n-1,\alpha/2} n^{-1/2}S < \mu < \overline{X} + t_{n-1,\alpha/2} n^{-1/2}S \right\} = 1 - \alpha.$$

In other words,

$$\left(\overline{X} - t_{n-1,\alpha/2}n^{-1/2}S, \overline{X} + t_{n-1,\alpha/2}n^{-1/2}S\right) \qquad (9.2.10)$$

is a $(1 - \alpha)$ two-sided confidence interval estimator for μ. ▲

Example 9.2.9 Normal Variance: Suppose that $X_1, ..., X_n$ are iid $N(\mu, \sigma^2)$ with both unknown parameters $\mu \in \Re$ and $\sigma \in \Re^+, n \geq 2$. Given some $\alpha \in (0, 1)$, we wish to construct a $(1-\alpha)$ two-sided confidence interval for σ^2. Let $\overline{X}$ be the sample mean and $S^2 = (n-1)^{-1}\Sigma_{i=1}^{n}(X_i - \overline{X})^2$ be the sample variance. The statistic $T \equiv (\overline{X}, S)$ is minimal sufficient for (μ, σ). Here, the distributions of the X's belong to the location-scale family. The pivot $U = (n-1)S^2/\sigma^2$ has the Chi-square distribution with $(n-1)$ degrees of freedom. Recall that $\chi^2_{\nu,\gamma}$ is the upper $100\gamma\%$ point of the Chi-square distribution with ν degrees of freedom.

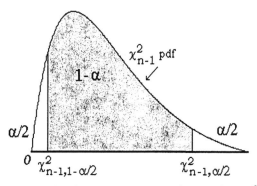

Figure 9.2.6. The Area on the Right (or Left) of $\chi^2_{n-1,\alpha/2}$
(or $\chi^2_{n-1,1-\alpha/2}$) Is $\alpha/2$

The distribution of S^2 belongs to the scale family. So, we can claim that $P\{\chi^2_{n-1,1-\alpha/2} < U < \chi^2_{n-1,\alpha/2}\} = 1 - \alpha$. See the Figure 9.2.6. Thus, we claim that

$$P_{\mu,\sigma}\left\{(\chi^2_{n-1,\alpha/2})^{-1}(n-1)S^2 < \sigma^2 < (\chi^2_{n-1,1-\alpha/2})^{-1}(n-1)S^2\right\}$$
$$= 1 - \alpha.$$

In other words,

$$\left((\chi^2_{n-1,\alpha/2})^{-1}(n-1)S^2, (\chi^2_{n-1,1-\alpha/2})^{-1}(n-1)S^2\right) \qquad (9.2.11)$$

is a $(1 - \alpha)$ two-sided confidence interval estimator for σ^2. ▲

Example 9.2.10 Joint Confidence Intervals for the Normal Mean and Variance: Suppose that $X_1, ..., X_n$ are iid $N(\mu, \sigma^2)$ with both unknown parameters $\mu \in \Re$ and $\sigma \in \Re^+, n \geq 2$. Given some $\alpha \in (0, 1)$, we wish to construct $(1 - \alpha)$ joint two-sided confidence intervals for both μ and σ^2. Let $\overline{X}$ be the sample mean and $S^2 = (n-1)^{-1}\Sigma_{i=1}^n (X_i - \overline{X})^2$ be the sample variance. The statistic $T \equiv (\overline{X}, S)$ is minimal sufficient for (μ, σ).

From the Example 9.2.8, we claim that

$$J_1 = \left(\overline{X} - t_{n-1,\gamma/2} n^{-1/2} S, \overline{X} + t_{n-1,\gamma/2} n^{-1/2} S \right) \qquad (9.2.12)$$

is a $(1 - \gamma)$ confidence interval for μ for any fixed $\gamma \in (0, 1)$. Similarly, from the Example 9.2.9, we claim that

$$J_2 = \left((\chi^2_{n-1,\delta/2})^{-1}(n-1)S^2, (\chi^2_{n-1,1-\delta/2})^{-1}(n-1)S^2 \right) \qquad (9.2.13)$$

is a $(1 - \delta)$ confidence interval for σ^2 for any $\delta \in (0, 1)$. Now, we can write

$$
\begin{aligned}
&P_{\mu,\sigma}\{\mu \in J_1 \cap \sigma^2 \in J_2\} \\
&\geq P_{\mu,\sigma}\{\mu \in J_1\} + P_{\mu,\sigma}\{\sigma^2 \in J_2\} - 1, \text{ using} \\
&\qquad \text{the Bonferroni inequality (Theorem 3.9.10)} \\
&= 1 - \gamma - \delta.
\end{aligned}
\qquad (9.2.14)
$$

Now, if we choose $0 < \gamma, \delta < 1$ so that $\gamma + \delta = \alpha$, then we can think of $\{J_1, J_2\}$ as the two-sided *joint confidence intervals* for the unknown parameters μ, σ^2 respectively with the *joint confidence coefficient* at least $(1 - \alpha)$. Customarily, we pick $\gamma = \delta = \frac{1}{2}\alpha$. ▲

> One will find more closely related problems on joint confidence intervals in the Exercise 9.2.7 and Exercises 9.2.11-9.2.12.

9.2.3   The Interpretation of a Confidence Coefficient

Next, let us explain in general how we *interpret the confidence coefficient or the coverage probability* defined by (9.1.1). Consider the confidence interval J for θ. Once we observe a particular data $\mathbf{X} = \mathbf{x}$, a two-sided confidence interval *estimate* of θ is going to be $(T_L(\mathbf{x}), T_U(\mathbf{x}))$, a *fixed* subinterval of the real line. Note that there is nothing random about this observed interval estimate $(T_L(\mathbf{x}), T_U(\mathbf{x}))$ and recall that the parameter θ is unknown $(\in \Theta)$ but it is a *fixed entity*. The interpretation of the phrase "$(1 - \alpha)$ confidence" simply means this: Suppose hypothetically that we keep observing different data $\mathbf{X} = \mathbf{x}_1, \mathbf{x}_2, \mathbf{x}_3, ...$ for a long time, and we keep constructing the corresponding observed confidence interval estimates $(T_L(\mathbf{x}_1), T_U(\mathbf{x}_1)), (T_L(\mathbf{x}_2), T_U(\mathbf{x}_2)), (T_L(\mathbf{x}_3), T_U(\mathbf{x}_3)), \dots$. In the long run, out of all these intervals constructed, approximately $100(1-\alpha)\%$ would include the unknown value of the parameter θ. This goes hand in hand with the relative frequency interpretation of probability calculations explained in Chapter 1.

In a *frequentist paradigm*, one *does not* talk about the *probability* of a fixed interval estimate including or not including the unknown value of θ.

Example 9.2.11 (Example 9.2.1 Continued) Suppose that $X_1, \dots, X_n$ are iid $N(\mu, \sigma^2)$ with the unknown parameter $\mu \in \Re$. We assume that $\sigma \in \Re^+$ is known. We fix $\alpha = .05$ so that $J = \left(\overline{X} - 1.645 \frac{\sigma}{\sqrt{n}}, \infty\right)$ will be a 95% lower confidence interval for μ. Using the MINITAB Release 12.1, we generated a normal population with $\mu = 5$ and $\sigma = 1$. First, we considered $n = 10$. In the ith replication, we obtained the value of the sample mean $\overline{x}_i$ and computed the lower end point $\overline{x}_i - 1.645 \frac{1}{\sqrt{10}}$ of the observed confidence interval, $i = 1, \dots, k$ with $k = 100, 200, 500$. Then, *approximately* 5% of the total number (k) of intervals so constructed can be expected not to include the true value $\mu = 5$. Next, we repeated the simulated exercise when $n = 20$. The following table summarizes the findings.

Table 9.2.1. Number of Intervals Not Including the True Value
$\mu = 5$ Out of k Simulated Confidence Intervals

	$k = 100$	$k = 200$	$k = 500$
$n = 10$	4	9	21
$n = 20$	4	8	22

The number of simulations (k) is not particularly very high in this example. Yet, we notice about four or five percent non-coverage which is what one may expect. Among k constructed intervals, if n_k denotes the number of intervals which *do not* include the true value of μ, then we can claim that $\lim\limits_{k \to \infty} \frac{n_k}{k} = .05$. ▲

9.2.4   Ideas of Accuracy Measures

In the two Examples 9.2.7-9.2.8 we used the equal tail percentage points of the standard normal and the Student's t distributions. Here, both pivotal distributions were symmetric about the origin. Both the standard normal and the Student's t_{n-1} pdf's obviously integrate to $(1 - \alpha)$ respectively on the intervals $(-z_{\alpha/2}, z_{\alpha/2})$ and $(-t_{n-1,\alpha/2}, t_{n-1,\alpha/2})$. But, for these two distributions, if there existed any *asymmetric* (around zero) and *shorter interval* with the same coverage probability $(1 - \alpha)$, then we should have instead mimicked that in order to arrive at the proposed confidence intervals.

Let us focus on the Example 9.2.7 and explain the case in point. Suppose that Z is the standard normal random variable. Now, we have $P\{-z_{\alpha/2} < Z < z_{\alpha/2}\} = 1 - \alpha$. Suppose that one can determine two other positive numbers a, b such that $P\{-a < Z < b\} = 1 - \alpha$ and at the same time $2z_{\alpha/2} > b + a$. That is, the two intervals $(-z_{\alpha/2}, z_{\alpha/2})$ and $(-a, b)$ have the same coverage probability $(1 - \alpha)$, but $(-a, b)$ is a shorter interval. In that case, instead of the solution proposed in (9.2.9), we should have suggested $\left(\overline{X} - an^{-1/2}\sigma, \overline{X} + bn^{-1/2}\sigma\right)$ as the confidence interval for μ. One may ask: Was the solution in (9.2.9) proposed because the interval $(-z_{\alpha/2}, z_{\alpha/2})$ happened to be the shortest one with $(1 - \alpha)$ coverage in a standard normal distribution? The answer is: yes, that is so. In the case of the Example 9.2.8, the situation is not quite the same but it remains similar. This will become clear if one contrasts the two Examples 9.2.12-9.2.13. For the record, let us prove the Theorem 9.2.1 first.

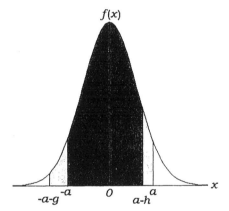

Figure 9.2.7. The Area Between $-a$ and a Is $(1 - \alpha)$. The Area Between $-(a + g)$ and $(a - h)$ Is $(1 - \alpha)$

Theorem 9.2.1 *Suppose that X is a continuous random variable having a unimodal pdf $f(x)$ with the support $\Re$. Assume that $f(x)$ is symmetric around $x = 0$, that is $f(-x) = f(x)$ for all $x > 0$. Let $P(-a < X < a) = 1 - \alpha$ for some $0 < \alpha < \frac{1}{2}$. Now, suppose that the positive numbers g, h are such that one has $P(-a - g < X < a - h) = 1 - \alpha$. Then, the interval $(-a - g, a - h)$ must be wider than the interval $(-a, a)$.*

Proof It will suffice to show that $g > h$. The mode of $f(x)$ must be at

$x = 0$. We may assume that $a > h$ since we have $0 < \alpha < \frac{1}{2}$. Since we have $P(-a < X < a) = P(-a - g < X < a - h)$, one obviously, has

$$\int_{x=-a-g}^{-a} f(x)dx = \int_{x=a-h}^{a} f(x)dx. \tag{9.2.15}$$

But, the pdf $f(x)$ is symmetric about $x = 0$ and $f(x)$ is assumed positive for all $x \in \Re$. The Figure 9.2.7 describes a situation like this. Hence, the integrals in (9.2.15) can be equal if and only if $\int_{x=a}^{a+g} f(x)dx = \int_{x=a-h}^{a} f(x)dx$ which can happen as long as the interval $(a - h, a)$ is shorter than the interval $(a, a + g)$. The result then follows. The details are left out as an exercise. ∎

Remark 9.2.1 The Theorem 9.2.1 proved that among all $(1 - \alpha)$ intervals going to the left of $(-a, a)$, the interval $(-a, a)$ was the shortest one. Since, $f(x)$ is assumed symmetric about $x = 0$, it also follows from this result that among all $(1 - \alpha)$ intervals going to the right of $(-a, a)$, the interval $(-a, a)$ is again the shortest one.

Example 9.2.12 (Example 9.2.7 Continued) The length L_n of the confidence interval from (9.2.9) amounts to $2z_{\alpha/2}n^{-1/2}\sigma$ which is the shortest width among all $(1 - \alpha)$ confidence intervals for the unknown mean μ in a $N(\mu, \sigma^2)$ population with $\sigma(> 0)$ known. The Theorem 9.2.1 immediately applies. Also observe that the *shortest width*, namely $2z_{\alpha/2}n^{-1/2}\sigma$, is a fixed number here. ▲

Example 9.2.13 (Example 9.2.8 Continued) The length L_n of the confidence interval from (9.2.10) amounts to $2t_{n-1,\alpha/2}n^{-1/2}S$ which is a random variable to begin with. So, we do not discuss whether the confidence interval from (9.2.10) has the shortest width among all $(1 - \alpha)$ confidence intervals for the unknown mean μ. The width L_n is not even a fixed number! Observe that $E[L_n] = 2t_{n-1,\alpha/2}\frac{\sigma}{\sqrt{n(n-1)}}E[\sqrt{Y}]$ where Y has the χ_{n-1}^2 distribution. But, the expression for $E[\sqrt{Y}]$ is a function of n only. Now, Theorem 9.2.1 directly implies that the confidence interval from (9.2.10) has the *shortest expected width* among all $(1 - \alpha)$ confidence intervals for μ in a $N(\mu, \sigma^2)$ population when $\sigma^2(> 0)$ is also unknown. ▲

Both examples handled the location parameter estimation problems and we could claim the *optimality* properties (*shortest width* or *shortest expected width*) associated with the proposed $(1 - \alpha)$ confidence intervals. Since the pivotal pdf's were symmetric about zero and unimodal, these intervals had each tail area probability $\frac{1}{2}\alpha$. Recall the Figures 9.2.4-9.2.5.

In the location parameter case, even if the pivotal pdf is *skewed* but *unimodal*, a suitable concept of "optimality" of $(1 - \alpha)$ confidence intervals can be formulated. The corresponding result will coincide with Theorem

9.2.1 in the symmetric case. But, any "optimality" property in the skewed unimodal case does not easily translate into something nice and simple under the scale parameter scenario. For *skewed* pivotal distributions, in general, we can not claim useful and attractive optimality properties when a $(1 - \alpha)$ confidence interval is constructed with the tail area probability $\frac{1}{2}\alpha$ on both sides.

> *Convention*: For standard pivotal distributions such as Normal, Student's t, Chi-square and F, we customarily assign the tail area probability $\frac{1}{2}\alpha$ on both sides in order to construct a $100(1 - \alpha)\%$ confidence interval.

9.2.5 Using Confidence Intervals in the Tests of Hypothesis

Let us think of testing a null hypothesis $H_0 : \theta = \theta_0$ against an alternative hypothesis $H_1 : \theta > \theta_0$ (or $H_1 : \theta < \theta_0$ or $H_1 : \theta \neq \theta_0$). Depending on the nature of the alternative hypothesis, the rejection region $\mathcal{R}$ respectively becomes upper or lower or two-sided. The reader has observed that a confidence interval for θ can also be upper or lower or two-sided.

Suppose that one has constructed a $(1 - \alpha)$ *lower* confidence interval estimator $J_1 = (T_L(\mathbf{X}), \infty)$ for θ. Then any null hypothesis $H_0 : \theta = \theta_0$ will be rejected at level α in favor of the alternative hypothesis $H_1 : \theta > \theta_0$ if and only if θ_0 falls outside the confidence interval J_1.

Suppose that one has constructed a $(1 - \alpha)$ *upper* confidence interval estimator $J_2 = (-\infty, T_U(\mathbf{X}))$ for θ. Then any null hypothesis $H_0 : \theta = \theta_0$ will be rejected at level α in favor of the alternative hypothesis $H_1 : \theta < \theta_0$ if and only if θ_0 falls outside the confidence interval J_2.

Suppose that one has constructed a $(1 - \alpha)$ *two-sided* confidence interval estimator $J_3 = (T_L(\mathbf{X}), T_U(\mathbf{X}))$ for θ. Then any null hypothesis $H_0 : \theta = \theta_0$ will be rejected at level α in favor of the alternative hypothesis $H_1 : \theta \neq \theta_0$ if and only if θ_0 falls outside the confidence interval J_3.

> Once we have a $(1 - \alpha)$ confidence interval J, it is clear that any null hypothesis $H_0 : \theta = \theta_0$ will be rejected at level α as long as $\theta_0 \notin J$. But, rejection of H_0 leads to acceptance of H_1 whose nature depends upon whether J is upper-, lower- or two-sided.

> Sample size determination is crucial in practice if we wish to have a preassigned "size" of a confidence region. See Chapter 13.

9.3 Two-Sample Problems

Tests of hypotheses for the equality of means or variances of two independent populations need machineries which are different in flavor from the theory of MP or UMP tests developed in Chapter 8. These topics are delegated to Chapter 11. So, in order to construct confidence intervals for the difference of means (or location parameters) and the ratio of variances (or the scale parameters), we avoid approaching the problems through the inversion of test procedures. Instead, we focus only on the *pivotal techniques*.

Here, the basic principles remain same as in Section 9.2.2. For an unknown real valued parametric function $\kappa(\boldsymbol{\theta})$, suppose that we have a point estimator $\widehat{\kappa(\boldsymbol{\theta})}$ which is based on a (minimal) sufficient statistic $\mathbf{T}$ for $\boldsymbol{\theta}$. Often, the exact distribution of $\{\widehat{\kappa(\boldsymbol{\theta})} - \kappa(\boldsymbol{\theta})\}/\tau$ will become free from $\boldsymbol{\theta}$ for all $\boldsymbol{\theta} \in \Theta$ with some $\tau(> 0)$.

If τ happens to be known, then we obtain two suitable numbers a and b such that $P_{\boldsymbol{\theta}}[a < \{\widehat{\kappa(\boldsymbol{\theta})} - \kappa(\boldsymbol{\theta})\}/\tau < b] = 1 - \alpha$. This will lead to a $(1 - \alpha)$ two-sided confidence interval for $\kappa(\boldsymbol{\theta})$.

If τ happens to be unknown, then we may estimate it by $\hat{\tau}$ and use the new pivot $\{\widehat{\kappa(\boldsymbol{\theta})} - \kappa(\boldsymbol{\theta})\}/\hat{\tau}$ instead. If the exact distribution of $\{\widehat{\kappa(\boldsymbol{\theta})} - \kappa(\boldsymbol{\theta})\}/\hat{\tau}$ become free from $\boldsymbol{\theta}$ for all $\boldsymbol{\theta} \in \Theta$, then again we obtain two suitable numbers a and b such that $P_{\boldsymbol{\theta}}[a < \{\widehat{\kappa(\boldsymbol{\theta})} - \kappa(\boldsymbol{\theta})\}/\hat{\tau} < b] = 1 - \alpha$. Once more this will lead to a $(1 - \alpha)$ two-sided confidence interval for $\kappa(\boldsymbol{\theta})$. This is the way we will handle the location parameter cases.

In a scale parameter case, we may proceed with a pivot $\widehat{\kappa(\boldsymbol{\theta})}/\kappa(\boldsymbol{\theta})$ whose distribution will often remain the same for all $\boldsymbol{\theta} \in \Theta$. Then, we obtain two suitable numbers a and b such that $P_{\boldsymbol{\theta}}[a < \widehat{\kappa(\boldsymbol{\theta})}/\kappa(\boldsymbol{\theta}) < b] = 1 - \alpha$. This will lead to a $(1 - \alpha)$ two-sided confidence interval for $\kappa(\boldsymbol{\theta})$.

9.3.1 *Comparing the Location Parameters*

Examples include estimation of the difference of (i) the means of two independent normal populations, (ii) the location parameters of two independent negative exponential populations, and (iii) the means of a bivariate normal population.

Example 9.3.1 Difference of Normal Means with a Common Unknown Variance: Recall the Example 4.5.2 as needed. Suppose that the random variables $X_{i1}, ..., X_{in_i}$ are iid $N(\mu_i, \sigma^2), i = 1, 2$, and that the X_{1j}'s are independent of the X_{2j}'s. We assume that all three parameters are unknown and $\boldsymbol{\theta} = (\mu_1, \mu_2, \sigma) \in \Re \times \Re \times \Re^+$. With fixed $\alpha \in (0, 1)$, we wish to construct a $(1-\alpha)$ two-sided confidence interval for $\mu_1 - \mu_2 (= \kappa(\boldsymbol{\theta}))$

based on the sufficient statistics for $\boldsymbol{\theta}$. With $n_i \geq 2$, let us denote

$$\overline{X}_i = n_i^{-1}\Sigma_{j=1}^{n_i}X_{ij}, \ S_i^2 = (n_i - 1)^{-1}\Sigma_{j=1}^{n_i}(X_{ij} - \overline{X}_i)^2$$
$$S_P^2 = (n_1 + n_2 - 2)^{-1}\left\{(n_1 - 1)S_1^2 + (n_2 - 1)S_2^2\right\} \tag{9.3.1}$$

for $i = 1, 2$. Here, S_P^2 is the *pooled sample variance*.

Based on (4.5.8), we define the pivot

$$U = \{n_1^{-1} + n_2^{-1}\}^{-\frac{1}{2}}[(\overline{X}_1 - \overline{X}_2) - (\mu_1 - \mu_2)]S_P^{-1} \tag{9.3.2}$$

which has the Student's t_ν distribution with $\nu = (n_1 + n_2 - 2)$ degrees of freedom. Now, we have $P\{-t_{\nu,\alpha/2} < U < t_{\nu,\alpha/2}\} = 1 - \alpha$ where $t_{\nu,\alpha/2}$ is the upper $100(1 - \frac{1}{2}\alpha)\%$ point of the Student's t distribution with ν degrees of freedom. Thus, we claim that

$$P_\theta[(\overline{X}_1 - \overline{X}_2) - t_{\nu,\alpha/2}S_P\{n_1^{-1} + n_2^{-1}\}^{1/2} < \mu_1 - \mu_2$$
$$< (\overline{X}_1 - \overline{X}_2) + t_{\nu,\alpha/2}S_P\{n_1^{-1} + n_2^{-1}\}^{1/2}] = 1 - \alpha.$$

Now, writing $W = \overline{X}_1 - \overline{X}_2$, we have

$$\left(W - t_{\nu,\alpha/2}\{n_1^{-1} + n_2^{-1}\}^{1/2}S_P, W + t_{\nu,\alpha/2}\{n_1^{-1} + n_2^{-1}\}^{1/2}S_P\right) \tag{9.3.3}$$

as our $(1 - \alpha)$ two-sided confidence interval for $\mu_1 - \mu_2$. ▲

Example 9.3.2 Difference of Negative Exponential Locations with a Common Unknown Scale: Suppose that the random variables $X_{i1}, ..., X_{in}$ are iid having the common pdf $f(x; \mu_i, \sigma), i = 1, 2$, where we denote $f(x; \mu, \sigma) = \sigma^{-1}exp\{-(x - \mu)/\sigma\}I(x > \mu)$. Also let the X_{1j}'s be independent of the X_{2j}'s. We assume that all three parameters are unknown and $\boldsymbol{\theta} = (\mu_1, \mu_2, \sigma) \in \Re \times \Re \times \Re^+$. With fixed $\alpha \in (0, 1)$, we wish to construct a $(1 - \alpha)$ two-sided confidence interval for $\mu_1 - \mu_2(= \kappa(\boldsymbol{\theta}))$ based on the sufficient statistics for $\boldsymbol{\theta}$. With $n \geq 2$, let us denote

$$X_i^{(1)} = \min_{1 \leq j \leq n} X_{ij}, \ W_i = (n - 1)^{-1}\Sigma_{j=1}^n(X_{ij} - X_i^{(1)}),$$
$$W_P = \frac{1}{2}\{W_1 + W_2\} \tag{9.3.4}$$

for $i = 1, 2$.

Here, W_P is the *pooled* estimator of σ. It is easy to verify that W_P estimates σ unbiasedly too. It is also easy to verify the following claims:

$$V[W_1] = V[W_2] = (n - 1)^{-1}\sigma^2 \text{ and } V[W_P] = \frac{1}{2}(n - 1)^{-1}\sigma^2.$$

That is, $V[W_P] < V[W_i], i = 1, 2$ so that the pooled estimator W_P is indeed a better unbiased estimator of σ than either $W_i, i = 1, 2$.

From the Example 4.4.12, recall that $2(n-1)W_i/\sigma$ is distributed as $\chi^2_{2n-2}, i = 1, 2$, and these are independent. Using the reproductive property of independent Chi-squares (Theorem 4.3.2, part (iii)) we can claim that $4(n-1)W_P\sigma^{-1} = \{2(n-1)W_1 + 2(n-1)W_2\}\sigma^{-1}$ which has a Chi-square distribution with $4(n-1)$ degrees of freedom. Also, $(X_1^{(1)}, X_2^{(1)})$ and W_P are independent. Hence, we may use the following pivot

$$U = n[(X_1^{(1)} - X_2^{(1)}) - (\mu_1 - \mu_2)]W_P^{-1}. \qquad (9.3.5)$$

Now, let us look into the distribution of U. We know that

$$Y_i = n(X_i^{(1)} - \mu_i)/\sigma, i = 1, 2, \text{ are iid standard exponential.}$$

Hence, the pdf of $Q = Y_1 - Y_2$ would be given by $\frac{1}{2}e^{-|q|}I(q \in \Re)$ so that the random variable $|Q|$ has the standard exponential distribution. Refer to the Exercise 4.3.4, part (ii). In other words, $2|Q|$ is distributed as χ^2_2. Also Q, W_P are independently distributed. Now, we rewrite the expression from (9.3.5) as

$$|U| = [2|Q|\sigma^{-1}/2] \div [4(n-1)W_P\sigma^{-1}/(4n-4)],$$

that is, the pivotal distribution of $|U|$ is given by $F_{2,4n-4}$. Let $F_{2,4n-4,\alpha}$ be the upper $100\alpha\%$ point of the F distribution with 2 and $(4n-4)$ degrees of freedom. See the Figure 9.3.1. We can say that $P\{-F_{2,4n-4,\alpha} < U < F_{2,4n-4,\alpha}\} = 1 - \alpha$ and claim that

$$P_\theta\left\{(\mu_1 - \mu_2) \in [(X_1^{(1)} - X_2^{(1)}) \pm F_{2,4n-4,\alpha}n^{-1}W_P]\right\} = 1 - \alpha.$$

In other words,

$$\left((X_1^{(1)} - X_2^{(1)}) - F_{2,4n-4,\alpha}n^{-1}W_P, (X_1^{(1)} - X_2^{(1)}) + F_{2,4n-4,\alpha}n^{-1}W_P\right)$$
$$(9.3.6)$$

is a $(1 - \alpha)$ two-sided confidence interval estimator for $(\mu_1 - \mu_2)$. ▲

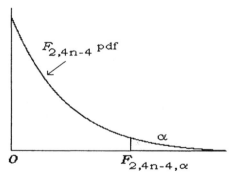

Figure 9.3.1. The Shaded Area on the Right of $F_{2,4n-4,\alpha}$ Is α

Confidence interval estimation of $(\mu_1 - \mu_2)/\sigma$ in the two Examples 9.3.1-9.3.2 using the Bonferroni inequality are left as Exercises 9.3.3 and 9.3.5. Look at the Example 9.2.10 as needed.

Example 9.3.3 The Paired Difference t Method: Sometimes the two populations may be assumed normal but they may be dependent. In a large establishment, for example, suppose that X_{1j}, X_{2j} respectively denote the job performance score before and after going through a week-long job enhancement program for the jth employee, $j = 1, ..., n(\geq 2)$. We assume that these employees are selected randomly and independently of each other. We wish to compare the average "before and after" job performance scores in the population. Here, observe that X_{1j}, X_{2j} are dependent random variables. The methodology from the Example 9.3.1 will not apply here.

Suppose that the pairs of random variables (X_{1j}, X_{2j}) are iid bivariate normal, $N_2(\mu_1, \mu_2, \sigma_1^2, \sigma_2^2, \rho), j = 1, ..., n(\geq 2)$. Let all five parameters be unknown, $(\mu_i, \sigma_i) \in \Re \times \Re^+, i = 1, 2$ and $-1 < \rho < 1$. With fixed $\alpha \in (0, 1)$, we wish to construct a $(1 - \alpha)$ two-sided confidence interval for $\mu_1 - \mu_2$ based on the sufficient statistics for $\boldsymbol{\theta}(= (\mu_1, \mu_2, \sigma_1, \sigma_2, \rho))$. Let us denote

$$Y_j = X_{1j} - X_{2j}, j = 1, ..., n, \text{ and } \overline{Y} = n^{-1}\Sigma_{j=1}^n Y_j,$$
$$S^2 = (n - 1)^{-1}\Sigma_{j=1}^n(Y_j - \overline{Y})^2. \tag{9.3.7}$$

In (9.3.7), observe that $Y_1, ..., Y_n$ are iid $N(\mu_1 - \mu_2, \sigma^2)$ where $\sigma^2 = \sigma_1^2 + \sigma_2^2 - 2\rho\sigma_1\sigma_2$. Since both mean $\mu_1 - \mu_2$ and variance σ^2 of the common normal distribution of the Y's are unknown, the original two-sample problem reduces to a one-sample problem (Example 9.2.8) in terms of the random samples on the Y's.

We consider the pivot

$$U = \sqrt{n}[\overline{Y} - (\mu_1 - \mu_2)]/S,$$

which has the Student's t distribution with $(n - 1)$ degrees of freedom. As before, let $t_{n-1,\alpha/2}$ be the upper $100(1 - \frac{1}{2}\alpha)\%$ point of the Student's t distribution with $(n - 1)$ degrees of freedom. Thus, along the lines of the Example 9.2.8, we propose

$$\left(\overline{X} - t_{n-1,\alpha/2}n^{-1/2}S, \overline{X} + t_{n-1,\alpha/2}n^{-1/2}S\right) \tag{9.3.8}$$

as our $(1 - \alpha)$ two-sided confidence interval estimator for $(\mu_1 - \mu_2)$. ▲

9.3.2 Comparing the Scale Parameters

The examples include estimation of the ratio of (i) the variances of two independent normal populations, (ii) the scale parameters of two independent negative exponential populations, and (iii) the scale parameters of two independent uniform populations.

Example 9.3.4 Ratio of Normal Variances: Recall the Example 4.5.3 as needed. Suppose that the random variables $X_{i1}, ..., X_{in_i}$ are iid $N(\mu_i, \sigma_i^2), n_i \geq 2, i = 1, 2$, and that the X_{1j}'s are independent of the X_{2j}'s. We assume that all four parameters are unknown and $(\mu_i, \sigma_i) \in \Re \times \Re^+, i = 1, 2$. With fixed $\alpha \in (0, 1)$, we wish to construct a $(1 - \alpha)$ two-sided confidence interval for σ_1^2/σ_2^2 based on the sufficient statistics for $\theta(= (\mu_1, \mu_2, \sigma_1, \sigma_2))$. Let us denote

$$\overline{X}_i = n_i^{-1}\Sigma_{j=1}^{n_i} X_{ij}, \quad S_i^2 = (n_i - 1)^{-1}\Sigma_{j=1}^{n_i}(X_{ij} - \overline{X}_i)^2 \qquad (9.3.9)$$

for $i = 1, 2$ and consider the pivot

$$U = [S_1^2/\sigma_1^2] \div [S_2^2/\sigma_2^2]. \qquad (9.3.10)$$

It should be clear that U is distributed as F_{n_1-1, n_2-1} since $(n_i - 1)S_i^2/\sigma_i^2$ is distributed as $\chi_{n_i-1}^2, i = 1, 2$, and these are also independent. As before, let us denote the upper $100(\alpha/2)\%$ point of the F_{n_1-1, n_2-1} distribution by $F_{n_1-1, n_2-1, \alpha/2}$. See the Figure 9.3.2.

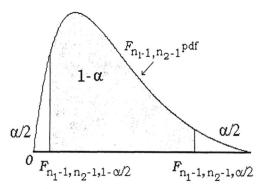

Figure 9.3.2. Area on the Right (or Left) of $F_{n_1-1, n_2-1, \alpha/2}$
(or $F_{n_1-1, n_2-1, 1-\alpha/2}$) Is $\alpha/2$

Thus, we can write $P\{F_{n_1-1, n_2-1, 1-\alpha/2} < U < F_{n_1-1, n_2-1, \alpha/2}\} = 1 - \alpha$ and claim that

$$P_\theta \left\{ F_{n_1-1, n_2-1, \alpha/2}^{-1}[S_1^2/S_2^2] < \sigma_1^2/\sigma_2^2 < F_{n_1-1, n_2-1, 1-\alpha/2}^{-1}[S_1^2/S_2^2] \right\}$$

$$= 1 - \alpha.$$

Hence, we conclude that

$$\left(F^{-1}_{n_1-1,n_2-1,\alpha/2}[S_1^2/S_2^2], F^{-1}_{n_1-1,n_2-1,1-\alpha/2}[S_1^2/S_2^2]\right) \qquad (9.3.11)$$

is a $(1 - \alpha)$ two-sided confidence interval estimator for the variance ratio σ_1^2/σ_2^2. By taking the square root throughout, it follows immediately from (9.3.11) that

$$\left(F^{-1/2}_{n_1-1,n_2-1,\alpha/2}[S_1/S_2], F^{-1/2}_{n_1-1,n_2-1,1-\alpha/2}[S_1/S_2]\right) \qquad (9.3.12)$$

is a $(1 - \alpha)$ two-sided confidence interval estimator for the ratio σ_1/σ_2. ▲

For the F distribution, one has the following relationship:
$$F_{\nu_1,\nu_2,1-\alpha} = F^{-1}_{\nu_2,\nu_1,\alpha} \text{ for any } \alpha \in (0,1).$$
It helps to find the lower α percentage points of F_{ν_1,ν_2} from the reciprocal of the upper α percentage points of F_{ν_2,ν_1}.

$\qquad (9.3.13)$

Example 9.3.5 Ratio of Negative Exponential Scales: Suppose that the random variables $X_{i1}, ..., X_{in_i}$ are iid having the common pdf $f(x; \mu_i, \sigma_i), n_i \geq 2, i = 1, 2$, where $f(x; \mu, \sigma) = \sigma^{-1}exp\{-(x - \mu)/\sigma\}I(x > \mu)$. Also let the X_{1j}'s be independent of the X_{2j}'s. Here we assume that all four parameters are unknown and $(\mu_i, \sigma_i) \in \Re \times \Re^+, i = 1, 2$. With fixed $\alpha \in (0, 1)$, we wish to construct a $(1 - \alpha)$ two-sided confidence interval for σ_1/σ_2 based on the sufficient statistics for $\boldsymbol{\theta}(= (\mu_1, \mu_2, \sigma_1, \sigma_2))$. We denote

$$X_i^{(1)} = \min_{1 \leq j \leq n_i} X_{ij}, \quad W_i = (n_i - 1)^{-1}\Sigma_{j=1}^{n_i}(X_{ij} - X_i^{(1)}) \qquad (9.3.14)$$

for $i = 1, 2$ and consider the pivot

$$U = [W_1/\sigma_1] \div [W_2/\sigma_2]. \qquad (9.3.15)$$

It is clear that U is distributed as $F_{2n_1-2,2n_2-2}$ since $2(n_i - 1)W_i/\sigma_i$ is distributed as $\chi^2_{2n_i-2}, i = 1, 2$, and these are also independent. As before, let us denote the upper $100(\alpha/2)\%$ point of the $F_{2n_1-2,2n_2-2}$ distribution by $F_{2n_1-2,2n_2-2,\alpha/2}$. Thus, we can write $P\{F_{2n_1-2,2n_2-2,1-\alpha/2} < U < F_{2n_1-2,2n_2-2,\alpha/2}\} = 1 - \alpha$ and hence claim that

$$P_\theta \left\{ F^{-1}_{2n_1-2,2n_2-2,\alpha/2}[W_1/W_2] < \sigma_1/\sigma_2 < F^{-1}_{n_1-1,n_2-1,1-\alpha/2}[W_1/W_2] \right\}$$

$$= 1 - \alpha.$$

This leads to the following conclusion:

$$\left(F^{-1}_{2n_1-2,2n_2-2,\alpha/2}[W_1/W_2], F^{-1}_{2n_1-2,2n_2-2,1-\alpha/2}[W_1/W_2] \right) \qquad (9.3.16)$$

is a $(1-\alpha)$ two-sided confidence interval estimator for σ_1/σ_2. ▲

Example 9.3.6 Ratio of Uniform Scales: Suppose that the random variables $X_{i1}, ..., X_{in}$ are iid Uniform$(0, \theta_i), i = 1, 2$. Also let the X_{1j}'s be independent of the X_{2j}'s. We assume that both the parameters are unknown and $\boldsymbol{\theta} = (\theta_1, \theta_2) \in \Re^+ \times \Re^+$. With fixed $\alpha \in (0, 1)$, we wish to construct a $(1 - \alpha)$ two-sided confidence interval for θ_1/θ_2 based on the sufficient statistics for (θ_1, θ_2). Let us denote $X_i^{(n)} = \max_{1 \leq j \leq n} X_{ij}$ for $i = 1, 2$ and we consider the following pivot:

$$U = [X_1^{(n)}/\theta_1] \div [X_2^{(n)}/\theta_2]. \qquad (9.3.17)$$

It should be clear that the distribution function of $X_i^{(n)}/\theta_i$ is simply t^n for $0 < t < 1$ and zero otherwise, $i = 1, 2$. Thus, one has

$$-nlog\{X_i^{(n)}/\theta_i\}, i = 1, 2,$$

distributed as iid standard exponential random variable. Recall the Example 4.2.5. We use the Exercise 4.3.4, part (ii) and claim that

$$W = -nlog(U) = -nlog\left[\{X_1^{(n)}/\theta_1\} \div \{X_2^{(n)}/\theta_2\} \right]$$

has the pdf $\frac{1}{2}e^{-|w|}I(w \in \Re)$. Then, we proceed to solve for $a(> 0)$ such that $P\{|-nlog(U)| < a\} = 1-\alpha$. In other words, we need $\int_{w=a}^{\infty} \frac{1}{2}e^{-|w|}dw = \frac{1}{2}\alpha$ which leads to the expression $a = log(1/\alpha)$. Then, we can claim that $P\left\{ -a < -nlog\left[\{X_1^{(n)}\theta_2\}\{X_2^{(n)}\theta_1\} \right] < a \right\} = 1 - \alpha$. Hence, one has:

$$P_\theta \left\{ e^{-a/n}[X_1^{(n)}/X_2^{(n)}] < \theta_1/\theta_2 < e^{a/n}[X_1^{(n)}/X_2^{(n)}] \right\} = 1 - \alpha,$$

which leads to the following conclusion:

$$\left(e^{-a/n}[X_1^{(n)}/X_2^{(n)}], e^{a/n}[X_1^{(n)}/X_2^{(n)}] \right) \qquad (9.3.18)$$

is a $(1-\alpha)$ two-sided confidence interval estimator for the ratio θ_1/θ_2. Look at the closely related Exercise 9.3.11. ▲

9.4   Multiple Comparisons

We first *briefly* describe some confidence region problems for the mean vector of a p-dimensional multivariate normal distribution when the p.d. dispersion matrix (i) is known and (ii) is of the form $\sigma^2 H$ with a known $p \times p$ matrix H but σ is unknown. Next, we compare the mean of a control with the means of independent treatments followed by the analogous comparisons among the variances. Here, one encounters important applications of the multivariate normal, t and F distributions which were introduced earlier in Section 4.6.

9.4.1   Estimating a Multivariate Normal Mean Vector

Example 9.4.1 Suppose that $\mathbf{X}_1, ..., \mathbf{X}_n$ are iid p-dimensional multivariate normal, $N_p(\boldsymbol{\mu}, \boldsymbol{\Sigma})$, random variables. Let us assume that the dispersion matrix $\boldsymbol{\Sigma}$ is p.d. and known. With given $\alpha \in (0, 1)$, we wish to derive a $(1 - \alpha)$ confidence region for the mean vector $\boldsymbol{\mu}$. Towards this end, from the Theorem 4.6.1, part (ii), let us recall that

$$n(\overline{\mathbf{X}} - \boldsymbol{\mu})' \boldsymbol{\Sigma}^{-1}(\overline{\mathbf{X}} - \boldsymbol{\mu}) \text{ is distributed as } \chi_p^2, \text{ where}$$
$$\overline{\mathbf{X}}(= n^{-1}\Sigma_{i=1}^n \mathbf{X}_i) \text{ is the sample mean vector.} \tag{9.4.1}$$

Now, consider the non-negative expression $n(\overline{\mathbf{X}} - \boldsymbol{\mu})' \boldsymbol{\Sigma}^{-1}(\overline{\mathbf{X}} - \boldsymbol{\mu})$ as the pivot and denote

$$U = n(\overline{\mathbf{X}} - \boldsymbol{\mu})' \boldsymbol{\Sigma}^{-1}(\overline{\mathbf{X}} - \boldsymbol{\mu}).$$

We write $\chi_{p,\alpha}^2$ for the upper $100\alpha\%$ point of the χ_p^2 distribution, that is $P\{U < \chi_{p,\alpha}^2\} = 1 - \alpha$. Thus, we define a p-dimensional confidence region $\mathcal{Q}$ for $\boldsymbol{\mu}$ as follows:

$$\mathcal{Q} = \{\boldsymbol{\omega} \in \Re^p : n(\overline{\mathbf{X}} - \boldsymbol{\omega})' \boldsymbol{\Sigma}^{-1}(\overline{\mathbf{X}} - \boldsymbol{\omega}) < \chi_{p,\alpha}^2\}. \tag{9.4.2}$$

One can immediately claim that

$$P_{\boldsymbol{\mu},\boldsymbol{\Sigma}}\{\boldsymbol{\mu} \in \mathcal{Q}\} = P_{\boldsymbol{\mu},\boldsymbol{\Sigma}}\{n(\overline{\mathbf{X}} - \boldsymbol{\mu})' \boldsymbol{\Sigma}^{-1}(\overline{\mathbf{X}} - \boldsymbol{\mu}) < \chi_{p,\alpha}^2\} = 1 - \alpha,$$

and hence, $\mathcal{Q}$ is a $(1 - \alpha)$ confidence region for the mean vector $\boldsymbol{\mu}$. Geometrically, the confidence region $\mathcal{Q}$ will be a p-dimensional ellipsoid having its center at the point $\overline{\mathbf{X}}$, the sample mean vector. ▲

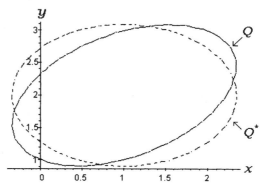

Figure 9.4.1. The Elliptic Confidence Regions $\mathcal{Q}$ from (9.4.3)
and $\mathcal{Q}^*$ from (9.4.4)

Example 9.4.2 (Example 9.4.1 Continued) Suppose that $\mathbf{X}_1, ..., \mathbf{X}_{10}$ are
iid 2-dimensional normal, $N_2(\boldsymbol{\mu}, \boldsymbol{\Sigma})$, random variables with $\boldsymbol{\Sigma} = \begin{pmatrix} 3 & 1 \\ 1 & 2 \end{pmatrix}$.
We fix $\alpha = .05$, that is we require a 95% confidence region for $\boldsymbol{\mu}$. Now, one
has $\chi^2_{2,.05} = -2log(.05) = 5.9915$. Suppose also that the observed value of
$\overline{\mathbf{X}}'$ is $(1, 2)$. Then, the confidence region from (9.4.2) simplifies to

$$\mathcal{Q} = \{(x, y) \in \Re^2 : 2(x - 1)^2 + 3(y - 2)^2 \\ -2(x - 1)(y - 2) < 2.99575\}, \tag{9.4.3}$$

which should be elliptic with its center at the point $(1, 2)$. The Figure 9.4.1
gives a picture (solid curve) of the region $\mathcal{Q}$ which is the inner disk of the
ellipse. The horizontal (x) and vertical (y) axis respectively correspond to
μ_1 and μ_2.

Instead, if we had $\boldsymbol{\Sigma} = \begin{pmatrix} 3 & 0 \\ 0 & 2 \end{pmatrix}$, then one can check that a 95%
confidence region for $\boldsymbol{\mu}$ will also turn out to be elliptic with its center at
the point $(1, 2)$. Let us denote

$$\mathcal{Q}^* = \{(x, y) \in \Re^2 : \tfrac{10}{3}(x - 1)^2 + 5(y - 2)^2 < 5.9915\}. \tag{9.4.4}$$

The Figure 9.4.1 gives a picture (dotted curve) of the region $\mathcal{Q}^*$ which is
the inner disk of the ellipse. ▲

Example 9.4.3 Suppose that $\mathbf{X}_1, ..., \mathbf{X}_n$ are iid p-dimensional multivariate normal, $N_p(\boldsymbol{\mu}, \boldsymbol{\Sigma})$, random variables with $n \geq 2$. Let us assume that
the dispersion matrix $\boldsymbol{\Sigma} = \sigma^2 H$ where H is a p.d. and known matrix but
the scale multiplier $\sigma^2(\in \Re^+)$ is unknown. With fixed $\alpha \in (0, 1)$, we wish
to derive a $(1 - \alpha)$ confidence region for the mean vector $\boldsymbol{\mu}$.

Let us denote

$$\overline{\mathbf{X}} = n^{-1}\Sigma_{i=1}^{n}\mathbf{X}_i \text{ and } S^2 = (pn - p)^{-1}\Sigma_{i=1}^{n}(\mathbf{X}_i - \overline{\mathbf{X}})'H^{-1}(\mathbf{X}_i - \overline{\mathbf{X}}),$$

which respectively estimate μ and σ^2. Recall from Theorem 4.6.1, part (ii) that

$$n(\overline{\mathbf{X}} - \boldsymbol{\mu})'H^{-1}(\overline{\mathbf{X}} - \boldsymbol{\mu})/\sigma^2 \text{ is distributed as } \chi_p^2. \qquad (9.4.5)$$

One can show easily that

$$(pn - p)S^2/\sigma^2 \text{ is distributed as } \chi_{p(n-1)}^2, \text{ and that the mean} \\ \text{vector } \overline{\mathbf{X}} \text{ and } S^2 \text{ are independently distributed.} \qquad (9.4.6)$$

Combining (9.4.5) and (9.4.6) we define the pivot

$$U = n(\overline{\mathbf{X}} - \boldsymbol{\mu})'H^{-1}(\overline{\mathbf{X}} - \boldsymbol{\mu})/(pS^2). \qquad (9.4.7)$$

But, we note that we can rewrite U as

$$[n(\overline{\mathbf{X}} - \boldsymbol{\mu})'H^{-1}(\overline{\mathbf{X}} - \boldsymbol{\mu})/(p\sigma^2)] \div [\{(pn - p)S^2/\sigma^2\}/(pn - p)], \qquad (9.4.8)$$

so that U has the $F_{p,pn-p}$ distribution. We write $F_{p,pn-p,\alpha}$ for the upper $100\alpha\%$ point of the $F_{p,pn-p}$ distribution, that is $P\{U < F_{p,pn-p,\alpha}\} = 1 - \alpha$. Thus, we define a p-dimensional confidence region $\mathcal{Q}$ for μ as follows:

$$\mathcal{Q} = \{\boldsymbol{\omega} \in \Re^p : n(\overline{\mathbf{X}} - \boldsymbol{\omega})'H^{-1}(\overline{\mathbf{X}} - \boldsymbol{\omega}) < F_{p,pn-p,\alpha}pS^2\}. \qquad (9.4.9)$$

Thus, we can immediately claim that

$$P_{\mu,\sigma}\{\boldsymbol{\mu} \in \mathcal{Q}\} = P_{\mu,\sigma}\left([\{n(\overline{\mathbf{X}} - \boldsymbol{\mu})'H^{-1}(\overline{\mathbf{X}} - \boldsymbol{\mu})\}/\{pS^2\}] < F_{p,pn-p,\alpha}\right)$$
$$= 1 - \alpha,$$

and hence $\mathcal{Q}$ is a $(1 - \alpha)$ confidence region for the mean vector μ. Again, geometrically the confidence region $\mathcal{Q}$ will be a p-dimensional ellipsoid with its center at the point $\overline{\mathbf{X}}$. ▲

9.4.2   Comparing the Means

Example 9.4.4 Suppose that $X_{i1}, ..., X_{in}$ are iid random samples from the $N(\mu_i, \sigma^2)$ population, $i = 0, 1, ..., p$. The observations $X_{01}, ..., X_{0n}$ refer to a control with its mean μ_0 whereas $X_{i1}, ..., X_{in}$ are the observations from the ith treatment, $i = 1, ..., p$. Let us also assume that all the observations from the treatments and control are independent and that all the parameters are unknown.

The problem is one of comparing the treatment means $\mu_1, ..., \mu_p$ with the control mean μ_0 by constructing a $(1 - \alpha)$ *joint* confidence region for estimating the parametric function $(\mu_1 - \mu_0, ..., \mu_p - \mu_0)$. Let us denote

$$\mathbf{Y}'_i = (X_{1i} - X_{0i}, X_{2i} - X_{0i}, ..., X_{pi} - X_{0i}), i = 1, ..., n. \qquad (9.4.10)$$

The random variables $\mathbf{Y}_1, ..., \mathbf{Y}_n$ are obviously iid. Observe that any linear function of $\mathbf{Y}_i$ is a linear function of the independent normal variables $X_{1i}, X_{2i}, ..., X_{pi}$ and X_{0i}. Thus, $\mathbf{Y}_1, ..., \mathbf{Y}_n$ are iid p-dimensional normal variables. The common distribution is given by $N_p(\boldsymbol{\mu}, \boldsymbol{\Sigma})$ where

$$\boldsymbol{\mu}' = (\mu_1 - \mu_0, ..., \mu_p - \mu_0)$$

and $\boldsymbol{\Sigma} = \sigma^2 H$ where

$$H_{p \times p} = (h_{ij}), h_{ii} = 2, h_{ij} = 1, i \neq j = 1, ..., p. \qquad (9.4.11)$$

Let us denote $\overline{X}_i = n^{-1} \Sigma_{j=1}^n X_{ij}, \overline{\mathbf{Y}} = (\overline{X}_1 - \overline{X}_0, \overline{X}_2 - \overline{X}_0, ..., \overline{X}_p - \overline{X}_0), i = 0, 1, ..., p$. The customary unbiased estimator of σ^2 is the pooled sample variance,

$$S^2 = \{(p+1)(n-1)\}^{-1} \Sigma_{i=0}^p \Sigma_{j=1}^n (X_{ij} - \overline{X}_i)^2,$$

and it is known that $(p+1)(n-1)S^2/\sigma^2$ is distributed as χ_ν^2 with the degree of freedom $\nu = (p+1)(n-1)$. Also, S^2 and $(\overline{X}_0, \overline{X}_1, ..., \overline{X}_p)$ are independently distributed, and hence S^2 and $\overline{\mathbf{Y}}$ are independently distributed. Next, let us consider the pivot

$$\begin{aligned} \mathbf{U} &= \sqrt{n}(\overline{\mathbf{Y}} - \boldsymbol{\mu})/(\sqrt{2}S) \\ &= \sqrt{n} \left(\overline{Y}_1 - \mu_1 + \mu_0, ..., \overline{Y}_p - \mu_p + \mu_0 \right) /(\sqrt{2}S). \end{aligned} \qquad (9.4.12)$$

But, observe that $\sqrt{n}(\overline{\mathbf{Y}} - \boldsymbol{\mu})/(\sqrt{2}\sigma)$ is distributed as $N_p(\mathbf{0}, \boldsymbol{\Sigma})$ where the correlation matrix

$$\boldsymbol{\Sigma}_{p \times p} = (\sigma_{ij}), \sigma_{ii} = 1, \sigma_{ij} = \tfrac{1}{2}, i \neq j = 1, ..., p.$$

Thus, the pivot $\mathbf{U}$ has a multivariate t distribution, $Mt_p(\nu = (p+1)(n-1), \boldsymbol{\Sigma})$, defined in Section 4.6.2. Naturally, we have the case of equicorrelation ρ where $\rho = \tfrac{1}{2}$. So, we may determine a positive number $h \equiv h_{\nu,\alpha}$ which is the upper equicoordinate $100\alpha\%$ point of the $Mt_p(\nu, \boldsymbol{\Sigma})$ distribution in the following sense:

$$\begin{aligned} P_{\mu,\sigma} &\left\{ \sqrt{n} \left| (\overline{Y}_i - \mu_i + \mu_0) \right| /(\sqrt{2}S) < h \text{ for all } i = 1, ..., p \right\} \\ &= 1 - \alpha. \end{aligned} \qquad (9.4.13)$$

The tables from Cornish (1954, 1962), Dunnett (1955), Dunnett and Sobel (1954, 1955), Krishnaiah and Armitage (1966), and Hochberg and Tamhane (1987) provide the values of "h" in various situations. For example, if $\alpha = .05$ and $p = 2$, then from Table 5 of Hochberg and Tamhane (1987) we can read off $h = 2.66, 2.54$ respectively when $n = 4, 5$. Next, we simply rephrase (9.4.13) to make the following joint statements:

$$\mu_i - \mu_0 \in \left(\overline{X}_i - \overline{X}_0 - hn^{-1/2}[\sqrt{2}S], \overline{X}_i - \overline{X}_0 + hn^{-1/2}[\sqrt{2}S]\right)$$

$$\text{for all } i = 1, ..., p.$$

$$(9.4.14)$$

The simultaneous confidence intervals given by (9.4.14) jointly have $100(1-\alpha)\%$ confidence. ▲

Example 9.4.5 Suppose that $X_{i1}, ..., X_{in}$ are iid random samples from the $N(\mu_i, \sigma^2)$ population, $i = 1, ..., 4$. The observations $X_{i1}, ..., X_{in}$ refer to the ith treatment, $i = 1, ..., 4$. Let us assume that all the observations from the treatments are independent and that all the parameters are unknown.

Consider, for example, the problem of jointly estimating the parameters $\theta_1 = \mu_1 - \mu_2, \theta_2 = \mu_2 + \mu_3 - 2\mu_4$ by means of a simultaneous confidence region. How should one proceed? Let us denote

$$\mathbf{Y}'_i = (X_{1i} - X_{2i}, X_{2i} + X_{3i} - 2X_{4i}), i = 1, ..., n. \qquad (9.4.15)$$

The random variables $\mathbf{Y}_1, ..., \mathbf{Y}_n$ are obviously iid. Also observe that any linear function of $\mathbf{Y}_i$ is a linear function of the independent normal variables $X_{1i}, ..., X_{4i}$. Thus, $\mathbf{Y}_1, ..., \mathbf{Y}_n$ are iid 2-dimensional normal variables. The common distribution is given by $N_2(\boldsymbol{\theta}, \boldsymbol{\Sigma})$ where $\boldsymbol{\theta}' = (\theta_1, \theta_2)$ and

$$\boldsymbol{\Sigma} = \sigma^2 H \text{ with } H_{p \times p} = \begin{pmatrix} 2 & 1 \\ 1 & 6 \end{pmatrix}. \qquad (9.4.16)$$

Now, along the lines of the Example 9.4.3, one can construct a $100(1-\alpha)\%$ joint elliptic confidence region for the parameters θ_1, θ_2. On the other hand, one may proceed along the lines of the Example 9.4.4 to derive a $100(1-\alpha)\%$ joint confidence intervals for the parameters θ_1, θ_2. The details are left out as Exercise 9.4.1. ▲

9.4.3 Comparing the Variances

Example 9.4.6 Suppose that $X_{i1}, ..., X_{in}$ are iid random samples from the $N(\mu_i, \sigma_i^2)$ population, $i = 0, 1, ..., p$. The observations $X_{01}, ..., X_{0n}$ refer to a control population with its mean μ_0 and variance σ_0^2 whereas $X_{i1}, ..., X_{in}$

refer to the ith treatment, $i = 1, ..., p$. Let us also assume that the all observations from the treatments and control are independent.

The problem is one of comparing the treatment variances $\sigma_1^2, ..., \sigma_p^2$ with the control variance σ_0^2 by constructing a joint confidence region of the parametric function $(\sigma_1^2/\sigma_0^2, ..., \sigma_p^2/\sigma_0^2)$. Let us denote $S_i^2 = (n-1)^{-1}\Sigma_{j=1}^n (X_{ij} - \overline{X}_i)^2$, the ith sample variance which estimates $\sigma_i^2, i = 0, 1, ..., p$. These sample variances are all independent and also $(n-1)S_i^2/\sigma_i^2$ is distributed as $\chi_{n-1}^2, i = 0, 1, ..., p$. Next, let us consider the pivot

$$\mathbf{U} = \left([S_1^2\sigma_0^2]/[S_0^2\sigma_1^2], [S_2^2\sigma_0^2]/[S_0^2\sigma_2^2], ..., [S_p^2\sigma_0^2]/[S_0^2\sigma_p^2]\right), \qquad (9.4.17)$$

which has a multivariate F distribution, $MF_p(\nu_0, \nu_1, ..., \nu_p)$ with $\nu_0 = \nu_1 = ... = \nu_p = n - 1$, defined in Section 4.6.3. With $\sigma' = (\sigma_0, \sigma_1, ..., \sigma_p)$, one may find a positive number $b \equiv b_{p,n,\alpha}$ such that

$$P_\sigma \left\{ [S_i^2\sigma_0^2]/[S_0^2\sigma_i^2] < b \text{ for all } i = 1, ..., p \right\} = 1 - \alpha. \qquad (9.4.18)$$

The Tables from Finney (1941) and Armitage and Krishnaiah (1964) will provide the b values for different choices of n and α. Next, we may rephrase (9.4.18) to make the following joint statements:

$$\sigma_i^2/\sigma_0^2 \in \left(S_i^2/[bS_0^2], \infty\right) \text{ for all } i = 1, ..., p. \qquad (9.4.19)$$

The simultaneous confidence intervals given by (9.4.19) jointly have $100(1 - \alpha)\%$ confidence. ▲

Example 9.4.7 (Example 9.4.6 Continued) From the pivot and its distribution described by (9.4.17), it should be clear how one may proceed to obtain the simultaneous $(1 - \alpha)$ two-sided confidence intervals for the variance ratios $\sigma_i^2/\sigma_0^2, i = 1, ..., p$. We need to find two positive numbers a and b, $a < b$, such that

$$P\left\{ [S_i^2\sigma_0^2]/[S_0^2\sigma_i^2] < b \text{ for all } i = 1, ..., p\right\} = 1 - \tfrac{1}{2}\alpha,$$
$$P\left\{ [S_i^2\sigma_0^2]/[S_0^2\sigma_i^2] < a \text{ for all } i = 1, ..., p\right\} = \tfrac{1}{2}\alpha,$$

so that with $\sigma' = (\sigma_0, \sigma_1, ..., \sigma_p)$ we have

$$P_\sigma \left\{ a < [S_i^2\sigma_0^2]/[S_0^2\sigma_i^2] < b \text{ for all } i = 1, ..., p \right\} = 1 - \alpha. \qquad (9.4.20)$$

Using (9.4.20), one can obviously determine simultaneous $(1 - \alpha)$ two-sided confidence intervals for all the variance ratios $\sigma_i^2/\sigma_0^2, i = 1, ..., p$. We leave the details out as Exercise 9.4.4.

But, the numerical determination of the numbers a and b is not so simple. The following result of Hewett and Bulgren (1971) may help:

$$P\left\{a < [S_i^2\sigma_0^2]/[S_0^2\sigma_i^2] < b \text{ for all } i = 1, ..., p\right\}$$
$$\geq [P\{a < F_{n-1,n-1} < b\}]^p . \qquad (9.4.21)$$

Now, equating $P\{a < F_{n-1,n-1} < b\}$ with $(1 - \alpha)^{1/p}$, we may determine approximate choices of a and b corresponding to equal tails $(= \frac{1}{2}(1 - (1 - \alpha)^{1/p}))$ of the $F_{n-1,n-1}$ distribution. This approximation works well when $n \leq 21$.

9.5   Exercises and Complements

9.1.1 Suppose that X_1, X_2 are iid with the common exponential pdf $f(x; \theta) = \theta^{-1}exp\{-x/\theta\}I(x > 0)$ where $\theta(> 0)$ is the unknown parameter. We are given $\alpha \in (0, 1)$.
   (i)   Based on X_1 alone, find an appropriate upper (lower) $(1 - \alpha)$ confidence interval for θ;
   (ii)   Based on X_1, X_2, find an appropriate upper (lower) and two-sided $(1 - \alpha)$ confidence interval for θ.
Give comments and compare the different confidence intervals.

9.1.2 Let X have the Laplace pdf $f(x; \theta) = \frac{1}{2}exp\{-|x - \theta|\}I(x \in \Re)$ where $\theta(\in \Re)$ is the unknown parameter. We are given $\alpha \in (0, 1)$. Based on X, find an appropriate upper (lower) and two-sided $(1 - \alpha)$ confidence interval for θ.

9.1.3 Let X have the Cauchy pdf $f(x; \theta) = \frac{1}{\pi}\{1 + (x - \theta)^2\}^{-1}I(x \in \Re)$ where $\theta(\in \Re)$ is the unknown parameter. We are given $\alpha \in (0, 1)$. Based on X, find an appropriate upper (lower) and two-sided $(1 - \alpha)$ confidence interval for θ.

9.1.4 Suppose that X has the Laplace pdf $f(x; \theta) = \frac{1}{2\theta}exp\{-|x|/\theta\}I(x \in \Re)$ where $\theta(\in \Re^+)$ is the unknown parameter. We are given $\alpha \in (0, 1)$. Based on X, find an appropriate upper (lower) and two-sided $(1 - \alpha)$ confidence interval for θ.

9.2.1 Suppose that X has $N(0, \sigma^2)$ distribution where $\sigma(\in \Re^+)$ is the unknown parameter. Consider the confidence interval $J = (|X|, 10|X|)$ for the parameter σ.
   (i)   Find the confidence coefficient associated with the interval J;
   (ii)   What is the expected length of the interval J?

9.2.2 Suppose that X has its pdf $f(x; \theta) = 2(\theta - x)\theta^{-2}I(0 < x < \theta)$ where $\theta(\in \Re^+)$ is the unknown parameter. We are given $\alpha \in (0, 1)$. Consider the pivot $U = X/\theta$ and derive a $(1 - \alpha)$ two-sided confidence interval for θ.

9.2.3 Suppose that $X_1, ..., X_n$ are iid Gamma(a, b) where $a(> 0)$ is known but $b(> 0)$ is assumed unknown. We are given $\alpha \in (0, 1)$. Find an appropriate pivot based on the minimal sufficient statistic and derive a two-sided $(1 - \alpha)$ confidence interval for b.

9.2.4 Suppose that $X_1, ..., X_n$ are iid $N(\mu, \sigma^2)$ where $\mu(\in \Re)$ is known but $\sigma(\in \Re^+)$ is assumed unknown. We are given $\alpha \in (0, 1)$. We wish to obtain a $(1 - \alpha)$ confidence interval for σ.

   (i) Find both upper and lower confidence intervals by inverting appropriate UMP level α tests;

   (ii) Find a two-sided confidence interval by considering an appropriate pivot based only on the minimal sufficient statistic.

9.2.5 Let $X_1, ..., X_n$ be iid with the common negative exponential pdf $f(x; \sigma) = \sigma^{-1}exp\{-(x - \theta)/\sigma\}I(x > \theta)$. We suppose that $\theta(\in \Re)$ is known but $\sigma(\in \Re^+)$ is unknown. We are given $\alpha \in (0, 1)$. We wish to obtain a $(1 - \alpha)$ confidence interval for σ.

   (i) Find both upper and lower confidence intervals by inverting appropriate UMP level α tests;

   (ii) Find a two-sided confidence interval by considering an appropriate pivot based only on the minimal sufficient statistic.

{*Hint*: The statistic $\Sigma_{i=1}^n(X_i - \theta)$ is minimal sufficient for σ.}

9.2.6 Let $X_1, ..., X_n$ be iid with the common negative exponential pdf $f(x; \theta, \sigma) = \sigma^{-1}exp\{-(x - \theta)/\sigma\}I(x > \theta)$. We suppose that both the parameters $\theta(\in \Re)$ and $\sigma(\in \Re^+)$ are unknown. We are given $\alpha \in (0, 1)$. Find a $(1 - \alpha)$ two-sided confidence interval for σ by considering an appropriate pivot based only on the minimal sufficient statistics.{*Hint*: Can a pivot be constructed from the statistic $\Sigma_{i=1}^n(X_i - X_{n:1})$ where $X_{n:1} = \min_{1 \leq i \leq n} X_i$, the smallest order statistic?}

9.2.7 Let $X_1, ..., X_n$ be iid with the common negative exponential pdf $f(x; \theta, \sigma) = \sigma^{-1}exp\{-(x - \theta)/\sigma\}I(x > \theta)$. We suppose that both the parameters $\theta(\in \Re)$ and $\sigma(\in \Re^+)$ are unknown. We are given $\alpha \in (0, 1)$. Derive the joint $(1 - \alpha)$ two-sided confidence intervals for θ and σ based only on the minimal sufficient statistics. {*Hint*: Proceed along the lines of the Example 9.2.10.}

9.2.8 Let $X_1, ..., X_n$ be iid Uniform$(-\theta, \theta)$ where $\theta(\in \Re^+)$ is assumed unknown. We are given $\alpha \in (0, 1)$. Derive a $(1 - \alpha)$ two-sided confidence interval for θ based only on the minimal sufficient statistics. {*Hint*: Can one justify working with $|X_{n:n}|$ to come up with a suitable pivot?}

9.2.9 Let $X_1, ..., X_n$ be iid having the Rayleigh distribution with the common pdf $f(x; \theta) = 2\theta^{-1}x\,exp(-x^2/\theta)I(x > 0)$ where $\theta(> 0)$ is the unknown parameter. We are given $\alpha \in (0, 1)$. First find both upper and lower $(1 - \alpha)$ confidence intervals for θ by inverting appropriate UMP level α tests. Next, consider an appropriate pivot and hence determine a $(1 - \alpha)$ two-sided confidence interval for θ based only on the minimal sufficient statistic.

9.2.10 Let $X_1, ..., X_n$ be iid having the Weibull distribution with the common pdf $f(x; a) = a^{-1}bx^{b-1}exp(-x^b/a)I(x > 0)$ where $a(> 0)$ is an unknown parameter but $b(> 0)$ is assumed known. We are given $\alpha \in (0, 1)$. First find both upper and lower $(1 - \alpha)$ confidence intervals for a by inverting appropriate UMP level α tests. Next, consider an appropriate pivot and hence determine a $(1 - \alpha)$ two-sided confidence interval for a based only on the minimal sufficient statistic.

9.2.11 (Example 9.2.10 Continued) Suppose that $X_1, ..., X_n$ are iid $N(\mu, \sigma^2)$ with both unknown parameters $\mu \in \Re$ and $\sigma \in \Re^+, n \geq 2$. Given $\alpha \in (0, 1)$, we had found $(1 - \alpha)$ joint confidence intervals J_1, J_2 for μ and σ^2. From this, derive a confidence interval for each parametric function given below with the confidence coefficient at least $(1 - \alpha)$.

(i) $\mu + \sigma$;   (ii) $\mu + \sigma^2$;   (iii) μ/σ;   (iv) μ/σ^2.

9.2.12 (Exercise 9.2.7 Continued) Let $X_1, ..., X_n$ be iid with the common negative exponential pdf $f(x; \theta, \sigma) = \sigma^{-1}exp\{-(x - \theta)/\sigma\}I(x > \theta)$. We suppose that both the parameters $\theta(\in \Re)$ and $\sigma(\in \Re^+)$ are unknown. We are given $\alpha \in (0, 1)$. In the Exercise 9.2.7, the joint $(1 - \alpha)$ two-sided confidence intervals for θ and σ were derived. From this, derive a confidence interval for each parametric function given below with the confidence coefficient at least $(1 - \alpha)$.

(i) $\theta + \sigma$;   (ii) $\theta + \sigma^2$;   (iii) θ/σ;   (iv) θ/σ^2.

9.2.13 A soda dispensing machine automatically fills the soda cans. The actual amount of fill must not vary too much from the target (12 fluid ounces) because the overfill will add extra cost to the manufacturer while the underfill will generate complaints from the customers. A random sample of 15 cans gave a standard deviation of .008 ounces. Assuming a normal distribution for the fill, estimate the true population variance with the help of a 95% two-sided confidence interval.

9.2.14 Ten automobiles of the same make and model were tested by drivers with similar road habits, and the gas mileage for each was recorded over a week. The summary results were $\bar{x} = 22$ miles per gallon and $s = 3.5$ miles per gallon. Construct a 90% two-sided confidence interval for the true average (μ) gas mileage per gallon. Assume a normal distribution for the gas mileage.

9.2.15 The waiting time (in minutes) at a bus stop is believed to have an exponential distribution with mean $\theta(> 0)$. The waiting times on ten occasions were recorded as follows:

$$6.2 \quad 5.8 \quad 4.5 \quad 6.1 \quad 4.6 \quad 4.8 \quad 5.3 \quad 5.0 \quad 3.8 \quad 4.0$$

(*i*)   Construct a 95% two-sided confidence interval for the true average waiting time;

(*ii*)   Construct a 95% two-sided confidence interval for the true variance of the waiting time;

(*iii*)   At 5% level, is there sufficient evidence to justify the claim that the average waiting time exceeds 5 minutes?

9.2.16 Consider a normal population with unknown mean μ and $\sigma = 25$. How large a sample size n is needed to estimate μ within 5 units with 99% confidence?

9.2.17 (Exercise 9.2.10 Continued) In the laboratory, an experiment was conducted to look into the average number of days a variety of weed takes to germinate. Twelve seeds of this variety of weed were planted on a dish. From the moment the seeds were planted, the time (days) to germination was recorded for each seed. The observed data follows:

$$4.39 \quad 6.04 \quad 6.43 \quad 6.98 \quad 2.61 \quad 5.87$$
$$2.73 \quad 7.74 \quad 5.31 \quad 3.27 \quad 4.36 \quad 4.61$$

The team's expert in areas of soil and weed sciences believed that the time to germination had a Weibull distribution with its pdf

$$f(x;a) = a^{-1}bx^{b-1}exp(-x^b/a)I(x > 0),$$

where $a(> 0)$ is unknown but $b = 3$. Find a 90% two-sided confidence interval for the parameter a depending on the minimal sufficient statistic.

9.2.18 (Exercise 9.2.17 Continued) Before the lab experiment was conducted, the team's expert in areas of soil and weed sciences believed that the average time to germination was 3.8 days. Use the confidence interval found in the Exercise 9.2.17 to test the expert's belief. What are the respective null and alternative hypotheses? What is the level of the test?

9.3.1 (Example 9.3.1 Continued) Suppose that the random variables $X_{i1}, ..., X_{in_i}$ are iid $N(\mu_i, \sigma_i^2), i = 1, 2$, and that the X_{1j}'s are independent of the X_{2j}'s. Here we assume that the means μ_1, μ_2 are unknown but σ_1, σ_2 are known, $(\mu_i, \sigma_i) \in \Re \times \Re^+, i = 1, 2$. With fixed $\alpha \in (0, 1)$, construct a

$(1-\alpha)$ two-sided confidence interval for $\mu_1 - \mu_2$ based on sufficient statistics for (μ_1, μ_2). Is the interval shortest among all $(1 - \alpha)$ two-sided confidence intervals for $\mu_1 - \mu_2$ depending only on the sufficient statistics for (μ_1, μ_2)?

9.3.2 (Example 9.3.1 Continued) Suppose that the random variables $X_{i1}, ..., X_{in_i}$ are iid $N(\mu_i, k_i\sigma^2), n_i \geq 2, i = 1, 2$, and that the X_{1j}'s are independent of the X_{2j}'s. Here we assume that all three parameters μ_1, μ_2, σ are unknown but k_1, k_2 are positive and known, $(\mu_1, \mu_2, \sigma) \in \Re \times \Re \times \Re^+$. With fixed $\alpha \in (0, 1)$, construct a $(1 - \alpha)$ two-sided confidence interval for $\mu_1 - \mu_2$ based on sufficient statistics for (μ_1, μ_2, σ). Is the interval shortest on the average among all $(1 - \alpha)$ two-sided confidence intervals for $\mu_1 - \mu_2$ depending on sufficient statistics for (μ_1, μ_2, σ)?{*Hint*: Show that the sufficient statistic is $\left(\Sigma_{j=1}^{n_1} X_{1j}, \Sigma_{j=1}^{n_2} X_{2j}, \frac{1}{k_1}\Sigma_{j=1}^{n_1} X_{1j}^2 + \frac{1}{k_2}\Sigma_{j=1}^{n_2} X_{2j}^2\right)$. Start creating a pivot by standardizing $\overline{X}_1 - \overline{X}_2$ with the help of an analog of the pooled sample variance. For the second part, refer to the Example 9.2.13.}

9.3.3 (Example 9.3.1 Continued) Suppose that the random variables $X_{i1}, ..., X_{in_i}$ are iid $N(\mu_i, \sigma^2), n_i \geq 2, i = 1, 2$, and that the X_{1j}'s are independent of the X_{2j}'s. Here we assume that all three parameters μ_1, μ_2, σ are unknown, $(\mu_1, \mu_2, \sigma) \in \Re \times \Re \times \Re^+$. With fixed $\alpha \in (0, 1)$, construct a $(1 - \alpha)$ two-sided confidence interval for $(\mu_1 - \mu_2)/\sigma$ based on sufficient statistics for (μ_1, μ_2, σ). {*Hint*: Combine the separate estimation problems for $(\mu_1 - \mu_2)$ and σ via the Bonferroni inequality.}

9.3.4 (Example 9.3.2 Continued) Suppose that the random variables $X_{i1}, ..., X_{in}$ are iid having the common pdf $f(x; \mu_i, \sigma_i), i = 1, 2$, where we denote $f(x; \mu, \sigma) = \sigma^{-1} exp\{-(x - \mu)/\sigma\}I(x > \mu)$. Also let the X_{1j}'s be independent of the X_{2j}'s. Here we assume that the location parameters μ_1, μ_2 are unknown but σ_1, σ_2 are known, $(\mu_i, \sigma_i) \in \Re \times \Re^+, i = 1, 2$. With fixed $\alpha \in (0, 1)$, construct a $(1-\alpha)$ two-sided confidence interval for $\mu_1 - \mu_2$ based on sufficient statistics for (μ_1, μ_2).

9.3.5 (Example 9.3.2 Continued) Suppose that the random variables $X_{i1}, ..., X_{in}$ are iid having the common pdf $f(x; \mu_i, \sigma), i = 1, 2$ and $n \geq 2$, where we denote $f(x; \mu, \sigma) = \sigma^{-1} exp\{-(x - \mu)/\sigma\}I(x > \mu)$. Also let the X_{1j}'s be independent of the X_{2j}'s. Here we assume that all three parameters are unknown, $(\mu_1, \mu_2, \sigma) \in \Re \times \Re \times \Re^+$. With fixed $\alpha \in (0, 1)$, construct a $(1 - \alpha)$ two-sided confidence interval for $(\mu_1 - \mu_2)/\sigma$ based on sufficient statistics for (μ_1, μ_2, σ). {*Hint*: Combine the separate estimation problems for $(\mu_1 - \mu_2)$ and σ via the Bonferroni inequality.}

9.3.6 Two types of cars were compared for their braking distances. Test runs were made for each car in a driving range. Once a car reached the stable speed of 60 miles per hour, the brakes were applied. The distance (feet) each car travelled from the moment the brakes were applied to the moment the car came to a complete stop was recorded. The summary statistics are shown below:

Car	Sample Size	$\bar{x}$	s
Make A	$n_A = 12$	37.1	3.1
Make B	$n_B = 10$	39.6	4.3

Construct a 95% two-sided confidence interval for $\mu_B - \mu_A$ based on sufficient statistics. Assume that the elapsed times are distributed as $N(\mu_A, \sigma^2)$ and $N(\mu_B, \sigma^2)$ respectively for the Make A and B cars with all parameters unknown. Is the interval shortest on the average among all 95% two-sided confidence intervals for $\mu_B - \mu_A$ depending on sufficient statistics for (μ_A, μ_B, σ)?

9.3.7 In an experiment in nutritional sciences, a principal investigator considered 8 overweight men with comparable backgrounds which *included* eating habits, family traits, health condition and job related stress. A study was conducted to estimate the average weight reduction for overweight men following a regimen involving nutritional diet and exercise. The technician weighed in each individual before they entered this program. At the conclusion of the two-month long study, each individual was weighed. The data follows:

ID# of Individual	Weight (x_1, pounds) Before Study	Weight (x_2, pounds) After Study
1	235	220
2	189	175
3	156	150
4	172	160
5	165	169
6	180	170
7	170	173
8	195	180

Here, it is not reasonable to assume that in the relevant target population, X_1 and X_2 are independent. The nutritional scientist believed that the assumption of a bivariate normal distribution would be more realistic. Obtain a 95% two-sided confidence interval for $\mu_1 - \mu_2$ where $E(X_i) = \mu_i, i = 1, 2$.

9.3.8 (Exercise 9.3.7 Continued) Suppose that the scientist believed before running this experiment that on the average the weight would go down by at least 10 pounds when such overweight men went through their regimen of diet and exercise for a period of two months. Can the 95% confidence interval found in the Exercise 9.3.7 be used to test the scientist's belief at 5% level? If not, then find an appropriate 95% confidence interval for $\mu_1 - \mu_2$ in the Exercise 9.3.7 and use that to test the hypothesis at 5% level.

9.3.9 Two neighboring towns wanted to compare the variations in the time (minutes) to finish a 5k-run among the first place winners during each town's festivities such as the heritage day, peach festival, memorial day, and other town-wide events. The following data was collected recently by the town officials:

Town A (x_A):   18   20   17   22   19   18   20   18   17
Town B (x_B):   20   17   25   24   18   23

Assume that the performances of these first place winners are independent and that the first place winning times are normally distributed within each town's festivities. Obtain a 90% two-sided confidence interval for σ_A/σ_B based on sufficient statistics for $(\mu_A, \mu_B, \sigma_A, \sigma_B)$. Can we conclude at 10% level that $\sigma_A = \sigma_B$?

9.3.10 Suppose that the random variables $X_{i1}, ..., X_{in_i}$ are iid having the common pdf $f(x; \theta_i), i = 1, 2$, where we denote the exponential pdf $f(x; a) = a^{-1} exp\{-x/a\} I(x > 0), a \in \Re^+$. Also, let the X_{1j}'s be independent of the X_{2j}'s. We assume that both the parameters θ_1, θ_2 are unknown, $(\theta_1, \theta_2) \in \Re^+ \times \Re^+$. With fixed $\alpha \in (0, 1)$, derive a $(1 - \alpha)$ two-sided confidence interval for θ_2/θ_1 based on sufficient statistics for (θ_1, θ_2). {*Hint:* Consider creating a pivot out of $\Sigma_{j=1}^{n_2} X_{2j}/\Sigma_{j=1}^{n_1} X_{1j}$.}

9.3.11 (Example 9.3.6 Continued) Suppose that the random variables $X_{i1}, ..., X_{in}$ are iid Uniform$(0, \theta_i), i = 1, 2$. Also, let the X_{1j}'s be independent of the X_{2j}'s. We assume that both the parameters are unknown, $(\theta_1, \theta_2) \in \Re^+ \times \Re^+$. With fixed $\alpha \in (0, 1)$, derive a $(1 - \alpha)$ two-sided confidence interval for $\theta_1/(\theta_1 + \theta_2)$ based on sufficient statistics for (θ_1, θ_2).

9.4.1 (Example 9.4.5 Continued) Suppose that $X_{i1}, ..., X_{in}$ are iid random samples from the $N(\mu_i, \sigma^2)$ population, $i = 1, ..., 4$. The observations $X_{i1}, ..., X_{in}$ refer to the ith treatment, $i = 1, ..., 4$. Let us assume that the treatments are independent too and that all the parameters are unknown. With fixed $\alpha \in (0, 1)$, derive the $(1 - \alpha)$ joint confidence intervals for estimating the parameters $\theta_1 = \mu_1 - \mu_2, \theta_2 = \mu_2 + \mu_3 - 2\mu_4$.

9.4.2 (Example 9.4.4 Continued) Suppose that $X_{i1}, ..., X_{in}$ are iid random samples from the $N(\mu_i, \sigma^2)$ population, $i = 1, ..., 5$. The observations $X_{i1}, ..., X_{in}$ refer to the ith treatment, $i = 1, ..., 5$. Let us assume that the treatments are independent too and that all the parameters are unknown. With fixed $\alpha \in (0, 1)$, derive the $(1-\alpha)$ joint ellipsoidal confidence region for estimating the parameters $\theta_1 = \mu_1 - \mu_2, \theta_2 = \mu_2 - \mu_3$ and $\theta_3 = \mu_3 + \mu_4 - 2\mu_5$.

9.4.3 (Exercise 9.4.2 Continued) Suppose that $X_{i1}, ..., X_{in}$ are iid random samples from the $N(\mu_i, \sigma^2)$ population, $i = 1, ..., 5$. The observations $X_{i1}, ..., X_{in}$ refer to the ith treatment, $i = 1, ..., 5$. Let us assume that the treatments are independent too and that all the parameters are unknown. With fixed $\alpha \in (0, 1)$, derive the $(1 - \alpha)$ joint confidence intervals for estimating the parameters $\theta_1 = \mu_1 - \mu_2, \theta_2 = \mu_2 - \mu_3$ and $\theta_3 = \mu_3 + \mu_4 - 2\mu_5$.

9.4.4 Use the equation (9.4.20) to propose the simultaneous $(1 - \alpha)$ two-sided confidence intervals for the variance ratios.

9.4.5 Let $X_{i1}, ..., X_{in}$ be iid having the common pdf $f(x; \mu_i, \sigma_i), n \geq 2$, where $f(x; \mu, \sigma) = \sigma^{-1} exp\{-(x - \mu)/\sigma\}I(x > \mu), (\mu_i, \sigma_i) \in \Re \times \Re^+$, $i = 0, 1, ..., p$. The observations $X_{01}, ..., X_{0n}$ refer to a control whereas $X_{i1}, ..., X_{in}$ refer to the ith treatment, $i = 1, ..., p$. Suppose that all the observations are independent and all the parameters are unknown. Start with an appropriate p-dimensional pivot depending on the sufficient statistics and then, with fixed $\alpha \in (0, 1)$, derive the simultaneous $(1 - \alpha)$ two-sided confidence intervals for the ratios $\sigma_i/\sigma_0, i = 1, ..., p$. {*Hint*: Proceed along the lines of the Examples 9.4.6-9.4.7.}

9.4.6 (Exercise 9.3.10 Continued) Suppose that the random variables $X_{i1}, ..., X_{in}$ are iid having the common pdf $f(x; \theta_i)$ where we denote the exponential pdf $f(x; a) = a^{-1} exp\{-x/a\}I(x > 0), \theta_i > 0, i = 0, 1, ..., p$. The observations $X_{01}, ..., X_{0n}$ refer to a control whereas $X_{i1}, ..., X_{in}$ refer to the ith treatment, $i = 1, ..., p$. Suppose that all the observations are independent and all the parameters are unknown. Start with an appropriate p-dimensional pivot depending on the sufficient statistics and then, with fixed $\alpha \in (0, 1)$, derive the simultaneous $(1 - \alpha)$ two-sided confidence intervals for the ratios $\theta_i/\theta_0, i = 1, ..., p$. {*Hint*: Proceed along the lines of the Examples 9.4.6-9.4.7.}

10

Bayesian Methods

10.1 Introduction

The nature of what we are going to discuss in this chapter is conceptually very different from anything we had included in the previous chapters. Thus far, we have developed methodologies from the point of view of a frequentist. We started with a random sample $X_1, ..., X_n$ from a population having the pmf or pdf $f(x; \vartheta)$ where $x \in \mathcal{X}$ and $\vartheta \in \Theta$. The unknown parameter ϑ is assumed fixed. A frequentist's inference procedures depended on the *likelihood function* denoted earlier by $L(\vartheta) = \Pi_{i=1}^n f(x_i; \vartheta)$ where ϑ is unknown but fixed.

In the Bayesian approach, the experimenter believes from the very beginning that the unknown parameter ϑ *is a random variable* having its own probability distribution on the space Θ. Now that ϑ is assumed *random*, the likelihood function will be same as $L(\theta)$ given that $\vartheta = \theta$. Let us denote the pmf or pdf of ϑ by $h(\theta)$ at the point $\vartheta = \theta$ which is called the *prior distribution* of ϑ.

> In previous chapters, we wrote $f(x; \theta)$ for the pmf or pdf of X.
> Now, $f(x; \theta)$ denotes the *conditional* pmf or pdf of X given $\vartheta = \theta$.

> The unknown parameter ϑ is assumed a random variable with its
> own distribution $h(\theta)$ when $\vartheta = \theta \in \Theta$, the parameter space.

The prior distribution $h(\theta)$ often reflects an experimenter's *subjective belief* regarding which ϑ values are more (or less) likely when one considers the whole parameter space Θ. The prior distribution is ideally fixed before the data gathering begins. An experimenter may utilize related expertise and other knowledge in order to come up with a realistic prior distribution $h(\theta)$.

The Bayesian paradigm requires one to follow along these lines: perform all statistical inferences and analysis after combining the information about ϑ contained in the collected data as evidenced by the likelihood function $L(\theta)$ given that $\vartheta = \theta$, as well as that from the prior distribution $h(\theta)$. One combines the evidences about ϑ derived from both the prior distribution and the likelihood function by means of the Bayes's Theorem (Theorem 1.4.3). After combining the information from these two sources, we come up with what is known as a *posterior distribution*. All Bayesian inferences are then guided by this posterior distribution. This approach which originated from Bayes's Theorem was due to Rev. Thomas Bayes (1783). A strong theoretical foundation evolved due to many fundamental contributions of de Finetti (1937), Savage (1954), and Jeffreys (1957), among others. The contributions of L. J. Savage were beautifully synthesized in Lindley (1980).

From the very beginning, R. A. Fisher was vehemently opposed to anything Bayesian. Illuminating accounts of Fisher's philosophical arguments as well as his interactions with some of the key Bayesian researchers of his time can be found in the biography written by his daughter, Joan Fisher Box (1978). Some interesting exchanges between R. A. Fisher and H. Jeffreys as well as L. J. Savage are included in the edited volume of Bennett (1990). The articles of Buehler (1980), Lane (1980) and Wallace (1980) gave important perspectives on possible connections between Fisher's fiducial inference and the Bayesian doctrine.

A branch of statistics referred to as "decision theory" provides a formal structure to find *optimal decision rules* whenever possible. Abraham Wald gave the foundation to this important area during the early to mid 1940's. Wald's (1950) book, *Statistical Decision Functions*, is considered a classic in this area. Berger (1985) treated modern decision theory with much emphasis on Bayesian arguments. Ferguson (1967), on the other hand, gave a more balanced view of the area. The titles on the cover of these two books clearly emphasize this distinction.

In Section 10.2, we give a formal discussion of the prior and posterior distributions. The Section 10.3 first introduces the concept of *conjugate priors* and then posterior distributions are derived in some standard cases when ϑ has a conjugate prior. In this section, we also include an example where the assumed prior for ϑ is not chosen from a conjugate family. In Section 10.4, we develop the point estimation problems under the *squared error loss* function and introduce the *Bayes estimator* for ϑ. In the same vein, Section 10.5.1 develops interval estimation problems. These are customarily referred to as *credible interval estimators* of ϑ. Section 10.5.2 highlights important conceptual differences between a credible interval estimator and a confidence interval estimator. Section 10.6 briefly touches upon the concept of a Bayes test of hypotheses whereas Section 10.7 gives some examples of Bayes estimation under non-conjugate priors.

We can not present a full-blown discussion of the Bayes theory at this level. It is our hope that the readers will get a taste of the underlying basic principles from a brief exposure to this otherwise vast area.

10.2   Prior and Posterior Distributions

The unknown parameter ϑ itself is assumed to be a random variable having its pmf or pdf $h(\theta)$ on the space Θ. We say that $h(\theta)$ is the *prior distribution* of ϑ. In this chapter, the parameter ϑ will represent a continuous *real valued* variable on Θ which will customarily be a subinterval of the real line $\Re$. Thus we will commonly refer to $h(\theta)$ as the prior pdf of ϑ.

The evidence about ϑ derived from the prior pdf is then combined with that obtained from the likelihood function by means of the Bayes's Theorem (Theorem 1.4.3). Observe that the likelihood function is the *conditional* joint pmf or pdf of $\mathbf{X} = (X_1, ..., X_n)$ given that $\vartheta = \theta$. Suppose that the statistic T is (minimal) sufficient for θ in the likelihood function of the observed data $\mathbf{x}$ given that $\vartheta = \theta$. The (minimal) sufficient statistic T will be frequently real valued and we will work with its pmf or pdf $g(t; \theta)$, given that $\vartheta = \theta$, with $t \in \mathcal{T}$ where $\mathcal{T}$ is an appropriate subset of $\Re$.

For the uniformity of notation, however, we will treat T as a continuous variable and hence the associated probabilities and expectations will be written in the form of integrals over the space $\mathcal{T}$. It should be understood that integrals will be interpreted as the appropriate sums when T is a discrete variable instead.

The *joint pdf* of T and ϑ is then given by

$$g(t; \theta)h(\theta) \text{ for all } t \in \mathcal{T} \text{ and } \theta \in \Theta. \tag{10.2.1}$$

The *marginal pdf* of T will be obtained by integrating the joint pdf from (10.2.1) with respect to θ. In other words, the marginal pdf of T can be written as

$$m(t) = \int_{\theta \in \Theta} g(t; \theta)h(\theta)d\theta \text{ for all } t \in \mathcal{T}. \tag{10.2.2}$$

Thus, we can obtain the conditional pdf of ϑ given that $T = t$ as follows:

$$k(\theta; t) \equiv k(\theta \mid T = t) = g(t; \theta)h(\theta)/m(t) \text{ for all } t \in \mathcal{T}$$
$$\text{and } \theta \in \Theta \text{ such that } m(t) > 0. \tag{10.2.3}$$

In passing, let us remark that the expression for $k(\theta; t)$ follows directly by applying the Bayes's Theorem (Theorem 1.4.3). In the statement of the Bayes's Theorem, simply replace $P(A_j \mid B), P(A_j), P(B \mid A_i)$ and $\Sigma_{i=1}^{k}$ by $k(\theta; t), h(\theta), g(t; \theta)$ and $\int_{\theta \in \Theta}$ respectively.

Definition 10.2.1 *The conditional pdf* $k(\theta; t)$ *of* ϑ *given that* $T = t$ *is called the posterior distribution of* ϑ.

Under the Bayesian paradigm, after having the likelihood function and the prior, the posterior pdf $k(\theta; t)$ of ϑ epitomizes how one combines the information about ϑ obtained from two separate sources, the prior knowledge and the collected data. The tractability of the final analytical expression of $k(\theta; t)$ largely depends on how easy or difficult it is to obtain the expression of $m(t)$. In some cases, the marginal distribution of T and the posterior distribution can be evaluated only numerically.

Example 10.2.1 Let $X_1, ..., X_n$ be iid Bernoulli(θ) given that $\vartheta = \theta$ where ϑ is the unknown probability of success, $0 < \vartheta < 1$. Given $\vartheta = \theta$, the statistic $T = \Sigma_{i=1}^n X_i$ is minimal sufficient for θ, and one has

$$g(t; \theta) = \binom{n}{t}\theta^t(1-\theta)^{n-t} \text{ for } t \in \mathcal{T}, \qquad (10.2.4)$$

where $\mathcal{T} = \{0, 1, ..., n\}$. Now, suppose that the prior distribution of ϑ on the space $\Theta = (0, 1)$ is taken to be Uniform$(0, 1)$, that is $h(\theta) = I(0 < \theta < 1)$. From (10.2.2), for $t \in \mathcal{T}$, we then obtain the marginal pmf of T as follows:

$$
\begin{aligned}
m(t) &= \int_{\theta=0}^{1} \binom{n}{t}\theta^t(1-\theta)^{n-t}d\theta \\
&= \binom{n}{t}\int_{\theta=0}^{1}\theta^t(1-\theta)^{n-t}d\theta \\
&= \binom{n}{t}b(t+1, n-t+1) \text{ where } b(\alpha, \beta) \text{ is} \\
&\qquad \text{the Beta function from (1.6.25).}
\end{aligned} \qquad (10.2.5)
$$

Thus, for any fixed value $t \in \mathcal{T}$, using (10.2.3) and (10.2.5), the posterior pdf of ϑ given the data $T = t$ can be expressed as

$$
\begin{aligned}
&k(\theta; t) \\
&= \left\{\binom{n}{t}\theta^t(1-\theta)^{n-t}h(\theta)\right\} / \left\{\binom{n}{t}b(t+1, n-t+1)\right\} \qquad (10.2.6) \\
&= [b(t+1, n-t+1)]^{-1}\theta^t(1-\theta)^{n-t}, \text{ for all } \theta \in (0, 1).
\end{aligned}
$$

That is, the posterior distribution of the success probability ϑ is described by the Beta$(t+1, n-t+1)$ distribution. ▲

> In principle, one may carry out similar analysis even if the unknown parameter ϑ is vector valued. But, in order to keep the presentation simple, we do not include any such example here. Look at the Exercise 10.3.8 for a taste.

Example 10.2.2 (Example 10.2.1 Continued) Let $X_1, ..., X_{10}$ be iid Bernoulli(θ) given that $\vartheta = \theta$ where ϑ is the unknown probability of success, $0 < \vartheta < 1$. The statistic $T = \Sigma_{i=1}^{10} X_i$ is minimal sufficient for θ given that $\vartheta = \theta$. Assume that the prior distribution of ϑ on the space $\Theta = (0, 1)$ is Uniform$(0, 1)$. Suppose that we have observed the particular value $T = 7$,

the number of successes out of ten Bernoulli trials. From (10.2.6) we know
that the posterior pdf of the success probability ϑ is that of the Beta(8, 4)
distribution. This posterior density has been plotted as a solid curve in the
Figure 10.2.1. This curve is skewed to the left. If we had observed the data
$T = 5$ instead, then the posterior distribution of the success probability
ϑ will be described by the Beta(6, 6) distribution. This posterior density
has been plotted as a dashed curve in the same Figure 10.2.1. This curve
is symmetric about $\theta = .5$. The Figure 10.2.1 clearly shows that under the
same *uniform* prior

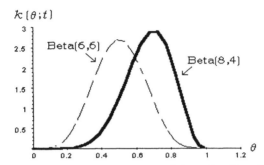

Figure 10.2.1. Posterior PDF's of ϑ Under Two Prior
Distributions of ϑ When $t = 7$

regarding the success probability ϑ but with different observed data, the
shape of the posterior distribution changes one's perception of what values
of ϑ are more (or less) probable. ▲

10.3 The Conjugate Priors

If the prior $h(\theta)$ is such that the integral in the equation (10.2.2) can
not be analytically found, then it will be nearly impossible to derive a
clean expression for the posterior pdf $k(\theta; t)$. In the case of many likelihood
functions, we may postulate a special type of prior $h(\theta)$ so that we can
achieve technical simplicity.

Definition 10.3.1 *Suppose that the prior pdf $h(\theta)$ for the unknown
parameter ϑ belongs to a particular family of distributions, $\mathcal{P}$. Then, $h(\theta)$
is called a conjugate prior for ϑ if and only if the posterior pdf $k(\theta; t)$ also
belongs to the same family $\mathcal{P}$.*

What it means is this: if $h(\theta)$ is chosen, for example, from the family of
beta distributions, then $k(\theta; t)$ must correspond to the pdf of an appropriate

beta distribution. Similarly, for conjugacy, if $h(\theta)$ is chosen from the family of normal or gamma distributions, then $k(\theta; t)$ must correspond to the pdf of an appropriate normal or gamma distribution respectively. The property of conjugacy demands that $h(\theta)$ and $k(\theta; t)$ must belong to the same family of distributions.

Example 10.3.1 (Example 10.2.1 Continued) Suppose that we have the random variables $X_1, ..., X_n$ which are iid Bernoulli(θ) given that $\vartheta = \theta$, where ϑ is the unknown probability of success, $0 < \vartheta < 1$. Given that $\vartheta = \theta$, the statistic $T = \Sigma_{i=1}^{n} X_i$ is minimal sufficient for θ, and recall that one has $g(t; \theta) = \binom{n}{t} \theta^t (1 - \theta)^{n-t}$ for $t \in \mathcal{T} = \{0, 1, ..., n\}$.

In the expression for $g(t; \theta)$, carefully look at the part which depends on θ, namely $\theta^t (1 - \theta)^{n-t}$. It resembles a beta pdf without the normalizing constant. Hence, we suppose that the prior distribution of ϑ on the space $\Theta = (0, 1)$ is Beta(α, β) where $\alpha (> 0)$ and $\beta (> 0)$ are known numbers. From (10.2.2), for $t \in \mathcal{T}$, we then obtain the marginal pmf of T as follows:

$$m(t) = \binom{n}{t} \frac{b(t + \alpha, n + \beta - t)}{b(\alpha, \beta)} \text{ where } b(\alpha, \beta)$$
$$\text{is the Beta function from (1.6.25).} \tag{10.3.1}$$

Now, using (10.2.3) and (10.3.1), the posterior pdf of ϑ given the data $T = t$ simplifies to

$$k(\theta; t) = [b(t + \alpha, n - t + \beta)]^{-1} \theta^{t + \alpha - 1} (1 - \theta)^{n + \beta - t - 1},$$
$$\text{for all } \theta \in (0, 1), \tag{10.3.2}$$

and fixed values $t \in \mathcal{T}$. In other words, the posterior pdf of the success probability ϑ is same as that for the Beta($t + \alpha, n - t + \beta$) distribution.

> We started with the beta prior and ended up with a beta posterior. Here, the beta pdf for ϑ is the conjugate prior for ϑ.

In the Example 10.2.1, the uniform prior was actually the Beta(1, 1) distribution. The posterior found in (10.3.2) agrees fully with that given by (10.2.6) when $\alpha = \beta = 1$. ▲

Example 10.3.2 Let $X_1, ..., X_n$ be iid Poisson(θ) given that $\vartheta = \theta$, where $\vartheta (> 0)$ is the unknown population mean. Given that $\vartheta = \theta$, the statistic $T = \Sigma_{i=1}^{n} X_i$ is minimal sufficient for θ, and recall that one has $g(t; \theta) = e^{-n\theta} (n\theta)^t / t!$ for $t \in \mathcal{T} = \{0, 1, 2, ...\}$.

In the expression of $g(t; \theta)$, look at the part which depends on θ, namely $e^{-n\theta} \theta^t$. It resembles a gamma pdf without the normalizing constant. Hence, we suppose that the prior distribution of ϑ on the space $\Theta = (0, \infty)$ is

Gamma(α, β) where $\alpha(> 0)$ and $\beta(> 0)$ are known numbers. From (10.2.2), for $t \in \mathcal{T}$, we then obtain the marginal pmf of T as follows:

$$
\begin{aligned}
m(t) &= \frac{n^t}{t! \beta^\alpha \Gamma(\alpha)} \int_{\theta=0}^{\infty} e^{-\theta(n\beta+1)/\beta} \theta^{\alpha+t-1} d\theta \\
&= \frac{n^t \beta^t \Gamma(\alpha+t)}{t! (n\beta+1)^{\alpha+t} \Gamma(\alpha)}.
\end{aligned}
\tag{10.3.3}
$$

Now, using (10.2.3) and (10.3.3), the posterior pdf of ϑ given the data $T = t$ simplifies to

$$
k(\theta; t) = \frac{\{(n\beta+1)/\beta\}^{\alpha+t}}{\Gamma(\alpha+t)} e^{-\theta(n\beta+1)/\beta} \theta^{\alpha+t-1} \text{ for all } \theta \in (0, \infty), \tag{10.3.4}
$$

and fixed values $t \in \mathcal{T}$. In other words, the posterior pdf of ϑ is the same as that for the Gamma$(t + \alpha, \beta(n\beta + 1)^{-1})$ distribution.

> We plugged in a gamma prior and ended up with a gamma posterior.

In this example, observe that the gamma pdf for ϑ is the conjugate prior for ϑ. ▲

Example 10.3.3 (Example 10.3.1 Continued) Suppose that we have the random variables $X_1, ..., X_n$ which are iid Bernoulli(θ) given that $\vartheta = \theta$, where ϑ is the unknown probability of success, $0 < \vartheta < 1$. As before, consider the statistic $T = \Sigma_{i=1}^n X_i$ which is minimal sufficient for θ given that $\vartheta = \theta$. Let us assume that the prior distribution of ϑ on the space $\Theta = (0, 1)$ is described as Beta(α, β) where $\alpha(> 0)$ and $\beta(> 0)$ are known numbers. In order to find the posterior distribution of ϑ, there is no real need to determine $m(t)$ first. The joint distribution of (ϑ, T) is given by

$$
\binom{n}{t} \frac{1}{b(\alpha, \beta)} \theta^t (1 - \theta)^{n-t} \theta^{\alpha-1} (1 - \theta)^{\beta-1} \propto \theta^{\alpha+t-1} (1 - \theta)^{\beta+n-t-1} \tag{10.3.5}
$$

for $0 < \theta < 1, t = 0, 1, ..., n$. Now, upon close examination of the rhs of (10.3.5), we realize that it does resemble a beta density without its normalizing constant. Hence, the posterior pdf of the success probability ϑ is going to be that of the Beta$(t + \alpha, n - t + \beta)$ distribution. ▲

Example 10.3.4 Let $X_1, ..., X_n$ be iid Poisson(θ) given that $\vartheta = \theta$, where $\vartheta(> 0)$ is the unknown population mean. As before, consider the statistic $T = \Sigma_{i=1}^n X_i$ which is minimal sufficient for θ given that $\vartheta = \theta$. Let us suppose that the prior distribution of ϑ on the space $\Theta = (0, \infty)$ is Gamma(α, β) where $\alpha(> 0)$ and $\beta(> 0)$ are known numbers. In order to find the posterior distribution of ϑ, again there is no real need to determine $m(t)$ first. The joint distribution of (ϑ, T) is given by

$$
\frac{n^t}{t!} \frac{1}{\beta^\alpha \Gamma(\alpha)} e^{-n\theta} \theta^t e^{-\theta/\beta} \theta^{\alpha-1} \propto e^{-(n+\frac{1}{\beta})\theta} \theta^{\alpha+t-1} \tag{10.3.6}
$$

for $0 < \theta < \infty$. Now, upon close examination of the rhs of (10.3.6), we realize that it does resemble a gamma density without its normalizing constant. Hence, the posterior pdf of ϑ is same as that for the Gamma$(t + \alpha, \beta(n\beta + 1)^{-1})$ distribution. ▲

> For this kind of analysis to go through, the observations $X_1, ..., X_n$ do not necessarily have to be iid to begin with. Look at the Exercises 10.3.1, 10.3.7, 10.4.2, 10.5.1 and 10.5.8.

Example 10.3.5 Let $X_1, ..., X_n$ be iid $N(\theta, \sigma^2)$ given that $\vartheta = \theta$, where $\vartheta(\in \Re)$ is the unknown population mean and $\sigma(> 0)$ is assumed known. Consider the statistic $T = \Sigma_{i=1}^{n} X_i$ which is minimal sufficient for θ given that $\vartheta = \theta$. Let us suppose that the prior distribution of ϑ on the space $\Theta = \Re$ is $N(\tau, \delta^2)$ where $\tau(\in \Re)$ and $\delta(> 0)$ are known numbers. In order to find the posterior distribution of ϑ, again there is no real need to determine $m(t)$ first. The joint distribution of (ϑ, T) is *proportional* to

$$
exp[\theta\{(t/\sigma^2) + (\tau/\delta^2)\}] \; exp[-\theta^2\{(n/\sigma^2) + (1/\delta^2)\}/2]
$$
$$
\propto exp\left\{-\tfrac{1}{2}\left(\tfrac{n}{\sigma^2} + \tfrac{1}{\delta^2}\right)\left[\theta - \left(\tfrac{t}{\sigma^2} + \tfrac{\tau}{\delta^2}\right)\left(\tfrac{n}{\sigma^2} + \tfrac{1}{\delta^2}\right)^{-1}\right]\right\}, \tag{10.3.7}
$$

for $\theta \in \Re$. Now, upon close examination of the last expression in (10.3.7), we realize that it does resemble a normal density without its normalizing constant.

> We started with a normal prior and ended up in a normal posterior.

With $\mu = \left(\tfrac{t}{\sigma^2} + \tfrac{\tau}{\delta^2}\right)\left(\tfrac{n}{\sigma^2} + \tfrac{1}{\delta^2}\right)^{-1}$ and $\sigma_0^2 = \left(\tfrac{n}{\sigma^2} + \tfrac{1}{\delta^2}\right)^{-1}$, the posterior distribution of ϑ turns out to be $N(\mu, \sigma_0^2)$. In this example again, observe that the normal pdf for ϑ is the conjugate prior for ϑ. ▲

> A conjugate prior may not be reasonable in every problem. Look at the Examples 10.6.1-10.6.4.

> In situations where the domain space $\mathcal{T}$ itself depends on the unknown parameter ϑ, one needs to be very careful in the determination of the posterior distribution. Look at the Exercises 10.3.4-10.3.6, 10.4.5-10.4.7 and 10.5.4-10.5.7.

The prior $h(\theta)$ used in the analysis reflects the experimenter's *subjective belief* regarding which ϑ-subsets are more (or less) likely. An experimenter may utilize hosts of related expertise to arrive at a realistic prior distribution $h(\theta)$. In the end, all types of Bayesian inferences follow from the posterior pdf $k(\theta; t)$ which is directly affected by the choice of $h(\theta)$. This is why it is of paramount importance that the prior pmf or pdf $h(\theta)$ is fixed in advance of data collection so that both the evidences regarding ϑ obtained from the likelihood function and the prior distribution remain useful and credible.

10.4   Point Estimation

In this section, we explore briefly how one approaches the point estimation problems of an unknown parameter ϑ under a particular *loss function*. Recall that the data consists of a random sample $\mathbf{X} = (X_1, ..., X_n)$ given that $\vartheta = \theta$. Suppose that a real valued statistic T is (minimal) sufficient for θ given that $\vartheta = \theta$. Let $\mathcal{T}$ denote the domain of t. As before, instead of considering the likelihood function itself, we will only consider the pmf or pdf $g(t; \theta)$ of the sufficient statistic T at the point $T = t$ given that $\vartheta = \theta$, for all $t \in \mathcal{T}$. Let $h(\theta)$ be the *prior distribution* of $\vartheta, \theta \in \Theta$.

An arbitrary point estimator of ϑ may be denoted by $\delta \equiv \delta(T)$ which takes the value $\delta(t)$ when one observes $T = t, t \in \mathcal{T}$. Suppose that the *loss* in estimating ϑ by the estimator $\delta(T)$ is given by

$$L^*(\vartheta, \delta) \equiv L^*(\vartheta, \delta(T)) = [\delta(T) - \vartheta]^2, \qquad (10.4.1)$$

which is referred to as the *squared error loss*.

The mean squared error (MSE) discussed in Section 7.3.1 will correspond to the weighted average of the loss function from (10.4.1) with respect to the weights assigned by the pmf or pdf $g(t; \theta)$. In other words, this average is actually a conditional average given that $\vartheta = \theta$. Let us define, conditionally given that $\vartheta = \theta$, the *risk function* associated with the estimator δ:

$$R^*(\theta, \delta) \equiv E_{T|\vartheta=\theta}[L^*(\theta, \delta)] = \int_{\mathcal{T}} L^*(\theta, \delta(t))g(t; \theta)dt. \qquad (10.4.2)$$

This is the *frequentist risk* which was referred to as MSE_δ in the Section 7.3. In Chapter 7, we saw examples of estimators δ_1 and δ_2 with risk functions $R^*(\theta, \delta_i), i = 1, 2$ where $R^*(\theta, \delta_1) > R^*(\theta, \delta_2)$ for some parameter values θ whereas $R^*(\theta, \delta_1) \leq R^*(\theta, \delta_2)$ for other parameter values θ. In other words, by comparing the two frequentist risk functions of δ_1 and δ_2, in some situations one may not be able to judge which estimator is decisively superior.

The prior $h(\theta)$ sets a sense of preference and priority of some values of ϑ over other values of ϑ. In a situation like this, while comparing two estimators δ_1 and δ_2, one may consider averaging the associated frequentist risks $R^*(\theta, \delta_i), i = 1, 2$, with respect to the prior $h(\theta)$ and then check to see which weighted average is smaller. The estimator with the smaller average risk should be the preferred estimator.

So, let us define the *Bayesian risk* (as opposed to the frequentist risk)

$$r^*(\vartheta, \delta) \equiv E_\vartheta[R^*(\vartheta, \delta)] = \int_\Theta R^*(\theta, \delta)h(\theta)d\theta. \qquad (10.4.3)$$

Suppose that $\mathcal{D}$ is the class of all estimators of ϑ whose Bayesian risks are finite. Now, the *best estimator* under the Bayesian paradigm will be δ^* from $\mathcal{D}$ such that

$$r^*(\vartheta, \delta^*) = \underset{\delta \in \mathcal{D}}{Inf} \, r^*(\vartheta, \delta). \qquad (10.4.4)$$

Such an estimator will be called the *Bayes estimator* of ϑ. In many standard problems, the Bayes estimator δ^* happens to be unique.

Let us suppose that we are going to consider only those estimators δ and prior $h(\theta)$ so that both $R^*(\theta, \delta)$ and $r^*(\vartheta, \delta)$ are finite, $\theta \in \Theta$.

Theorem 10.4.1 *The Bayes estimator $\delta^* \equiv \delta^*(T)$ is to be determined in such a way that the posterior risk of $\delta^*(t)$ is the least possible, that is*

$$\int_\Theta L^*(\theta, \delta^*(t))k(\theta; t)d\theta = \underset{\delta \in \mathcal{D}}{Inf} \int_\Theta L^*(\theta, \delta(t))k(\theta; t)d\theta \qquad (10.4.5)$$

for all possible observed data $t \in \mathcal{T}$.

Proof Assuming that $m(t) > 0$, let us express the Bayesian risk in the following form:

$$\begin{aligned} r^*(\vartheta, \delta) &= \int_\Theta \int_\mathcal{T} L^*(\theta, \delta(t))g(t; \theta)h(\theta)dtd\theta \\ &= \int_\mathcal{T} \left[\int_\Theta L^*(\theta, \delta(t))k(t; \theta)d\theta\right] m(t)dt. \end{aligned} \qquad (10.4.6)$$

In the last step, we used the relation $g(t; \theta)h(\theta) = k(t; \theta)m(t)$ and the fact that the order of the double integral $\int_\Theta \int_\mathcal{T}$ can be changed to $\int_\mathcal{T} \int_\Theta$ because the integrands are non-negative. The interchanging of the order of the integrals is allowed here in view of a result known as Fubini's Theorem which is stated as Exercise 10.4.10 for the reference.

Now, suppose that we have observed the data $T = t$. Then, the Bayes estimate $\delta^*(t)$ must be the one associated with the $\underset{\delta \in \mathcal{D}}{Inf} \int_\Theta L^*(\theta, \delta(t))k(\theta; t)d\theta$, that is the smallest posterior risk. The proof is complete. ∎

An attractive feature of the Bayes estimator δ^* is this: Having observed $T = t$, we can explicitly determine $\delta^*(t)$ by implementing the process of minimizing the posterior risk as stated in (10.4.5). In the case of the squared error loss function, the determination of the Bayes estimator happens to be very simple.

Theorem 10.4.2 *In the case of the squared error loss function, the Bayes estimate $\delta^* \equiv \delta^*(t)$ is the mean of the posterior distribution $k(\theta;t)$, that is*

$$\delta^*(t) = \int_\Theta \theta k(\theta;t)d\theta \equiv E_{\theta|T=t}[\vartheta] \qquad (10.4.7)$$

for all possible observed data $t \in T$.

Proof In order to determine the estimator $\delta^*(T)$, we need to minimize $\int_\Theta L^*(\theta,\delta(t))k(\theta;t)d\theta$ with respect to δ, for every fixed $t \in T$. Let us now rewrite

$$\begin{aligned}\int_\Theta L^*(\theta,\delta(t))k(\theta;t)d\theta &= \int_\Theta [\theta^2 - 2\theta\delta + \delta^2]k(\theta;t)d\theta \\ &= a(t) - 2\delta E_{\theta|T=t}[\vartheta] + \delta^2,\end{aligned} \qquad (10.4.8)$$

where we denote $a(t) = \int_\Theta \theta^2 k(\theta;t)d\theta$. In (10.4.8) we used the fact that $\int_\Theta k(\theta;t)d\theta = 1$ because $k(\theta;t)$ is a probability distribution on Θ. Now, we look at the expression $a(t) - 2\delta E_{\theta|T=t}[\vartheta] + \delta^2$ as a function of $\delta \equiv \delta(t)$ and wish to minimize this with respect to δ. One can accomplish this task easily. We leave out the details as an exercise. ∎

Example 10.4.1 (Example 10.3.1 Continued) Suppose that we have the random variables $X_1, ..., X_n$ which are iid Bernoulli(θ) given that $\vartheta = \theta$, where ϑ is the unknown probability of success, $0 < \vartheta < 1$. Given that $\vartheta = \theta$, the statistic $T = \Sigma_{i=1}^n X_i$ is minimal sufficient for θ. Suppose that the prior distribution of ϑ is Beta(α, β) where $\alpha(> 0)$ and $\beta(> 0)$ are known numbers. From (10.3.2), recall that the posterior distribution of ϑ given the data $T = t$ happens to be Beta($t + \alpha, n - t + \beta$) for $t \in T = \{0, 1, ..., n\}$. In view of the Theorem 10.4.2, under the squared error loss function, the Bayes estimator of ϑ would be the mean of the posterior distribution, namely, the mean of the Beta($t + \alpha, n - t + \beta$) distribution. One can check easily that the mean of this beta distribution simplifies to $(t + \alpha)/(\alpha + \beta + n)$ so that we can write:

$$\text{The Bayes estimator of } \vartheta \text{ is } \widehat{\vartheta}_B = \frac{(t + \alpha)}{(\alpha + \beta + n)}. \qquad (10.4.9)$$

Now, we can rewrite the Bayes estimator as follows:

$$\widehat{\vartheta}_B = \frac{n(t/n) + (\alpha + \beta)[\alpha/(\alpha + \beta)]}{(\alpha + \beta + n)} = \frac{n\overline{x} + (\alpha + \beta)[\alpha/(\alpha + \beta)]}{n + \alpha + \beta} \qquad (10.4.10)$$

From the likelihood function, that is given $\vartheta = \theta$, the maximum likelihood estimate or the UMVUE of θ would be $\overline{x}$, whereas the mean of the prior distribution is $\alpha/(\alpha+\beta)$. The Bayes estimator is an appropriate weighted average of $\overline{x}$ and $\alpha/(\alpha+\beta)$. If the sample size n is large, then the classical estimator $\overline{x}$ gets more weight than the mean of the prior belief, namely $\alpha/(\alpha+\beta)$ so that the observed data is valued more. For small sample sizes, $\overline{x}$ gets less weight and thus the prior information is valued more. That is, for large sample sizes, the sample evidence is weighed in more, whereas if the sample size is small, the prior mean of ϑ is trusted more. ▲

Example 10.4.2 (Example 10.3.5 Continued) Let $X_1, ..., X_n$ be iid $N(\theta, \sigma^2)$ given that $\vartheta = \theta$, where $\vartheta(\in \Re)$ is the unknown population mean and $\sigma(> 0)$ is assumed known. Consider the statistic $T = \Sigma_{i=1}^n X_i$ which is minimal sufficient for θ given that $\vartheta = \theta$. Let us suppose that the prior distribution of ϑ is $N(\tau, \delta^2)$ where $\tau(\in \Re)$ and $\beta(> 0)$ are known numbers. From the Example 10.3.5, recall that the posterior distribution of ϑ is $N(\mu, \sigma_0^2)$ where $\mu = \left(\frac{t}{\sigma^2} + \frac{\tau}{\delta^2}\right)\left(\frac{n}{\sigma^2} + \frac{1}{\delta^2}\right)^{-1}$ and $\sigma_0^2 = \left(\frac{n}{\sigma^2} + \frac{1}{\delta^2}\right)^{-1}$. In view of the Theorem 10.4.2, under the squared error loss function, the Bayes estimator of ϑ would be the mean of the posterior distribution, namely, the $N(\mu, \sigma_0^2)$ distribution. In other words, we can write:

The Bayes estimator of ϑ is $\widehat{\vartheta}_B = \left(\frac{t}{\sigma^2} + \frac{\tau}{\delta^2}\right) / \left(\frac{n}{\sigma^2} + \frac{1}{\delta^2}\right).$ (10.4.11)

Now, we can rewrite the Bayes estimator as follows:

$$\widehat{\vartheta}_B = \left(\frac{n}{\sigma^2}\overline{x} + \frac{1}{\delta^2}\tau\right) / \left(\frac{n}{\sigma^2} + \frac{1}{\delta^2}\right).$$ (10.4.12)

From the likelihood function, that is given $\vartheta = \theta$, the maximum likelihood estimate or the UMVUE of θ would be $\overline{x}$, whereas the mean of the prior distribution is τ. The Bayes estimate is an appropriate weighted average of $\overline{x}$ and τ. If $\frac{n}{\sigma^2}$ is larger than $\frac{1}{\delta^2}$, that is if the classical estimate $\overline{x}$ is more reliable (smaller variance) than the prior mean τ, then the sample mean $\overline{x}$ gets more weight than the prior belief. When $\frac{n}{\sigma^2}$ is smaller than $\frac{1}{\delta^2}$, the prior mean of ϑ is trusted more than the sample mean $\overline{x}$. ▲

10.5   Credible Intervals

Let us go back to the basic notions of Section 10.2. After combining the information about ϑ gathered separately from two sources, namely the observed data $\mathbf{X} = \mathbf{x}$ and the prior distribution $h(\theta)$, suppose that we have obtained the posterior pmf or pdf $k(\theta; t)$ of ϑ given $T = t$. Here, the statistic T is assumed to be (minimal) sufficient for θ, given that $\vartheta = \theta$.

Definition 10.5.1 *With fixed α, $0 < \alpha < 1$, a subset Θ^* of the parameter space Θ is called a $100(1 - \alpha)\%$ credible set for the unknown parameter ϑ*

if and only if the posterior probability of the subset Θ^* *is at least* $(1 - \alpha)$, *that is*

$$\int_{\Theta^*} k(\theta; t) d\theta \geq 1 - \alpha. \qquad (10.5.1)$$

In some situations, a $100(1 - \alpha)\%$ credible set $\Theta^*(\subset \Theta)$ for the unknown parameter ϑ may not be a subinterval of Θ. The geometric structure of Θ^* will largely depend upon the nature of the prior pdf $h(\theta)$ on Θ. When Θ^* is an interval, we naturally refer to it as a $100(1 - \alpha)\%$ *credible interval* for ϑ.

10.5.1   Highest Posterior Density

It is conceivable that one will have to choose a $100(1 - \alpha)\%$ credible set Θ^* from a long list of competing credible sets for ϑ. Naturally, we should include only those values of ϑ in a credible set which are "very likely" according to the posterior distribution. Also, the credible set should be "small" because we do not like to remain too uncertain about the value of ϑ. The concept of the *highest posterior density* (HPD) credible set combines these two ideas.

Definition 10.5.2 *With some fixed* α, $0 < \alpha < 1$, *a subset* Θ^* *of the parameter space* Θ *is called a* $100(1 - \alpha)\%$ *highest posterior density (HPD) credible set for the unknown parameter* ϑ *if and only if the subset* Θ^* *has the following form:*

$$\Theta^* = \{\theta \in \Theta : k(\theta; t) \geq a\}, \qquad (10.5.2)$$

where $a \equiv a(\alpha, t)$ *is the largest constant such that the posterior probability of* Θ *is at least* $(1 - \alpha)$.

Now, suppose that the posterior density $k(\theta; t)$ is *unimodal*. In this case, a $100(1 - \alpha)\%$ HPD credible set happens to be an interval. If we assume that the posterior pdf is *symmetric* about its finite mean (which would coincide with the its median too), then the HPD $100(1 - \alpha)\%$ credible interval Θ^* will be the *shortest* as well as *equal tailed and symmetric* about the posterior mean. In order to verify this conclusion, one may apply the Theorem 9.2.1 in the context of the posterior pdf $k(\theta; t)$.

Example 10.5.1 (Examples 10.3.5 and 10.4.2 Continued) Let $X_1, ..., X_n$ be iid $N(\theta, \sigma^2)$ given that $\vartheta = \theta$ where $\vartheta(\in \Re)$ is the unknown population mean and $\sigma(> 0)$ is assumed known. Consider the statistic $T = \Sigma_{i=1}^n X_i$ which is minimal sufficient for θ given that $\vartheta = \theta$. Let us suppose that the prior distribution of ϑ is $N(\tau, \delta^2)$ where $\tau(\in \Re)$ and $\delta(> 0)$ are known numbers. From the Example 10.3.5, recall that the posterior distribution of ϑ is $N(\mu, \sigma_0^2)$ where $\mu = \left(\frac{t}{\sigma^2} + \frac{\tau}{\delta^2}\right)\left(\frac{n}{\sigma^2} + \frac{1}{\delta^2}\right)^{-1}$ and $\sigma_0^2 = \left(\frac{n}{\sigma^2} + \frac{1}{\delta^2}\right)^{-1}$. Also, from the Example 10.4.2, recall that under the *squared error loss function*, the Bayes estimate of ϑ would be the mean of the posterior distribution, namely $\widehat{\vartheta}_B = \left(\frac{n}{\sigma^2}\overline{x} + \frac{1}{\delta^2}\tau\right) / \left(\frac{n}{\sigma^2} + \frac{1}{\delta^2}\right)$. Let $z_{\alpha/2}$ stands for the upper $100(\alpha/2)\%$ point of the standard normal distribution. The posterior distribution $N(\mu, \sigma_0^2)$ being symmetric about μ, the HPD $100(1-\alpha)\%$ credible interval Θ^* will become

$$\left\{\left(\frac{n}{\sigma^2}\overline{x} + \frac{1}{\delta^2}\tau\right) / \left(\frac{n}{\sigma^2} + \frac{1}{\delta^2}\right)\right\} \pm z_{\alpha/2}\left(\frac{n}{\sigma^2} + \frac{1}{\delta^2}\right)^{-1/2}. \qquad (10.5.3)$$

This interval is centered at the posterior mean $\widehat{\vartheta}_B$ and it stretches either way by $z_{\alpha/2}$ times the posterior standard deviation σ_0. ▲

Example 10.5.2 (Example 10.5.1 Continued) In an elementary statistics course with large enrollment, the instructor postulated that the midterm examination score (X) should be distributed normally with an unknown mean θ and variance 30, given that $\vartheta = \theta$. The instructor assumed the prior distribution $N(73, 20)$ for ϑ and then looked at the midterm examination scores of $n = 20$ randomly selected students. The observed data follows:

$$\begin{array}{cccccccccc}
85 & 78 & 87 & 92 & 66 & 59 & 88 & 61 & 59 & 78 \\
82 & 72 & 75 & 79 & 63 & 67 & 69 & 77 & 73 & 81
\end{array}$$

One then has $\overline{x} = 74.55$. We wish to construct a 95% HPD credible interval for the population average score ϑ so that we have $z_{\alpha/2} = 1.96$. The posterior mean and variance are respectively

$$\widehat{\vartheta}_B = \left(\tfrac{20}{30}(74.55) + \tfrac{1}{20}(73)\right) / \left(\tfrac{20}{30} + \tfrac{1}{20}\right) \approx 74.442,$$
$$\sigma_0^2 = \left(\tfrac{20}{30} + \tfrac{1}{20}\right)^{-1} = 1.3953.$$

From (10.5.3), we claim that the 95% HPD credible interval for ϑ would be

$$74.442 \pm 1.96\sqrt{1.3953} \approx 74.442 \pm 2.315 \qquad (10.5.4)$$

which will be approximately the interval $(72.127, 76.757)$. ▲

Example 10.5.3 (Example 10.4.1 Continued) Suppose that we have the random variables $X_1, ..., X_n$ which are iid Bernoulli(θ) given that $\vartheta = \theta$ where ϑ is the unknown probability of success, $0 < \vartheta < 1$. Given that $\vartheta = \theta$, the statistic $T = \Sigma_{i=1}^n X_i$ is minimal sufficient for θ. Suppose that the prior distribution of ϑ is Beta(α, β) where $\alpha(> 0)$ and $\beta(> 0)$ are known numbers. From (10.3.2), recall that the posterior distribution of ϑ is Beta$(t + \alpha, n - t + \beta)$ for $t \in \mathcal{T} = \{0, 1, ..., n\}$. Using the Definition 10.5.2, one can immediately write down the HPD $100(1 - \alpha)\%$ credible interval Θ^* for ϑ: Let us denote

$$\Theta^* = \{\theta \in (0,1) : \tfrac{1}{b(t+\alpha, n-t+\beta)}\theta^{t+\alpha-1}(1-\theta)^{n-t+\beta-1} > a\} \qquad (10.5.5)$$

with the largest positive number a such that $P\{\Theta^* \mid T = t\} \geq 1 - \alpha$. ▲

Example 10.5.4 (Example 10.5.3 Continued) In order to gather information about a new pain-killer, it was administered to $n = 10$ comparable patients suffering from the same type of headache. Each patient was checked after one-half hour to see if the pain was gone and we found seven patients reporting that their pain vanished. It is postulated that we have observed the random variables $X_1, ..., X_{10}$ which are iid Bernoulli(θ) given that $\vartheta = \theta$ where $0 < \vartheta < 1$ is the unknown probability of pain relief for each patient, $0 < \vartheta < 1$. Suppose that the prior distribution of ϑ was fixed in advance of data collection as Beta($2, 6$). Now, we have observed seven patients who got the pain relief. From (10.5.5), one can immediately write down the form of the HPD 95% credible interval Θ^* for ϑ as

$$\Theta^* = \{\theta \in (0,1) : \tfrac{1}{b(9,9)}\theta^8(1-\theta)^8 > a\} \qquad (10.5.6)$$

with the largest positive number a such that $P\{\Theta^* \mid T = t\} \geq 1 - \alpha$. Let us rewrite (10.5.6) as

$$\Theta^* \equiv \Theta^*(b) = \{\theta \in (0,1) : \theta(1-\theta) > b\}, b > 0. \qquad (10.5.7)$$

We will choose the positive number b in (10.5.7) such that $P\{\Theta^*(b) \mid T = t\} = 1 - \alpha$. We can simplify (10.5.7) and equivalently express, for $0 < b < \tfrac{1}{4}$,

$$\Theta^*(b) = \left\{\theta \in (0,1) : \tfrac{1}{2} - \sqrt{(\tfrac{1}{4} - b)} < \theta < \tfrac{1}{2} + \sqrt{(\tfrac{1}{4} - b)}\right\}. \qquad (10.5.8)$$

Now, the *posterior probability content* of the credible interval $\Theta^*(b)$ is given by

$$q(b) = \int_{\Theta^*(b)} \tfrac{1}{b(9,9)}\theta^8(1-\theta)^8, 0 < b < \tfrac{1}{4}. \qquad (10.5.9)$$

We have given a plot of the function $q(b)$ in the Figure 10.5.1. From this figure, we can guess that the posterior probability of the credible interval

$\Theta^*(b)$ is going to be 95% when we choose $b \approx .2$.

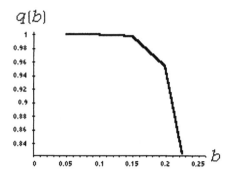

Figure 10.5.1. Plot of the Probability Content $q(b)$ from (10.5.9)

With the help of MAPLE, we numerically solved the equation $q(b) = .95$ and found that $b \approx .20077$. In other words, (10.5.8) will lead to the HPD 95% credible interval for ϑ which is $(.27812, .72188)$. ▲

10.5.2   Contrasting with the Confidence Intervals

Now, we emphasize important conceptual differences in the notions of a $100(1-\alpha)\%$ *confidence interval* and a $100(1-\alpha)\%$ (HPD) *credible interval* for an unknown parameter ϑ. In either case, one starts with an appropriate likelihood function. In the case of a confidence interval, the unknown parameter ϑ is assumed fixed. In the case of a credible interval, however, the unknown parameter ϑ is assumed to be a random variable having some pdf $h(\theta)$ on Θ. Under the frequentist paradigm, we talk about the unknown parameter ϑ belonging to a constructed confidence interval, only from the point of view of limiting relative frequency idea. Recall the interpretation given in Section 9.2.3. On the other hand, under the Bayesian paradigm we *can evaluate* the posterior probability that ϑ belongs to a credible interval by integrating the posterior pdf on the interval. In this sense, the confidence coefficient may be regarded as a measure of *initial accuracy* which is fixed in advance by the experimenter whereas the posterior probability content of a credible interval may be regarded as the *final accuracy*. The latter conclusion may look more appealing in some applications but we may also remind the reader that a HPD credible interval arises at the expense of *assuming* that (i) the unknown parameter ϑ is a random variable *and* (ii) ϑ has a known prior distribution $h(\theta)$ for $\theta \in \Theta$.

Example 10.5.5 (Example 10.5.2 Continued) With the sample of size $n = 20$ midterm examination scores, we had $\bar{x} = 74.55$ and $\sigma^2 = 30$. So a 95% confidence interval for the population mean ϑ would be $74.55 \pm 1.96(\sqrt{30/20})$ which will lead to the interval $(72.15, 76.95)$. It is not meaningful for a frequentist to talk about the chance that the unknown parameter ϑ belongs or does not belong to the interval $(72.15, 76.95)$ because ϑ is not a random variable. In order to make any such statement meaningful, the phrase "chance" would have to be tied in with some sense of randomness about ϑ in the first place! A frequentist will add, however, that in a large collection of such confidence intervals constructed in the same fashion with $n = 20$, approximately 95% of the intervals will include the unknown value of ϑ.

On the other hand, the 95% HPD credible interval for ϑ turned out to be $(72.127, 76.757)$. Here, one should have no confusion about the interpretation of this particular interval's "probability" content. The posterior pdf puts exactly 95% posterior probability on this interval. This statement is *verifiable* because the posterior is completely known once the observed value of the sample mean becomes available.

In this example, one may be tempted to say that the credible interval is more accurate than the confidence interval because it is shorter of the two. But, one should be very careful to pass any swift judgement. Before one compares the two intervals, one should ask: Are these two interval estimates comparable at all, and if they are, then in what sense are they comparable? This sounds like a very simple question, but it defies any simple answer. We may mention that philosophical differences between the frequentist and Bayesian schools of thought remains deep-rooted from the days of Fisher and Jeffreys. ▲

10.6   Tests of Hypotheses

We would like to recall that a concept of the p-value associated with the Neyman-Pearson's frequentist theory of tests of hypotheses was briefly discussed in Section 8.4.1. In a Bayesian framework, one may argue that making a decision to reject or accept H_0 based on the p-value is not very reasonable. Berger et al. (1997) gave an interesting exposition. A *Bayes test* of hypotheses is then formulated as follows.

Let us consider iid real valued random variables $X_1, ..., X_n$ having a common pmf or pdf $f(x; \vartheta)$ where the unknown parameter ϑ takes the values $\theta \in \Theta(\subseteq \Re)$. We wish to test a null hypothesis $H_0 : \vartheta \in \Theta_0$ against an alternative hypothesis $H_1 : \vartheta \in \Theta_1$ where $\Theta_0 \cap \Theta_1 = \varphi, \Theta_0 \cup \Theta_1 \subseteq \Theta$. Given $\vartheta = \theta$, we first write down the likelihood function $L(\theta) = \Pi_{i=1}^n f(x_i; \theta), \theta \in \Theta$ and obtain the (minimal) sufficient statistic $T \equiv T(\mathbf{X})$ which is assumed real valued. Let us continue to denote the prior of ϑ and the pmf or pdf of T given $\vartheta = \theta$ by $h(\theta)$ and $g(t; \theta)$ respectively.

Next, one obtains the posterior distribution $k(\theta; t)$ as in (10.2.3) and proceeds to evaluate the posterior probability of the null and alternative spaces Θ_0, Θ_1 respectively as follows:

$$\alpha_0 = \int_{\theta \in \Theta_0} k(\theta; t) d\theta, \alpha_1 = \int_{\theta \in \Theta_1} k(\theta; t) d\theta \text{ for } t \in \mathcal{T}. \tag{10.6.1}$$

We assume that both α_0, α_1 are positive and view these as the *posterior evidences* in favor of H_0, H_1 respectively.

A *Bayes test* will reject H_0 if and only if $\alpha_0 < \alpha_1$. $\tag{10.6.2}$

We refrain from giving more details. One may refer to Ferguson (1967) and Berger (1985).

Example 10.6.1 (Example 10.3.5 Continued) Let $X_1, ..., X_5$ be iid $N(\theta, 4)$ given that $\vartheta = \theta$ where $\vartheta (\in \Re)$ is the unknown population mean. Let us suppose that the prior distribution of ϑ on the space $\Theta = \Re$ is $N(3, 1)$. We wish to test a null hypothesis $H_0 : \vartheta < 1$ against an alternative hypothesis $H_1 : \vartheta > 2.8$. Consider the statistic $T = \Sigma_{i=1}^5 X_i$ which is minimal sufficient for θ given that $\vartheta = \theta$. Suppose that the observed value of T is $t = 6.5$.

With $\mu = \left(\frac{6.5}{4} + 3\right)\left(\frac{5}{4} + 1\right)^{-1} \approx 2.0556$ and $\sigma_0^2 = \left(\frac{5}{4} + 1\right)^{-1} \approx .44444$, the posterior distribution of ϑ turns out to be $N(\mu, \sigma_0^2)$. Let us denote a random variable Y which is distributed as $N(\mu, \sigma_0^2)$ and $Z = (Y - \mu)/\sigma_0$.

Now, from (10.6.1), we have $\alpha_0 = P(Y < 1) = P(Z < -1.5834) \approx .056665$ and $\alpha_1 = P(Y > 2.8) = P(Z > 1.1166) \approx .13208$. Thus, the Bayes test from (10.6.2) will reject H_0. ▲

So far, we have relied heavily upon conjugate priors. But, in some situations, a conjugate prior may not be available or may not seem very appealing. The set of examples in the next section will highlight a few such scenarios.

10.7   Examples with Non-Conjugate Priors

The following examples exploit some specific non-conjugate priors. Certainly these are *not* the only choices of such priors. Even though the chosen priors are non-conjugate, we are able to derive analytically the posterior and Bayes estimate.

Example 10.7.1 Let X be $N(\theta, 1)$ given that $\vartheta = \theta$ where ϑ is the unknown population mean. Consider the statistic $T = X$ which is minimal sufficient for θ given that $\vartheta = \theta$. We have been told that ϑ is positive and so there is no point in assuming a normal prior for the parameter ϑ. For simplicity, let us suppose that the prior distribution of ϑ on the space $\Theta = \Re^+$ is exponential with known mean $\alpha^{-1}(> 0)$. Now, let us first proceed to determine the marginal pdf $m(x)$. The joint distribution of (ϑ, X) is given by

$$
\begin{aligned}
&\alpha(\sqrt{2\pi})^{-1} exp\{-\tfrac{1}{2}(x^2 - 2\theta x + \theta^2) - \alpha\theta\} \\
&= \alpha(\sqrt{2\pi})^{-1} exp\{\tfrac{1}{2}\alpha^2 - \alpha x\} \times \\
&\quad exp\{-\tfrac{1}{2}[\theta - (x - \alpha)]^2\} \text{ for } x \in \Re, \theta \in \Re^+.
\end{aligned}
\tag{10.7.1}
$$

Recall the standard normal pdf $\phi(y) = (\sqrt{2\pi})^{-1} exp\{-\tfrac{1}{2}y^2\}$ and the df $\Phi(y) = \int_{-\infty}^{y} \phi(u)du$, with $y \in \Re$. Thus, for all $x \in \Re$, the marginal pdf of X can be written as

$$
\begin{aligned}
m(x) \\
&= \alpha(\sqrt{2\pi})^{-1} exp\{\tfrac{1}{2}\alpha^2 - \alpha x\} \int_{\theta=0}^{\infty} \phi(\theta - [x - \alpha])d\theta \\
&= \alpha(\sqrt{2\pi})^{-1} exp\{\tfrac{1}{2}\alpha^2 - \alpha x\}\{1 - \Phi(-[x - \alpha])\} \\
&= \alpha(\sqrt{2\pi})^{-1} exp\{\tfrac{1}{2}\alpha^2 - \alpha x\}\Phi([x - \alpha]).
\end{aligned}
\tag{10.7.2}
$$

Now, combining (10.7.1)-(10.7.2), we obtain the posterior pdf of ϑ given that $X = x$ as follows: For all $x \in \Re, \theta \in \Re^+$,

$$
\begin{aligned}
k(\theta; x) \\
&= \{\Phi([x - \alpha])\}^{-1}(\sqrt{2\pi})^{-1} exp\{-\tfrac{1}{2}[\theta - (x - \alpha)]^2\} \\
&= \{\Phi([x - \alpha])\}^{-1}\phi(\theta - [x - \alpha]).
\end{aligned}
\tag{10.7.3}
$$

In this example, the chosen prior is *not* a conjugate one and yet we have been able to derive the expression of the posterior pdf in a closed form. In the Exercise 10.7.1 we ask the reader to check by integration that $k(\theta; x)$ is indeed a pdf on the positive half of the real line. Also refer to the Exercises 10.7.2-10.7.3. ▲

Example 10.7.2 (Example 10.7.1 Continued) Let X be $N(\theta, 1)$ given that $\vartheta = \theta$ where ϑ is the unknown population mean. Consider the statistic $T = X$ which is minimal sufficient for θ given that $\vartheta = \theta$. We were told that ϑ was positive and we took the exponential distribution with the known mean $\alpha^{-1}(> 0)$ as our prior for ϑ.

Let us recall the standard normal pdf $\phi(y) = (\sqrt{2\pi})^{-1} exp\{-\tfrac{1}{2}y^2\}$ and the df $\Phi(y) = \int_{-\infty}^{y} \phi(u)du$, for $y \in \Re$. Now, from the Example 10.7.1, we know that the posterior distribution of ϑ is given by $k(\theta; x) = \{\Phi([x - \alpha])\}^{-1}\phi(\theta - [x - \alpha])$ for $\theta > 0, -\infty < x < \infty$.

In view of the Theorem 10.4.2, under the squared error loss function, the Bayes estimate of ϑ would be the mean of the posterior distribution. In other words, one can write the Bayes estimate $\widehat{\vartheta}_B$ of ϑ as

$$
\begin{aligned}
\int_{\theta=0}^{\infty} & \theta k(\theta; x) d\theta \\
&= \{\Phi([x-\alpha])\}^{-1} \int_{\theta=0}^{\infty} \theta \phi(\theta - [x-\alpha]) d\theta \\
&= \{\Phi([x-\alpha])\}^{-1} \int_{\theta=0}^{\infty} [\theta - (x-\alpha)] \phi(\theta - [x-\alpha]) d\theta \qquad (10.7.4) \\
&\quad + \{\Phi([x-\alpha])\}^{-1} (x-\alpha) \int_{\theta=0}^{\infty} \phi(\theta - [x-\alpha]) d\theta \\
&= I_1 + I_2, \text{ say.}
\end{aligned}
$$

It is easy to check that $I_2 = x - \alpha$. Next, we rewrite I_1 as

$$
\begin{aligned}
\{\Phi([x-\alpha])\}^{-1} & \int_{y=-(x-\alpha)}^{\infty} y \phi(y) dy \\
&= (\sqrt{2\pi})^{-1} \{\Phi([x-\alpha])\}^{-1} \int_{y=(x-\alpha)^2/2}^{\infty} e^{-u} du \qquad (10.7.5) \\
&= \{\phi([x-\alpha])/\Phi([x-\alpha])\}.
\end{aligned}
$$

Now, combining (10.7.4)-(10.7.5), we get

$$
\widehat{\vartheta}_B = [x-\alpha] + \frac{\phi([x-\alpha])}{\Phi([x-\alpha])} \qquad (10.7.6)
$$

as the Bayes estimate of ϑ. Also refer to the Exercises 10.7.2-10.7.3. ▲

Example 10.7.3 (Example 10.7.2 Continued) Let X be $N(\theta, 1)$ given that $\vartheta = \theta$ where ϑ is the unknown population mean. Consider the statistic $T = X$ which is minimal sufficient for θ given that $\vartheta = \theta$. We were told that ϑ was positive and we supposed that the prior distribution of ϑ was exponential with the known mean $\alpha^{-1}(> 0)$. Under the squared error loss function, the Bayes estimate of ϑ turned out to be $\widehat{\vartheta}_B$ given by (10.7.6). Observe that the posterior distribution's support is $\Re^+$ and thus $\widehat{\vartheta}_B$, the mean of this posterior distribution, must be positive. But, notice that the observed data x may be larger or smaller than α. Can we check that the expression of $\widehat{\vartheta}_B$ will always lead to a positive estimate? The answer will be in the *affirmative* if we verify the following claim:

$$
p(x) = x\Phi(x) + \phi(x) > 0 \text{ for all } x \in \Re. \qquad (10.7.7)
$$

Note that $\frac{d}{dx}\Phi(x) = \phi(x)$ and $\frac{d}{dx}\phi(x) = -x\phi(x)$ for all $x \in \Re$. Thus, one has $\frac{d}{dx}p(x) = \Phi(x)$ which is positive for all $x \in \Re$ so that the function $p(x) \uparrow$ in x. Next, let us consider the behavior of $p(x)$ when x is near negative infinity. By appealing to L'Hôpital's rule from (1.6.29), we observe that $\lim_{x \to -\infty} x\Phi(x) = -\lim_{x \to -\infty} x^2\phi(x) = 0$. Hence, $\lim_{x \to -\infty} p(x) = 0$. Combine this with the fact that $p(x)$ is increasing in $x(\in \Re)$ to validate (10.7.7). In other words, the expression of $\widehat{\vartheta}_B$ is always positive. Also refer to the Exercises 10.7.2-10.7.3. ▲

In the three previous examples, we had worked with the normal likelihood function given $\vartheta = \theta$. But, we were told that ϑ was positive and hence we were forced to put a prior only on $(0, \infty)$. We experimented with the exponential prior for ϑ which happened to be skewed to the right. In the case of a normal likelihood, some situations may instead call for a non-conjugate *but symmetric* prior for ϑ. Look at the next example.

Example 10.7.4 Let X be $N(\theta, 1)$ given that $\vartheta = \theta$ where $\vartheta(\in \Re)$ is the unknown population mean. Consider the statistic $T = X$ which is minimal sufficient for θ given that $\vartheta = \theta$. We have been told that (i) ϑ is an arbitrary real number, (ii) ϑ is likely to be distributed symmetrically around zero, and (iii) the prior probability around zero is little more than what it is likely with a normal prior. With the $N(0, 1)$ prior on ϑ, we note that the prior probability of the interval $(-.01, .01)$ amounts to 7.9787×10^{-3} whereas with the Laplace prior pdf $h(\theta) = \frac{1}{2}e^{-|\theta|}I(\theta \in \Re)$, the prior probability of the same interval amounts to 9.9502×10^{-3}. So, let us start with the Laplace prior $h(\theta)$ for ϑ. From the apriori description of ϑ alone, it does not follow however that the chosen prior distribution is the only viable candidate. But, let us begin some preliminary investigation anyway.

One can verify that the marginal pdf $m(x)$ of X can be written as

$$
\begin{aligned}
&\tfrac{1}{2}exp\{-x^2/2\}[\{1 - \Phi(x+1)\}exp\{(x+1)^2/2\} \\
&+\Phi(x-1)exp\{(x-1)^2/2\}] \text{ for all } x \in \Re .
\end{aligned}
\tag{10.7.8}
$$

Next, one may verify that the posterior pdf of ϑ given that $X = x$ can be expressed as follows: For all $x \in \Re$ and $\theta \in \Re$,

$$
k(\theta; x) = \begin{cases} [b(x)]^{-1}exp\{\tfrac{1}{2}(x+1)^2\}\phi([\theta + (x+1)]) \text{ if } \theta \leq 0 \\ [b(x)]^{-1}exp\{\tfrac{1}{2}(x-1)^2\}\phi([\theta - (x-1)]) \text{ if } \theta > 0, \end{cases}
\tag{10.7.9}
$$

where $b(x) = \{1 - \Phi(x+1)\}exp\{(x+1)^2/2\}+\Phi(x-1)exp\{(x-1)^2/2\}$. The details are left out as the Exercise 10.7.4. Also refer to the Exercises 10.7.5-10.7.6. ▲

10.8   Exercises and Complements

10.2.1 Suppose that $X_1, ..., X_n$ are iid with the common pdf $f(x; \theta) = \theta e^{-\theta x}I(x > 0)$ given that $\vartheta = \theta$ where $\vartheta(> 0)$ is the unknown parameter.

Assume the prior density $h(\theta) = \alpha e^{-\alpha\theta} I(\theta > 0)$ where $\alpha(> 0)$ is known. Derive the posterior pdf of ϑ given that $T = t$ where $T = \Sigma_{i=1}^n X_i, t > 0$.

10.2.2 (Example 10.2.1 Continued) Suppose that $X_1, ..., X_5$ are iid with the common pdf $f(x; \theta) = \theta e^{-\theta x} I(x > 0)$ given that $\vartheta = \theta$ where $\vartheta(> 0)$ is the unknown parameter. Assume the prior density $h(\theta) = \frac{1}{8} e^{-\theta/8} I(\theta > 0)$. Draw and compare the posterior pdf's of ϑ given that $\Sigma_{i=1}^5 X_i = t$ with $t = 15, 40, 50$.

10.2.3 Suppose that $X_1, ..., X_n$ are iid Poisson(θ) given that $\vartheta = \theta$ where $\vartheta(> 0)$ is the unknown population mean. Assume the prior density $h(\theta) = e^{-\theta} I(\theta > 0)$. Derive the posterior pdf of ϑ given that $T = t$ where $T = \Sigma_{i=1}^n X_i, t = 0, 1, 2, ...$.

10.2.4 (Example 10.2.3 Continued) Suppose that $X_1, ..., X_{10}$ are iid Poisson(θ) given that $\vartheta = \theta$ where $\vartheta(> 0)$ is the unknown population mean. Assume the prior density $h(\theta) = \frac{1}{4} e^{-\theta/4} I(\theta > 0)$. Draw and compare the posterior pdf's of ϑ given that $\Sigma_{i=1}^{10} X_i = t$ with $t = 30, 40, 50$.

10.3.1 Let X_1, X_2 be independent, X_1 be distributed as $N(\theta, 1)$ and X_2 be distributed as $N(2\theta, 3)$ given that $\vartheta = \theta$ where $\vartheta(\in \Re)$ is the unknown parameter. Consider the minimal sufficient statistic T for θ given that $\vartheta = \theta$. Let us suppose that the prior distribution of ϑ on the space $\Theta = \Re$ is $N(5, \tau^2)$ where $\tau(> 0)$ is a known number. Derive the posterior distribution of ϑ given that $T = t, t \in \Re$. {*Hint*: Given $\vartheta = \theta$, is the statistic T normally distributed?}

10.3.2 Suppose that $X_1, ..., X_n$ are iid Exponential(θ) given that $\vartheta = \theta$ where the parameter $\vartheta(> 0)$ is unknown. We say that ϑ has the *inverted gamma prior*, denoted by IGamma(α, β), whenever the prior pdf is given by

$$h(\theta) = \{\beta^\alpha \Gamma(\alpha) \theta^{\alpha+1}\}^{-1} exp\{-1/(\theta\beta)\} I(\theta > 0),$$

where α, β are known positive numbers. Denote the sufficient statistic $T = \Sigma_{i=1}^n X_i$ given that $\vartheta = \theta$.

(i)   Show that the prior distribution of ϑ is IGamma(α, β) if and only if ϑ^{-1} has the Gamma(α, β) distribution;

(ii)   Show that the posterior distribution of ϑ given $T = t$ turns out to be IGamma$\left(n + \alpha, \{t + \beta^{-1}\}^{-1}\right)$.

10.3.3 Suppose that $X_1, ..., X_n$ are iid $N(0, \theta^2)$ given that $\vartheta = \theta$ where the parameter $\vartheta(> 0)$ is unknown. Assume that ϑ^2 has the inverted gamma prior IGamma(α, β), where α, β are known positive numbers. Denote the sufficient statistic $T = \Sigma_{i=1}^n X_i^2$ given that $\vartheta = \theta$. Show that the posterior distribution of ϑ^2 given $T = t$ is an appropriate inverted gamma distribution.

10.3.4 Suppose that $X_1, ..., X_n$ are iid Uniform$(0, \theta)$ given that $\vartheta = \theta$ where the parameter $\vartheta(> 0)$ is unknown. We say that ϑ has the *Pareto* prior, denoted by Pareto(α, β), when the prior pdf is given by

$$h(\theta) = \beta \alpha^\beta \theta^{-(\beta+1)} I(\alpha < \theta < \infty),$$

where α, β are known positive numbers. Denote the sufficient statistic $T = X_{n:n}$, the largest order statistic, given that $\vartheta = \theta$. Show that the posterior distribution of ϑ given $T = t$ turns out to be Pareto$(\max(t, \alpha), n + \beta)$.

10.3.5 (Exercise 10.3.4 Continued) Let $X_1, ..., X_n$ be iid Uniform$(0, a\theta)$ given that $\vartheta = \theta$ where the parameter $\vartheta(> 0)$ is unknown, but $a(> 0)$ is assumed known. Suppose that ϑ has the Pareto(α, β) prior where α, β are known positive numbers. Denote the sufficient statistic $T = X_{n:n}$, the largest order statistic, given that $\vartheta = \theta$. Show that the posterior distribution of ϑ given $T = t$ is an appropriate Pareto distribution.

10.3.6 Let $X_1, ..., X_n$ be iid Uniform$(-\theta, \theta)$ given that $\vartheta = \theta$ where the parameter $\vartheta(> 0)$ is unknown. Suppose that ϑ has the Pareto(α, β) prior where α, β are known positive numbers. Denote the minimal sufficient statistic $T = |X|_{n:n}$, the largest order statistic among $|X_1|, ..., |X_n|$, given that $\vartheta = \theta$. Show that the posterior distribution of ϑ given $T = t$ is an appropriate Pareto distribution. {*Hint*: Can this problem be reduced to the Exercise 10.3.4?}

10.3.7 Let X_1, X_2, X_3 be independent, X_1 be distributed as $N(\theta, 1)$, X_2 be distributed as $N(2\theta, 3)$, and X_3 be distributed as $N(\theta, 3)$ given that $\vartheta = \theta$ where $\vartheta(\in \Re)$ is the unknown parameter. Consider the minimal sufficient statistic T for θ given that $\vartheta = \theta$. Let us suppose that the prior distribution of ϑ on the space $\Theta = \Re$ is $N(2, \tau^2)$ where $\tau(> 0)$ is a known number. Derive the posterior distribution of ϑ given that $T = t, t \in \Re$. {*Hint*: Given $\vartheta = \theta$, is the statistic T normally distributed?}

10.3.8 We denote $\boldsymbol{\vartheta} = (\vartheta_1, \vartheta_2^2), \boldsymbol{\theta} = (\theta_1, \theta_2^2)$ and suppose that $X_1, ..., X_n$ are iid $N(\theta_1, \theta_2^2)$ given that $\boldsymbol{\vartheta} = \boldsymbol{\theta}$ where the parameters $\vartheta_1(\in \Re), \vartheta_2(\in \Re^+)$ are assumed both unknown. Consider the minimal sufficient statistic $\mathbf{T} = (\overline{X}, S^2)$ for $\boldsymbol{\theta}$ given that $\boldsymbol{\vartheta} = \boldsymbol{\theta}$. We are given the joint prior distribution of $\vartheta_1, \vartheta_2^2$ as follows:

$$h(\theta_1, \theta_2^2) = h_{1|2}(\theta_1) h_2(\theta_2^2),$$

where $h_{1|2}(\theta_1)$ stands for the pdf of the $N(\mu, \theta_2^2)$ distribution and $h_2(\theta_2^2)$ stands for the pdf of the IGamma$(2, 1)$ distribution. The form of the inverted gamma pdf was given in the Exercise 10.3.3.

(i)   Given that $\vartheta = \theta$, write down the joint pdf of $\overline{X}, S^2$ by taking the product of the two separate pdf's because these are independent, and this will serve as the likelihood function;

(ii)   Multiply the joint pdf from part (i) with the joint prior pdf $h(\theta_1, \theta_2^2)$. Then, combine the terms involving θ_1 (and θ_2^2) and the terms involving only θ_2^2. From this, conclude that the joint posterior pdf $k(\boldsymbol{\theta}; \mathbf{t})$ of $\boldsymbol{\vartheta}$ given that $\overline{X} = \overline{x}, S^2 = s^2$, that is $\mathbf{t} = (\overline{x}, s^2)$ can be viewed as $k_{1|\vartheta_2^2 = \theta_2^2, \mathbf{t}}(\theta_1) k_{2| \mathbf{t}}(\theta_2^2)$. Here, $k_{1|\vartheta_2^2 = \theta_2^2, \mathbf{t}}(\theta_1)$ stands for the pdf (in the variable θ_1) of the $N(\mu_0, \sigma_0^2)$ distribution with $\mu_0 = (\mu + n\overline{x})/(n+1), \sigma_0^2 = \theta_2^2/(n+1)$, and $k_{2| \mathbf{t}}(\theta_2^2)$ stands for the pdf (in the variable θ_2^2) of the IGamma(α_0, β_0) with $\alpha_0 = 2 + \frac{1}{2}n, \beta_0^{-1} = 1 + \frac{1}{2}(n-1)s^2 + \frac{1}{2}(n+1)^{-1}n(\overline{x} - \mu)^2$;

(iii)   From the joint posterior pdf of $\boldsymbol{\vartheta}$ given that $\overline{X} = \overline{x}, S^2 = s^2$, integrate out θ_1 and thus show that the marginal posterior pdf of θ_2^2 is the same as that of the IGamma(α_0, β_0) distribution as in part (ii);

(iv)   From the joint posterior pdf of $\boldsymbol{\vartheta}$ given that $\overline{X} = \overline{x}, S^2 = s^2$, integrate out θ_2^2 and thus show that the marginal posterior pdf of θ_1 is the same as that of an appropriate Student's t distribution.

10.4.1 (Example 10.3.2 Continued) Let $X_1, ..., X_n$ be iid Poisson(θ) given that $\vartheta = \theta$ where $\vartheta(> 0)$ is the unknown population mean. We suppose that the prior distribution of ϑ on the space $\Theta = (0, \infty)$ is Gamma(α, β) where $\alpha(> 0)$ and $\beta(> 0)$ are known numbers. Under the squared error loss function, derive the expression of the Bayes estimate $\widehat{\vartheta}_B$ for ϑ. {*Hint*: Use the form of the posterior from (10.3.3) and then appeal to the Theorem 10.4.2.}

10.4.2 (Exercise 10.3.1 Continued) Let X_1, X_2 be independent, X_1 be distributed as $N(\theta, 1)$ and X_2 be distributed as $N(2\theta, 3)$ given that $\vartheta = \theta$ where $\vartheta(\in \Re)$ is the unknown parameter. Let us suppose that the prior distribution of ϑ on the space $\Theta = \Re$ is $N(5, \tau^2)$ where $\tau(> 0)$ is a known number. Under the squared error loss function, derive the expression of the Bayes estimate $\widehat{\vartheta}_B$ for ϑ. {*Hint*: Use the form of the posterior from Exercise 10.3.1 and then appeal to the Theorem 10.4.2.}

10.4.3 (Exercise 10.3.2 Continued) Let $X_1, ..., X_n$ be iid Exponential(θ) given that $\vartheta = \theta$ where the parameter $\vartheta(> 0)$ is unknown. Assume that ϑ has the inverted gamma prior IGamma(α, β) where α, β are known positive numbers. Under the squared error loss function, derive the expression of the Bayes estimate $\widehat{\vartheta}_B$ for ϑ. {*Hint*: Use the form of the posterior from Exercise 10.3.2 and then appeal to the Theorem 10.4.2.}

10.4.4 (Exercise 10.3.3 Continued) Let $X_1, ..., X_n$ be iid $N(0, \theta^2)$ given that $\vartheta = \theta$ where the parameter $\vartheta(> 0)$ is unknown. Assume that ϑ^2 has the inverted gamma prior IGamma(α, β) where α, β are known positive numbers. Under the squared error loss function, derive the expression of the Bayes estimate $\widehat{\vartheta}_B$ for ϑ. {*Hint*: Use the form of the posterior from Exercise 10.3.3 and then appeal to the Theorem 10.4.2.}

10.4.5 (Exercise 10.3.4 Continued) Let $X_1, ..., X_n$ be iid Uniform$(0, \theta)$ given that $\vartheta = \theta$ where the parameter $\vartheta(> 0)$ is unknown. Suppose that ϑ has the Pareto(α, β) prior pdf $h(\theta) = \beta \alpha^\beta \theta^{-(\beta+1)} I(\alpha < \theta < \infty)$ where α, β are known positive numbers. Under the squared error loss function, derive the expression of the Bayes estimate $\widehat{\vartheta}_B$ for ϑ. {*Hint*: Use the form of the posterior from Exercise 10.3.4 and then appeal to the Theorem 10.4.2.}

10.4.6 (Exercise 10.3.5 Continued) Let $X_1, ..., X_n$ be iid Uniform$(0, a\theta)$ given that $\vartheta = \theta$ where the parameter $\vartheta(> 0)$ is unknown, but $a(> 0)$ is assumed known. Suppose that ϑ has the Pareto(α, β) prior where α, β are known positive numbers. Under the squared error loss function, derive the expression of the Bayes estimate $\widehat{\vartheta}_B$ for ϑ. {*Hint*: Use the form of the posterior from Exercise 10.3.5 and then appeal to the Theorem 10.4.2.}

10.4.7 (Exercise 10.3.6 Continued) Let $X_1, ..., X_n$ be iid Uniform$(-\theta, \theta)$ given that $\vartheta = \theta$ where the parameter $\vartheta(> 0)$ is unknown. Suppose that ϑ has the Pareto(α, β) prior where α, β are known positive numbers. Under the squared error loss function, derive the expression of the Bayes estimate $\widehat{\vartheta}_B$ for ϑ. {*Hint*: Use the form of the posterior from Exercise 10.3.6 and then appeal to the Theorem 10.4.2.}

10.4.8 A soda dispensing machine is set up so that it automatically fills the soda cans. The actual amount of fill must not vary too much from the target (12 fl. ounces) because the overfill will add extra cost to the manufacturer while the underfill will generate complaints from the customers hampering the image of the company. A random sample of the fills for 15 cans gave $\Sigma_{i=1}^{15}(x_i - 12)^2 = 0.14$. Assuming a $N(12, \theta^2)$ distribution for the actual fills given that $\vartheta = \theta(> 0)$ and the IGamma$(10, 10)$ prior for ϑ^2, obtain the Bayes estimate of the true population variance under the squared error loss.

10.4.9 Ten automobiles of the same make and model were driven by the drivers with similar road habits and the gas mileage for each was recorded over a week. The summary results included $\overline{x} = 22$ miles per gallon. Assume a $N(\theta, 4)$ distribution for the actual gas mileage given that $\vartheta = \theta(\in \Re)$ and the $N(20, 6)$ prior for ϑ. Construct a 90% HPD credible interval for the true average gas mileage per gallon, ϑ. Give the Bayes estimate of the true average gas mileage per gallon ϑ under the squared error loss.

10.4.10 (Fubini's Theorem) Suppose that a function of two real variables $g(x_1, x_2)$ is either non-negative or integrable on the space $\mathcal{X} = \mathcal{X}_1 \times \mathcal{X}_2 (\subseteq \Re \times \Re)$. That is, for all $(x_1, x_2) \in \mathcal{X}$, the function $g(x_1, x_2)$ is either non-negative or integrable. Then, show that the order of the (two-dimensional) integrals can be interchanged, that is, one can write

$$\int_{\mathcal{X}_2} \int_{\mathcal{X}_1} g(x_1, x_2) dx_1 dx_2 = \int_{\mathcal{X}_2} \left\{ \int_{\mathcal{X}_1} g(x_1, x_2) dx_1 \right\} dx_2$$
$$= \int_{\mathcal{X}_1} \left\{ \int_{\mathcal{X}_2} g(x_1, x_2) dx_2 \right\} dx_1.$$

{*Note*: This is a hard result to verify. It is stated here for reference purposes and completeness. Fubini's Theorem was used in the proof of the Theorem 10.4.1 for changing the order of two integrals.}

10.5.1 (Exercise 10.4.2 Continued) Let X_1, X_2 be independent, X_1 be distributed as $N(\theta, 1)$ and X_2 be distributed as $N(2\theta, 3)$ given that $\vartheta = \theta$ where $\vartheta(\in \Re)$ is the unknown parameter. Let us suppose that the prior distribution of ϑ on the space $\Theta = \Re$ is $N(5, \tau^2)$ where $\tau(> 0)$ is a known number. With fixed $0 < \alpha < 1$, derive the $100(1 - \alpha)\%$ HPD credible interval for ϑ. {*Hint*: Use the form of the posterior from the Exercise 10.3.1.}

10.5.2 (Exercise 10.4.3 Continued) Let $X_1, ..., X_n$ be iid Exponential(θ) given that $\vartheta = \theta$ where the parameter $\vartheta(> 0)$ is unknown. Suppose that ϑ has the IGamma(1, 1) prior pdf $h(\theta) = \theta^{-2} exp\{-1/\theta\} I(\theta > 0)$. With fixed $0 < \alpha < 1$, derive the $100(1 - \alpha)\%$ HPD credible interval for ϑ. {*Hint*: Use the form of the posterior from the Exercise 10.3.2.}

10.5.3 (Exercise 10.4.4 Continued) Let $X_1, ..., X_n$ be iid $N(0, \theta^2)$ given that $\vartheta = \theta$ where the parameter $\vartheta(> 0)$ is unknown. Suppose that ϑ^2 has the IGamma(1, 1). With fixed $0 < \alpha < 1$, derive the $100(1 - \alpha)\%$ HPD credible interval for ϑ^2. {*Hint*: Use the form of the posterior from the Exercise 10.3.3.}

10.5.4 (Exercise 10.4.5 Continued) Let $X_1, ..., X_n$ be iid Uniform$(0, \theta)$ given that $\vartheta = \theta$ where the parameter $\vartheta(> 0)$ is unknown. Suppose that ϑ has the Pareto(τ, β) prior, that is the prior pdf is given by $h(\theta) = \beta\tau^\beta \theta^{-(\beta+1)} I(\tau < \theta < \infty)$ where τ, β are known positive numbers. With fixed $0 < \alpha < 1$, derive the $100(1-\alpha)\%$ HPD credible interval for ϑ. {*Hint*: Use the form of the posterior from the Exercise 10.3.4.}

10.5.5 (Exercise 10.4.6 Continued) Let $X_1, ..., X_n$ be iid Uniform$(0, a\theta)$ given that $\vartheta = \theta$ where the parameter $\vartheta(> 0)$ is unknown, but $a(> 0)$ is assumed known. Suppose that ϑ has the Pareto(τ, β) prior where τ, β are known positive numbers. With fixed $0 < \alpha < 1$, derive the $100(1 - \alpha)\%$ HPD credible interval for ϑ. {*Hint*: Use the form of the posterior from the Exercise 10.3.5.}

10.5.6 (Exercise 10.4.7 Continued) Let $X_1, ..., X_n$ be iid Uniform$(-\theta, \theta)$ given that $\vartheta = \theta$ where the parameter $\vartheta (> 0)$ is unknown. Suppose that ϑ has the Pareto(τ, β) prior where τ, β are known positive numbers. With fixed $0 < \alpha < 1$, derive the $100(1 - \alpha)\%$ HPD credible interval for ϑ. Under the squared error loss function, derive the expression of the Bayes estimate $\widehat{\vartheta}_B$ for ϑ. {*Hint*: Use the form of the posterior from the Exercise 10.3.6.}

10.5.7 Suppose that X has its pdf $f(x; \theta) = 2\theta^{-2}(\theta - x)I(0 < x < \theta)$ given that $\vartheta = \theta$ where the parameter $\vartheta (> 0)$ is unknown. Suppose that ϑ has the Gamma$(3, 2)$ distribution.

(*i*) Show that the posterior pdf $k(\theta; x) = \frac{1}{4}(\theta - x)e^{-(\theta-x)/2} \times$
       $I(0 < x < \theta)$;

(*ii*) With fixed $0 < \alpha < 1$, derive the $100(1 - \alpha)\%$ HPD credible
        interval for ϑ.

10.5.8 (Exercise 10.3.7 Continued) Let X_1, X_2, X_3 be independent, X_1 be distributed as $N(\theta, 1)$, X_2 be distributed as $N(2\theta, 3)$, and X_3 be distributed as $N(\theta, 3)$ given that $\vartheta = \theta$ where $\vartheta (\in \Re)$ is the unknown parameter. Consider the minimal sufficient statistic T for θ given that $\vartheta = \theta$. Let us suppose that the prior distribution of ϑ on the space $\Theta = \Re$ is $N(2, 9)$. Suppose that the following data has been observed:

$$x_1 = 3.5, x_2 = 8.2 \text{ and } x_3 = 4.1.$$

Derive the 95% HPD credible interval for ϑ. Under the squared error loss function, derive the expression of the Bayes estimate $\widehat{\vartheta}_B$ for ϑ. {*Hint*: First find the posterior along the lines of the Exercise 10.3.7.}

10.6.1 (Example 10.6.1 Continued) Let $X_1, ..., X_{10}$ be iid $N(\theta, 4.5)$ given that $\vartheta = \theta$ where $\vartheta (\in \Re)$ is the unknown parameter. Let us suppose that the prior distribution of ϑ on the space $\Theta = \Re$ is $N(4, 2)$. Consider the statistic $T = \Sigma_{i=1}^{10} X_i$ which is minimal sufficient for θ given that $\vartheta = \theta$. Suppose that the observed value of T is $t = 48.5$. Use the Bayes test from (10.6.2) to choose between a null hypothesis $H_0 : \vartheta < 3$ against an alternative hypothesis $H_1 : \vartheta \geq 5$.

10.6.2 (Example 10.3.3 Continued) Let X_1, X_2, X_3 be iid Bernoulli(θ) given that $\vartheta = \theta$ where ϑ is the unknown probability of success, $0 < \vartheta < 1$. Let us assume that the prior distribution of ϑ on the space $\Theta = (0, 1)$ is described as Beta$(2, 4)$. Consider the statistic $T = \Sigma_{i=1}^3 X_i$ which is minimal sufficient for θ given that $\vartheta = \theta$. Suppose that the observed value of T is $t = 2$. Use the Bayes test from (10.6.2) to choose between a null hypothesis

$H_0 : \vartheta \leq .3$ against an alternative hypothesis $H_1 : \vartheta \geq .5$. {*Hint*: Use integration by parts when evaluating α_0, α_1 from (10.6.1). Review (1.6.28) as needed.}

10.6.3 (Example 10.3.4 Continued) Let $X_1, ..., X_4$ be iid Poisson(θ) given that $\vartheta = \theta$ where $\vartheta(> 0)$ is the unknown population mean. Let us suppose that the prior distribution of ϑ on the space $\Theta = (0, \infty)$ is Gamma($2, 3$). Consider the statistic $T = \Sigma_{i=1}^4 X_i$ which is minimal sufficient for θ given that $\vartheta = \theta$. Suppose that the observed value of T is $t = 1$. Use the Bayes test from (10.6.2) to choose between a null hypothesis $H_0 : \vartheta \leq 1.5$ against an alternative hypothesis $H_1 : \vartheta \geq 2.5$. {*Hint*: Use integration by parts when evaluating α_0, α_1 from (10.6.1). Review (1.6.28) as needed.}

10.7.1 (Example 10.7.1 Continued) Using the definition of a pdf, show that the function $k(\theta; x)$ given in (10.7.3) is indeed a probability density function of ϑ.

10.7.2 Let $X_1, ..., X_n$ be iid $N(\theta, 1)$ given that $\vartheta = \theta$ where $\vartheta(\in \Re^+)$ is the unknown parameter. Let us suppose that the prior pdf of ϑ on the space $\Theta = \Re^+$ is given by $h(\theta) = \alpha e^{-\alpha\theta} I(\theta > 0)$ where $\alpha(> 0)$ is a known number. Consider $T = \overline{X}$, the sample mean, which is minimal sufficient for θ given that $\vartheta = \theta$.

(i)   Derive the marginal pdf $m(t)$ of T for $t \in \Re$;

(ii)   Derive the posterior pdf $k(\theta; t)$ for ϑ given that $T = t$, for $\theta > 0$ and $-\infty < t < \infty$;

(iii)   Under the squared error loss function, derive the Bayes estimate $\widehat{\vartheta}_B$ and show, as in the Example 10.7.3, that $\widehat{\vartheta}_B$ can take only positive values.

{*Hint*: Proceed along the lines of the Examples 10.7.1-10.7.2.}

10.7.3 (Exercise 10.7.2 Continued) Let $X_1, ..., X_{10}$ be iid $N(\theta, 1)$ given that $\vartheta = \theta$ where $\vartheta(\in \Re^+)$ is the unknown parameter. Let us suppose that the prior pdf of ϑ on the space $\Theta = \Re^+$ is given by $h(\theta) = \alpha e^{-\alpha\theta} I(\theta > 0)$ where $\alpha(> 0)$ is a known number. Suppose that the observed value of the sample mean was 5.92. Plot the posterior pdf $k(\theta; t)$ assigning the values $\alpha = \frac{1}{2}, \frac{1}{4}$ and $\frac{1}{8}$. Comment on the empirical behaviors of the corresponding posterior pdf and the Bayes estimates of ϑ.

10.7.4 (Example 10.7.4 Continued) Verify the expressions of $m(x)$ and $k(\theta; x)$ given by (10.7.8) and (10.7.9) respectively. Under the squared error loss function, derive the form of the Bayes estimate $\widehat{\vartheta}_B$.

10.7.5 Let $X_1, ..., X_n$ be iid $N(\theta, 1)$ given that $\vartheta = \theta$ where $\vartheta(\in \Re)$ is the unknown parameter. Let us suppose that the prior pdf of ϑ on the space $\Theta = \Re$ is given by $h(\theta) = \frac{1}{2}\alpha e^{-\alpha|\theta|} I(\theta \in \Re)$ where $\alpha(> 0)$ is a known number. Consider $T = \overline{X}$, the sample mean, which is minimal sufficient for

θ given that $\vartheta = \theta$.

- (i)   Derive the marginal pdf $m(t)$ of T for $t \in \Re$;
- (ii)   Derive the posterior pdf $k(\theta; t)$ for ϑ given that $T = t$, for $\theta > 0$ and $-\infty < t < \infty$;
- (iii)   Under the squared error loss function, derive the Bayes estimate $\widehat{\vartheta}_B$ and show, as in the Example 10.7.3, that $\widehat{\vartheta}_B$ can take only positive values.

{*Hint*: Proceed along the Example 10.7.4.}

10.7.6 (Exercise 10.7.5 Continued) Let $X_1, ..., X_{10}$ be iid $N(\theta, 1)$ given that $\vartheta = \theta$ where $\vartheta(\in \Re)$ is the unknown parameter. Let us suppose that the prior pdf of ϑ on the space $\Theta = \Re$ is given by $h(\theta) = \frac{1}{2}\alpha e^{-\alpha|\theta|}I(\theta \in \Re)$ where $\alpha(> 0)$ is a known number. Suppose that the observed value of the sample mean was 5.92. Plot the posterior pdf $k(\theta; t)$ assigning the values $\alpha = \frac{1}{2}, \frac{1}{4}$ and $\frac{1}{8}$. Comment on the empirical behaviors of the corresponding posterior pdf and the Bayes estimates of ϑ.

11

Likelihood Ratio and Other Tests

11.1 Introduction

In Chapter 8, a theory of UMP level α tests was developed for a simple null hypothesis against a lower- or upper-sided alternative hypothesis. But, we have mentioned that even in a one-parameter problem, sometimes a UMP level α test does not exist when choosing between a simple null hypothesis against a two-sided alternative hypothesis. Recall the situation from Section 8.5.1 in the case of testing the mean of a normal distribution with known variance when the alternative hypothesis was two-sided. One may also recall the Exercises 8.5.2, 8.5.4 and 8.5.5 in this context. In these situations, the *likelihood ratio tests* provide useful methodologies. This general approach to construct test procedures for composite null and alternative hypotheses was developed by Neyman and Pearson (1928a,b,1933a,b).

We start with iid real valued random variables $X_1, ..., X_n$ having a common pdf $f(x; \boldsymbol{\theta})$ where the unknown parameter $\boldsymbol{\theta}$ consists of $p(\geq 1)$ components $(\theta_1, ..., \theta_p) \in \Theta(\subseteq \Re^p)$. We wish to test

a null hypothesis		an alternative hypothesis
$H_0 : \boldsymbol{\theta} \in \Theta_0$	versus	$H_1 : \boldsymbol{\theta} \in \Theta_1$

with a given level α where $\Theta_1 = \Theta - \Theta_0, 0 < \alpha < 1$. First, let us write down the likelihood function:

$$L(\boldsymbol{\theta}) = \Pi_{i=1}^n f(x_i; \boldsymbol{\theta}), \boldsymbol{\theta} \in \Theta. \qquad (11.1.1)$$

Then, we look at $\underset{\boldsymbol{\theta} \in \Theta_0}{Sup} L(\boldsymbol{\theta})$ which is interpreted as the best evidence in favor of the null hypothesis H_0. On the other hand, $\underset{\boldsymbol{\theta} \in \Theta}{Sup} L(\boldsymbol{\theta})$ is interpreted as the overall best evidence in favor of $\boldsymbol{\theta}$ without regard to any restrictions. Now, the likelihood ratio (LR) *test statistic* is defined as

$$\Lambda = \underset{\boldsymbol{\theta} \in \Theta_0}{Sup} L(\boldsymbol{\theta}) / \underset{\boldsymbol{\theta} \in \Theta}{Sup} L(\boldsymbol{\theta}), \qquad (11.1.2)$$

whereas the LR test is implemented as follows:

$$\text{Reject } H_0 \text{ if and only if } \Lambda \text{ is "small" } (< k). \qquad (11.1.3)$$

Note that small values of Λ are associated with the small values of $\underset{\theta \in \Theta_0}{Sup} \; L(\boldsymbol{\theta})$ relative to $\underset{\theta \in \Theta}{Sup} \; L(\boldsymbol{\theta})$. If the best evidence in favor of the null hypothesis appears weak, then the null hypothesis is rejected. That is, we reject H_0 for significantly small values of Λ.

It is easy to see that one must have $0 < \Lambda < 1$ because in the definition of Λ, the supremum in the numerator (denominator) is taken over a smaller (larger) set $\Theta_0(\Theta)$. The cut-off number $k \in (0, 1)$ has to be chosen in such a way that the LR test from (11.1.3) has the required level α.

For simplicity, we will handle only a special kind of null hypothesis. Let us test

$$H_0 : \theta_1 = \theta_1^* \text{ versus } H_1 : \theta_1 \neq \theta_1^*,$$

where θ_1^* is a known and fixed value of the (sub-)parameter θ_1. One may be tempted to say that the hypothesis H_0 is a simple null hypothesis. But, actually it may not be so. Even though H_0 specifies a fixed value for a single component of $\boldsymbol{\theta}$, observe that the other components of $\boldsymbol{\theta}$ remain unknown and arbitrary.

How should we evaluate $\underset{\theta \in \Theta_0}{Sup} \; L(\boldsymbol{\theta})$? First, in the expression of $L(\boldsymbol{\theta})$, we must plug in the value θ_1^* in the place of θ_1. Then, we maximize the likelihood function $L(\theta_1^*, \theta_2, ..., \theta_p)$ with respect to the (sub-)parameters $\theta_2, ..., \theta_p$ by substituting their respective MLE's when we know that $\theta_1 = \theta_1^*$. On the other hand, the $\underset{\theta \in \Theta}{Sup} \; L(\boldsymbol{\theta})$ is found by plugging in the MLE's of all the components $\theta_1, ..., \theta_p$ in the likelihood function. In the following sections, we highlight these step by step derivations in a variety of situations.

Section 11.2 introduces LR tests for the mean and variance of a normal population. In Section 11.3, we discuss LR tests for comparing the means and variances of two independent normal populations. In Section 11.4, under the assumption of bivariate normality, test procedures are given for the population correlation coefficient ρ and for comparing the means as well as variances.

11.2   One-Sample Problems

We focus on a single normal population and some LR tests associated with it. With fixed $\alpha \in (0, 1)$, first a level α LR test is derived for a specified population mean against the two-sided alternative hypothesis, and come up with the customary two-sided Z-test (t-test) when the population variance is assumed known (unknown). Next, we obtain a level α LR test for a specified population variance against the two-sided alternative hypothesis, and come up with the customary two-sided χ^2-test assuming that the population mean is unknown.

11.2.1 LR Test for the Mean

Suppose that $X_1, ..., X_n$ are iid observations from the $N(\mu, \sigma^2)$ population where $\mu \in \Re, \sigma \in \Re^+$. We assume that μ is unknown. Given $\alpha \in (0, 1)$, consider choosing between a null hypothesis $H_0 : \mu = \mu_0$ and a two-sided alternative hypothesis $H_1 : \mu \neq \mu_0$ with level α where μ_0 is a fixed real number. We address the cases involving known σ or unknown σ separately. As usual, $\overline{X} = n^{-1}\Sigma_{i=1}^{n}X_i$ and $S^2 = (n-1)^{-1}\Sigma_{i=1}^{n}(X_i - \overline{X})^2$ respectively denote the sample mean and variance.

Variance Known

Since σ is known, we have $\theta = \mu, \Theta_0 = \{\mu_0\}$ and $\Theta = \Re$. In this case, H_0 is a simple null hypothesis. The likelihood function is given by

$$L(\mu) = \{\sigma\sqrt{2\pi}\}^{-n} exp\{-\tfrac{1}{2\sigma^2}\Sigma_{i=1}^{n}(x_i - \mu)^2\} \text{ for all } \mu \in \Re. \qquad (11.2.1)$$

Observe that

$$\underset{\mu\in\Theta_0}{Sup}\ L(\mu) = \{\sigma\sqrt{2\pi}\}^{-n} exp\{-\tfrac{1}{2\sigma^2}\Sigma_{i=1}^{n}(x_i - \mu_0)^2\}, \qquad (11.2.2)$$

since Θ_0 has the single element μ_0. On the other hand, one has

$$\underset{\mu\in\Theta}{Sup}\ L(\mu) = \{\sigma\sqrt{2\pi}\}^{-n} exp\{-\tfrac{1}{2\sigma^2}\Sigma_{i=1}^{n}(x_i - \overline{x})^2\}, \text{ since } \overline{x} \text{ is}$$
$$\text{the MLE of } \mu, \text{ so that } \overline{x} \text{ maximizes } L(\mu) \text{ over } \Theta. \qquad (11.2.3)$$

Note that for any real number c, we can write

$$\Sigma_{i=1}^{n}(x_i - c)^2 = \Sigma_{i=1}^{n}(x_i - \overline{x})^2 + n(\overline{x} - c)^2, \qquad (11.2.4)$$

and hence by combining (11.2.2)-(11.2.3) we obtain the likelihood ratio

$$\begin{aligned}
\Lambda &= exp\{-\tfrac{1}{2\sigma^2}\left[\Sigma_{i=1}^{n}(x_i - \mu_0)^2 - \Sigma_{i=1}^{n}(x_i - \overline{x})^2\right]\} \\
&= exp\{-\tfrac{1}{2\sigma^2}\left[n(\overline{x} - \mu_0)^2\right]\}, \text{ in view of (11.2.4)}.
\end{aligned} \qquad (11.2.5)$$

Intuitively speaking, one may be inclined to reject H_0 if and only if $\underset{\mu\in\Theta_0}{Sup} L(\mu)$ is small or $\underset{\mu\in\Theta}{Sup} L(\mu)$ is large. But for some data, $\underset{\mu\in\Theta_0}{Sup} L(\mu)$ and

$Sup_{\mu\in\Theta} L(\mu)$ may be small or large at the same time. Thus, *more formally*, one rejects H_0 if and only if Λ is small. Thus, we decide as follows:

$$\text{Reject } H_0 \text{ if and only if } n(\overline{X} - \mu_0)^2/\sigma^2 > k, \qquad (11.2.6)$$

where $k(> 0)$ is a *generic constant*. That is, we reject H_0 if and only if $\overline{X} - \mu_0$ is too large $(> \sqrt{k})$ or too small $(< -\sqrt{k})$. The implementable form of the level α LR test will look like this:

$$\text{Reject } H_0 \text{ if and only if } \left|\sqrt{n}(\overline{X} - \mu_0)/\sigma\right| > z_{\alpha/2}. \qquad (11.2.7)$$

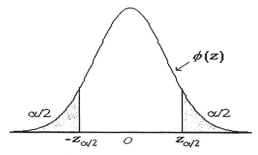

Figure 11.2.1. Two-Sided Standard Normal Rejection Region

See the Figure 11.2.1. The level of the two-sided Z-test (11.2.7) can be evaluated as follows:

$$
\begin{aligned}
&P\{\text{Reject } H_0 \text{ when } \mu = \mu_0\} \\
&= P\{\left|\sqrt{n}(\overline{X} - \mu_0)/\sigma\right| > z_{\alpha/2} \text{ when } \mu = \mu_0\}
\end{aligned} \qquad (11.2.8)
$$

which is α since $\sqrt{n}(\overline{X} - \mu_0)/\sigma$ is distributed as $N(0,1)$ if $\mu = \mu_0$.

From Section 8.5.1, recall that no UMP test exists for this problem.

Variance Unknown

We have $\boldsymbol{\theta} = (\mu, \sigma^2), \Theta_0 = \{(\mu_0, \sigma^2) : \mu_0 \text{ is fixed, } \sigma \in \Re^+\}$ and $\Theta = \{(\mu, \sigma^2) : \mu \in \Re, \sigma \in \Re^+\}$. In this case, H_0 is *not* a simple null hypothesis. We assume that the sample size n is at least two. The likelihood function is given by

$$
\begin{aligned}
L(\mu, \sigma^2) = \{\sigma\sqrt{2\pi}\}^{-n} exp\{-\Sigma_{i=1}^n (x_i - \mu)^2/(2\sigma^2)\} \\
\text{for all } (\mu, \sigma^2) \in \Re \times \Re^+.
\end{aligned} \qquad (11.2.9)
$$

Observe that

$$\underset{\mu\in\Theta_0}{Sup}\ L(\mu,\sigma^2)$$

$$= \underset{\mu\in\Theta_0}{Sup}\{\sigma\sqrt{2\pi}\}^{-n}\,exp\{-\Sigma_{i=1}^n(x_i-\mu_0)^2/(2\sigma^2)\}$$

$$= \{\widehat{\sigma}\sqrt{2\pi}\}^{-n}\,exp\{-\Sigma_{i=1}^n(x_i-\mu_0)^2/(2\widehat{\sigma}^2)\},\ \text{since } \widehat{\sigma}^2 = \qquad (11.2.10)$$

$$n^{-1}\Sigma_{i=1}^n(x_i-\mu_0)^2 \text{ is the MLE of } \sigma^2 \text{ if } \mu=\mu_0$$

$$= \{\widehat{\sigma}\sqrt{2\pi}\}^{-n}\,exp\{-n/2\}.$$

On the other hand, one has

$$\underset{(\mu,\sigma^2)\in\Theta}{Sup}\ L(\mu,\sigma^2)$$

$$= \{\widehat{\sigma}^*\sqrt{2\pi}\}^{-n}\ exp\{-\Sigma_{i=1}^n(x_i-\overline{x})^2/(2\widehat{\sigma}^{*2})\},$$

$$\text{since } \overline{x},\widehat{\sigma}^{*2} = n^{-1}\Sigma_{i=1}^n(x_i-\overline{x})^2 \text{ are the} \qquad (11.2.11)$$

$$\text{MLE's of } \mu \text{ and } \sigma^2 \text{ respectively}$$

$$= \{\widehat{\sigma}^*\sqrt{2\pi}\}^{-n}\,exp\{-n/2\}.$$

Now, we combine (11.2.4) and (11.2.10)-(11.2.11) to express the likelihood ratio

$$\Lambda = \{\widehat{\sigma}^{*2}/\widehat{\sigma}^2\}^{n/2} = \left[1 + \frac{n(\overline{x}-\mu_0)^2}{\Sigma_{i=1}^n(x_i-\overline{x})^2}\right]^{-n/2}. \qquad (11.2.12)$$

Now, we reject H_0 if and only if Λ is small. Thus, we decide as follows:

$$\text{Reject } H_0 \text{ if and only if } n(\overline{X}-\mu_0)^2/\Sigma_{i=1}^n(X_i-\overline{X})^2 > k, \qquad (11.2.13)$$

where $k(>0)$ is a *generic constant*. That is, we reject H_0 if and only if $\overline{X}-\mu_0$ when properly scaled becomes too large or too small. The implementable form of the level α LR test would then look like this:

$$\text{Reject } H_0 \text{ if and only if } \left|\sqrt{n}(\overline{X}-\mu_0)/S\right| > t_{n-1,\alpha/2}. \qquad (11.2.14)$$

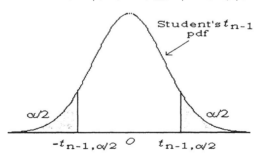

Figure 11.2.2. Two-Sided Student's t_{n-1} Rejection Region

See the Figure 11.2.2. The level of the two-sided t test (11.2.14) can be evaluated as follows:

$$P\{\text{Reject } H_0 \text{ when } \mu = \mu_0\}$$
$$= P\{|\sqrt{n}(\overline{X} - \mu_0)/S| > t_{n-1,\alpha/2} \text{ when } \mu = \mu_0\}, \qquad (11.2.15)$$

which is α since $\sqrt{n}(\overline{X} - \mu_0)/S$ has the Student's t distribution with $n - 1$ degrees of freedom if $\mu = \mu_0$.

In principle, one may think of a LR test as long as one starts with the likelihood function. The observations do not need to be iid. Look at the Exercises 11.2.6 and 11.2.9.

Example 11.2.1 In a recent meeting of the association for the commuting students at a college campus, an issue came up regarding the weekly average commuting distance (μ). A question was raised whether the weekly average commuting distance was 340 miles. Ten randomly selected commuters were asked about how much (X) each had driven to and from campus in the immediately preceding week. The data follows:

$$351.9, 357.5, 360.1, 370.4, 323.6, 332.1, 346.6, 355.5, 351.0, 348.4$$

One obtains $\overline{x} = 349.71$ miles and $s = 13.4987$ miles. Assume normality for the weekly driving distances. We may like to test $H_0 : \mu = 340$ against $H_1 : \mu \neq 340$ at the 10% level. From (11.2.14), we have the observed value of the test statistic:

$$t_{calc} = \sqrt{n}(\overline{x} - \mu_0)/s = \sqrt{10}(349.71 - 340)/13.4987 \approx 2.275.$$

With $\alpha = .10$ and 9 degrees of freedom, one has $t_{9,.05} = 1.8331$. Since $|t_{calc}|$ exceeds $t_{9,.05}$, we reject the null hypothesis at the 10% level. In other words, at the 10% level, we conclude that the average commuting distance per week is significantly different from 340 miles. ▲

11.2.2 LR Test for the Variance

Suppose that $X_1, ..., X_n$ are iid observations from the $N(\mu, \sigma^2)$ population where $\mu \in \Re, \sigma \in \Re^+$. We assume that both μ and σ are unknown. Given $\alpha \in (0, 1)$, we wish to find a level α LR test for choosing between a null hypothesis $H_0 : \sigma = \sigma_0$ and a two-sided alternative hypothesis $H_1 : \sigma \neq \sigma_0$ where σ_0 is a fixed positive real number. In this case, H_0 is *not* a simple null hypothesis. As usual, we denote the sample mean $\overline{X} = n^{-1}\Sigma_{i=1}^{n}X_i$ and the sample variance $S^2 = (n-1)^{-1}\Sigma_{i=1}^{n}(X_i - \overline{X})^2$ for $n \geq 2$. We have

$\theta = (\mu, \sigma^2), \Theta_0 = \{(\mu, \sigma_0^2) : \mu \in \Re, \sigma_0 \text{ is fixed}\}$ and $\Theta = \{(\mu, \sigma^2) : \mu \in \Re, \sigma \in \Re^+\}$.

The likelihood function is again given by

$$L(\mu, \sigma^2) = \{\sigma\sqrt{2\pi}\}^{-n} exp\{-\Sigma_{i=1}^n (x_i - \mu)^2/(2\sigma^2)\}$$
$$\text{for all } (\mu, \sigma^2) \in \Re \times \Re^+. \tag{11.2.16}$$

Now, observe that

$$\begin{aligned} \underset{(\mu,\sigma^2)\in\Theta_0}{Sup} & L(\mu, \sigma^2) \\ &= \underset{\mu\in\Re}{Sup}\{\sigma_0\sqrt{2\pi}\}^{-n} exp\{-\Sigma_{i=1}^n (x_i - \mu)^2/(2\sigma_0^2)\} \\ &= \{\sigma_0\sqrt{2\pi}\}^{-n} \ exp\{-\Sigma_{i=1}^n (x_i - \overline{x})^2/(2\sigma_0^2)\}, \\ & \text{since } \overline{x} \text{ is the MLE of } \mu \text{ if } \sigma = \sigma_0. \end{aligned} \tag{11.2.17}$$

On the other hand, one has $\underset{(\mu,\sigma^2)\in\Theta}{Sup} L(\mu, \sigma^2) = \{\widehat{\sigma}^*\sqrt{2\pi}\}^{-n} exp\{-n/2\}$ from (11.2.11) where $\widehat{\sigma}^{*2} = n^{-1}\Sigma_{i=1}^n (x_i - \overline{x})^2$. Now, we combine this with (11.2.17) to obtain the likelihood ratio

$$\Lambda = \left[\left(\widehat{\sigma}^{*2}/\sigma_0^2\right) exp\left\{-\left(\widehat{\sigma}^{*2}/\sigma_0^2\right) + 1\right\} \right]^{n/2}. \tag{11.2.18}$$

Now, one rejects H_0 if and only if Λ is small. Thus, we decide as follows:

$$\text{Reject } H_0 \text{ if and only if } \left(\widehat{\sigma}^{*2}/\sigma_0^2\right) exp\left\{1 - (\widehat{\sigma}^{*2}/\sigma_0^2)\right\} < k, \tag{11.2.19}$$

where $k(> 0)$ is a *generic constant*.

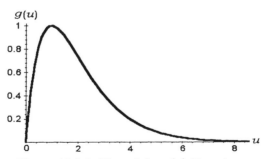

Figure 11.2.3. Plot of the $g(u)$ Function

In order to express the LR test in an implementable form, we proceed as follows: Consider the function $g(u) = ue^{1-u}$ for $u > 0$ and investigate its behavior to check when it is small $(< k)$. We note that $g(1) = 1$ and $g'(u) = \{(1 - u)/u\}g(u)$ which is positive (negative) when $u < 1$ ($u > 1$). Hence, the function $g(u)$ is strictly increasing (decreasing) on the left (right) hand side of $u = 1$. Thus, $g(u)$ is going to be "small" for both very small and very large values of $u(> 0)$. This feature is also clear from the plot of the function $g(u)$ given in the Figure 11.2.1. Thus, we rewrite the LR test (11.2.19) as follows:

Reject H_0 if and only if $\widehat{\sigma}^{*2}/\sigma_0^2 < a$ or $\widehat{\sigma}^{*2}/\sigma_0^2 > b$

for $0 < a < b < \infty$, (11.2.20)

as long a, b as are chosen so that the test has level α.

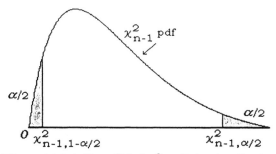

Figure 11.2.4. Two-Sided χ^2_{n-1} Rejection Region

Recall that $(n - 1)S^2/\sigma_0^2 = \Sigma_{i=1}^n (X_i - \overline{X})^2/\sigma_0^2 = n\widehat{\sigma}^{*2}/\sigma_0^2$ has a Chi-square distribution with $n - 1$ degrees of freedom if $\sigma = \sigma_0$ and hence a level α LR test can be expressed as follows:

Reject H_0 if and only if $(n - 1)S^2/\sigma_0^2 > \chi^2_{n-1,\alpha/2}$ or

$(n - 1)S^2/\sigma_0^2 < \chi^2_{n-1,1-\alpha/2}$. (11.2.21)

That is, we reject H_0 if and only if S^2/σ_0^2 when properly scaled becomes too large or too small. See the Figure 11.2.4. The case when μ is known has been left as the Exercises 11.2.2-11.2.3. Also look at the related Exercise 11.2.4.

Recall from the Exercise 8.5.5 that no UMP level α test exists for testing H_0 versus H_1 even if μ is known.

Example 11.2.2 In a dart-game, the goal is to throw a dart and hit the bull's eye at the center. After the dart lands on the board, its distance (X)

from the center is measured in inches. A player made 7 attempts to hit the bull's eye and the observed x values were recorded as follows:

$$2.5, 1.2, 3.0, 2.3, 4.4, 0.8, 1.6$$

Assume a normal distribution for X. We wish to test $H_0 : \sigma = 1$ against $H_1 : \sigma \neq 1$ at 5% level. One obtains $\bar{x} = 2.2571$ inches and $s = 1.2164$ inches. From (11.2.21), we have the observed value of the test statistic:

$$\chi^2_{calc} = (n-1)s^2/\sigma_0^2 = 6(1.2164)^2 \approx 8.878.$$

With $\alpha = .05$ and 6 degrees of freedom, one has $\chi^2_{6,.975} = 1.2373$ and $\chi^2_{6,.025} = 14.449$. Since χ^2_{calc} lies between the two numbers 1.2373 and 14.449, we accept H_0 or conclude that there is not enough evidence to reject H_0 at 5% level. ▲

Example 11.2.3 A preliminary mathematics screening test was given to a group of twenty applicants for the position of actuary. This group's test scores (X) gave $\bar{x} = 81.26$ and $s = 15.39$. Assume a normal distribution for X. The administrator wished to test $H_0 : \sigma = 12$ against $H_1 : \sigma \neq 12$ at 10% level. From (11.2.21), we have the observed value of the test statistic:

$$\chi^2_{calc} = (n-1)s^2/\sigma_0^2 = 19(15.39)^2/12^2 \approx 31.251.$$

With $\alpha = .10$ and 19 degrees of freedom, one has $\chi^2_{19,.95} = 10.117$ and $\chi^2_{19,.05} = 30.144$. Since χ^2_{calc} lies outside of the interval $(10.117, 30.144)$, one should reject at 10% level. ▲

11.3   Two-Sample Problems

We focus on two independent normal populations and some associated likelihood ratio tests. With fixed $\alpha \in (0, 1)$, first a level α LR test is derived for the equality of means against a two-sided alternative hypothesis when the common population variance is unknown and come up with the customary two-sided t-test which uses the pooled sample variance. Next, we derive a level α LR test for the equality of variances against a two-sided alternative hypothesis when the population means are unknown and come up with the customary two-sided F-test.

11.3.1   Comparing the Means

Suppose that the random variables $X_{i1}, ..., X_{in_i}$ are iid $N(\mu_i, \sigma^2), i = 1, 2$, and that the X_{1j}'s are independent of the X_{2j}'s. We assume that all three parameters are unknown and $\theta = (\mu_1, \mu_2, \sigma) \in \Re \times \Re \times \Re^+$. Given $\alpha \in (0,1)$, we wish to find a level α LR test for choosing between a null hypothesis $H_0 : \mu_1 = \mu_2$ and a two-sided alternative hypothesis $H_1 : \mu_1 \neq \mu_2$. With $n_i \geq 2$, let us denote

$$\overline{X}_i = n_i^{-1} \Sigma_{j=1}^{n_i} X_{ij}, \; S_i^2 = (n_i - 1)^{-1} \Sigma_{j=1}^{n_i} (X_{ij} - \overline{X}_i)^2$$
$$S_P^2 = (n_1 + n_2 - 2)^{-1} \left\{ (n_1 - 1) S_1^2 + (n_2 - 1) S_2^2 \right\}$$

$$(11.3.1)$$

for $i = 1, 2$. Here, S_P^2 is the pooled estimator of σ^2.

Since H_0 specifies that the two means are same, we have $\Theta_0 = \{(\mu, \mu, \sigma^2) : \mu \in \Re, \sigma \in \Re^+\}$, and $\Theta = \{(\mu_1, \mu_2, \sigma^2) : \mu_1 \in \Re, \mu_2 \in \Re, \sigma \in \Re^+\}$. The likelihood function is given by

$$L(\mu_1, \mu_2, \sigma^2) = \{\sigma\sqrt{2\pi}\}^{-(n_1+n_2)} exp\left\{ -(2\sigma^2)^{-1} \Sigma_{i=1}^2 \Sigma_{j=1}^{n_i} (x_{ij} - \mu_i)^2 \right\}$$

for all $(\mu_1, \mu_2, \sigma^2) \in \Re \times \Re \times \Re^+$.

$$(11.3.2)$$

Thus, we can write

$$\underset{(\mu_1,\mu_2,\sigma^2)\in\Theta_0}{Sup} L(\mu_1, \mu_2, \sigma^2)$$
$$= \underset{-\infty<\mu<\infty, 0<\sigma^2<\infty}{Sup} \{\sigma\sqrt{2\pi}\}^{-(n_1+n_2)} exp\{-\Sigma_{i=1}^2 \Sigma_{j=1}^{n_i} (x_{ij} - \mu)^2 / (2\sigma^2)\}.$$

$$(11.3.3)$$

One should check that the maximum likelihood estimates of μ, σ^2 obtained from this restricted likelihood function turns out to be

$$\widehat{\mu} = (n_1 \overline{x}_1 + n_2 \overline{x}_2)/(n_1 + n_2), \widehat{\sigma^2} = \Sigma_{i=1}^2 \Sigma_{j=1}^{n_i} (x_{ij} - \widehat{\mu})^2 / (n_1 + n_2).$$

$$(11.3.4)$$

Hence, from (11.3.3)-(11.3.4) we have

$$\underset{(\mu_1,\mu_2,\sigma^2)\in\Theta_0}{Sup} L(\mu_1, \mu_2, \sigma^2) = \{\widehat{\sigma}\sqrt{2\pi}\}^{-(n_1+n_2)} exp\{-(n_1 + n_2)/2\}.$$

$$(11.3.5)$$

On the other hand, one has

$$\underset{(\mu_1,\mu_2,\sigma^2)\in\Theta}{Sup} L(\mu_1, \mu_2, \sigma^2)$$
$$= \{\widehat{\sigma}^*\sqrt{2\pi}\}^{-(n_1+n_2)} exp\{-\Sigma_{i=1}^2 \Sigma_{j=1}^{n_i} (x_{ij} - \overline{x}_i)^2 / (2\widehat{\sigma}^{*2})\},$$
$$\text{since } \overline{x}_1, \overline{x}_2 \text{ and } \widehat{\sigma}^{*2} = (n_1 + n_2)^{-1} \Sigma_{i=1}^2 \Sigma_{j=1}^{n_i} (x_{ij} - \overline{x}_i)^2$$
$$\text{are the MLE's of } \mu_1, \mu_2, \sigma^2$$
$$= \{\widehat{\sigma}^*\sqrt{2\pi}\}^{-(n_1+n_2)} exp\{-(n_1 + n_2)/2\}.$$

$$(11.3.6)$$

Now, we combine (11.3.5)-(11.3.6) to express the likelihood ratio as

$$\Lambda = \{\widehat{\sigma}^{*2}/\widehat{\sigma}^2\}^{(n_1+n_2)/2} = \left[\frac{\Sigma_{i=1}^2 \Sigma_{j=1}^{n_i}(x_{ij} - \overline{x}_i)^2}{\Sigma_{i=1}^2 \Sigma_{j=1}^{n_i}(x_{ij} - \widehat{\mu})^2}\right]^{(n_1+n_2)/2} \qquad (11.3.7)$$

with $\widehat{\mu}$ from (11.3.4) and reject H_0 if and only if Λ is small. Thus, we decide as follows:

$$\text{Reject } H_0 \text{ if and only if } \frac{\Sigma_{i=1}^2 \Sigma_{j=1}^{n_i}(X_{ij} - \overline{X}_i)^2}{\Sigma_{i=1}^2 \Sigma_{j=1}^{n_i}(X_{ij} - \widehat{\mu})^2} < k, \qquad (11.3.8)$$

where $k(> 0)$ is a *generic constant*. Again, let us utilize (11.2.4) and write

$$\begin{aligned}\Sigma_{i=1}^2 \Sigma_{j=1}^{n_i}(x_{ij} - \widehat{\mu})^2 &= \Sigma_{i=1}^2 \Sigma_{j=1}^{n_i}(x_{ij} - \overline{x}_i)^2 + \Sigma_{i=1}^2 n_i(\overline{x}_i - \widehat{\mu})^2 \\ &= \Sigma_{i=1}^2 \Sigma_{j=1}^{n_i}(x_{ij} - \overline{x}_i)^2 + \left(\frac{n_1 n_2}{n_1 + n_2}\right)(\overline{x}_1 - \overline{x}_2)^2,\end{aligned}$$

which implies that

$$\frac{\Sigma_{i=1}^2 \Sigma_{j=1}^{n_i}(X_{ij} - \widehat{\mu})^2}{\Sigma_{i=1}^2 \Sigma_{j=1}^{n_i}(X_{ij} - \overline{X}_i)^2} = 1 + \left(\frac{n_1 n_2}{n_1 + n_2}\right)\frac{(\overline{X}_1 - \overline{X}_2)^2}{\Sigma_{i=1}^2 \Sigma_{j=1}^{n_i}(X_{ij} - \overline{X}_i)^2}. \qquad (11.3.9)$$

In other words, the "small" values of $\Sigma_{i=1}^2 \Sigma_{j=1}^{n_i}(X_{ij} - \overline{X}_i)^2/\Sigma_{i=1}^2 \Sigma_{j=1}^{n_i}(X_{ij} - \widehat{\mu})^2$ will correspond to the "large" values of $(\overline{X}_1 - \overline{X}_2)^2/\Sigma_{i=1}^2 \Sigma_{j=1}^{n_i}(X_{ij} - \overline{X}_i)^2$. Thus, we can rewrite the test (11.3.8) as follows:

$$\text{Reject } H_0 \text{ if and only if } \frac{|\overline{X}_1 - \overline{X}_2|}{\sqrt{\Sigma_{i=1}^2 \Sigma_{j=1}^{n_i}(X_{ij} - \overline{X}_i)^2}} > k, \qquad (11.3.10)$$

where $k(> 0)$ is an appropriate number.

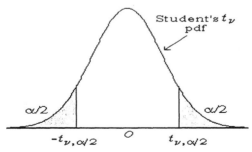

Figure 11.3.1. Two-Sided Student's t_ν Rejection Region
Where the Degree of Freedom Is $\nu = n_1 + n_2 - 2$

From Example 4.5.2 recall that $\{n_1^{-1} + n_2^{-1}\}^{-1/2}(\overline{X}_1 - \overline{X}_2)S_P^{-1}$ has the Student's t distribution with $n_1 + n_2 - 2$ degrees of freedom. Thus, in view of (11.3.10), the implementable form of a level α LR test would be:

Reject H_0 if and only if

$$\{n_1^{-1} + n_2^{-1}\}^{-1/2} \left|\overline{X}_1 - \overline{X}_2\right| S_P^{-1} > t_{n_1+n_2-2,\alpha/2}. \qquad (11.3.11)$$

See the Figure 11.3.1. Note that the LR test rejects H_0 when $\overline{X}_1 - \overline{X}_2$ is sizably different from zero with proper scaling.

> Look at the Exercises 11.3.2-11.3.3 for a LR test of the equality of means in the case of known variances.

> Look at the Exercises 11.3.4-11.3.6 for a LR test to choose between $H_0 : \mu_1 - \mu_2 = D$ versus $H_1 : \mu_1 - \mu_2 \neq D$.

Example 11.3.1 Weekly salaries (in dollars) of two typical high-school seniors, Lisa and Mike, earned during last summer are given below:

Lisa: $234.26, 237.18, 238.16, 259.53, 242.76, 237.81, 250.95, 277.83$

Mike: $187.73, 206.08, 176.71, 213.69, 224.34, 235.24$

Assume independent normal distributions with unknown average weekly salaries, μ_L for Lisa and μ_M for Mike, but with common unknown variance σ^2. At 5% level we wish to test whether the average weekly salaries are same for these two students, that is we have to test $H_0 : \mu_L = \mu_M$ versus $H_1 : \mu_L \neq \mu_M$. One has

$$\overline{x}_L = 247.31 \quad s_L^2 = 223.0918 \quad \overline{x}_M = 207.20 \quad s_M^2 = 491.1681$$

so that the pooled sample variance

$$s_P^2 = \frac{7(223.0918) + 5(491.1681)}{8 + 6 - 2} \approx 334.79.$$

From (11.3.11), we find the observed value of the test statistic:

$$t_{calc} = \{n_L^{-1} + n_M^{-1}\}^{-1/2} \left|\overline{x}_L - \overline{x}_M\right| s_P^{-1} = \frac{|247.31 - 207.20|}{\sqrt{(\frac{1}{8} + \frac{1}{6})(334.79)}} \approx 4.059.$$

With $\alpha = .05$ and 12 degrees of freedom, we have $t_{12,.025} = 2.1788$. Since $|t_{calc}|$ exceeds $t_{12,.025}$, we reject the null hypothesis at 5% level. At 5% level, we conclude that the average weekly salaries of Lisa and Mike were significantly different. ▲

11.3.2 Comparing the Variances

Suppose that the random variables $X_{i1}, ..., X_{in_i}$ are iid $N(\mu_i, \sigma_i^2), n_i \geq 2, i = 1, 2$, and that the X_{1j}'s are independent of the X_{2j}'s. We assume that all four parameters are unknown, $(\mu_i, \sigma_i) \in \Re \times \Re^+, i = 1, 2$. With fixed $\alpha \in (0, 1)$, we wish to construct a level α LR test for choosing between a null hypothesis $H_0 : \sigma_1 = \sigma_2$ and a two-sided alternative hypothesis $H_1 : \sigma_1 \neq \sigma_2$. Let us denote $\boldsymbol{\theta} = (\mu_1, \mu_2, \sigma_1^2, \sigma_2^2)$ and for $i = 1, 2$,

$$\overline{X}_i = n_i^{-1}\Sigma_{j=1}^{n_i}X_{ij}, \ S_i^2 = (n_i - 1)^{-1}\Sigma_{j=1}^{n_i}(X_{ij} - \overline{X}_i)^2. \qquad (11.3.12)$$

Since H_0 specifies that the two variances are same, we can write $\Theta_0 = \{(\mu_1, \mu_2, \sigma^2, \sigma^2) : \mu_1 \in \Re, \mu_2 \in \Re, \sigma \in \Re^+\}$, and $\Theta = \{(\mu_1, \mu_2, \sigma_1^2, \sigma_2^2) : \mu_1 \in \Re, \mu_2 \in \Re, \sigma_1 \in \Re^+, \sigma_2 \in \Re^+\}$. The likelihood function is given by

$$L(\mu_1, \mu_2, \sigma_1^2, \sigma_2^2)$$
$$= \{\sqrt{2\pi}\}^{-(n_1+n_2)}(\sigma_1^{n_1}\sigma_2^{n_2})^{-1}exp[-\{\Sigma_{j=1}^{n_1}(x_{1j} - \mu_1)^2/(2\sigma_1^2)\}$$
$$-\{\Sigma_{j=1}^{n_2}(x_{2j} - \mu_2)^2/(2\sigma_2^2)\}]$$

$$(11.3.13)$$

for all $(\mu_1, \mu_2, \sigma_1^2, \sigma_2^2) \in \Re \times \Re \times \Re^+ \times \Re^+$. Thus, we can write

$$\underset{(\mu_1, \mu_2, \sigma^2, \sigma^2) \in \Theta_0}{Sup} L(\mu_1, \mu_2, \sigma_1^2, \sigma_2^2)$$
$$= \underset{\mu_1, \mu_2, \sigma^2}{Sup} \{\sigma\sqrt{2\pi}\}^{-(n_1+n_2)} exp\{-\frac{1}{2\sigma^2}\Sigma_{i=1}^2\Sigma_{j=1}^{n_i}(x_{ij} - \mu_i)^2\} \qquad (11.3.14)$$
$$\text{for } -\infty < \mu_1, \mu_2 < \infty, 0 < \sigma^2 < \infty.$$

Now, consider the restricted likelihood function from (11.3.14). One should check that the maximum likelihood estimates of μ_1, μ_2, σ^2 obtained from this restricted likelihood function turns out to be

$$\widehat{\mu}_1 = \overline{x}_1, \widehat{\mu}_2 = \overline{x}_2, \widehat{\sigma^2} = \Sigma_{i=1}^2\Sigma_{j=1}^{n_i}(x_{ij} - \overline{x}_i)^2/(n_1 + n_2). \qquad (11.3.15)$$

Hence, from (11.3.14)-(11.3.15) we have

$$\underset{(\mu_1, \mu_2, \sigma^2, \sigma^2) \in \Theta_0}{Sup} L(\mu_1, \mu_2, \sigma_1^2, \sigma_2^2) = \{\widehat{\sigma}\sqrt{2\pi}\}^{-(n_1+n_2)} exp\{-(n_1 + n_2)/2\}.$$

$$(11.3.16)$$

On the other hand, one has

$$\underset{(\mu_1,\mu_2,\sigma_1^2,\sigma_2^2)\in\Theta}{Sup} L(\mu_1,\mu_2,\sigma_1^2,\sigma_2^2)$$

$$= \{\sqrt{2\pi}\}^{-(n_1+n_2)}(\hat{\sigma}_1^{n_1}\hat{\sigma}_2^{n_2})^{-1}exp\{-\tfrac{1}{2}\Sigma_{i=1}^2\tfrac{1}{\hat{\sigma}_i^2}\Sigma_{j=1}^{n_i}(x_{ij}-\overline{x}_i)^2\},$$

$$\text{since } \overline{x}_i, \hat{\sigma}_i^2 = n_i^{-1}\Sigma_{j=1}^{n_i}(x_{ij}-\overline{x}_i)^2 \text{ are the MLE's}$$

$$\text{of } \mu_i, \sigma_i^2, i = 1, 2$$

$$= \{\sqrt{2\pi}\}^{-(n_1+n_2)}(\hat{\sigma}_1^{n_1}\hat{\sigma}_2^{n_2})^{-1}exp\{-(n_1+n_2)/2\}. \tag{11.3.17}$$

Now, we combine (11.3.16)-(11.3.17) to express the likelihood ratio as

$$\Lambda = \frac{\hat{\sigma}_1^{n_1}\hat{\sigma}_2^{n_2}}{\hat{\sigma}^{(n_1+n_2)}} = \frac{aS_1^{n_1}S_2^{n_2}}{[S_1^2+bS_2^2]^{(n_1+n_2)/2}}, \tag{11.3.18}$$

where $a \equiv a(n_1,n_2), b \equiv b(n_1,n_2)$ are positive numbers which depend on n_1, n_2 only. Now, one rejects H_0 if and only if Λ is small. Thus, we decide as follows:

$$\text{Reject } H_0 \text{ if and only if } \frac{S_1^{n_1}S_2^{n_2}}{[S_1^2+bS_2^2]^{(n_1+n_2)/2}} < k,$$

or equivalently,

$$\text{Reject } H_0 \text{ if and only if } \frac{(S_1^2/S_2^2)^{n_1/2}}{[(S_1^2/S_2^2)+b]^{(n_1+n_2)/2}} < k, \tag{11.3.19}$$

where $k(> 0)$ is a *generic constant*.

In order to express the LR test in an implementable form, we proceed as follows: Consider the function $g(u) = u^{n_1/2}(u+b)^{-(n_1+n_2)/2}$ for $u > 0$ and investigate its behavior in order to check when it is small ($< k$). Note that $g'(u) = \tfrac{1}{2}u^{(n_1-2)/2}(u+b)^{-(n_1+n_2+2)/2}\{n_1b - n_2u\}$ which is positive (negative) when $u < (>)n_1b/n_2$. Hence, $g(u)$ is strictly increasing (decreasing) on the left (right) hand side of $u = n_1b/n_2$. Thus, $g(u)$ is going to be "small" for both very small or very large values of $u(> 0)$.

Next, we rewrite the LR test (11.3.19) as follows:

$$\text{Reject } H_0 \text{ if and only if } S_1^2/S_2^2 < c \text{ or } S_1^2/S_2^2 > d$$

$$\text{for } 0 < c < d < \infty, \tag{11.3.20}$$

where the numbers c, d are chosen in such a way that the test has level α.

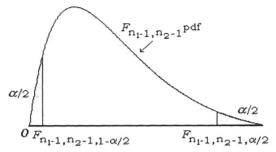

Figure 11.3.2. Two-Sided F_{n_1-1, n_2-1} Rejection Region

Recall that S_1^2/S_2^2 has the F distribution with the degrees of freedom $n_1 - 1, n_2 - 1$ when $\sigma_1 = \sigma_2$, that is under the null hypothesis H_0. Hence, a level α LR test can be written as follows:

$$\text{Reject } H_0 \text{ if and only if } S_1^2/S_2^2 > F_{n_1-1, n_2-1, \alpha/2} \text{ or}$$
$$S_1^2/S_2^2 < F_{n_1-1, n_2-1, 1-\alpha/2}. \tag{11.3.21}$$

Thus, this test rejects H_0 when S_1^2/S_2^2 is sizably different from one. See the Figure 11.3.2.

> Look at the Exercises 11.3.9-11.3.10 for a LR test of the equality of variances in the case of known means.

> Look at the Exercise 11.3.11 for a LR test to choose between $H_0 : \sigma_1/\sigma_2 = D(> 0)$ versus $H_1 : \sigma_1/\sigma_2 \neq D$.

Example 11.3.2 Over a period of 6 consecutive days, the opening prices (dollars) of two well known stocks were observed and recorded as follows:

$$\text{Stock \#1: } 39.09, 39.70, 41.77, 38.96, 41.42, 42.26$$
$$\text{Stock \#2: } 42.33, 39.16, 42.10, 40.92, 46.47, 45.02$$

Let us make a naive assumption that the two stock prices went up or down independently during the period under study. The stock prices gave rise to

$$\overline{x}_1 = 40.53 \quad s_1^2 = 2.11 \quad \overline{x}_2 = 42.67 \quad s_2^2 = 7.16$$

and the question we want to address is whether the variabilities in the opening prices of the two stocks are the same at 10% level. That is, we want to test $H_0 : \sigma_1 = \sigma_2$ versus the two-sided alternative hypothesis

$H_1 : \sigma_1 \neq \sigma_2$. Assume normality. From (11.3.21), we find the observed value of the test statistic:

$$F_{calc} = s_1^2/s_2^2 = \tfrac{2.11}{7.16} \approx .295.$$

With $\alpha = .10$, one has $F_{5,5,.05} = 5.0503$ and $F_{5,5,.95} = \tfrac{1}{5.0503} \approx .19801$. Since F_{calc} lies between the two numbers $.19801$ and 5.0503, we conclude at 10% level that the two stock prices were equally variable during the six days under investigation. ▲

> The problem of testing the equality of means of two independent normal populations with unknown and unequal variances is hard. It is referred to as the *Behrens-Fisher problem*. For some ideas and references, look at both the Exercises 11.3.15 and 13.2.10.

11.4   Bivariate Normal Observations

We have discussed LR tests to check the equality of means of two *independent* normal populations. In some situations, however, the two normal populations may be dependent. Recall the Example 9.3.3. Different test procedures are used in practice in order to handle such problems.

Suppose that the pairs of random variables (X_{1i}, X_{2i}) are iid bivariate normal, $N_2(\mu_1, \mu_2, \sigma_1^2, \sigma_2^2, \rho), i = 1, ..., n(\geq 2)$. Here we assume that all five parameters are unknown, $(\mu_l, \sigma_l) \in \Re \times \Re^+, l = 1, 2$ and $-1 < \rho < 1$. Test procedures are summarized for the population correlation coefficient ρ and for comparing the means μ_1, μ_2 as well as the variances σ_1^2, σ_2^2.

11.4.1   Comparing the Means: The Paired Difference t Method

With fixed $\alpha \in (0,1)$, we wish to find a level α test for a null hypothesis $H_0 : \mu_1 = \mu_2$ against the upper-, lower-, or two-sided alternative hypothesis H_1. The methodology from the Section 11.3.1 will not apply here. Let us denote

$$Y_i = X_{1i} - X_{2i}, i = 1, ..., n, \text{ and } \overline{Y} = n^{-1}\Sigma_{i=1}^n Y_i,$$
$$S^2 = (n-1)^{-1}\Sigma_{i=1}^n(Y_i - \overline{Y})^2. \tag{11.4.1}$$

Observe that $Y_1, ..., Y_n$ are iid $N(\mu_1 - \mu_2, \sigma^2)$ where $\sigma^2 = \sigma_1^2 + \sigma_2^2 - 2\rho\sigma_1\sigma_2$. Since the mean $\mu_1 - \mu_2$ and the variance σ^2 of the common normal distribution of the Y's are unknown, the two-sample problem on hand is reduced to a one-sample problem in terms of the pivot

$$U = \frac{n^{1/2}[\overline{Y} - (\mu_1 - \mu_2)]}{S} \underset{H_0}{\overset{\text{Under}}{\Longrightarrow}} U_{calc} = \frac{n^{1/2}\overline{Y}}{S}. \tag{11.4.2}$$

Under the null hypothesis $H_0 : \mu_1 = \mu_2$, the statistic U has the Student's t distribution with $n - 1$ degrees of freedom.

Upper-Sided Alternative Hypothesis

We test $H_0 : \mu_1 = \mu_2$ versus $H_1 : \mu_1 > \mu_2$. See the Figure 11.4.1. Along the lines of Section 8.4, we can propose the following upper-sided level α test:

Reject H_0 in favor of $H_1 : \mu_1 > \mu_2$ if and only if $U_{calc} > t_{n-1,\alpha}$.

$$\tag{11.4.3}$$

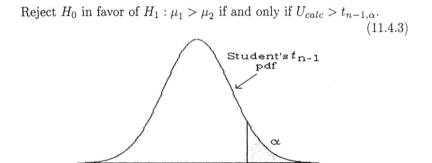

Figure 11.4.1. Upper-Sided Student's t_{n-1} Rejection Region

Lower-Sided Alternative Hypothesis

We test $H_0 : \mu_1 = \mu_2$ versus $H_1 : \mu_1 < \mu_2$. See the Figure 11.4.2. Along the lines of Section 8.4, we can propose the following lower-sided level α test:

Reject H_0 in favor of $H_1 : \mu_1 < \mu_2$ if and only if $U_{calc} < -t_{n-1,\alpha}$.

$$\tag{11.4.4}$$

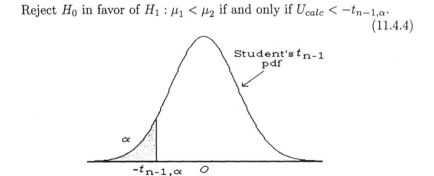

Figure 11.4.2. Lower-Sided Student's t_{n-1} Rejection Region

Two-Sided Alternative Hypothesis

We test $H_0 : \mu_1 = \mu_2$ versus $H_1 : \mu_1 \neq \mu_2$. See the Figure 11.4.3. Along the lines of Section 11.2.1, we can propose the following two-sided level α test:

Reject H_0 in favor of $H_1 : \mu_1 \neq \mu_2$ if and only if

$$U_{calc} < -t_{n-1,\alpha/2} \text{ or } U_{calc} > t_{n-1,\alpha/2}.$$ (11.4.5)

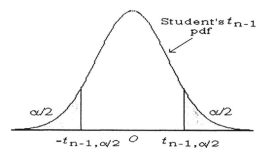

Figure 11.4.3. Two-Sided Student's t_{n-1} Rejection Region

Example 11.4.1 In a large establishment, suppose that X_{1i}, X_{2i} respectively denote the job performance score before and after going through a week-long job training program for the ith employee, $i = 1, ..., n (\geq 2)$. We may assume that these employees are picked randomly and independently of each other and want to compare the average job performance scores in the population, before and after the training. Eight employee's job performance scores (out of 100 points) were recorded as follows.

ID #	X_1	X_2	Y $X_1 - X_2$	ID #	X_1	X_2	Y $X_1 - X_2$
1	70	80	-10	5	89	94	-5
2	85	83	2	6	78	86	-8
3	67	75	-8	7	63	69	-6
4	74	80	-6	8	82	78	4

Assume a bivariate normal distribution for (X_1, X_2). The question we wish to address is whether the training program has been effective. That is, we want to test $H_0 : \mu_1 = \mu_2$ versus $H_1 : \mu_1 < \mu_2$, say, at the 1% level. From the 8 observed values of Y, we get the sample mean and variance $\bar{y} = -4.625, s^2 = 24.8393$. From (11.4.4), under H_0 we find the observed value of the test statistic:

$$u_{calc} = \frac{\sqrt{n}\bar{y}}{s} = \frac{(-4.625)\sqrt{8}}{\sqrt{24.8393}} \approx -2.624.$$

With $\alpha = .01$ and 7 degrees of freedom, one has $t_{7,.01} = 2.9980$. But, since u_{calc} does not go below $-t_{7,.01}$, we do not reject the null hypothesis at 1% level. In other words, at 1% level, we conclude that the job training has not been effective. ▲

11.4.2 LR Test for the Correlation Coefficient

With fixed $\alpha \in (0,1)$, we wish to construct a level α LR test for a null hypothesis $H_0 : \rho = 0$ against a *two-sided alternative* hypothesis $H_1 : \rho \neq 0$. We denote $\boldsymbol{\theta} = (\mu_1, \mu_2, \sigma_1^2, \sigma_2^2, \rho)$ and write $\Theta_0 = \{(\mu_1, \mu_2, \sigma_1^2, \sigma_2^2, 0) : \mu_1 \in \Re, \mu_2 \in \Re, \sigma_1 \in \Re^+, \sigma_2 \in \Re^+\}$, and $\Theta = \{(\mu_1, \mu_2, \sigma_1^2, \sigma_2^2, \rho) : \mu_1 \in \Re, \mu_2 \in \Re, \sigma_1 \in \Re^+, \sigma_2 \in \Re^+, \rho \in (-1,1)\}$. The likelihood function is given by

$$L(\mu_1, \mu_2, \sigma_1^2, \sigma_2^2, \rho)$$

$$= \{2\pi\sigma_1\sigma_2\sqrt{(1-\rho^2)}\}^{-n} exp \left\{ -\frac{1}{2(1-\rho^2)} \left[\frac{\Sigma_{i=1}^n (x_{1i} - \mu_1)^2}{\sigma_1^2} \right. \right. \qquad (11.4.6)$$

$$\left. \left. + \frac{\Sigma_{i=1}^n (x_{2i} - \mu_2)^2}{\sigma_2^2} - \frac{2\rho\Sigma_{i=1}^n (x_{1i} - \mu_1)(x_{2i} - \mu_2)}{\sigma_1\sigma_2} \right] \right\}$$

for all $\boldsymbol{\theta} \in \Theta$. We leave it as Exercise 11.4.5 to show that the MLE's for $\mu_1, \mu_2, \sigma_1^2, \sigma_2^2$ and ρ are respectively given by $\overline{x}_1, \overline{x}_2, u_1^2 = n^{-1}\Sigma_{i=1}^n (x_{1i} - \overline{x}_1)^2, u_2^2 = n^{-1}\Sigma_{i=1}^n (x_{2i} - \overline{x}_2)^2$ and $r = n^{-1}\Sigma_{i=1}^n (x_{1i} - \overline{x}_1)(x_{2i} - \overline{x}_2)/(u_1 u_2)$. These stand for the customary sample means, sample variances (not unbiased), and the sample correlation coefficient. Hence, from (11.4.6) one has

$$\underset{\boldsymbol{\theta} \in \Theta}{Sup} \, L(\mu_1, \mu_2, \sigma_1^2, \sigma_2^2, \rho) = \{2\pi u_1 u_2 \sqrt{(1-r^2)}\}^{-n} exp\{-n\}. \qquad (11.4.7)$$

Under the null hypothesis, that is when $\rho = 0$, the likelihood function happens to be

$$L(\mu_1, \mu_2, \sigma_1^2, \sigma_2^2)$$

$$= \{2\pi\sigma_1\sigma_2\}^{-n} exp \left\{ -\frac{1}{2} \left[\frac{\Sigma_{i=1}^n (x_{1i} - \mu_1)^2}{\sigma_1^2} + \frac{\Sigma_{i=1}^n (x_{2i} - \mu_2)^2}{\sigma_2^2} \right] \right\}.$$

$$(11.4.8)$$

We leave it as the Exercise 11.4.4 to show that the MLE's for μ_1, μ_2, σ_1^2 and σ_2^2 are respectively given by $\overline{x}_1, \overline{x}_2, u_1^2 = n^{-1}\Sigma_{i=1}^n (x_{1i} - \overline{x}_1)^2$ and $u_2^2 = n^{-1}\Sigma_{i=1}^n (x_{2i} - \overline{x}_2)^2$. These again stand for the customary sample means and sample variances (not unbiased). Hence, from (11.4.8) one has

$$\underset{\boldsymbol{\theta} \in \Theta_0}{Sup} \, L(\mu_1, \mu_2, \sigma_1^2, \sigma_2^2) = \{2\pi u_1 u_2\}^{-n} exp\{-n\}. \qquad (11.4.9)$$

Now, one rejects H_0 if and only if Λ is small. Thus, we decide as follows:

$$\text{Reject } H_0 \text{ if and only if } (1 - r^2)^{n/2} < k,$$

or equivalently

$$\text{Reject } H_0 \text{ if and only if } r^2/(1 - r^2) > k, \qquad (11.4.10)$$

where $k(> 0)$ is a *generic constant*. Note that $r^2/(1 - r^2)$ is a one-to-one function of r^2. It is easy to see that this test rejects H_0 when r is sizably different from zero.

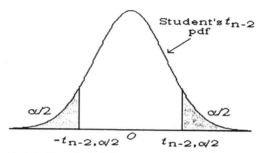

Figure 11.4.4. Two-Sided Student's t_{n-2} Rejection Region

Now, recall from (4.6.10) that $r\sqrt{n-2}/\sqrt{1-r^2}$ has the Student's t distribution with $n-2$ degrees of freedom when $\rho = 0$. This is why we assumed that n was at least three. The derivation of this sampling distribution was one of the earliest fundamental contributions of Fisher (1915). See the Figure 11.4.4. Now, from (11.4.10), a level α LR test can be expressed as follows:

$$\text{Reject } H_0 \text{ if and only if } \left| \frac{r\sqrt{n-2}}{\sqrt{1-r^2}} \right| > t_{n-2,\alpha/2}. \qquad (11.4.11)$$

Upper-Sided Alternative Hypothesis

We test $H_0 : \rho = 0$ versus $H_1 : \rho > 0$. See the Figure 11.4.5. Along the lines of (11.4.11), we can propose the following upper-sided level α test:

$$\text{Reject } H_0 \text{ in favor of } H_1 : \rho > 0 \text{ if and only if } \frac{r\sqrt{n-2}}{\sqrt{1-r^2}} > t_{n-2,\alpha}. \qquad (11.4.12)$$

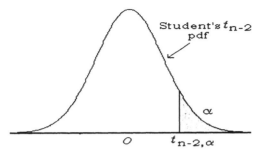

Figure 11.4.5. Upper-Sided Student's t_{n-2} Rejection Region

Lower-Sided Alternative Hypothesis

We test $H_0 : \rho = 0$ versus $H_1 : \rho < 0$. See the Figure 11.4.6. Along the lines of (11.4.11), one can also propose the following lower-sided level α test:

Reject H_0 in favor of $H_1 : \mu_1 < \mu_2$ if and only if $\dfrac{r\sqrt{n-2}}{\sqrt{1-r^2}} < -t_{n-2,\alpha}$.

$$(11.4.13)$$

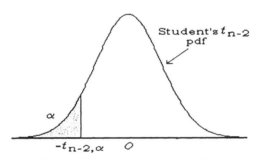

Figure 11.4.6. Lower-Sided Student's t_{n-2} Rejection Region

Example 11.4.2 (Example 11.4.1 Continued) Consider the data on (X_1, X_2) from Example 11.4.1 for 8 employees on their job performance scores before and after the training. Assuming a bivariate normal distribution for (X_1, X_2), we wish to test whether the job performance scores before and after the training are correlated. At 10% level, first we may want to test $H_0 : \rho = 0$ versus $H_1 : \rho \neq 0$. For the observed data, one should check that $r = .837257$. From (11.4.11), we find the observed value of the test statistic:

$$t_{calc} = \frac{r\sqrt{n-2}}{\sqrt{1-r^2}} = \frac{.837257\sqrt{6}}{\sqrt{1-(.837257)^2}} \approx 3.7506.$$

With $\alpha = .10$ and 6 degrees of freedom, one has $t_{6,.05} = 1.9432$. Since $|t_{calc}|$ exceeds $t_{6,.05}$, we reject the null hypothesis at 10% level and conclude that the job performance scores before and after training appear to be significantly correlated. ▲

11.4.3 Test for the Variances

With fixed $\alpha \in (0,1)$, we wish to construct a level α test for the null hypothesis $H_0 : \sigma_1 = \sigma_2$ against the upper-, lower-, or two-sided alternative hypothesis H_1. The methodology from Section 11.3.2 does not apply. Let us denote

$$Y_{1i} = X_{1i} + X_{2i}, Y_{2i} = X_{1i} - X_{2i}, i = 1, ..., n,$$
$$\overline{Y}_1 = n^{-1}\Sigma_{i=1}^n Y_{1i}, \overline{Y}_2 = n^{-1}\Sigma_{i=1}^n Y_{2i}$$
$$S_1^2 = (n-1)^{-1}\Sigma_{i=1}^n (Y_{1i} - \overline{Y}_1)^2, S_2^2 = (n-1)^{-1}\Sigma_{i=1}^n (Y_{2i} - \overline{Y}_2)^2$$
$$r^* = (n-1)^{-1}\Sigma_{i=1}^n (Y_{1i} - \overline{Y}_1)(Y_{2i} - \overline{Y}_2)/(S_1 S_2).$$
$$(11.4.14)$$

Observe that (Y_{1i}, Y_{2i}) are iid bivariate normal, $N_2(\nu_1, \nu_2, \tau_1^2, \tau_2^2, \rho^*), i = 1, ..., n (\geq 3)$ where $\nu_1 = \mu_1 + \mu_2, \nu_2 = \mu_1 - \mu_2, \tau_1^2 = \sigma_1^2 + \sigma_2^2 + 2\rho\sigma_1\sigma_2, \tau_2^2 = \sigma_1^2 + \sigma_2^2 - 2\rho\sigma_1\sigma_2$ and $Cov(Y_{1i}, Y_{2i}) = \sigma_1^2 - \sigma_2^2$ so that $\rho^* = (\sigma_1^2 - \sigma_2^2)/(\tau_1\tau_2)$. Of course all the parameters $\nu_1, \nu_2, \tau_1^2, \tau_2^2, \rho^*$ are unknown, $(\nu_l, \tau_l) \in \Re \times \Re^+, l = 1, 2$ and $-1 < \rho^* < 1$.

Now, it is clear that testing the original null hypothesis $H_0 : \sigma_1 = \sigma_2$ is equivalent to testing a null hypothesis $H_0 : \rho^* = 0$ whereas the upper-, lower-, or two-sided alternative hypothesis regarding σ_1, σ_2 will translate into an upper-, lower-, or two-sided alternative hypothesis regarding ρ^*. So, a level α test procedure can be derived by mimicking the proposed methodologies from (11.4.11)-(11.4.13) once r is replaced by the new sample correlation coefficient r^* obtained from the transformed data $(Y_{1i}, Y_{2i}), i = 1, ..., n (\geq 3)$.

Upper-Sided Alternative Hypothesis

We test $H_0 : \sigma_1 = \sigma_2$ versus $H_1 : \sigma_1 > \sigma_2$. See the Figure11.4.5. Along the lines of (11.4.12), we can propose the following upper-sided level α test:

Reject H_0 in favor of $H_1 : \sigma_1 > \sigma_2$ if and only if $\dfrac{r^*\sqrt{n-2}}{\sqrt{1-r^{*2}}} > t_{n-2,\alpha}$.

$$(11.4.15)$$

Lower-Sided Alternative Hypothesis

We test $H_0 : \sigma_1 = \sigma_2$ versus $H_1 : \sigma_1 < \sigma_2$. See the Figure 11.4.6. Along the lines of (11.4.13), we can propose the following lower-sided level α test.

Reject H_0 in favor of $H_1 : \sigma_1 < \sigma_2$ if and only if $\dfrac{r^*\sqrt{n-2}}{\sqrt{1-r^{*2}}} < -t_{n-2,\alpha}.$

$$(11.4.16)$$

Two-Sided Alternative Hypothesis

We test $H_0 : \sigma_1 = \sigma_2$ versus $H_1 : \sigma_1 \neq \sigma_2$. See the Figure 11.4.4. Along the lines of (11.4.11), we can propose the following two-sided level α test.

Reject H_0 in favor of $H_1 : \sigma_1 \neq \sigma_2$ if and only if $\left|\dfrac{r^*\sqrt{n-2}}{\sqrt{1-r^{*2}}}\right| > t_{n-2,\alpha/2}.$

$$(11.4.17)$$

Example 11.4.3 (Example 11.4.1 Continued) Consider that data on (X_1, X_2) from the Example 11.4.1 for the 8 employees on their job performance scores before and after the training. Assuming the bivariate normal distribution for (X_1, X_2), we may like to test whether the job performance scores after the training are less variable than those taken before the training. At the 1% level, we may want to test $H_0 : \sigma_1 = \sigma_2$ versus $H_1 : \sigma_1 > \sigma_2$ or equivalently test $H_0 : \rho^* = 0$ versus $H_1 : \rho^* > 0$ where ρ^* is the population correlation coefficient between Y_1, Y_2. For the observed data, one has

ID #	Y_1 $X_1 + X_2$	Y_2 $X_1 - X_2$	ID #	Y_1 $X_1 + X_2$	Y_2 $X_1 - X_2$
1	150	−10	5	183	−5
2	168	2	6	164	−8
3	142	−8	7	132	−6
4	154	−6	8	160	4

One should check that the sample correlation coefficient between Y_1, Y_2 is $r^* = .34641$. From (11.4.15), we find the observed value of the test statistic:

$$t_{calc} = \frac{r^*\sqrt{n-2}}{\sqrt{1-r^{*2}}} = \frac{.34641\sqrt{6}}{\sqrt{1-(.34641)^2}} \approx .905.$$

With $\alpha = .01$ and 6 degrees of freedom, one has $t_{6,.01} = 3.1427$. Since t_{calc} does not exceed $t_{6,.01}$, we do not reject the null hypothesis at 1% level and conclude that the variabilities in the job performance scores before and after the training appear to be same. ▲

11.5   Exercises and Complements

11.2.1 Verify the result given in (11.2.4).

11.2.2 Suppose that $X_1, ..., X_n$ are iid $N(0, \sigma^2)$ where $\sigma(> 0)$ is the unknown parameter. With preassigned $\alpha \in (0, 1)$, derive a level α LR test for the null hypothesis $H_0 : \sigma^2 = \sigma_0^2(> 0)$ against an alternative hypothesis $H_1 : \sigma^2 \neq \sigma_0^2$ in the implementable form. {*Note*: Recall from the Exercise 8.5.5 that no UMP level α test exists for testing H_0 versus H_1}.

11.2.3 Suppose that $X_1, ..., X_n$ are iid $N(\mu, \sigma^2)$ where $\sigma(\in \Re^+)$ is the unknown parameter but $\mu(\in \Re)$ is assumed known. With preassigned $\alpha \in (0, 1)$, derive a level α LR test for a null hypothesis $H_0 : \sigma^2 = \sigma_0^2(> 0)$ against an alternative hypothesis $H_1 : \sigma^2 \neq \sigma_0^2$ in the implementable form. {*Note*: Recall from the Exercise 8.5.5 that no UMP level α test exists for testing H_0 versus H_1}.

11.2.4 Suppose that X_1, X_2 are iid $N(\mu, \sigma^2)$ where $\mu(\in \Re), \sigma(\in \Re^+)$ are both assumed unknown parameters. With preassigned $\alpha \in (0, 1)$, reconsider the level α LR test from (11.2.21) for choosing between a null hypothesis $H_0 : \sigma^2 = \sigma_0^2(> 0)$ against an alternative hypothesis $H_1 : \sigma^2 \neq \sigma_0^2$. Show that the same test can be expressed as follows: Reject H_0 if and only if $|X_1 - X_2| > \sqrt{2}\sigma_0 z_{\alpha/2}$.

11.2.5 Suppose that $X_1, ..., X_n$ are iid having the common exponential pdf $f(x; \theta) = \theta^{-1} exp\{-x/\theta\} I(x > 0)$ where $\theta(> 0)$ is assumed unknown. With preassigned $\alpha \in (0, 1)$, derive a level α LR test for a null hypothesis $H_0 : \theta = \theta_0(> 0)$ against an alternative hypothesis $H_1 : \theta \neq \theta_0$ in the implementable form. {*Note*: Recall from the Exercise 8.5.4 that no UMP level α test exists for testing H_0 versus H_1}.

11.2.6 Suppose that X_1 and X_2 are independent random variables respectively distributed as $N(\mu, \sigma^2), N(3\mu, 2\sigma^2)$ where $\mu \in \Re$ is the unknown parameter and $\sigma \in \Re^+$ is assumed known. With preassigned $\alpha \in (0, 1)$, derive a level α LR test for $H_0 : \mu = \mu_0$ versus $H_1 : \mu \neq \mu_0$ where $\mu_0(\in \Re)$ is a fixed number, in the implementable form. {*Hint*: Write down the likelihood function along the line of (8.3.31) and then proceed directly as in Section 11.2.1.}

11.2.7 Suppose that $X_1, ..., X_n$ are iid having the Rayleigh distribution with the common pdf $f(x; \theta) = 2\theta^{-1} x exp(-x^2/\theta) I(x > 0)$ where $\theta(> 0)$ is the unknown parameter. With preassigned $\alpha \in (0, 1)$, derive a level α LR test for $H_0 : \theta = \theta_0$ versus $H_1 : \theta \neq \theta_0$ where $\theta_0(\in \Re^+)$ is a fixed number, in the implementable form.

11.2.8 Suppose that $X_1, ..., X_n$ are iid having the Weibull distribution with the common pdf $f(x; a) = a^{-1} b x^{b-1} exp(-x^b/a) I(x > 0)$ where $a(> 0)$ is an unknown parameter but $b(> 0)$ is assumed known. With preassigned $\alpha \in (0, 1)$, derive a level α LR test for $H_0 : a = a_0$ versus $H_1 : a \neq a_0$ where a_0 is a positive number, in the implementable form.

11.2.9 Let us denote $\mathbf{X}' = (X_1, X_2)$ where $\mathbf{X}$ is assumed to have the bivariate normal distribution, $N_2(\mu, \mu, 1, 1, \frac{1}{\sqrt{2}})$. Here, we consider $\mu(\in \Re)$ as the unknown parameter. With preassigned $\alpha \in (0, 1)$, derive a level α LR test for $H_0 : \mu = \mu_0$ versus $H_1 : \mu \neq \mu_0$ where μ_0 is a real number, in the implementable form.{*Hint*: Write down the likelihood function along the line of (8.3.33) and proceed accordingly.}

11.2.10 Let $X_1, ..., X_n$ be iid positive random variables having the common lognormal pdf $f(x; \mu) = \{x\sigma\sqrt{2\pi}\}^{-1} exp\{-[log(x) - \mu]^2/(2\sigma^2)\}$ with $x > 0, -\infty < \mu < \infty, 0 < \sigma < \infty$. Here, μ is the only unknown parameter. With preassigned $\alpha \in (0, 1)$, derive a level α LR test for $H_0 : \mu = \mu_0$ versus $H_1 : \mu \neq \mu_0$ where μ_0 is a real number, in the implementable form.

11.2.11 Suppose that $X_1, ..., X_n$ are iid having the common Laplace pdf $f(x; b) = \frac{1}{2}b^{-1} exp(-|x - a|/b)I(x \in \Re)$ where $b(> 0)$ is an unknown parameter but $a(\in \Re)$ is assumed known. With preassigned $\alpha \in (0, 1)$, derive a level α LR test for a null hypothesis $H_0 : b = b_0(> 0)$ against the alternative hypothesis $H_1 : b \neq b_0$ in the implementable form. {*Hint*: Use the Exercise 11.2.4.}

11.2.12 Suppose that $X_1, ..., X_n$ are iid having the common negative exponential pdf $f(x; \mu) = \sigma^{-1} exp\{-(x - \mu)/\sigma\}I(x > \mu)$ where $\mu(\in \Re)$ is an unknown parameter but $\sigma(\in \Re^+)$ is assumed known. With preassigned $\alpha \in (0, 1)$, derive a level α LR test for $H_0 : \mu = \mu_0$ versus $H_1 : \mu \neq \mu_0$ where μ_0 is a real number, in the implementable form.

11.2.13 Suppose that $X_1, ..., X_n$ are iid random variables having the common pdf $f(x; \theta) = \theta^{-1}x^{(1-\theta)/\theta}I(0 < x < 1)$ with $0 < \theta < \infty$. Here, θ is the unknown parameter. With preassigned $\alpha \in (0, 1)$, derive a level α LR test for $H_0 : \theta = \theta_0$ versus $H_1 : \theta \neq \theta_0$ where θ_0 is a positive number, in the implementable form.

11.2.14 A soda dispensing machine automatically fills the soda cans. The actual amount of fill must not vary too much from the target (12 fl. ounces) because the overfill will add extra cost to the manufacturer while the underfill will generate complaints from the customers. A random sample of the fills for 15 cans gave a standard deviation of .008 ounces. Assuming a normal distribution for the fills, test at 5% level whether the true population standard deviation is different from .01 ounces.

11.2.15 Ten automobiles of the same make and model were used by drivers with similar road habits and the gas mileage for each was recorded over a week. The summary results were $\bar{x} = 22$ miles per gallon and $s = 3.5$ miles per gallon. Test at 10% level whether the true average gas mileage per gallon is any different from 20.

11.2.16 The waiting time (in minutes) for a passenger at a bus stop is assumed to have an exponential distribution with mean $\theta (> 0)$. The waiting times for ten passengers were recorded as follows in one afternoon:

$$6.2 \quad 5.8 \quad 4.5 \quad 6.1 \quad 4.6 \quad 4.8 \quad 5.3 \quad 5.0 \quad 3.8 \quad 4.0$$

Test at 5% level whether the true average waiting time is any different from 5.3 minutes. {*Hint:* Use the test from Exercise 11.2.5.}

11.3.1 Verify the expressions of the MLE's in (11.3.4).

11.3.2 Let the random variables $X_{i1}, ..., X_{in_i}$ be iid $N(\mu_i, \sigma_i^2), i = 1, 2$, and that the X_{1j}'s be independent of the X_{2j}'s. Here we assume that μ_1, μ_2 are unknown and $(\mu_1, \mu_2) \in \Re \times \Re$ but $(\sigma_1, \sigma_2) \in \Re^+ \times \Re^+$ are assumed known. With preassigned $\alpha \in (0, 1)$, derive a level α LR test for $H_0 : \mu_1 = \mu_2$ versus $H_1 : \mu_1 \neq \mu_2$ in the implementable form. {*Hint:* The LR test rejects H_0 if and only if $|\overline{X}_1 - \overline{X}_2| > z_{\alpha/2}\sqrt{n_1^{-1}\sigma_1^2 + n_1^{-1}\sigma_1^2}$.}

11.3.3 Let the random variables $X_{i1}, ..., X_{in_i}$ be iid $N(\mu_i, \sigma^2), i = 1, 2$, and that the X_{1j}'s be independent of the X_{2j}'s. Here we assume that μ_1, μ_2 are unknown and $(\mu_1, \mu_2) \in \Re \times \Re$ but $\sigma \in \Re^+$ is assumed known. With preassigned $\alpha \in (0, 1)$, derive a level α LR test for $H_0 : \mu_1 = \mu_2$ versus $H_1 : \mu_1 \neq \mu_2$ in the implementable form. {*Hint:* The LR test rejects H_0 if and only if $|\overline{X}_1 - \overline{X}_2| > \sigma z_{\alpha/2}\sqrt{(n_1^{-1} + n_1^{-1})}$.}

11.3.4 Let the random variables $X_{i1}, ..., X_{in_i}$ be iid $N(\mu_i, \sigma^2), i = 1, 2$, and that the X_{1j}'s be independent of the X_{2j}'s. Here we assume that μ_1, μ_2, σ are all unknown and $(\mu_1, \mu_2) \in \Re \times \Re, \sigma \in \Re^+$. With preassigned $\alpha \in (0, 1)$ and a real number D, show that a level α LR test for $H_0 : \mu_1 - \mu_2 = D$ versus $H_1 : \mu_1 - \mu_2 \neq D$ would reject H_0 if and only if $\{n_1^{-1} + n_2^{-1}\}^{-\frac{1}{2}} |\overline{X}_1 - \overline{X}_2 - D| S_P^{-1} > t_{n_1+n_2-2,\alpha/2}$. {*Hint:* Repeat the techniques from Section 11.3.1.}

11.3.5 (Exercise 11.3.2 Continued) Let the random variables $X_{i1}, ..., X_{in_i}$ be iid $N(\mu_i, \sigma_i^2), i = 1, 2$, and that the X_{1j}'s be independent of the X_{2j}'s. Here we assume that μ_1, μ_2 are unknown and $(\mu_1, \mu_2) \in \Re \times \Re$ but $(\sigma_1, \sigma_2) \in \Re^+ \times \Re^+$ are assumed known. With preassigned $\alpha \in (0, 1)$ and a real number D, derive a level α LR test for $H_0 : \mu_1 - \mu_2 = D$ versus $H_1 : \mu_1 - \mu_2 \neq D$ in the implementable form. {*Hint:* The LR test rejects H_0 if and only if $|\overline{X}_1 - \overline{X}_2 - D| > z_{\alpha/2}\sqrt{n_1^{-1}\sigma_1^2 + n_1^{-1}\sigma_1^2}$.}

11.3.6 (Exercise 11.3.3 Continued) Let the random variables $X_{i1}, ..., X_{in_i}$ be iid $N(\mu_i, \sigma^2), i = 1, 2$, and that the X_{1j}'s be independent of the X_{2j}'s. Here we assume that μ_1, μ_2 are unknown and $(\mu_1, \mu_2) \in \Re \times \Re$ but $\sigma \in \Re^+$ is assumed known. With preassigned $\alpha \in (0, 1)$ and a real number D, derive a level α LR test for $H_0 : \mu_1 - \mu_2 = D$ versus $H_1 : \mu_1 - \mu_2 \neq D$

in the implementable form. {*Hint*: The LR test rejects H_0 if and only if $|\overline{X}_1 - \overline{X}_2 - D| > \sigma z_{\alpha/2}\sqrt{n_1^{-1} + n_1^{-1}}$.}

11.3.7 Two types of cars were compared for the braking distances. Test runs were made for each car in a driving range. Once a car reached the stable speed of 60 miles per hour, the brakes were applied. The distance (feet) each car travelled from the moment the brakes were applied to the moment the car came to a complete stop was recorded. The summary statistics are shown below:

Car	Sample Size	$\overline{x}$	s
Make A	$n_A = 12$	37.1	3.1
Make B	$n_B = 10$	39.6	4.3

Assume that the elapsed times are distributed as $N(\mu_A, \sigma^2)$ and $N(\mu_B, \sigma^2)$ respectively for the make A and B cars with all parameters unknown. Test at 5% level whether the average braking distances of the two makes are significantly different.

11.3.8 Verify the expressions of the MLE's in (11.3.15).

11.3.9 Let the random variables $X_{i1}, ..., X_{in_i}$ be iid $N(0, \sigma_i^2), i = 1, 2$, and that the X_{1j}'s be independent of the X_{2j}'s. Here we assume that $(\sigma_1, \sigma_2) \in \Re^+ \times \Re^+$ are unknown. With preassigned $\alpha \in (0, 1)$, derive a level α LR test for $H_0 : \sigma_1 = \sigma_2$ versus $H_1 : \sigma_1 \neq \sigma_2$ in the implementable form.

11.3.10 Let the random variables $X_{i1}, ..., X_{in_i}$ be iid $N(\mu_i, \sigma_i^2), i = 1, 2$, and that the X_{1j}'s be independent of the X_{2j}'s. Here we assume that $(\mu_1, \mu_2) \in \Re \times \Re$ are known but $(\sigma_1, \sigma_2) \in \Re^+ \times \Re^+$ are unknown. With preassigned $\alpha \in (0, 1)$, derive a level α LR test for $H_0 : \sigma_1 = \sigma_2$ versus $H_1 : \sigma_1 \neq \sigma_2$ in the implementable form.

11.3.11 Let the random variables $X_{i1}, ..., X_{in_i}$ be iid $N(\mu_i, \sigma_i^2), i = 1, 2$, and that the X_{1j}'s be independent of the X_{2j}'s. Here we assume that $\mu_1, \mu_2, \sigma_1, \sigma_2$ are all unknown and $(\mu_1, \mu_2) \in \Re^2, (\sigma_1, \sigma_2) \in \Re^{+2}$. With preassigned $\alpha \in (0, 1)$ and a positive number D, show that the level α LR test for $H_0 : \sigma_1/\sigma_2 = D$ versus $H_1 : \sigma_1/\sigma_2 \neq D$, would reject H_0 if and only if $S_1^2/(D^2 S_2^2) > F_{n_1-1,n_2-1,\alpha/2}$ or $S_1^2/(D^2 S_2^2) < F_{n_1-1,n_2-1,1-\alpha/2}$. {*Hint*: Repeat the techniques from Section 11.3.2.}

11.3.12 Let the random variables $X_{i1}, ..., X_{in_i}$ be iid Exponential$(\theta_i), i = 1, 2$, and that the X_{1j}'s be independent of the X_{2j}'s. Here we assume that θ_1, θ_2 are unknown and $(\theta_1, \theta_2) \in \Re^+ \times \Re^+$. With preassigned $\alpha \in (0, 1)$, derive a level α LR test for $H_0 : \theta_1 = \theta_2$ versus $H_1 : \theta_1 \neq \theta_2$ in the implementable form. {*Hint*: Proceed along Section 11.3.2. With $0 < c < d < \infty$, a LR test rejects H_0 if and only if $\overline{X}_1/\overline{X}_2 < c$ or $> d$.}

11.3.13 Two neighboring towns wanted to compare the variations in the time (minutes) to finish a 5k-run among the first place winners during each town's festivities such as the heritage day, peach festival, memorial day, and other town-wide events. The following data was collected recently by these two towns:

Town A (x_A):   18   20   17   22   19   18   20   18   17
Town B (x_B):   20   17   25   24   18   23

Assume that the performances of the first place winners are independent and that the first place winning times are normally distributed within each town. Then, test at 1% level whether the two town's officials may assume $\sigma_A = \sigma_B$.

11.3.14 Let the random variables $X_{i1}, ..., X_{in_i}$ be iid $N(\mu_i, \sigma^2), n_i \geq 2, i = 1, 2, 3$ and that the X_{1j}'s, X_{2j}'s and X_{3j}'s be all independent. Here we assume that $\mu_1, \mu_2, \mu_3, \sigma$ are all unknown and $(\mu_1, \mu_2, \mu_3) \in \Re^3, \sigma \in \Re^+$. With preassigned $\alpha \in (0, 1)$, show that a level α LR test for $H_0 : \mu_1 + \mu_2 = 2\mu_3$ versus $H_1 : \mu_1 + \mu_2 \neq 2\mu_3$ would reject H_0 if and only if $\{n_1^{-1} + n_2^{-1} + n_3^{-1}\}^{-\frac{1}{2}} |\overline{X}_1 + \overline{X}_2 - 2\overline{X}_3| S_P^{-1} > t_{n_1+n_2+n_3-3,\alpha/2}$ where S_P^2 is understood to be the *corresponding* pooled sample variance based on $n_1 + n_2 + n_3 - 3$ degrees of freedom. {*Hint*: Repeat the techniques from Section 11.3.1. Under H_0, while writing down the likelihood function, keep μ_1, μ_2 but replace μ_3 by $\frac{1}{2}(\mu_1 + \mu_2)$ and maximize the likelihood function with respect to μ_1, μ_2 only.}

11.3.15 (Behrens-Fisher problem) Let $X_{i1}, ..., X_{in_i}$ be iid $N(\mu_i, \sigma_i^2)$ random variables, $n_i \geq 2, i = 1, 2$, and that the X_{1j}'s be independent of the X_{2j}'s. Here we assume that $\mu_1, \mu_2, \sigma_1, \sigma_2$ are all unknown and $(\mu_1, \mu_2) \in \Re^2, (\sigma_1, \sigma_2) \in \Re^{+2}, \sigma_1 \neq \sigma_2$. Let $\overline{X}_i, S_i^2$ respectively be the sample mean and variance, $i = 1, 2$. With preassigned $\alpha \in (0, 1)$, we wish to have a level α test for $H_0 : \mu_1 = \mu_2$ versus $H_1 : \mu_1 \neq \mu_2$ in the implementable form. It may be natural to use the test statistic U_{calc} where $U_{calc} = \{\overline{X}_1 - \overline{X}_2\}/\sqrt{n_1^{-1}S_1^2 + n_2^{-1}S_2^2}$. Under H_0, the statistic U_{calc} has approximately a Student's $t_{\widehat{\nu}}$ distribution with $\widehat{\nu} = \{n_1^{-1}S_1^2 + n_2^{-1}S_2^2\}^2 \{(n_1^3 - n_1^2)^{-1}S_1^4 + (n_2^3 - n_2^2)^{-1}S_2^4\}$. Obtain a two-sided approximate t test based on U_{calc}. {*Hint*: This is referred to as the *Behrens-Fisher problem*. Its development, originated from Behrens (1929) and Fisher (1935,1939), is historically rich. Satterthwaite (1946) obtained the approximate distribution of U_{calc} under H_0, by matching "moments." There is a related confidence interval estimation problem for the ratio μ_1/μ_2 which is referred to as the

Fieller-Creasy problem. Refer to Creasy (1954) and Fieller (1954). Both these problems were elegantly reviewed by Kendall and Stuart (1979) and Wallace (1980). In the Exercise 13.2.10, we have given the two-stage sampling technique of Chapman (1950) for constructing a *fixed-width confidence interval* for $\mu_1 - \mu_2$ with the *exact confidence coefficient* at least $1 - \alpha$.}

11.4.1 A nutritional science project had involved 8 overweight men of comparable background which *included* eating habits, family traits, health condition and job related stress. An experiment was conducted to study the average reduction in weight for overweight men following a particular regimen of nutritional diet and exercise. The technician weighed in each individual before they were to enter this program. At the conclusion of the study which took two months, each individual was weighed in again. It was believed that the assumption of a bivariate normal distribution would be reasonable to use for (X_1, X_2).

Test at 5% level whether the true average weights taken before and after going through the regimen are significantly different. At 10% level, is it possible to test whether the true average weight taken after going through the regimen is significantly lower than the true average weight taken before going through the regimen? The observed data is given in the adjoining table.

ID# of Individual	Weight (x_1, pounds) Before Study	Weight (x_2, pounds) After Study
1	235	220
2	189	175
3	156	150
4	172	160
5	165	169
6	180	170
7	170	173
8	195	180

{*Hint*: Use the methodology from Section 11.4.1.}

11.4.2 Suppose that the pairs of random variables (X_{1i}, X_{2i}) are iid bivariate normal, $N_2(\mu_1, \mu_2, \sigma_1^2, \sigma_2^2, \rho), i = 1, ..., n(\geq 2)$. Here we assume that μ_1, μ_2 are unknown but $\sigma_1^2, \sigma_2^2, \rho$ are known where $(\mu_l, \sigma_l) \in \Re \times \Re^+, l = 1, 2$ and $-1 < \rho < 1$. With fixed $\alpha \in (0, 1)$, construct a level α test for a null hypothesis $H_0 : \mu_1 = \mu_2$ against a two-sided alternative hypothesis $H_1 : \mu_1 \neq \mu_2$ in the implementable form. {*Hint*: Improvise with the methodology from Section 11.4.1.}

11.4.3 Suppose that the pairs of random variables (X_{1i}, X_{2i}) are iid bivariate normal, $N_2(\mu_1, \mu_2, \sigma^2, \sigma^2, \rho), i = 1, ..., n(\geq 2)$. Here we assume that all the parameters μ_1, μ_2, σ^2 and ρ are unknown where $(\mu_1, \mu_2) \in \Re \times \Re, \sigma \in \Re^+$ and $-1 < \rho < 1$. With fixed $\alpha \in (0, 1)$, construct a level α test for a null hypothesis $H_0 : \mu_1 = \mu_2$ against a two-sided alternative hypothesis $H_1 : \mu_1 \neq \mu_2$ in the implementable form. {$Hint$: Improvise with the methodology from Section 11.4.1. Is it possible to have the t test based on $2(n-1)$ degrees of freedom?}

11.4.4 Suppose that the pairs of random variables (X_{1i}, X_{2i}) are iid bivariate normal, $N_2(\mu_1, \mu_2, \sigma_1^2, \sigma_2^2, 0), i = 1, ..., n(\geq 2)$. Here we assume that all the parameters μ_1, μ_2, σ_1^2 and σ_2^2 are unknown where $(\mu_l, \sigma_l) \in \Re \times \Re^+, l = 1, 2$. Show that the MLE's for μ_l, σ_l^2 are respectively given by $\overline{X}_l$ and $U_l^2 = n^{-1}\Sigma_{i=1}^n (X_{li} - \overline{X}_l)^2, l = 1, 2$. {$Hint$: Consider the likelihood function from (11.4.8) and then proceed by taking its natural logarithm, followed by its partial differentiation.}

11.4.5 Suppose that the pairs of random variables (X_{1i}, X_{2i}) are iid bivariate normal, $N_2(\mu_1, \mu_2, \sigma_1^2, \sigma_2^2, \rho), i = 1, ..., n(\geq 2)$. Here we assume that all the parameters $\mu_1, \mu_2, \sigma_1^2, \sigma_2^2$ and ρ are unknown where $(\mu_l, \sigma_l) \in \Re \times \Re^+, l = 1, 2, -1 < \rho < 1$. Consider the likelihood function from (11.4.6) and then proceed by taking its natural logarithm, followed by its partial differentiation.

(i)   Simultaneously solve $\partial log L/\partial \mu_l = 0$ to show that $\overline{X}_l$ is the MLE of $\mu_l, l = 1, 2$;

(ii)   Simultaneously solve $\partial log L/\partial \sigma_l^2 = 0, \partial log L/\partial \rho = 0, l = 1, 2$ and show that the MLE's for σ_l^2 and ρ are respectively $U_l^2 = n^{-1}\Sigma_{i=1}^n (X_{li} - \overline{X}_l)^2$ and the sample correlation coefficient, $r = n^{-1}\Sigma_{i=1}^n (X_{1i} - \overline{X}_1)(X_{2i} - \overline{X}_2)/(U_1 U_2), l = 1, 2$.

11.4.6 A researcher wanted to study whether the proficiency in two specific courses, sophomore history (X_1) and calculus (X_2), were correlated. From the large pool of sophomores enrolled in the two courses, ten students were randomly picked and their midterm grades in the two courses were recorded. The data is given below.

Student Number:	1	2	3	4	5	6	7	8	9	10
History Score (X_1):	80	75	68	78	80	70	82	74	72	77
Calculus Score (X_2):	90	85	72	92	78	87	73	87	74	85

Assume that (X_1, X_2) has a bivariate normal distribution in the population. Test whether ρ_{X_1, X_2} can be assumed zero with $\alpha = .10$.

11.4.7 In what follows, the data on systolic blood pressure (X_1) and age (X_2) for a sample of 10 women of similar health conditions are given.

ID #	X_1	X_2	ID #	X_1	X_2
1	122	41	6	144	44
2	148	52	7	138	51
3	146	54	8	138	56
4	162	60	9	145	49
5	135	45	10	144	58

At 5% level, test whether the population correlation coefficient ρ_{X_1,X_2} is significantly different from zero. Assume the bivariate normality of (X_1, X_2).

11.4.8 The strength of the right and left grips, denoted by X_1 and X_2 respectively, were checked for 12 auto accident victims during routine therapeutic exams at a rehab center. The observed values of X_1 and X_2 are both coded scores between zero and ten. Here, a low (high) value indicates significant weakness (strength) in the grip. Assume the bivariate normality of (X_1, X_2). The data is given in the enclosed table.

ID #	X_1	X_2	ID #	X_1	X_2	ID #	X_1	X_2
1	6.2	6.8	5	3.7	2.8	9	7.7	6.4
2	5.3	4.9	6	5.4	6.2	10	4.9	7.8
3	7.1	7.6	7	5.0	5.8	11	6.5	6.5
4	7.8	6.9	8	8.2	7.9	12	5.2	6.0

(i)   At 1% level, test whether the mean grip strengths in the two arms are same;

(ii)   At 5% level, test whether the variabilities of the strengths in the two arms are same;

(iii) At 1% level, test if the right and left hand grip's strengths are significantly correlated.

12

Large-Sample Inference

12.1 Introduction

In the previous chapters we gave different approaches of statistical inference. Those methods were meant to deliver predominantly exact answers, whatever be the sample size n, large or small. Now, we summarize *approximate* confidence interval and test procedures which are meant to work when the sample size n is large. We emphasize that these methods allow us to construct confidence intervals with *approximate confidence coefficient* $1 - \alpha$ or construct tests with *approximate level* α.

Section 12.2 gives some useful large-sample properties of a *maximum likelihood estimator* (MLE). In Section 12.3, we introduce large-sample confidence interval and test procedures for (i) the mean μ of a population having an unknown distribution, (ii) the success probability p in the Bernoulli distribution, and (iii) the mean λ of a Poisson distribution. The *variance stabilizing transformation* is introduced in Section 12.4 and we first exhibit the two customary transformations $\sin^{-1}(\sqrt{p})$ and $\sqrt{\lambda}$ used respectively in the case of a Bernoulli(p) and Poisson(λ) population. Section 12.4.3 includes Fisher's $\tanh^{-1}(\rho)$ transformation in the context of the correlation coefficient ρ in a bivariate normal population.

12.2 The Maximum Likelihood Estimation

In this section, we provide a brief introductory discussion of some of the useful large sample properties of the MLE. Consider random variables $X_1, ..., X_n$ which are iid with a common pmf or pdf $f(x; \theta)$ where $x \in \mathcal{X} \subseteq \Re$ and $\theta \in \Theta \subseteq \Re$. Having observed the data $\mathbf{X} = \mathbf{x}$, recall that the *likelihood function* is given by

$$L(\theta) = \Pi_{i=1}^{n} f(x_i; \theta). \qquad (12.2.1)$$

We denote it as a function of θ alone because the observed data $\mathbf{x} = (x_1, ..., x_n)$ is held fixed.

We make some standing assumptions. These requirements are, in spirit, similar to those used in the derivation of Cramér-Rao inequality.

A1: The expressions $\frac{\partial}{\partial\theta}[log f(x;\theta)]$, $\frac{\partial^2}{\partial\theta^2}[log f(x;\theta)]$, and $\frac{\partial^3}{\partial\theta^3}[log f(x;\theta)]$ are assumed finite for all $x \in \mathcal{X}$ and for all θ in an interval around the true unknown value of θ.

A2: Consider the three integrals $\int \frac{\partial}{\partial\theta}[f(x;\theta)]dx$, $\int \frac{\partial^2}{\partial\theta^2}[f(x;\theta)]dx$ and $\int \left\{\frac{\partial}{\partial\theta}[f(x;\theta)]\right\}^2 dx$. The first two integrals amount to zero whereas the third integral is positive for the true unknown value of θ.

A3: For every θ in an interval around the true unknown value of θ, $\left|\frac{\partial^3}{\partial\theta^3}[log f(x;\theta)]\right| < a(x)$ such that $E_\theta[a(X_1)] < b$ where b is a constant which is independent of θ.

The assumptions A1-A3 are routinely satisfied by many standard distributions, for example, binomial, Poisson, normal and exponential. In order to find the MLE for θ, one frequently takes the derivative of the likelihood function and then solve the *likelihood equation*:

$$\frac{\partial}{\partial\theta}[L(\theta)] = 0. \qquad (12.2.2)$$

One will be tempted to ask: Does this equation necessarily have any solution? If so, is the solution unique? The assumptions A1-A3 will guarantee that we can answer both questions in the affirmative. For the record, we state the following results:

> The likelihood equation from (12.2.2) has a solution, denoted by $\widehat{\theta}_n \equiv \widehat{\theta}_n(\mathbf{x})$, such that $\widehat{\theta}_n(\mathbf{X})$ is a consistent estimator for θ.

$$(12.2.3)$$

In other words, the MLE of θ will stay close to the unknown but true value of θ with high probability when the sample size n is sufficiently large.

> Let $\widehat{\theta}_n(\mathbf{x})$ be a solution of (12.2.2) where $\widehat{\theta}_n(\mathbf{X})$ is consistent for θ. Then, $\sqrt{n}[\widehat{\theta}_n(\mathbf{X}) - \theta] \xrightarrow{\mathcal{L}} N(0, 1/\mathcal{I}(\theta))$ as $n \to \infty$ where θ is the true value. Here, $\mathcal{I}(\theta)$ is the Fisher information in a single observation.

$$(12.2.4)$$

In other words, a properly normalized version of the MLE of θ will converge (in distribution) to a standard normal variable when the sample size n is large.

In a variety of situations, the asymptotic variance of the MLE $\widehat{\theta}_n(\mathbf{X})$ coincides with $1/\mathcal{I}(\theta)$. One may recall that $1/\mathcal{I}(\theta)$ is the Cramér-Rao lower bound (CRLB) for the variance of unbiased estimators of θ. What we are claiming then is this: In many routine problems, the variance of $\widehat{\theta}_n(\mathbf{X})$, the MLE of θ, has asymptotically the smallest possible value. This phenomenon was referred to as the *asymptotic efficiency* property of the MLE by Fisher (1922,1925a).

Fisher (1922,1925a) gave the foundations of the likelihood theory. Cramér (1946a, Chapter 33) pioneered the mathematical treatments under full generality with assumptions along the lines of A1-A3. One may note that the assumptions A1-A3 give merely *sufficient* conditions for the stated large sample properties of an MLE.

Many researchers have derived properties of MLE's comparable to those stated in (12.2.3)-(12.2.4) under less restrictive assumptions. The reader may look at some of the early investigations due to Cramér (1946a,b), Neyman (1949), Wald (1949a), LeCam (1953,1956), Kallianpur and Rao (1955) and Bahadur (1958). This is by no means an exhaustive list. In order to gain historical as well as technical perspectives, one may consult Cramér (1946a), LeCam (1986a,b), LeCam and Yang (1990), and Sen and Singer (1993). Admittedly, any detailed analysis is way beyond the scope of this book.

We should add, however, that in general an MLE $\widehat{\theta}$ may not be consistent for θ. Some examples of inconsistent MLE's were constructed by Neyman and Scott (1948) and Basu (1955b). But, under mild regularity conditions, the result (12.2.3) shows that an MLE $\widehat{\theta}$ is indeed consistent for the parameter θ it is supposed to estimate in the first place. This result, in conjunction with the invariance property (Theorem 7.2.1), make the MLE's very appealing in practice.

Example 12.2.1 Suppose that $X_1, ..., X_n$ are iid with the common Cauchy pdf $f(x; \theta) = \pi^{-1}\{1 + (x - \theta)^2\}^{-1}I(x \in \Re)$ where $\theta(\in \Re)$ is the unknown parameter. The likelihood function from (12.2.1) is then given by

$$L(\theta) = \frac{1}{\pi^n} \prod_{i=1}^n \frac{1}{\{1 + (x_i - \theta)^2\}}.$$

Now, the MLE of θ is a solution of the equation $\frac{\partial}{\partial\theta}[log\{L(\theta)\}] = 0$ which amounts to solving the equation

$$\sum_{i=1}^n \frac{(x_i - \theta)}{\{1 + (x_i - \theta)^2\}} = 0 \qquad (12.2.5)$$

for θ. Having observed the data $x_1, ..., x_n$, one would solve the equation (12.2.5) numerically with an iterative approach. Unfortunately, *no* analytical expression for the MLE $\widehat{\theta}_n$ is available.

We leave it as Exercise 12.2.1 to check that the Fisher information $\mathcal{I}(\theta)$ in a single observation is 2. Then, we invoke the result from (12.2.4) to conclude that

$$\sqrt{n}[\widehat{\theta}_n(\mathbf{X}) - \theta] \xrightarrow{\mathcal{L}} N(0, \tfrac{1}{2}) \text{ as } n \to \infty, \qquad (12.2.6)$$

even though we do not have any explicit expression for $\widehat{\theta}_n(\mathbf{X})$. ▲

Example 12.2.2 (Example 12.2.1 Continued) Consider the enclosed data from a Cauchy population with pdf $\pi^{-1}\{1 + (x - \theta)^2\}^{-1} I(x \in \Re), \theta \in \Re$. The sample size is 30 which is regarded large for practical purposes. In order to find the maximum likelihood estimate $\widehat{\theta}$ of θ, we solved the likelihood equation (12.2.5) numerically. We accomplished this with the help of MAPLE. For the observed data, the solution turns out to be $\widehat{\theta} = 1.4617$. That is, the MLE of θ is 1.4617 and its *approximate variance* is $2n^{-1}$ with $n = 30$.

2.23730	2.69720	$-.50220$	$-.11113$	2.13320
0.04727	0.51153	2.57160	1.48200	$-.88506$
2.16940	-27.410	-1.1656	1.78830	28.0480
-1.7565	-3.8039	3.21210	5.03620	2.60930
2.77440	2.66690	$-.23249$	8.71200	1.95450
0.87362	-10.305	3.03110	2.47850	1.03120

In view of (12.2.6), an *approximate* 99% confidence interval for θ is constructed as $\widehat{\theta} \pm 2.58 \frac{\sqrt{2}}{\sqrt{30}}$ which simplifies to $1.4617 \pm .66615$. We may add that the data was simulated with $\theta = 2$ and the true value of θ does belong to the confidence interval. ▲

Example 12.2.3 (Example 12.2.2 Continued) Suppose that a random sample of size $n = 30$ is drawn from a Cauchy population having its pdf $f(x; \theta) = \pi^{-1}\{1 + (x - \theta)^2\}^{-1} I(x \in \Re)$ where $\theta (\in \Re)$ is the unknown parameter. We are told that the maximum likelihood estimate of θ is 5.84. We wish to test a null hypothesis $H_0 : \theta = 5$ against an alternative hypothesis $H_1 : \theta \neq 5$ with *approximate* 1% level. We argue that approximately $z_{calc} = \sqrt{30}(5.84 - 5)/\sqrt{2} \approx 3.2533$ but $z_{\alpha/2} = 2.58$. Since $|z_{calc}|$ exceeds $z_{\alpha/2}$, we reject H_0 at the *approximate* level α. That is, at the *approximate* level 1%, we have enough evidence to claim that the true value of θ is different from 5. ▲

> Exercises 12.2.3-12.2 4 are devoted to a Logistic distribution with the location parameter θ. The situation is similar to what we have encountered in the case of the Cauchy population.

12.3 Confidence Intervals and Tests of Hypothesis

We start with *one-sample* problems and give confidence intervals and tests for the unknown mean μ of an arbitrary population having a finite variance. Next, such methodologies are discussed for the success probability p in Bernoulli trials and the mean λ in a Poisson distribution. Then, these techniques are extended for analogous *two-sample* problems. These methodologies rely heavily upon the central limit theorem (CLT) introduced in Chapter 5.

12.3.1 The Distribution-Free Population Mean

Suppose that $X_1, ..., X_n$ are iid real valued random variables from a population with the pmf or pdf $f(x; \boldsymbol{\theta})$ where the parameter vector $\boldsymbol{\theta}$ is unknown or the functional form of f is unknown. Let us assume that the variance $\sigma^2 \equiv \sigma^2(\boldsymbol{\theta})$ of the population is finite and positive which in turn implies that the mean $\mu \equiv \mu(\boldsymbol{\theta})$ is also finite. Suppose that μ and σ are both unknown. Let $\overline{X}_n = n^{-1}\Sigma_{i=1}^n X_i, S_n^2 = (n-1)^{-1}\Sigma_{i=1}^n(X_i - \overline{X}_n)^2$ be respectively the sample mean and variance for $n \geq 2$. We can apply the CLT and write: As $n \to \infty$,

$$\sqrt{n}(\overline{X}_n - \mu)/\sigma \overset{\pounds}{\to} N(0,1) \text{ and } \sqrt{n}(\overline{X}_n - \mu)/S_n \overset{\pounds}{\to} N(0,1). \quad (12.3.1)$$

From (12.3.1) what we conclude is this: For large sample size n, we should be able to use the random variable $\sqrt{n}(\overline{X}_n - \mu)/S_n$ as an *approximate pivot* since its asymptotic distribution is $N(0,1)$ which is free from both μ, σ.

Confidence Intervals for the Population Mean

Let us first pay attention to the **one-sample** problems. With preassigned $\alpha \in (0,1)$, we claim that

$$P_{\boldsymbol{\theta}}\{|\sqrt{n}(\overline{X}_n - \mu)/S_n| < z_{\alpha/2}\} \approx 1 - \alpha, \quad (12.3.2)$$

which leads to the confidence interval

$$[\,\overline{X}_n \pm z_{\alpha/2}n^{-1/2}S_n] \text{ for the population mean } \mu \quad (12.3.3)$$

with *approximate confidence coefficient* $1 - \alpha$. Recall that $z_{\alpha/2}$ is the upper $100(\alpha/2)\%$ point of the standard normal distribution. See the Figure 12.3.1.

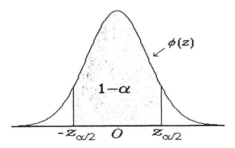

Figure 12.3.1. Standard Normal PDF: The Shaded Area Between $-z_{\alpha/2}$ and $z_{\alpha/2}$ is $1 - \alpha$

In Chapter 9 we had a similar confidence interval for the unknown mean with *exact confidence coefficient* $1 - \alpha$. That methodology worked for all sample sizes, large or small, but at the expense of the assumption that the population distribution was normal. The present confidence interval from (12.3.3) works for large n, *whatever be the population distribution* as long as σ is positive and finite. We do realize that the associated confidence coefficient is claimed only *approximately* $1 - \alpha$.

Next, let us briefly discuss the **two-sample** problems. Suppose that the random variables $X_{i1}, ..., X_{in_i}$ are iid from the ith population having a common pmf or pdf $f_i(x_i; \boldsymbol{\theta}_i)$ where the parameter vector $\boldsymbol{\theta}_i$ is unknown or the functional form of f_i itself is unknown, $i = 1, 2$. Let us assume that the variance $\sigma_i^2 \equiv \sigma_i^2(\boldsymbol{\theta}_i)$ of the ith population is finite and positive which in turn implies that its mean $\mu_i \equiv \mu_i(\boldsymbol{\theta}_i)$ is also finite, $i = 1, 2$. Assume that μ_i, σ_i^2 are unknown, $i = 1, 2$. Let us also suppose that the X_{1j}'s are independent of the X_{2j}'s and denote the sample mean $\overline{X}_{in_i} = n_i^{-1} \Sigma_{j=1}^{n_i} X_{ij}$ and sample variance $S_{in_i}^2 = (n_i - 1)^{-1} \Sigma_{j=1}^{n_i} (X_{ij} - \overline{X}_i)^2, n_i \geq 2, i = 1, 2$. In this case, one can invoke the following form of the CLT: If $n_1 \to \infty, n_2 \to \infty$ such that $n_1/n_2 \to \delta$ for some $0 < \delta < \infty$, then

$$\frac{(\overline{X}_{1n_1} - \overline{X}_{2n_2}) - (\mu_1 - \mu_2)}{\sqrt{\sigma_1^2 n_1^{-1} + \sigma_2^2 n_2^{-1}}} \xrightarrow{\pounds} N(0, 1). \tag{12.3.4}$$

Thus, using Slutsky's Theorem, we can immediately conclude: If $n_1 \to \infty, n_2 \to \infty$ such that $n_1/n_2 \to \delta$ for some $0 < \delta < \infty$, then

$$\frac{(\overline{X}_{1n_1} - \overline{X}_{2n_2}) - (\mu_1 - \mu_2)}{\sqrt{S_{1n_1}^2 n_1^{-1} + S_{2n_2}^2 n_2^{-1}}} \xrightarrow{\pounds} N(0, 1). \tag{12.3.5}$$

For large sample sizes n_1 and n_2, we should be able to use the random variable $\{(\overline{X}_{1n_1} - \overline{X}_{2n_2}) - (\mu_1 - \mu_2)\}/\sqrt{S_{1n_1}^2 n_1^{-1} + S_{2n_2}^2 n_2^{-1}}$ as an *approximate pivot* since its asymptotic distribution is $N(0, 1)$ which is free from μ_1, μ_2, σ_1 and σ_2. With preassigned $\alpha \in (0, 1)$, we claim that

$$P_{\boldsymbol{\theta}_1, \boldsymbol{\theta}_2} \left\{ \left| \frac{(\overline{X}_{1n_1} - \overline{X}_{2n_2}) - (\mu_1 - \mu_2)}{\sqrt{S_{1n_1}^2 n_1^{-1} + S_{2n_2}^2 n_2^{-1}}} \right| < z_{\alpha/2} \right\} \approx 1 - \alpha, \tag{12.3.6}$$

which leads to the confidence interval

$$[(\overline{X}_{1n_1} - \overline{X}_{2n_2}) \pm z_{\alpha/2} \sqrt{S_{1n_1}^2 n_1^{-1} + S_{2n_2}^2 n_2^{-1}}] \text{ for } (\mu_1 - \mu_2) \tag{12.3.7}$$

with *approximate confidence coefficient* $1 - \alpha$.

Tests of Hypotheses for the Population Mean

Let us first pay attention to the **one-sample** problems. With preassigned $\alpha \in (0, 1)$, we wish to test a null hypothesis $H_0 : \mu = \mu_0$ at an *approximate* level α where μ_0 is a specified real number.. The alternative hypotheses will be given shortly. One considers the test statistic

$$U = \sqrt{n}(\overline{X}_n - \mu_0)/S_n, \text{ approximately distributed as } N(0, 1) \quad (12.3.8)$$

for large n when $\mu = \mu_0$.

Upper-Sided Alternative Hypothesis

Here we test $H_0 : \mu = \mu_0$ versus $H_1 : \mu > \mu_0$. Along the lines of Section 8.4, we can propose the following upper-sided *approximate* level α test:

Reject H_0 in favor of $H_1 : \mu > \mu_0$ if and only if $U > z_\alpha$. $\quad (12.3.9)$

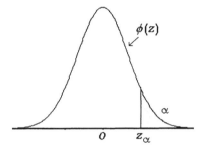

Figure 12.3.2. Upper-Sided Standard Normal Rejection Region

Lower-Sided Alternative Hypothesis

Here we test $H_0 : \mu = \mu_0$ versus $H_1 : \mu < \mu_0$. Along the lines of Section 8.4, we can propose the following lower-sided *approximate* level α test:

Reject H_0 in favor of $H_1 : \mu < \mu_0$ if and only if $U < -z_\alpha$. $\quad (12.3.10)$

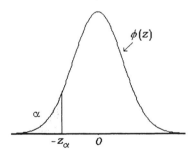

Figure 12.3.3. Lower-Sided Standard Normal Rejection Region

Two-Sided Alternative Hypothesis

Here we test $H_0 : \mu = \mu_0$ versus $H_1 : \mu \neq \mu_0$. Along the lines of Section 11.2.1, we can propose the following two-sided *approximate* level α test:

Reject H_0 in favor of $H_1 : \mu \neq \mu_0$ if and only if $|U| > z_{\alpha/2}$.   (12.3.11)

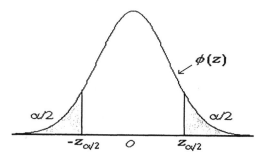

Figure 12.3.4. Two-Sided Standard Normal Rejection Region

Next, let us briefly address the **two-sample** problems. With preassigned $\alpha \in (0,1)$, we wish to test a null hypothesis $H_0 : \mu_1 = \mu_2$ with an *approximate* level α. The specific alternative hypotheses will be given shortly. Using (12.3.5), one considers the test statistic

$$W = \frac{(\overline{X}_{1n_1} - \overline{X}_{2n_2})}{\sqrt{S_{1n_1}^2 n_1^{-1} + S_{2n_2}^2 n_2^{-1}}}, \text{ approximately distributed as } N(0,1)$$

(12.3.12)

for large n_1, n_2 when $\mu_1 = \mu_2$.   ·

Upper-Sided Alternative Hypothesis

Here we test $H_0 : \mu_1 = \mu_2$ versus $H_1 : \mu_1 > \mu_2$. See, for example, the Figure 12.3.2. Along the lines of Section 8.4, we can propose the following upper-sided *approximate* level α test:

Reject H_0 in favor of $H_1 : \mu_1 > \mu_2$ if and only if $W > z_\alpha$.   (12.3.13)

Lower-Sided Alternative Hypothesis

Here we test $H_0 : \mu_1 = \mu_2$ versus $H_1 : \mu_1 < \mu_2$. See, for example, the Figure 12.3.3. Along the lines of Section 8.4, we can propose the following lower-sided *approximate* level α test:

Reject H_0 in favor of $H_1 : \mu_1 < \mu_2$ if and only if $W < -z_\alpha$.   (12.3.14)

Two-Sided Alternative Hypothesis

Here we test $H_0 : \mu_1 = \mu_2$ versus $H_1 : \mu_1 \neq \mu_2$. See, for example, the Figure 12.3.4. Along the lines of Section 11.2.1, we can propose the following two-sided *approximate* level α test:

Reject H_0 in favor of $H_1 : \mu_1 \neq \mu_2$ if and only if $|W| > z_{\alpha/2}$.

$$(12.3.15)$$

> In practice, a sample of size **thirty or more** is considered **large**.
> Refer to pages 258-261.

Example 12.3.1 The management of a rehab center for patients with acute illnesses wants to estimate the average number of days its patients typically spends with them. A random sample of $40(= n)$ patient's records gave $\bar{x}_n = 20.7$ days and $s_n = 5.6$ days. The administrator of the center now wants to look at a 90% confidence interval for the average (μ) number of days of stay. We do not have any specifics regarding the population distribution of the number of days these patients spend in the center. But, the sample size n is large, and thus we invoke (12.3.3). An *approximate* 90% confidence interval for μ would be $20.7 \pm 1.645(\frac{1}{\sqrt{40}})5.6$, that is $(19.2435, 22.1565)$. The administrator would conclude that the average stay at the center is between 19.2 and 22.2 days with approximate 90% confidence. ▲

Example 12.3.2 A local department store wanted to estimate the average account balance owed by the store's typical charge card holders. From a list of charge card accounts, the store manager drew a random sample of $50(= n)$ accounts for closer inspection. The 50 accounts with charge card balance gave $\bar{x}_n = \$105.28$ and $s_n = \$12.67$. The manager wished to look at a 95% confidence interval for the average (μ) amount owed per card. Again we do not have any specifics regarding the population distribution for the amount owed per card. But, the sample size n is large, and thus we invoke (12.3.3). An *approximate* 95% confidence interval for μ would be $105.28 \pm 1.96(\frac{1}{\sqrt{50}})12.67$, that is $(101.77, 108.79)$. The manager concluded that the average amount owed per charge card at the store was between $\$101.77$ and $\$108.79$ with approximate 95% confidence. ▲

Example 12.3.3 Two instructors were teaching different but large sections of the first-year undergraduate statistics service course from the same text book. After the mid-term examinations in both sections, the accompanied data was compiled from the *randomly selected* students' grades. Based on this data, is it possible to conclude at an *approximate* 1% level that the students in the second section outscored the students in the first section on the average? In Section #i, suppose that the unknown population average exam grade is $\mu_i, i = 1, 2$. We wish to test $H_0 : \mu_1 = \mu_2$ versus $H_1 : \mu_1 < \mu_2$ at an approximate 1% level.

Instructor #1	Instructor #2
$n_1 = 55$	$n_2 = 40$
$\overline{x}_{1n_1} = 72.4$	$\overline{x}_{2n_2} = 74.4$
$s_{1n_1} = 5.29$	$s_{2n_2} = 3.55$

Now, we have

$$w_{calc} = \frac{(\overline{x}_{1n_1} - \overline{x}_{2n_2})}{\sqrt{s_{1n_1}^2 n_1^{-1} + s_{2n_2}^2 n_2^{-1}}} = \frac{72.4 - 74.4}{\sqrt{\{(5.29)^2/55\} + \{(3.55)^2/40\}}}$$

$$= \frac{-2.0}{\sqrt{.5088 + .31506}} \approx -2.2035.$$

We find that $z_{.01} = 2.33$ and hence in view of (12.3.14), we fail to reject H_0 at an approximate 1% level since $w_{calc} > -z_{.01}$. In other words, the students' average mid-term exam performances in the two sections appear to be the same at the *approximately* 1% level. ▲

Remark 12.3.1 In general, suppose that one has a consistent estimator $\widehat{\theta}_n$ for θ such that $\sqrt{n}(\widehat{\theta}_n - \theta) \xrightarrow{\mathcal{L}} N(0, b^2(\theta))$ as $n \to \infty$, with some continuous function $b(\theta) > 0$, and for all $\theta \in \Theta$. Then, using Slutsky's Theorem, we can claim that $\widehat{\theta}_n \pm z_{\alpha/2} n^{-1/2} b(\widehat{\theta}_n)$ is an *approximate* $1 - \alpha$ confidence interval for θ if n is large. Look at the following example.

Example 12.3.4 In (12.3.3) we have given an *approximate* $1 - \alpha$ confidence interval for μ if n is large when the population distribution is unspecified. Now, suppose that we wish to construct an *approximate* $1 - \alpha$ confidence interval for μ^2. Using Mann-Wald Theorem from Chapter 5, we claim that $\sqrt{n}(\overline{X}_n^2 - \mu^2) \xrightarrow{\mathcal{L}} N(0, 4\mu^2\sigma^2)$ as $n \to \infty$. Refer to Example 5.3.7. Thus, an *approximate* $1 - \alpha$ confidence interval for μ^2 can be constructed as $\overline{X}_n^2 \pm 2z_{\alpha/2} n^{-1/2} |\overline{X}_n| S_n$ when n is large. ▲

12.3.2   The Binomial Proportion

Suppose that $X_1, ..., X_n$ are iid Bernoulli(p) random variables where $0 < p < 1$ is the unknown parameter. The minimal sufficient estimator of p is the sample mean $\overline{X}_n$ which is same as the sample proportion $\widehat{p}_n$ of the number of 1's, that is the proportion of successes in n independent replications. We can immediately apply the CLT and write:

$$\sqrt{n}(\widehat{p}_n - p)/\sqrt{\widehat{p}_n(1 - \widehat{p}_n)} \xrightarrow{\mathcal{L}} N(0, 1) \text{ as } n \to \infty. \tag{12.3.16}$$

We use $\sqrt{n}(\widehat{p}_n - p)/\sqrt{\widehat{p}_n(1 - \widehat{p}_n)}$ as an *approximate pivot* since its asymptotic distribution is $N(0,1)$ which is free from p.

Confidence Intervals for the Success Probability

Let us first pay attention to the **one-sample** problems. With preassigned $\alpha \in (0,1)$, we claim that

$$P_p\left\{\left|\sqrt{n}(\widehat{p}_n - p)/\sqrt{\widehat{p}_n(1 - \widehat{p}_n)}\right| < z_{\alpha/2}\right\} \approx 1 - \alpha, \qquad (12.3.17)$$

which leads to the confidence interval

$$\left[\widehat{p}_n \pm z_{\alpha/2}n^{-1/2}\sqrt{\widehat{p}_n(1 - \widehat{p}_n)}\right] \text{ for } p \qquad (12.3.18)$$

with *approximate confidence coefficient* $1 - \alpha$. Recall that $z_{\alpha/2}$ is the upper $100(\alpha/2)\%$ point of the standard normal distribution. See, for example, the Figure 12.3.1.

A different confidence interval for p is given in Exercise 12.3.8.

Next, let us briefly discuss the **two-sample** problems. Suppose that the random variables $X_{i1}, ..., X_{in_i}$ are iid from the ith population having the Bernoulli(p_i) distribution where $0 < p_i < 1$ is unknown, $i = 1,2$. We suppose that the X_{1j}'s are independent of the X_{2j}'s and denote the sample mean $\widehat{p}_{in_i} \equiv \overline{X}_{in_i}$ obtained from the ith population, $i = 1,2$. In this case, one invokes the following CLT: If $n_1 \to \infty, n_2 \to \infty$ such that $n_1/n_2 \to \delta$ for some $0 < \delta < \infty$, then

$$\frac{(\widehat{p}_{1n_1} - \widehat{p}_{2n_2}) - (p_1 - p_2)}{\sqrt{\widehat{p}_{1n_1}(1 - \widehat{p}_{1n_1})n_1^{-1} + \widehat{p}_{2n_2}(1 - \widehat{p}_{2n_2})n_2^{-1}}} \xrightarrow{\pounds} N(0,1). \qquad (12.3.19)$$

For large sample sizes n_1 and n_2, we should be able to use the random variable $\{(\widehat{p}_{1n_1} - \widehat{p}_{2n_2}) - (p_1 - p_2)\}/\sqrt{\widehat{p}_{1n_1}(1 - \widehat{p}_{1n_1})n_1^{-1} + \widehat{p}_{2n_2}(1 - \widehat{p}_{2n_2})n_2^{-1}}$ as an *approximate pivot* since its asymptotic distribution is $N(0,1)$ which is free from p_1 and p_2. With preassigned $\alpha \in (0,1)$, for large n_1 and n_2, we claim that

$$P_{p_1,p_2}\left\{\left|\frac{(\widehat{p}_{1n_1} - \widehat{p}_{2n_2}) - (p_1 - p_2)}{\sqrt{\widehat{p}_{1n_1}(1 - \widehat{p}_{1n_1})n_1^{-1} + \widehat{p}_{2n_2}(1 - \widehat{p}_{2n_2})n_2^{-1}}}\right| < z_{\alpha/2}\right\} \approx 1 - \alpha,$$
$$(12.3.20)$$

which leads to the confidence interval

$$[(\widehat{p}_{1n_1} - \widehat{p}_{2n_2}) \pm z_{\alpha/2}\sqrt{\widehat{p}_{1n_1}(1 - \widehat{p}_{1n_1})n_1^{-1} + \widehat{p}_{2n_2}(1 - \widehat{p}_{2n_2})n_2^{-1}}]$$
$$(12.3.21)$$

for $(p_1 - p_2)$ with *approximate confidence coefficient* $1 - \alpha$.

Tests of Hypotheses for the Success Probability

Let us first pay attention to the **one-sample** problems. With preassigned $\alpha \in (0,1)$, we test a null hypothesis $H_0 : p = p_0$ with an *approximate level* α. The specific alternative hypotheses will be given shortly. Here, $0 < p_0 < 1$ is a fixed. We consider the test statistic

$$U = \sqrt{n}(\widehat{p}_n - p_0)/\sqrt{p_0(1 - p_0)}, \text{ approximately distributed as } N(0,1)$$
$$(12.3.22)$$

for large n when $p = p_0$.

Upper-Sided Alternative Hypothesis

Here we test $H_0 : p = p_0$ versus $H_1 : p > p_0$. See, for example, the Figure 12.3.2. We can propose the following upper-sided *approximate* level α test:

Reject H_0 in favor of $H_1 : p > p_0$ if and only if $U > z_\alpha$.   (12.3.23)

Lower-Sided Alternative Hypothesis

Here we to test $H_0 : p = p_0$ versus $H_1 : p < p_0$. See, for example, the Figure 12.3.3. We can propose the following lower sided *approximate* level α test:

Reject H_0 in favor of $H_1 : p < p_0$ if and only if $U < -z_\alpha$.   (12.3.24)

Two-Sided Alternative Hypothesis

Here we test $H_0 : p = p_0$ versus $H_1 : p < p_0$. See, for example, the Figure 12.3.4. We can propose the following two sided *approximate* level α test:

Reject H_0 in favor of $H_1 : p \neq p_0$ if and only if $|U| > z_{\alpha/2}$.   (12.3.25)

Remark 12.3.2 In the case of the upper- or lower-sided alternative hypothesis, one should in fact prefer to use the corresponding randomized UMP tests discussed in Chapter 8 instead of the proposed approximate tests given by (12.3.23) and (12.3.24). On the other hand, when n is large, finding the cut-off points for implementing the UMP test may become too cumbersome. Hence, the approximate tests in proposed (12.3.23) and (12.3.24) are also used in practice whenever n is large.

Next, let us briefly address the **two-sample** problems. With preassigned $\alpha \in (0,1)$, we test a null hypothesis $H_0 : p_1 = p_2$ with the *approximate level* α. The specific alternative hypotheses will be given shortly. Under the null hypothesis H_0 we have $p_1 = p_2 = p$ where $0 < p < 1$ is unknown and we can write the variance of $(\widehat{p}_{1n_1} - \widehat{p}_{2n_2})$ as $p(1-p)(n_1^{-1} + n_2^{-1})$. Now, the common p can be estimated by the *pooled estimator*, namely,

$$\widehat{p}_{\mathbf{n}} = (n_1 \widehat{p}_{1n_1} + n_2 \widehat{p}_{2n_2})/(n_1 + n_2) \text{ where } \mathbf{n} = (n_1, n_2). \qquad (12.3.26)$$

Thus, we consider the test statistic

$$W = \frac{(\widehat{p}_{1n_1} - \widehat{p}_{2n_2})}{\sqrt{\widehat{p}_{\mathbf{n}}(1 - \widehat{p}_{\mathbf{n}})\{n_1^{-1} + n_2^{-1}\}}}, \text{ approximately distributed as } N(0,1)$$

$$(12.3.27)$$

for large n_1, n_2 when $p_1 = p_2$.

Upper-Sided Alternative Hypothesis

Here we test $H_0 : p_1 = p_2$ versus $H_1 : p_1 > p_2$. See, for example, the Figure 12.3.2. We can propose the following upper-sided *approximate* level α test:

Reject H_0 in favor of $H_1 : p_1 > p_2$ if and only if $W > z_\alpha$.   (12.3.28)

Lower-Sided Alternative Hypothesis

Here we test $H_0 : p_1 = p_2$ versus $H_1 : p_1 < p_2$. See, for example, the Figure 12.3.3. We can propose the following lower-sided *approximate* level α test:

Reject H_0 in favor of $H_1 : p_1 < p_2$ if and only if $W < -z_\alpha$.   (12.3.29)

Two-Sided Alternative Hypothesis

Here we test $H_0 : p_1 = p_2$ versus $H_1 : p_1 \neq p_2$. See, for example, the Figure 12.3.4. We can propose the following two-sided *approximate* level α test:.

Reject H_0 in favor of $H_1 : p_1 \neq p_2$ if and only if $|W| > z_{\alpha/2}$.   (12.3.30)

| In practice, a sample of size **thirty or more** is considered **large**. |

Example 12.3.5 The members of a town council wished to estimate the percentage of voters in favor of computerizing the local library's present cataloging system. A committee obtained a random sample of 300 voters, of whom 175 indicated that they favored the proposed computerization. Let p denote the proportion of voters in the town who favored the proposed computerization. Then, $\widehat{p} = \frac{175}{300} = .58333$. Since $n = 300$ is large, in view of (12.3.18), an approximate 90% confidence interval for p will be

$\widehat{p} \pm 1.645\sqrt{\widehat{p}(1-\widehat{p})/300}$ which simplifies to $.58333 \pm .04682$. The committee concluded with *approximate* 90% confidence that between 53.7% and 63% of the voters in the town favored the proposed computerization project. ▲

Example 12.3.6 The manager of a local chain of fitness centers claimed that 65% of their present members would renew membership for the next year. A close examination of a random sample of 100 present members showed that 55 renewed membership for the next year. At 1% level, is there sufficient evidence that the manager is off the mark? Let p denote the proportion of present members who would renew membership. The manager's null hypothesis is going to be $H_0 : p = .65$ but what should be the appropriate alternative hypothesis? It is reasonable to assume that the manager does not believe that p is likely to be larger than .65? Because if he did, he would have claimed a higher value of p to bolster the image. So, we formulate the alternative hypothesis as $H_1 : p < .65$. Now, we have $\widehat{p} = 55/100 = .55$ and $z_{.01} = 2.33$. Since $n = 100$ is large, in view of (12.3.24), we obtain

$$u_{calc} = \sqrt{100}(.55 - .65)/\sqrt{.65(1 - .65)} \approx -2.097.$$

At an *approximate* 1% level, we do not reject H_0 since $u_{calc} > -z_{.01}$, that is we do not have sufficient evidence to claim that the manager's belief is entirely wrong. ▲

Example 12.3.7 The two locations of a department store wanted to compare proportions of their satisfied customers. The store at location #1 found that out of 200 randomly picked customers, 155 expressed satisfaction, whereas the store at location #2 found that out of 150 randomly picked customers, 100 expressed satisfaction. Let p_i stand for the population proportion of satisfied customers for the store at location #$i, i = 1, 2$. The question we want to address is whether the proportions of satisfied customers at the two locations are same at 5% level. We consider testing $H_0 : p_1 = p_2$ versus $H_1 : p_1 \neq p_2$ with $\alpha = .05$ so that $z_{.025} = 1.96$. Since the sample sizes are large, we apply the test procedure from (12.3.30). From (12.3.26), observe that $\widehat{p} = (155 + 100)/(200 + 150) \approx .72857$. Thus, in view of (12.3.27), one has

$$w_{calc} = (.775 - .66667)/\sqrt{(.72857)(.27143)(.005 + .00667)} \approx 2.255,$$

which exceeds $z_{.025}$. Thus, we reject $H_0 : p_1 = p_2$ in favor of $H_1 : p_1 \neq p_2$ at an *approximate* 5% level. In other words, the proportions of satisfied customers at the two locations appear to be significantly different at an *approximate* 5% level. ▲

12.3.3   The Poisson Mean

Suppose that $X_1, ..., X_n$ are iid Poisson(λ) where $0 < \lambda < \infty$ is the unknown parameter. The minimal sufficient estimator of λ is the sample mean $\overline{X}_n$ which is denoted by $\widehat{\lambda}_n$. We immediately apply the CLT to write:

$$\sqrt{n}(\widehat{\lambda}_n - \lambda)/\sqrt{\widehat{\lambda}_n} \xrightarrow{\mathcal{L}} N(0, 1) \text{ as } n \to \infty. \tag{12.3.31}$$

We will use $\sqrt{n}(\widehat{\lambda}_n - \lambda)/\sqrt{\widehat{\lambda}_n}$ as an *approximate pivot* since its asymptotic distribution is $N(0, 1)$ which is free from λ.

Confidence Intervals for the Mean

Let us first pay attention to the **one-sample** problems. With preassigned $\alpha \in (0, 1)$, we claim that

$$P_\lambda \left\{ \left| \sqrt{n}(\widehat{\lambda}_n - \lambda)/\sqrt{\widehat{\lambda}_n} \right| < z_{\alpha/2} \right\} \approx 1 - \alpha, \tag{12.3.32}$$

which leads to the confidence interval

$$\left[\widehat{\lambda}_n \pm z_{\alpha/2} n^{-1/2} \sqrt{\widehat{\lambda}_n} \right] \text{ for } \lambda \tag{12.3.33}$$

with an *approximate confidence coefficient* $1 - \alpha$. Recall that $z_{\alpha/2}$ is the upper $100(\alpha/2)\%$ point of the standard normal distribution. See, for example, the Figure 12.3.1.

A different confidence interval for λ is given in the Exercise 12.3.14.

Next, let us briefly discuss the **two-sample** problems. Suppose that the random variables $X_{i1}, ..., X_{in_i}$ are iid from the ith population having the Poisson(λ_i) distribution where $0 < \lambda_i < \infty$ is unknown, $i = 1, 2$. We suppose that the X_{1j}'s are independent of the X_{2j}'s and denote the sample mean $\widehat{\lambda}_{in_i} = \overline{X}_{in_i}$ obtained from the ith population, $i = 1, 2$. In this case, one invokes the following CLT: If $n_1 \to \infty, n_2 \to \infty$ such that $n_1/n_2 \to \delta$ for some $0 < \delta < \infty$, then

$$\frac{(\widehat{\lambda}_{1n_1} - \widehat{\lambda}_{2n_2}) - (\lambda_1 - \lambda_2)}{\sqrt{\widehat{\lambda}_{1n_1} n_1^{-1} + \widehat{\lambda}_{2n_2} n_2^{-1}}} \xrightarrow{\mathcal{L}} N(0, 1). \tag{12.3.34}$$

For large sample sizes n_1 and n_2, we should use the random variable $\{(\widehat{\lambda}_{1n_1} - \widehat{\lambda}_{2n_2}) - (\lambda_1 - \lambda_2)\}/\sqrt{\widehat{\lambda}_{1n_1} n_1^{-1} + \widehat{\lambda}_{2n_2} n_2^{-1}}$ as an *approximate pivot* since its asymptotic distribution is $N(0,1)$ which is free from λ_1, λ_2. With preassigned $\alpha \in (0,1)$, we claim that

$$P_{\lambda_1,\lambda_2}\left\{\left|\frac{(\widehat{\lambda}_{1n_1} - \widehat{\lambda}_{2n_2}) - (\lambda_1 - \lambda_2)}{\sqrt{\widehat{\lambda}_{1n_1}n_1^{-1} + \widehat{\lambda}_{2n_2}n_2^{-1}}}\right| < z_{\alpha/2}\right\} \approx 1 - \alpha, \qquad (12.3.35)$$

which leads to the confidence interval

$$\left[(\widehat{\lambda}_{1n_1} - \widehat{\lambda}_{2n_2}) \pm z_{\alpha/2}\sqrt{\widehat{\lambda}_{1n_1}n_1^{-1} + \widehat{\lambda}_{2n_2}n_2^{-1}}\right] \qquad (12.3.36)$$

for $(\lambda_1 - \lambda_2)$ with an *approximate confidence coefficient* $1 - \alpha$.

Tests of Hypotheses for the Mean

Let us first pay attention to the **one-sample** problems. With preassigned $\alpha \in (0,1)$, we test a null hypothesis $H_0 : \lambda = \lambda_0$ with an *approximate level* α where $\lambda_0(> 0)$ is a fixed number. The specific alternative hypotheses will be given shortly. We consider the test statistic

$$U = \sqrt{n}(\widehat{\lambda}_n - \lambda_0)/\sqrt{\lambda_0}, \text{ approximately distributed as } N(0,1) \quad (12.3.37)$$

for large n when $\lambda = \lambda_0$.

Upper-Sided Alternative Hypothesis

Here we test $H_0 : \lambda = \lambda_0$ versus $H_1 : \lambda > \lambda_0$. See, for example, the Figure 12.3.2. We can propose the following upper-sided *approximate* level α test:

Reject H_0 in favor of $H_1 : \lambda > \lambda_0$ if and only if $U > z_\alpha$. (12.3.38)

Lower-Sided Alternative Hypothesis

Here we test $H_0 : \lambda = \lambda_0$ versus $H_1 : \lambda < \lambda_0$. See, for example, the Figure 12.3.3. We can propose the following lower-sided *approximate* level α test:

Reject H_0 in favor of $H_1 : \lambda < \lambda_0$ if and only if $U < -z_\alpha$. (12.3.39)

Two-Sided Alternative Hypothesis

Here we test $H_0 : \lambda = \lambda_0$ versus $H_1 : \lambda \neq \lambda_0$. See, for example, the Figure 12.3.4. We can propose the following two-sided *approximate* level α test:

Reject H_0 in favor of $H_1 : \lambda \neq \lambda_0$ if and only if $|U| > z_{\alpha/2}$. (12.3.40)

The Remark 12.3.2 again holds in essence in the case of the Poisson problem. Also, one can easily tackle the associated two-sample problems in the Poisson case. We leave out the details as exercises.

> In practice, a sample of size **thirty or more** is considered **large.**

Exercise 12.3.8 Customer arrivals at a train station's ticket counter have a Poisson distribution with mean λ per hour. The station manager wished to check whether the rate of customer arrivals is 40 per hour or more at 10% level. The data was collected for 120 hours and we found that $5,000$ customers arrived at the ticket counter during these 120 hours. With $\alpha = .10$ so that $z_{.10} = 1.28$, we wish to test $H_0 : \lambda = 40$ versus $H_1 : \lambda > 40$. We have $\widehat{\lambda} = 5000/120 \approx 41.6667$. Now, in view of (12.3.38), we obtain $u_{calc} = \sqrt{120}(41.6667 - 40)/\sqrt{40} \approx 2.8868$ which is larger than $z_{.10}$. Thus, we reject $H_0 : \lambda = 40$ in favor of $H_1 : \lambda > 40$ at *approximate* 10% level. That is, it appears that customers arrive at the ticket counter at a rate exceeding 40 per hour. ▲

12.4   The Variance Stabilizing Transformations

Suppose that we have a sequence of real valued statistics $\{T_n; n \geq 1\}$ such that $n^{1/2}(T_n - \theta) \xrightarrow{\pounds} N(0, \sigma^2)$ as $n \to \infty$. Then, the Mann-Wald Theorem from Section 5.3.2 will let us conclude that

$$n^{1/2}[g(T_n) - g(\theta)] \xrightarrow{\pounds} N(0, [\sigma g'(\theta)]^2) \text{ as } n \to \infty, \qquad (12.4.1)$$

if $g(.)$ is a continuous real valued function and $g'(\theta)$ is finite and nonzero. When σ^2 itself involves θ, one may like to determine an appropriate function $g(.)$ such that for large n, the *approximate variance* of the associated transformed statistic $g(T_n)$ becomes free from the unknown parameter θ. Such a function $g(.)$ is called a *variance stabilizing transformation.*

In the case of both the Bernoulli and Poisson problems, discussed in the previous section, the relevant statistic T_n happened to be the sample mean $\overline{X}_n$ and its variance was given by $p(1 - p)/n$ or λ/n as the case may be. In what follows, we first exhibit the variance stabilizing transformation $g(.)$ in the case of the Bernoulli and Poisson problems. We also include the famous *arctan transformation* due to Fisher, obtained in the context of asymptotic distribution of the sample correlation coefficient in a bivariate normal population.

12.4.1   The Binomial Proportion

Let us go back to Section 12.3.2. Suppose that $X_1, ..., X_n$ are iid Bernoulli(p) random variables where $0 < p < 1$ is the unknown parameter. The minimal sufficient estimator of p is the sample mean $\overline{X}_n$ which is same as the sample proportion $\widehat{p}_n$ of successes in n independent runs. One has, $E_p[\widehat{p}_n] = p$ and $V_p[\widehat{p}_n] = p(1-p)/n$. Even though $\sqrt{n}\,(\widehat{p}_n - p)\,/\sqrt{p(1-p)} \overset{\pounds}{\to} N(0,1)$ as $n \to \infty$, it will be hard to use $\sqrt{n}(\widehat{p}_n - p)/\sqrt{p(1-p)}$ as a pivot to construct tests and confidence intervals for p. Look at the Exercise 12.3.8. Since the normalizing constant in the denominator depends on the unknown parameter p, the "power" calculations will be awkward too.

We invoke Mann-Wald Theorem from (12.4.1) and require a suitable function $g(.)$ such that the asymptotic variance of $\sqrt{n}[g(\widehat{p}_n) - g(p)]$ becomes free from p. In other words, we want to have

$$g'(p)\sqrt{p(1-p)} = k, \text{ a constant,} \qquad (12.4.2)$$

that is $g(p) = \int \frac{k}{\sqrt{p(1-p)}} dp$. We substitute $p = \sin^2(\theta)$ to write

$$g(p) = 2k \int d\theta = 2k \sin^{-1}(\sqrt{p}) + \text{ constant.} \qquad (12.4.3)$$

From (12.4.3), it is clear that the transformation $\sin^{-1}\left(\sqrt{\widehat{p}_n}\right)$ should be looked at carefully. Let us now consider the asymptotic distribution of

$$\sqrt{n}\left[\sin^{-1}\left(\sqrt{\widehat{p}_n}\right) - \sin^{-1}\left(\sqrt{p}\right)\right].$$

We apply (12.4.1) with $g(p) = \sin^{-1}(\sqrt{p})$ to claim that

$$\sqrt{n}\left[\sin^{-1}\left(\sqrt{\widehat{p}_n}\right) - \sin^{-1}\left(\sqrt{p}\right)\right] \overset{\pounds}{\to} N(0, \tfrac{1}{4}) \text{ as } n \to \infty,$$

since we have $g'(p) = \frac{1}{\sqrt{1-(\sqrt{p})^2}} \frac{1}{2\sqrt{p}} = \frac{1}{2\sqrt{p(1-p)}}$. That is, for large n, we can consider the pivot

$$2\sqrt{n}\left[\sin^{-1}\left(\sqrt{\widehat{p}_n}\right) - \sin^{-1}\left(\sqrt{p}\right)\right], \text{ which is approximately } N(0,1).$$
$$(12.4.4)$$

In the literature, for moderate values of n, the following fine-tuned approximation is widely used:

$$2\sqrt{n}\left[\sin^{-1}\left(\sqrt{\{\widehat{p}_n + \tfrac{3}{8n}\}\{1 + \tfrac{3}{4n}\}^{-1}}\right) - \sin^{-1}\left(\sqrt{p}\right)\right]. \qquad (12.4.5)$$

See Johnson and Kotz (1969, Chapter 3, Section 8.1) for a variety of other related approximations.

For large n, one may use either (12.4.4) or (12.4.5) to derive an *approximate* $100(1 - \alpha)\%$ confidence interval for p. Also, in order to test the null hypothesis $H_0 : p = p_0$, for large n, one may use the test statistic

$$Z_{calc} = 2\sqrt{n} \left[\sin^{-1}\left(\sqrt{\widehat{p}_n}\right) - \sin^{-1}\left(\sqrt{p_0}\right)\right] \qquad (12.4.6)$$

to come up with an *approximate* level α test against an appropriate alternative hypothesis. The details are left out for brevity.

Example 12.4.1 (Example 12.3.5 Continued) Let p denote the proportion of voters in the town who favored the proposed computerization. We wish to test a null hypothesis $H_0 : p = .60$ against an alternative hypothesis $H_1 : p < .60$. In a random sample of 300 voters, 175 indicated that they favored the proposed computerization. Now, $\widehat{p} = \frac{175}{300} = .58333$ and we have $z_{calc} = 2\sqrt{300} \left[\sin^{-1}\left(\sqrt{.58333}\right) - \sin^{-1}\left(\sqrt{.60}\right)\right] \approx -.588$ in view of (12.4.6), since $n = 300$ is large. Now, one has $z_{.01} = 2.33$ and thus at 1% level, we fail to reject H_0. In other words, we conclude that at an *approximate* 1% level, we do not have enough evidence to validate the claim that less than sixty percent of the voters are in favor of computerizing the library's present cataloging system. ▲

Example 12.4.2 (Example 12.3.6 Continued) The manager of a local chain of fitness centers claimed that 65% of their present members would renew memberships for the next year. A random sample of 100 present members showed that 55 have renewed memberships for the next year. At 1% level, is there sufficient evidence to claim that the manager is off the mark? Let p denote the proportion of present members who would renew membership. The null hypothesis is $H_0 : p = .65$ and the alternative hypothesis is $H_1 : p < .65$. We have $\widehat{p} = 55/100 = .55$ and $z_{.01} = 2.33$. Since $n = 100$ is large, in view of (12.4.6), we obtain $z_{calc} = 2\sqrt{100} \left[\sin^{-1}\left(\sqrt{.55}\right) - \sin^{-1}\left(\sqrt{.65}\right)\right] \approx -2.0453$. At an *approximate* 1% level, we do not reject H_0 since $z_{calc} > -z_{.01}$. So, we do not have sufficient evidence to claim that the manager's belief is entirely wrong. ▲

Example 12.4.3 (Examples 12.3.5 and 12.4.1 Continued) In a random sample of 300 voters, 175 indicated that they favored the proposed computerization. Let p denote the proportion of voters in the whole town who favor the proposed computerization. Then, $\widehat{p} = \frac{175}{300} = .58333$. Since $n = 300$ is large, in view of (12.4.4), an *approximate* 90% confidence interval for $\sin^{-1}\left(\sqrt{p}\right)$ will be $\sin^{-1}\left(\sqrt{.58333}\right) \pm 1.645 \left(\frac{1}{2\sqrt{300}}\right)$ which reduces to $.86912 \pm .04749$. We conclude that $\sin^{-1}\left(\sqrt{p}\right)$ lies between .82163 and

.91661, and hence p lies between $.5362$ and $.6297$. We report that between 53.62% and 62.97% of the voters in the town favor the proposed computerization. ▲

In Examples 12.4.1-12.4.3, the sample size n was so large that the results obtained after variance stabilization looked nearly identical to what was found in Section 12.3.1.

Example 12.4.4 (Examples 12.3.5 and 12.4.1 Continued) Suppose that in a random sample of 30 voters, 18 indicated that they favored the proposed computerization. Let p denote the proportion of voters in the town who favored the proposed computerization. Then, $\widehat{p} = \frac{18}{30} = .6$. Since $n = 30$ is large, in view of (12.4.4), an *approximate* 90% confidence interval for $\sin^{-1}\left(\sqrt{p}\right)$ will be $\sin^{-1}\left(\sqrt{.6}\right) \pm 1.645 \left(\frac{1}{2\sqrt{30}}\right)$ which reduces to $.88608 \pm .15017$. In other words, we conclude that $\sin^{-1}\left(\sqrt{p}\right)$ lies between $.73591$ and 1.03625, and hence p lies between $.45059$ and $.74046$. We report that among the voters, between 45.06% and 74.05% favor the proposed computerization.

On the other hand, in view of (12.3.18), an *approximate* 90% confidence interval for p will be $\widehat{p} \pm 1.645\sqrt{\widehat{p}(1-\widehat{p})/30}$ which simplifies to $.6 \pm .14713$. We conclude that among the voters, between 45.29% and 74.71% in the town favor the proposed computerization. Now, one can feel some difference in the two approximations. ▲

For moderate values of n, an *approximate* $100(1-\alpha)\%$ confidence interval for p based on (12.4.4) is expected to fare a little better than that based on (12.3.18). An *approximate* $100(1-\alpha)\%$ confidence interval for p would be (b_L, b_U) where $b_L = \sin^2\left\{\sin^{-1}\left(\sqrt{\widehat{p}_n}\right) - z_{\alpha/2}\left(\frac{1}{2\sqrt{n}}\right)\right\}, b_U = \sin^2\left\{\sin^{-1}\left(\sqrt{\widehat{p}_n}\right) + z_{\alpha/2}\left(\frac{1}{2\sqrt{n}}\right)\right\}$ which are based on (12.4.4). Another *approximate* $100(1-\alpha)\%$ confidence interval for p would be (a_L, a_U) where $a_L = \widehat{p}_n - z_{\alpha/2}\sqrt{\frac{\widehat{p}_n(1-\widehat{p}_n)}{n}}, a_U = \widehat{p}_n + z_{\alpha/2}\sqrt{\frac{\widehat{p}_n(1-\widehat{p}_n)}{n}}$ which are based on (12.3.18). Since a_L, b_L may fall below zero or a_U, b_U may exceed unity, instead of the intervals (a_L, a_U) and (b_L, b_U) we may respectively consider the confidence interval estimators

$$(\max(a_L, 0), \min(a_U, 1)) \text{ and } (\max(b_L, 0), \min(b_U, 1)) \text{ for } p.$$

In Table 12.4.1, we exhibit these two respective intervals based on simulated data from the Bernoulli(p) population with 10 replications when

$\alpha = .05, p = .6, .7$ and $n = 30$.

Table 12.4.1. Comparison of Ten Approximate 95% Confidence
Intervals Based on (12.3.18) and (12.4.4)

#	$p = .6$ and $z \equiv z_{.025} = 1.96$				$p = .7$ and $z \equiv z_{.025} = 1.96$			
	a_L	a_U	b_L	b_U	a_L	a_U	b_L	b_U
1	.3893	.7440	.3889	.7360	.5751	.8916	.5637	.8734
2	.4980	.8354	.4910	.8212	.4609	.8058	.4561	.7937
3	.3211	.6789	.3249	.6751	.4980	.8354	.4910	.8212
4	.4980	.8354	.4910	.8212	.6153	.9180	.6016	.8980
5	.3893	.7440	.3889	.7360	.6153	.9180	.6016	.8980
6	.3211	.6789	.3249	.6751	.5751	.8916	.5637	.8734
7	.3548	.7119	.3565	.7060	.6569	.9431	.6409	.9211
8	.4247	.7753	.4221	.7653	.6153	.9180	.6016	.8980
9	.4609	.8058	.4561	.7937	.6153	.9180	.6016	.8980
10	.4609	.8058	.4561	.7937	.4980	.8354	.4910	.8212

Based on this exercise, it appears that the intervals (b_L, b_U) obtained via variance stabilizing transformation are on the whole a little "tighter" than the intervals (a_L, a_U). The reader may perform large-scale simulations in order to compare two types of confidence intervals for a wide range of values of p and n.

12.4.2   The Poisson Mean

Suppose that $X_1, ..., X_n$ are iid Poisson(λ) random variables where $0 < \lambda < \infty$ is the unknown parameter. The minimal sufficient estimator of λ is $\overline{X}_n$, denoted by $\widehat{\lambda}_n$. One has $E_p[\widehat{\lambda}_n] = \lambda$ and $V_p[\widehat{\lambda}_n] = \lambda/n$. Even though $\sqrt{n}(\widehat{\lambda}_n - \lambda)/\sqrt{\lambda} \overset{\mathcal{L}}{\to} N(0,1)$ as $n \to \infty$, it will be hard to use $\sqrt{n}(\widehat{\lambda}_n - \lambda)/\sqrt{\lambda}$ as a pivot o construct tests and confidence intervals for λ. See the Exercise 12.3.14. Since the normalizing constant in the denominator depends on the unknown parameter λ, the "power" calculations will be awkward too.

We may invoke Mann-Wald Theorem from (12.4.1) and require a suitable function $g(.)$ such that the asymptotic variance of $\sqrt{n}\left[g(\widehat{\lambda}_n) - g(\lambda)\right]$ becomes free from λ. In other words, we want to have

$$g'(\lambda)\sqrt{\lambda} = k, \text{ a constant,} \tag{12.4.7}$$

that is

$$g(\lambda) = \int k\lambda^{-1/2} d\lambda = 2k\sqrt{\lambda} + \text{constant.} \tag{12.4.8}$$

From (12.4.8), it is clear that we should look at the transformation $\sqrt{\widehat{\lambda}_n}$ carefully and consider the asymptotic distribution of $\sqrt{n}\left[\sqrt{\widehat{\lambda}_n} - \sqrt{\lambda}\right]$. In view of (12.4.1) with $g(\lambda) = \sqrt{\lambda}, g'(\lambda) = \frac{1}{2\sqrt{\lambda}}$, we can claim that

$$\sqrt{n}\left[\sqrt{\widehat{\lambda}_n} - \sqrt{\lambda}\right] \xrightarrow{\pounds} N(0, \tfrac{1}{4}) \text{ as } n \to \infty.$$

That is, for large n, we consider the pivot

$$2\sqrt{n}\left[\sqrt{\widehat{\lambda}_n} - \sqrt{\lambda}\right], \text{ which is approximately } N(0,1). \qquad (12.4.9)$$

See Johnson and Kotz (1969, Chapter 4, Section 7) for a variety of other related approximations.

For large n, one may use (12.4.9) to derive an *approximate* $100(1-\alpha)\%$ confidence interval for λ. Also, in order to test a null hypothesis $H_0 : \lambda = \lambda_0$, for large n, one may use the test statistic

$$Z_{calc} = 2\sqrt{n}\left[\sqrt{\widehat{\lambda}_n} - \sqrt{\lambda_0}\right] \qquad (12.4.10)$$

to come up with an *approximate* level α test against an appropriate alternative hypothesis. The details are left out for brevity.

Example 12.4.5 A manufacturer of $3.5''$ diskettes measured the quality of its product by counting the number of missing pulse (X) on each. We are told that X follows a Poisson(λ) distribution where $\lambda(> 0)$ is the unknown average number of missing pulse per diskette and that in order to stay competitive, the parameter λ should not exceed .009. A random sample of 1000 diskettes were tested which gave rise to a sample mean $\overline{x} = .0162$. Is there sufficient evidence to reject $H_0 : \lambda = .009$ in favor of $H_1 : \lambda > .009$ *approximately* at 5% level? In view of (12.4.10), we obtain

$$z_{calc} = 2\sqrt{1000}\left[\sqrt{.0162} - \sqrt{.009}\right] \approx 2.0499,$$

which exceeds $z_{.05} = 1.645$. Hence, we reject H_0 *approximately* at 5% level. Thus, at an *approximate* 5% level, there is sufficient evidence that the defective rate of diskettes is not meeting the set standard. ▲

12.4.3   The Correlation Coefficient

Suppose that the pairs of random variables (X_{1i}, X_{2i}) are iid bivariate normal, $N_2(\mu_1, \mu_2, \sigma_1^2, \sigma_2^2, \rho), i = 1, ..., n(\geq 4)$. It will be clear later from

(12.4.19) why we have assumed that the sample size is at least four. Let all five parameters be unknown with $(\mu_l, \sigma_l) \in \Re \times \Re^+, l = 1, 2$ and $-1 < \rho < 1$. In Section 11.4.2, a level α likelihood ratio test was given for $H_0 : \rho = 0$ against $H_1 : \rho \neq 0$. In Exercise 11.4.5, one verified that the MLE's for $\mu_1, \mu_2, \sigma_1^2, \sigma_2^2$ and ρ were respectively given by

$$\overline{x}_{1n}, \overline{x}_{2n}, u_{1n}^2 = n^{-1}\Sigma_{i=1}^n(x_{1i} - \overline{x}_{1n})^2, u_{2n}^2 = n^{-1}\Sigma_{i=1}^n(x_{2i} - \overline{x}_{2n})^2, \text{ and}$$

$$r_n \equiv r = n^{-1}\Sigma_{i=1}^n(x_{1i} - \overline{x}_{1n})(x_{2i} - \overline{x}_{2n})/(u_{1n}u_{2n}).$$

These stand for the customary sample means, sample variances (not unbiased), and the sample correlation coefficient.

The level α LR test for $H_0 : \rho = 0$ against $H_1 : \rho \neq 0$ is this:

$$\text{Reject } H_0 \text{ if and only if } \left| \frac{r_n\sqrt{n-2}}{\sqrt{1-r_n^2}} \right| > t_{n-2,\alpha/2}. \tag{12.4.11}$$

We recall that under H_0, the pivot $r_n\sqrt{n-2}/\sqrt{1-r_n^2}$ has the Student's t distribution with $n-2$ degrees of freedom.

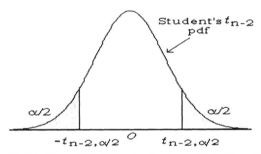

Figure 12.4.1. Two-Sided Student's t_{n-2} Rejection Region

But, now suppose that we wish to construct an *approximate* $100(1-\alpha)\%$ confidence interval for ρ. In this case, we need to work with the *non-null* distribution of the sample correlation coefficient r_n. The *exact* distribution of r_n, when $\rho \neq 0$, was found with ingenious geometric techniques by Fisher (1915). That exact distribution being very complicated, Fisher (1915) proceeded to derive the following asymptotic distribution when $\rho \neq 0$:

$$\sqrt{n}(r_n - \rho) \xrightarrow{\pounds} N\left(0, (1-\rho^2)^2\right) \text{ as } n \to \infty. \tag{12.4.12}$$

For a proof of (12.4.12), one may look at Sen and Singer (1993, pp. 134-136) among other sources.

Again, one should realize that a variance stabilizing transformation may be useful here. We invoke Mann-Wald Theorem from (12.4.1) and require a suitable function $g(.)$ such that the asymptotic variance of $\sqrt{n}\,[g(r_n) - g(\rho)]$ becomes free from ρ. In other words, we want to have

$$g'(\rho)(1 - \rho^2) = k, \text{ a constant}, \qquad (12.4.13)$$

that is $g(\rho) = k \int \frac{1}{(1-\rho^2)} d\rho$. Hence, we rewrite

$$g(\rho) = \tfrac{1}{2}k \int \left\{ \tfrac{1}{1-\rho} + \tfrac{1}{1+\rho} \right\} d\rho = \tfrac{1}{2}k\log\left\{ \tfrac{1+\rho}{1-\rho} \right\} + \text{constant}. \qquad (12.4.14)$$

From (12.4.14), it is clear that we should look at the transformations

$$U_n = \frac{1}{2}\log\left\{ \frac{1+r_n}{1-r_n} \right\} \text{ and } \xi = \frac{1}{2}\log\left\{ \frac{1+\rho}{1-\rho} \right\}, \qquad (12.4.15)$$

and consider the asymptotic distribution of $\sqrt{n}[U_n - \xi]$.

Now, in view of (12.4.1) we claim that

$$\sqrt{n}[U_n - \xi] \xrightarrow{\mathcal{L}} N(0,1) \text{ as } n \to \infty, \qquad (12.4.16)$$

since with $g(\rho) = \tfrac{1}{2}\log\left\{ \tfrac{1+\rho}{1-\rho} \right\}$, one has $g'(\rho) = \frac{1}{1-\rho^2}$. That is, for large n, we should consider the pivot

$$\sqrt{n}[U_n - \xi], \text{ which is approximately } N(0,1), \qquad (12.4.17)$$

where U_n and ξ were defined in (12.4.15).

One may verify that the transformations given in (12.4.15) can be *equivalently* stated as

$$U_n = \tanh^{-1}(r_n) \text{ and } \xi = \tanh^{-1}(\rho), \qquad (12.4.18)$$

which are referred to as Fisher's Z transformations introduced in 1925.

Fisher obtained the first four moments of $\tanh^{-1}(r_n)$ which were later updated by Gayen (1951). It turns out that the variance of $\tanh^{-1}(r_n)$ is approximated better by $\frac{1}{n-3}$ rather than $\frac{1}{n}$ when n is moderately large. Hence, in applications, we use the pivot

$$\sqrt{n-3}\left[\tanh^{-1}(r_n) - \tanh^{-1}(\rho)\right], \text{ which is approximately } N(0,1),$$
$$(12.4.19)$$

whatever be $\rho, -1 < \rho < 1$. We took n at least four in order to make $\sqrt{n-3}$ meaningful.

For large n, we use (12.4.19) to derive an *approximate* $100(1 - \alpha)\%$ confidence interval for ρ. Also, to test a null hypothesis $H_0 : \rho = \rho_0$, for large n, one uses the test statistic

$$Z_{calc} = \sqrt{n-3}\left[\tanh^{-1}(r_n) - \tanh^{-1}(\rho_0)\right] \qquad (12.4.20)$$

and comes up with an *approximate* level α test against an appropriate alternative hypothesis. The details are left out for brevity.

Example 12.4.6 (Examples 11.4.1-11.4.2 Continued) Consider the variables (X_1, X_2) from Example 11.4.1 for employees on their job performance scores before and after the training. Assuming a bivariate normal distribution for (X_1, X_2), we want to test $H_0 : \rho = 0.7$ versus $H_1 : \rho > 0.7$ at an *approximate* 5% level. From a set of observed data for 85 employees, we found $r_n = .757$. Thus, we have $\tanh^{-1}(r) = .98915, \tanh^{-1}(\rho_0) = .8673$, and hence from (12.4.20), we obtain:

$$z_{calc} = \sqrt{(85-3)}[.98915 - .8673] \approx 1.1034.$$

Now, with $\alpha = ..05$, one has $z_{.05} = 1.645$. Since z_{calc} does not exceed $z_{.05}$, we do not reject the null hypothesis at an *approximate* 5% level. We conclude that the correlation coefficient ρ between the job performance scores does not appear to be significantly larger than 0.7. ▲

Example 12.4.7 A psychologist wanted to study the relationship between the age (X_1, in years) and the average television viewing time per day (X_2, in minutes) in a large group of healthy children between the ages 4 and 12 years. From the study group, 42 children were randomly picked. For each child, the television viewing time was recorded every day during the two preceding weeks and X_2 is the average of the 14 recorded numbers. From this set of 42 observations on the pair of random variables (X_1, X_2), we found $r_n = .338$. Assuming a bivariate normal distribution for (X_1, X_2), we wish to obtain an *approximate* 95% confidence interval for ρ, the population correlation coefficient. From (12.4.19), we claim that $\tanh^{-1}(r_n) \pm \frac{1.96}{\sqrt{39}}$ is an *approximate* 95% confidence interval for $\tanh^{-1}(\rho)$. We have $\tanh^{-1}(r_n) = .35183$ and so the interval $(0.03798, 0.66568)$ for $\tanh^{-1}(\rho)$ has an *approximate* 95% confidence. In other words, the interval $(.03796, .58213)$ for ρ has an *approximate* 95% confidence. ▲

12.5   Exercises and Complements

12.2.1 (Example 12.2.1 Continued) Suppose that a random variable X has the Cauchy pdf $f(x; \theta) = \pi^{-1}\{1 + (x - \theta)^2\}^{-1}I(x \in \Re)$ where $\theta(\in \Re)$ is the unknown parameter. Show that the Fisher information, $\mathcal{I}(\theta) = 2$. {*Hint*: With $y = x - \theta$, first write $\mathcal{I}(\theta) = \pi^{-1}\int_{-\infty}^{\infty} 4y^2(1+y^2)^{-3}dy$. Evaluate the integral by substituting $y = \tan(z)$.}

12.2.2 Suppose that $\widehat{\theta}_n \equiv \widehat{\theta}_n(\mathbf{X})$ is the MLE of $\theta(\in \Re - \{0\})$ which satisfies both (12.2.3)-(12.2.4). Then, find the asymptotic distribution of $\sqrt{n}\left[\frac{\widehat{\theta}_n(\mathbf{X})}{\theta} - 1\right]$ as $n \to \infty$.

12.2.3 Suppose that a random variable X has the Logistic distribution with its pdf $f(x;\theta) = e^{-(x-\theta)}\{1 + e^{-(x-\theta)}\}^{-2}I(x \in \Re)$ where $\theta(\in \Re)$ is the unknown parameter. Show that the Fisher information, $\mathcal{I}(\theta) = 1/3$. {*Hint*: With $y = x - \theta$, express $\mathcal{I}(\theta)$ as

$$\int_{-\infty}^{\infty}[e^{-y}\{1 + e^{-y}\}^{-2} + 4e^{-3y}\{1 + e^{-y}\}^{-4} - 4e^{-2y}\{1 + e^{-y}\}^{-3}]dy.$$

Evaluate this integral by substituting $u = \{1 + e^{-y}\}^{-1}$.}

12.2.4 (Exercise 12.3.3 Continued) Suppose that $X_1, ..., X_n$ are iid random variables having the common Logistic pdf $f(x;\theta) = e^{-(x-\theta)}\{1 + e^{-(x-\theta)}\}^{-2}I(x \in \Re)$ where $\theta(\in \Re)$ is the unknown parameter.

(*i*)   Write down the likelihood equation (12.2.2) in this situation. Is it possible to obtain an analytical expression of the MLE, $\widehat{\theta}_n \equiv \widehat{\theta}_n(\mathbf{X})$ where $\mathbf{X} = (X_1, ..., X_n)$? If not, how could one find the estimate $\widehat{\theta}_n(\mathbf{x})$ for θ having obtained the data $\mathbf{x}$?

(*ii*)   Show that the MLE, $\widehat{\theta}_n$ is consistent for θ;

(*iii*)   Show that asymptotically (as $n \to \infty$), $\sqrt{n}\{\widehat{\theta}_n(\mathbf{X}) - \theta\}$ is distributed as $N(0, 3)$. {*Hint*: Use Exercise 12.2.3.}

12.3.1 A coffee dispensing machine automatically fills the cups placed underneath. The average (μ) amount of fill must not vary too much from the target (4 fl. ounces) because the overfill will add extra cost to the manufacturer while the underfill will generate complaints from the customers. A random sample of fills for 35 cups gave $\overline{x} = 3.98$ ounces and $S = .008$ ounces. Test a null hypothesis $H_0 : \mu = 4$ versus an alternative hypothesis $H_1 : \mu \neq 4$ at an *approximate* 1% level.

12.3.2 (Exercise 12.3.1 Continued) Obtain an *approximate* 95% confidence interval for the average (μ) amount of fill per cup.

12.3.3 Sixty automobiles of the same make and model were tested by drivers with similar road habits, and the gas mileage for each was recorded over a week. The summary results were $\overline{x} = 19.28$ miles per gallon and $s = 2.53$ miles per gallon. Construct an *approximate* 90% confidence interval for the true average (μ) gas mileage per gallon.

12.3.4 A company has been experimenting with a new liquid diet. The investigator wants to test whether the average (μ) weight loss for individuals on this diet is more that five pounds over the initial two-week period. Fifty individuals with similar age, height, weight, and metabolic structure were started on this diet. After two weeks, the weight reduction for each individual was recorded which gave the sample average 5.78 pounds and standard deviation 1.12 pounds. Test a null hypothesis $H_0 : \mu = 5$ versus an alternative hypothesis $H_1 : \mu > 5$ at an *approximate* 5% level.

12.3.5 A receptionist in a medical clinic thought that a patient's average (μ) waiting time to see one of the doctors exceeded twenty minutes. A random sample of sixty patients visiting the clinic during a week gave the average waiting time $\bar{x} = 28.65$ minutes and standard deviation $s = 2.23$ minutes. Can the receptionist's feeling regarding the average waiting time be validated at an *approximate* 5% level?

12.3.6 (Exercise 12.3.5 Continued) Obtain an *approximate* 95% confidence interval for the average (μ) waiting time for patients visiting the clinic.

12.3.7 A large tire manufacturing company has two factories A and B. It is believed that the employees in Factory A are paid less monthly salaries on the average than the employees in Factory B even though these employees had nearly similar jobs and job-related performances. The local union of employees in Factory A randomly sampled 35 employees from each factory and recorded each individual's monthly gross salary. The summary data follows:

Factory A	Factory B
$n_1 = 35$	$n_2 = 35$
$\bar{x}_{1n_1} = \$2854.72$	$\bar{x}_{2n_2} = \$3168.27$
$s_{1n_1} = \$105.29$	$s_{2n_2} = \$53.55$

At an *approximate* 5% level, test whether the data validates the belief that employees in Factory A are paid less on the average than employees in Factory B.

12.3.8 Suppose that $X_1, ..., X_n$ are iid Bernoulli(p) random variables where $0 < p < 1$ is the unknown parameter. The minimal sufficient estimator of p is the sample proportion $\widehat{p}_n$ of successes in n independent replications. One observes that $\sqrt{n}(\widehat{p}_n - p)/\sqrt{p(1-p)}$ is asymptotically standard normal. Hence, for large n, we have

$$P_p\left\{ \left| \sqrt{n}(\widehat{p}_n - p)/\sqrt{p(1-p)} \right| < z_{\alpha/2} \right\} \approx 1 - \alpha$$
$$\Rightarrow P_p\{p : n(\widehat{p}_n - p)^2 < [p(1-p)]z_{\alpha/2}^2\} \approx 1 - \alpha.$$

Now, solve the quadratic equation in p to derive an alternative (to that given in (12.3.18)) *approximate* $100(1 - \alpha)$% confidence interval for p.

12.3.9 A critic of the insurance industry claimed that less than 30% of the women in the work-force in a city carried employer-provided health benefits. In a random sample of 2000 working women in the city, we observed 20% with employer-provided health benefits. At an *approximate* 5% level, test the validity of the critic's claim.

12.3.10 A die has been rolled 100 times independently and the face 6 came up 25 times. Suppose that p stands for the probability of a six in a single trial. At an *approximate* 5% level, test a null hypothesis $H_0 : p = \frac{1}{6}$ versus an alternative hypothesis $H_1 : p \neq \frac{1}{6}$.

12.3.11 (Exercise 12.3.10 Continued) A die has been rolled 1000 times independently and the face 6 came up 250 times. Suppose that p stands for the probability of a six in a single trial. At an *approximate* 5% level, test a null hypothesis $H_0 : p = \frac{1}{6}$ versus an alternative hypothesis $H_1 : p \neq \frac{1}{6}$. Also, obtain an *approximate* 99% confidence interval for p.

12.3.12 We looked at two specific brands (A and B) of refrigerators in the market and we were interested to compare the percentages (p_A and p_B) requiring service calls during warranty. We found that out of 200 brand A refrigerators, 15 required servicing, whereas out of 100 brand B refrigerators, 6 required servicing during the warranty. At an *approximate* 5% level, test a null hypothesis $H_0 : p_A = p_B$ versus an alternative hypothesis $H_1 : p_A \neq p_B$.

12.3.13 (Exercise 12.3.12 Continued) Obtain an *approximate* 90% confidence interval for $p_A - p_B$.

12.3.14 Suppose that $X_1, ..., X_n$ are iid Poisson(λ) random variables where $0 < \lambda < \infty$ is the unknown parameter. The minimal sufficient estimator of λ is the sample mean $\overline{X}_n$ which is denoted by $\widehat{\lambda}_n$. One observes immediately that $\sqrt{n}(\widehat{\lambda}_n - \lambda)/\sqrt{\lambda}$ is also asymptotically standard normal. Hence, for large n, we have $P_\lambda \left\{ \left| \sqrt{n}(\widehat{\lambda}_n - \lambda)/\sqrt{\lambda} \right| < z_{\alpha/2} \right\} \approx 1 - \alpha$ so that $P_\lambda\{\lambda : n(\widehat{\lambda}_n - \lambda)^2 < \lambda z_{\alpha/2}^2\} \approx 1 - \alpha$. Now, solve the quadratic equation in λ to derive an alternative (to that given in (12.3.33)) *approximate* $100(1 - \alpha)$% confidence interval for λ.

12.3.15 Derive an *approximate* $100(1-\alpha)$% confidence interval for $\lambda_1 - \lambda_2$ in the two-sample situation for independent Poisson distributions.

12.3.16 Derive tests for the null hypothesis $H_0 : \lambda_1 = \lambda_2$ in the two-sample situation for independent Poisson distributions with *approximate* level α, when the alternative hypothesis is either upper-, lower-, or two-sided respectively.

12.3.17 (**Sign Test**) Suppose that $X_1, ..., X_n$ are iid continuous random variables with a common pdf $f(x; \theta)$ which is symmetric around $x = \theta$ where $x \in \Re, \theta \in \Re$. That is, the parameter θ is the population median assumed unknown, and f is assumed unknown too. The problem is to test

$H_0 : \theta = 0$ versus $H_1 : \theta > 0$ with given level $\alpha \in (0,1)$. Define $Y_i = I(X_i > 0)$, that is $Y_i = 1$ or 0 according as $X_i > 0$ or $X_i \leq 0, i = 1, ..., n$.

(i)   Show that $Y_1, ..., Y_n$ are iid Bernoulli(p) where $p = \frac{1}{2}$ or $p > \frac{1}{2}$ according as $\theta = 0$ or $\theta > 0$ respectively whatever be f;

(ii)   Argue that one can equivalently test $H_0 : p = \frac{1}{2}$ versus $H_1 : p > \frac{1}{2}$ with the help of the observations $Y_1, ..., Y_n$;

(iii) Propose an appropriate level α test from Section 12.3.1.

{*Note*: The associated Bernoulli parameter test is called a "Sign Test" because $\Sigma_{i=1}^n Y_i$ counts the number of positive observations among the X's. A sign test is called *nonparametric* or *distribution-free* since the methodology depending on Y's does not use an explicit form of the function f.}

12.3.18 (Comparison of a Sign Test with a Z Test in a Normal Distribution) Suppose that $X_1, ..., X_n$ are iid $N(\mu, \sigma^2)$ random variables where $\mu \in \Re$ is assumed unknown, but $\sigma \in \Re^+$ is known. The problem is to test $H_0 : \mu = 0$ versus $H_1 : \mu > 0$ with given level $\alpha \in (0,1)$. Define $Y_i = I(X_i > 0)$, that is $Y_i = 1$ or 0 according as $X_i > 0$ or $X_i \leq 0, i = 1, ..., n$. The UMP level α test is expressed as follows:

Test #1: Reject H_0 if and only if $\sqrt{n}\overline{X}_n/\sigma > z_\alpha$.

Whereas a level α sign test from Exercise 12.3.17 will look like this for large n:

Test #2: Reject H_0 if and only if $2\sqrt{n}(\overline{Y}_n - \frac{1}{2}) > z_\alpha$.

Now, address the following problems.

(i)   Compare the powers of Test #1 and Test #2 when $n = 30$, $40, 50, \alpha = .05$ and $\mu = .03, .02, .01$ (which are close to $\mu = 0$);

(ii)   Let us fix $n = 50$. Evaluate the sample sizes n_1, n_2, n_3 such that the Test #2 with sample size $n = n_1, n = n_2, n = n_3$ has respectively the same power as that of Test #1, used with $n = 50$, at $\mu = .03, .02, .01$;

(iii) How does n_1, n_2, n_3 compare with $n = 50$? Any comments?

{*Note*: Such investigations eventually lead to a notion called the *Bahadur efficiency*. Refer to Bahadur (1971).}

12.4.1 Suppose that $X_1, ..., X_n$ are iid with the common $N(\mu, \sigma^2)$ distribution where the parameters $\mu \in \Re, \sigma \in \Re^+$. Thus, $\sqrt{n}(S_n^2 - \sigma^2)$ is asymptotically (as $n \to \infty$) distributed as $N(0, 2\sigma^4)$ where S_n^2 is the sample variance. Find the variance stabilizing transformation of S_n^2.

12.4.2 Suppose that $X_1, ..., X_n$ are iid with the common exponential distribution having the mean $\beta(> 0)$. Thus, $\sqrt{n}(\overline{X}_n - \beta)$ is asymptotically (as $n \to \infty$) distributed as $N(0, \beta^2)$ where $\overline{X}_n$ is the sample mean. Find the variance stabilizing transformation of $\overline{X}_n$.

12.4.3 Suppose that $X_1, ..., X_n$ are iid with the common $N(\theta, \theta^2)$ distribution where the parameter $\theta \in \Re^+$. Let us denote $T_n = n^{-1}\Sigma_{i=1}^n X_i^2$. Derive the asymptotic (as $n \to \infty$) distribution $\sqrt{n}(T_n - 2\theta^2)$. Find the variance stabilizing transformation of T_n.

12.4.4 Verify the result given by (12.4.18).

12.4.5 A researcher wanted to study the strength of the correlation coefficient (ρ) between the proficiency in the two specific courses, namely first-year physics (X_1) and calculus (X_2), in a college campus. From the large pool of first-year students enrolled in these two courses, thirty eight students were randomly picked and their midterm grades in the two courses were recorded. From this data, we obtained the sample correlation coefficient $r = .663$. Assume that (X_1, X_2) has a bivariate normal distribution in the population. Test whether ρ can be assumed to exceed 0.5 with an *approximate* level $\alpha = .10$.

12.4.6 The data on systolic blood pressure (X_1) and age (X_2) for a sample of 45 men of similar health conditions was collected. From this data, we obtained the sample correlation coefficient $r = .385$. Obtain an approximate 95% confidence interval for the population correlation coefficient ρ. Assume bivariate normality of (X_1, X_2) in the population.

12.4.7 The strength of the right and left grips, denoted by X_1 and X_2 respectively, were tested for 120 auto accident victims during routine therapeutic exams in a rehab center. The observed values of X_1 and X_2 were both coded between zero and ten, a low (high) value indicating significant weakness (strength) in the grip. Assume bivariate normality of (X_1, X_2) in the population. The data gave rise to the sample correlation coefficient $r = .605$. Test whether ρ can be assumed 0.5 with an *approximate* level $\alpha = .10$.

13

Sample Size Determination: Two-Stage Procedures

13.1 Introduction

Consider a sequence of iid observations $X_1, X_2, \ldots$ from a $N(\mu, \sigma^2)$ population where both parameters $\mu \in \Re, \sigma \in \Re^+$ are unknown. Having recorded a fixed number of observations $X_1, \ldots, X_n, n \geq 2$, and with preassigned $\alpha \in (0, 1)$, the customary $100(1 - \alpha)\%$ confidence interval for the population mean μ is given by

$$J_n = \left[\overline{X}_n \pm t_{n-1,\alpha/2} n^{-1/2} S_n \right]. \tag{13.1.1}$$

Here, $\overline{X}_n = n^{-1} \Sigma_{i=1}^{n} X_i$, $S_n^2 = (n-1)^{-1} \Sigma_{i=1}^{n} (X_i - \overline{X}_n)^2$ and $t_{n-1,\alpha/2}$ stands for the upper $100(\alpha/2)\%$ point of the Student's t distribution with $(n-1)$ degrees of freedom.

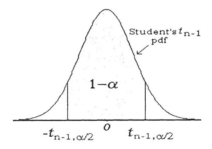

Figure 13.1.1. The Area on the Right (or Left) of $t_{n-1,\alpha/2}$ (or $-t_{n-1,\alpha/2}$) Is $\alpha/2$

The width of the confidence interval J_n is given by

$$W_n = 2t_{n-1,\alpha/2} n^{-1/2} S_n. \tag{13.1.2}$$

> Note that the width W_n is a *random variable* even though n is held fixed. It is so because S_n is a random variable.

Now, recall that the support for the random variable S_n is the whole positive half of the real line, that is S_n has a positive density at the point s if and only if $s > 0$.

| How wide can the customary confidence interval J_n be? |

First, one should realize that $P\{W_n > k\} > 0$ for any fixed $k > 0$, that is there is no guarantee at all that the width W_n is necessarily going to be "small". In Table 13.1.1, we present a summary from a simple simulated exercise. Using MINITAB Release 12.1, we generated n random samples from a $N(0, \sigma^2)$ population where we let $n = 5, 10, \alpha = .05$ and $\sigma = 1, 2, 5$. Since the distribution of the sample standard deviation S_n, and hence that of W_n, is free from the parameter μ, we fixed the value $\mu = 0$ in this illustration. With a fixed pair of values of n and σ, we gathered a random sample of size n, one hundred times independently, thereby obtaining the confidence interval J_n for the parameter μ each time. Consequently, for a fixed pair of values of n and σ, we came up with one hundred values of the random variable W_n, the width of the actually constructed 95% confidence interval for μ. Table 13.1.1 lists some descriptive statistics found from these one hundred observed values of W_n.

Table 13.1.1. Simulated Description of 95% Confidence Interval's Width W_n Given by (13.1.2): 100 Replications

	n = 5
	minimum = 2.841, maximum = 24.721, mean = 11.696,
$\sigma = 5$	standard deviation = 4.091, median = 11.541
	n = 10
	minimum = 2.570, maximum = 10.606, mean = 6.661,
	standard deviation = 1.669, median = 6.536
	n = 5
	minimum = 1.932, maximum = 9.320, mean = 4.918,
$\sigma = 2$	standard deviation = 1.633, median = 4.889
	n = 10
	minimum = 1.486, maximum = 5.791, mean = 2.814,
	standard deviation = 0.701, median = 2.803
	n = 5
	minimum = 0.541, maximum = 5.117, mean = 2.245,
$\sigma = 1$	standard deviation = 0.863, median = 2.209
	n = 10
	minimum = 0.607, maximum = 2.478, mean = 1.394,
	standard deviation = 0.344, median = 1.412

From Table 13.1.1, it is obvious that for a fixed value of σ, the random width W_n goes down when n increases and for a fixed value of n, the same goes down when σ decreases. But, it may feel awkward when we realize that for any fixed pair of n and σ, an experimenter may get stuck with a confidence interval J_n whose observed width w_n is relatively "large". Hence, we may like to construct a confidence interval J for μ such that

(i)   the length of J is $2d$ where $d(> 0)$ is preassigned, **and**

(ii) $P_{\mu,\sigma^2}\{\mu \in J\} \geq 1 - \alpha$ where $0 < \alpha < 1$ is preassigned,   (13.1.3)

for all μ and σ^2.

> An interval estimator J satisfying **both** requirements in (13.1.3) is called a *fixed-width confidence interval* for μ.

Dantzig (1940) had shown that no matter what fixed-sample-size procedure one uses, that is regardless of what the fixed sample size n is, one can not find a fixed-width confidence interval estimator for μ from the random sample $X_1, ..., X_n$. In other words, we *can not simultaneously control* the width of the confidence interval for μ and its confidence coefficient if the sample size n is held fixed in advance. A general non-existential result along this line is found in Ghosh et al. (1997, Section 3.7).

We can, however, design an experiment by determining the appropriate sample size as a positive integer valued random variable to solve this problem. Stein (1945,1949) gave an ingenious method of determining the required sample size by sampling in two steps. This method of data collection is called *two-stage sampling* or *double sampling*.

This sampling technique falls in the area of *Sequential Analysis*. Abraham Wald was the leading architect who laid the foundation of sequential sampling, an approach that was essential to curtail quality control inspections of shipments of weapons during the World War. In the early 1940's, Wald and his collaborators created the mathematical foundation in this area. Wald's classic book on this subject, *Sequential Analysis*, appeared in 1947. In its introduction, Wald credited P. C. Mahalanobis by noting that some of the key ideas from Mahalanobis (1940) may be considered the "forerunner" of sequential analysis. Mahalanobis (1940) introduced the fundamental idea of taking pilot samples before conducting large-scale surveys. The method of *two-stage sampling* or *double sampling* relies heavily on pilot samples.

The area of sequential analysis is both vast and rich. Sequential sampling strategies are applied in areas including (i) clinical trials [Armitage (1973), Whitehead (1986,1991)], (ii) reliability, life-tests and quality control [Basu

(1991)], and (iii) survival analysis [Gardiner and Susarla (1991), Gardiner et. al. (1986)].

To achieve a balanced perspective, the reader may browse through the books of Wald (1947), Bechhofer et al. (1968), Ghosh (1970), Armitage (1973), Sen (1981), Woodroofe (1982), Siegmund (1985), Whitehead (1983), Mukhopadhyay and Solanky (1994), Balakrishnan and Basu (1995), and Ghosh et al. (1997) as well as the *Handbook of Sequential Analysis* edited by Ghosh and Sen (1991). The *Sequential Analysis* journal provides the most up-to-date account in this important area. We hope that the cited sources will guide interested readers to consider many other important and relevant references.

There are important reasons why we include such topics to end this book. One point to note is that many results derived earlier come together here as important tools and machineries needed for technical analyses. But, there is another point to note. From previous chapters, a reader may get an impression that all reasonable statistical problems can be handled by the classical fixed-sample-size analyses and we show otherwise.

We recall Example 7.3.2 where we showed that it was impossible to find any unbiased estimator of p^{-1} when the data consisted of iid Bernoulli(p) observations $X_1, ..., X_n$ with predetermined n. But, in Example 7.3.3 we showed how inverse binomial sampling method could provide an unbiased estimator of p^{-1}, a parametric function which is very important in statistical applications in ecology.

In the present chapter, we touch upon two statistical problems with very realistic goals which can not be solved if the sample size is fixed in advance, but these problems can be solved with the help of two-stage sampling strategies. The emphasis is to show that there is a wide world out there beyond the classical fixed-sample-size analysis.

> Neither a *fixed-width* confidence interval satisfying both requirements in (13.1.3) nor the *bounded risk* point estimation problem can be solved by any fixed-sample-size method. A Stein type *two-stage* methodology provides an exact solution for both problems.

In Section 13.2, we give the construction of a fixed-width confidence interval for the unknown mean of a normal population whose variance is completely unknown. This is accomplished (Theorem 13.2.2) by the Stein two-stage methodology. In Section 13.3, we briefly address a point estimation problem by requiring a preassigned upper bound on the mean square error.

13.2   The Fixed-Width Confidence Interval

Let us begin with a $N(\mu, \sigma^2)$ population where $\mu(\in \Re), \sigma(\in \Re^+)$ are assumed unknown. We wish to construct a *fixed-width confidence interval* $\mathcal{J}$ for μ such that both requirements in (13.1.3) are met. That is, the length of $\mathcal{J}$ has to be $2d$ and $P_{\mu,\sigma^2}\{\mu \in \mathcal{J}\} \geq 1 - \alpha$ for all μ, σ^2 where $d(> 0), 0 < \alpha < 1$ are preassigned.

Having recorded the observations $X_1, ..., X_n$, let us closely examine the fixed-width confidence interval

$$\mathcal{J}_n = [\overline{X}_n - d, \overline{X}_n + d] \text{ for the population mean } \mu. \tag{13.2.1}$$

One may recall that $\phi(.)$ and $\Phi(.)$ respectively stand for the pdf and df of a standard normal random variable, that is $\phi(z) = \{\sqrt{2\pi}\}^{-1} exp\{-z^2/2\}$ and $\Phi(z) = \int_{-\infty}^{z} \phi(y)dy, z \in \Re$. Next, we have

$$P_{\mu,\sigma^2}\{\mu \in \mathcal{J}_n\} = P_{\mu,\sigma^2}\{|\overline{X}_n - \mu| \leq d\} = 2\Phi(\sqrt{n}d/\sigma) - 1 \geq 1 - \alpha, \tag{13.2.2}$$

if and only if

$$2\Phi(\sqrt{n}d/\sigma) - 1 \geq 2\Phi(z_{\alpha/2}) - 1 \Leftrightarrow \sqrt{n}d/\sigma \geq z_{\alpha/2}$$
$$\Leftrightarrow n \geq z_{\alpha/2}^2 \sigma^2/d^2 = C, \text{ say.} \tag{13.2.3}$$

Had the population variance σ^2 been known, we would have taken $n = \langle C \rangle + 1$ and constructed the confidence interval $\mathcal{J}_n$.

> The notation $\langle u \rangle$ stands for the largest integer $< u$.

For example, if somehow we knew that $C = 27.02$ or 23 or 102.801, then n will respectively be 28 or 23 or 103.

But, the magnitude of C is unknown because σ^2 is unknown. We really do not know how many observations to take so that the fixed-width confidence interval $\mathcal{J}_n$ from (13.2.1) will work. In this literature, the sample size C, pretending that it is an integer, is referred to as the *optimal fixed sample size* required to solve the problem, had σ^2 been known.

13.2.1   Stein's Sampling Methodology

Since the *optimal fixed sample size* C is unknown, Stein (1945,1949) formulated the following two-stage sampling strategy. One starts with a preliminary set of observations $X_1, ..., X_m$ where the pilot sample size $m(\geq 2)$ is predetermined. This step is referred to as the *first stage* of sampling.

The pilot observations are utilized to estimate the population variance σ^2 by the corresponding sample variance S_m^2. Next, let us define a *stopping variable*,

$$N = \max\left\{m, \left\langle t_{m-1,\alpha/2}^2 S_m^2/d^2\right\rangle + 1\right\}, \qquad (13.2.4)$$

which is an observable positive integer valued random variable. In the expression of C, we have replaced σ^2 and $z_{\alpha/2}^2$ respectively by the sample variance S_m^2 and $t_{m-1,\alpha/2}^2$. Thus, the expression $t_{m-1,\alpha/2}^2 S_m^2/d^2$, used in the definition of N, is viewed as an estimator of C and hence we view N as an estimator of C. Observe that the stopping variable N depends upon d, α, m and the first-stage observations $X_1, ..., X_m$ through S_m^2.

If $N = m$, then we like to think that we have started with too many pilot observations and hence we do not take any other observations in the second stage. But, if $N > m$, then we sample the difference $(N - m)$ in the second stage to come up with $(N - m)$ new observations $X_{m+1}, ..., X_N$. After combining observations from the two-stages, the data consists of $X_1, ..., X_N$ whether any observations are drawn in the second stage or not.

Next, on the basis of N and $X_1, ..., X_N$, one finds the sample mean $\overline{X}_N(= N^{-1}\Sigma_{i=1}^N X_i)$ and proposes the fixed-width confidence interval

$$\mathcal{J}_N = [\overline{X}_N - d, \overline{X}_N + d] \text{ for the population mean } \mu. \qquad (13.2.5)$$

This is what is referred to as the Stein two-stage sampling methodology.

13.2.2   Some Interesting Properties

We should check whether the confidence coefficient associated with the interval estimator $\mathcal{J}_N$ is at least $1-\alpha$, the preassigned goal. Theorem 13.2.2 settles this query in the affirmative. We first summarize some preliminary but interesting results.

Theorem 13.2.1 *For the stopping variable N defined by (13.2.4), for all fixed $\mu \in \Re, \sigma \in \Re^+, d \in \Re^+$ and $\alpha \in (0,1)$, one has:*

(i) $P_{\mu,\sigma^2}\{N < \infty\} = 1$;

(ii) $t_{m-1,\alpha/2}^2 \sigma^2 d^{-2} \le E_{\mu,\sigma^2}[N] \le m + t_{m-1,\alpha/2}^2 \sigma^2 d^{-2}$;

(iii) $Q \equiv N^{1/2}(\overline{X}_N - \mu)/\sigma$ *is distributed as $N(0,1)$. Also, the random variables Q and S_m^2 are independent*;

(iv) $N^{1/2}(\overline{X}_N - \mu)/S_m$ *is distributed as the Student's t random variable with $m - 1$ degrees of freedom*;

(v) $\overline{X}_N$ *is an unbiased estimator of μ with its variance given by $\sigma^2 E_{\mu,\sigma^2}[1/N]$.*

Proof From the definition of N in (13.2.4), one can verify that

$$t^2_{m-1,\alpha/2}S^2_m d^{-2} \leq N \leq m + t^2_{m-1,\alpha/2}S^2_m d^{-2}, \qquad (13.2.6)$$

which is referred to as the *basic inequality*. Now, let us proceed with the proof.

(i) Observe that one has

$$
\begin{aligned}
P_{\mu,\sigma^2}\{N = \infty\} \\
&= \lim_{n\to\infty} P_{\mu,\sigma^2}\{N > n\} \\
&\leq \lim_{n\to\infty} P_{\mu,\sigma^2}\{m + t^2_{m-1,\alpha/2}S^2_m d^{-2} > n\} \text{ in view of} \\
&\quad \text{the basic inequality (13.2.6)} \\
&\leq \lim_{n\to\infty} P_{\mu,\sigma^2}\{Y > (n-m)(m-1)d^2 t^2_{m-1,\alpha/2}\sigma^{-2}\}, \text{ where} \\
&\quad Y = (m-1)S^2_m \sigma^{-2}, \text{ which is distributed as } \chi^2_{m-1}.
\end{aligned}
\qquad (13.2.7)
$$

But, with $a_n = (n-m)(m-1)d^2 t^2_{m-1,\alpha/2}\sigma^{-2}$ for $n \geq m$, we can write

$$
\begin{aligned}
P_{\mu,\sigma^2}\{Y > a_n\} \\
&= \left\{2^{(m-1)/2}\Gamma(\tfrac{1}{2}(m-1))\right\}^{-1} \int_{y=a_n}^{\infty} e^{-y/2} y^{(m-3)/2} dy \\
&\leq \left\{2^{(m-1)/2}\Gamma(\tfrac{1}{2}(m-1))\right\}^{-1} e^{-a_n/2} \int_{y=a_n}^{\infty} y^{(m-3)/2} dy, \\
&\quad \text{since } e^{-y/2} \leq e^{-a_n/2} \text{ for all } y \in (a_n, \infty) \\
&= 2\left\{2^{(m-1)/2}(m-1)\Gamma(\tfrac{1}{2}(m-1))\right\}^{-1} e^{-a_n/2} a_n^{(m-1)/2}.
\end{aligned}
\qquad (13.2.8)
$$

Now, $\lim_{n\to\infty} a_n = \infty$ and one can verify using L'Hôpital's rule from (1.6.29) that $\lim_{n\to\infty} e^{-a_n/2} a_n^{(m-1)/2} = 0$. Hence, from (13.2.8) we conclude that $\lim_{n\to\infty} P_{\mu,\sigma^2}\{Y > a_n\} = 0$. Then, combining this with (13.2.7), we conclude that $P_{\mu,\sigma^2}\{N = \infty\} = 0$ so that one has $P_{\mu,\sigma^2}\{N < \infty\} = 1$.

(ii) We take expectations throughout the basic inequality in (13.2.6) and use the fact that $E_{\mu,\sigma^2}(S^2_m) = \sigma^2$. We leave the details out as Exercise 13.2.1.

(iii) For any fixed $x \in \Re$, let us write

$$
\begin{aligned}
P_{\mu,\sigma^2}\{N^{1/2}(\overline{X}_N - \mu)/\sigma \leq x\} \\
&= \Sigma_{n=m}^{\infty} P_{\mu,\sigma^2}\{N^{1/2}(\overline{X}_N - \mu)/\sigma \leq x \cap N = n\} \\
&= \Sigma_{n=m}^{\infty} P_{\mu,\sigma^2}\{n^{1/2}(\overline{X}_n - \mu)/\sigma \leq x \cap N = n\} \\
&= \Sigma_{n=m}^{\infty} P_{\mu,\sigma^2}\{n^{1/2}(\overline{X}_n - \mu)/\sigma \leq x\} P_{\mu,\sigma^2}\{N = n\}.
\end{aligned}
\qquad (13.2.9)
$$

The last step in (13.2.9) is valid because (a) the random variable $I(N = n)$ is determined only by S_m^2, (b) $\overline{X}_n$ and S_m^2 are independent for all $n \geq m$, and hence the two random variables $\overline{X}_n$ and $I(N = n)$ are independent for all $n \geq m$. Refer to Exercise 13.2.2. Now, we can rewrite (13.2.9) as

$$
\begin{aligned}
P_{\mu,\sigma^2}&\{N^{1/2}(\overline{X}_N - \mu)/\sigma \leq x\} \\
&= \Sigma_{n=m}^{\infty}\Phi(x)P_{\mu,\sigma^2}\{N = n\}, \text{ since } \Phi(.) \text{ is the df of} \\
&\quad n^{1/2}(\overline{X}_n - \mu)/\sigma \text{ for all fixed } n \geq m \\
&= \Phi(x), \text{ since } \Sigma_{n=m}^{\infty}P_{\mu,\sigma^2}\{N = n\} = P_{\mu,\sigma^2}(N < \infty) = 1,
\end{aligned}
\tag{13.2.10}
$$

in view part (i). Hence, $N^{1/2}(\overline{X}_N - \mu)/\sigma$ is distributed as $N(0,1)$.

Observe that given S_m^2, we have a fixed value of N. Hence, given S_m^2, we claim that the conditional distribution of $N^{1/2}(\overline{X}_N - \mu)/\sigma$ is $N(0,1)$. But, this conditional distribution does not involve the given value of the conditioning variable S_m^2. Thus, $N^{1/2}(\overline{X}_N-\mu)/\sigma$ and S_m^2 are independently distributed.

(iv) We combine facts from part (iii) and Definition 4.5.1 to conclude that $N^{1/2}(\overline{X}_N - \mu)/S_m$ is distributed as Student's t with $m - 1$ degrees of freedom.

(v) By the Theorem 3.3.1, part (i), we can express $E_{\mu,\sigma^2}[\overline{X}_N]$ as

$$
\begin{aligned}
\Sigma_{n=m}^{\infty}&E_{\mu,\sigma^2}[\overline{X}_N \mid N = n]P_{\mu,\sigma^2}(N = n) \\
&= \Sigma_{n=m}^{\infty}E_{\mu,\sigma^2}[\overline{X}_n \mid N = n]P_{\mu,\sigma^2}(N = n) \\
&= \Sigma_{n=m}^{\infty}E_{\mu,\sigma^2}[\overline{X}_n]P_{\mu,\sigma^2}(N = n), \text{ since } \overline{X}_n \text{ and } I(N = n) \\
&\quad \text{are independent for all } n \geq m \\
&= \mu\Sigma_{n=m}^{\infty}P_{\mu,\sigma^2}(N = n),
\end{aligned}
\tag{13.2.11}
$$

which reduces to μ since $P_{\mu,\sigma^2}(N < \infty) = 1$. In the same fashion, one can verify that $E_{\mu,\sigma^2}[\overline{X}_N^2] = \mu^2 + \sigma^2 E_{\mu,\sigma^2}[1/N]$ and the expression of $V_{\mu,\sigma^2}[\overline{X}_N]$ can be found easily. The details are left out as Exercise 13.2.3. ∎

Now we are in a position to state a fundamental result which is due to Stein (1945,1949).

Theorem 13.2.2 *For the stopping variable N defined by (13.2.4) and the fixed-width confidence interval $\mathcal{J}_N = [\overline{X}_N - d, \overline{X}_N + d]$ for the population mean μ, we have:*

$$
P_{\mu,\sigma^2}\{\mu \in \mathcal{J}_N\} \geq 1 - \alpha
$$

for all fixed $\mu \in \Re, \sigma \in \Re^+, d \in \Re^+$ and $\alpha \in (0,1)$.

Proof Let us express the confidence coefficient associated with the fixed-width confidence interval $\mathcal{J}_N$ as

$$
\begin{aligned}
P_{\mu,\sigma^2}\{\mu \in \mathcal{J}_N\} &= P_{\mu,\sigma^2}\{|\overline{X}_N - \mu| \le d\} \\
&= P_{\mu,\sigma^2}\{N^{1/2}|\overline{X}_N - \mu|/S_m \le dN^{1/2}/S_m\}.
\end{aligned}
$$
(13.2.12)

But, from the basic inequality (13.2.6) for N, we see that $dN^{1/2}/S_m \ge t_{m-1,\alpha/2}$ w.p.1, and hence the event $N^{1/2}|\overline{X}_N - \mu|/S_m \le t_{m-1,\alpha/2}$ implies the event $N^{1/2}|\overline{X}_N - \mu|/S_m \le dN^{1/2}/S_m$. That is, the set A where $N^{1/2}|\overline{X}_N - \mu|/S_m \le t_{m-1,\alpha/2}$ holds is a subset of the set B where $N^{1/2}|\overline{X}_N - \mu|/S_m \le dN^{1/2}/S_m$ holds. Thus, we can claim that

$$
\begin{aligned}
&P_{\mu,\sigma^2}\{N^{1/2}|\overline{X}_N - \mu|/S_m \le dN^{1/2}/S_m\} \\
&\ge P_{\mu,\sigma^2}\{N^{1/2}|\overline{X}_N - \mu|/S_m \le t_{m-1,\alpha/2}\}.
\end{aligned}
$$
(13.2.13)

Next, one combines (13.2.12)-(13.2.13) to write

$$
P_{\mu,\sigma^2}\{\mu \in \mathcal{J}_N\} \ge P_{\mu,\sigma^2}\{N^{1/2}|\overline{X}_N - \mu|/S_m \le t_{m-1,\alpha/2}\} = 1 - \alpha,
$$
(13.2.14)

since $t_{m-1,\alpha/2}$ is the upper $100(\alpha/2)\%$ point of the Student's t distribution with df $m - 1$. In (13.2.14), we have used the result from Theorem 13.2.1, part (iv). ∎

Example 13.2.1 (Simulation) With the help of a computer, we generated a normal population with $\mu = 2$ and $\sigma = 5$. Let us pretend that we do not know μ, σ. We consider $\alpha = .05$, that is we require a 95% fixed-width confidence interval for μ. We also fixed some d values (given in the accompanying Table 13.2.1) and took $m = 5, 10$. The two-stage estimation procedure (13.2.4)-(13.2.5) is implemented as follows:

For the ith *independent replication*, we start with a *new* set of observations $x_{i1}, ..., x_{im}$ from the computer-generated population and obtain $\overline{x}_{im} = m^{-1}\Sigma_{j=1}^m x_{ij}, s_{im}^2 = (m-1)^{-1}\Sigma_{j=1}^m(x_{ij} - \overline{x}_{im})^2$ which lead to an *observed* value, $N = n_i$. Then, the pilot data $x_{i1}, ..., x_{im}$ is updated in order to arrive at the final set of data $x_{i1}, ..., x_{in_i}, i = 1, ..., 5000(= k$, say). Then, we compute $\overline{n} = k^{-1}\Sigma_{i=1}^k n_i, s^2(\overline{n}) = \{k(k-1)\}^{-1}\Sigma_{i=1}^k(n_i - \overline{n})^2$ which are respectively the unbiased estimates for $E(N)$ and its variance. During the ith independent replication, we also record $p_i = 1$ (or 0) according as $\mu = 2 \in (\notin)$ to the *observed* fixed-width confidence interval $\mathcal{J}_{n_i} = [\overline{x}_{in_i} \pm d]$ where $\overline{x}_{in_i} = n_i^{-1}\Sigma_{j=1}^{n_i} x_{ij}, i = 1, ..., k$. Then, we obtain $\overline{p} = k^{-1}\Sigma_{i=1}^k p_i, s^2(\overline{p}) = (k-1)^{-1}\overline{p}(1 - \overline{p})$ which are respectively the unbiased estimates for the coverage probability and its variance. The Table 13.2.1 summarizes our findings.

From this table, some features are immediately noticed. First, the $\bar{p}$ values have always exceeded the target confidence coefficient 0.95 except in one case. In the case $m = 10, C = 50$ we have $\bar{p} = .9458$ with its estimated standard error .0037. Since k is very large, an approximate 95% confidence interval for the true coverage probability (p) in this case can be taken to be $\bar{p} \pm 1.96s(\bar{p})$ which amounts to (.9336, .9531). Note that the target coverage probability, that is 0.95, lies between the two confidence bounds 0.9336 and 0.9531.

Table 13.2.1. Moderate Sample Performance of Stein's
Two-Stage Procedure (13.2.4)-(13.2.5) with $\alpha = .05$:
5000 Replications

d	C	$\bar{n}$	$s(\bar{n})$	$\bar{p}$	$s(\bar{p})$
$m = 5, z_{.025} = 1.96, t_{4,.025} = 2.776$					
1.789	30	62.32	0.825	0.9882	0.0041
1.386	50	103.03	1.023	0.9788	0.0043
0.980	100	199.20	1.765	0.9867	0.0041
$m = 10, z_{.025} = 1.96, t_{9,.025} = 2.262$					
1.789	30	40.48	0.264	0.9577	0.0029
1.386	50	67.22	0.416	0.9458	0.0037
0.980	100	135.16	0.915	0.9524	0.0032

Next, we observe that the $\bar{n}$ values are sizably larger than the corresponding C values, a feature which validates the fact that the Stein procedure over-samples significantly. But, the extent of over-sampling falls substantially when we choose $m = 10$ instead of $m = 5$. Compared with C, over-sampling amounts to approximately 100% and 33% respectively when $m = 5$ and $m = 10$. We may add that the ratio $t_{m-1,.025}^2/z_{.025}^2$ approximately reduces to 2.006 and 1.332 respectively when $m = 5$ and $m = 10$. Since $C = z_{\alpha/2}^2 \sigma^2/d^2$ and $E(N) \approx t_{m-1,\alpha/2}^2 \sigma^2/d^2$, the amount of over-sampling we have experienced should be quite expected. More elaborate exercises with computer simulations are left as class projects. ▲

The Theorem 13.2.2 shows that the two-stage procedure of Stein solved a fundamental problem of statistical inference which could not be tackled by any fixed-sample-size methodology.

On the other hand, there are several important issues one can raise in this context. The choice of m, the pilot sample size, plays a crucial role in the performance of the two-stage estimation procedure (13.2.4)-(13.2.5).

Moshman (1958) discussed a couple of approaches to determine a reasonable choice of the pilot sample size m. But, whether one selects a small or large value of m, it is known that the final sample size N tends to significantly overestimate C on the average. We may add that even when C is very large, the final sample size N tends to significantly overestimate C on the average. Starr (1966) systematically investigated these other related issues.

But, there is no denying of the fact that the Stein procedure *solved* a fundamental problem of statistical inference which could not otherwise be solved by any fixed-sample-size design. One may refer to Cox (1952), Chatterjee (1991), Mukhopadhyay (1980,1991), Ghosh and Mukhopadhyay (1976,1981) and Ghosh et al. (1997, Chapter 6), among other sources, to appreciate a fuller picture of different kinds of developments in this vein.

If one can assume that the unknown parameter σ has a known lower bound $\sigma_L(> 0)$, then the choice of $m(\approx z_{\alpha/2}^2 \sigma_L^2/d^2)$ becomes clear. In this situation, Stein's two-stage fixed-width confidence interval procedure, when appropriately modified, becomes very "efficient" as shown recently by Mukhopadhyay and Duggan (1997,1999).

The exact confidence coefficient for Stein's two-stage confidence interval $\mathcal{J}_N$ from (13.2.5) is given in the Exercise 13.2.4.

The *two-sample* fixed-width confidence problem for the difference of means of independent normal populations with unknown and unequal variances is left as Exercise 13.2.10. This is referred to as the *Behrens-Fisher problem*. For ideas and references, look at both the Exercises 11.3.15 and 13.2.10.

13.3   The Bounded Risk Point Estimation

Again, we begin with a $N(\mu,\sigma^2)$ population where $\mu \in \Re, \sigma \in \Re^+$ are assumed unknown. Having recorded a fixed number of observations $X_1, ..., X_n$, the customary point estimator for μ is the sample mean $\overline{X}_n(= n^{-1}\Sigma_{i=1}^n X_i)$. As before, let us denote $S_n^2 = (n-1)^{-1}\Sigma_{i=1}^n (X_i - \overline{X}_n)^2$ for $n \geq 2$, and us suppose that the loss incurred in estimating μ by $\overline{X}_n$ is measured by the *squared error loss function*,

$$L(\mu,\overline{X}_n) = a[\overline{X}_n - \mu]^2, \text{ with some fixed and known } a(> 0). \quad (13.3.1)$$

The associated *risk* is given by

$$R_n(\mu,\overline{X}_n) = E_{\mu,\sigma^2}[L(\mu,\overline{X}_n)] = a\sigma^2/n, \quad (13.3.2)$$

which remains unknown since σ^2 is assumed unknown. We recall that this is the *frequentist risk* discussed earlier in Section 10.4.

It is clear that based on the data $X_1, ..., X_n$ we may estimate $R_n(\mu, \overline{X}_n)$ by $\widehat{R_n} = aS_n^2/n$. In Table 13.3.1, we present a summary from a simple simulated exercise. Using MINITAB Release 12.1, we generated n random samples from a $N(0, \sigma^2)$ population where we let $a = 1, n = 5, 10$ and $\sigma = 5, 2, 1$. Since the distribution of the sample variance S_n^2, and hence that of $\widehat{R_n}$, is free from the parameter μ, we fixed the value $\mu = 0$ in this illustration. With a fixed pair of values of n and σ, we gathered a random sample of size n, one hundred times independently, thereby obtaining an observed value of $\widehat{R_n}$ each time. Consequently, for a fixed pair of values of n and σ, we came up with one hundred values of the random variable $\widehat{R_n}$. Table 13.3.1 lists some descriptive statistics derived from these one hundred observed values of $\widehat{R_n}$.

Table 13.3.1. Simulated Description of the Estimated
Values $\widehat{R_n}$: 100 Replications

	n = 5
	minimum = .232, maximum = 16.324, mean = 4.534
$\sigma = 5$	standard deviation = 3.247, median = 3.701
	n = 10
	minimum = .457, maximum = 9.217, mean = 2.597
	standard deviation = 1.443, median = 2.579
	n = 5
	minimum = .0569, maximum = 3.3891, mean = .7772
$\sigma = 2$	standard deviation = .6145, median = .6578
	n = 10
	minimum = .1179, maximum = 1.0275, mean = .3927
	standard deviation = .1732, median = .3608
	n = 5
	minimum = .0135, maximum = .7787, mean = .1904
$\sigma = 1$	standard deviation = .1281, median = .1684
	n = 10
	minimum = .02843, maximum = .24950, mean = .10113
	standard deviation = .05247, median = .09293

From Table 13.3.1, it is obvious that for a fixed value of σ, the estimated risk $\widehat{R_n}$ goes down when n increases and for a fixed value of n, the same goes down when σ decreases. But, it may feel awkward when we realize that for any fixed pair n and σ, an experimenter may get stuck with an estimated risk which is relatively "large".

Hence, we may like to estimate μ by $\overline{X}_n$ in such a way that the risk,

$$R_n(\mu, \overline{X}_n) \leq \omega \text{ for preassigned } \omega(> 0). \tag{13.3.3}$$

This formulation is described as the *bounded risk* point estimation approach. The requirement in (13.3.3) will guarantee that the risk does not exceed a preassigned number ω. It is a very useful goal to achieve, particularly when ω is chosen small.

But, there is a problem here. Even such a simplistic goal *can not be attained* by any fixed-sample-size procedure. Refer to a general non-existential result found in Ghosh et al. (1997, Section 3.7). In Section 13.3.1, we discuss a Stein type two-stage sampling technique for this problem. In Section 13.3.2, we show that the bounded risk point estimation problem can be solved by adopting the proposed methodology.

13.3.1 The Sampling Methodology

The risk, $R_n(\mu, \overline{X}_n)$ is bounded from above by ω, a preassigned positive number, if and only if the sample size n is the smallest integer $\geq a\sigma^2/\omega = n^*$, say. Had σ been known, we would take a sample of size $n = \langle n^* \rangle + 1$ and estimate μ by $\overline{X}_n$ obtained from the observations $X_1, ..., X_n$. But, the fact is that n^* is unknown since σ remains unknown. Since no fixed-sample-size solution exists, we propose an appropriate two-stage procedure for this problem. This methodology was discussed in Rao (1973, pp. 486-488). One may refer to Ghosh and Mukhopadhyay (1976) for more details.

One starts with initial observations $X_1, ..., X_m$ where the pilot sample size $m(\geq 4)$ is predetermined. The basic idea is to begin the experiment with such m which is hopefully "small" compared with n^*. In real life though, one starts with some reasonably chosen m, but it is impossible to check whether m is small compared with n^* because n^* is unknown in the first place.

In any case, let the experiment begin with a pilot sample size $m(\geq 4)$ and the observations $X_1, ..., X_m$. The data gathered from the *first stage* of sampling are then utilized to estimate the population variance σ^2 by the sample variance S_m^2. Let us denote

$$b_m = \frac{a(m-1)}{(m-3)} \text{ for } m \geq 4 \tag{13.3.4}$$

and define the stopping variable

$$N = \max\left\{m, \langle b_m S_m^2/\omega \rangle + 1\right\}, \tag{13.3.5}$$

which is a positive integer valued random variable. Observe that N depends on a, ω, m and the first-stage observations $X_1, ..., X_m$ through S_m^2. The expression $b_m S_m^2 / \omega$ used in the definition of N is viewed as an estimator of n^*, the optimal fixed-sample size had σ^2 been known. Since σ^2 is unknown, we have replaced it by the sample variance S_m^2 and a by b_m in the expression of n^*.

If $N = m$, then perhaps we have started with too many pilot observations to begin with and hence we do not see a need to take any other observations in the second stage. But, if $N > m$, then we sample the difference $(N - m)$ in the second stage which means that we take $(N - m)$ new observations $X_{m+1}, ..., X_N$ in the second stage. After combining the observations obtained from the two-stages, the data consists of $X_1, ..., X_N$ whether any observations are drawn in the second stage or not, that is the stopping variable N is the final sample size.

On the basis of N and $X_1, ..., X_N$, we obtain the sample mean, $\overline{X}_N = N^{-1} \Sigma_{i=1}^N X_i$ and propose to estimate the population mean μ by $\overline{X}_N$.

13.3.2 Some Interesting Properties

We first summarize some preliminary but yet interesting results.

Theorem 13.3.1 *For the stopping variable N defined by (13.3.5), for all fixed $\mu \in \Re, \sigma \in \Re^+, a \in \Re^+$ and $\omega \in \Re^+$, one has:*

(i) $P_{\mu,\sigma^2}\{N < \infty\} = 1$;

(ii) $b_m \sigma^2 \omega^{-1} \le E_{\mu,\sigma^2}[N] \le m + b_m \sigma^2 \omega^{-1}$;

(iii) $Q \equiv N^{1/2}(\overline{X}_N - \mu)/\sigma$ *is distributed as* $N(0, 1)$. *Also, the random variables Q and S_m^2 are independent;*

(iv) $N^{1/2}(\overline{X}_N - \mu)/S_m$ *is distributed as the Student's t random variable with $m - 1$ degrees of freedom;*

(v) $\overline{X}_N$ *is an unbiased estimator of μ.*

Its proof is similar to that of Theorem 13.2.1 and so the details are left out as Exercise 13.3.1.

Now, let us prove a fundamental result in this connection which shows that the bounded risk point estimation problem for the mean of a normal population, having an unknown variance, can be solved with the help of a properly designed two-stage sampling strategy.

Theorem 13.3.2 *For the stopping variable N defined by (13.3.5) and the proposed estimator $\overline{X}_N$ for the population mean μ, we have the associated risk:*

$$E_{\mu,\sigma^2}[L(\mu, \overline{X}_N)] \le \omega$$

for all fixed $\mu \in \Re, \sigma \in \Re^+, a \in \Re^+$ and $\omega \in \Re^+$.

Proof Let us first express the risk function $E_{\mu,\sigma^2}[L(\mu, \overline{X}_N)]$ associated with $\overline{X}_N$ as follows:

$$
\begin{aligned}
aE_{\mu,\sigma^2}&[(\overline{X}_N - \mu)^2] \\
&= aE\left\{E_{\mu,\sigma^2}[(\overline{X}_N - \mu)^2 \mid N]\right\}, \text{ Theorem 3.3.1, part (i)} \\
&= a\Sigma_{n=m}^{\infty} E_{\mu,\sigma^2}[(\overline{X}_N - \mu)^2 \mid N = n]P_{\mu,\sigma^2}(N = n) \qquad (13.3.6) \\
&= a\Sigma_{n=m}^{\infty} E_{\mu,\sigma^2}[(\overline{X}_n - \mu)^2]P_{\mu,\sigma^2}(N = n), \text{ since } \overline{X}_n \text{ and} \\
&\quad I(N = n) \text{ are independent for all } n \geq m .
\end{aligned}
$$

Now, since $E_{\mu,\sigma^2}[(\overline{X}_n - \mu)^2] = \sigma^2 n^{-1}$ for every fixed n and $P_{\mu,\sigma^2}(N < \infty) = 1$, from (13.3.6) we obtain

$$E_{\mu,\sigma^2}[L(\mu, \overline{X}_N)] = a\sigma^2 E_{\mu,\sigma^2}[1/N]. \qquad (13.3.7)$$

Next, from the definition of N in (13.3.5), we note that $N \geq b_m S_m^2/\omega$ w.p.1. Hence, we can write:

$$N^{-1} \leq \{\omega/b_m\}\{1/S_m^2\} = \{\omega(m-1)/b_m\}\sigma^{-2}\{1/Y\} \text{ w.p.1} \qquad (13.3.8)$$

where $Y = (m - 1)S_m^2/\sigma^2$ which is distributed as χ_{m-1}^2. Now, we combine the facts $E\{1/Y\} = \{2^{(m-3)/2}\Gamma(\frac{1}{2}(m - 3))\}\{2^{(m-1)/2}\Gamma(\frac{1}{2}(m - 1))\}^{-1}$ and $\Gamma(\frac{1}{2}(m - 1)) = (\frac{1}{2}(m - 3))\Gamma(\frac{1}{2}(m - 3))$ with (13.3.7)-(13.3.8) to obtain:

$$
\begin{aligned}
E_{\mu,\sigma^2}&[L(\mu, \overline{X}_N)] \\
&\leq a\sigma^2\{\omega(m - 1)/b_m\}\sigma^{-2}E\{1/Y\} \\
&= \{a\omega(m - 1)/b_m\}\{2^{(m-3)/2}\Gamma(\tfrac{m-3}{2})\}\{2^{(m-1)/2}\Gamma(\tfrac{m-1}{2})\}^{-1} \qquad (13.3.9) \\
&= \{a\omega(m - 1)\}/\{(m - 3)b_m\} \\
&= \omega, \text{ since we had chosen } b_m = a(m - 1)/(m - 3).
\end{aligned}
$$

Thus, the proof is complete. ∎

Again, the choice of m, the pilot sample size, plays a crucial role in the performance of the two-stage estimation procedure (13.3.5). Whether one selects a "small" or "large" value of m, the final sample size N will tend to overestimate n^* on the average. But, there is no denying of the fact that a Stein type two-stage procedure has *solved* a fundamental problem of statistical inference which could not otherwise be solved by any fixed-sample-size strategy.

If one can assume that the unknown parameter σ has a known lower bound $\sigma_L(> 0)$, then the choice of $m(\approx a\sigma_L^2/\omega)$ becomes quite apparent. In this situation, a Stein type two-stage bounded risk point estimation procedure, when appropriately modified, becomes very "efficient" as shown recently by Mukhopadhyay and Duggan (1999).

13.4   Exercises and Complements

13.2.1 Use the basic inequality from (13.2.6) to prove Theorem 13.2.1, part (ii).

13.2.2 Let $X_1, ..., X_n$ be iid $N(\mu, \sigma^2)$ random variables. Use Helmert's transformation from Chapter 4 (Example 4.4.9) to show that the sample mean $\overline{X}_n$ and the sample variance S_m^2 are distributed independently for all fixed $n \geq m (\geq 2)$. This result has been utilized in the proof of Theorem 13.2.1, part (iii).

13.2.3 For the two-stage procedure (13.2.4), show that $V_{\mu,\sigma^2}[\overline{X}_N] = \sigma^2 E_{\mu,\sigma^2}[1/N]$. Also evaluate $E_{\mu,\sigma^2}[(\overline{X}_N - \mu)^k]$ for $k = 3, 4$. {*Hint*: Use the fact that the two random variables $\overline{X}_n$ and $I(N = n)$ are independent for all $n \geq m$.}

13.2.4 Consider the two-stage fixed-width confidence interval $\mathcal{J}_N$ from (13.2.5) for the unknown population mean μ. Show that

$$P_{\mu,\sigma^2}[\mu \in \mathcal{J}_N] = 2E_{\mu,\sigma^2}[\Phi(N^{1/2}d/\sigma)] - 1.$$

{*Hint*: Use the fact that the two random variables $\overline{X}_n$ and $I(N = n)$ are independent for all $n \geq m$.}

13.2.5 Let $X_1, ..., X_n$ be iid random variables with the common pdf $f(x; \mu, \sigma) = \sigma^{-1} exp\{-(x - \mu)/\sigma\}I(x > \mu)$ where both the parameters $\mu(\in \Re)$ and $\sigma(\in \Re^+)$ are assumed unknown. Recall that the MLE for μ is $X_{n:1}$, the smallest order statistic. Having two preassigned numbers $d(> 0)$ and $\alpha \in (0, 1)$, suppose that we consider a *fixed-width* confidence interval $\mathcal{J}_n = [X_{n:1} - d, X_{n:1}]$ for μ. Note that for the upper limit in the interval $\mathcal{J}_n$, there may not be any point in taking $X_{n:1} + d$, because $X_{n:1} > \mu$ w.p.1. In other words, $X_{n:1}$ is a "natural" upper limit for μ. We also require that the confidence coefficient be at least $1 - \alpha$.

(i)   Show that $P_{\mu,\sigma}\{\mu \in \mathcal{J}_n\} = 1 - e^{-nd/\sigma}$;
(ii) Hence, show that $P_{\mu,\sigma}\{\mu \in \mathcal{J}_n\} \geq 1 - \alpha$ if and only if n is the smallest integer $\geq a\sigma/d = C$, say, where $a = -\log(\alpha)$. The optimal fixed sample size, had the scale parameter σ been known, is then C, assuming that C is an integer.

{*Hint*: Refer to the Example 4.4.12.}

13.2.6 (Exercise 13.2.5 Continued) Whatever be the sample size n, a fixed-width confidence interval $\mathcal{J}_n$ given in Exercise 13.2.5 can not be constructed such that $P_{\mu,\sigma}\{\mu \in \mathcal{J}_n\} \geq 1 - \alpha$ for all fixed d, α, μ and σ. Ghurye (1958) proposed the following Stein type two-stage sampling strategy. Let the experiment begin with the sample size $m(\geq 2)$ and initial observations $X_1, ..., X_m$. This is the *first stage* of sampling. These pilot observations are then utilized to estimate the scale parameter σ by $T_m = (m-1)^{-1}\Sigma_{i=1}^m(X_i - X_{m:1})$. Next, define the stopping variable,

$$N = \max\left\{m, \langle F_{2,2m-2,\alpha}T_m/d\rangle + 1\right\},$$

which is a positive integer valued random variable. Here, N estimates C. Recall that $F_{2,2m-2,\alpha}$ is the upper $100\alpha\%$ point of the $F_{2,2m-2}$ distribution. This procedure is implemented along the lines of Stein's two-stage scheme. Finally, having obtained N and $X_1, ..., X_N$, we construct the fixed-width confidence interval $\mathcal{J}_N = [X_{N:1} - d, X_{N:1}]$ for μ.

(i) Write down the corresponding basic inequality along the lines of (13.2.6). Hence, show that $d^{-1}F_{2,2m-2,\alpha}\sigma \le E_{\mu,\sigma}[N] \le m$ $+d^{-1}F_{2,2m-2,\alpha}\sigma$;

(ii) Show that $X_{n:1}$ and $I(N = n)$ are independent for all $n \ge m$;

(iii) Show that $Q \equiv N(X_{N:1} - \mu)/\sigma$ is distributed as the standard exponential random variable.

{*Hint*: From Chapter 4, recall that $n(X_{n:1} - \mu)/\sigma$ and $2(m-1)T_m/\sigma$ are respectively distributed as the standard exponential and χ^2_{2m-2} random variables, and these are also independent.}

13.2.7 (Exercise 13.2.6 Continued) Show that $P_{\mu,\sigma}\{\mu \in \mathcal{J}_N\} \ge 1-\alpha$ for all fixed d, α, μ and σ. This result was proved in Ghurye (1958). {*Hint*: First show that $N(X_{N:1} - \mu)/T_m$ is distributed as $F_{2,2m-2}$ and then proceed as in the proof of Theorem 13.2.2.}

13.2.8 (Exercise 13.2.6 Continued) Verify the following expressions.

(i) $E_{\mu,\sigma}[X_{N:1}] = \mu + \sigma E_{\mu,\sigma}[1/N]$;

(ii) $E_{\mu,\sigma}[(X_{N:1} - \mu)^2] = 2\sigma^2 E_{\mu,\sigma}[1/N^2]$;

(iii) $P_{\mu,\sigma}\{\mu \in \mathcal{J}_N\} = 1 - E_{\mu,\sigma}[e^{-Nd/\sigma}]$.

{*Hint*: Use the fact that the two random variables $X_{n:1}$ and $I(N = n)$ are independent for all $n \ge m$.}

13.2.9 Let the random variables $X_{i1}, X_{i2}, ...$ be iid $N(\mu_i, \sigma^2), i = 1, 2$, and that the X_{1j}'s be independent of the X_{2j}'s. We assume that all three parameters are unknown and $(\mu_1, \mu_2, \sigma) \in \Re \times \Re \times \Re^+$. With preassigned $d(> 0)$ and $\alpha \in (0,1)$, we wish to construct a $100(1-\alpha)\%$ fixed-width confidence interval for $\mu_1 - \mu_2(= \mu$, say). Having recorded $X_{i1}, ..., X_{in}$ with $n \ge 2$, let us denote $\overline{X}_{in} = n^{-1}\Sigma_{j=1}^n X_{ij}, S_{in}^2 = (n-1)^{-1}\Sigma_{j=1}^n(X_{ij} - \overline{X}_{in})^2$ for $i = 1, 2$ and the pooled sample variance $S_{Pn}^2 = \frac{1}{2}\left\{S_1^2 + S_2^2\right\}$. Consider the confidence interval $\mathcal{J}_n = [\overline{X}_{1n} - \overline{X}_{2n} - d, \overline{X}_{1n} - \overline{X}_{2n} + d]$ for $\mu_1 - \mu_2$. Let us also denote $\boldsymbol{\theta} = (\mu_1, \mu_2, \sigma)$.

(i) Obtain the expression for $P_{\boldsymbol{\theta}}\{\mu \in \mathcal{J}_n\}$ and evaluate the optimal fixed sample size C, had σ been known, such that with $n = \langle C\rangle$ $+1$, one has $P_{\boldsymbol{\theta}}\{\mu \in \mathcal{J}_n\} \ge 1 - \alpha$;

(*ii*)   Propose an appropriate Stein type two-stage stopping variable
N for which one will be able to conclude that $P_\theta\{\mu \in \mathcal{J}_N\}$
$\geq 1 - \alpha$ for all fixed θ, d, α. Here, N should be defined using
the pooled sample variance S_{Pm}^2 obtained from pilot samples.

{*Note*: Analogous problems for the negative exponential populations
were discussed by Mauromaustakos (1984) and Mukhopadhyay and Mauromoustakos (1987).}

13.2.10 (Behrens-Fisher problem) Let $X_{i1}, ..., X_{in_i}$ be iid $N(\mu_i, \sigma_i^2)$
random variables, $n_i \geq 2, i = 1, 2$, and that the X_{1j}'s be independent of the
X_{2j}'s. Here we assume that $\mu_1, \mu_2, \sigma_1, \sigma_2$ are all unknown and $(\mu_1, \mu_2) \in$
$\mathfrak{R}^2, (\sigma_1, \sigma_2) \in \mathfrak{R}^{+2}, \sigma_1 \neq \sigma_2$. Let $\overline{X}_{in_i}, S_{in_i}^2$ respectively be the sample mean
and variance, $i = 1, 2$. With preassigned numbers $\alpha \in (0, 1)$ and $d(> 0)$,
we wish to construct a confidence interval $\mathcal{J}$ for $\mu_1 - \mu_2$ such that (i) the
length of J is $2d$ as well as (ii) $P_\theta\{\mu_1 - \mu_2 \in \mathcal{J}\} \geq 1 - \alpha$ for all fixed
$\theta = (\mu_1, \mu_2, \sigma_1, \sigma_2)$ and d, α.

(*i*)   Let us denote a fixed-width confidence interval $\mathcal{J}_\mathbf{n} = [\{\overline{X}_{1n_1} -$
$\overline{X}_{2n_2}\} \pm d]$ for $\mu_1 - \mu_2$ where $\mathbf{n} = (n_1, n_2)$. Find the expression
for $P_\theta\{\mu_1 - \mu_2 \in \mathcal{J}_\mathbf{n}\}$ with fixed $\mathbf{n}$;

(*ii*)   Define h such that $P\{W_1 - W_2 \leq h\} = 1 - \frac{1}{2}\alpha$, where W_1, W_2
are iid Student's t_{m-1} variables with some fixed $m \geq 2$. Let m
be the pilot sample size from both populations. We define $N_i =$
$\max\left\{m, \langle h^2 S_{im}^2/d^2\rangle + 1\right\}$, and sample the difference $N_i - m$
in the second stage from the ith population, $i = 1, 2$. Now, with
$\mathbf{N} = (n_1, n_2)$, we propose the fixed-width confidence interval
$\mathcal{J}_\mathbf{N} = [\{\overline{X}_{1n_1} - \overline{X}_{2n_2}\} \pm d]$ for $\mu_1 - \mu_2$. Show that $P_\theta\{\mu_1 - \mu_2$
$\in \mathcal{J}_\mathbf{N}\} \geq 1 - \alpha$ for all fixed $\theta = (\mu_1, \mu_2, \sigma_1, \sigma_2)$ and d, α.

{*Hint*: It is a hard problem. This *exact and elegant solution* is due to
Chapman (1950). In the Exercise 11.3.15, we cited the original papers
of Behrens (1929), Fisher (1935,1939), Creasy (1954) and Fieller (1954).
Ghosh (1975) tabulated the h values for a range of values of m and α.
The analogous two-sample problem for the negative exponential distributions was developed in Mukhopadhyay and Hamdy (1984). Section 6.7.1
in Ghosh et al. (1997) gave a detailed account of the related sequential
literature.}

13.3.1 Prove Theorem 13.3.1 along the lines of the proof of Theorem
13.2.1.

13.3.2 Let us consider a sequence of iid observations $X_1, X_2, ...$ from a
$N(\mu, \sigma^2)$ population where the parameters $\mu \in \mathfrak{R}$ and $\sigma \in \mathfrak{R}^+$ are both
assumed unknown. Having recorded the observations $X_1, ..., X_n$, the cus-
tomary point estimator for the population mean μ is the sample mean

$\overline{X}_n(= n^{-1}\Sigma_{i=1}^n X_i)$. We denote $S_n^2 = (n-1)^{-1}\Sigma_{i=1}^n (X_i - \overline{X}_n)^2$ for $n \geq 2$ and $\boldsymbol{\theta} = (\mu, \sigma^2)$. Suppose that the loss incurred in estimating μ by $\overline{X}_n$ is given by the function $L(\mu, \overline{X}_n) = a|\overline{X}_n - \mu|^t$ where a and t are fixed and known positive numbers.

(i) Derive the associated fixed-sample-size risk $R_n(\mu, \overline{X}_n)$ which is given by $E_{\boldsymbol{\theta}}[L(\mu, \overline{X}_n)]$;

(ii) The experimenter wants to bound the risk $R_n(\mu, \overline{X}_n)$ from above by ω, a preassigned positive number. Find an expression of n^*, the optimal fixed-sample-size, if σ^2 were known.

13.3.3 (Exercise 13.3.2 Continued) For the *bounded risk* point estimation problem formulated in Exercise 13.3.2, propose an appropriate two-stage stopping variable N along the lines of (13.3.5) and estimate μ by the sample mean $\overline{X}_N$. Find a proper choice of "b_m" required in the definition of N by proceeding along (13.3.9) in order to claim that $E_{\boldsymbol{\theta}}[L(\mu, \overline{X}_N)] \leq \omega$ for all $\boldsymbol{\theta}, a$ and t. Is it true that $b_m > 1$ for all positive numbers t? Show that $\overline{X}_N$ is unbiased for μ.

13.3.4 Let the random variables $X_{i1}, X_{i2}, ...$ be iid $N(\mu_i, \sigma^2), i = 1, 2$, and that the X_{1j}'s be independent of the X_{2j}'s. We assume that all three parameters are unknown and $(\mu_1, \mu_2, \sigma^2) \in \Re \times \Re \times \Re^+$. Having recorded $X_{i1}, ..., X_{in}$ with $n \geq 2$, let us denote $\overline{X}_{in} = n^{-1}\Sigma_{j=1}^n X_{ij}, S_{in}^2 = (n-1)^{-1}\Sigma_{j=1}^n (X_{ij} - \overline{X}_{in})^2$ for $i = 1, 2$ and the pooled sample variance $S_{Pn}^2 = \frac{1}{2}\{S_1^2 + S_2^2\}$. The customary estimator of $\mu_1 - \mu_2$ is taken to be $\overline{X}_{1n} - \overline{X}_{2n}$. Let us also denote $\boldsymbol{\theta} = (\mu_1, \mu_2, \sigma^2)$. Suppose that the loss incurred in estimating $\mu_1 - \mu_2$ by $\overline{X}_{1n} - \overline{X}_{2n}$ is given by the function $L(\mu_1 - \mu_2, \overline{X}_{1n} - \overline{X}_{2n}) = a|\{\overline{X}_{1n} - \overline{X}_{2n}\} - \{\mu_1 - \mu_2\}|^t$ where a and t are fixed and known positive numbers.

(i) Derive the fixed-sample-size risk $R_n(\mu_1 - \mu_2, \overline{X}_{1n} - \overline{X}_{2n})$ which is given by $E_{\boldsymbol{\theta}}[L(\mu_1 - \mu_2, \overline{X}_{1n} - \overline{X}_{2n})]$;

(ii) The experimenter wants to bound $R_n(\mu_1 - \mu_2, \overline{X}_{1n} - \overline{X}_{2n})$ from above by ω, a preassigned positive number. Find the the expression of n^*, the optimal fixed-sample-size, had σ^2 been known.

13.3.5 (Exercise 13.3.4 Continued) For the *bounded risk* point estimation problem formulated in Exercise 13.3.4, propose an appropriate two-stage stopping variable N along the lines of (13.3.5) depending on S_{Pm}^2 and estimate $\mu_1 - \mu_2$ by $\overline{X}_{1N} - \overline{X}_{2N}$. Find a proper choice of "b_m" required in the definition of N by proceeding along (13.3.9) in order to claim that $E_{\boldsymbol{\theta}}[L(\mu_1 - \mu_2, \overline{X}_{1N} - \overline{X}_{2N})] \leq \omega$ for all $\boldsymbol{\theta}, a$ and t. Is it true that $b_m > 1$ for all positive numbers t? Show that $\overline{X}_{1N} - \overline{X}_{2N}$ is unbiased for $\mu_1 - \mu_2$.

13.3.6 Let $X_1, X_2, \ldots$ be iid random variables with the common pdf $f(x; \mu, \sigma) = \sigma^{-1} exp\{-(x - \mu)/\sigma\} I(x > \mu)$ where both the parameters $\mu(\in \Re)$ and $\sigma(\in \Re^+)$ are assumed unknown. Having recorded $X_1, \ldots, X_n$ for $n \geq 2$, we estimate μ and σ by $X_{n:1}$ and $T_n = (n-1)^{-1} \Sigma_{i=1}^n (X_i - X_{n:1})$. Let us denote $\theta = (\mu, \sigma)$. Suppose that the loss incurred in estimating μ by $X_{n:1}$ is given by the function $L(\mu, X_{n:1}) = a(X_{n:1} - \mu)^t$ where a and t are fixed and known positive numbers.

    (i)   Derive the associated fixed-sample-size risk $R_n(\mu, X_{n:1})$ which is given by $E_\theta[L(\mu, X_{n:1})]$;

    (ii)   The experimenter wants to have the risk $R_n(\mu, X_{n:1})$ bounded from above by ω, a preassigned positive number. Find the expression of n^*, the optimal fixed-sample-size, had the scale parameter σ been known.

13.3.7 (Exercise 13.3.6 Continued) For the *bounded risk* point estimation problem formulated in Exercise 13.3.6, propose an appropriate two-stage stopping variable N along the lines of (13.3.5) and estimate μ by the smallest order statistic $X_{N:1}$. Find the proper choice of "b_m" required in the definition of N by proceeding along (13.3.9) in order to claim that $E_\theta[L(\mu, X_{N:1})] \leq \omega$ for all θ, a and t. Is it true that $b_m > 1$ for all positive numbers t? Obtain the expressions for $E_\theta[X_{N:1}]$ and $V_\theta[X_{N:1}]$.

13.3.8 Let the random variables $X_{i1}, X_{i2}, \ldots$ be iid having the common pdf $f(x; \mu_i, \sigma), i = 1, 2$, where $f(x; \mu, \sigma) = \sigma^{-1} exp\{-(x - \mu)/\sigma\} I(x > \mu)$. Also let the X_{1j}'s be independent of the X_{2j}'s. We assume that all three parameters are unknown and $(\mu_1, \mu_2, \sigma) \in \Re \times \Re \times \Re^+$. Having recorded $X_{i1}, \ldots, X_{in}$ with $n \geq 2$, let us denote $X_{n:1}^{(i)} = \min_{1 \leq j \leq n} X_{ij}$, $U_{in} = (n-1)^{-1} \Sigma_{j=1}^n (X_{ij} - X_{n:1}^{(i)}), i = 1, 2$, and the pooled estimator $U_{Pn} = \frac{1}{2}\{U_{1n} + U_{2n}\}$ for the scale parameter σ. The customary estimator of $\mu_1 - \mu_2$ is $X_{n:1}^{(1)} - X_{n:1}^{(2)}$. Let us denote $\theta = (\mu_1, \mu_2, \sigma)$. Suppose that the loss incurred in estimating $\mu_1 - \mu_2$ by $X_{n:1}^{(1)} - X_{n:1}^{(2)}$ is given by the function

$$L\left(\mu_1 - \mu_2, X_{n:1}^{(1)} - X_{n:1}^{(2)}\right) = a \left|\left\{X_{n:1}^{(1)} - X_{n:1}^{(2)}\right\} - \{\mu_1 - \mu_2\}\right|^t$$ where a and t are fixed and known positive numbers.

    (i)   Derive the associated fixed-sample-size risk $R_n\left(\mu_1 - \mu_2, X_{n:1}^{(1)}\right.$ $\left. -X_{n:1}^{(2)}\right)$ which is given by $E_\theta\left[L\left(\mu_1 - \mu_2, X_{n:1}^{(1)} - X_{n:1}^{(2)}\right)\right]$;

    (ii)   The experimenter wants to have the risk $R_n\left(\mu_1 - \mu_2, X_{n:1}^{(1)}\right.$ $\left. -X_{n:1}^{(2)}\right)$ bounded from above by ω, a preassigned positive number. Find the expression of n^*, the optimal fixed-sample -size, had σ been known.

13.3.9 (Exercise 13.3.8 Continued) For the *bounded risk* point estimation problem formulated in Exercise 13.3.8, propose an appropriate two-stage stopping variable N along the lines of (13.3.5) depending upon U_{Pm} and estimate $\mu_1 - \mu_2$ by $X_{N:1}^{(1)} - X_{N:1}^{(2)}$. Find the proper choice of "b_m" required in the definition of N by proceeding along (13.3.9) in order to claim that $E_\theta \left[L \left(\mu_1 - \mu_2, X_{N:1}^{(1)} - X_{N:1}^{(2)} \right) \right] \leq \omega$ for all θ, a and t. Is it true that $b_m > 1$ for all positive numbers t? Obtain the expressions for $E_\theta \left[X_{N:1}^{(1)} - X_{N:1}^{(2)} \right]$ and $V_\theta \left[X_{N:1}^{(1)} - X_{N:1}^{(2)} \right]$.

14

Appendix

The first section includes a summary of frequently used abbreviations and some notation. The Section 14.2 provides brief biographical notes on some of the luminaries mentioned in the book. In the end, a few standard statistical tables are included. These tables were prepared with the help of MAPLE.

14.1 Abbreviations and Notation

Some of the abbreviations and notation are summarized here.

$\mathcal{X}$	space of the values x for the random variable X
$\Re$	real line; $(-\infty, \infty)$
$\Re^+$	positive half of $\Re$; $(0, \infty)$
$\in$; $x \in A$	belongs to; x belongs to the set A
$A(\subset) \subseteq B$	A is a (proper) subset of B
▲	denotes end of an example
■	denotes end of proof in theorems
◆	denotes end of proof of lengthy subcases
φ	empty set
lhs	left hand side
rhs	right hand side
$log(a), a > 0$	natural (base e) logarithm of a
pmf; PMF	probability mass function
pdf; PDF	probability density function
df; DF	distribution function
cdf; CDF	cumulative distribution function
mgf; MGF	moment generating function
iid	independent and identically distributed
$\Pi_{i=1}^n a_i$	product of $a_1, a_2, ..., a_n$
w.p. 1	with probability one
$\mathbf{x}, \mathbf{X}, \boldsymbol{\theta}$	vectors of dimension more than one
$a \equiv b$	a, b are equivalent
$\Gamma(\alpha), \alpha > 0$	gamma function; $\int_0^\infty e^{-x} x^{\alpha-1} dx$

591

$\text{beta}(\alpha, \beta), \alpha > 0, \beta > 0$	beta function; $\Gamma(\alpha)\Gamma(\beta)/\Gamma(\alpha + \beta)$
$[f(x)]_{x=a}^{x=b}$	stands for $f(b) - f(a)$
$\frac{dg(t)}{dt}$ or $dg(t)/dt$	first derivative of $g(t)$ with respect to t
$\frac{d^2g(t)}{dt^2}$ or $d^2g(t)/dt^2$	second derivative of $g(t)$ with respect to t
$\frac{\partial g(s,t)}{\partial t}$ or $\partial g(s,t)/\partial t$	first partial derivative of $g(s,t)$ with respect to t
$\frac{\partial^2 g(s,t)}{\partial t^2}$ or $\partial^2 g(s,t)/\partial t^2$	second partial derivative of $g(s,t)$ with respect to t
$\frac{\partial^2 g(s,t)}{\partial s \partial t}$ or $\partial^2 g(s,t)/\partial s \partial t$	$\frac{\partial}{\partial s}\{\partial g(s,t)/\partial t\}$
$I(A)$ or I_A	indicator function of the set A
$\mathcal{I}(\theta)$	Fisher information about θ
$X_n \xrightarrow{P} a$	X_n converges to a in probability as $n \to \infty$
$X_n \xrightarrow{\mathcal{L}} X$	X_n converges to X in law (or distribution) as $n \to \infty$
$a(x) \backsim b(x)$	$a(x)/b(x) \to 1$ as $x \to \infty$
χ_ν^2	Chi-square distribution with ν degrees of freedom
$\phi(.), \Phi(.)$	standard normal pdf and cdf respectively
$\overline{X}$ or $\overline{X}_n$	sample mean from $X_1, ..., X_n$
S^2 or S_n^2	sample variance from $X_1, ..., X_n$; divisor $n - 1$
$X_{n:i}$	i^{th} order statistic in $X_{n:1} \le X_{n:2} \le ... \le X_{n:n}$
$det(A)$	determinant of a square matrix A
A'	transposed matrix obtained from $A_{m \times n}$
$I_{k \times k}$	identity matrix of order $k \times k$
WLLN	weak law of large numbers
SLLN	strong law of large numbers
CLT	central limit theorem
$\langle u \rangle$	largest integer $< u$
$a \approx b$	a and b are approximately same
p.d.	positive definite
n.d.	negative definite
p.s.d.	positive semi definite
$B(\theta)$	bias of an estimator
MSE	mean squared error
MLE	maximum likelihood estimator
CRLB	Cramér-Rao lower bound
UMVUE	uniformly minimum variance unbiased estimator
(U)MP test	(uniformly) most powerful test

(U)MPU test	(uniformly) most powerful unbiased test
LR test	likelihood ratio test
$a \propto b$	a and b are proportional
z_α	upper $100\alpha\%$ point of $N(0,1)$
$\chi^2_{\nu,\alpha}$	upper $100\alpha\%$ point of χ^2_ν
$t_{\nu,\alpha}$	upper $100\alpha\%$ point of t_ν
$F_{\upsilon_1,\nu_2,\alpha}$	upper $100\alpha\%$ point of F_{υ_1,ν_2}

14.2   A Celebration of Statistics: Selected Biographical Notes

In this section, we add brief biographical notes for selected luminaries in statistics mentioned in this textbook. Each biography is prepared with the help of available published materials. In each case, we have cited the original sources.

The aim is to let the readers appreciate that the field of statistics has developed, and it continues to flourish, because it attracted some of the best minds of the twentieth century. What we present here is sketchy. The space is limited. *The list is certainly not exhaustive.*

The journal *Statistical Science*, published by the Institute of Mathematical Statistics, routinely prints conversation articles of eminent statisticians. These articles are filled with historical materials. A broader sense of history will emerge from the *additional* conversation articles in *Statistical Science:* **H. Akaike** [Findley and Parzen (1995)], **T. W. Anderson** [DeGroot (1986b)], **G. A. Barnard** [DeGroot (1988)], **M. Bartlett** [Olkin (1989)], **H. Bergström** [Råde (1997)], **H. Chernoff** [Bather(1996)], **D. R. Cox** [Reid (1994)], **H. Daniels** [Whittle (1993)], **F. N. David** [Laird (1989)], **J. Doob** [Snell (1997)], **D. J. Finney** [MacNeill (1993)], **J. Gani** [Heyde(1995)], **B. V. Gnedenko** [Singpurwalla and Smith (1992)], **I. J. Good** [Banks (1996)], **S. S. Gupta** [McDonald (1998)], **M. Hansen** [Olkin (1987)], **D. Kendall** [Bingham (1996)], **L. Kish** [Frankel and King (1996)], **C. C. Li** [Chen and Tai (1998)], **D. V. Lindley** [Smith (1995)], **S. K. Mitra** [Mukhopadhyay (1997)], **Y. V. Prokhorov** [Shepp (1992)], **F. Proschan** [Hollander and Marshall (1995)], **I. R. Savage** [Sampson and Spencer (1999)], **E. Seiden** [Samuel-Cahn (1992)], **H. Solomon** [Switzer (1992)], **G. Watson** [Beran and Fisher (1998)]. Bellhouse and Genest

(1999) as well as the discussants included an account of the rich history of statistics in Canada.

The interview articles of **C. W. Dunnett** [Liaison (1993)] and **I. Olkin** [Press(1989)] are filled with historical remarks about colleagues, lives and careers. These articles were included in the edited *Festschrift Volumes* respectively An interview of **H. Robbins** [Page (1989)] appeared in *"Herbert Robbins: Selected Papers"* (T. L. Lai and D. Siegmund eds.) which is a delight to read.

Morris H. DeGroot was born on **June 8, 1931** in Scranton, Pennsylvania. The dedication and memorial article in *Statist. Sci.* (1991, **6**, 4-14), *Biography of Morris H. DeGroot*, portrays the life and work of DeGroot, and it is rich in history. DeGroot had a key role in creating the new journal, *Statistical Science*, and he served as its Executive Editor during its formative years. DeGroot was also responsible for several early interview articles. DeGroot died on **November 2, 1989**.

The Archives of Films of Distinguished Statisticians in the American Statistical Association has taped lectures and interviews of some of the eminent statisticians, including: T. W. Anderson, D. Blackwell, R. C. Bose, G. Box, H. Chernoff, W. G. Cochran, D. R. Cox, H. Cramér, E. W. Deming, F. Graybill, I. J. Good, M. Hansen, R. V. Hogg, S. Hunter, O. Kempthorne, E. L. Lehmann, J. Neyman, F. Mosteller, G. Noether, I. Olkin, E. J. G. Pitman, C. R. Rao, H. Robbins, E. L. Scott, J. W. Tukey, and M. Zelen. These are wonderful films, filled with fascinating stories. A local Chapter of the American Statistical Association (ASA) or other scholarly societies may consider renting some of the films from the ASA's Archives for their meetings.

The edited volume of Pearson and Kendall (1970) includes very valuable selected articles and hence it will undoubtedly serve as an excellent resource. One may also consult the monographs and articles of Aldrich (1997), Anderson (1996), Craig (1986), David (1998), Edwards (1997a,b), Fienberg (1992,1997), Ghosh et al. (1999), Halmos (1985), Hogg (1986), LeCam (1986b), Shafer (1986), Stigler (1980,1986,1989,1991,1996) and Zabell (1989) for a better appreciation of the international statistical heritage.

With due respect to *all* contributors, both past and present, in our young field of statistics, we present the *selected* biographies. We earnestly hope that the lives and careers of these fascinating individuals, *and* many others who unfortunately could not be mentioned here in detail, will continue to inspire, nurture and touch the souls of the future generations of statisticians.

LET THE CELEBRATION BEGIN

R. R. Bahadur: Raghu Raj Bahadur was born in Delhi, India on **April**

30, 1924. He completed his school and college education in Delhi. He received Ph.D. in statistics from the University of North Carolina at Chapel Hill. He was in the first batch of Ph.D. advisees of Herbert Robbins.

During 1956-1961, Bahadur was a professor in the Research and Training School of the Indian Statistical Institute, Calcutta. In 1961, he left the Institute to join University of Chicago, a position he held ever since.

Bahadur's contributions on large deviation theory, sufficiency, MLE, comparisons of tests, sample quantiles, sequential decisions, transitive sufficiency, are truly noteworthy. He was a master in his unique approach to unveil the inner beauty in some of the hardest problems. Anything he published is considered a jewel by many statisticians. The phrases *Bahadur Efficiency*, *Bahadur Slope* and *Bahadur Representation of Quantiles* have become household words in statistics. At the 1974 inauguration ceremony of the Delhi campus of the Indian Statistical Institute, Jerzy Neyman referred to Bahadur as the brightest of the two hundred and fifty stars of Indian origin shining in U.S.A. in the field of statistics.

In September, 1969, Bahadur gave the NSF-CBMS lectures in the Department of Statistics of Florida State University, which led to his monograph, *Some Limit Theorems in Statistics* (1971, SIAM). This monograph is a classic in large sample theory. An international symposium was arranged in Delhi in December 1988 to honor the memory of R. C. Bose. Bahadur edited the symposium volume, *Probability, Statistics and Design of Experiments* (1990, Wiley Eastern).

Bahadur received many honors in U.S.A. and India. He received Guggenheim Fellowship in 1968. He was an elected Fellow of the Indian National Science Academy and the Institute of Mathematical Statistics. Bahadur became President (1974) of the Institute of Mathematical Statistics and gave the Wald Lecture. Bahadur was very friendly, unassuming and always approachable. He cared about the people around him, students and colleagues alike.

Bahadur died on **June 7, 1997** after long illness. In the obituary of Bahadur, Stigler (1997) wrote: "He was extremely modest in demeanor and uncomfortable when being honored, ..." Stigler's article and Bahadur's own little write-up, *Remarks on Transitive Sufficiency*, which was included in Ghosh et al. (1992) provide more details on this one of a kind statistician's life and work.

D. Basu: Debabrata Basu was born on **July 5, 1924** in Dhaka, India (now Bangladesh). His father was the Head of the Department of Mathematics at Dhaka University. He went through the undergraduate and Masters' degree programs at this university. During this period, he was charmed by the lectures of the famous number theorist, T. Vijayraghavan. Basu moved to Calcutta in 1948.

In 1950, Basu enrolled as the first research scholar under the guidance of C. R. Rao in Indian Statistical Institute (ISI) and completed his Ph.D. thesis within two years. In 1953, he went to University of California, Berkeley, on a Fulbright fellowship where he came in close contact with Jerzy Neyman. In 1955, when R. A. Fisher visited ISI, Basu got the first-hand taste of Fisherian inference. With his uncanny ability to produce counter-examples, quite often spontaneously, Basu started raising fundamental philosophical issues. He felt uncomfortable with Fisherian doctrine as well as the Neyman-Pearsonian approach. Eventually, he became an ardent Bayesian. He vigorously gave voice to the staunchest criticisms of many frequentist ideas.

Basu always gave inspiring lectures. He travelled extensively all over the world. During 1976-1986, he remained as a professor in the department of statistics in Florida State University, Tallahassee. Currently, he is a Professor Emeritus in ISI and Florida State University.

Basu is an elected Fellow of both the Institute of Mathematical Statistics and the American Statistical Association. He is highly regarded for his penetrating essays on the foundations of statistical inference. Ghosh (1988) edited a special volume containing most of Basu's critical essays on *Statistical Information and Likelihood*. He also made fundamental contributions on inferences for finite population sampling. A Festschrift volume, *Current Issues in Statistical Inference: Essays in Honor of D. Basu*, was presented to him on his 65th birthday. This special volume was edited by Ghosh and Pathak (1992).

Basu's article about himself, *Learning from Counterexamples: At the Indian Statistical Institute in the Early Fifties*, was included in Ghosh et al. (1992). It mentions many interesting stories and encounters with Fisher, Neyman, Wald and Rao. Basu has been a dedicated bridge player and he loves gardening. Unfortunately, his health has been failing for sometime.

D. Blackwell: David Blackwell was born on **April 24, 1919**, in Centralia, Illinois. He got into the University of Illinois and received the A.B. (1938), A.M. (1939), and Ph.D. (1941) degrees, all in mathematics.

Blackwell entered college to become an elementary school teacher. He came in contact with J. L. Doob at the University of Illinois. Blackwell wrote his Ph.D. thesis under the guidance of Doob, but the thesis topic was neither in the area of statistics nor probability.

After receiving the Ph.D. degree in mathematics, Blackwell was not sure whether he would be interested in statistics. According to his own admission [DeGroot (1986a)], during the period 1941-1944, it was very hard for him to land a suitable job. He accepted a faculty position in mathematics at

Howard University, Washington D.C., and in 1944 he met A. Girshick when Blackwell went to attend a Washington chapter meeting of the American Statistical Association where Girshick gave a lecture on sequential analysis. Blackwell was impressed.

In the meantime, Blackwell came across the works of A. Wald and his collaborators at Columbia University. In the early 1950's, he started thinking about a Bayesian formulation of sequential experiments. Lengthy collaborations between Blackwell and Girshick followed which led to the Blackwell-Girshick classic, *Theory of Games and Statistical Decisions* (1954, Wiley). Blackwell also wrote an elementary textbook, *Basic Statistics* (1969, McGraw-Hill).

Blackwell stayed with the faculty in Howard University for ten years and in 1954, he joined the department of statistics at the University of California, Berkeley, after some persuasion from J. Neyman. He has been in Berkeley ever since.

Blackwell has made fundamental contributions in several areas including sufficiency and information, game theory and optimal statistical decisions. He became President (1955) of the Institute of Mathematical Statistics, President of the Bernoulli Society, and Vice President of the American Statistical Association, the American Mathematical Society, and the International Statistical Institute. He has delivered the Fisher Lecture and has received prestigious awards and honorary degrees. The interview article [DeGroot (1986a)] with Blackwell gives a much broader perspective of his life and career. Blackwell is a Professor Emeritus at Berkeley.

H. Cramér: Harald Cramér was born on **September 25, 1893**, in Stockholm, Sweden. He studied mathematics and chemistry in Stockholm University where he received the Ph.D. (1917) degree in pure mathematics. His research career spanned over seven decades while his first publication appeared in 1913 which dealt with a problem on biochemistry of yeast. In 1920, he published on the theory of prime numbers.

During 1917-1929, Cramér was in the faculty of mathematics in Stockholm University, where he became a professor of mathematical statistics and actuarial science in 1929 and continued in that position until 1958. He became the President of Stockholm University in 1950 and the Chancellor of the entire Swedish University System in 1958. He retired from the Swedish University System in 1961, but his professional career marched on for the next two decades. He travelled extensively, including multiple visits to the University of North Carolina at Chapel Hill, Princeton, Copenhagen, Paris, Helsinki, and the Indian Statistical Institute in Calcutta.

Among Cramér's many path-breaking contributions, lies his seminal book, *Mathematical Methods of Statistics* (1945 from Stockholm; 1946 from Princeton University Press). During the next several decades, this book stood as the gold standard of what mathematical statistics ought to be. The *Mathematical Methods of Statistics* has been translated in other languages including Russian, Polish, Spanish, and Japanese. Cramér also wrote other important books and monographs including the one on stochastic processes, co-authored with R. Leadbetter (1967, Wiley).

Cramér's contributions in Markov and other stochastic processes, probability theory, and large sample theory, in particular, have been legendary in how they influenced the research methods for a very long time. Cramér's (1942) paper on harmonic analysis has been included in the *Breakthroughs in Statistics, Volume I* [Johnson and Kotz (1992)]. The *Cramér-Rao inequality* and the *Cramér-Rao Lower Bound* are two household phrases.

Cramér had close professional ties with H. Hotelling, R. A. Fisher, A. N. Kolmogorov, P. Lévy, W. Feller, J. Neyman, P. C. Mahalanobis, D. G. Kendall, J. L. Doob, and C. R. Rao, among others. He was a superb lecturer and a great story-teller. He could grab any audience's attention as soon as he walked into a lecture hall. He was always charming, friendly and unassuming.

Cramér received many honors including the Guy Medal in Gold from the Royal Statistical Society in 1972. He received several honorary degrees including a D.Sc. degree from the Indian Statistical Institute.

Cramér's (1976) lengthy article beautifully portrayed his recollections on the development of probability theory. (D. G.) Kendall (1983) wrote a charming tribute to Cramér on the celebration of his 90th birthday. Wegman (1986) detailed Cramér's many personal recollections. In the interview article of H. Bergström [Råde (1997)], some of Cramér's major contributions are mentioned.

Cramér died on **October 5, 1985**. The legacy of his long, vigorous and distinguished career continues to nurture the growth of mathematical statistics.

B. de Finetti: Bruno de Finetti was born on **June 13, 1906** in Innsbruck, Austria. He first joined the faculty of mathematics at the University of Milan, but eventually settled down in Rome. He was one of the strongest proponents of the *subjectivistic interpretation of probability* which is the life-line of the Bayesian doctrine. His books, de Finetti (1972,1974), on subjective probability are master pieces. de Finetti introduced the concept of *exchangeability* which profoundly influenced the modern Bayesian school. de Finetti's (1937) article has been included in the *Breakthroughs in Statistics Volume I* [Johnson and Kotz (1992)]. His autobiographical piece, *Probability and My Life,* included in Gani (1982), is filled with fascinating stories. de Finetti passed away in Rome on **July 20, 1985**.

W. Feller: William Feller was born on **July 7, 1906**, in Zagreb, Yugoslavia (now Croatia). After finishing Masters degree there in 1925, he worked in the University of Göttingen during 1925-1928 where he received Ph.D. degree in 1926 at a tender age of twenty. Feller knew David Hilbert and Richard Courant from Göttingen. He spent 1934-1939 at the University of Stockholm where he came to know Harald Cramér.

In 1939, Feller joined Brown University at Providence as an associate professor. He became the first Executive Editor of *Mathematical Reviews* and continuously gave extraordinary service to the profession for six long years in that position. In 1944, Feller became a U.S. citizen. Then, he moved to Cornell University and finally in 1950, he joined Princeton University as the prestigious Eugene Huggins Professor.

Feller's research covered many areas including calculus, geometry, functional analysis, and probability theory. He was keenly interested in mathematical genetics. Some of his best known papers in probability theory made fundamental contributions in and around the *classical limit theorems*. These included the *Central Limit Theorem,* the *Law of the Iterated Logarithm*, and the *Renewal Theory*. Feller's versatility as a first-rate mathematician was recognized and highly respected by eminent mathematicians including F. Hausdorff, A. N. Kolmogorov, G. H. Hardy, J. E. Littlewood, A. Y. Khinchine, P. Lévy and P. Erdös. During 1950-1962, Feller went on to break new grounds on the theory of diffusion which was earlier developed by Kolmogorov.

Feller's seminal works, *An Introduction to Probability Theory and Its Applications, Volume 1* (1950; third edition 1968) and *Volume 2* (1956; second edition in 1971), both published by Wiley, are still regarded as classics in probability theory.

Feller received many honors. He was a member of the U.S. National Academy of Sciences, a past Governor of Mathematical Association of America, and President (1946) of the Institute of Mathematical Statistics.

Feller was to receive the U.S. National Medal of Science, the highest honor an American scientist can receive, in a White House ceremony on February 16, 1970. But the world of mathematics was stunned in grief to learn that William Feller passed away on **January 14, 1970**, in New York.

By action of the Council of the Institute of Mathematical Statistics, the 1970 volume of the *Annals of Mathematical Statistics* was dedicated to the memory of William Feller. Its opening article "William Feller 1906-1970 " prepared by the Editors of the *Annals of Mathematical Statistics*, described the life and career of this one of a kind giant among mathematicians.

R. A. Fisher: Ronald Aylmer Fisher was born on **February 17, 1890** in North London. When he was six years old, Fisher studied astronomy and later he took to mathematics. During early upbringing in a private school, he developed extraordinary skills in geometrical ideas. His uncanny depth in geometry first became vivid in his derivation [Fisher (1915)] of the distribution of the sample correlation coefficient.

Fisher went to Cambridge in 1909 to study mathematics and physics where he came across K. Pearson's (1903) paper on the theory of evolution. He was very influenced by this paper and throughout his life, Fisher remained deeply committed to both genetics and evolution. Fisher was regarded as a first-rate geneticist of his time. He received B.A. from Cambridge in 1912, having passed the Mathematical Tripos Part II as a Wrangler. On the outbreak of war, Fisher volunteered for the military services in August, 1914, but he was disappointed by his rejection on account of poor eye sight.

Fisher received both M.A. (1920) and Sc.D. (1926) degrees from Cambridge. During 1919-1933, he was a statistician in the Rothamsted Experimental Station, Harpenden. During 1933-1943, he was the Galton Professor of Eugenics in the University College, London. In the University College, he became the Editor of Annals of Eugenics and during 1943-1957, he was the Arthur Balfour Professor of Genetics in Cambridge. During 1959-1962, Fisher lived in South Australia as a Research Fellow at the CSIRO Division of Mathematical Statistics, University of Adelaide.

The fundamental concepts of likelihood, sufficiency, ancillarity, conditionality, maximum likelihood estimation, consistency, and efficiency were fully developed by Fisher from ground zero. He had built a logical theory of scientific inference based on the information gathered from data. In the areas of correlations, partial correlations, directional data, multivariate analysis, discriminant analysis, factor analysis, principal components, design of experiments, modeling, anthropology, genetics, for example, Fisher gave the foundations. Anderson (1996) reviewed Fisher's fundamental contributions in the area of multivariate analysis. Efron (1998) explained Fisher's influence and legacy in the twenty first century.

Fisher (1930) developed the notion of *fiducial probability* distributions for unknown parameters which were supposedly generated by inverting the distribution of appropriate pivotal random variables. He was very fond of fiducial probability and applied this concept vigorously whenever he had an opportunity to do so, for example in Fisher (1935,1939). The articles of Buehler (1980), Lane (1980) and Wallace (1980) provide important perspectives of fiducial inference.

Fisher was an ardent frequentist. On many occasions, he fought tooth and nail to defend *against* any hint or allegation from the Bayesians that somewhere he tacitly used the Bayes's Theorem without mentioning it.

The Bayesians argued that Fisher's theory of fiducial probability was not valid without fully subscribing to the Bayesian principles. But, Fisher vehemently defended his position that he never relied upon Bayesian arguments to develop the concept of fiducial probability. He never caved in.

Fisher had close ties with W. S. Gosset who is perhaps better known under the pseudonym "Student" than under his own name. Gosset graduated (1899) with a first class degree in chemistry from New College in Oxford, then joined the famous Guinness Brewery in Dublin as a brewer, and stayed with this brewing firm for all his life, ultimately becoming the Head Brewer in a new installation operated by the Guinness family at Park Royal, London in 1935. Gosset needed and developed statistical methods for small sample sizes which he would then apply immediately to ascertain relationships between the key ingredients in beer. Gosset's path-breaking 1908 paper gave the foundation of the t-distribution where he derived the probable error of a correlation coefficient in 1908 and made several forceful conjectures, most of these being proven true later by Fisher (1915). For both Gosset and Fisher, making scientific inferences and using small-sample statistics for experimental data went hand-in-hand on a daily basis. The theory of statistics was *essential* for applying statistics. There was no line drawn between the two. Fisher created the foundation of this discipline. Rao (1992a) called Fisher "The founder of modern statistics." Fisher regarded Gosset very highly and he once described Gosset as "Faraday of statistics." Box (1987) pointed out various connections and collaborations among Guinness, Gosset and Fisher.

Fisher's first book, *Statistical Methods for Research Workers*, was out in 1925. *The Genetic Theory of Natural Selection*, first appeared in 1930. *The Design of Experiments*, appeared in 1935. *The Statistical Tables for Biological, Agricultural and Medical Research* of R. A. Fisher and F. Yates was out in 1938. Fisher's three other books, *The Theory of Inbreeding, Contributions to Mathematical Statistics*, followed by *Statistical Methods and Scientific Inference*, respectively appeared in 1949, 1950 and 1956. These books have served as landmarks in the history of the statistical science.

Fisher received many honors. The list includes, Fellow of the Royal Society (1929), Honorary Fellow of the Indian Statistical Institute (1937), Royal Medal (1938), Darwin Medal (1948), and Copley Medal (1955) of the Royal Society, Guy Medal in Gold of the Royal Statistical Society (1946). He became a Foreign Associate of the U.S. National Academy of Sciences (1948). He was Knighted in 1952.

An extra-ordinary body of now well-known developments in modern statistical theory and methods originated in the work of Fisher. One encounters Fisherian thoughts and approaches practically in all aspects of statistics. Fisher (1922) has been included in the *Breakthroughs in Statistics, Volume I* [Johnson and Kotz (1992)]. Also, Fisher (1925b,1926) have been included in the *Breakthroughs in Statistics, Volume II* [Johnson and Kotz (1993)].

Early in life, Fisher got into some conflicts with Karl Pearson on statistical issues. For example, Fisher criticized (K.) Pearson's *method of moments* as he pushed forward the MLE's. (K.) Pearson did not take these criticisms lightly. One will notice that in the team formed for the cooperative project [Soper et al. (1917)] studying the distribution of the sample correlation coefficient, under the leadership of (K.) Pearson, the young Fisher was not included. This happened in spite of the fact that Fisher was right there and he earned quite some fame for his brilliant paper of 1915 on the distribution of the sample correlation coefficient. Fisher felt hurt as he was left out of this project. He never published again in *Biometrika* with (K.) Pearson as its Editor. Refer to DasGupta (1980) for some historical details.

After Karl Pearson resigned from Galton Chair in 1933, his department was split into two separate departments. R. A. Fisher was appointed as new Galton Professor, and (E. S.) Pearson became the Reader and Head of a *separate* statistics department. Fisher did not like this arrangement very much.

Neyman and (E. S.) Pearson (1928a,b) approached inference problems to build statistical tools for experimenters to choose between two classes of models. These culminated later into path-breaking contributions, Neyman and (E. S.) Pearson (1333a,b). Fisher criticized the Neyman-Pearson approach claiming that his estimation theory, along with the likelihood and sufficiency, was quite adequate, and that the work of Neyman and Pearson on testing of hypotheses was misguided. This controversy between Fisher and Neyman-Pearson was never settled. Some of the letters exchanged between Neyman and Fisher have been included in Bennett (1990). In the 1961 paper, *Silver jubilee of my dispute with Fisher*, Neyman sounded more conciliatory.

Fisher's work was compiled as, *Collected Papers of R. A. Fisher, Volumes 1-5*, edited by J. H. Bennett (in 1971-1974) at the University of Adelaide. An interesting feature of this compilation is that many articles were reprinted with updated commentaries from Fisher himself.

Fisher's daughter, Joan Fisher Box, wrote (1978) a fascinating biography, *R. A. Fisher: The Life of a Scientist*. In Bennett's (1990) edited volume containing many of Fisher's correspondences, one will discover, for example, interesting exchanges between him and G. A. Barnard, W. G. Cochran, H. Hotelling, H. Jeffreys, P. C. Mahalanobis, J. Neyman, C. R. Rao, L. J.

Savage, S. S. Wilks and J. Tukey. The article of Savage (1976), the volume edited by Fienberg and Hinkley (1980), and the articles of Barnard (1992), Karlin (1992), Rao (1992a) honoring Fisher's centenary, deserve attention.

D. Basu's article, *Learning from Counterexamples: At the Indian Statistical Institute in the Early Fifties*, about himself [Ghosh et al. (1992)] and the interview article of Robbins [Page (1989)] mention interesting stories about Fisher. Both (M. G.) Kendall (1963) and Neyman (1967) also gave much historical perspectives.

Fisher travelled almost everywhere. At the request of P. C. Mahalanobis, for example, he frequently visited the Indian Statistical Institute. Sometimes, Fisher went to plead with the President as well as the Prime Minister of India when Mahalanobis faced serious threat of cuts in Federal funding. Fisher visited University of North Carolina at Chapel Hill and Iowa State University at Ames a couple of times each. Many of these visits are colorfully portrayed in his biography [Box (1978)].

Hinkley (1980b) opened with the sentence, "R. A. Fisher was without question a genius." In the published interview article [Folks (1995)], O. Kempthorne mentioned giving a seminar at Ames with the title, "R. A. Fisher $= 1.4$ of Gauss." Fisher remained vigorously productive throughout his life. He was always invigorating, totally engulfed in the creation of new ideas, and forcefully challenging the status-quo in science and society.

He died peacefully on **July 29, 1962** in Adelaide, South Australia. E. A. Cornish, in his address [reprinted in Bennett (1990), pp. xvi-xviii] at Fisher's funeral on August 2, 1962 in Adelaide, had said, "He must go down in history as one of the great men of this our twentieth century." Nobody should ever argue with this assertion.

A. N. Kolmogorov: Andrei Nikolaevitch Kolmogorov was born on **April 25, 1903** during a journey to his mother's home. His mother died during childbirth. His father perished during the offensive by Dynikin in 1919. An aunt brought up Kolmogorov in a village near Volga.

Kolmogorov went to Moscow University in 1920 to study mathematics. In 1922, he constructed the first example of an integrable function whose Fourier series diverged almost everywhere, and then gave the example of an integrable function whose Fourier series diverged *everywhere*. He became an instant international celebrity. Kolmogorov wrote his first paper, jointly with Khinchine, which included the famous *Three Series Theorem* and *Kolmogorov Inequality*.

Kolmogorov found a job as a secondary school teacher. He then became a doctoral student under the supervision of N. N. Luzin. During this apprenticeship, one was normally required to complete fourteen different courses in mathematics but one could substitute any final examination by submitting some independent project in the assigned field. Before the final examinations, Kolmogorov submitted fourteen original research papers written in the required fields. Buried in these papers, there were new discoveries which included some of his fundamental contributions on the *Strong Law of Large Numbers*, the *Central Limit Theorem*, and the *Law of the Iterated Logarithm*. His 1931 paper, *Markov Processes with Continuous States in Continuous Time*, laid the foundation of the stochastic diffusion theory which continues to stand as a landmark, just like many of Kolmogorov's other discoveries. We recall that during 1950-1962, W. Feller broke new grounds on the theory of diffusion and Kolmogorov was delighted.

Kolmogorov's monograph, *Foundations of the Theory of Probability*, first appeared in German in 1933 whose English translation became available in 1950. This immediately became the manifesto in the axiomatic development in probability. The world recognizes Kolmogorov as the guiding light and a true pioneer in probability theory. His 1933 paper on the empirical determination of a distribution has been included in the *Breakthroughs in Statistics, Volume II* [Johnson and Kotz (1993)]. Kolmogorov also wrote other important books and monographs including, *Introductory Real Analysis*, coauthored with S. V. Fomin (1968 Russian edition, 1970 English edition by Prentice-Hall).

Kolmogorov travelled extensively and enjoyed exchanging ideas with colleagues. He energetically listened to anyone around him, young and experienced alike. He had ties with H. Cramér, D. G. Kendall, W. Feller, M. Fréchet, J. Hadamard, P. Lévy, E. B. Dynkin, J. L. Doob, B. V. Gnedenko, A. Rényi, J. Neyman, P. C. Mahalanobis, J. B. S. Haldane, Yu. V. Linnik and C. R. Rao, among others.

Kolmogorov received many honors, including honorary D.Sc. degrees from the University of Paris, University of Stockholm, Indian Statistical Institute, and Hungary. He became a foreign member of the Polish Academy of Sciences and the GDR Academy of Sciences, as well as a honorary member of the London Mathematical Society, the Royal Statistical Society, the Romanian Academy, and the Hungarian Academy. In 1967, he became a member of the U. S. National Academy of Sciences.

The two articles, (D. G.) Kendall (1991) and Shiryaev (1991), provide fascinating commentaries on the life and accomplishments of Kolmogorov. The article of Shiryaev, the first Ph.D. student and later a close associate of Kolmogorov, is particularly rich in its portrayal of the human elements. Kolmogorov enjoyed telling stories and jokes. He also loved mountaineering, skiing, hiking and swimming. The July 1989 issue of the *Annals of Probability* and the September 1990 issue of the *Annals of Statistics*, both published by the Institute of Mathematical Statistics, devoted a total of

195 pages to Kolmogorov's life and contributions, plus a list of his 518 publications.

He constantly strived toward the development of a rigorous curriculum in mathematics in schools. He loved teaching, playing, and listening to the young children in the secondary school where he taught mathematics for many decades. Subsequently, this school has been named after Kolmogorov. He passed away on **October 20, 1987**.

E. L. Lehmann: Erich L. Lehmann was born on **November 20, 1917**, in Strasbourg, France. He was raised in Germany and he is of Jewish ancestry. When the Nazis came to power in 1933, his family decided to settle in Switzerland where he attended high school. Originally he had his mind set on studying the German literature. But, his father suggested the route of mathematics instead. In his interview article [DeGroot (1986c)], Lehmann said that he did not really know what mathematics might entail, but he agreed with his father anyway. In the article, he mentioned that he did not argue with his parents on the choice of a career path. He went to study mathematics in Zurich.

The war broke out in Europe. Lehmann arrived in New York in 1940 and went on to receive the M.A. (1942) and Ph.D. (1946) degrees in mathematics from the University of California, Berkeley. Since then, he has been in the faculty at Berkeley. For his Ph.D. thesis, related to the UMPU tests, Lehmann got advice from J. Neyman, G. Pólya and P. L. Hsu. The interview [DeGroot (1986c)] and his two articles, Lehmann (1993,1997), tell many stories about his life and career.

Lehmann nurtured and developed ideas within the context of the theory of tests of hypothesis in ways which at times had been different from Neyman's original approach. Neyman did not approve such departure from his theory. Lehmann started preparing the manuscript for the book, *Testing Statistical Hypotheses*. Neyman demanded to see the manuscript before it was published! Since then, Lehmann was not asked to teach a course on testing of statistical hypotheses at Berkeley as long as Neyman was around. These episodes caused unhappy strains in the relationship between Lehmann and Neyman. One may refer to DeGroot (1986c, pp. 246-247).

The 1959 (Wiley) classic, *Testing Statistical Hypotheses*, followed by its second edition (1986, Wiley), has been on the "must reading list" practically in all statistics Ph.D. programs. Lehmann's famous *Notes on the Theory of Estimation* has been available in Berkeley since 1951. The *Theory of Point Estimation* first appeared in 1983 (Wiley), a second edition (coauthored with G. Casella) of which came out in 1998 (Springer-Verlag). Lehmann also wrote few other books including the *Nonparametrics: Statistical Methods Based on Ranks* (1975, Holden-Day).

Lehmann's contributions, particularly on sufficiency, minimal sufficiency, completeness, unbiased tests, nonparametric inference, have been far reaching. The papers of Lehmann and Scheffé (1950, 1955, 1956, *Sankhyā*) on completeness, similar regions and unbiased estimation have profoundly influenced the research and the understanding of statistical inference. He was Editor of the Annals of Mathematical Statistics during 1953-1955. He received many honors, including President (1961) of the Institute of Mathematical Statistics, the Guggenheim Fellowship awarded three times, and elected (1978) member of the U. S. National Academy of Sciences. In February 1985, Lehmann received a honorary doctorate degree from the University of Leiden. Lehmann stays active in research and he frequently travels to conferences.

P. C. Mahalanobis: Prasanta Chandra Mahalanobis, without whom the Indian statistical movement would probably never have been initiated, was born on **June 29, 1893** in Calcutta. His parents' family-friends and relatives were in the forefront of the nineteenth century awakening in India. After receiving B.Sc. (Honors, 1912) degree in Physics from the University of Calcutta, Mahalanobis went for a casual visit to London in the summer of 1913. He was awarded senior scholarships from King's College where he studied physics. Upon his return to India, he joined as a lecturer of physics in Presidency College, Calcutta in 1915. Mahalanobis was an eminent physicist of his time.

Because of his analytical mind, he was often called upon by influential family-friends to help analyze data for important government as well as university related projects and reports. Mahalanobis started reading some of the old issues of *Biometrika*, the copies of which he happened to bring along from his trip to Cambridge.

Within the Physics Department in Presidency College, Mahalanobis initiated a Statistical Laboratory in the late 1920's. This Statistical Laboratory eventually grew up to be the famed Indian Statistical Institute (ISI) on April 28, 1932. Mahalanobis's energy and vision led to a phenomenal growth of a newly created discipline in India. Other brilliant individuals joined hands with him including, not in any particular order, S. N. Roy, R. C. Bose, S. S. Bose, C. R. Rao, K. Kishen, K. R. Nair, D. B. Lahiri, G. Kallianpur, D. Basu, R. R. Bahadur, S. R. S. Varadhan, S. K. Mitra, G. P. Patil, J. Sethuraman and R. G. Laha.

In 1933, *Sankhyā: The Indian Journal of Statistics* was started with Mahalanobis as its Editor. He continued to edit the journal until his death. Due to his initiative and involvement, ISI played a key role in formulating India's second five-year economic plan after independence. The large-scale surveys were initiated and the National Sample Survey Organization was created. National Statistical Service was established and Mahalanobis succeeded in obtaining recognition of Statistics as a discipline, separate from mathematics, by the Indian Science Congress. Apart from these and host of other accomplishments of *national importance*, Mahalanobis also wrote five books and 210 papers.

Mahalanobis (1932) first constructed statistical tables for the F distribution since these were directly needed for the analyses of variance in field experiments. Fisher, on the other hand, popularized tables for $\frac{1}{2} log(F)$, namely Fisher's Z. Snedecor's (1934) book apparently overlooked Mahalanobis's earlier contribution and independently provided the F tables saying simply that they were computed from Fisher's table.

Mahalanobis's early papers on meteorology, anthropology, economic planning, flood control, psychology, multivariate analysis, design of experiments and sample surveys are particularly noteworthy. He was interested in all aspects of science, including biology, geology, genetics, botany, and speech recognition. He was also a literary figure and critic of his time.

In large-scale surveys, Mahalanobis (1940) pioneered the idea of gathering samples in successive stages to increase efficiency. The method of using pilot samples before running the main survey was developed in the same paper. Wald (1947, p. 2) commented that this fundamental contribution may be regarded as a forerunner of Sequential Analysis. *Mahalanobis's D^2*, a measure of distance between two populations, is a fundamental concept in multivariate analysis.

Figure 14.2.1. P. C. Mahalanobis Centennial Postage Stamp
Issued by the Government of India, 1993

Mahalanobis received the highest honors awarded by the Government of India including the title *Padmabibhusan,* which means that the awardee adds glory even to the flower Lotus. During the centennial year 1993, the Government of India issued a *postage stamp* with Mahalanobis's picture on it. This is how the masses of India celebrated statistics and quantitative literacy!

Mahalanobis became President (1950) of the Indian Science Congress. Many other academies from India and all over the world bestowed honors upon him, including the Fellowship of the Royal Society, the Foreign Membership of the U.S.S.R. Academy of Sciences, and the Statistical Advisor to the Cabinet of Indian Government since 1949, Gold Medal from the Czechoslovak Academy of Sciences. He received a number of honorary degrees including a Honorary Ph.D. (1956) from the University of Calcutta and the Deshikottama (Honorary D.Litt., 1961) from Visva Bharati, the University founded by Tagore, the Indian poet and Nobel Laureate.

Mahalanobis was instrumental in opening up the continuous flow of scientific exchanges and visits, particularly with U.S.A. and many East European Countries. He was able to create a vigorous statistical infrastructure in India at all imaginable levels. He was a true ambassador of statistical science.

Additional details can be found in a series of commemorative articles which appeared in *Sankhyā, Series A* and *B*, 1973. The published interviews of Rao [DeGroot (1987)] and Mitra [Mukhopadhyay (1997)] as well as the essays written by Rao (1992b) and Hansen (1987) would be very informative. One may also look at the entry [Rao (1968)] on *P. C. Mahalanobis* in the *International Encyclopedia of Statistics*. The recent article of Ghosh et al. (1999) also gives many historical details on Mahalanobis's contributions.

In June,1972, Mahalanobis was hospitalized for some surgery. He was recovering well. On June 27, like every other day, from his hospital bed he cleared some files and wrote letters. He gave dictations on some policy matters regarding the Institute's affairs and its *future*. Perhaps he felt that the end was near! This superbly powerful, dynamic and productive visionary's life came to a halt in Calcutta on **June 28, 1972**, exactly one day shy of his 79th birthday.

J. Neyman: Jerzy Neyman was born on **April 16, 1894** in Bendery, Russia. His parents were Polish. He received early education from governesses, alternately French and German, and this helped in his proficiency in several languages. When he was twelve, his father passed away and his family moved, ultimately settling in Kharkov.

Neyman entered University of Kharkov in 1912 to study physics and mathematics. S. N. Bernstein, a Russian probabilist, was among his teachers. Bernstein mentioned Karl Pearson's (1892) monograph, *Grammar of*

Science, to Neyman. This monograph had a life-long influence on Neyman.

Very early in college, Neyman was keenly interested in the theory of Lebesgue measure. But, life-style was uncertain and difficult during those days of wars. Poland and Russia started fighting about territories. Because of his Polish background, Neyman was arrested as an enemy alien and he spent a few weeks in jail. But, he was needed in the University to teach, and hence he was ultimately let go.

At the age of 27, Neyman first had the opportunity to visit Poland in an exchange of prisoners of war and in Warsaw, he met the Polish mathematician W. Sierpiński who encouraged him and thought highly of his research. Ultimately Sierpiński helped Neyman to get a job as a statistician in the National Institute of Agriculture. Here, he was to take meteorological observations and help with agricultural experiments.

Neyman's doctoral thesis (1923) in the University of Warsaw had dealt with probabilistic considerations in agricultural trials. In 1924, he went to the University College London for a year to study under Karl Pearson where he met three other statisticians: R. A. Fisher, E. S. Pearson, and W. S. Gosset ("Student"). Some of Neyman's papers were known to the people at the University College because their English translations were already available. Neyman spent the following year in Paris with a Rockefeller Fellowship, and became acquainted with H. Lebesgue, J. Hadamard, and E. Borel.

In late 1920's, Neyman focused more upon mathematical statistics as well as applications in economics, insurance, biology and industry. His collaboration with E. S. Pearson started around 1925. Since early 1920's, (E. S.) Pearson began developing his own philosophy of statistical methods and inference. He also started to appreciate and build practical statistical models, particularly useful for industrial applications. During 1928-1938, extensive collaborations took place between (E. S.) Pearson and Neyman.

Neyman and (E. S.) Pearson (1928a,b) approached inference problems to build statistical tools for experimenters to choose between two classes of models. These culminated later into path-breaking contributions, Neyman and (E. S.) Pearson (1333a,b). In the latter papers, likelihood ratio tests in the multi-parameter cases were fully developed. Neyman and Pearson (1933a) has been included in the *Breakthroughs in Statistics, Volume I* [Johnson and Kotz (1992)]. Neyman-Pearson's formulation of *optimal tests* ultimately evolved into *optimal decision functions* for more general statistical problems in the hands of Wald (1949b,1950).

Fisher criticized the Neyman-Pearson approach claiming that his estimation theory, along with the likelihood and sufficiency, was quite adequate, and that the work of Neyman and Pearson on testing of hypotheses was misguided. This controversy between Fisher and Neyman-Pearson was never settled. Some of the letters exchanged between Neyman and Fisher have been included in Bennett (1990). In his 1961 paper, *Silver jubilee of my dispute with Fisher*, Jerzy Neyman sounded more conciliatory.

Neyman made fundamental contributions in survey sampling in 1934 and later developed the notion of estimation via confidence sets in 1937. Neyman (1934) has been included in the *Breakthroughs in Statistics, Volume II* [Johnson and Kotz (1993)]. In all three fundamental areas of contributions, namely tests of hypotheses-survey sampling-confidence sets, Neyman characteristically started by formulating the problem from what used to be very vague concepts and then slowly building a logical structure within which one could carry out scientific inferences.

His research arena was too vast for us to discuss fully. His contributions have been far reaching. Many of his discoveries have become a part of the statistical folklore. The phrases like *Neyman-Pearson Lemma, Neyman Structure, Optimal $C(\alpha)$ Tests, Neyman Factorization, Neyman Accuracy*, are parts of statistical vocabulary.

Neyman came to visit U.S.A. for six weeks in 1937, lecturing in universities and the Department of Agriculture, Washington D.C., and at that time he was invited to join the University of California at Berkeley as a professor of statistics to build a separate statistical unit. The offer came at a time when he was mentally ready to move away from the uncertainty in Europe and the oppression of Hitler. Neyman arrived in Berkeley in August, 1938, and the rest is history.

In Berkeley, Neyman built the prestigious Statistical Laboratory and Department of Statistics and he was the key attraction. He initiated and continued to organize the Berkeley Symposia where the brightest researchers assembled to exchange ideas. Many path-breaking discoveries during 1950's-1970's appeared in the highly acclaimed *Proceedings of the Berkeley Symposium in Statistics and Probability*. Neyman actively edited the Berkeley Symposia volumes, sometimes with the help of his younger colleagues.

Neyman travelled extensively. When he did not travel, he hosted distinguished visitors at Berkeley. He had close ties with the leaders of centers of learning in other countries. For example, Neyman had close ties with E. S. Pearson, H. Cramér, A. N. Kolmogorov and P. C. Mahalanobis, and they were very friendly. On a number of occasions, he visited (E. S.) Pearson in the University College, Cramér in Stockholm, and Mahalanobis in the Indian Statistical Institute. Interesting details regarding Neyman's visits to India can be found in a series of Mahalanobis commemorative articles which had appeared in *Sankhyā, Series A* and *B*, 1973.

Neyman's smiling face was unmistakable whenever he walked into a room full of people. He encouraged bright and upcoming researchers. He was very approachable and he often said, "I like young people." Neyman had a very strong personality, but he also had a big heart.

Neyman received many honors and awards for his monumental contributions to statistical science. He was elected to the U.S. National Academy of Sciences. He became a foreign member of both the Polish and the Swedish Academies of Science and a Fellow of the Royal Society. He received the Guy Medal in Gold from the Royal Statistical Society. He was awarded several honorary doctorate degrees, for example, from the University of Chicago, the University of California at Berkeley, the University of Stockholm, the University of Warsaw, and the Indian Statistical Institute.

In 1968, Neyman was awarded the highest honor an American scientist could receive. He was awarded the U.S. National Medal of Science from the White House.

The article of LeCam and Lehmann (1974), prepared in celebration of Neyman's 80th birthday, described and appraised the impact of his contributions. The biographical book of Reid (1982), *Neyman from Life,* portrays a masterful account of his life and work. One may also look at the entry [Scott (1985)] on *Jerzy Neyman,* included in the *Encyclopedia of Statistical Sciences.* Neyman died on **August 5, 1981** in Berkeley, California.

E. S. Pearson: Egon Sharpe Pearson was born on **August 11, 1895** in Hampstead, England. His father was Karl Pearson. Egon Pearson finished his school education in Oxford. He moved to Trinity College, Cambridge, and obtained First Class in Part I of Mathematical Tripos. Due to ill-health and time-out needed for war service, his B.A. degree from Cambridge was delayed and it was completed in 1920.

The *Biometrika*'s first issue came out in October 1901 when it was founded by W. F. R. Weldon, F. Galton, and K. Pearson. From a very young age, Egon Pearson saw how this journal was nurtured, and he grew emotionally attached to it. In 1921, he joined his father's internationally famous department at the University College London as a lecturer of statistics.

Egon Pearson became the Assistant Editor of *Biometrika* in 1924. After Karl Pearson resigned from the Galton Chair in 1933, his department was split into two separate departments and R. A. Fisher was appointed as the new Galton Professor whereas Egon Pearson became the Reader and Head of a *separate* statistics department. Fisher did not like this arrangement very much. Egon Pearson became a professor in 1935 and the Managing Editor of *Biometrika* in 1936. He remained associated with the University College London until his retirement in 1960.

Since early 1920's, Egon Pearson started developing his own philosophy of statistical methods and inference. He began to appreciate and build practical statistical models, particularly useful for industrial applications. On either count, philosophically he started drifting away from his "father's footsteps." It was quite some challenge for the younger Pearson to form his own strong ideological base and research interests under the "shadow" of Karl Pearson. During 1928-1938, extensive collaborations took place between Egon Pearson and Jerzy Neyman whom he first met in 1924 when Neyman came to visit the Galton Laboratory.

Neyman and (E. S.) Pearson (1928a,b) approached inference problems to build statistical tools for experimenters to choose between two classes of models. These culminated later into path-breaking contributions, Neyman and (E. S.) Pearson (1333a,b). In the latter papers, likelihood ratio tests in the multi-parameter cases were fully developed. Fisher criticized the Neyman-Pearson approach claiming that his estimation theory, along with the likelihood and sufficiency, was quite adequate, and that the work of Neyman and Pearson on testing of hypotheses was misguided. This controversy between Fisher and Neyman-Pearson was never settled.

Neyman and (E. S.) Pearson (1933a) has been included in the *Breakthroughs in Statistics, Volume I* [Johnson and Kotz (1992)]. Egon Pearson's (1966) own account of his collaborations with Neyman was recorded in the article, *The Neyman-Pearson Story: 1926-1934*. Egon Pearson's interests and enthusiasm in editing statistical tables have been particularly noteworthy. With H. O. Hartley, the revised *Biometrika Tables for Statisticians, Volume 1* and *2*, were respectively published (Cambridge University Press) in 1954 and 1972. Some of these tables are indispensable even today.

Egon Pearson received many honors. In 1935, he received the Weldon Prize and Medal, and became President (1955) of the Royal Statistical Society for a two-year period. In 1955, he was awarded the Guy Medal in Gold by the Royal Statistical Society.

E. S. Pearson died on **June 12, 1980** in Midhurst, Sussex. The opening articles of Bartlett (1981) and Tippett (1981) in *Biometrika* detailed Egon Pearson's productive life and career. The obituary article which appeared in the *J. Roy. Statist. Soc. , Ser. A* (1981, 144, pp. 270-271) is also very informative. The article, *Egon S. Pearson (August 11, 1895-June 12, 1980): An appreciation*, written by Jerzy Neyman (1981) ended with the following sentence: "The memories of cooperation with Egon during the decade 1928-1938 are very dear to me."

K. Pearson: Karl Pearson was born on **March 27, 1857** in London. He attended University College School and the Kings College, Cambridge as a scholar. He got his degree in 1879 securing the Third Wrangler position in Mathematical Tripos. He was appointed to the Chair of Applied

Mathematics at the University College London in 1885 where he stayed throughout his career. In 1911, he moved to the Chair of Eugenics which was newly established.

Karl Pearson was an established mathematician with interests in other fields, including physics, law, genetics, history of religion and literature. Around 1885, he started formulating problems arising from observational studies and systematically developed analysis of data. W. F. R. Weldon and F. Galton influenced his thinking process significantly. At one time, he was consumed by Galton's *Natural Inheritance* from 1889.

Karl Pearson published an astonishingly large volume of original papers in genetics, evolution, biometry, eugenics, anthropology, astronomy, and other areas. His contributions on moments, correlations, association, system of frequency curves, probable errors of moments and product moments, and Chi-square goodness-of-fit, among his other discoveries, have become a part of the folklore in statistics. Stigler (1989) throws more light on the invention of correlation. He created the preliminary core of statistics. He was the nucleus in the movement of systematic statistical thinking when it was essentially unheard of. (K.) Pearson (1900) has been included in the *Breakthroughs in Statistics, Volume II* [Johnson and Kotz (1993)].

Weldon, Galton and (K.) Pearson started the journal, *Biometrika*, whose first issue came out in October 1901. Karl Pearson edited the journal until 1936. He resigned from the Galton Chair in 1933 and his department was split into two *separate* departments. His son, Egon Pearson became the Reader and Head of a *separate* statistics department whereas R. A. Fisher was appointed as the new Galton Professor.

K. Pearson received many honors, including the election (1896) to the Royal Society and was awarded the Darwin Medal in 1898. Egon Pearson's (1938) monograph, *Karl Pearson: An Appreciation of Some Aspects of his Life and Work*, is fascinating to read. One may also look at the entry [David (1968)] on *Karl Pearson*, included in the *International Encyclopedia of Statistics* for more information. Karl Pearson died on **April 27, 1936** in Coldharbour, Surrey.

C. R. Rao: Calyampudi Radhakrishna Rao was born on **September 10, 1920**, in Karnataka, India. He received M.A. (1940) in mathematics from the Andhra University, another M.A. (1943) in statistics from the University of Calcutta, Ph.D. (1948) and Sc.D. (1965) from the University of Cambridge. R. A. Fisher was Rao's adviser for his Ph.D. thesis in Cambridge. Fisher told Rao to find his own research problem. Rao worked by himself on discriminant analysis and classification problems related to his work on anthropology at the museum in King's College. Rao (1992b) recalled that when he showed his completed thesis to Fisher, he was told simply that "The problem was worth investigating." Rao's life, career, and accomplishments have been documented in an interview article [DeGroot (1987)]. Rao's (1992b) article, *Statistics as a Last Resort*, has many interesting stories and important historical details.

After securing M.A. degree in mathematics, Rao started looking for a job without much success. The war broke out. Eventually he came to Calcutta to be interviewed for a position of "a mathematician for the army service unit" and there he met an individual who was sent to Calcutta to receive "training in statistics" from the Indian Statistical Institute (ISI), founded earlier by P. C. Mahalanobis. He was told that statistics was a subject for the future. Rao went to visit ISI and he immediately decided to join to get "training in statistics." Mahalanobis admitted him to the Training Section of ISI starting January 1, 1941. The rest has been history.

After Rao returned to Calcutta in August, 1948 from Cambridge with a Ph.D. degree, he became a full professor in ISI. For the next thirty years, Rao was a key figure for nurturing the Institute's programs, goals, and aspirations. He has advised Ph.D. dissertations of many distinguished statisticians and probabilists, including D. Basu, V. S. Varadarajan, S. R. S. Varadhan, K. R. Parthasarathy, and DesRaj.

During 1949-1963, Rao was Head of the Division of Theoretical Statistics in ISI. He became Director of the Research and Training School in ISI in 1963. In 1976, he gave up his administrative position in ISI, but continued as the Jawaharlal Nehru Professor, a special chair created for Rao by the Prime Minister of India, Indira Gandhi. After the death of Mahalanobis, the Chief of ISI, Rao held the position of the Director and Secretary of ISI.

In the summer of 1978, Rao came for a casual visit to the University of Pittsburgh and he was invited to give a university-wide popular lecture. The day after his lecture, he was offered a position which was totally "unexpected" as Rao (1992b) recalled. He started a new career at the University of Pittsburgh in 1979. In 1988, he moved to the University of Pennsylvania as Professor and holder of the Eberly Chair in Statistics. He is also Director of the Center for Multivariate Analysis. He has been the Editor or Co-Editor of *Sankhyā, The Indian Journal of Statistics* for many years since 1964.

Rao has made legendary contributions practically in all areas of statistics. Many contributions, for example, on sufficiency, information, maximum likelihood, estimation, tests, multivariate analysis, discriminant analysis, linear models, linear algebra, generalized inverses of matrices, MINQE theory, design of experiments and combinatorics are path-breaking. Through the last five decades, many of his discoveries have been incorporated as standard material in courses and curriculum in mathematics, statistics, probability, electrical engineering, medical statistics, to name a few. The *Cramér-Rao inequality*, *Cramér-Rao lower bound*, *Rao-Blackwellization*, and *Rao's Score*, for example, have become household phrases in these and other fields. Rao's (1945) paper has been included in the *Breakthroughs in Statistics, Volume I* [Johnson and Kotz (1992)].

Rao also made penetrating discoveries in nonparametric inference, higher order efficiencies, estimating equations, principal component and factor analysis. His book, *Linear Statistical Inference*, essentially took off from where Cramér's (1946a) classic text had left. This phenomenally successful work first appeared in 1965 (Wiley), with its second edition out in 1973 (Wiley), and it has been translated in many languages including Russian, German, Japanese, Czech, Polish, and Chinese.

In applied areas, for example, in anthropometry, psychometry, econometrics, quality control, genetics and statistical ecology, Rao's contributions have been equally far-reaching and deep. His book, *Advanced Statistical Methods in Biometric Research,* was published in 1952 (Wiley) and it included the most modern statistical analyses of the day. He introduced and developed a comprehensive theory of what are customarily called the *weighted distributions* which takes into account the visibility bias in sampling from wild-life populations, for example. The methodology of weighted distributions has played fundamental roles in areas including statistical ecology and environmental monitoring.

In mathematics also, Rao is considered a leader. We may simply mention the areas, for example, the characterization problems in probability, differential geometry, and linear algebra. He has been in the forefront of these areas over many decades. Much of the theory of the generalized inverses of matrices was developed, unified and made popular by Rao himself and jointly with S. K. Mitra and other colleagues. The monograph, *Generalized Inverse of Matrices and its Applications*, jointly written by Rao and Mitra (1971, Wiley) has been a landmark in this area. Rao's another monograph, *Characterization Problems of Mathematical Statistics*, written jointly with A. Kagan and Yu. V. Linnik (1972 Russian edition, 1973 Wiley edition) was very highly acclaimed.

The recent article of Ghosh et al. (1999) supplies many details on Rao's phenomenal contributions. He has written or coauthored more than one dozen books, over three hundred research papers, and edited many special volumes. For his legendary contributions, Rao has received many honors from all over the world. He has received more than a dozen honorary doctorate degrees from prestigious Universities and Institutes spread all over the globe.

Rao is a Fellow of the Royal Society and King's College, a Fellow of the

Indian National Science Academy, a Honorary Member of the International Statistical Institute, to name a few. He became President (1977) of the Institute of Mathematical Statistics. He is an elected member of the U. S. National Academy of Sciences.

Rao is very modest and has a charming personality. He is a captivating story-teller. His hobby is photography. At Penn State, Rao continues to keep himself busy, always working-guiding-sharing ideas with colleagues as well as students, and travelling both far and near.

L. J. Savage: Leonard Jimmie Savage was born in Detroit, Michigan on November 20, 1917. He made phenomenal contributions in the area of foundations of statistics, particularly in Bayesian inference. Savage's work has bolstered the Bayesian movement many fold over a period of several decades. Savage (1976) gave a penetrating overview of the body of works of R. A. Fisher on statistical inference. His 1954 monograph, *The Foundations of Statistics*, has been a landmark in statistics. Lindley (1980) beautifully synthesized Savage's contributions in statistics and probability.

Savage died in New Haven, Connecticut on **November 1, 1971**. Savage's (1981) volume of articles will attest to his profound influence and legacy in statistics and subjective probability theory. Many well-known personalities, including F. J. Anscombe, D. A. Berry, B. de Finetti, L. Dubins, S. E. Fienberg, M. Friedman, W. Kruskal, D. V. Lindley, F. Mosteller, W. A. Wallis and A. Zellner, have written about Savage and his contributions. These and other references are included in Savage (1981). Sampson and Spencer's (1999) interview article on I. Richard Savage, the brother of L. J. Savage, gives many personal as well as professional accounts of Jimmie Savage's life and work.

H. Scheffé: Henry Scheffé was born on **April 11, 1907**, in New York City. His parents were German.

In 1925, Scheffé became a student at the Polytechnic Institute of Brooklyn and afterward he worked as a technical assistant at Bell Telephone Laboratories. In 1928, he went to attend the University of Wisconsin to study mathematics. He had taken one statistics related course during this period. Scheffé received B.A. (1931) in mathematics with high honors, followed by Ph.D. (1936), both from the University of Wisconsin. His Ph.D. thesis dealt with asymptotic solutions of differential equations (1936, *Trans. Amer. Math. Soc.*, **40**, 127-154).

After finishing the Ph.D. degree, Scheffé taught in a number of places. In 1941, he joined Princeton University with the expectation of doing research in a more fertile area such as mathematical statistics instead of mathematics. There, he came to know T. W. Anderson, W. G. Cochran, W. J. Dixon, A. M. Mood, F. Mosteller, J. Tukey and S. S. Wilks, among others. Up until 1948, he also taught at Syracuse University and UCLA. During this period, he spent part of his time as a Guggenheim Fellow at the University of California, Berkeley, where he began collaborations with E. L. Lehmann. During 1948-1953, he was an associate professor of mathematical statistics at Columbia University.

In 1953, Scheffé moved to the University of California, Berkeley, as a professor of statistics and remained there until retirement in 1974. During 1965-1968, he became the Chairman of the statistics department at Berkeley.

Scheffé wrote the authoritative book, *The Analysis of Variance*, which was published in 1959 (Wiley). The papers of Lehmann and Scheffé (1950, 1955, 1956, *Sankhyā*) on completeness, similar regions and unbiased estimation have had a tremendous influence on research in statistics. Scheffé's fundamental contributions in the area of multiple comparisons have provided essential tools used by statisticians everywhere. His contributions in the area of the Behrens-Fisher problem are also particularly noteworthy.

Scheffé received many honors. He was President (1954) of the Institute of Mathematical Statistics, Vice President (1954-1956) of the American Statistical Association, and he received the Fulbright Research Award (1962-1963).

Scheffé enjoyed reading novels. He was fond of bicycling, swimming, snorkeling and backpacking. He loved nature and enjoyed travelling. Later in life, he learned to play the recorder and played chamber music with friends. After his retirement from Berkeley in 1974, he joined the University of Indiana in Bloomington for three years as a professor of mathematics. He returned to Berkeley in June, 1977 to start the process of revising his book, *The Analysis of Variance*. Daniel and Lehmann (1979) noted that this was not to be - Scheffé died on **July 5, 1977** from the injuries he sustained in a bicycle accident earlier that day.

By action of the Council of the Institute of Mathematical Statistics, the 1979 volume of the *Annals of Statistics* was dedicated to the memory of Henry Scheffé. Its opening article, prepared by Daniel and Lehmann (1979) detailed his life and career.

C. Stein: Charles Stein was born on **March 22, 1920**, in Brooklyn, New York. He received B.S. (1940) in mathematics from the University of Chicago. He served in the U. S. Army Air Force during 1942-1946 and became a Captain. He earned Ph.D. (1947) in mathematical statistics from Columbia University, and joined the faculty of Statistical Laboratory at the University of California, Berkeley.

In 1949-1950, Stein was a National Research Council Fellow in Paris. He was an associate professor at the University of Chicago during 1951-

1953. Since 1953, he has been a professor in the department of statistics at Stanford University.

In an interview article [DeGroot (1986d)], Stein told that he had always intended to be a mathematician. After studying some works of Wald, he published the landmark paper [Stein (1945)] on a two-stage sampling strategy for testing the mean of a normal population whose power did not depend on the unknown population variance.

Stein received encouragement from senior researchers including A. Wald, J. Neyman and K. J. Arrow. He started to generalize some of Wald's work on most stringent tests. In the interview article [DeGroot (1986d)], Stein mentioned that G. Hunt pointed it out to him that what he was doing was group theory. He did not realize this in the beginning. Eventually, *Hunt-Stein Theorem* became a household phrase within the realm of invariance where Stein made fundamental contributions.

Stein's best known result is perhaps his proof of the *inadmissibility* [Stein (1956)] of the sample mean vector $\overline{\mathbf{X}}$ as an estimator of the mean vector μ in the $N_p(\mu, I)$ population when the dimension p is three or higher, under the quadratic loss function. Later, James and Stein (1961) gave an explicit estimator which had its risk smaller than that of $\overline{\mathbf{X}}$, for all μ, under the quadratic loss function. The dominating estimator has come to be known as the *James-Stein* estimator. In this area, one frequently encounters a special identity which is called the *Stein Identity*. James and Stein (1961) has been included in the *Breakthroughs in Statistics, Volume I* [Johnson and Kotz (1992)]. Berger (1985) gave an elegant exposition of this problem and its generalizations. DeGroot (1986d) portrayed a delightful account of Stein's life and career. Stein remains busy and active in research.

"Student" (W. S. Gosset): William Sealy Gosset was born in **1876,** the oldest child of a Colonel in the Royal Engineers. He was a pioneer in the development of statistical methods for design and analysis of experiments. He is perhaps better known under the pseudonym "Student" than under his own name. In most of his papers, he preferred to use the pseudonym "Student" instead of his given name.

Following his father's footsteps, Gosset entered the Royal Military Academy, Woollwich, to become a Royal Engineer himself. But, he was rejected on account of poor eyesight. He graduated (1899) with a first class degree in chemistry from New College in Oxford, and then joined the famous Guinness Brewery in Dublin as a brewer. He stayed with this brewing firm for all his life, ultimately becoming the Head Brewer in a new installation operated by the Guinness family at Park Royal, London in 1935.

Gosset needed and developed statistical methods for small sample sizes which he would then apply immediately to ascertain relationships between the key ingredients in beer. His path-breaking 1908 paper, included in the *Breakthroughs in Statistics, Volume II* [Johnson and Kotz (1993)], gave the foundation of the t-distribution. Gosset derived the probable error of a correlation coefficient in 1908 and made several forceful conjectures, most of these being proven true later by Fisher (1915). Fisher once described Gosset as "Faraday of statistics."

The monographs and articles of (E. S.) Pearson (1938), Reid (1982), David (1968), Barnard (1992), Karlin (1992), Rao (1992a) would provide more details on the life and work of Gosset. Box's (1978) fascinating biography, *R. A. Fisher: The Life of a Scientist*, points out early diverse interactions among (K.) Pearson, Gosset, and Fisher. Box (1987) portrayed the connections and collaborations among Guinness, Gosset and Fisher. Additionally, both (M. G.) Kendall (1963) and Neyman (1967) presented much historical perspectives. One may also take a look at the entry [Irwin (1968)], *William Sealy Gosset*, included in the *International Encyclopedia of Statistics* for more information. Gosset died in **October, 1937.**

A. Wald: Abraham Wald was born on **October 31, 1902**, in Cluj, Romania. He was raised in an intellectual environment within his family. But shockingly, his parents, sisters, one brother who was considered intellectually gifted, their spouses and children, and other relatives, perished in German crematoria and concentration camps. As the son of an orthodox Jew, he faced many obstacles and was barred to pursue many interests as a child and young adult.

After graduating from the University of Cluj, Wald was finally admitted to study mathematics at the University of Vienna. He came to know K. Menger and K. Hahn in Vienna. Initially, Wald began to work on geometry. One of his proofs was later incorporated into the seventh edition of David Hilbert's classic text, *Grundlagen der Geometrie*. But, it was clear that there was practically no prospect of getting an academic position in Austria in spite of Wald's brilliance.

Menger advised him to move into some area of applied mathematics. He met O. Morgenstern and started working on mathematical economics. He made fundamental contributions in econometrics through a series of papers and a book. The three brilliant scholars, O. Morgenstern, J. von Neumann and A. Wald later emigrated to U.S.A., and in 1944, von Neuman and Morgenstern's famous book, *Theory of Games and Economic Behavior*, appeared. These collaborations led to a fuller growth of statistical decision theory, culminating into Wald's (1950) classic book, *Statistical Decision Functions*.

Wald came to U.S.A. in the summer of 1938 as a Fellow of the Cowles Commission for Research in Economics. Hotelling arranged for Wald's release from the Fellowship of the Cowles Commission to enable him to accept a Fellowship from the Carnegie Corporation and join Columbia University. Wald quickly learned statistical theory by attending lectures of Hotelling at Columbia. In Columbia, Wald had a long period of collaborations with J. Wolfowitz. Together, they wrote a number of fundamental papers in sequential analysis and multiple decision theory.

Wald's lecture notes on the topics he taught at Columbia were lucid but rigorous and challenging. These notes have taught and given inspirations to the next several generations of aspiring statisticians. During this period, he wrote a number of important papers including the fundamental one on estimation and testing [Wald (1939)], even before he came to know the details of statistical theory. Wald gave the foundation to the theory of statistical decision functions. This deep area is probably one of his greatest contributions. The sequential analysis is another vast area where he made legendary contributions. Wald's (1947) book, *Sequential Analysis*, has always been regarded as a classic. The two papers, Wald (1945,1949b), have been included in the *Breakthroughs in Statistics, Volume I* [Johnson and Kotz (1992)].

At the invitation of Mahalanobis, in November, 1950, Wald left for a visit to the Indian Statistical Institute, Calcutta, and meet colleagues and students there. After the official obligations of the trip was taken care of, Wald along with his wife left on **December 13, 1950**, on an Air India flight for a visit to the picturesque southern India. Unfortunately, the plane got lost in the fog and crashed into the peaks of the Nilgiri mountains. All passengers aboard that flight, including Wald and his wife, perished. This unfortunate accident shook the world of mathematics, statistics and economics. The world lost a true genius at a tender age of 48.

D. Basu's article, *Learning from Counterexamples: At the Indian Statistical Institute in the Early Fifties*, included in Ghosh et al. (1992), recalls interesting stories about Wald at the time of his visit to the Institute in November-December, 1950. These are important stories because very little is otherwise known about what Wald was doing or thinking during the last few days of his life.

Wald became President (1948) of the Institute of Mathematical Statistics. At the Abraham Wald Memorial Session held on September 7, 1951, during the Minneapolis meeting of the Institute of Mathematical Statistics, J. Wolfowitz had said, "The personal loss will be felt by his numerous friends, but all must mourn for the statistical discoveries yet unmade which were buried in the flaming wreckage on a mountain side in South India and which will slowly and painfully have to be made by others." By action of the Council of the Institute of Mathematical Statistics, the 1952 volume of the *Annals of Mathematical Statistics* was dedicated to the memory of Abraham Wald. Its opening articles, prepared by Wolfowitz (1952), Menger (1952), and Tintner (1952) detailed the life and career of Wald.

Epilogue

It will be safe to say that we have already included too many citations in this section. Yet, it is possible, however, that we have left out the references of some important historical volumes or documents. But, for a reader who is just beginning to get interested in the *history of statistics*, a very long list of references will probably do more harm than good. As a starter, the following narrower list will hopefully serve a reader better:

R. A. Fisher, The Life of a Scientist, by Joan Fisher Box (1978)
Neyman from Life, by Constance Reid (1982)
The Making of Statisticians, edited by Joe Gani (1982)
American Contributions to Mathematical Statistics in the Nineteenth Century, edited by Steve Stigler (1980)
The History of Statistics: The Measurement of Uncertainty Before 1900, by Steve Stigler (1986)
Glimpses of India's Statistical Heritage, edited by J. K. Ghosh, S. K. Mitra, and K. R. Parthasarathy (1992)

| **Enjoy and Celebrate Statistics by Learning Its History** |

14.3   Selected Statistical Tables

This section provides some of the standard statistical tables. These were prepared with the help of MAPLE.

Tables 14.3.1a-14.3.1b correspond to the distribution function of a standard normal distribution. Tables 14.3.2, 14.3.3 and 14.3.4 respectively correspond to the percentage points of the Chi-square, Student's t, and F distribution. One may look at Lindley and Scott (1995) or other sources for more extensive sets of tables like these.

14.3.1   The Standard Normal Distribution Function

In the Tables 14.3.1a and 14.3.1b, the first column and row respectively designate the "first" and "second" decimal points of z. Let us suppose that

Z stands for the standard normal random variable. Look at the Figures 14.3.1-14.3.2.

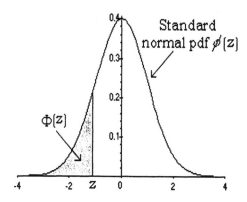

Figure 14.3.1. $N(0,1)$ DF $\Phi(z)$ Is the Shaded Area When $z < 0$

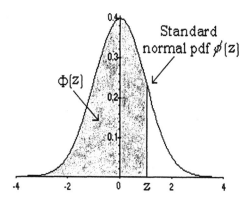

Figure 14.3.2. $N(0,1)$ DF $\Phi(z)$ Is the Shaded Area When $z > 0$

Now, consider the following examples:

$$
\begin{aligned}
P\{Z < 2\} \quad &= \Phi(2) = .97725; \\
P\{Z < 2.1\} \quad &= \Phi(2.1) = .98214; \\
P\{Z < 1.21\} \quad &= \Phi(1.21) = .88686; \\
P\{Z < 1.28\} \quad &= \Phi(1.28) = .89973; \\
P\{Z < -2\} \quad &= 1 - P\{Z < 2\} = 1 - \Phi(2) \\
&= 1 - .97725 = .02275;
\end{aligned}
$$

$$P\{Z < -1.21\} \qquad = 1 - P\{Z < 1.21\} = 1 - \Phi(1.21)$$
$$= 1 - .88686 = .11314;$$

$$P\{0 < Z < 2.1\} \qquad = \Phi(2.1) - \Phi(0) = .98214 - .5$$
$$= .48214;$$

$$P\{-1.21 < Z < 2.1\} \quad = \Phi(2.1) - \Phi(-1.21)$$
$$= .98214 - .11314 = .869;$$

$$P\{Z < -1.28\} \qquad = 1 - P\{Z < 1.28\} = 1 - \Phi(1.28)$$
$$= 1 - .89973 = .10027;$$

$$P\{-1.28 < Z < 2.36\} \quad = \Phi(2.36) - \Phi(-1.28)$$
$$= .99086 - .10027 = .89059.$$

Table 14.3.1a. Values of the $N(0,1)$ Distribution
Function: $\Phi(z) = \int_{-\infty}^{z} (\sqrt{2\pi})^{-1} e^{-x^2/2} dx$

z	.00	.01	.02	.03	.04
0.0	.50000	.50399	.50798	.51197	.51595
0.1	.53983	.54380	.54776	.55172	.55567
0.2	.57926	.58317	.58706	.59095	.59483
0.3	.61791	.62172	.62552	.62930	.63307
0.4	.65542	.65910	.66276	.66640	.67003
0.5	.69146	.69497	.69847	.70194	.70540
0.6	.72575	.72907	.73237	.73565	.73891
0.7	.75804	.76115	.76424	.76730	.77035
0.8	.78814	.79103	.79389	.79673	.79955
0.9	.81594	.81859	.82121	.82381	.82639
1.0	.84134	.84375	.84614	.84849	.85083
1.1	.86433	.86650	.86864	.87076	.87286
1.2	.88493	.88686	.88877	.89065	.89251
1.3	.90320	.90490	.90658	.90824	.90988
1.4	.91924	.92073	.92220	.92364	.92507
1.5	.93319	.93448	.93574	.93699	.93822
1.6	.94520	.94630	.94738	.94845	.94950
1.7	.95543	.95637	.95728	.95818	.95907
1.8	.96407	.96485	.96562	.96638	.96712
1.9	.97128	.97193	.97257	.97320	.97381
2.0	.97725	.97778	.97831	.97882	.97932
2.1	.98214	.98257	.98300	.98341	.98382
2.2	.98610	.98645	.98679	.98713	.98745
2.3	.98928	.98956	.98983	.99010	.99036
2.4	.99180	.99202	.99224	.99245	.99266
2.5	.99379	.99396	.99413	.99430	.99446
2.6	.99534	.99547	.99560	.99573	.99585
2.7	.99653	.99664	.99674	.99683	.99693
2.8	.99744	.99752	.99760	.99767	.99774
2.9	.99813	.99819	.99825	.99831	.99836
3.0	.99865	.99869	.99874	.99878	.99882
3.1	.99903	.99906	.99910	.99913	.99916
3.2	.99931	.99934	.99936	.99938	.99940
3.3	.99952	.99953	.99955	.99957	.99958
3.4	.99966	.99968	.99969	.99970	.99971
3.5	.99977	.99978	.99978	.99979	.99980

Table 14.3.1b. Values of the N(0,1) Distribution
Function: $\Phi(z) = \int_{-\infty}^{z}(\sqrt{2\pi})^{-1}e^{-x^2/2}dx$

z	.05	.06	.07	.08	.09
0.0	.51994	.52392	.52790	.53188	.53586
0.1	.55962	.56356	.56749	.57142	.57535
0.2	.59871	.60257	.60642	.61026	.61409
0.3	.63683	.64058	.64431	.64803	.65173
0.4	.67364	.67724	.68082	.68439	.68793
0.5	.70884	.71226	.71566	.71904	.72240
0.6	.74215	.74537	.74857	.75175	.75490
0.7	.77337	.77637	.77935	.78230	.78524
0.8	.80234	.80511	.80785	.81057	.81327
0.9	.82894	.83147	.83398	.83646	.83891
1.0	.85314	.85543	.85769	.85993	.86214
1.1	.87493	.87698	.87900	.88100	.88298
1.2	.89435	.89617	.89796	.89973	.90147
1.3	.91149	.91309	.91466	.91621	.91774
1.4	.92647	.92785	.92922	.93056	.93189
1.5	.93943	.94062	.94179	.94295	.94408
1.6	.95053	.95154	.95254	.95352	.95449
1.7	.95994	.96080	.96164	.96246	.96327
1.8	.96784	.96856	.96926	.96995	.97062
1.9	.97441	.97500	.97558	.97615	.97670
2.0	.97982	.98030	.98077	.98124	.98169
2.1	.98422	.98461	.98500	.98537	.98574
2.2	.98778	.98809	.98840	.98870	.98899
2.3	.99061	.99086	.99111	.99134	.99158
2.4	.99286	.99305	.99324	.99343	.99361
2.5	.99461	.99477	.99492	.99506	.99520
2.6	.99598	.99609	.99621	.99632	.99643
2.7	.99702	.99711	.99720	.99728	.99736
2.8	.99781	.99788	.99795	.99801	.99807
2.9	.99841	.99846	.99851	.99856	.99861
3.0	.99886	.99889	.99893	.99896	.99900
3.1	.99918	.99921	.99924	.99926	.99929
3.2	.99942	.99944	.99946	.99948	.99950
3.3	.99960	.99961	.99962	.99964	.99965
3.4	.99972	.99973	.99974	.99975	.99976
3.5	.99981	.99981	.99982	.99983	.99983

14.3.2   Percentage Points of the Chi-Square Distribution

The Table 14.3.2 provides the lower $100\gamma\%$ point $\chi^2_{\nu,1-\gamma}$ for the χ^2_ν distribution for different values of ν and γ. See the Figure 14.3.3.

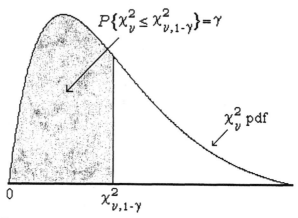

$$P\{\chi^2_\nu \le \chi^2_{\nu,1-\gamma}\} = \gamma$$

χ^2_ν pdf

0

$\chi^2_{\nu,1-\gamma}$

Figure 14.3.3. The Shaded Area Is the Probability γ

Suppose that X has the χ^2_ν distribution with some appropriate ν. Now, consider the following examples:

With $\nu = 3, P\{X \le .1149\}$ $\qquad = .01 \Rightarrow \chi^2_{3,.99} = .1149;$

With $\nu = 10, P\{X \le 18.307\}$ $\qquad = .95 \Rightarrow \chi^2_{10,.05} = 18.307;$

With $\nu = 15, P\{6.2621 \le X \le a\}$ $\quad = .925 \Rightarrow a = 24.996.$

Remark 14.3.1 The percentage points of the χ^2_1 distribution can be easily found using the standard normal Table 14.3.1. Observe that

$$\Phi\left(\sqrt{\chi^2_{1,1-\gamma}}\right) = \tfrac{1}{2}(1+\gamma).$$

For example, when $\gamma = .95$, one has $\Phi\left(\sqrt{\chi^2_{1,.05}}\right) = .975$. Thus, from the Table 14.3.1b we find that $\sqrt{\chi^2_{1,.05}} = 1.96$, that is $\chi^2_{1,.05} = (1.96)^2 = 3.8416$. In the Chi-square Table 14.3.2, we have $\chi^2_{1,.05} = 3.8415$.

Remark 14.3.2 The percentage points of the χ^2_2 distribution can be easily found using direct integrals of the exponential pdf. From (5.4.16), observe that

$$\chi^2_{2,1-\gamma} = 2log(1/(1-\gamma)).$$

For example, when $\gamma = .95$, one has $\chi^2_{2,.05} = 2log(1/.05) \approx 5.991465$. In the Chi-square Table 14.3.2, we have $\chi^2_{2,.05} = 5.9915$.

Table 14.3.2. Lower $100\gamma\%$ Point $\chi^2_{\nu,1-\gamma}$ for the χ^2_ν

Distribution: $P(\chi^2_\nu \leq \chi^2_{\nu,1-\gamma}) = \gamma$

ν	$\gamma = .01$	$\gamma = .025$	$\gamma = .05$	$\gamma = .95$	$\gamma = .975$	$\gamma = .99$
1	.00016	.00098	.00393	3.8415	5.0239	6.6349
2	.02010	.05064	.10259	5.9915	7.3778	9.2103
3	.11490	.21570	.35185	7.8147	9.3484	11.345
4	.29711	.48442	.71072	9.4877	11.143	13.277
5	.55430	.83121	1.1455	11.070	12.833	15.086
6	.87209	1.2373	1.6354	12.592	14.449	16.812
7	1.2390	1.5643	2.1673	14.067	16.013	18.475
8	1.6465	2.1797	2.7326	15.507	17.535	20.090
9	2.0879	2.7004	3.3251	16.919	19.023	21.666
10	2.5582	3.2470	3.9403	18.307	20.483	23.209
11	3.0535	3.8157	4.5748	19.675	21.920	24.725
12	3.5706	4.4038	5.2260	21.026	23.337	26.217
13	4.1069	5.0088	5.8919	22.362	24.736	27.688
14	4.6604	5.6287	6.5706	23.685	26.119	29.141
15	5.2293	6.2621	7.2609	24.996	27.488	30.578
16	5.8122	6.9077	7.9616	26.296	28.845	32.000
17	6.4078	7.5642	8.6718	27.587	30.191	33.409
18	7.0149	8.2307	9.3905	28.869	31.526	34.805
19	7.6327	8.9065	10.117	30.144	32.852	36.191
20	8.2604	9.5908	10.851	31.410	34.170	37.566
21	8.8972	10.283	11.591	32.671	35.479	38.932
22	9.5425	10.982	12.338	33.924	36.781	40.289
23	10.196	11.689	13.091	35.172	38.076	41.638
24	10.856	12.401	13.848	36.415	39.364	42.980
25	11.524	13.120	14.611	37.652	40.646	44.314
26	12.198	13.844	15.379	35.563	41.923	45.642
27	12.879	14.573	16.151	40.113	43.195	46.963
28	13.565	15.308	16.928	41.337	44.461	48.278
29	14.256	16.047	17.708	42.550	45.731	49.590
30	14.953	16.791	18.493	43.780	46.980	50.890

14.3.3   Percentage Points of the Student's t Distribution

The Table 14.3.3 provides the lower $100\gamma\%$ point $t_{\nu,1-\gamma}$ for the Student's t_{ν} distribution for different values of ν and γ. See the Figure 14.3.4.

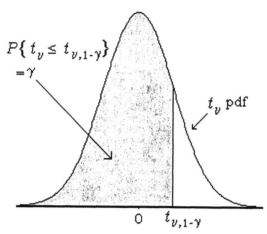

Figure 14.3.4. The Shaded Area Is the Probability γ

Suppose that X has the Student's t_{ν} distribution with some appropriate ν. Now, consider the following examples:

With $\nu = 4, P\{X \leq 2.1318\}$ $= .95 \Rightarrow t_{4,.05} = 2.1318;$

With $\nu = 10, P\{X \leq -2.7638\}$ $= .01 \Rightarrow t_{10,.99} = -2.7638;$

With $\nu = 15, P\{-1.7531 \leq t_{15} \leq a\}$ $= .94 \Rightarrow a = 2.6025.$

Remark 14.3.3 The percentage points of the t_1 distribution can be easily found by directly integrating its pdf which happens to be the same as the Cauchy pdf. Observe that

$$\gamma = \tfrac{1}{\pi} \int_{-\infty}^{t_{\nu,1-\gamma}} (1+x^2)^{-1} dx = \tfrac{1}{2} + \tfrac{1}{\pi} \arctan(t_{\nu,1-\gamma})$$

$$\Rightarrow t_{\nu,1-\gamma} = \tan\left(\pi(\gamma - \tfrac{1}{2})\right).$$

For example, when $\gamma = .95$, one has $t_{1,.05} = \tan\left(\pi(.95 - \tfrac{1}{2})\right) \approx 6.3138$. In the t Table 14.3.3, we have $t_{1,.05} = 6.314$.

Table 14.3.3. Lower $100\gamma\%$ Point $t_{\nu,1-\gamma}$ for the Student's t_ν
Distribution: $P(t_\nu \leq t_{\nu,1-\gamma}) = \gamma$

ν	$\gamma = .90$	$\gamma = .95$	$\gamma = .975$	$\gamma = .99$	$\gamma = .995$
1	3.0777	6.3140	12.706	31.821	63.657
2	1.8856	2.9200	4.3027	6.9646	9.9248
3	1.6377	2.3534	3.1824	4.5407	5.8409
4	1.5332	2.1318	2.7764	3.7469	4.6041
5	1.4759	2.0150	2.5706	3.3649	4.0321
6	1.4398	1.9432	2.4469	3.1427	3.7074
7	1.4149	1.8946	2.3646	2.9980	3.4995
8	1.3968	1.8595	2.3060	2.8965	3.3554
9	1.3830	1.8331	2.2622	2.8214	3.2498
10	1.3722	1.8125	2.2281	2.7638	3.1693
11	1.3634	1.7959	2.2010	2.7181	3.1058
12	1.3562	1.7823	2.1788	2.6810	3.0545
13	1.3502	1.7709	2.1604	2.6503	3.0123
14	1.3450	1.7613	2.1448	2.6245	2.9768
15	1.3406	1.7531	2.1314	2.6025	2.9467
16	1.3368	1.7459	2.1199	2.5835	2.9208
17	1.3334	1.7396	2.1098	2.5669	2.8982
18	1.3304	1.7341	2.1009	2.5524	2.8784
19	1.3277	1.7291	2.0930	2.5395	2.8609
20	1.3253	1.7247	2.0860	2.5280	2.8453
21	1.3232	1.7207	2.0796	2.5176	2.8314
22	1.3212	1.7171	2.0739	2.5083	2.8188
23	1.3195	1.7139	2.0687	2.4999	2.8073
24	1.3178	1.7109	2.0639	2.4922	2.7969
25	1.3163	1.7081	2.0595	2.4851	2.7874
26	1.3150	1.7056	2.0555	2.4786	2.7787
27	1.3137	1.7033	2.0518	2.4727	2.7707
28	1.3125	1.7011	2.0484	2.4671	2.7633
29	1.3114	1.6991	2.0452	2.4620	2.7564
30	1.3104	1.6973	2.0423	2.4573	2.7500
35	1.3062	1.6896	2.0301	2.4377	2.7238
40	1.3031	1.6839	2.0211	2.4233	2.7045
45	1.3006	1.6794	2.0141	2.4121	2.6896
50	1.2987	1.6759	2.0086	2.4033	2.6778
100	1.2901	1.6602	1.9840	2.3642	2.6259

14.3.4   Percentage Points of the F Distribution

The Table 14.3.4 provides the lower $100\gamma\%$ point $F_{\nu_1,\nu_2,1-\gamma}$ for the F_{ν_1,ν_2} distribution for different values of ν_1, ν_2 and γ. See the Figure 14.3.5.

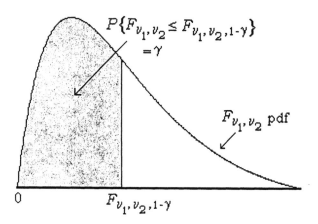

Figure 14.3.5. The Shaded Area Is the Probability γ

Suppose that X has the F_{ν_1,ν_2} distribution with some appropriate ν_1, ν_2. Now, consider the following examples:

With $\nu_1 = 5, \nu_2 = 8, P\{X \leq 3.6875\} \quad = .95 \Rightarrow F_{5,8,.05} = 3.6875;$

With $\nu_1 = 10, \nu_2 = 5, P\{X \leq 10.051\} \quad = .99 \Rightarrow F_{10,5,.01} = 10.051.$

In confidence interval and testing of hypothesis problems, one may need the values of $F_{\nu_1,\nu_2,1-\gamma}$ where γ is small. In such situations, one needs to recall the fact that if X has the F_{ν_1,ν_2} distribution, then $Y = 1/X$ has the F_{ν_2,ν_1} distribution. So, suppose that for some small value of γ, we wish to find the positive number b such that $P\{X \leq b\} = \delta$. But, observe that $P\{Y \geq 1/b\} = P\{X \leq b\} = \delta$, so that $P\{Y \leq 1/b\} = 1 - \delta$. Now, one can obtain the value of $a(= 1/b)$ from the Table 14.3.4 with $\gamma = 1 - \delta$ and *the degrees of freedom reversed*.

Look at the following example. We wish to determine the positive number b such that with $\nu_1 = 5, \nu_2 = 8, P\{X \leq b\} = .05$. In the Table 14.3.4, for $F_{8,5}$ we find the number 4.8183 which corresponds to $\gamma = .95 \; (= 1 - .05)$. That is, $b = \frac{1}{4.8183} \approx .20754$.

Remark 14.3.4 For the F distribution, computing its percentage points when the numerator degree of freedom is two can be fairly painless. We had actually shown a simple way to accomplish this in Chapter 5. One can apply

(5.4.17) with $\alpha = 1 - \gamma$ and verify that

$$F_{2,\nu_2,1-\gamma} = \tfrac{1}{2}\nu_2 \left\{ (1-\gamma)^{-2/\nu_2} - 1 \right\}.$$

For example, when $\gamma = .90, .95, .975, .99$ and $.995$, this formula provides the values $F_{2,2,1-\gamma} = 9.0, 19.0, 39.0, 99.0$ and 199.0 respectively which match with the corresponding entries in the Table 14.3.4. Also, for example, this formula provides the values $F_{2,5,.05} = 5.7863$ and $F_{2,8,.025} = 6.0596$ whereas the corresponding entries in the Table 14.3.4 are 5.7861 and 6.0595 respectively.

Table 14.3.4. Lower $100\gamma\%$ Point $F_{\nu_1,\nu_2,1-\gamma}$ for the F_{ν_1,ν_2} Distribution: $P(F_{\nu_1,\nu_2} \leq F_{\nu_1,\nu_2,1-\gamma}) = \gamma$ Where ν_1, ν_2 are Respectively the Numerator and Denominator Degrees of Freedom

ν_1	γ	ν_2						
		1	2	5	8	10	15	20
	.90	39.863	8.5263	4.0604	3.4579	3.2850	3.0732	2.9747
	.95	161.45	18.513	6.6079	5.3177	4.9646	4.5431	4.3512
1	.975	647.79	38.506	10.007	7.5709	6.9367	6.1995	6.1995
	.99	4052.2	98.503	16.258	11.259	10.044	8.6831	8.0960
	.995	16211	198.50	22.785	14.688	12.826	10.798	9.9439
	.90	49.500	9.0000	3.7797	3.1131	2.9245	2.6952	2.5893
	.95	199.50	19.000	5.7861	4.4590	4.1028	3.6823	3.4928
2	.975	799.50	39.000	8.4336	6.0595	5.4564	4.765	4.4613
	.99	4999.5	99.000	13.274	8.6491	7.5594	6.3589	5.8489
	.995	20000	199.00	18.314	11.042	9.4270	7.7008	6.9865
	.90	57.240	9.2926	3.453	2.7264	2.5216	2.273	2.1582
	.95	230.16	19.296	5.0503	3.6875	3.3258	2.9013	2.7109
5	.975	921.85	39.298	7.1464	4.8173	4.2361	3.5764	3.2891
	.99	5763.6	99.299	10.967	6.6318	5.6363	4.5556	4.1027
	.995	23056	199.30	14.940	8.3018	6.8724	5.3721	4.7616
	.90	59.439	9.3668	3.3393	2.5893	2.3772	2.1185	1.9985
	.95	238.88	19.371	4.8183	3.4381	3.0717	2.6408	2.4471
8	.975	956.66	39.373	6.7572	4.4333	3.8549	3.1987	2.9128
	.99	5981.1	99.374	10.289	6.0289	5.0567	4.0045	3.5644
	.995	23925	199.37	13.961	7.4959	6.1159	4.6744	4.0900
	.90	60.195	9.3916	3.2974	2.538	2.3226	2.0593	1.9367
	.95	241.88	19.396	4.7351	3.3472	2.9782	2.5437	2.3479
10	.975	968.63	39.398	6.6192	4.2951	3.7168	3.0602	2.7737
	.99	6055.8	99.399	10.051	5.8143	4.8491	3.8049	3.3682
	.995	24224	199.40	13.618	7.2106	5.8467	4.4235	3.8470
	.90	61.220	9.4247	3.2380	2.4642	2.2435	1.9722	1.8449
	.95	245.95	19.429	4.6188	3.2184	2.8450	2.4034	2.2033
15	.975	984.87	39.431	6.4277	4.1012	3.5217	2.8621	2.5731
	.99	6157.3	99.433	9.7222	5.5151	4.5581	3.5222	3.0880
	.995	24630	199.43	13.146	6.8143	5.4707	4.0698	3.5020
	.90	61.740	9.4413	3.2067	2.4246	2.2007	1.9243	1.7938
	.95	248.01	19.446	4.5581	3.1503	2.7740	2.3275	2.1242
20	.975	993.10	39.448	6.3286	3.9995	3.4185	2.7559	2.4645
	.99	6208.7	99.449	9.5526	5.3591	4.4054	3.3719	2.9377
	.995	24836	199.45	12.903	6.6082	5.274	3.8826	3.3178

References

Abramowitz, M. and Stegun, I.A. (1972). *Handbook of Mathematical Functions* (edited volume). Dover Publications, Inc., New York.

Aitken, A.C. and Silverstone, H. (1942). On the estimation of statistical parameters. *Proc. Roy. Soc. Edinburgh, Ser. A*, **61**, 186-194.

Aldrich, J. (1997). R. A. Fisher and the making of maximum likelihood 1912-1922. *Statist. Sci.*, **3**, 162-176.

Anderson, T.W. (1996). R. A. Fisher and multivariate analysis. *Statist. Sci.*, **11**, 20-34.

Anscombe, F.J. (1949). The statistical analysis of insect counts based on the negative binomial distribution. *Biometrika*, **5**, 165-173.

Armitage, J.V. and Krishnaiah, P.R. (1964). Tables for Studentized largest chi-square and their applications. *Report ARL 64-188,* Aerospace research Laboratories, Wright-Patterson Air Force Base, Ohio.

Armitage, P. (1973). *Sequential Medical Trials*, second edition. Blackwell Scientific Publications, Oxford.

Bahadur, R.R. (1954). Sufficiency and statistical decision functions. *Ann. Math. Statist.*, **25**, 423-462.

Bahadur, R.R. (1957). On unbiased estimates of uniformly minimum variance. *Sankhyā*, **18**, 211-224.

Bahadur, R.R. (1958). Examples of inconsistency of maximum likelihood estimates. *Sankh-yā*, **20**, 207-210.

Bahadur, R.R. (1971). *Some Limit Theorems in Statistics*. NSF-CBMS Monograph No. 4. Society for Industrial and Applied Mathematics, Philadelphia.

Bahadur, R.R. (1990). *Probability Statistics and Design of Experiments* (edited volume). Festschrift in Honor of the Late Professor R. C. Bose. Wiley Eastern Ltd., New Delhi.

Balakrishnan, N. and Basu, A.P. (1995). *The Exponential Distributions: Theory, Methods and Applications* (edited volume). Gordon and Breach, Amsterdam.

Banks, D.L. (1996). A conversation with I. J. Good. *Statist. Sci.*, **11**, 1-19.

Barankin, E.W. and Maitra, A. (1963). Generalization of the Fisher-Darmois-Koopman-Pitman theorem on sufficient statistics. *Sankhyā, Ser. A*, **25**, 217-244.

Barnard, G. (1992). Review of "Statistical Inference and Analysis: Selected Correspondence of R. A. Fisher (Edited by J.H. Bennett)". *Statist. Sci.*, **7**, 5-12.

Barndorff-Nielsen, G. (1978). *Information and Exponential Families in Statistical Theory*. John Wiley & Sons, Inc., New York.

Bartlett, M.S. (1981). Egon Sharpe Pearson, 1895-1980. *Biometrika*, **68**, 1-12.

Basu, A.P. (1991). Sequential methods in reliability and life testing. *Handbook of Sequential Analysis* (B.K. Ghosh and P.K. Sen, eds.), *Chapter 25*, pp. 581-592. Marcel Dekker, Inc., New York.

Basu, D. (1955a). On statistics independent of a complete sufficient statistic. *Sankhyā*, **15**, 377-380.

Basu, D. (1955b). An inconsistency of the method of maximum likelihood. *Ann. Math. Statist.*, **26**, 144-145.

Basu, D. (1958). On statistics independent of sufficient statistics. *Sankhyā*, **20**, 223-226.

Basu, D. (1964). Recovery of ancillary information. *Contributions to Statistics, the 70th Birthday Festschrift Volume Presented to P. C. Mahalanobis.* Pergamon Press, Oxford.

Bather, J. (1996). A conversation with Herman Chernoff. *Statist. Sci.*, **11**, 335-350.

Bayes, T. (1783). An essay towards solving a problem in the doctrine of chances. *Phil. Trans. Roy. Soc.*, **53**, 370-418.

Bechhofer, R.E., Kiefer, J. and Sobel, M. (1968). *Sequential Identification and Ranking Procedures.* University of Chicago Press, Chicago.

Behrens, W.V. (1929). Ein betrag zur fehlerberechnung bei weniger beobachtungen. *Landwirtschaftliche Jahrbücher*, **68**, 807-837.

Bellhouse, D.R. and Genest, C. (1999). A history of the Statistical Society of Canada: The formative years. *Statist. Sci.*, **14**, 80-125.

Bennett, J.H. (1971-1974). *Collected Works of R. A. Fisher, Volumes 1-5* (edited volumes). University of Adelaide, South Australia.

Bennett, J.H. (1990). *Statistical Inference and Analysis: Selected Correspondence of R. A. Fisher* (edited volume). Oxford University Press, Oxford.

Beran, R.J. and Fisher, N.I. (1998). A conversation with Geoff Watson. *Statist. Sci.*, **13**, 75-93.

Berger, J.O. (1985). *Statistical Decision Theory and Bayesian Analysis*, second edition. Springer-Verlag, Inc., New York.

Berger, J.O., Boukai, B. and Wang, Y. (1997). Unified frequentist and Bayesian testing of a precise hypothesis (with discussions by D.V. Lindley, T.A. Louis and D. V. Hinkley). *Statist. Sci.*, **12**, 133-160.

Bingham, N.H. (1996). A conversation with David Kendall. *Statist. Sci.*, **11**, 159-188.

Blackwell, D. (1947). Conditional expectation and unbiased sequential estimation. *Ann. Math. Statist.*, **18**, 105-110.

Blackwell, D. and Girshick, M.A. (1954). *Theory of Games and Statistical*

Decisions. John Wiley & Sons, Inc., New York. Reprinted (1979) by the Dover Publications, Inc., New York.

Box, J.F. (1978). *R. A. Fisher, The Life of a Scientist*. John Wiley & Sons, Inc., New York.

Box, J.F. (1987). Guinness, Gosset, Fisher, and small samples. *Statist. Sci.*, **2**, 45-52.

Brown, L.D. (1964). Sufficient statistics in the case of independent random variables. *Ann. Math. Statist.*, **35**, 1456-1474.

Buehler, R. (1980). Fiducial inference. *R. A. Fisher: An Appreciation* (S.E. Fienberg and D.V. Hinkley, eds.), pp. 109-118. Springer-Verlag, Inc., New York.

Chapman, D.G. (1950). Some two-sample tests. *Ann. Math. Statist.*, **21**, 601-606.

Chatterjee, S.K. (1991). Two-stage and multistage procedures. *Handbook of Sequential Analysis* (B.K. Ghosh and P.K. Sen, eds.), *Chapter 2*, pp. 21-45. Marcel Dekker, Inc., New York.

Chen, T.T. and Tai, J.J. (1998). A conversation with C. C. Li. *Statist. Sci.*, **13**, 378-387.

Chung, K.L. (1974). *A Course in Probability Theory*. Academic Press, New York.

Cornish, E.A. (1954). The multivariate *t*-distribution associated with a set of normal sample deviates. *Austr. J. Physics*, **7**, 531-542.

Cornish, E.A. (1962). The multivariate *t*-distribution associated with the general multivariate normal distribution. *Technical Paper No. 13*, Div. Math. Statist., CSIRO, Australia.

Cox, D.R. (1952). Estimation by double sampling. *Biometrika*, **39**, 217-227.

Craig, C.C. (1986). Early days in statistics at Michigan. *Statist. Sci.*, **1**, 292-293.

Cramér, H. (1942). On harmonic analysis in certain functional spaces. *Arkiv Mat. Astron. Fysik*, **28B**. Reprinted in *Breakthroughs in Statistics Volume I* (S. Kotz and N. L. Johnson, eds.), 1992. Springer-Verlag, Inc., New York.

Cramér, H. (1946a). *Mathematical Methods of Statistics*. Princeton University Press, Princeton.

Cramér, H. (1946b). A contribution to the theory of statistical estimation. *Skand. Akt. Tidskr.*, **29**, 85-94.

Cramér, H. (1976). Half a century with probability theory: Some personal recollections. *Ann. Probab.*, **4**, 509-546.

Creasy, M.A. (1954). Limits for the ratios of means. *J. Roy. Statist. Soc., Ser. B*, **16**, 186-194.

Cressie, N., Davis, A.S., Folks, J.L., and Policello II, G.E. (1981). The moment-generating function and negative integer moments. *Amer. Statist.*, **35**, 148-150.

Daniel, C. and Lehmann, E.L. (1979). Henry Scheffé 1907-1977. *Ann. Statist.*, **7**, 1149-1161.

Dantzig, G.B. (1940). On the non-existence of tests of Student's hypothesis having power functions independent of σ. *Ann. Math. Statist.*, **11**, 186-192.

Darmois, G. (1945). Sur les lois limites de la dispersion de certaines estimations. *Rev. Int. Statist. Inst.*, **13**, 9-15.

DasGupta, S. (1980). Distributions of the correlation coefficient. *R. A. Fisher: An Appreciation* (S.E. Fienberg and D.V. Hinkley, eds.), pp. 9-16. Springer-Verlag, Inc., New York.

DasGupta, S. and Perlman, M. (1974). Power of the noncentral F-test: Effect of additional variates on Hotelling's T^2-test. *J. Amer. Statist. Assoc.*, **69**, 174-180.

David, F.N. (1968). Karl Pearson. *International Encyclopedia of Statistics* (W.H. Kruskal and J.M. Tanur eds.), **1**, pp. 653-655. Collier Macmillan Publishing Co., Inc., London. Reprinted (1978) by the Free Press, New York.

David, H.A. (1998). Statistics in U.S. universities in 1933 and the establishment of the statistical laboratory at Iowa State. *Statist. Sci.*, **13**, 66-74.

de Finetti, B. (1937). Foresight: Its logical laws, its subjective sources. Translated and reprinted, *Studies in Subjective Probability* (H. Kyberg and H. Smokler, eds.), 1964, pp. 93-158. John Wiley & Sons, Inc., New York. Reprinted in *Breakthroughs in Statistics Volume I* (S. Kotz and N. L. Johnson, eds.), 1992. Springer-Verlag, Inc., New York.

de Finetti, B. (1972). *Probability, Induction, and Statistics.* John Wiley & Sons, Inc., New York.

de Finetti, B. (1974). *Theory of Probability, Volumes 1 and 2.* John Wiley & Sons, Inc., New York.

DeGroot, M.H. (1986a). A conversation with David Blackwell. *Statist. Sci.*, **1**, 40-53.

DeGroot, M.H. (1986b). A conversation with T. W. Anderson. *Statist. Sci.*, **1**, 97-105.

DeGroot, M.H. (1986c). A conversation with Erich L. Lehmann. *Statist. Sci.*, **1**, 243-258.

DeGroot, M.H. (1986d). A conversation with Charles Stein. *Statist. Sci.*, **1**, 454-462.

DeGroot, M.H. (1987). A conversation with C. R. Rao. *Statist. Sci.*, **2**, 53-67.

DeGroot, M.H. (1988). A conversation with George A. Barnard. *Statist. Sci.*, **3**, 196-212.

Dunnett, C.W. (1955). A multiple comparison procedure for comparing several treatments with a control. *J. Amer. Statist. Assoc.*, **50**, 1096-1121.

Dunnett, C.W. and Sobel, M. (1954). A bivariate generalization of Student's *t*-distribution, with tables for certain special cases. *Biometrika*, **41**, 153-169.

Dunnett, C.W. and Sobel, M. (1955). Approximations to the probability integral and certain percentage points of a multivariate analogue of Student's *t* distribution. *Biometrika*, **42**, 258-260.

Edwards, A.W.F. (1997a). Three early papers on efficient estimation. *Statist. Sci.*, **12**, 35-47.

Edwards, A.W.F. (1997b). What did Fisher mean by "Inverse Probability" in 1912-1922? *Statist. Sci.*, **12**, 177-184.

Efron, B.F (1975). Defining the curvature of a statistical problem (with applications to second order efficiency). *Ann. Statist.*, **3**, 1189-1242.

Efron, B.F. (1978). The geometry of exponential families. *Ann. Statist.*, **6**, 367-376.

Efron, B.F. (1998). R. A. Fisher in the 21st century (with discussions by D.R. Cox, R. Kass, O. Barndorff-Nielsen, D.V. Hinkley, D.A.S. Fraser and A.P. Dempster). *Statist. Sci.*, **13**, 95-122.

Feller, W. (1968). *An Introduction to Probability and Its Applications, Volume 1*, third edition. John Wiley & Sons, Inc., New York.

Feller, W. (1971). *An Introduction to Probability and Its Applications, Volume 2*, second edition. John Wiley & Sons, Inc., New York.

Ferguson, T.S. (1967). *Mathematical Statistics*. Academic Press, Inc., New York.

Fieller, E.C. (1954). Some problems in interval estimation. *J. Roy. Statist. Soc., Ser. B*, **16**, 175-185.

Fienberg, S.E. (1992). A brief history of statistics in three and one-half chapters: A review essay. *Statist. Sci.*, **7**, 208-225.

Fienberg, S.E. (1997). Introduction to R. A. Fisher on inverse probability and likelihood. *Statist. Sci.*, **12**, 161.

Fienberg, S.E. and Hinkley, D.V. (1980). *R. A. Fisher: An Appreciation* (edited volume). Lecture Notes in Statistics No. 1. Springer-Verlag, Inc., New York.

Findley, D.F. and Parzen, E. (1995). A conversation with Hirottugu Akaike. *Statist. Sci.*, **10**, 104-117.

Finney, D.J. (1941). The distribution of the ratio of estimates of the two variances in a sample from a normal bivariate population. *Biometrika*, **30**, 190-192.

Fisher, R.A. (1912). On an absolute criterion for fitting frequency curves. *Messeng. Math.*, **42**, 155-160.

Fisher, R.A. (1915). Frequency distribution of the values of the correlation coefficients in samples from an indefinitely large population. *Biometrika*, **10**, 507-521.

Fisher, R.A. (1920). A mathematical examination of the methods of determining the accuracy of an observation by the mean error, and by the mean square error. *Monthly Notices Roy. Astronom. Soc.*, **80**, 758-770.

Fisher, R.A. (1921). On the "probable error" of a coefficient of correlation deduced from a small sample. *Metron*, **1**, 3-32.

Fisher, R.A. (1922). On the mathematical foundations of theoretical statistics. *Phil. Trans. Roy. Soc.*, **A222**, 309-368. Reprinted in *Breakthroughs in Statistics Volume I* (S. Kotz and N. L. Johnson, eds.), 1992. Springer-Verlag, New York.

Fisher, R.A. (1925a). Theory of statistical estimation. *Proc. Camb. Phil. Soc.*, **22**, 700-725.

Fisher, R.A. (1925b). *Statistical Methods for Research Workers*. Oliver and Boyd, Edinburgh. Reprinted (1973) by Hafner, New York. Reprinted in *Breakthroughs in Statistics Volume II* (S. Kotz and N. L. Johnson, eds.), 1993. Springer-Verlag, Inc., New York.

Fisher, R.A. (1926). The arrangement of field experiments. *J. Min. Agric. G. Br.*, **33**, 503-515. Reprinted in *Breakthroughs in Statistics Volume II* (S. Kotz and N. L. Johnson, eds.), 1993. Springer-Verlag, Inc., New York.

Fisher, R.A. (1928). Moments and product moments of sampling distributions. *Proc. London Math. Soc.*, **30**, 199-238.

Fisher, R.A. (1930). Inverse probability. *Proc. Camb. Phil. Soc.*, **26**, 528-535.

Fisher, R.A. (1934). Two new properties of mathematical likelihood. *Proc. Roy. Soc., Ser. A*, **144**, 285-307.

Fisher, R.A. (1935). The fiducial argument in statistical inference. *Ann. Eugenics*, **6**, 391-398.

Fisher, R.A. (1939). The comparison of samples with possibly unequal variances. *Ann. Eugenics*, **9**, 174-180.

Fisher, R.A. (1956). *Statistical Methods and Scientific Inference*. Oliver and Boyd, Edinburgh and London.

Folks, J.L. (1981). *Ideas of Statistics*. John Wiley & Sons, Inc., New York.

Folks, J.L. (1995). A conversation with Oscar Kempthorne. *Statist. Sci.*, **10**, 321-336.

Frankel, M. and King, B. (1996). A conversation with Leslie Kish. *Statist. Sci.*, **11**, 65-87.

Fréchet, M. (1943). Sur l'extension de certaines evaluations statistiques de petits echantillons. *Rev. Int. Inst. Statist.*, **11**, 182-205.

Gani, J. (1982). *The Making of Statisticians* (edited volume). Springer-Verlag, Inc., New York.

Gardiner, J.C. and Susarla, V. (1991). Time-sequential estimation. *Hand-*

book of Sequential Analysis (B.K. Ghosh and P.K. Sen, eds.), *Chapter 27*, pp. 613-631. Marcel Dekker, Inc., New York.

Gardiner, J.C., Susarla, V. and van Ryzin, J. (1986). Time-sequential estimation of the exponential mean under random withdrawals. *Ann. Statist.*, **14**, 607-618.

Gauss, C.F. (1821). *Theoria combinationis observationum erroribus minimis obnoxiae.* Its English translation is available in Gauss's publications (1803-1826).

Gayen, A.K. (1951). The frequency distribution of the product moment correlation coefficient in random samples of any size drawn from non-normal universes. *Biometrika*, **38**, 219-247.

Ghosh, B.K. (1970). *Sequential Tests of Statistical Hypotheses.* Addison-Wesley, Reading.

Ghosh, B.K. (1973). Some monotonicity theorems for χ^2, F and t distributions with applications. *J. Roy. Statist. Soc., Ser. B*, **35**, 480-492.

Ghosh, B.K. (1975). On the distribution of the difference of two t-variables. *J. Amer. Statist. Assoc.*, **70**, 463-467.

Ghosh, B.K. and Sen, P.K. (1991). *Handbook of Sequential Analysis* (edited volume). Marcel Dekker, Inc., New York.

Ghosh, J.K. (1988). *Statistical Information and Likelihood: A Collection of Critical Essays by Dr. D. Basu* (edited volume). Lecture Notes in Statistics No. 45. Springer-Verlag, Inc., New York.

Ghosh, J.K., Maiti, P., Rao, T.J., and Sinha, B.K. (1999). Evolution of statistics in India. *International Statist. Rev.*, **67**, 13-34.

Ghosh, J.K., Mitra, S.K., and Parthasarathy, K.R. (1992). *Glimpses of India's Statistical Heritage.* Wiley Eastern, Ltd., New Delhi.

Ghosh, M. and Mukhopadhyay, N. (1976). On two fundamental problems of sequential estimation. *Sankhyā, Ser. A*, **38**, 203-218.

Ghosh, M. and Mukhopadhyay, N. (1981). Consistency and asymptotic efficiency of two-stage and sequential estimation procedures. *Sankhyā, Ser. A*, **43**, 220-227.

Ghosh, M., Mukhopadhyay, N. and Sen, P.K. (1997). *Sequential Estimation.* John Wiley & Sons, Inc., New York.

Ghosh M. and Pathak, P.K. (1992). *Current Issues in Statistical Inference: Essays in Honor of D. Basu* (edited volume). Institute of Mathematical Statistics Lecture Notes-Monograph Series, U.S.A.

Ghurye, S.G. (1958). Note on sufficient statistics and two-stage procedures. *Ann. Math. Statist.*, **29**, 155-166.

Gleser, L.J. and Healy, J.D. (1976). Estimating the mean of a normal distribution with known coefficient of variation. *J. Amer. Statist. Assoc.*, **71**, 977-981.

Grams, W.F. and Serfling, R.J. (1973). Convergence rates for U-statistics and related statistics. *Ann. Statist.*, **1**, 153-160.

Haldane, J.B.S. (1956). The estimation and significance of the logarithm of a ratio of frequencies. *Ann. Human Genetics*, **20**, 309-311.

Halmos, P.R. (1946). The theory of unbiased estimation. *Ann. Math. Statist.*, **17**, 34-43.

Halmos, P.R. (1985). *I want to Be a Mathematician: An Automathography*. Springer-Verlag, Inc., New York.

Halmos, P.R. and Savage, L.J. (1949). Application of the Radon-Nikodym theorem to the theory of sufficient statistics. *Ann. Math. Statist.*, **20**, 225-241.

Hansen, M.H. (1987). Some history and reminiscences on survey sampling. *Statist. Sci.*, **2**, 180-190.

Hewett, J. and Bulgren, W.G. (1971). Inequalities for some multivariate f-distributions with applications. *Technometrics*, **13**, 397-402.

Heyde, C. (1995). A conversation with Joe Gani. *Statist. Sci.*, **10**, 214-230.

Hinkley, D.V. (1980a). Fisher's development of conditional inference. *R. A. Fisher: An Appreciation* (S.E. Fienberg and D.V. Hinkley, eds.), pp. 101-108. Springer-Verlag, Inc., New York.

Hinkley, D.V. (1980b). R. A. Fisher: Some introductory remarks. *R. A. Fisher: An Appreciation* (S.E. Fienberg and D.V. Hinkley, eds.), pp. 1-5. Springer-Verlag, Inc., New York.

Hipp, C. (1974). Sufficient statistics and exponential families. *Ann. Statist.*, **2**, 1283-1292.

Hochberg, Y. and Tamhane, A.C. (1987). *Multiple Comparison Procedures*. John Wiley & Sons, Inc., New York.

Hoeffding, W. (1948). A class of statistics with asymptotically normal distributions. *Ann. Math. Statist.*, **19**, 293-325.

Hogg, R.V. (1986). On the origins of the Institute of Mathematical Statistics. *Statist. Sci.*, **10**, 285-291.

Hollander, M. and Marshall, A.W. (1995). A conversation with Frank Proschan. *Statist. Sci.*, **1**, 118-133.

Irwin, J.O. (1968). William Sealy Gosset. *International Encyclopedia of Statistics* (W.H. Kruskal and J.M. Tanur eds.), **1**, pp. 409-413. Collier Macmillan Publishing Co., Inc., London. Reprinted (1978) by the Free Press, New York.

James, W. and Stein, C. (1961). Estimation with quadratic loss. *Proc. Fourth Berkeley Symp. Math. Statist. Probab.*, **1**, 361-379. University of California Press, Berkeley.

Jeffreys, H. (1957). *Scientific Inference*. Cambridge University Press, London.

Johnson, N.L. and Kotz, S. (1969). *Distributions in Statistics: Discrete Distributions*. John Wiley & Sons, Inc., New York.

Johnson, N.L. and Kotz, S. (1970). *Distributions in Statistics: Continuous Univariate Distributions-2.* John Wiley & Sons, Inc., New York.

Johnson, N.L. and Kotz, S. (1972). *Distributions in Statistics: Continuous Multivariate Distributions.* John Wiley & Sons, Inc., New York.

Johnson, N.L. and Kotz, S. (1992). *Breakthroughs in Statistics Volume I* (edited volume). Springer-Verlag, Inc., New York.

Johnson, N.L. and Kotz, S. (1993). *Breakthroughs in Statistics Volume II* (edited volume). Springer-Verlag, Inc., New York.

Kallianpur, G. and Rao, C.R. (1955). On Fisher's lower bound to asymptotic variance of a consistent estimate. *Sankhyā*, **15**, 331-342.

Karlin, S. (1992). R. A. Fisher and evolutionary theory. *Statist. Sci.*, **7**, 5-12.

Karlin, S. and Rubin, H. (1956). The theory of decision procedures for distributions with monotone likelihood ratio. *Ann. Math. Statist.*, **27**, 272-299.

Kendall, D.G. (1983). A tribute to Harald Cramér. *J. Roy. Statist. Soc., Ser. A*, **146**, 211-212.

Kendall, D.G. (1991). Kolmogorov as I remember him. *Statist. Sci.*, **6**, 303-312.

Kendall, M.G. (1963). Ronald Aylmer Fisher, 1890-1962. *Biometrika*, **50**, 1-16.

Kendall, M.G. and Stuart, A. (1979). *The Advanced Theory of Statistics, Volume II: Inference and Relationship*, fourth edition. Macmillan, New York.

Khan, R.A. (1968). A note on estimating the mean of a normal distribution with known coefficient of variation. *J. Amer. Statist. Assoc.*, **63**, 1039-1041.

Kimbal, A.W. (1951). On dependent tests of significance in the analysis of variance. *Ann. Math. Statist.*, **22**, 600-602.

Kolmogorov, A.N. (1933). On the empirical determination of a distribution. *G. 1st. Ital. Attuari*, **4**, 83-91. Reprinted in *Breakthroughs in Statistics Volume II* (S. Kotz and N. L. Johnson, eds.), 1993. Springer-Verlag, Inc., New York.

Kolmogorov, A.N. (1950a). Unbiased estimates (in Russian). *Izvestia Acad. Nauk. USSR*, **14**, 303-326. (Amer. Math. Soc. Translations No. 90).

Kolmogorov, A.N. (1950b). *Foundations of the Theory of Probability* (German edition, 1933). Chelsea, New York.

Krishnaiah, P.R. and Armitage, J.V. (1966). Tables for multivariate *t*-distribution. *Sankhyā, Ser. B*, **28**, 31-56.

Laird, N.M. (1989). A conversation with F. N. David. *Statist. Sci.*, **4**, 235-246.

Lane, D.A. (1980). Fisher, Jeffreys, and the nature of inference. *R. A.*

Fisher: An Appreciation (S.E. Fienberg and D.V. Hinkley, eds.), pp. 148-160. Springer-Verlag, Inc., New York.

LeCam, L. (1953). On some asymptotic properties of maximum likelihood estimates and related Bayes's estimates. *Univ. California Publ. Statist.*, **1**, 277-330.

LeCam, L. (1956). On the asymptotic theory of estimation and testing hypotheses. *Proc. Third Berkeley Symp. Math. Statist. Probab.*, **1**, 129-156. University of California Press, Berkeley.

LeCam, L. (1986a). *Asymptotic Methods in Statistical Decision Theory.* Springer-Verlag, Inc., New York.

LeCam, L. (1986b). The central limit theorem around 1935 (with discussions by H.F. Trotter, J.L. Doob and D. Pollard). *Statist. Sci.*, **1**, 78-96.

LeCam, L. and Lehmann, E.L. (1974). J Neyman: On the occasion of his 80th birthday. *Ann. Statist.*, **2**, vii-xiii.

LeCam, L. and Yang, G.C. (1990). *Asymptotics in Statistics: Some Basic Concepts.* Springer-Verlag, Inc., New York.

Lehmann, E.L. (1951). *Notes on the Theory of Estimation.* University of California Press, Berkeley.

Lehmann, E.L. (1983). *Theory of Point Estimation.* John Wiley & Sons, Inc., New York.

Lehmann, E.L. (1986). *Testing Statistical Hypotheses*, second edition. John Wiley & Sons, Inc., New York.

Lehmann, E.L. (1993). Mentors and early collaborators: Reminiscences from the years 1940-1956 with an epilogue. *Statist. Sci.*, **8**, 331-341.

Lehmann, E.L. (1997). Testing Statistical Hypotheses: The story of a book. *Statist. Sci.*, **12**, 48-52.

Lehmann, E.L. and Casella, G. (1998). *Theory of Point Estimation*, second edition. Springer-Verlag, Inc., New York.

Lehmann, E.L. and Scheffé, H. (1950). Completeness, similar regions and unbiased estimation-Part I. *Sankhyā*, **10**, 305-340.

Lehmann, E.L. and Scheffé, H. (1955). Completeness, similar regions and unbiased estimation-Part II. *Sankhyā*, **15**, 219-236.

Lehmann, E.L. and Scheffé, H. (1956). Corrigenda: Completeness, similar regions and unbiased estimation-Part I. *Sankhyā*, **17**, 250.

Liaison (1993). A conversation with Charles W. Dunnett. *Multiple Comparisons, Selection, and Applications in Biometry, A Festschrift in Honor of Charles W. Dunnett* (F. Hoppe, ed.), pp. 1-10. Marcel Dekker, Inc., New York.

Lindley, D.V. (1980). L. J. Savage - His work in probability and statistics. *Ann. Statist.*, **8**, 1-24.

Lindley, D.V. and Scott, W.F. (1995). *New Cambridge Statistical Tables*, second edition. Cambridge University Press, Cambridge.

Lukacs, E. (1960). *Characteristic Functions.* Charles Griffin & Co., Ltd.,

London.

MacNeill, I. (1993). A conversation with David J. Finney. *Statist. Sci.*, **8**, 187-201.

Mahalanobis, P.C. (1932). Auxiliary tables for Fisher's Z-test in analysis of variance. (Statistical notes for agricultural workers, No. 3). *Indian J. Agric. Sci.*, **2**, 679-693.

Mahalanobis, P.C. (1940). A sample survey of acreage under jute in Bengal, with discussion on planning of experiments. *Proc. Second Indian Statist. Conference.* Statistical Publishing Society, Calcutta.

Mauromaustakos, A. (1984). A three-stage estimation for the negative exponential. *Masters thesis*, Department of Statist., Oklahoma State Univ., Stillwater.

McDonald, G.C. (1998). A conversation with Shanti Gupta. *Statist. Sci.*, **13**, 291-305.

Menger, K. (1952). The formative years of Abraham wald and his work in geometry. *Ann. Math. Statist.*, **32**, 14-20.

Moshman, J. (1958). A method for selecting the size of the initial sample in Stein's two-sample procedure. *Ann. Math. Statist.*, **29**, 667-671.

Mukhopadhyay, N. (1980). A consistent and asymptotically efficient two-stage procedure to construct fixed-width confidence interval for the mean. *Metrika*, **27**, 281-284.

Mukhopadhyay, N. (1991). Sequential point estimation. *Handbook of Sequential Analysis* (B.K. Ghosh and P.K. Sen, eds.), *Chapter 10*, pp. 245-267. Marcel Dekker, Inc., New York.

Mukhopadhyay, N. (1997). A conversation with Sujit Kumar Mitra. *Statist. Sci.*, **12**, 61-75.

Mukhopadhyay, N. and Duggan, W.T. (1997). Can a two-stage procedure enjoy second-order properties? *Sankhyā, Ser. A*, **59**, 435-448.

Mukhopadhyay, N. and Duggan, W.T. (1999). On a two-stage procedure having second-order properties with applications. *Ann. Inst. Statist. Math*, **51**, in press.

Mukhopadhyay, N. and Hamdy, H.I. (1984). On estimating the difference of location parameters of two negative exponential distributions. *Canad. J. Statist.*, **12**, 67-76.

Mukhopadhyay, N. and Mauromoustakos, A. (1987). Three-stage estimation for the negative exponential distributions. *Metrika*, **34**, 83-93.

Mukhopadhyay, N. and Solanky, T.K.S. (1994). *Multistage Selection and Ranking Procedures.* Marcel Dekker, Inc., New York.

Neyman, J. (1934). On two different aspects of the representative method. *J. Roy. Statist. Soc.*, **97**, 558-625. Reprinted in *Breakthroughs in Statistics Volume II* (S. Kotz and N. L. Johnson, eds.), 1993. Springer-Verlag, Inc., New York.

Neyman, J. (1935a). Sur un teorema concernente le cosidette statistiche sufficienti. *Giorn. 1st. Ital. Att.*, **6**, 320-334.

Neyman, J. (1935b). On the problem of confidence intervals. *Ann. Math. Statist.*, **6**, 111-116.

Neyman, J. (1937). Outline of a theory of statistical estimation based on the classical theory of probability. *Phil. Trans. Roy. Soc., Ser. A*, **236**, 333-380.

Neyman, J. (1949). Contribution to the theory of the χ^2 test. *Proc. First Berkeley Symp. Math. Statist. Probab.*, **1**, 239-273. University of California Press, Berkeley.

Neyman, J. (1961). Silver jubilee of my dispute with Fisher. *J. Oper. Res. Soc. (Japan)*, **3**, 145-154.

Neyman, J. (1967). R. A. Fisher: An appreciation. *Science*, **156**, 1456-1460.

Neyman, J. (1981). Egon S. Pearson (August 11, 1895-June 12, 1980): An appreciation. *Ann. Statist.*, **9**, 1-2.

Neyman, J. and Pearson, E.S. (1928a). On the use and interpretation of certain test criteria for purposes of statistical inference, Part I. *Biometrika*, **20A**, 175–240.

Neyman, J. and Pearson, E.S. (1928b). On the use and interpretation of certain test criteria for purposes of statistical inference, Part II. *Biometrika*, **20A**, 263-294.

Neyman, J. and Pearson, E.S. (1933a). On the problem of the most efficient tests of statistical hypotheses. *Phil. Trans. Roy. Soc., Ser. A*, **231**, 289-337. Reprinted in *Breakthroughs in Statistics Volume I* (S. Kotz and N. L. Johnson, eds.), 1992. Springer-Verlag, Inc., New York.

Neyman, J. and Pearson, E.S. (1933b). The testing of statistical hypotheses in relation to probabilities *a priori*. *Proc. Camb. Phil. Soc.*, **24**, 492-510.

Neyman, J. and Scott, E.L. (1948). Consistent estimates based on partially consistent observations. *Econometrica*, **16**, 1-32.

Olkin, I. (1987). A conversation with Morris Hansen. *Statist. Sci.*, **2**, 162-179.

Olkin, I. (1989). A conversation with Maurice Bartlett. *Statist. Sci.*, **4**, 151-163.

Page, W. (1989). An interview of Herbert Robbins. *Herbert Robbins: Selected Papers* (T.L. Lai and D. Siegmund, eds.), pp. xix-xli. Springer-Verlag, Inc., New York. Reprinted from the *College Mathematics Journal* (1984), **15**.

Pearson, E.S. (1938). *Karl Pearson: An Appreciation of Some Aspects of His Life and Work*. Cambridge University Press.

Pearson, E.S. (1966). The Neyman-Pearson story: 1926-1934. *Research Papers in Statistics: Festschrift for J. Neyman* (F.N. David, ed.). John

Wiley & Sons, Inc., New York.

Pearson, E.S. and Kendall, M.G. (1970). *Studies in the History of Statistics and Probability* (edited volume). Charles Griffin & Co., Ltd., London.

Pearson, K. (1900). On the criterion that a given system of deviations from the probable in the case of a correlated system of variables is such that it can be reasonably supposed to have arisen from random sampling. *Phil. Mag.*, **5:50**, 157-172. Reprinted in *Breakthroughs in Statistics Volume II* (S. Kotz and N. L. Johnson, eds.), 1993. Springer-Verlag, Inc., New York.

Pearson, K. (1902). On the systematic fitting of curves to observations and measurements - part I. *Biometrika*, **1**, 265-303.

Pearson, K. (1903). Mathematical contributions to the theory of evolution XII: On a generalized theory of alternative inheritance, with a special reference to Mendel's Laws. *Phil. Trans.*, **203A**, 53-87.

Petersen, C.G.L. (1896). The yearly immigration of young plaice into the Limfjord from the German Sea. *Rep. Danish Bio. Statist.*, **6**, 1-48.

Press, S.J. (1989). A conversation with Ingram Olkin. *Contributions to Probability and Statistics, Essays in Honor of Ingram Olkin* (L.J. Gleser, M.D. Perlman, S.J. Press, and A.R. Sampson, eds.), pp. 7-33. Springer-Verlag, Inc., New York.

Råde, L. (1997). A conversation with Harald Bergström. *Statist. Sci.*, **12**, 53-60.

Ramachandran, B.R. (1967). *Advanced Theory of Characteristic Functions*. Statistical Publishing Society, Calcutta.

Rao, C.R. (1945). Information and accuracy attainable in the estimation of statistical parameters. *Bull. Calcutta Math. Soc.*, **37**, 81-91. Reprinted in *Breakthroughs in Statistics Volume I* (S. Kotz and N. L. Johnson, eds.), 1992. Springer-Verlag, Inc., New York.

Rao, C.R. (1947). Minimum variance and the estimation of several parameters. *Proc. Camb. Phil. Soc.*, **43**, 280-283.

Rao, C.R. (1968). P.C. Mahalanobis. *International Encyclopedia of Statistics* (W.H. Kruskal and J.M. Tanur, eds.), **1**, pp. 571-576. Collier Macmillan Publishing Co., Inc., London. Reprinted (1978) by the Free Press, New York.

Rao, C.R. (1973). *Linear Statistical Inference and Its Applications*, second edition. John Wiley & Sons, Inc., New York.

Rao, C.R. (1992a). R. A. Fisher: The founder of modern statistics. *Statist. Sci.*, **7**, 5-12.

Rao, C.R. (1992b). Statistics as a last resort. *Glimpses of India's Statistical Heritage* (Ghosh, J.K., Mitra, S.K., and Parthasarathy, K.R., eds.), pp. 153-213. Wiley Eastern, Ltd., New Delhi.

Reid, C. (1982). *Neyman from Life*. Springer-Verlag, Inc., New York.

Reid, N. (1994). A conversation with Sir David Cox. *Statist. Sci.*, **9**,

439-455.

Reid, N. (1995). The roles of conditioning in inference (with discussions by G. Casella, A.P. Dawid, T.J. DiCiccio, V.P. Godambe, C. Goutis, B. Li, B.C. Lindsay, P. McCullagh, L.A. Ryan, T.A. Severini and M.T. Wells). *Statist. Sci.*, **10**, 138-157.

Sampson, A.R. and Spencer, B. (1999). A conversation with I. Richard Savage. *Statist. Sci.*, **14**, 126-148.

Samuel-Cahn, E. (1992). A conversation with Esther Seiden. *Statist. Sci.*, **7**, 339-357.

Satterthwaite, F.E. (1946). An approximate distribution of estimates of variance components. *Biometrics Bulletin*, **2**, 110-114.

Savage, L.J. (1954). *The Foundations of Statistics*. John Wiley & Sons, Inc., New York. Reprinted (1972) by Dover Publications, Inc., New York.

Savage, L.J. (1976). On rereading R. A. Fisher. *Ann. Statist.*, **4**, 441-500.

Savage, L.J. (1981). *The Writings of Leonard Jimmie Savage - A Memorial Selection*. Amer. Statist. Assoc. and Inst. Math. Statist., Washington, D.C.

Scheffé, H. and Tukey, J.W. (1944). A formula for sample sizes for population tolerance limits. *Ann. Math. Statist.*, **15**, 217.

Scott, E.L. (1985). Jerzy Neyman. *Encyclopedia of Statistical Sciences*, **6** (S. Kotz and N.L. Johnson, eds.), pp. 215-223. John Wiley & Sons, Inc., New York.

Sen, P.K. (1981). *Sequential Nonparametrics*. John Wiley & Sons, Inc., New York.

Sen, P.K. and Ghosh, M. (1981). Sequential point estimation of estimable parameters based on U-statistics.. *Sankhyā, Ser. A*, **43**, 331-344.

Sen, P.K. and Singer, J.O. (1993). *Large Sample Methods in Statistics*. Chapman & Hall, Inc., New York.

Serfling, R.J. (1980). *Approximation Theorems of Mathematical Statistics*. John Wiley & Sons, Inc., New York.

Shafer, G. (1986). Savage revisited (with discussions by D.V. Lindley, A.P. Dawid, P.C. Fishburn, R.M. Dawes and J.W. Pratt). *Statist. Sci.*, **1**, 463-501.

Shepp, L. (1992). A conversation with Yuri Vasilyevich Prokhorov. *Statist. Sci.*, **7**, 123-130.

Shiryaev, A.N. (1991). Everything about Kolmogorov was unusual ... . *Statist. Sci.*, **6**, 313-318.

Siegmund, D. (1985). *Sequential Analysis: Tests and Confidence Intervals*. Springer-Verlag, Inc., New York.

Singpurwalla, N.D. and Smith, R.L. (1992). A conversation with Boris Vladimirovich Gnedenko. *Statist. Sci.*, **7**, 273-283.

Smith, A. (1995). A conversation with Dennis Lindley. *Statist. Sci.*, **10**, 305-319.

Snedecor, G.W. (1934). *Calculation and Interpretation of Analysis of Variance and Covariance*. Collegiate Press, Ames.

Snell, J.L. (1997). A conversation with Joe Doob. *Statist. Sci.*, **12**, 301-311.

Soper, H.E. (1913). On the probable error of the correlation coefficient to a second approximation. *Biometrika*, **9**, 91-115.

Soper, H.E., Young, A.W., Cave, B.M., Lee, A. and Pearson, K. (1917). On the distribution of the correlation coefficient in small samples. A cooperative study. *Biometrika*, **11**, 328-413.

Starr, N. (1966). The performance of a sequential procedure for fixed-width interval estimate. *Ann. Math. Statist.*, **37**, 36-50.

Stein, C. (1945). A two sample test for a linear hypothesis whose power is independent of the variance. *Ann. Math. Statist.*, **16**, 243-258.

Stein, C. (1949). Some problems in sequential estimation (abstract). *Econometrica*, **17**, 77-78.

Stein, C. (1956). Inadmissibility of the usual estimator of the mean of a multivariate normal distribution. *Proc. Third Berkeley Symp. Math. Statist. Probab.*, **1**, 197-206. University of California Press, Berkeley.

Stigler, S.M. (1980). *American Contributions to Mathematical Statistics in the Nineteenth Century* (edited volume). Arno Press, New York.

Stigler, S.M. (1986). *The History of Statistics: The Measurement of Uncertainty Before 1900*. Harvard University Press, Cambridge.

Stigler, S.M. (1989). Francis Galton's account of the invention of correlation. *Statist. Sci.*, **4**, 77-86.

Stigler, S.M. (1991). Stochastic simulation in the nineteenth century. *Statist. Sci.*, **6**, 89-97.

Stigler, S.M. (1996). The history of statistics in 1933. *Statist. Sci.*, **11**, 244-252.

Stigler, S.M. (1997). Raghu Raj Bahadur 1924-1997. *Inst. Math. Statist. Bul.*, **26**, 357-358.

"Student"(W.S. Gosset) (1908). The probable error of a mean. *Biometrika*, **6**, 1-25. Reprinted in *Breakthroughs in Statistics Volume II* (S. Kotz and N. L. Johnson, eds.), 1993. Springer-Verlag, Inc., New York.

Switzer, P. (1992). A conversation with Herbert Solomon. *Statist. Sci.*, **7**, 388-401.

Tintner, G. (1952). Abraham Wald's contributions to econometrics. *Ann. Math. Statist.*, **32**, 21-28.

Tippett, L.H.C. (1981). Egon Sharpe Pearson, 1895-1980: Appreciation by L.H.C. Tippett. *Biometrika*, **68**, 7-12.

Tong, Y.L. (1990). *The Multivariate Normal Distribution*. Springer-Verlag, Inc., New York.

Unni, K. (1978). The theory of estimation in algebraic and analytic exponential families with applications to variance components models. *Unpublished Ph.D. Dissertation*, Indian Statistical Institute, Calcutta.

Wald, A. (1939). Contributions to the theory of statistical estimation and testing hypotheses. *Ann. Math. Statist.*, **10**, 299-326.

Wald, A. (1945). Sequential tests of statistical hypotheses. *Ann. Math. Statist.*, **16**, 117-196. Reprinted in *Breakthroughs in Statistics Volume I* (S. Kotz and N. L. Johnson, eds.), 1992. Springer-Verlag, Inc., New York.

Wald, A. (1947). *Sequential Analysis*. John Wiley & Sons, Inc., New York. Reprinted (1973) by Dover Publications, Inc., New York.

Wald, A. (1949a). Note on the consistency of maximum likelihood estimate. *Ann. Math. Statist.*, **20**, 595-601.

Wald, A. (1949b). Statistical decision functions. *Ann. Math. Statist.*, **29**, 165-205. Reprinted in *Breakthroughs in Statistics Volume I* (S. Kotz and N. L. Johnson, eds.), 1992. Springer-Verlag, Inc., New York.

Wald, A. (1950). *Statistical Decision Functions*. John Wiley & Sons, Inc., New York.

Wallace, D.L. (1980). The Behrens-Fisher and Fieller-Creasy problems. *R. A. Fisher: An Appreciation* (S.E. Fienberg and D.V. Hinkley, eds.), pp. 119-147. Springer-Verlag, Inc., New York.

Wegman, E.J. (1986). Some personal recollections of Harald Cramér on the development of statistics and probability. *Statist. Sci.*, **1**, 528-535.

Whitehead, J. (1983). *The Design and Analysis of Sequential Clinical Trials*. Ellis Horwood, Chichester.

Whitehead, J. (1986). On the bias of maximum likelihood estimation following a sequential test. *Biometrika*, **73**, 573-581.

Whitehead, J. (1991). Sequential methods in clinical trials. *Handbook of Sequential Analysis* (B.K. Ghosh and P.K. Sen, eds.), *Chapter 26*, pp. 593-611. Marcel Dekker, Inc., New York.

Whittle, P. (1993). A conversation with Henry Daniels. *Statist. Sci.*, **8**, 342-353.

Wolfowitz, J. (1952). Abraham Wald, 1902-1950. *Ann. Math. Statist.*, **32**, 1-13.

Woodroofe, M. (1982). *Nonlinear Renewal Theory in Sequential Analysis*. NSF-CBMS Monograph No. 39. Society for Industrial and Applied Mathematics, Philadelphia.

Zabell, S. (1989). R. A. Fisher on the history of inverse probability (with discussions by R.L. Plackett and G.A. Barnard). *Statist. Sci.*, **4**, 247-263.

Zehna, P.W. (1966). Invariance of maximum likelihood estimators. *Ann. Math. Statist.*, **37**, 744.

Zinger, A.A. (1958). The independence of quasi-polynomial statistics and analytical properties of distributions. *Theory Probab. Appl.*, **3**, 247-265.

Index

A

B

I

P

R